MATHEMATICS FOR HIGH SCHOOL TEACHERS

AN ADVANCED PERSPECTIVE

Zalman Usiskin
The University of Chicago

Anthony Peressini
University of Illinois at Urbana-Champaign

Elena Anne Marchisotto
California State University, Northridge

Dick Stanley
University of California, Berkeley

Pearson Education, Inc.
Upper Saddle River, New Jersey 07458

Library of Congress Cataloging-in-Publication Data
Usiskin, Zalman
Mathematics for high school teachers: an advanced perspective
/ Zalman Usiskin, Anthony Peressini, Elena Anne Marchisotto, Dick Stanley
p. cm.
Includes bibliographical references and index.
ISBN 0-13-044941-5 (hc)
1. Mathematics. 2. Mathematics—Study and teaching (Secondary)
I. Usiskin, Zalman.
QA39.3 .M38 2003 CIP
510'.71'2—dc21 2002072760

Editor in Chief/Acquisitions Editor: *Sally Yagan*
Production Editor: *Lynn Savino Wendel*
Vice President/Director of Production and Manufacturing: *David W. Riccardi*
Senior Managing Editor: *Linda Mihatov Behrens*
Assistant Managing Editor: *Bayani Mendoza De Leon*
Executive Managing Editor: *Kathleen Schiaparelli*
Manufacturing Buyer: *Michael Bell*
Manufacturing Manager: *Trudy Pisciotti*
Marketing Manager: *Krista M. Bettino*
Marketing Assistants: *Rachel Beckman, Christine Bayeux*
Editorial Assistant: *Joanne Wendelken*
Art Director: *John Christiana*
Interior and Cover Designer: *Maureen Eide*
Art Editor: *Thomas Benfatti*
Creative Director: *Carole Anson*
Director of Creative Services: *Paul Belfanti*
Cover Photo Credits: *David Witonsky, Susan Chang*
Art Studio: *Laserwords*

Pearson Education, Inc.
Upper Saddle River, New Jersey 07458

Printed in the United States of America

10 9 8 7 6 5 4 3

ISBN 0-13-044941-5

Pearson Education LTD., *London*
Pearson Education Australia PTY, Limited, *Sydney*
Pearson Education Singapore, Pte. Ltd.
Pearson Education North Asia Ltd., *Hong Kong*
Pearson Education Canada, Ltd., *Toronto*
Pearson Educación de Mexico, S.A. de C.V.
Pearson Education—Japan, *Tokyo*
Pearson Education Malaysia, Pte. Ltd.

CONTENTS

NUMBER SYSTEM STRUCTURES 245

PART II GEOMETRY WITH CONNECTIONS TO ALGEBRA AND ANALYSIS

CONGRUENCE 275

PREFACE

Mathematics for High School Teachers–An Advanced Perspective is intended as a text for mathematics courses for prospective or experienced secondary school mathematics teachers and all others who wish to examine high school mathematics from a higher point of view.

Preliminary versions of the book have been used in a variety of ways, ranging from junior and senior (capstone) or graduate mathematics courses for pre-service secondary mathematics education majors to graduate professional development courses for teachers. Some courses included both undergraduate and graduate students and practicing teachers with good success.

There is enough material in this book for at least a full year (two semesters) of study under normal conditions, even if only about half of the problems are assigned. With a few exceptions, the chapters are relatively independent and an instructor may choose from them. However, some chapters contain more sophisticated content than others. Here are four possible sequences for a full semester's work:

Algebra emphasis: Chapters 1–6
Geometry emphasis: Chapters 1, 7–11
Introductory emphasis: Chapters 1, 3, 4, 7, 8, 10
More advanced emphasis: Chapters 1, 2, 5, 6, 9, 11.

In each sequence we suggest beginning with Chapter 1 so that students are aware of the features of this book and of some of the differences between it and other mathematics texts they may have used. More information and suggestions in this regard can be found in the Instructor's Notes. Additional instructional resources are also at the web site http://www.prenhall.com/usiskin.

The presentation assumes the student has had at least one year of calculus and a post-calculus mathematics course (such as real analysis, linear algebra, or abstract algebra) in which proofs were required and algebraic structures were discussed. The term "from an advanced standpoint" is taken to mean that the text examines high school mathematical ideas from a perspective appropriate for college mathematics majors, and makes use of the kind of mathematical knowledge and sophistication the student is gaining or has gained in other courses.

Two basic characteristics of *Mathematics for High School Teachers–An Advanced Perspective*, taken together, distinguish courses taught from this book from many current courses. First, the material is rooted in the core mathematical content and problems of high school mathematics courses before calculus. Specifically, the development emanates from the major concepts found in high school mathematics: numbers, algebra, geometry, and functions. Second, the concepts and problems are treated from a mathematically advanced standpoint, and differ considerably from materials designed for high school students.

The authors feel that the mathematical content in this book lies in an area of mathematics that is of great benefit to all those interested in mathematics at the secondary school level, but is rarely seen by them. Specifically, we have endeavored to include:

1. analyses of alternate definitions, language, and approaches to mathematical ideas
2. extensions and generalizations of familiar theorems

3. discussions of the historical contexts in which concepts arose and have changed over time
4. applications of the mathematics in a wide range of settings
5. analyses of common problems of high school mathematics from a deeper mathematical level
6. demonstrations of alternate ways of approaching problems, including ways with and without calculator and computer technology
7. connections between ideas that may have been studied separately in different courses
8. relationships of ideas studied in school to ideas students may encounter in later study.

There are many reasons why we believe a teacher or other person interested in high school mathematics should have this knowledge. Here are a few. Knowing alternate approaches helps in making decisions regarding curriculum, selection of materials, and lesson plans. Being able to connect, extend, and relate mathematical ideas to each other and to the mathematics a student may take later helps in designing courses and responding to student questions. Having a sense of history and the stories behind the mathematics can make lessons more interesting and engaging for both teacher and student. Encountering the richness of the mathematics that is studied at the high school level helps us to understand why some students are turned on by that mathematics, while others have difficulty with it.

ACKNOWLEDGMENTS

Mathematics for High School Teachers–An Advanced Perspective was developed and tested under a grant to the University of California at Berkeley from the Stuart Foundation of San Francisco for a project entitled High School Mathematics from an Advanced Standpoint. Original conceptualizers of the project were Dick Stanley and Zalman Usiskin. We are extremely grateful for the support of the Stuart Foundation throughout the development and to Phil Daro for his advice in obtaining this support.

We wish to thank the principal investigators of the project in the Department of Mathematics at the University of California at Berkeley, Calvin C. Moore and Hung-Hsi Wu, and the Professional Development Program for providing the facilities and administrative backup necessary for a project of this kind.

We were assisted in planning by an advisory board consisting of the following individuals: Gordon Bushaw, Central Kitsap School District (WA); Duane DeTemple, Washington State University; Jacqueline M. Dewar, Loyola Marymount University; Wade Ellis, Jr., West Valley College; Scott Farrand, California State University, Sacramento; Brenda Hull, Mt. San Jacinto High School (CA); James King, University of Washington; Bill Kring, Eisenhower High School, Yakima (WA); Carolyn R. Mahoney, Elizabeth City State University (NC); and Harris Shultz, California State University, Fullerton. We thank these individuals for their help and guidance.

It takes skill, courage, flexibility, and sometimes quite a bit of tolerance to pilot materials that have never before been taught. We are grateful to the following individuals who piloted the materials at their institutions and gave us their reactions and comments: Alice Artzt, Queens College of the City of New York; Irene Bloom, Arizona State University; Janet Caldwell, Rowan University; E. Graham Evans, University of Illinois at Urbana-Champaign; Mary Garner, Kennesaw State University; Emiliano Gómez, University of California at Berkeley: Abbe Herzig, Rutgers University; Lynne Ipiña, University of Wyoming; Michael Keynes, University of California at Berkeley; James Madden, Louisiana State University; Tom McGannon, St. Xavier University; Frank Neubrander, Louisiana State University; Anthony Phillips, State University of New York at Stony Brook; Lew Romagnano, The Metropolitan State College of Denver; Wendy Sanchez, Kennesaw State University; Robin Sue Sanders, Buffalo State College; Angelo Segalla, California State University, Long Beach; Alan Sultan, Queens College of the City of New York; and Gordon Woodward, University of Nebraska, Lincoln.

We asked several mathematicians to review and comment on the penultimate version of the manuscript. We benefited greatly from the careful reading, detailed comments, and helpful suggestions of each of them, to whom we give special thanks: Richard Askey, University of Wisconsin; Susanna Epp, DePaul University; Robin Hartshorne, University of California at Berkeley; Reuben Hersh, University of New Mexico (emeritus); Roger Howe, Yale University; and Alan Tucker, State University of New York at Stony Brook.

Our project was greatly assisted by the help of office staff and students, including: Judie Welch, Manya Raman, Lily Storm, Francisco Rios, and Krysten Morganti at Berkeley; Mari Flores, Kaveh Shamsa, Richard Fine, Martin Tippens Jr., and James Castro at Northridge; Amanda C. Jones and Dan Willms at Chicago. We also wish to thank several colleagues for their reactions and advice: Bruce Reznick, Bruce Berndt, and John Wetzel of the University of Illinois at Urbana-Champaign, Patrick Callahan of the University of California Office of the President, and James T. Smith of San Francisco State University.

The editing, production, and testing of the materials was centered at the University of Chicago School Mathematics Project (UCSMP) at the University of Chicago. Susan Chang, Margaret Liput, and Carol Siegel were of great assistance throughout. The final editing of the manuscript and of the answers to all questions was done with great patience and skill by David Witonsky of UCSMP.

We wish to emphasize that even with all this help and advice from so many individuals, the authors alone are responsible for the content herein. These materials have gone through many trial versions, and we have endeavored to make each version better and freer from errors than the previous. We hope to receive additional comments from instructors and students who use the materials to help us improve future versions and to identify those aspects of the materials that are found to be most appealing.

Work on a manuscript such as this one, which takes countless hours over many years and goes well beyond normal hours of work, tests the patience of family and friends. Without the support of our spouses, Karen Usiskin, Joan Peressini, and Joe Marchisotto, we would not have been able to undertake this venture. We owe more to them than any words could express.

ZALMAN USISKIN
ANTHONY PERESSINI
ELENA ANNE MARCHISOTTO
DICK STANLEY

Chapter 1

WHAT IS MEANT BY "AN ADVANCED PERSPECTIVE"?

This book contains mathematics that the authors believe is particularly suitable for teachers and others who wish a deeper knowledge and understanding of the mathematics taught in high schools. A glance at the table of contents reveals that most of the core topics of a typical high school mathematics program are considered here, and a closer review of the book's content shows that the topical coverage is actually much broader.

Our objective is to discuss content of high school mathematics from an advanced perspective. This perspective takes into account not only the many interconnections among high school mathematics topics, but also their relationship to college-level mathematics. This perspective includes a deeper analysis of problems and concepts drawn from high school mathematics to reveal important new mathematical insights and understandings. This advanced perspective is also mindful of the historical and conceptual evolution of mathematical theory and school mathematics. The goal of this advanced perspective is to encourage you to explore mathematical ideas in depth and to explain mathematical ideas with clarity and precision. In this way, we hope to help you to develop a deeper understanding of high school mathematics and a new appreciation of its beauty, its logical structure, and its applicability.

In this chapter, we discuss several specific examples that illustrate some of the features of our advanced perspective on high school mathematics that are the backbone of this book and that you may not have seen described as we envision them. We refer to these features as *concept analysis*, *problem analysis*, and *mathematical connections*.

An Example of Concept Analysis: Parallelism

Concept analysis involves tracing the origins and applications of a concept, looking at the different ways in which it appears both within and outside mathematics, and examining the various representations and definitions used to describe it and their consequences. Here we look at the concept of *parallelism* or, more simply, *parallel*.

Exploring alternate definitions

What does *parallel* mean? It seems possible to answer this question merely by invoking a definition. The most common definition found in schoolbooks applies parallelism to lines and defines "parallel" lines as lines in the same plane that do not intersect. But this is not the only possible definition, and the idea of *parallel* is quite a bit broader. In James and James's *Mathematics Dictionary*, the word *parallel* is defined as "equidistant apart". This dictionary then describes parallel curves, parallel lines, parallel rays, parallel planes, parallel surfaces, parallel vectors, parallels of latitude, etc. For *parallel rays*, the entry states that sometimes it is required that they point in the same direction. So under some definitions, opposite rays would be parallel and under other definitions they would not be parallel.

Included among the dictionary descriptions of parallel objects are the following four different characterizations:

parallel objects:
- are equidistant apart
- do not intersect
- go in the same direction
- can be obtained from each other by a translation.

The meaning of each of these characterizations for lines in the plane is straightforward, and any one of them could be (and has been!) used as the definition. However, these characterizations are not logically equivalent. A line is parallel to itself under the last two of these definitions but not under the second and only under the first if zero distance between lines is allowed.

Mathematical definitions are formulated within a mathematical system so that deductions can be made from them. The theorems that are deduced add to our understanding of the idea and can sometimes become alternate definitions. However, these abstract formulations may not precisely represent the intuitive ideas that created the need for the definition in the first place. By choosing any one definition without considering the others, we may lose sight of the other possibilities. Knowledge of the variety of possibilities can assist teachers in knowing why students have trouble both in using their intuition and in applying the abstractions.

Examining instances and applications

Figure 1

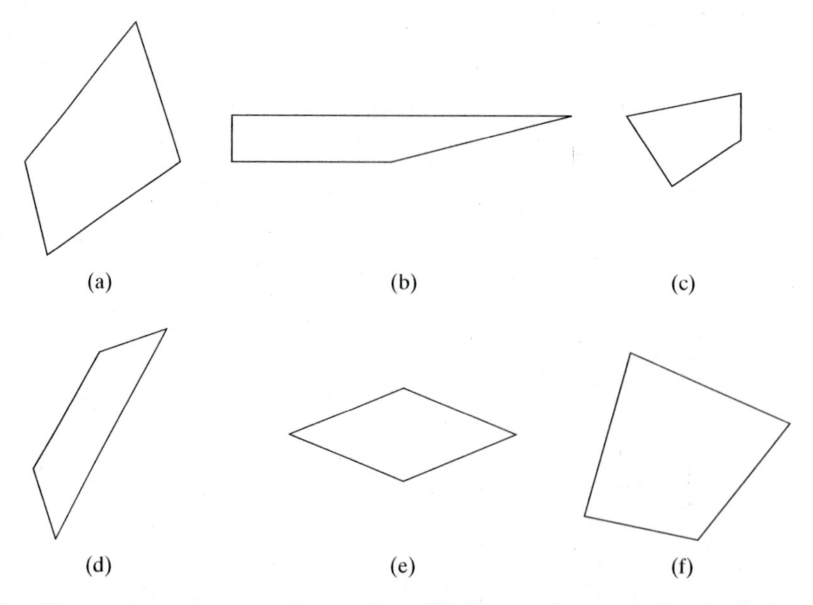

Different instances of "parallel" may involve different intuitions. For example, if different quadrilaterals are drawn as in Figure 1, how would you visually decide which has a pair of parallel sides?

The answer might depend on the particular quadrilateral. You might visually extend the opposite sides in your mind and decide if they would intersect. You might look to see if the opposite sides go in the same direction. You might visualize the (perpendicular) distance between the sides to see if it is constant.

Figure 2

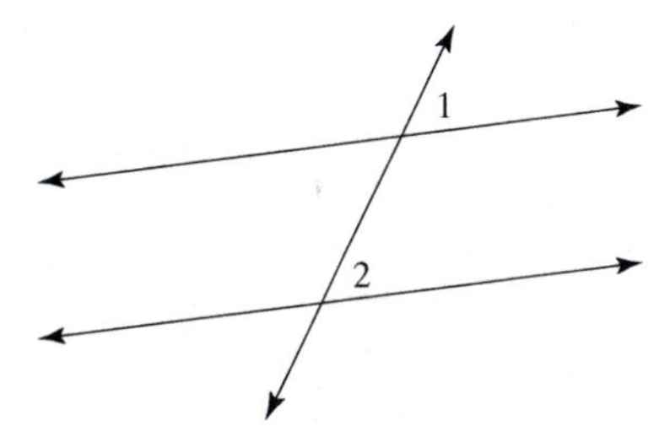

As another example, corresponding angles 1 and 2 in Figure 2, formed by two lines cut by a transversal, might seem to have the same measure because their sides (parallel rays!) go in the same directions, or because the sides are translation images of one another, rather than because the corresponding sides are equidistant or do not intersect.

No analysis of a concept is complete without some examination of its applications. Such an analysis may tell you the origins of the idea and also may suggest some extensions. The ancients might have used the tracks of chariots as their example of parallel lines. Today, railroad tracks are often presented. This example applies when the tracks are straight but may not be true when the tracks curve unless we have appropriately defined parallel curves. When we say that two streets are parallel, we usually mean parallel in the sense of being equidistant on a plane or going in the same direction. Yet north–south streets, if extended, would intersect at both the north and south poles. Examples of parallelism on a smaller scale are easier to find: lines on a sheet of lined paper, opposite sides of trapezoids and many other geometric figures, and so on.

The concept of parallel also applies to objects in space. For example, two planes or two lines in space may be parallel to one another. We may speak of parallel walls of a room. Also, when we describe locations on Earth's surface (modeled as the surface of a sphere) we often speak of "parallels of latitude".

Generalizations

Figure 3

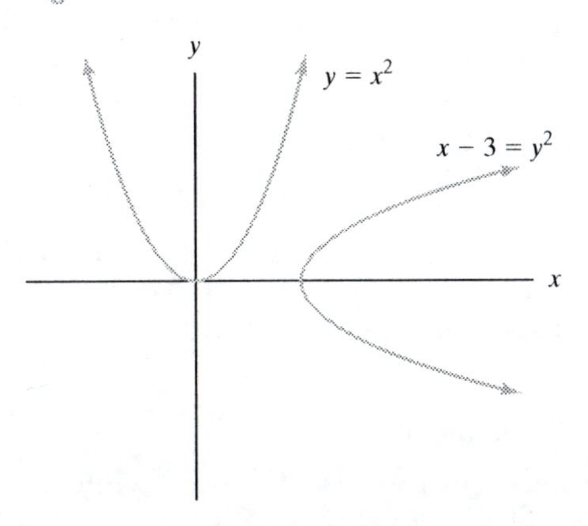

Figure 4

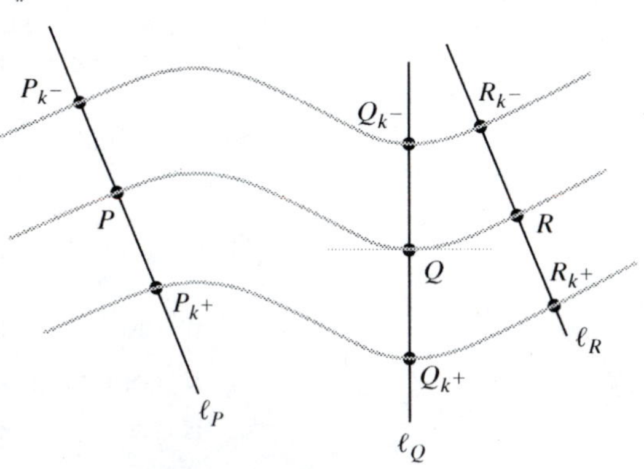

It is instructive to examine instances of the word *parallel* in contexts more general than that of line segments or lines or planes to see how these characterizations of parallel generalize. Consider again the idea of parallel curves in a plane. In this more general setting, we ask whether any of the four characterizations of parallel lines mentioned above apply easily or with little modification. To define parallel curves merely as "curves that do not intersect" would seem to be insufficient, for parabolas such as the graphs of $y = x^2$ and $x - 3 = y^2$ (Figure 3) would be parallel.

To define parallel curves as "curves that are equidistant" requires some meaning for the distance between curves. One possible meaning is that two curves are equidistant if and only if there is a 1-1 correspondence between them such that there is a segment of fixed length that joins corresponding points and this segment is perpendicular to the tangents to the curves at those points. This definition is quite restrictive but applies to parallel lines and to concentric circles. Another possible meaning was given by Gottfried Leibniz in 1692. Let C be a smooth (differentiable) curve that does not intersect itself. Leibniz considered the curves at a distance $k > 0$ from C to be parallel to C. What are these curves? For each point P on C, let ℓ_P be the line perpendicular to the tangent to C at P, as shown in Figure 4. There are two points on ℓ_P whose distance from P is k; we call them P_k^+ and P_k^-. The set of all points constructed in this way forms the two curves.

To define parallel curves as "curves that go in the same direction", we might assign a meaning to the term *going in the same direction* by saying that the tangent lines at corresponding points are parallel. The curves of Figure 5 could then be parallel, but so too could be the curves of Figure 6.

Figure 5

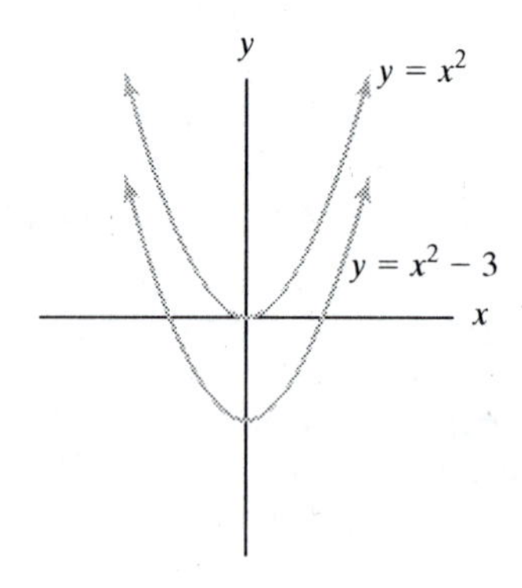

Figure 6

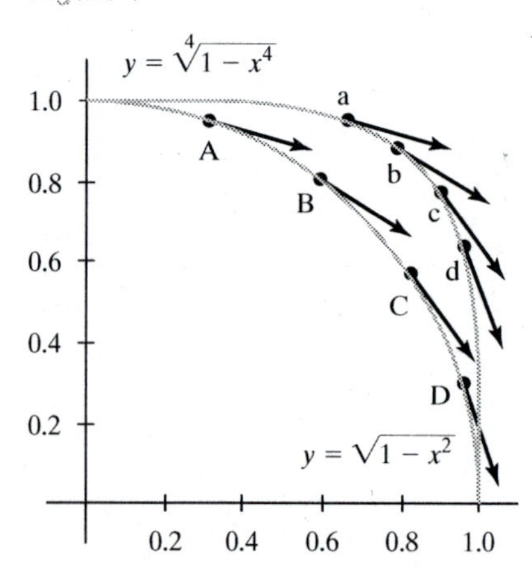

To define two curves as parallel if "one curve is the translation image of another" (as in Figure 5) has the difficulty that some parallel curves, such as the graphs of the sine and cosine functions, would intersect. Thus, different descriptions of a mathematical concept that are equivalent in one context may not be equivalent or even easily definable in a more general context.

Before reading on, answer the following question.

Question 1: Consider the following instances of objects that may seem to be "parallel" and give your opinion of which (if any) of the four characterizations of the *idea* of parallel seem to apply most naturally.

a. coincident lines

b. two concentric circles in the plane

c. the graphs of the equations $y = x^2$ and $y = x^2 + 1$ in the xy-plane.

Questions in the reading, like the one identified above, are meant for you to consider before reading on. If the question has a correct answer (i.e., it is not asking for an opinion) and no answer is given in the reading, then an answer is supplied after the Problems for that section.

Your analysis of Question 1 should reveal that the answer to the question of whether the objects in the pair are parallel depends on the definition of parallel that you choose to adopt.

History

The concept of parallelism has played an important role in the history of geometry due to the influence of the *Elements* of Euclid. In this work, Euclid took as a postulate (in modern language) "If two lines are cut by a transversal and the sum of the measures of the interior angles on the same side of the transversal is less than 180°, then the lines intersect on that side of the transversal." In Chapter 7 of this book, we discuss Euclid's *Elements*, and in Chapter 11, we examine this postulate in some detail. Here, it suffices to say that this postulate, which is called *Euclid's parallel postulate*, clearly represents the notion of parallel as "do not intersect". Euclid's parallel postulate motivated a number of mathematicians to attempt to prove it from the other postulates. As a direct outgrowth of the futility of attempts over two thousand years, non-Euclidean geometries were discovered. These geometries changed the ways in which mathematicians viewed geometry in particular and mathematics in general, and had a profound impact on how we think of axiom systems today.

Concept analysis includes the type of examination of alternate definitions, instances, and generalizations of a mathematical idea that we have outlined here for the "parallel" concept. The results of concept analysis often make us realize that mathematics is not as rigid as it is sometimes made out to be, and it may suggest changes in our formalization and interpretation of the idea. Many of the chapters of this book might be thought of as extended concept analyses of the important ideas of high school mathematics: function, equation, congruence, distance, etc. But within these concept analyses are other kinds of analyses, to which we now turn.

An Example of Problem Analysis: Matching an Average

Problem analysis involves more than finding different ways of solving a problem. It includes looking at a problem after it has been solved and examining what has been done. Will the method of solution work for other problems? Can we extend the problem? And so forth.

The problem: The average test-grade problem

To illustrate, we begin with a typical problem found in algebra books.

> Jane has an average of 87 after 4 tests. What score does she need on the fifth test to average 90 for all five tests?

Solving the problem

To answer this question, the algebra student is expected to let a variable such as x stand for Jane's score on the 5th test and to solve $\frac{87 \cdot 4 + x}{5} = 90$. But many students will use arithmetic, working somewhat as follows. To average 90 points on 5 tests means to have 450 points. Jane has 348 points, so that she needs 102 points. The algebra not only verifies this answer but mimics the arithmetic. To solve the equation, we multiply both sides by 5 and then subtract $87 \cdot 4$. Students who do the problem arithmetically will naturally wonder why algebra is needed when they see that the algebra merely repeats the arithmetic, and they are right because algebra is not needed for this problem.

Generalizing the problem

This problem means little if it ends here. First we interpret the answer. If the tests contain at most 100 points, then Jane cannot average 90 for all five tests. We might ask: Given her current situation, what is the highest average score she can attain? What is the lowest average score she can attain? These questions lead us to delve into the relationship between Jane's score on the 5th test and her average. Then we cannot avoid algebra.

Let A equal Jane's average for all 5 tests. $A = \frac{87 \cdot 4 + x}{5}$. A is a function of x, so we can write $A = f(x) = \frac{87 \cdot 4 + x}{5} = \frac{1}{5}x + 69.6$, so f is a linear function with slope $\frac{1}{5}$. We can use the equation $A = \frac{87 \cdot 4 + x}{5}$ to determine what Jane needs to obtain any given average, not merely the single average of 90 presented in the original problem, by solving for x: $x = 5A - 87 \cdot 4$.

Representing the problem

A geometric picture can add new insight into an algebraic problem. A graph of $A = \frac{348 + x}{5}$ over the interval $0 \leq x \leq 100$ shows all the possible solutions (Figure 7).

Figure 7

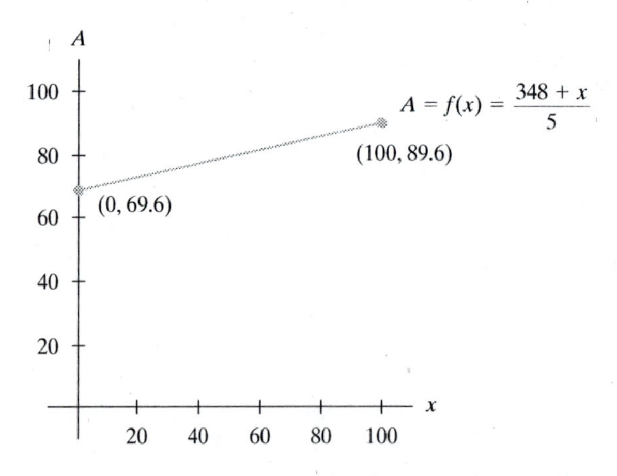

Jane's lowest possible average is 69.6, while her highest possible average is 89.6. Now we see the power of algebra to solve *an entire set of problems* at once.

The linear relationship between x and A shows that this is a constant-increase situation. Each point Jane earns on the 5th test adds the same amount to her average for the five tests. In fact, each point Jane scores on *any* of the tests contributes $\frac{1}{5}$ point to her average. Thus the problem analysis has shown us a different way to look at the problem.

Extending the problem: The Scoring Title Problem

In Jane's case, we sought a specific average. But now suppose that the average to be attained is not known. We illustrate this situation with actual data, not invented for instructional purposes. In April, 1998, the basketball players Michael Jordan and Shaquille (Shaq) O'Neal were vying for the season individual scoring title until the last game of the season. The scoring title is won by the player with the highest average number of points per game, calculated by dividing the total number of points by the number of games the player has played. Before the last game, Jordan had scored 2313 points in 81 games, for an average of $\frac{2313}{81} \approx 28.6 \frac{\text{points}}{\text{game}}$ (customarily, averages in newspapers are rounded to the nearest tenth). Shaq had scored 1666 points in 59 games, for an average of $\frac{1666}{59} \approx 28.2 \frac{\text{points}}{\text{game}}$. No one else had a chance to win the title. The Scoring Title Problem can be stated as follows:

> Given the above information, with what numbers of points in their final games does Shaquille O'Neal win the scoring title over Michael Jordan?

It is easy to try out various scenarios. You could estimate how many points Jordan and Shaq will score in their last game, and then calculate, on the basis of those guesses, who would win the scoring title. These calculations require only arithmetic. If there were only a couple of possibilities for the numbers of points scored, this arithmetic would give you rather quickly all of the possible outcomes. But there are many possibilities: Each player might reasonably score any number of points from 10 to 60.

For this reason it is useful to consider all the scenarios at one time, and algebra is needed. How can we use algebra to analyze what will happen in the final game? We show two ways, one using equations and inequalities, the other using functions. You will see that each gives a different and useful perspective on the situation. Seeing both together helps us understand the situation better, and also gives good insight into the relative role of graphs of functions and equations.

Question 2: If Jordan scores more points in the last game than Shaq, will he necessarily win?

Question 3: If Shaq scores more points in the last game than Jordan, will he necessarily win?

An equations/inequalities approach

If Jordan scores j points in his last game, then he has a total of $2313 + j$ points in 82 games, for an average of $\frac{2313 + j}{82} \frac{\text{points}}{\text{game}}$. Similarly, if s is the number of points Shaq scores in his last game, he will then have $1666 + s$ points in 60 games, for an average of $\frac{1666 + s}{60} \frac{\text{points}}{\text{game}}$. Consequently, Jordan wins the scoring title whenever

$$\frac{2313 + j}{82} > \frac{1666 + s}{60}.$$

On the other hand, Shaq wins the scoring title whenever

$$\frac{2313 + j}{82} < \frac{1666 + s}{60}.$$

Representing the extended problem

Where do we go now? First, we must ask what it means to *solve* inequalities like these. We cannot simply say j equals this and s equals that, because there are infinitely many solutions even if one of the variables is fixed. So we graph the boundary to the inequalities. That is the line with equation

$$\frac{2313 + j}{82} = \frac{1666 + s}{60}.$$

Solving for s in terms of j, $s = \frac{30j + 1084}{41}$. This line is graphed in Figure 8.

Figure 8

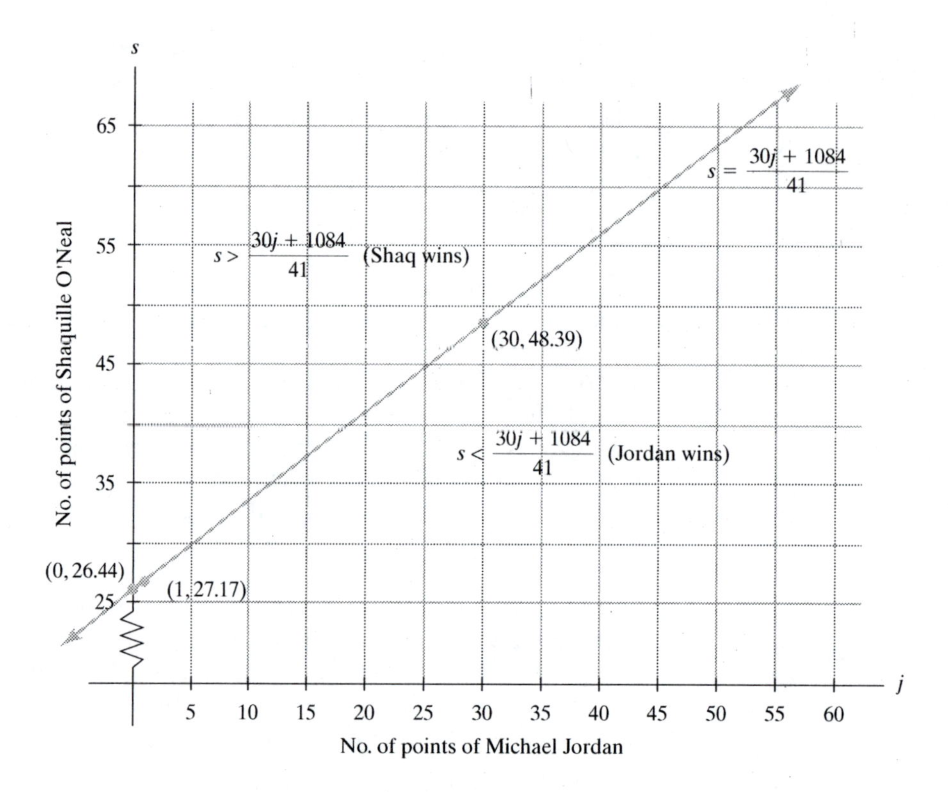

For $j \geq 0$, each point on the line with at least one integer coordinate has meaning. If Jordan scores 0 points, Shaq still needs to score over 26.44 points to win the title. That is, he needs 27 points or more. If Jordan scores 1 point, Shaq needs over 27.17 points—28 points or more—to win. If Jordan scores 30 points, Shaq needs 49 points or more. We see that the lattice points (the points with integer coordinates) above the line in the first quadrant or on the s-axis offer all the possible ways in which Shaq wins. The lattice points below the line in the first quadrant or on the j-axis show all the possible ways in which Jordan wins.

Question 4: Redraw the graph of Figure 8 with the axes intersecting at $(0, 0)$ and including the line $s = j$. On your graph, show how Shaq could in theory win the title while scoring fewer points than Jordan in the final game.

So, in theory, Shaq could have a lower average than Jordan before the final game, and also score fewer points than Jordan in the final game, yet still have a *higher* average overall. (This phenomenon, in which an overall average is in a different order than the two partial averages that comprise it, is known as Simpson's paradox.)

It happened that Jordan scored 44 points in his last game of the regular season. His game was over before Shaq's game started. When $j = 44$, $s = 58.63\ldots$, which meant that Shaq had to score 59 points in his last regular season game to win the title. Shaq would have needed a personal record for him to win the scoring title. Shaq scored 39 points, which is terrific scoring, but it was not enough to win the scoring title.

Question 5: Suppose Shaq had scored 39 points in his last game, and it had been played before Jordan's last game. Indicate (a) on the graph of Figure 8, and (b) by solving an equation or inequality, how many points Jordan would have needed to score to win the title.

A functions approach

We begin this approach by representing each player's average points per game for the whole season as a function of the points he scores in the final game.

Jordan: $$A = f(j) = \frac{2313 + j}{82}$$

Shaq: $$A = g(s) = \frac{1666 + s}{60}$$

The functions f and g are graphed in Figures 9a and 9b.

The answer to Question 5 is visualized in a different way in Figure 9 than it is in Figure 8. For any k, the x values where the graphs of f and g intersect the horizontal line $A = k$ are the numbers of points Jordan and Shaq need to score in the last game to obtain a season average of k. By drawing three sides of a rectangle, we see on the horizontal axis of Figure 9, the coordinates of the points that are on the line graphed in Figure 8. For example, Figure 9 pictures 18 as the answer to Question 5 and shows that after Jordan scored 44 points, Shaq would have needed to score 59 points in his last game to match Jordan's average.

There is another relationship between Figures 8 and 9. In Figure 9 we have graphed $f(j) = \frac{2313 + j}{82}$, and $g(s) = \frac{1666 + s}{60}$. The equation of the line graphed in Figure 8 signifies when the two averages are the same, that is, it is $f(j) = g(s)$. Since g is a linear function, g has an inverse, and $g^{-1} \circ f(j) = g^{-1} \circ g(s) = s$. That

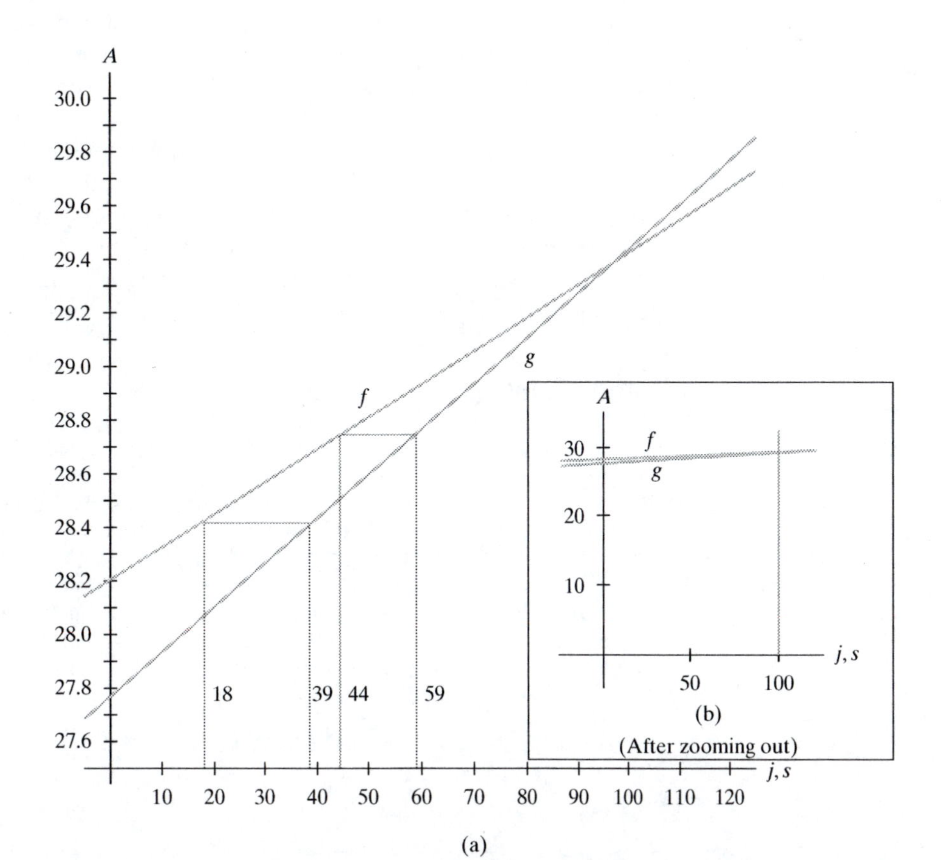

(a)

is, $s = g^{-1} \circ f(j)$. Thus the equation that is graphed in Figure 8, $s = \frac{30j + 1084}{41}$, is the equation of $g^{-1} \circ f$. We show this schematically.

$$\begin{array}{ccccccc} j & \xrightarrow{f} & f(j) & & g(s) & \xleftarrow{g} & s \\ \begin{pmatrix}\text{Points scored by Jordan}\\ \text{in last game of season}\end{pmatrix} & \xrightarrow{f} & \begin{pmatrix}\text{Jordan's season}\\ \text{average}\end{pmatrix} & & \begin{pmatrix}\text{Shaq's season}\\ \text{average}\end{pmatrix} & \xleftarrow{g} & \begin{pmatrix}\text{Points scored by Shaq}\\ \text{in last game of season}\end{pmatrix} \end{array}$$

If $f(j) = g(s)$, then $g^{-1} \circ f(j) = s$.

Notice the meaning of $g^{-1} \circ f$. The function f maps Jordan's last game points onto Jordan's season average. The function g maps the number of points Shaq gets in the last game onto his average for the season. If the season averages are to be equal, then g^{-1} is mapping Jordan's average onto the number of points Shaq needs to get that average. Thus the function $g^{-1} \circ f$ maps the number of points Jordan scores in the last game onto the number of points Shaq needs to have the same average as Jordan. In this way, function composition and function inverses both generalize and provide an explanation for the graphs in Figures 8 and 9.

Throughout this book, you will encounter extended analyses of problems like this one, in which we begin with a simple problem and examine its solutions and variants to deepen both the understanding of the problem and the understanding of the underlying mathematics.

An Example of Mathematical Connections: + and ·

Some mathematical connections relate different areas of mathematics, such as algebra to geometry. Others connect the mathematics studied before college to the mathematics studied in college-level courses, such as relating school algebra to college-level abstract algebra. Still others show analogies between concepts, such as between area and volume.

For our example in this chapter, we exhibit a set of corresponding properties of addition and multiplication. Let p, q, and r denote real numbers and a, b, and c denote corresponding positive real numbers. Throughout this example, the left column contains properties arising from addition while the right column contains corresponding properties arising from multiplication. We begin by thinking of $p + q$ as corresponding to ab.

Commutative group properties

	Addition in the Set of Real Numbers	**Multiplication in the Set of Positive Reals**
Closure	$p + q$ is real.	ab is real.
Commutativity	$p + q = q + p$	$ab = ba$
Associativity	$p + (q + r) = (p + q) + r$	$a(bc) = (ab)c$
Identity	There exists a number 0 such that for all p, $p + 0 = 0 + p = p$.	There exists a number 1 such that for all a, $a \cdot 1 = 1 \cdot a = a$.
Inverse	For all p, there exists a number $-p$ such that $p + -p = 0$.	For all a, there exists a number $\frac{1}{a}$ such that $a \cdot \frac{1}{a} = 1$.

For those of you who are familiar with abstract algebra, these correspondences are often described with the following statements:

1. The set **R** of real numbers with the operation + of addition is a *commutative group* $\langle \mathbf{R}, + \rangle$.
2. The set $\mathbf{R}^+$ of positive real numbers with the operation · of multiplication is a *commutative group* $\langle \mathbf{R}^+, \cdot \rangle$.

Multiples and powers

The correspondence between addition and multiplication by no means ends here. Let m and n be fixed nonnegative integers. We see that the common properties of integer powers (sometimes called "laws of exponents"), which emanate from multiplication, correspond to the distributive property and other properties of addition and multiplication. For instance, the fact that the zero power equals 1 (the multiplicative identity) corresponds to the fact that the zero multiple equals 0 (the additive identity).

Doubles/squares	$p + p = 2p$	$a \cdot a = a^2$
Triples/cubes	$p + p + p = 3p$	$a \cdot a \cdot a = a^3$
Multiples/powers	$\underbrace{p + p + \cdots + p}_{m \text{ terms}} = mp$	$\underbrace{a \cdot a \cdot \cdots \cdot a}_{m \text{ factors}} = a^m$
Sum of multiples/ product of powers	$mp + np = (m + n)p$	$a^m \cdot a^n = a^{m+n}$
Zero multiple/power	$0p = 0$	$a^0 = 1$
Multiple of sum/ power of product	$m(p + q) = mp + mq$	$(ab)^m = a^m \cdot b^m$
Multiple of multiple/ power of power	$m(np) = (mn)p$	$(a^n)^m = a^{mn}$

Question 6: What property of multiplication corresponds to $3(2x + 5y) = 6x + 15y$?

Inverses and inverse operations

Allow m and n to be negative integers. The first row in the table shows another way in which opposites correspond to reciprocals, and the rest follows. The third row shows that the *difference* in subtraction corresponds to the *quotient* in division.

m* = −1 *and inverse	$-1 \cdot p = -p$	$a^{-1} = \frac{1}{a}$
Inverse of inverse	$-(-p) = p$	$\frac{1}{\frac{1}{a}} = a$
Inverse operations	$p - q = p + (-q)$	$a \div b = \frac{a}{b} = a \cdot \frac{1}{b}$
One-step equations	$p + q = r$ if and only if $p = r - q$	$ab = c$ if and only if $a = \frac{c}{b}$

Operations on fractions

Once division has been introduced in the multiplication column, some operations with fractions can be handled. Multiplication and division of fractions correspond to addition and subtraction of differences. Here s corresponds to d.

Adding differences/ multiplying fractions	$(p - q) + (r - s) = (p + r) - (q + s)$	$\frac{a}{b} \cdot \frac{c}{d} = \frac{ac}{bd}$
Subtracting differences/ dividing fractions	$(p - q) - (r - s) = (p + s) - (q + r)$	$\frac{a}{b} \div \frac{c}{d} = \frac{ad}{bc}$

nth parts and nth roots

What multiple of p, when added to itself, gives p? The answer is $\frac{1}{2}p$. The corresponding question for multiplication is, What power of a, when multiplied by itself, gives a? We recognize the answer as a square root of a. To reveal the correspondence between mp and a^m, think of a square root of a as $a^{1/2}$. Furthermore, since the numbers in the multiplication column are positive real numbers, $a^{1/2}$ is the positive square root of a. We call $\frac{1}{n}p$ the nth part of p. The nth part in addition corresponds to the nth root in multiplication. Notice that radical notation disguises the correspondence.

Adding halves/ multiplying square roots	$\frac{1}{2}p + \frac{1}{2}q = \frac{1}{2}(p + q)$	$a^{1/2} \cdot b^{1/2} = (ab)^{1/2}$
nth part/nth root	$\underbrace{\frac{1}{n}p + \frac{1}{n}p + \ldots \frac{1}{n}p}_{n \text{ terms}} = p$	$\underbrace{a^{1/n} \cdot a^{1/n} \cdot \ldots \cdot a^{1/n}}_{n \text{ factors}} = a$
Rational multiples/ rational powers	$m(\frac{1}{n})p = (\frac{1}{n})mp = (\frac{m}{n})p$	$(a^{1/n})^m = (a^m)^{1/n} = a^{m/n}$ $(\sqrt[n]{a})^m = \sqrt[n]{a^m} = a^{m/n}$

Inequalities

In this correspondence between addition and multiplication, order is preserved. The preservation leads to corresponding inequalities. The numbers 0 and 1 play corresponding roles here just as they do as identities for the operations. Here they might be called *pivots*, since the sign of the inequality in each column can change when a number switches from one side of the pivot to the other. The familiar properties of multiplication of positive and negative numbers correspond to properties of powers with bases greater than or less than 1.

Order	$p < q$	$a < b$
Positive multiples/positive powers keep order	$m > 0$ and $p < q \Rightarrow mp < mq$	$m > 0$ and $a < b \Rightarrow a^m < b^m$
Negative multiples/negative powers reverse order	$m < 0$ and $p < q \Rightarrow mp > mq$	$m < 0$ and $a < b \Rightarrow a^m > b^m$
Multiples and pivots/ powers and pivots	$m > 0$ and $p > 0 \Rightarrow mp > 0$ $m > 0$ and $p < 0 \Rightarrow mp < 0$	$m > 0$ and $a > 1 \Rightarrow a^m > 1$ $m > 0$ and $a < 1 \Rightarrow a^m < 1$

Relating functions

There are also corresponding functions. The slope of the (nonconstant) linear function L, which can be any real number, corresponds to the base of the (nonconstant) exponential function E, which must be a positive number. Notice in the following table that m is the independent variable (not the usual x), q is the slope (not m), and p is the y-intercept. In the language of linear increase and exponential growth, m represents time, p and a represent the initial values, and q (the slope) and b (the base) indicate how fast the increase and growth in the functions L and E are taking place. If q is negative, there will be linear decrease. This corresponds to a value for the positive base b of the exponential function that is less than 1, yielding exponential decay.

Arithmetic sequences/ geometric sequences	$p, p+q, p+2q, \dots p+(m-1)q$	$a, ab, ab^2, \dots ab^{m-1}$
Linear functions/ exponential functions	$L: m \rightarrow p + mq$	$E: m \rightarrow ab^m$
Addition of linear functions/multiplication of exponential functions	If L_1 and L_2 are linear functions with slopes q_1 and q_2, then $L_1 + L_2$ is linear with slope $q_1 + q_2$.	If E_1 and E_2 are exponential functions with bases b_1 and b_2, then $E_1 \cdot E_2$ is exponential with base $b_1 b_2$.

We return to the correspondences between linear and exponential functions in Chapter 3.

Analyzing the correspondence

We started from

p corresponds to a

q corresponds to b,

and $p + q$ corresponds to ab.

We found that

$p - q$ corresponds to $\frac{a}{b}$,

and mq corresponds to b^m.

In the correspondences $q \rightarrow b$ and $mq \rightarrow b^m$, let $q = 1$. Then we have the correspondence $m \rightarrow b^m$, which is an exponential function. The correspondence the other way is the inverse of the exponential function, the logarithm function. Specifically, if p is replaced by $\log a$, q by $\log b$, r by $\log c$, and s by $\log d$, every property in the addition column is equivalent (in the sense of being true for the same values of a, b, c, and d) to the corresponding property in the multiplication column. In the language of abstract algebra, the group of real numbers with the operation of addition is *isomorphic* to the group of positive real numbers with the operation of multiplication, and any logarithm function with a positive base greater than 1 provides the (order-preserving) correspondence.

We have by no means exhausted the corresponding properties of addition and multiplication. You are asked to explore other corresponding properties of these operations in the problems. Still other correspondences are mentioned in various places later in this book.

Chapter 1 Problems

1. Four conceptions of parallelism are described in this chapter. Which conception seems to you to be most closely related to the use of "parallel" in each situation below, and why?
   a. the parallel opposite edges of a parallelepiped
   b. parallel bars in gymnastics
   c. In a plane, if two lines are perpendicular to the same line, then they are parallel.
   d. parallel electrical circuits
   e. parallel rays from the Sun
   f. parallel processing (with computers)

2. Suppose parallel curves are defined as Leibniz did. Show that a consequence of this definition, if applied to curves in 3-space, is that there can be two curves that are parallel where one curve is a straight line and the other is not.

3. a. Consider the statement, "In a plane, segments on fixed rays from a point cut off by any parallel lines are proportional." Illustrate this statement with a diagram, and describe its hypothesis and conclusion in terms of your diagram.
   b. Consider the statement, "In a plane, segments on fixed parallel lines cut off by rays from a point are proportional." Illustrate this statement with a diagram, and describe its hypothesis and conclusion in terms of your diagram.
   c. Is either (or both) of **a** or **b** true? If so, why? If not, why not?

4. Analyze the concept of *absolute value* in a way similar to that done in this chapter for the concept of *parallel*. You may find it convenient to organize your analysis in the following way.
   a. Define the absolute value of a real number both algebraically and geometrically.
   b. Give properties of absolute value that are easily understood using the algebraic definition, and properties of absolute value that are easily understood using the geometric definition.
   c. Discuss the graph of the absolute value function f, where $f(x) = |x|$ and x is a real number.
   d. A function f is **additive** if for all x and y in its domain, $f(x + y) = f(x) + f(y)$. A function f is **multiplicative** if for all x and y in its domain, $f(xy) = f(x) \cdot f(y)$. Is the absolute value function additive? Why or why not? Is the absolute value function multiplicative? Why or why not?
   e. Generalize absolute value to apply to complex numbers. How does this generalization relate to the algebraic and geometric conceptions of absolute value of real numbers?
   f. How is absolute value generalized to apply to vectors of any dimension? To which conceptions of absolute value is this generalization linked?
   g. Give an application of absolute value to a situation originating outside of mathematics.

5. Generalize the average test-grade problem of this section as follows.
   a. Suppose Jane has an average of G after 4 tests. What score does she need on the 5th test to average H for all five tests?
   b. Suppose Jane has an average of G after n tests. What score does she need on the $(n + 1)$st test to average H for all $n + 1$ tests?
   c. Suppose Jane has an average of G after n tests. What average score does she need on the next m tests to average H for all $m + n$ tests?
   d. Comment on the relative difficulty of answering parts **a–c** for typical high school students.

6. Consider the Shaq–Jordan problem described in this chapter. Recall that the equation $s = \frac{30j + 1084}{41}$ gives the values of s and j for which the season averages of Shaq and Jordan would be equal.
   a. Find positive integer values of s and j that satisfy this equation. (A method is given in Chapter 5, but you are not expected to use that method. You may use trial and error.)
   b. The graph of this equation is a line. What is the meaning of the slope and s-intercept of this line?
   c. Solve this equation for j. What is the meaning of the slope and j-intercept of the line that is the graph of the resulting equation?
   d. Give the smallest integer number of points for Shaq and a larger integer number of points for Jordan in the last game that would have resulted in Shaq's having the higher season average.

7. Examine the general case of the Jordan-Shaq problem with variables defined as in the following table.

	Points So Far in Season	Number of Games Played in Season	Points in Last Game	Average Points per Game for Season
Jordan	J	n_J	j	
Shaq	S	n_S	s	

   a. Copy the table and fill in the right column.
   b. The functions f and g in this chapter are linear functions. Interpret the meaning of the slope and intercept of each of these functions.
   c. Equating the function values $\frac{S + s}{n_S} = \frac{J + j}{n_J}$, we can solve for s in terms of j or solve for j in terms of s. Do both and interpret the meaning of the slope and intercept of these functions.

8. At the beginning of the 2002 Major League Baseball season, Sammy Sosa had hit 450 home runs in his career. The major league career record for home runs is Hank Aaron's 755. Suppose Sosa plays m more seasons.
   a. How many home runs must he average per season to surpass Hank Aaron's career total?

b. Generalize part **a** in some way and discuss your generalization.

9. Consider the following classic problem.

> A substance is 99% water. Some water evaporates, leaving a substance that is 98% water. How much of the water has evaporated?

a. *Getting the initial answer.* Solve the given problem.
b. *The classic status of the problem.* Based on your answer to part **a**, indicate why you think this is called a classic problem.
c. *A numerical approach.* In answering part **a** you may have used algebra, setting up unknowns and solving equations. Answer part **a** again, this time using no algebra, but only concrete, numerical reasoning. (For example, if the original substance had 1 unit of solid stuff and 99 units water, the evaporated substance still has 1 unit of solid stuff, so . . . .)
d. *A diagrammatic approach.* Solve the problem yet again, this time using a diagram as your basic reasoning tool. (For example, a simple rectangle can be divided into two regions representing the water and the solid in the original substance. A different rectangle can represent the evaporated substance . . . .)
e. *Generalizing the problem.* Solve the problem again, but this time replace the numerical values 99% and 98% in the statement of the problem with general parameters. (This is a first step to generalizing the problem.)
f. *A functions approach.* The algebraic solution from part **e** is fully general, yet it is not fully revealing about why evaporating half the water has lowered the proportion of water only about 1%. In a sense, this solution is too general to focus on the essentials of this problem. Solve the problem again, but this time keep the specific numerical value 1% of the drop in water content part of the solution, and express the proportion of water evaporated as a function of the original proportion of water in the substance.
g. *Another functions approach.* The approach outlined in the discussion of part **f** is not the only functions approach possible. You can also express the proportion of water as a function of the absolute amount of water in the substance, letting the fixed amount of solute S = 1. Try this.
h. Summarize what you have learned from working on this problem analysis.

10. Refer to the correspondence between addition and multiplication described in this chapter. Six properties are given below. Into which column would each fall, addition or multiplication? Write the corresponding property that would go in the other column. We occasionally use different letters for the variables so that you cannot rely on the letters as guides.

a. $\frac{x^m}{x^n} = x^{m-n}$
b. $a(b + c) = ab + ac$
c. $\sqrt[n]{p} \cdot \sqrt[n]{q} = \sqrt[n]{pq}$
d. The geometric mean of p and q is $\sqrt{pq}$.
e. $(p - q) - (r - s) = (p - q) + (s - r)$
f. $\frac{x^5yz^4}{xyz^6} = \frac{x^4}{z^2} = x^4z^{-2}$

11. From the distributive property $mx + nx = (m + n)x$, using group properties, we can deduce $0x = 0$ for all x. Here is a proof:

> By the distributive property, for all real x, m, and n,
> $$mx + nx = (m + n)x.$$
> So $$mx + 0x = (m + 0)x.$$
> Since 0 is the additive identity,
> $$mx + 0x = mx.$$
> So $0x$ is the additive identity. So $0x = 0$.

Refer to the correspondence between addition and multiplication described in this chapter. Give the corresponding properties and proof for the other column.

12. Five properties of inequality are given below. Follow the directions of Problem 10. Again, different letters may be used for the variables so that you cannot rely on the letters as guides.

a. $a < b \iff a + c < b + c$
b. If $a > 0$ and $b > 0$, then $a + b > 0$.
c. If $a < 0$ and $b < 0$, then $a + b < 0$.
d. If $x > 0$ and $m < 0$, then $mx < 0$.
e. If $x < y$ and $n > 0$, then $\frac{x}{n} < \frac{y}{n}$.

ANSWERS TO QUESTIONS (IN THE CHAPTER TEXT):

3. This will happen if the point (j, s) is above the line $s = \frac{30j + 1084}{41}$ graphed in Figure 8 yet below the line $s = j$. These lines intersect at about (98.5, 98.5). Thus it is possible that Jordan could score more points than Shaq and still lose the title, but unlikely, since Shaq would have to score more than 98 points.

4. The lattice points (j, s) in the shaded region of Figure 10 indicate situations in which Jordan scores more points than Shaq in the final game yet loses the scoring title.

Figure 10

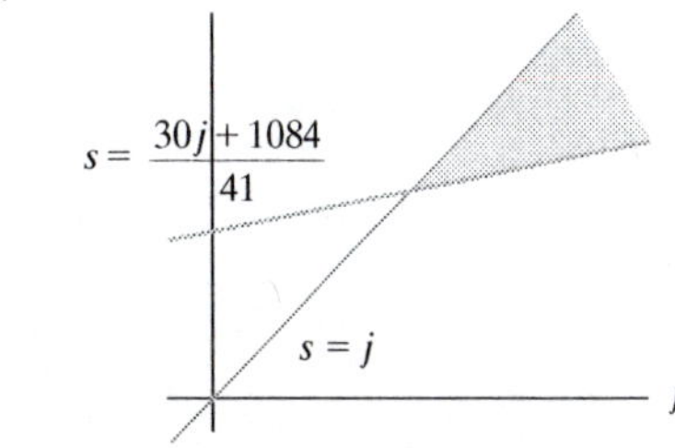

6. $(a^2b^5)^3 = a^6b^{15}$

Bibliography

Concept Analysis

Calinger, Ronald. *Vita Mathematica: Historical Research and Integration with Teaching*. Washington, DC: Mathematical Association of America, 1996.

This is a set of essays on the integration of the history of mathematics into mathematical teaching.

Freudenthal, Hans. *The Didactical Phenomenology of Mathematical Structures*. Boston: D. Reidel, 1983.

The entire book presents concept analysis of mathematical objects, together with some pedagogical implications.

Heath, Sir Thomas L. *The Thirteen Books of Euclid's Elements*. Translated from the text of Heiberg. Second edition. New York: Dover Publications, 1956.

Pages 202–220 of Part I of this classic three-part treatise contain an in-depth discussion of Euclid's parallel postulate and its influence on later mathematics.

James, Robert C., and Glenn James. *Mathematics Dictionary*. Fourth Edition. New York: Van Nostrand Reinhold, 1979.

Problem Analysis

Brown, Stephen I., and Marion I. Walter. *The Art of Problem Posing*. Philadelphia, PA: The Franklin Institute Press, 1983.

Brown and Walter offer suggestions for formulating questions about mathematical situations.

Polya, George. *How to Solve It*. Princeton, NJ: Princeton University Press, 1956.

Polya explicates his analysis of problem solving. The fourth step, "looking back", is most similar to problem analysis as used in this book.

Polya, George. *Induction and Analogy in Mathematics*. Volumes I and II. New York: Wiley, 1954.

These books are devoted to the ideas of problem analysis but at a higher mathematical level than this book.

Polya, George. *Mathematical Discovery*. Volumes I and II. New York: Wiley, 1962.

A continuation of the ideas found in Polya's earlier books.

Mathematical Connections

House, Peggy A., and Arthur F. Coxford, editors. *Connecting Mathematics across the Curriculum*, 1995 Yearbook of the National Council of Teachers of Mathematics. Reston, VA: NCTM, 1995.

This entire volume is filled with interesting ideas, with a very nice opening essay by Coxford titled "The Case for Connections."

National Council of Teachers of Mathematics. *Principles and Standards for School Mathematics*. Reston, VA: NCTM, 2000.

Pages 64–65 discuss connections broadly; pages 354–359 are devoted to connections in grades 9–12.

Usiskin, Zalman. "Some Corresponding Properties of Real Numbers and Implications for Teaching." *Educational Studies in Mathematics* 5 (1974), 279–290.

The isomorphism between $\langle \mathbf{R}, + \rangle$ and $\langle \mathbf{R}^+, \cdot \rangle$ is treated in detail.

Chapter 2

REAL NUMBERS AND COMPLEX NUMBERS

People have often co-opted words from outside mathematics to describe numbers. Consequently, students of mathematics are usually familiar with nonmathematical meanings of the words *natural, whole, real, complex, imaginary*, and *rational* before they encounter the same words as technical mathematics language. (The word *integer* is a notable exception of not having a common meaning outside mathematics.) Sometimes the mathematical terms are best understood in contrasting pairs. Thus *rational* is contrasted to *irrational, positive* to *negative*, and *real* to *imaginary*. In each case, the first of the pair has a meaning outside mathematics that conveys easier accessibility or greater utility, while the second of the pair evolved from a human tendency to view new things as strange, bad, or unreal.

The excess baggage of knowing nonmathematical meanings can affect how students view these numbers. Probably nowhere is this more pronounced than with the two terms *real* and *imaginary*. To mathematicians, an *imaginary number* is as real as a *real number*, and neither is imaginary in the nonmathematical sense of the word. But students are often influenced by these names to think that real numbers are the actual ones we work with, and to think that imaginary numbers are not numbers at all, but inventions of mathematicians to provide theoretical solutions to equations that have no utility outside mathematics.

What qualifies a mathematical object to be identified as some type of number? A simple answer is not as easy to obtain as it may first seem. Exactly what basic properties objects called "numbers" should possess can be a subject of debate. Some people would argue that telephone numbers are not numbers (in the mathematical sense) because the operations of addition and multiplication are not meaningful with them. Others would say that these are numbers whose use just happens not to employ arithmetic operations. We address that debate by introducing the idea of a *number system*. A **number system** is a set of objects together with operations (addition and multiplication and perhaps others) and relations (equality and perhaps order) that satisfy some predetermined properties (such as commutativity or the existence of an identity for an operation). With this distinction, we conclude that a telephone number such as 1-800-555-1212 is a number, but it is not a natural number because it does not possess the properties of natural numbers (for example, divisibility or the ability to be added). The Dewey Decimal library classification numbers for books also might be considered as numbers, but they are not rational numbers or real numbers because arithmetic operations are not meaningful with them.

In this chapter, we examine the numbers that make up rational, real, and complex number systems. Our approach is to start from their most familiar geometric representations, the real number line and the complex plane.

Unit 2.1 The Real Numbers

The natural numbers arose historically from the need to count. The extension of the system **N** of natural numbers to the system **Z** of integers was probably prompted by the need to maintain trade accounts. As early as 300 B.C., the Chinese and Indians used rod numerals (Figure 1).

Figure 1

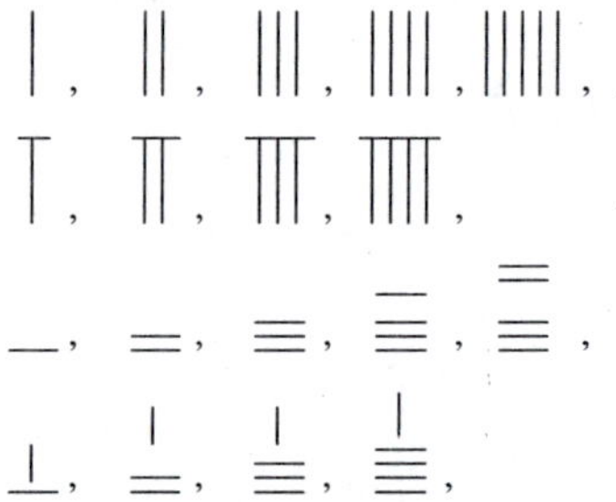

|, ||, |||, ||||, |||||, for 1 through 5,

for 6 through 9,

for 10, 20, 30, 40, and 50,

for 60, 70, 80, and 90

They used right to left positional notation for larger numbers (for example, 6221 was represented by ⊥ ||=|). They carried out commercial and governmental calculations by using rods of two different colors to distinguish between positive and negative numbers. Interestingly enough, red rods were used for positive numbers, and black rods for negative numbers, the opposite of later uses in Western countries! At first, zero was represented by an empty space in the numeral and later by a more conventional 0. These historical particulars aside, the mathematical importance of the extension from the natural numbers to the integers is that it extends the subtraction $a - b$, which is defined in **N** only for the case in which $a > b$, to arbitrary integers a and b. As a consequence, the equation $x + b = a$ is solvable in **Z** for all choices of a and b in **Z**.

The positive rational numbers were devised to help measure and compare the sizes of objects. The concept of "commensurability" of lengths was fundamental to the early development of geometry by the Greeks. The meaning of commensurability can be explained as follows: Suppose that A and B are two line segments. Then A and B are commensurable if there exist positive integers a and b such that $\frac{\text{length of } A}{\text{length of } B} = \frac{a}{b}$. The ancients used such ratios of positive integers (i.e., what evolved into today's positive rational numbers) not only to compare objects of different lengths but also different weights, areas, and so on.

The measurement of lengths, areas, and volumes of geometric objects became the primary objective of geometric analysis in antiquity. In the sixth century B.C., Pythagoras of Samos and his followers formalized and taught a philosophy of mathematics based on measurement and number ratios that had a deep influence on the evolution of mathematics in Greece for over one hundred years.

The Pythagoreans originally believed that the lengths of all segments in geometric objects are commensurable, and therefore the positive rational numbers are adequate for all measurement purposes. However, the Pythagoreans themselves discovered later that the length s of a side and the length d of a diagonal of a square were not commensurable; that is, they showed that there do not exist natural numbers m and n such that $\frac{d}{s} = \frac{m}{n}$ or, equivalently, $ms = nd$ (see Figure 2).

Figure 2

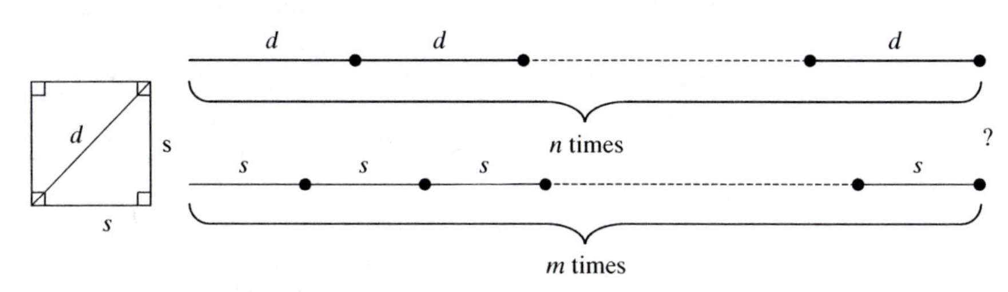

Question 1: Explain why, in this situation, $\frac{d}{s} = \frac{m}{n}$ is equivalent to $\sqrt{2} = \frac{m}{n}$.

As a result of this unexpected discovery, a major problem developed for the Greek geometers who were rigorously proving theorems. The validity of all theorems involving ratio and proportion was suddenly called into question. This discovery also raised the question of how the lengths of incommensurable segments could be compared. Eudoxus of Cnidus developed a new theory of proportions (360 B.C.), and this theory was incorporated in Euclid's *Elements* (c. 300 B.C.). The contributions of Eudoxus notwithstanding, the subsequent evolution of mathematics in Greece reflected a certain suspicion of measurement and number.

Because $d^2 = 2s^2$ for an isosceles right triangle by the Pythagorean Theorem, the statement that there do not exist natural numbers m and n such that $\frac{d}{s} = \frac{m}{n}$ is equivalent to the statement that there do not exist natural numbers m and n such that $\sqrt{2} = \frac{m}{n}$. In today's language, we say that $\sqrt{2}$ is not a rational number, that is, that $\sqrt{2}$ is an irrational number.

The incommensurability of the lengths of the sides and the diagonal of a square showed that, although the rational numbers arose from the need to measure objects, they are not adequate for that purpose. This inadequacy led very slowly to the use of the broader system of real numbers as the number system for measurement. Although irrational numbers were used in calculations and the solution of algebraic equations, the evolution of the real numbers as an algebraic system did not occur until the nineteenth century. At that time, the evolution of the theory of real functions from its roots in calculus finally necessitated a careful study of the foundations of analysis. This study was independently initiated by Bernhard Bolzano (1781–1848), a priest from Czechoslovakia, and the French mathematician Augustin-Louis Cauchy (pronounced Co′-shee) (1789–1857). The German mathematician Karl Weierstrass (1815–1897) sought to base analysis solidly on a number foundation without appealing to its connections with geometry. But it was the work of Richard Dedekind (1831–1916) and Georg Cantor (1845–1918) of Germany and Charles Méray (1835–1911) of France that finally established precise definitions for the real number system based on the system **Q** of rational numbers.

The comparatively slow development of the real numbers as a number system was due in part to the lack of a representation for irrational numbers that was conducive to calculation. The use of decimal representation by Simon Stevin (1548–1620)[1] was a

---

[1]Stevin did not invent decimal representation. In fact, a form of decimal representation was used by the ancient Chinese, and mathematical researchers of Stevin's time used it. However, the publication of Stevin's very popular book on decimal representation in 1585 resulted in widespread use of decimal representation among engineers, scientists, and other users of mathematics. The decimal point, to separate the integer and fractional parts of a real number, is first found in the 1619 posthumous publication of John Napier, *Mirifici Logarithmorum Canonis Constructio*, in which he describes how he constructed his tables of logarithms. (In Europe, a comma is still commonly used for this separation.)

great stride forward in this regard. However, another factor that contributed to the slow development of the theory of the real number system was that precise algebraic definitions of *real number* were not given until the years 1869–1879. Méray (1869) and Cantor (1879) constructed real numbers using sequences of rational numbers. Weierstrass (1872) defined real numbers in terms of infinite decimals. In the same year, Dedekind published a treatise in which the real numbers were defined on the basis of partitions of the set **Q** of rational numbers, partitions that are now known as Dedekind cuts. In that treatise, he also developed the algebraic structure of the real number system on the basis of these cuts.

Today there are two basic approaches that can be taken to the theory of real numbers. One is top-down, to assume that there exists a number system that has the properties of a complete ordered field. We take this approach in Chapter 6. The other approach is bottom-up, to construct the real numbers from the rational numbers as done by the mathematicians identified above. We do not carry out all the details of such a construction in this book. But we convey some aspects of the bottom-up approach in this chapter because it helps to illuminate the relationships among (1) the real numbers; (2) decimals, their most common representation; and (3) the geometry of the number line. In Chapter 6 we show how viewing the real numbers as an ordered field relate to (1), (2), and (3).

2.1.1 Rational numbers and irrational numbers

Rational numbers

When we think of rational numbers, we may think of *ratios*, the origin of their name.

Definition A number is **rational** if and only if[2] it can be written as the indicated quotient of two integers.

Numbers written as indicated quotients of two integers, such as $\frac{14}{3}$ or $\frac{-6}{15}$, are rational from the definition. As an immediate consequence of the definition, any integer k is a rational number, because it can be written as $\frac{k}{1}$, the quotient of two integers.

The indicated quotient of a divided by b may be denoted by a slash (a/b), a bar $(\frac{a}{b})$, or a division sign $(a \div b)$. In some countries, a colon $(a:b)$ is used. Either of a/b or $\frac{a}{b}$ is a **fraction**. The bar in $\frac{a}{b}$ also serves as a vinculum, a parenthetical grouping symbol, just as is found in the radical sign $\sqrt{}$. For this reason, $16 + 8/2 + 6$, which has no grouping symbol, equals $16 + 4 + 6$ or 26, while $\frac{16 + 8}{2 + 6}$ equals 3. To make the slash act like the fraction bar, parentheses need to be used: $(16 + 8)/(2 + 6) = 3$.

Although every fraction whose numerator and denominator are integers is rational, *a rational number is not the same as a fraction*. The definition of "rational number" does not require that a rational number *must* be written as a quotient of integers, only that it *can* be written that way. Consider the number one-half. It can be written as the fraction $\frac{1}{2}$, or in infinitely many other forms, including notations as diverse as $0.5, \frac{6}{12}, \sin 30°, \log \sqrt{10}, 2^{-1}, 64^{-1/6}$, and $\frac{\pi}{2\pi}$. Of these, only $\frac{6}{12}$ and $\frac{\pi}{2\pi}$ are fractions, and only $\frac{6}{12}$ is an indicated quotient of integers. So rational numbers are not determined by how they look, but by how they *can* look.

[2]Some mathematicians use "if" rather than "if and only if" when it is clear that a word is being defined. For further discussion, see Section 7.1.3.

The various ways of representing $\frac{1}{2}$ give sufficient evidence of the variety of forms in which rational numbers may be written and lead to a natural question: How do we know that -5.4322986, $\sqrt{12}\tan\frac{\pi}{3}$, and $\ln e^{9/8}$ are rational? (They all are.) There is no general procedure, only a requirement. We need to show that each is equal to a quotient of two integers.

Question 2: Write $\frac{\frac{4}{3}}{\frac{7}{2}}$, -5.4322986, $\sqrt{12}\,tan\,\frac{\pi}{3}$, and $\ln e^{9/8}$ as quotients of two integers to show that each is a rational number.

There are infinitely many ways to denote any particular positive rational number even if we insist on writing it as a quotient of two positive integers. For instance, $\frac{4}{3} = \frac{40}{30} = \frac{12}{9} = \frac{124}{93} = \cdots$. But only one of these fractions has the property that its numerator and denominator have no common integer factor greater than 1. This is the fraction that we pick when we say the rational number is **in lowest terms**. Two positive integers are **relatively prime** if they have no common integer factor greater than 1. Thus a fraction in lowest terms has a numerator and denominator that are relatively prime. A negative rational number such as $\frac{-7}{3}$ or $\frac{14}{-6}$ is in lowest terms if the absolute values of the numerator and denominator are relatively prime. Thus either $\frac{-7}{3}$ or $\frac{7}{-3}$ is in lowest terms. Some people prefer positive denominators and would consider $\frac{-7}{3}$ alone to be in lowest terms.

Operations on rational numbers

Why do we care whether a number is rational? One reason is that the algorithms we have for operations with fractions make rational numbers easy to add, subtract, multiply, and divide. For instance, it is almost always easier to multiply by sin 30° than to multiply by sin 40°, because the first of these equals $\frac{1}{2}$. And, as the following theorem shows, the sum, difference, product, and quotient of two rational numbers is itself a rational number (provided we do not divide by zero). [In the statement of part (b), $\mathbf{Q} - \{0\}$ means the set of rational numbers with 0 removed. In general, if B is a subset of A, then $\mathbf{A-B}$ is the set of elements in A that are not in B.]

Theorem 2.1

a. The set **Q** of rational numbers is closed under addition, subtraction, and multiplication.

b. The set $\mathbf{Q} - \{0\}$ of nonzero rational numbers is closed under division.

Proof:

a. We need show only that the sum, difference, and product of two arbitrary rationals is rational.

Let p and q be the two rational numbers. Since they are rational, there exist integers a, b, c, and d with $p = \frac{a}{b}$ and $q = \frac{c}{d}$, and $b \neq 0$ and $d \neq 0$. Then $p + q = \frac{a}{b} + \frac{c}{d} = \frac{ad + bc}{bd}$. Since the sum and product of two integers is an integer, both $ad + bc$ and bd are integers, and $bd \neq 0$ because neither b nor d is 0. Thus $p + q$ is the quotient of two integers and is rational.

The proofs of the rest of part (a) and part (b) are left to you. ┘

Estimating rational numbers

In 1978, the following multiple-choice problem was given to a random sample of 13-year-olds participating in the National Assessment of Educational Progress (NAEP).

Problem:	Estimate $\dfrac{12}{13} + \dfrac{7}{8}$.	
Choices:	A	1
	B	2
	C	19
	D	21
	E	I don't know

Only 24% selected the correct choice (B). We might conclude that 13-year-olds of that era did not know how to add fractions, but on the Second International Study of Mathematics Achievement in 1981, 84% of 8th graders correctly added two fractions with different denominators in a situation that was not multiple choice. Most researchers have concluded from these examples that the NAEP students tried to answer the above question by adding the fractions blindly without having any idea of the size of the numbers. But then, when they obtained a sum, they didn't know how to connect it with the choices!

Estimating a positive rational number $\frac{a}{b}$ to the nearest integer is easy if it is written as a *mixed number*. A **mixed number** is the sum of an integer and a fraction between 0 and 1, typically written with no space between them. For instance, the mixed number $32\frac{4}{5} = 32 + \frac{4}{5}$. In this case 32 is called the **integer part** of the rational number, and $\frac{4}{5}$ is its **fractional part**.

We can obtain an estimate of the size of a rational number by writing it as a mixed number. For instance, when a car is driven 453 miles between gasoline fill-ups, and it takes 16.8 gallons to fill the tank, the fuel efficiency in miles per gallon is $\frac{453}{16.8}$, a rational number that means little until we determine its integer part. Recall that $\lfloor t \rfloor$ denotes the greatest integer less than or equal to t. When we write a positive rational number t as a mixed fraction, the integer part is $\lfloor t \rfloor$.

To determine $\lfloor t \rfloor$ for a given positive number t, we divide. In the above case, we would divide 16.8 into 453. One way to do this is to multiply both divisor and dividend by 10 to obtain integers, and divide 168 into 4530. To obtain the integer part of the quotient, we may repeatedly subtract 168 from 4530, or we might use long division, or today most people would use a calculator. Regardless of the process, we find that $\lfloor \frac{4530}{168} \rfloor = 26$. This indicates that $\frac{4530}{168} = 26 + \frac{r}{168}$, where r is the remainder that needs to be determined. Multiplying both sides by 168, $4530 = 168 \cdot 26 + r$, from which $r = 162$. So $\frac{4530}{168} = 26\frac{162}{168} = 26\frac{27}{28}$.

The theorem about integers that guarantees a unique quotient and unique remainder in this process is called the *Division Algorithm*. We deduce the Division Algorithm from properties of natural numbers and discuss it in detail in Section 5.2.1 but state it here since we use it repeatedly in this chapter.

Theorem 5.3 **(Division Algorithm):** If a and b are integers and if $b > 0$, then there exist unique integers q and r such that $a = bq + r$ and $0 \le r < b$.

Dividing both sides by b, we obtain the form of the Division Algorithm that we used above.

Corollary (Alternate Form of Division Algorithm): If a and b are integers and if $b > 0$, then there exist unique integers q and r such that $\frac{a}{b} = q + \frac{r}{b}$, with $0 \le r < b$.

In some computer languages, q and r are denoted as **a div b** and **a mod b**. Notice in the corollary that since $r < b$, we have $\frac{r}{b} < 1$, and so $q = \lfloor \frac{a}{b} \rfloor$ if $a > 0$. In this way, the Division Algorithm explains why every rational number t is either an integer or between two consecutive integers, shows how those integers can be calculated, and also determines how to write t as a mixed number.

For instance, if $a = -164$ and $b = 5$, then we are looking for q and r with $-164 = 5q + r$ and $0 \leq r \leq 5$. $q = \lfloor \frac{-164}{5} \rfloor = \lfloor -32\frac{4}{5} \rfloor = -33$, from which $r = 1$. Thus $\frac{-164}{5}$ is between -33 and -32, and $\frac{-164}{5} = -32\frac{4}{5}$. So we say the integer part of $-32\frac{4}{5}$ is -33, and the fractional part is $\frac{1}{5}$. (Some people say the integer part of $-32\frac{4}{5}$ is -32, and the fractional part is $-\frac{4}{5}$.)

In the next section, we show how this process also enables us to write rational numbers as decimals. But now we turn to irrational numbers.

Irrational numbers

An **irrational number** is a real number that is not a rational number.

Today, irrational numbers are found throughout the study of middle school and high school mathematics. They include the common logarithms of all positive integers except the integer powers of 10 and the natural logarithms of all positive integers greater than 1; the sines, cosines, and tangents of all integer degrees except for some divisible by 15°; the square roots and cube roots of most of the positive integers; and, of course, π and all its rational multiples. They are lengths of segments in geometry, coefficients in formulas for area and volume of common figures, and values of some of the most important functions in mathematics. Pick the coefficients a, b, and c of the quadratic equation $ax^2 + bx + c = 0$ at random and you are more likely to have solutions that are not rational than to have rational solutions. Indeed, as we show in Section 2.1.3, there are far more irrational numbers than rational numbers.

The existence of irrational numbers

In this book, we give many proofs of the existence of irrational numbers. We show:

Square roots of positive integers are either positive integers or irrational. (Theorem 2.2)

Numbers represented by infinite nonrepeating decimals are irrational. (Section 2.1.2)

e is irrational. (Section 2.1.3)

The number of irrational numbers is infinite and not countable. (Section 2.1.4)

Roots of many polynomial equations are irrational. (Section 2.1.4)

We also discuss the irrationality of other specific numbers, such as π.

Our first proof shows that the square root of any positive integer that is not a perfect square is an irrational number. It relies on two facts: (1) Except for the order of the factors, every integer has a unique factorization into primes. This statement, known as the Fundamental Theorem of Arithmetic, is proved in Section 5.2.4. (2) If integers a and b are relatively prime (i.e., have no common integer factors >1), then a^2 and b^2 are relatively prime. Fact (2) follows quite quickly from (1).

Theorem 2.2 Let n be a positive integer. Then $\sqrt{n}$ is either an integer or it is irrational.

Proof: Obviously the proof has to take into account the meaning of "square root" and of "irrational". The proof is indirect. Suppose n is a positive integer, $\sqrt{n}$ is not an integer, and $\sqrt{n}$ is rational. Then there exist relatively prime integers a and

b with $\sqrt{n} = \frac{a}{b}$. Squaring both sides (using the definition of "square root"), $n = \frac{a^2}{b^2}$, from which $nb^2 = a^2$. Now if we factor a and b into primes, there are no common factors. So the factorizations of a^2 and b^2 have no common factors. Consequently, the factorizations of the equal numbers nb^2 and a^2 are different. Since two different factorizations of a^2 are impossible by the Fundamental Theorem of Arithmetic, the supposition must be false. So if $\sqrt{n}$ is not an integer, it must be irrational. ┘

As a result of Theorem 2.2, numbers such as $\sqrt{83}$ and $\sqrt{21}$ are irrational.

Theorem 2.2 can be considered to be a theorem about solutions to equations. It is equivalent to asserting that if p is not a perfect square, the polynomial equation $x^2 - p = 0$ has no rational solutions. This, in turn, is a special case of the *Rational Root Theorem* found in many textbooks.[3]

Once a number has been shown to be irrational, it can be used to produce many other irrational numbers.

Theorem 2.3 Let s be any nonzero rational number and v be any irrational number. Then $s + v$, $s - v$, sv, and $\frac{s}{v}$ are irrational.

Proof: The proof is indirect. Suppose s and v are as given and $s + v$ is rational. Then $(s + v) - s$ is rational from Theorem 2.1(a). But $(s + v) - s = v$, which is irrational. This contradiction shows that $s + v$ must be irrational. The other parts of the theorem can be proved in a similar fashion. ┘

Theorems 2.2 and 2.3 together show that such numbers as $2 + \sqrt{3}$ and $\dfrac{-14}{2 + \frac{3}{5}\sqrt{7}}$ are irrational.

Operations on irrational numbers

On occasion, irrational numbers are nicely related. For instance, the product of two square roots of positive integers is another square root of a positive integer, as can be seen from the identity $\sqrt{a} \cdot \sqrt{b} = \sqrt{ab}$. But this does not work for sums of square roots. Even as simple a sum as $\sqrt{2} + \sqrt{3}$ does not equal the square root of an integer. Relationships among irrationals are always tied to their origins. For instance, an identity such as $\log(xy) = \log x + \log y$ may involve irrational numbers, but its truth is traceable back to properties of logarithms, not to properties of irrational numbers.

Indeed, sums, differences, products, and quotients of irrational numbers may be rational or irrational. So the set **I** of irrational numbers is not closed under any of the operations of arithmetic.

---

[3]The Rational Root Theorem: If $p(x) = a_n x^n + a_{n-1}x^{n-1} + \cdots a_1 x + a_0$ is a polynomial with integer coefficients, and $\frac{a}{b}$ is a rational solution to $p(x) = 0$ in lowest terms, then a is a factor of a_0, and b is a factor of a_n.

2.1.1 Problems

1. Write each number as the quotient of two integers. If the number can be written as a mixed number, identify its integer part and fractional part in lowest terms.

a. 3.14159

b. $\frac{35789.22}{47.6}$

c. $\log_{10} \sqrt[3]{100}$

d. $-6\frac{3}{4}$

2. Give examples to show that the following statements are not true for all positive rational numbers x and y.

a. $\lfloor x + \frac{1}{10} \rfloor = \lfloor x \rfloor$

b. $\lfloor x \rfloor + \lfloor y \rfloor = \lfloor x + y \rfloor$

c. $\lfloor x \rfloor - \lfloor y \rfloor = \lfloor x - y \rfloor$

3. Expressions may look irrational yet still be rational.

a. Write $\sqrt{3+\sqrt{7}} - \sqrt{8-2\sqrt{7}}$ as the quotient of two integers.

b. Make up another example of the same type as part **a**.

c. Generalize parts **a** and **b**.

4. a. Prove Theorem 2.1(a) for subtraction.

b. Prove Theorem 2.1(a) for multiplication.

c. Prove Theorem 2.1(b).

5. a. Is the set of rational numbers closed under the operation of exponentiation? That is, for all rational numbers a and b, is a^b rational? Why or why not?

b. Is the set of irrational numbers closed under the operation of exponentiation?

6. Can you use the process in the proof of Theorem 2.2 to prove that the square root of 25 is irrational? If so, show how. If not, indicate where the proof breaks down.

7. Modify the proof of Theorem 2.2 to show that the positive nth root of any prime is irrational.

8. Prove or disprove the claim of a student that the sides of any right triangle can be written in the form $\sqrt{a}$, $\sqrt{b}$, and $\sqrt{a+b}$.

9. Let s be a nonzero rational number and v be irrational.

a. Prove that $s - v$ is irrational.

b. Prove that sv is irrational.

c. Prove that $\frac{s}{v}$ is irrational.

10. a. Give an example of two different irrational numbers whose sum is a rational number.

b. Give an example of two different irrational numbers v_1 and v_2 such that $\frac{v_1}{v_2}$ is rational.

11. a. Prove that if a, b, and c are integers and $\sqrt{b^2-4ac}$ is not an integer, then $\frac{-b \pm \sqrt{b^2-4ac}}{2a}$ is irrational.

b. Give a specific example of a quadratic equation whose solutions are proved irrational by applying the result of part **a**.

12. Although $\sqrt{2}+\sqrt{3}$ does not equal the square root of an integer, $\sqrt{27}+\sqrt{48}$ does.

a. What integer's square root equals $\sqrt{27}+\sqrt{48}$, and why?

b. Make up another example like $\sqrt{27}+\sqrt{48}$.

c. Find every set of different positive integers p, q, and r all less than 100 such that p and q are not perfect squares and $\sqrt{p}+\sqrt{q}=\sqrt{r}$.

13. a. Consider $\sqrt{n}$, where n is an integer and $1 \le n \le 10$. How many of these ten numbers are irrational?

b. An integer n is randomly chosen from 1 to k^2, where k is an integer. What is the probability that n is a perfect square? What is the probability that $\sqrt{n}$ is irrational?

ANSWERS TO QUESTIONS

1. Suppose m and n are natural numbers. Since $d^2 = 2s^2$, $\frac{d^2}{s^2} = 2$. Since d and s are (positive) lengths, $\sqrt{\frac{d^2}{s^2}} = \sqrt{2}$, from which $\frac{d}{s} = \sqrt{2}$. So if $\frac{d}{s} \neq \frac{m}{n}$, then $\sqrt{2} \neq \frac{m}{n}$. 2. $\frac{8}{21}$, $\frac{-54322986}{10000000}$, $\frac{6}{1}$, $\frac{9}{8}$

2.1.2 The number line and decimal representation of real numbers

In school algebra, real numbers are commonly described as numbers that can be represented by finite or infinite decimals. In geometry, they are introduced as numbers that are in one-to-one correspondence with the points on a line. In higher mathematics, real numbers may be defined in terms of rational numbers by least upper bounds, sums of infinite series, nested intervals, or Dedekind cuts. In this section, we connect these more advanced ideas to decimals and the number line. Our approach is to begin with the number line, describe rational numbers as series of the form $\sum_{i=-\infty}^{n} a_i b^i$ with $b = 10$, and then use the Nested Interval Property of the number line to obtain real numbers as decimals.

The number line

Although we often think of the arithmetic/algebra aspect of mathematics as being separate from the geometric aspect of mathematics, there is a fundamental interplay between the two. Arithmetic and algebra provide models for geometry, and vice versa. Geometry can be modeled algebraically through coordinates or vectors or complex numbers, and these algebraic models can lead to new geometric insights. In turn, geometric representations of real and complex numbers play important roles in our understanding of these numbers.

The *number line* or *real line* is a geometric model or representation of the system $\mathbf{R}$ of real numbers. In this model, we begin with a straight line and select two points on that line that represent the integers 0 and 1, as in Figure 3.

Figure 3

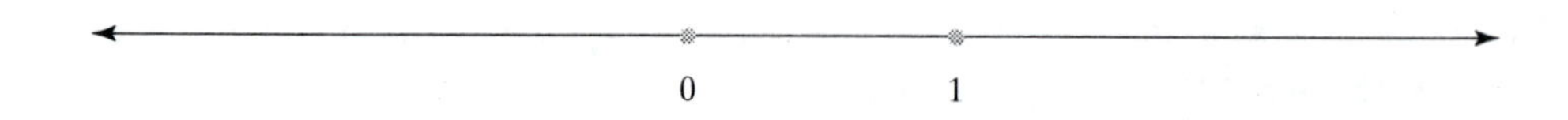

Then we represent the successive positive integers 2, 3, 4, ... by equally spaced points to the right of 1, and the successive negative integers by equally spaced points to the left of the point 0.

A positive rational number x that is not an integer can be represented by thinking of it as a mixed number. Suppose $x = \frac{a}{b}$, where a and b are integers; then $x = \frac{a}{b} = q + \frac{r}{b}$, where $q = \lfloor x \rfloor$ and $0 < r < b$. On the number line, x is represented by the point that divides the segment from q to $q + 1$ in the ratio $\frac{r}{b}$. For example, the rational number $\frac{18}{7} = 2 + \frac{4}{7}$. So $\frac{18}{7}$ is represented by the point between 2 and 3 on the number line that divides the segment from 2 to 3 in the ratio $\frac{4}{7}$. A negative rational number x is represented by the point on the other side of 0 at the same distance from 0 as the point representing $|x|$ (Figure 4).

Figure 4

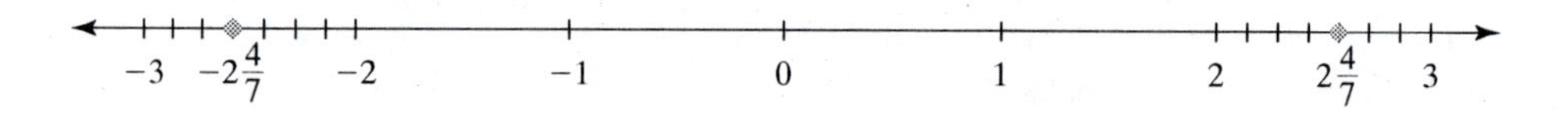

This number line model also allows us to represent irrational numbers by points on the number line. For example, we know that $\sqrt{2}$ is the length of either diagonal of a square with side length equal to 1. Consequently, the point on the real line corresponding to $\sqrt{2}$ can be constructed as the intersection of the real line and the arc centered at $(0, 0)$ and containing $(1, 1)$, as indicated in Figure 5.

Figure 5

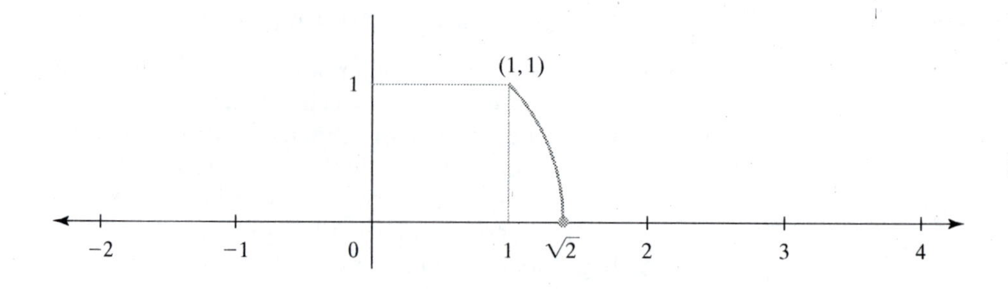

In this way, if an irrational number can be identified with a length, we can find the point on the number line corresponding to it.

This geometric description of the system $\mathbf{R}$ of real numbers is quite adequate for many purposes and is the basis of much of our intuition about real numbers. Most of us find this model to be natural and intuitive because of our experience with concrete scales and measuring devices such as thermometers and rulers. In this section and the next, we use this model of the real number system $x \in \mathbf{R}$ to describe its features. In Section 2.1.4, we outline how the system of real numbers can be defined without reference to the number line model and indicate why it was necessary to develop such definitions.

Intervals

An **interval** of numbers is a set containing all numbers between two given numbers together with one, both, or neither of the given numbers. Therefore, an interval can be modeled on the number line by a segment or ray with or without its endpoints. An interval is **closed** if it contains its endpoints and **open** if it does not contain them. The **length** of an interval with endpoints a and b and with $a < b$ is $b - a$. Interval notation and terminology are summarized in Table 1. (Read "$\in$" as "is an element of", $+\infty$ as "positive infinity", and $-\infty$ as "minus infinity".)

Table 1 Intervals

Name of interval	notation	inequality description	number line representation
finite, open	(a, b)	$a < x < b$	a b
finite, closed	$[a, b]$	$a \le x \le b$	a b
finite, half-open	$(a, b]$ $[a, b)$	$a < x \le b$ $a \le x < b$	a b a b
infinite, open	$(a, +\infty)$ $(-\infty, b)$	$a < x < +\infty$ $-\infty < x < b$	a b
infinite, closed	$[a, +\infty)$ $(-\infty, b]$	$a \le x < +\infty$ $-\infty < x \le b$	a b

Intervals can be used to describe the solution sets of equations and inequalities, the domains of functions, and bounds for estimates. For example, the solution set S of the inequality $x^2 - 6 \ge x$ is the set of all real numbers x with $x \ge 3$ or $x \le -2$, because

$$x^2 - 6 \ge x \iff x^2 - x - 6 \ge 0 \iff (x - 3)(x + 2) \ge 0 \iff x \ge 3 \text{ or } x \le -2.$$

(Read "$\iff$" as "if and only if.") S can be described in interval and in set notation.

$$S = (-\infty, -2] \cup [3, +\infty) = \{x \in \mathbf{R}: x \le -2 \text{ or } x \ge 3\}$$

As another example, the domain D of the tangent function can be described as

$$D = \left\{x \in R: x \ne (2k + 1)\left(\frac{\pi}{2}\right) \quad \text{for all integers } k\right\},$$

or as the union of all open intervals between successive odd multiples of $\frac{\pi}{2}$; that is,

$$D = \bigcup_{k \in Z}\left([2k - 1]\frac{\pi}{2}, [2k + 1]\frac{\pi}{2}\right).$$

As still another example, if the diameter of a rod must be within .05 mm of a desired value of 1.45 cm, then the diameter must lie on the closed interval [1.445 cm, 1.455 cm].

Apart from applications such as these, intervals are also used to define decimal representation and to describe some of the most basic properties of the real number system.

What is a decimal?

We can classify decimals into two categories: finite (terminating) decimals such as 2.25 and 0.4 and infinite (nonterminating) decimals such as $4.2\overline{3}$ or the decimal representation of $\pi = 3.141592\ldots$. We first define finite decimals. To formulate a proper definition, think about how you have used decimals in the past. A real number such as $3\frac{5}{8}$ is written in decimal notation as 3.625 because $3\frac{5}{8} = 3 + \frac{6}{10^1} + \frac{2}{10^2} + \frac{5}{10^3}$. The number 3.625 is a *finite decimal* representing $3\frac{5}{8}$. Here is a general definition.

Definition If a nonnegative real number x can be expressed as a (finite) sum of the form $x = D + \frac{d_1}{10^1} + \frac{d_2}{10^2} + \cdots + \frac{d_k}{10^k} = D + d_1 \cdot 10^{-1} + d_2 \cdot 10^{-2} + \cdots + d_k \cdot 10^{-k}$, where D and each d_n are nonnegative integers and $0 \le d_n \le 9$ for $n = 1, 2, \ldots, k$, then $D.d_1d_2\ldots d_k$ is the **finite decimal** representing x.

We say that D is the **integer part**, $.d_1d_2\ldots d_k$ is the **decimal part**, and d_i is the ***i*th decimal place** of the decimal.[4] The integer part is the greatest integer less than or equal to x.

If x is a negative real number and there is a finite decimal $D.d_1d_2\ldots d_k$ representing $-x$, then we write $-(D.d_1d_2\ldots d_k)$ or, more simply, $-D.d_1d_2\ldots d_k$ for the finite decimal representing x. In this case, the integer part is $-D - 1$. For example, if $x = -\frac{11}{8}$, then $x = -(1 + \frac{3}{8}) = -(1.375) = -(1 + \frac{3}{10} + \frac{7}{10^2} + \frac{5}{10^3})$. Thus -1.375 is the finite decimal representing x, with integer part -2, because $D = 1$ and so $-D - 1 = -2$.

Question 1: Give the values of D and the d_i for $98\frac{43}{3200}$.

Question 2:

a. Explain why every finite decimal represents a rational number.

b. Show by example that there are rational numbers that do not have finite decimal representations.

An **infinite decimal** is an infinite sequence $d = [D, d_1, d_2, d_3, \ldots, d_n, \ldots]$ of integers such that $0 \le d_k \le 9$ for all k in N. An infinite decimal d is usually written in the form

$$d = D.d_1d_2d_3\ldots d_n\ldots.$$

Every finite decimal $D.d_1d_2\ldots d_k$ can be regarded as an infinite decimal by identifying it with the infinite sequence $d = [D, d_1, d_2, \ldots, d_k, 0, 0, 0, \ldots, 0, \ldots]$. (See Problem 2 at the end of this section.)

How does a decimal determine a real number?

Think of the decimal for the number π, the circumference of a circle with unit diameter. If only the first two decimal places are known, that is, $\pi = 3.14\ldots$, then we know that π is in the closed interval $[3.14, 3.15]$, an interval of length 10^{-2}. Each succeeding decimal place places π in an interval of length $\frac{1}{10}$ the preceding interval. To five decimal places, $\pi = 3.14159\ldots$, which places π in the interval $[3.14159, 3.14160]$, an interval of length 10^{-5}. This interval is a subset of the preceding intervals and is said to be *nested* in each preceding interval.

---

[4] Historically, x was called a *decimal fraction*. Most books today avoid this vocabulary because the words "decimal" and "fraction" refer to representations of the number, not the number itself.

Definitions An interval I is **nested** in another interval J if and only if I is a subset of J, that is, $I \subseteq J$. A sequence $\{I_k\}$ of intervals is called a **nested sequence** if and only if $I_{k+1} \subseteq I_k$ for all k.

The determination of a single real number follows from the *Nested Interval Property* of the real numbers, which we assume.

Nested Interval Property: For any sequence of finite closed nested intervals, there is at least one point that belongs to all of them.

The Nested Interval Property is an assumed geometric property of the number line model of the real number system. From it, many important properties of the real numbers and real functions can be derived. These include the identification of the real number determined by a given decimal and rational approximations of real numbers. For instance, consider the nested sequence of closed intervals:

$$[3.1, 3.2], [3.14, 3.15], [3.141, 3.142], [3.1415, 3.1416], \ldots, [p_k, p_k + 10^{-k}], \ldots,$$

where p_k is the rational number whose finite decimal representation consists of the first k places of the decimal $\pi = 3.141592\ldots$. The Nested Interval Property asserts that there is at least one point that belongs to all of these intervals. Moreover, because the length of the kth interval in this sequence is 10^{-k}, at most one point can belong to all of these intervals. (If there were two points, the distance between them would be larger than 10^{-k} for some k, so they could not both be in all the $[p_k, p_k + 10^{-k}]$.) That unique point is the real number π.

More generally, if $\{I_k\}$ is any nested sequence of closed intervals with rational endpoints whose length decreases to 0, there is one and only one point that belongs to all of the intervals in the sequence. That point is the **real number determined by the sequence $\{I_k\}$**. There are many different nested sequence of closed intervals with rational endpoints that determine the same real number. For example, the nested sequences $\{I_k\}$ and $\{J_k\}$ defined by

$$I_k = [p_k - 10^{-k}, p_k + 10^{-k}] \quad \text{and} \quad J_k = \left[p_k, p_k + \frac{1}{k}\right]$$

also determine the real number π. The important thing about nested sequences of closed intervals with rational endpoints with lengths decreasing to 0 is that each such sequence corresponds to exactly one real number, and that number is represented on the number line by the one and only point common to all of the intervals.

Notice that every rational number x is a real number determined by such a sequence, for we need only take $I_n = [x, x + 10^{-n}]$ to obtain a sequence $\{I_k\}$ of nested closed intervals with rational endpoints that all contain x and only x.

Now consider the decimal $D.d_1d_2d_3\ldots d_n\ldots$ (finite or infinite). For each natural number k, define x_k to be the rational number represented by the finite decimal $D.d_1d_2d_3\ldots d_k$, and let $I_k = [x_k, x_k + 10^{-k}]$ be the closed interval with left endpoint x_k and length 10^{-k}. The closed intervals I_k are nested. Because the length of I_k is 10^{-k}, and 10^{-k} decreases to 0 as k increases, there is one and only one point x on the number line that belongs to all of the intervals I_k. This unique x is the **real number determined by the decimal** $D.d_1d_2d_3\ldots d_n\ldots$, and we say that $D.d_1d_2d_3\ldots d_n\ldots$ is a **decimal representation** of x.

EXAMPLE 1 Find the first five terms of a nested sequence $\{I_k\}$ of intervals for the finite decimal 3.625.

Solution Since we are given the decimal 3.625, the first five intervals of $\{I_k\}$ can be found merely by examining 3.625. They are

$$[3.6, 3.7], [3.62, 3.63], [3.625, 3.626], [3.6250, 3.6251], [3.62500, 3.62501].$$

The left endpoint of all of these intervals from I_3 onward is $3\frac{5}{8}$, which is the real number represented by a finite decimal 3.625.

EXAMPLE 2 Find the first five terms of the nested sequence $\{I_k\}$ of intervals for the decimal for $\sqrt{2}$.

Solution We know that $1 < \sqrt{2} < 2$. The first decimal place is the number d_1 from 0 to 9 such that

$$1 + \frac{d_1}{10} \le \sqrt{2} \le 1 + \frac{d_1 + 1}{10}.$$

Since d_1 can only be one of ten values, we can just try them all, and test by squaring the numbers. We find

$$1 + \frac{4}{10} \le \sqrt{2} \le 1 + \frac{5}{10}.$$

In decimal notation,

$$1.4 \le \sqrt{2} \le 1.5.$$

We proceed in the same way to find the second decimal place d_2, which satisfies

$$1 + \frac{4}{10} + \frac{d_2}{100} \le \sqrt{2} \le 1 + \frac{4}{10} + \frac{d_2 + 1}{100}.$$

Again there are only ten possible values and we find $d_2 = 1$ because

$$1 + \frac{4}{10} + \frac{1}{100} \le \sqrt{2} \le 1 + \frac{4}{10} + \frac{2}{100}.$$

That is,

$$1.41 \le \sqrt{2} \le 1.42.$$

Thus, for the decimal d representing $\sqrt{2}$, the first two intervals of $\{I_k\}$ are $[1.4, 1.5]$ and $[1.41, 1.42]$. With a calculator, we can find the next three intervals. They are $[1.414, 1.415]$, $[1.4142, 1.4143]$, and $[1.41421, 1.41422]$.

Question 3: Explain why the real number $\sqrt{2}$ is not an endpoint (left or right) of any of the intervals in the sequence $\{I_k\}$ whose first 5 terms are listed in Example 2.

The method of Example 2 enables us to find a decimal for any real number x that can be compared to rational numbers. Suppose $x > 0$, and let $D = \lfloor x \rfloor$. In Problem 7, you are asked to use the preceding constructions of d_1 and d_2 as the basis to prove that for each natural number k there is an integer d_k such that $0 \le d_k \le 9$ and

$$D + \frac{d_1}{10} + \cdots + \frac{d_k}{10^k} \le x < D + \frac{d_1}{10} + \cdots + \frac{d_k}{10^k} + \frac{1}{10^k}.$$

Then $D.d_1d_2d_3 \ldots d_k$ gives an approximation to the real number x **to k decimal places.**

If the given real number x is negative, then apply the preceding construction to the positive real number $y = -x$ to construct a decimal $D.d_1d_2d_3 \ldots d_k \ldots$ corresponding to y. Then $-D.d_1d_2d_3 \ldots d_k \ldots$ is defined to be the decimal representing x.

Can one real number have two different decimal representations?

The answer to this question is yes. Two different decimal representations exist for all rational numbers with finite decimals.

For example, the construction of a decimal representation for the number 1 one digit at a time, as in the preceding description, results in the decimal representation $1.0000\ldots00\ldots$. However, $.9999\ldots9\ldots$ is also a decimal representation of 1 because

$$0 + \frac{9}{10} + \cdots + \frac{9}{10^k} < 1 = 0 + \frac{9}{10} + \cdots + \frac{9}{10^k} + \frac{1}{10^k} \quad \text{for all } k \in \mathbf{N}.$$

(See Problem 6.) Similarly, $3.4999\ldots$ (with 9s repeating forever) $= 3.5000$ (with 0s repeating forever). In general, the reason two decimals exist for these numbers is that we have defined $d = [D, d_1, d_2, d_3, \ldots, d_k, \ldots]$ to be a decimal representation of a real number x provided that d and x satisfy the following inequalities:

$$D + \frac{d_1}{10} + \cdots + \frac{d_k}{10^k} \le x \le D + \frac{d_1}{10} + \cdots + \frac{d_k}{10^k} + \frac{1}{10^k} \quad \text{for all } k \in \mathbf{N}.$$

Consequently, if x is a decimal fraction represented by the finite decimal $d = D.d_1d_2\ldots d_m$, then

$$D + \frac{d_1}{10} + \cdots + \frac{d_m}{10^m} = x.$$

But then the infinite decimal $d = D.d_1d_2\ldots(d_m - 1)999\ldots9\ldots$ also represents x because

$$D + \frac{d_1}{10} + \cdots + \frac{d_m - 1}{10^m} < x = D + \frac{d_1}{10} + \cdots + \frac{d_m - 1}{10^m} + \frac{9}{10^{m+1}} + \cdots + \frac{9}{10^k} + \frac{1}{10^k} \quad \text{for all } k \ge m.$$

An alternate approach to decimal representation that results in a unique decimal representation for each real number is explored in Project 1 for this chapter.

It is possible to define operations of addition and multiplication for real numbers through their representation by sequences of nested closed intervals with rational endpoints. If $\{I_k\}$ and $\{J_k\}$ are the sequences of intervals that determine the real numbers x and y, we form a new sequence of nested intervals whose endpoints are the sums of the corresponding intervals for x and for y. This new sequence will contain a single real number we call the sum. For instance, here are the first five members of the sequences for $\sqrt{2}$, $\sqrt{3}$, and $\sqrt{2} + \sqrt{3}$.

$\sqrt{2}$	$\sqrt{3}$	$\sqrt{2} + \sqrt{3}$
$I_1 = [1.4, 1.5]$	$J_1 = [1.7, 1.8]$	$[3.1, 3.3]$
$I_2 = [1.41, 1.42]$	$J_2 = [1.73, 1.74]$	$[3.14, 3.16]$
$I_3 = [1.414, 1.415]$	$J_3 = [1.732, 1.733]$	$[3.146, 3.148]$
$I_4 = [1.4142, 1.4143]$	$J_4 = [1.7320, 1.7321]$	$[3.1462, 3.1464]$
$I_5 = [1.41421, 1.41422]$	$J_5 = [1.73205, 1.73206]$	$[3.14626, 3.14628]$

Although the intervals of the sum are not the same length as the intervals for $\sqrt{2}$ and $\sqrt{3}$, they are still nested and, since their lengths go to 0, there is only one number within all of them. In similar fashion, we can define multiplication of two real numbers. With these definitions, we can derive all the familiar properties of addition and multiplication of real numbers.

Repeating decimals

The decimal representing any rational number can be obtained by repeated application of the Division Algorithm. We illustrate this process by finding the decimal for $\frac{462}{13}$. The first application gives us 35, the integer part of this decimal. Each succeeding application uses 10 times the remainder from the previous step. We show the first four lines, which yield a quotient of 35.5384 and a remainder of $\frac{8}{130000}$.

$$462 = 13 \cdot 35 + 7$$

$$70 = 13 \cdot 5 + 5 \quad \Rightarrow \quad 7 = 13 \cdot \frac{5}{10} + \frac{5}{10} \quad \Rightarrow \quad 462 = 13\left(35 + \frac{5}{10}\right) + \frac{5}{10}$$

$$50 = 13 \cdot 3 + 11 \Rightarrow \frac{5}{10} = 13 \cdot \frac{3}{10^2} + \frac{11}{10^2} \Rightarrow 462 = 13\left(35 + \frac{5}{10} + \frac{3}{10^2}\right) + \frac{11}{10^2}$$

$$110 = 13 \cdot 8 + 6 \Rightarrow \frac{11}{10^2} = 13 \cdot \frac{8}{10^3} + \frac{6}{10^3} \Rightarrow 462 = 13\left(35 + \frac{5}{10} + \frac{3}{10^2} + \frac{8}{10^3}\right) + \frac{6}{10^3}$$

$$60 = 13 \cdot 4 + 8 \Rightarrow \frac{6}{10^3} = 13 \cdot \frac{4}{10^4} + \frac{8}{10^4} \Rightarrow 462 = 13\left(35 + \frac{5}{10} + \frac{3}{10^2} + \frac{8}{10^3} + \frac{4}{10^4}\right) + \frac{8}{10^4}$$

Because there are only 12 possible nonzero remainders, the cycle of quotients that begins 5384... must repeat after at most 12 steps.

Question 4: Carry out the next two applications of the Division Algorithm to show that the cycle of quotients for $\frac{462}{13}$ repeats after 6 steps.

Long division is a collapsed version of the aforementioned process. Here is the long division to find the decimal for $\frac{462}{13}$. (See Section 5.2.1 for another example relating long division and the Division Algorithm.) Compare each line with the steps of the calculations preceding Question 4. We show the part obtaining the integer 35 as one step even though most people would take two steps to get it. The final remainder 7 is equal to a remainder six steps earlier, so the cycle of quotients, 538461, will be repeated if the long division is continued. Therefore, $\frac{462}{13} = 35.538461538461\ldots538461\ldots$. We write this as $35.\overline{538461}$.

$$\begin{array}{r}
35.538461 \\
13\overline{)462.000000} \\
455 \\
\hline
7\,0 \\
6\,5 \\
\hline
50 \\
39 \\
\hline
110 \\
104 \\
\hline
60 \\
52 \\
\hline
80 \\
78 \\
\hline
20 \\
13 \\
\hline
7
\end{array}$$

More generally, if the decimal representation of a rational number $\frac{a}{b}$ does not terminate, then the decimal is **periodic** (or **repeating**); that is, there is a finite string $d_q d_{q+1} \ldots d_{q+p-1}$ of p digits in the decimal representation of that repeats forever from some point on. This is due to the fact that all of the remainders that occur in the Division Algorithm division procedure for constructing the decimal representation of $\frac{a}{b}$ must be positive integers less than b. Because there are only $b - 1$ such integers, the long division process must eventually cycle.

The shortest repeating string is called a **repetend**, and the length p of a repetend is called the **period** of the decimal. We have demonstrated the following theorem.

Theorem 2.4 Suppose $\frac{a}{b}$ is a rational number in lowest terms with $b > 0$ whose decimal representation is not terminating. Then $\frac{a}{b}$ is represented by a repeating decimal whose period is at most $b - 1$.

Sometimes the period of $\frac{a}{b}$ equals $b - 1$, as for $\frac{1}{7} = .\overline{142857}$. Sometimes the period of $\frac{a}{b}$ does not equal $b - 1$, as for $\frac{3}{11} = .\overline{27}$. In the next section, we explore the period of a periodic decimal representing a rational number.

2.1.2 Problems

1. Find decimals representing the rational numbers $\frac{21}{20}$ and $\frac{20}{21}$.

2. Suppose that x is a rational number represented by the finite decimal $D.d_1 d_2 \ldots d_k$.

a. Explain why an (infinite) decimal representing x is $D.d_1 d_2 \ldots d_k 0000 \ldots$.

b. What other infinite decimal represents x?

3. a. Give the first six digits in the decimal representation of $-\pi$.

b. Describe the first six intervals I_k in the construction of the decimal for $-\pi$.

4. Find the first three decimal places of $\sqrt{7}$ using only multiplication.

5. Find the decimal for each rational number by repeated applications of the Division Algorithm.

a. $\frac{817}{37}$ b. $\frac{6}{25}$ c. $\frac{46}{12}$

6. The following arguments are frequently used to convince students that $.9999\ldots = 1$:

Argument 1: Let $n = .99999\ldots$. Then $10n = 9.99999\ldots$, and so $9n = 10n - n = 9.0000\ldots$; therefore, $9n = 9$ and so $n = 1$.

Argument 2: We know that $\frac{1}{3} = .333333\ldots$, so $1 = 3(\frac{1}{3}) = 3(.33333333\ldots) = .99999999\ldots$.

Argument 3: By long division, you can see that

$$\frac{4}{9} = .44444\ldots \text{ and } \frac{5}{9} = .555555\ldots.$$

Therefore,

$$1 = \frac{4}{9} + \frac{5}{9} = .44444\ldots + .555555\ldots = .99999\ldots.$$

Argument 4: The decimal $.999999\ldots$ stands for the geometric series

$$\frac{9}{10} + \frac{9}{10^2} + \cdots + \frac{9}{10^n} + \ldots.$$

But when $|r| < 1$, the sum of the infinite geometric series $1 + r + r^2 + \cdots + r^n + \cdots$ is

$$\frac{1}{1-r}.$$

Therefore,

$$\begin{aligned} .9999\cdots &= \frac{9}{10} + \frac{9}{10^2} + \cdots + \frac{9}{10^n} + \cdots \\ &= \frac{9}{10}\left(1 + \left(\frac{1}{10}\right) + \left(\frac{1}{10}\right)^2 + \cdots \right. \\ &\quad \left. + \left(\frac{1}{10}\right)^{n-1} + \cdots\right) \\ &= \frac{9}{10}\left(\frac{1}{1 - \frac{1}{10}}\right) = 1. \end{aligned}$$

Give a justification for each step in these arguments.

7. Suppose that p and q are positive integers and that $a = \frac{p}{q}$. Explain why long division of p by q results in the decimal representation of a. (*Hint*: It is enough to explain why the decimal $d = [D, d_1, d_2, d_3, \ldots, d_k, \ldots]$ produced by long division satisfies

$$D + \frac{d_1}{10} + \cdots + \frac{d_k}{10^k} \le a < D + \frac{d_1}{10} + \cdots + \frac{d_k}{10^k} + \frac{1}{10^k}$$

for all k in N.)

*8. Prove that if x is a positive real number, then for each natural number k there is an integer d_k such that $0 \le d_k \le 9$ and such that

$$D + \frac{d_1}{10} + \cdots + \frac{d_k}{10^k} \le x < D + \frac{d_1}{10} + \cdots + \frac{d_k}{10^k} + \frac{1}{10^k}.$$

(*Hint*: Use the construction of d_1 and d_2 in this section as a guide for a proof by mathematical induction.)

ANSWERS TO QUESTIONS

1. $D = 98$, $d_1 = 0$, $d_2 = 1$, $d_3 = 3$, $d_4 = 4$, $d_5 = 3$, $d_6 = 7$, $d_7 = 8$.

2. a. Let $a = D.d_1d_2\ldots d_k$. Then $10^k \cdot a$ is an integer, call it n. Then $a = \frac{n}{10^k}$, which is a quotient of integers, so a is rational.

3. $\sqrt{2}$ is not a rational number, unlike all the endpoints, which are rational numbers.

4. $80 = 13 \cdot 6 + 2 \Rightarrow \frac{8}{10^4} = 13 \cdot \frac{6}{10^5} + \frac{2}{10^5} \Rightarrow 462 = 13\left(35 + \frac{5}{10} + \frac{3}{10^2} + \frac{8}{10^3} + \frac{4}{10^4} + \frac{6}{10^5}\right) + \frac{2}{10^5}$;

$20 = 13 \cdot 1 + 7 \Rightarrow \frac{20}{10^5} = 13 \cdot \frac{1}{10^6} + \frac{7}{10^6} \Rightarrow 462 = 13\left(35 + \frac{5}{10} + \frac{3}{10^2} + \frac{8}{10^3} + \frac{4}{10^4} + \frac{6}{10^5} + \frac{1}{10^6}\right) + \frac{7}{10^6}$.

Table 2

$\frac{1}{2} = 0.5$
$\frac{1}{3} = 0.\overline{3}$
$\frac{1}{4} = 0.25$
$\frac{1}{5} = 0.2$
$\frac{1}{6} = 0.1\overline{6}$
$\frac{1}{7} = 0.\overline{142857}$
$\frac{1}{8} = 0.125$
$\frac{1}{9} = 0.\overline{1}$
$\frac{1}{10} = 0.1$
$\frac{1}{11} = 0.\overline{09}$
$\frac{1}{12} = 0.08\overline{3}$
$\frac{1}{13} = 0.\overline{076923}$
$\frac{1}{14} = 0.0\overline{714285}$
$\frac{1}{15} = 0.0\overline{6}$
$\frac{1}{16} = 0.0625$

2.1.3 Periods of periodic decimals

The variety of types of decimals for rational numbers is illustrated in Table 2, which shows the decimal representations of the reciprocals of the integers 2 through 16.

Notice that six of these representations are finite decimals, while the remaining nine are periodic: Five have period 1, one has period 2, and three have period 6.

Also, for five of the nine periodic cases, the period starts right after the decimal point. In three of the other four, there is a delay of 1 digit before the period starts. In one periodic case, there is a delay of 2 digits.

What kind of pattern is there in the types of representations? More precisely, given an integer n, what can you predict about the decimal representation of $\frac{m}{n}$ if m and n are relatively prime positive integers?

The three types of decimals

Theorem 2.4 in the preceding section tells us that, if we divide 1 by an integer n, then after at most $n - 1$ steps the division process must either terminate or else start to repeat. This is true because there are only $n - 1$ possible nonzero remainders. If the process starts to repeat at some step, it can either start repeating from the beginning of the division process, or else from some intermediate point. This gives rise to three distinct types of decimal representations. Table 3 classifies the decimals from Table 2 into these types.

Table 3

Type of Decimal	Examples	General Form
terminating	$0.5, 0.25, 0.2, 0.125, 0.1, 0.0625$	$0.d_1d_2d_3\ldots d_t\ (d_t \neq 0)$
simple-periodic	$0.\overline{3}, 0.\overline{142857}, 0.\overline{1}, 0.\overline{09}, 0.\overline{076923}$	$0.\overline{d_1d_2d_3\ldots d_p}$
delayed-periodic	$0.1\overline{6}, 0.08\overline{3}, 0.0\overline{714285}, 0.0\overline{6}$	$0.d_1d_2d_3\ldots d_t\overline{d_{t+1}d_{t+2}d_{t+3}\ldots d_{t+p}}$

---

*An asterisk by a problem number indicates a problem perceived to be more difficult than other problems.

Question 1: Find the decimal representation of $\frac{1}{28}$ using the Division Algorithm or long division. Into which of the three categories does it fall? Explain the length of the repetend and the number of digits of the delay before the period starts in terms of the pattern of the remainders.

The examples in Tables 2 and 3 are all decimal representations of reciprocals of positive integers $\frac{1}{n}$. However, it is easy to extend what we learn from these reciprocals to decimal representations of any positive number $\frac{m}{n}$, where m and n are positive integers and $m < n$ because of the following result: *If* $\frac{m}{n}$ *is in lowest terms, the general form of the decimal representation of* $\frac{m}{n}$ *is the same as that of* $\frac{1}{n}$. (See Problem 2.)

By the *general form* of a decimal representation, we mean not only the type (terminating, simple-periodic, delayed-periodic), but also the length of the strings of digits in each part (the values of t and p in the notation of the second column of Table 3). The proof of this result will become apparent as we work through each type.

The big picture

Table 4 summarizes the results that are deduced in the remainder of this section. We need only consider rational numbers x between 0 and 1 written as $x = \frac{m}{n}$ in lowest terms because any other rational number is the sum of x and an integer, and the integer part of a rational number does not affect the general form of its decimal representation.

Table 4

Type of Decimal Representation	Form of Decimal Representation[5]	Rational Number $\frac{m}{n}$ in Lowest Terms	Equivalent Form of Rational Number
terminating	$0.d_1d_2d_3\dots d_t$ (t = max of r and s)	$\dfrac{m}{2^r \cdot 5^s}$ (Theorem 2.6)	$\dfrac{M}{10^t}$ (Theorem 2.5)
simple-periodic	$0.\overline{d_1d_2d_3\dots d_p}$	$\dfrac{m}{3^u \cdot 7^v \cdot 11^w \cdot \dots}$ (Theorem 2.8)	$\dfrac{M}{10^p - 1}$ (Theorem 2.7)
delayed-periodic	$0.d_1d_2\dots d_t\overline{d_{t+1}d_{t+2}\dots d_{t+p}}$ (t = max of r and s, $t > 0$)	$\dfrac{m}{2^r \cdot 5^s \cdot 3^u \cdot 7^v \cdot 11^w \cdot \dots}$ (Theorem 2.10)	$\dfrac{M}{10^t \cdot (10^p - 1)}$ (Theorem 2.9)

Terminating decimals

We begin with the simplest case: terminating decimals.

Suppose $x = \frac{d_1}{10^1} + \frac{d_2}{10^2} + \dots + \frac{d_t}{10^t}$ where $d_1, d_2, \dots, d_t$ are nonnegative integers, $d_t \neq 0$, and $d_k \leq 9$ for $k = 1, \dots, t$. Any such number is represented by the terminating decimal $.d_1d_2\dots d_{t-1}d_t$ with $d_t \neq 0$. Then $x = \frac{M}{10^t}$, where

$$M = 10^{t-1}d_1 + 10^{t-2}d_2 + \dots 10d_{t-1} + d_t.$$

For example, because $\frac{5}{8} = \frac{6}{10} + \frac{2}{10^2} + \frac{5}{10^3}$, $\frac{5}{8}$ is represented by the terminating decimal .625. Also, $\frac{5}{8}$ can be expressed as $\frac{5}{8} = \frac{625}{10^3}$, and

$$625 = 10^2 \cdot 6 + 10^1 \cdot 2 + 5.$$

[5]Throughout this section, strings of d_i such as $d_1d_2d_3\dots d_t$ stand for digits of a number (and not for multiplication).

Conversely, a fraction with a denominator that is a power of 10 has a terminating decimal representation that can be immediately written. For instance, $\frac{902}{10^8} = \frac{9 \cdot 10^2 + 2}{10^8} = \frac{9}{10^6} + \frac{2}{10^8} = .00000902$. You are asked to show this converse (see Problem 9).

Theorem 2.5 A number x between 0 and 1 has a terminating decimal representation $0.d_1d_2d_3 \ldots d_t$ (where $d_t \neq 0$) if and only if it can be represented in the form $x = \frac{M}{10^t}$ for some positive integer M that is not divisible by 10.

To use Theorem 2.5, we need to know whether or not a given rational number $\frac{m}{n}$ can be represented in the form $\frac{M}{10^t}$. For example, $\frac{1}{250}$ and $\frac{1}{64}$ can be represented in this form, but $\frac{1}{60}$ cannot.

Question 2: Explain why $\frac{1}{250}$ and $\frac{1}{64}$ have terminating decimals but $\frac{1}{60}$ does not.

There is a simple way of telling these cases apart.

Theorem 2.6 Suppose that $\frac{m}{n}$ is in lowest terms and that $m < n$. Then $\frac{m}{n}$ has a decimal representation that terminates after t digits if and only if $n = 2^r \cdot 5^s$ and t is the larger of r and s.

Proof: Since the theorem is an if-and-only-if statement, both directions of the implication must be proved.

($\Rightarrow$) Suppose $\frac{m}{n}$ has a decimal representation $D.d_1d_2d_3 \ldots d_t$. Then, by Theorem 2.5, $\frac{m}{n}$ can be represented in the form $\frac{M}{10^t} = \frac{M}{2^t \cdot 5^t}$ where M is not divisible by 10. Thus M may contain factors of 2 or 5 but not both. If we cancel these factors in M with corresponding factors in the denominator, we obtain a denominator $n = 2^r \cdot 5^s$, where t is the larger of r and s.

($\Leftarrow$) Suppose $n = 2^r \cdot 5^s$ and $r \geq s$. Then we can write $r = s + k$ for some nonnegative integer k, and so $\frac{m}{n} = \frac{m}{2^{s+k}5^s} = \frac{m \cdot 5^k}{2^{s+k}5^{s+k}} = \frac{m \cdot 5^k}{10^r}$, and $m \cdot 5^k$ is not divisible by 10 because $\frac{m}{n}$ is in lowest terms and n has factor of 2. This shows that the decimal representation of $\frac{m}{n}$ consists of the decimal representation of the integer $m \cdot 5^k$, but with the decimal point moved r digits to the left. The case $r < s$ is similar. ∎

Question 3: What fraction $\frac{m}{n}$ in lowest terms is represented by the terminating decimal 0.00056?

Question 4: Use Theorem 2.6 to find the number of digits in the decimal representation for $\frac{7}{80}$.

Decimals with simple-periodic representations

Theorem 2.6 shows that terminating decimals represent rational numbers in lowest terms whose denominators consist *only* of powers of 2 and 5. Now, we consider the decimal representations of those rational numbers in lowest terms whose denominators contain *no* powers of 2 and 5. Examples from Table 2 are written here.

$$\frac{1}{3} = 0.\overline{3} \qquad \frac{1}{7} = 0.\overline{142857} \qquad \frac{1}{9} = 0.\overline{1} \qquad \frac{1}{11} = 0.\overline{09} \qquad \frac{1}{13} = 0.\overline{076923}$$

It happens that these numbers all have *simple-periodic* decimal representations. (The periods for the five numbers just given are 1, 6, 1, 2, and 6, respectively.) Although simple to state, this property is not so easy to prove. We first demonstrate that *every rational $\frac{m}{n}$ that is represented by a simple-periodic decimal is equal to a fraction with a denominator consisting of all 9s*. For example, for the simple-periodic decimals from Table 2,

$$0.\overline{3} = \frac{3}{9} \qquad 0.\overline{1} = \frac{1}{9} \qquad 0.\overline{09} = \frac{9}{99} \qquad 0.\overline{076923} = \frac{76923}{999999} \qquad 0.\overline{142857} = \frac{142857}{999999}.$$

Notice not only that these simple-periodic decimals represent fractions with all 9s in the denominator, but also that the length of the period tells us how many 9s there are. It is easy to prove this.

Proof: Suppose $x = 0.\overline{d_1d_2d_3\ldots d_p}$. Multiply x by 10^p. Then subtract x to obtain $(10^p - 1)x = d_1d_2\ldots d_p$, a finite decimal. Then divide by $10^p - 1$ to obtain $x = \frac{d_1d_2d_3\ldots d_p}{10^p - 1}$; that is, x can be expressed as a fraction in which the numerator is the repetend of x and the denominator is the integer with p digits all equal to 9. ┘

Question 5: Represent $.\overline{314}$ in this form.

We have proved the $(\Rightarrow)$ direction of a statement whose converse is also true. That is, if a fraction can be written in a form (not necessarily lowest terms) with a denominator consisting of a string of 9s, then its decimal representation is simple-periodic. You can get a feel for this fact by working with actual examples using a calculator. The proof of the general statement is not difficult.

Theorem 2.7 A number x between 0 and 1 has a simple-periodic decimal representation $0.\overline{d_1d_2d_3\ldots d_p}$ if and only if x can be put in the form $\frac{M}{10^p - 1}$, where M is the integer $d_1d_2d_3\ldots d_p$.

Proof: Having proved the $(\Rightarrow)$ direction, we show the proof of the $(\Leftarrow)$ direction. Suppose $x = \frac{M}{10^p - 1}$ and M is the integer $d_1d_2d_3\ldots d_p$. Now let $\overline{M}$ be the periodic decimal $.\overline{d_1d_2d_3\ldots d_p}$. Then $10^p\overline{M} = M + \overline{M}$. So $\overline{M} = \frac{M}{10^p - 1}$. Consequently $\overline{M} = x$. ┘

Question 6: In the representation form guaranteed by this theorem, are M and $10^p - 1$ always relatively prime?

How do we apply Theorem 2.7? For example, given a rational number such as $\frac{1}{13}$, how do we know whether it can be put in a form $\frac{M}{10^p - 1}$ with all 9s in the denominator? We now are ready to prove the test we mentioned earlier.

Theorem 2.8 Suppose that $\frac{m}{n}$ is in lowest terms and that $m < n$. Then $\frac{m}{n}$ has a simple-periodic decimal representation if and only if 2 or 5 are not factors of n.

Proof:

$(\Rightarrow)$ If the rational number $\frac{m}{n}$ has a simple-periodic decimal representation, then by Theorem 2.7 it can be put in the form $\frac{M}{10^p - 1}$, and so $m \cdot (10^p - 1) = M \cdot n$. If n were to have a factor of 2 or 5, then m would also have that factor, since clearly $10^p - 1$ cannot. But this would contradict the fact that $\frac{m}{n}$ is in lowest terms. We conclude that n can have no factor of 2 or 5.

$(\Leftarrow)$ To prove the other direction of the if-and-only-if statement, suppose 2 and 5 are not factors of n. Then, by Theorem 2.6, $\frac{m}{n}$ cannot have a terminating decimal representation, so it will have a periodic decimal representation with a period p.

Consider the remainders r_k that occur in the long division of m by n with the numerator m regarded as the initial remainder r_0. We know that for some smallest nonnegative t, it is true that $r_{k+p} = r_k$ for all $k \geq t$. We must show that $t = 0$; that is, that the remainders begin to repeat after the first p steps.

Suppose to the contrary that $t > 0$. Then $r_{t+p} = r_t$ but $r_{t+p-1} \neq r_{t-1}$. The Division Algorithm implies that there exist integers q_t, q_{t+p}, r_t, and r_{t+p} with $0 < r_t < n$ and $0 < r_{t+p} < n$ such that

$$10r_{t-1} = nq_t + r_t$$
$$10r_{t+p-1} = nq_{t+p} + r_{t+p}$$

for suitable nonnegative integers q_t and q_{t+p}. By subtracting the first of these equations from the second and using the fact that $r_{t+p} = r_t$, we conclude that

$$10(r_{t+p-1} - r_{t-1}) = n(q_{t+p} - q_t).$$

Therefore, n is a divisor of $10(r_{t+p-1} - r_{t-1})$. Because n has no factors of 2 or 5, n must be a divisor of $r_{t+p-1} - r_{t-1}$. But the remainders r_{t+p-1} and r_{t-1} are positive integers less than n, so their difference $r_{t+p-1} - r_{t-1}$ is an integer between $-n$ and n. Because n is also a divisor of this integer, this integer must be 0. Thus, $r_{t+p-1} = r_{t-1}$, contrary to our supposition that $t > 0$. Therefore, the decimal representation of $\frac{m}{n}$ must be simple-periodic. ┘

Decimals with delayed-periodic representations

The characteristics of delayed-periodic decimals are a combination of the characteristics of the two previous cases (terminating decimals and simple-periodic decimals). Delayed-periodic decimals result when the denominator of $\frac{m}{n}$ in lowest terms has both 2's or 5's and some other prime factors as well. The highest power of the 2 or the 5 that divides n determines the length t of the delay before the repetend starts. But the period itself is determined only from an equal fraction that is found in basically the same way as for simple-periodic decimals.

Question 7: Consider $\frac{9}{28}$.

a. Give its decimal representation.

b. What is the length t of the delay?

c. What is the period p?

d. Express $\frac{9}{28}$ as a sum of the form $\frac{A}{10^t} + 10^{-t}\frac{B}{10^p - 1}$, where A and B are integers.

Theorem 2.9 A number x between 0 and 1 has a delayed-periodic decimal representation $0.d_1d_2\ldots d_t\overline{d_{t+1}d_{t+2}\ldots d_{t+p}}$ with period p and with t digits before the start of the repetend if and only if $x = \frac{M}{10^t(10^p - 1)}$, where the numerator is the integer $M = d_1d_2\ldots d_td_{t+1}d_{t+2}\ldots d_{t+p} - d_1d_2\ldots d_t$. (The denominator has a decimal representation consisting of p 9s followed by t 0s.)

Proof: Suppose x has the delayed-periodic decimal representation of the theorem. Multiply x by $10^t \cdot 10^p$ to obtain

$$10^t \cdot 10^p \cdot x = d_1d_2\ldots d_td_{t+1}d_{t+2}\ldots d_{t+p}.\overline{d_{t+1}d_{t+2}\ldots d_{t+p}}.$$

Then $10^t \cdot 10^p \cdot x - 10^t \cdot x = d_1d_2\ldots d_td_{t+1}d_{t+2}\ldots d_{t+p} - d_1d_2\ldots d_t = M$, where M is an integer. Solve for x to obtain the desired result: $x = \frac{M}{10^t(10^p - 1)}$.

To prove the other direction of the theorem, as the answer to Question 7 illustrates, $x = \frac{M}{10^t(10^p - 1)}$ can also be represented as a sum,

$$\frac{M}{10^t(10^p - 1)} = \frac{A}{10^t} + 10^{-t}\frac{B}{10^p - 1},$$

for suitable constants A and B. The result follows from this fact. The details are left for you.

Theorem 2.9 gives a necessary and sufficient condition for a fraction to have a delayed-periodic decimal representation.

Finally, we give the simple condition for predicting from the denominator n of a rational number $\frac{m}{n}$ in lowest terms those that have delayed-periodic forms. The proof follows from our earlier results.

Theorem 2.10 A rational number $\frac{m}{n}$ between 0 and 1 that is in lowest terms has a delayed-periodic decimal representation with a period that starts t digits after the decimal point if and only if $n = 2^r \cdot 5^s \cdot q$ for some integer q relatively prime to 2 and 5. Here r and s are not both 0, and t is the larger of r and s.

You might find it helpful to refer back to Table 4, which provides a summary of this section. Notice that the type of decimal representation for a rational number $\frac{m}{n}$ in lowest terms is determined entirely by the factors of n (see Problem 2).

2.1.3 Problems

1. Use the theorems of the text to predict the general form of the decimal representations of the reciprocals of the integers 17 to 41; that is, the type as well as the delay if the type is periodic.

2. Explain why, if $\frac{m}{n}$ is a rational number between 0 and 1, and if m and n are relatively prime, then the type of the decimal representation of $\frac{m}{n}$ is independent of m. (*Hint*: Consider the three types of decimal representations separately.)

3. a. The rational number $\frac{1}{19}$ has decimal representation $0.\overline{052631578947368421}$. This can be verified by a pencil and paper application of the Division Algorithm, but this is laborious. It would be nice to be able to do this on a calculator, but many calculators show only 6 or 7 digits. Find a method for using a calculator to "piece together" the full 18 digits of the period of $\frac{1}{19}$.
   b. Find the decimal representation of $\frac{12}{71}$.

4. Consider those *reciprocals of primes* that have *simple-periodic* decimal representations. Using the theorems of the section, prove that, of these:
   a. There is exactly 1 with period 1. What is it?
   b. There is exactly 1 with period 2. What is it?
   c. There is exactly 1 with period 3. What is it?
   d. There is exactly 1 with period 4. What is it?
   e. There are exactly 2 with period 5. What are they?
   f. There are exactly 2 with period 6. What are they?

5. Consider those *reciprocals of integers* that have *simple-periodic* decimal representations. Using the theorems of the section, prove that, of these:
   a. There are exactly 2 with period 1. What are they?
   b. There are exactly 3 with period 2. What are they?
   c. There are exactly 5 with period 3. What are they?

6. a. Find the decimal representations for $\frac{1}{27}$ and $\frac{1}{37}$.
   b. Explain the peculiar relationship between these decimals, and find other pairs of decimals with the same relationship.

7. Find the prime factorization of integers of the form $10^p - 1$ for $p = 1$ to 7. What is the relevance of these factorizations to the behavior of decimal representations?

8. Consider the decimal representations for $\frac{n}{7}$:

$$\frac{1}{7} = 0.\overline{142857}, \qquad \frac{2}{7} = 0.\overline{285714}, \qquad \text{etc.}$$

Find the other four representations. Notice that the digits of the period always appear in the same order. Is this peculiar to the denominator 7, or is it true of all prime denominators?

9. Prove the ($\Leftarrow$) direction of Theorem 2.5: If $x = \frac{M}{10^t}$ and M is an integer not divisible by 10, then x has a terminating decimal representation.

10. Another proof of the ($\Leftarrow$) implication for Theorem 2.8 is based on a theorem proved by Euler. Euler's theorem, which is a generalization of Fermat's Theorem (Theorem 6.4), states that if natural numbers a and n are relatively prime (have no common factors larger than 1), then the number $a^{\phi(n)} - 1$ has n as a factor. Here, $\phi(n)$ is the **Euler phi function**, defined as the number of natural numbers less than n and relatively prime to n. Since 10 and any number n with no factors of 2 or 5 are relatively prime, the theorem guarantees that $(10^{\phi(n)} - 1) = n \cdot M$ for some integer M. Hence we can write

$$\frac{m}{n} = \frac{m \cdot M}{n \cdot M} = \frac{m \cdot M}{10^{\phi(n)} - 1}$$

So Euler's Theorem provides the denominator of 9s. Theorem 2.7 now applies, showing that $\frac{m}{n}$ has a simple-periodic decimal representation.

Prove the following: The period p of the decimal representation of a rational number $\frac{m}{n}$ in lowest terms is a divisor of the number $\phi(n)$ of positive integers less than n that are relatively prime to n.

11. Consider $\frac{1}{21}$. You may need to refer to Problem 10.

a. Use the theorems of this section to explain why its decimal representation must be simple-periodic.

b. By computing $\phi(21)$, show that the period p of its decimal representation must divide 12, and hence be equal to 1, 2, 3, 4, 6, or 12.

c. Show that 21 must thus divide one of the numbers 99, 999, 9999, 999999, or 999999999999.

d. Show that the first of these numbers that 21 does divide is $10^6 - 1 = 999999$, and conclude that the period must be 6.

e. Compute the decimal for $\frac{1}{21}$ to verify your result.

12. Finish the proof of Theorem 2.9.

13. Prove Theorem 2.10.

ANSWERS TO QUESTIONS

1. $\frac{1}{28} = 0.03\overline{571428}$; delayed-periodic; remainders are 10, 16, 20, 4, 12, 8, 24, 16. The second appearance of 16 indicates a repetend 6 digits long; the first two quotients are 0 and 3, indicating a delay of 2 digits before the repetend.

2. $250 = 2^1 \cdot 5^3$, and $64 = 2^6 \cdot 5^0$, but $60 = 3 \cdot 2^2 \cdot 5^1$, which is not a product of 2's and 5's.

3. $\frac{56}{10^5} = \frac{7}{12500} = \frac{7}{2^2 5^5}$.

4. $\frac{7}{80} = \frac{7}{2^4 5^1}$. Hence $r = 4$, $s = 1$, and so $t = 4$. This means there will be 4 digits in the decimal representation of $\frac{7}{80}$. In fact, $\frac{7}{80} = 0.0875$.

5. $\frac{314}{999}$.

6. No. For example, the representation of this form for $0.\overline{3}$ is $\frac{3}{9}$.

7. a. $\frac{9}{28} = 0.32\overline{142857}$ b. 2

c. 6 d. $\frac{9}{28} = \frac{32}{10^2} + 10^{-2}\left(\frac{142857}{10^6 - 1}\right)$

2.1.4 The distributions of various types of real numbers

In this section, we consider the following question: How are the rational numbers and irrational numbers distributed among the real numbers? To answer this question, we first note that every real number is either an integer or lies between two integers.

Theorem 2.11 For any real number x, there is one and only one integer n such that $n \le x < n + 1$.

Proof: This number n is the greatest integer less than or equal to x, $\lfloor x \rfloor$. If a decimal representation of x is $D.d_1 d_2 d_3 \ldots$, then $\lfloor x \rfloor = D$ if x is positive, and $\lfloor x \rfloor = D - 1$ if x is negative. ┘

The integers are sprinkled through the real numbers in the sense that it is easy to find two real numbers between which there is no integer. But even if there is no integer between two real numbers x and y, there always is a rational number and an irrational number between them.

Theorem 2.12 For any pair of real numbers x and y with $x < y$, there is a rational number r and an irrational number s such that r and s are both in the open interval (x, y).

Proof: Suppose x has a decimal representation $D.d_1d_2d_3\ldots$ and y has a decimal representation $E.e_1e_2e_3\ldots$. (If either x or y has a finite decimal representation, then pick the representation with all 0s.) Since $x < y$, either $D < E$ (which we identify as the 0th decimal place) or there exists a smallest ith decimal place with $d_i < e_i$. There must be a d_j, with $j > i$, that is less than 9 (otherwise all digits would be 9's). Then the rational number $r = D.d_1d_2d_3\ldots(d_j + 1)$ is between x and y. This rational number has a decimal representation identical to the expansion for x for the first $j - 1$ decimal places and with a larger number in the jth place. Any number s obtained from r by adding a nonrepeating pattern of digits after the jth decimal place is irrational and also between x and y. ∎

Question 1: Identify a rational and an irrational number between 40.111 and 40.11101.

Algebraic Numbers

Even though every irrational number has a nonrepeating decimal representation, there is a sense in which we *know* some of these numbers. For instance, we know $\sqrt{56}$ as the positive solution to the equation $x^2 = 56$. With this property, we can perform operations with $\sqrt{56}$ and deduce other properties. This is an example of an *algebraic number*.

Definition A number r is **algebraic** if there is a nonzero polynomial $p(x)$ with integer coefficients such that r is a root of $p(x) = 0$.

The polynomial $p(x) = x^2 - 56$ suffices to show that $\sqrt{56}$ is algebraic.

Question 2: Show that $\sqrt[5]{\frac{11}{17}}$ is an algebraic number by identifying a polynomial equation $p(x) = 0$, where $p(x)$ has integer coefficients, that has $\sqrt[5]{\frac{11}{17}}$ as a solution.

Question 3: Some authors define a number r to be algebraic if it is the solution to a polynomial equation $p(x) = 0$, where $p(x)$ is a polynomial with *rational* coefficients. Is this definition equivalent to the one we have given? Why or why not?

An immediate consequence of the definition is the following theorem.

Theorem 2.13

a. Every rational number is an algebraic number.

b. Every kth root of a rational number r is an algebraic number.

Proof: The proof of each part requires finding an equation satisfied by the numbers under consideration, and that is left to you. (See Problems 3 and 4.) ∎

As you know, $\sin 30° = \sin\frac{\pi}{6} = \frac{1}{2}$, $\cos 30° = \cos\frac{\pi}{6} = \frac{\sqrt{3}}{2}$, and $\tan 45° = \tan\frac{\pi}{4} = 1$. Thus some values of trigonometric functions are algebraic numbers, and some are also rational. Notice that the arguments of each of these functions are integer degrees and rational multiples of π radians. The following general theorem tells us that some of the values of the trigonometric functions are algebraic and some other values are irrational. It assumes the arguments of the cosine function are in radians. Its proof can be found in Niven (1956).

Theorem 2.14

a. If r is rational, then $\cos(r\pi)$ is an algebraic number.

b. If r is rational and $r \neq 0$, then $\cos r$ is irrational.

The irrationality of π and e

Two of the most famous real numbers in mathematics are irrational. From the contrapositive of Theorem 2.14(b), it follows that π is irrational, because $\cos(\pi) = -1$. The first proof that π is irrational was given by Johann H. Lambert in 1761, using the theory of continued fractions. Here is a proof that e (the base of natural logarithms) is irrational.

Theorem 2.15 The base e of natural logarithms is an irrational number.

Proof: We begin with the Maclaurin series representation for e that is typically derived in calculus.

$$e = \sum_{n=0}^{\infty} \frac{1}{n!} = 1 + 1 + \frac{1}{2!} + \frac{1}{3!} + \cdots + \frac{1}{n!} + \cdots$$

Suppose that e were rational. Then $e = \frac{p}{q}$, where p and q are positive integers. Now

$$\begin{aligned} e - 1 - 1 - \frac{1}{2!} - \frac{1}{3!} - \cdots - \frac{1}{q!} \\ &= \frac{1}{(q+1)!} + \frac{1}{(q+2)!} + \cdots + \frac{1}{(q+n)!} + \cdots \\ &= \sum_{k=1}^{\infty} \frac{1}{(q+k)!} \end{aligned}$$

So

$$\begin{aligned} q!\left(e - 1 - 1 - \frac{1}{2!} - \frac{1}{3!} - \cdots - \frac{1}{q!}\right) &= \frac{1}{q+1} + \frac{1}{(q+1)(q+2)} + \cdots + \\ &\quad \frac{1}{(q+1)(q+2)\ldots(q+m)} + \cdots \\ &= \sum_{k=1}^{\infty} \frac{1}{(q+1)(q+2)\ldots(q+k)}. \end{aligned}$$

The left side is a positive integer because $\frac{q!}{k!}$ is an integer whenever $1 \leq k \leq q$. We now show that the right side must be less than 1. Because $\frac{1}{(q+1)(q+2)\ldots(q+k)} < \frac{1}{(q+1)^k}$ for each positive integer k, the right side is less than the sum of the geometric series $r + r^2 + r^3 + \cdots + r^n + \cdots = \frac{r}{1-r}$, where $r = \frac{1}{q+1}$. When $r = \frac{1}{q+1}, \frac{r}{1-r} = \frac{1}{q}$ and $\frac{1}{q} < 1$. This contradiction proves that e is not rational. ∎

Transcendental numbers

Theorems 2.13 to 2.15 raise the question whether any real numbers are not algebraic. The answer is that there are many nonalgebraic numbers. Such real numbers are called **transcendental**. In 1844, Joseph Liouville first showed that a particular number was transcendental. He proved that the number represented by the infinite decimal 0.1010010000001000..., where there are $n!$ 0's between the nth and $(n + 1)$st 1, is transcendental.

It has been found to be rather difficult to prove that particular numbers are transcendental. In 1873, Charles Hermite (1822–1901) first proved that e is transcendental using methods of higher algebra. A proof that e is transcendental, depending only on the tools of calculus, was given by Adolf Hurwitz in 1893 [see Niven (1956)].

In 1882, Ferdinand Lindemann (1852–1939) proved that π is transcendental using a line of thought similar to that of Hermite. In 1934, Theodor Schneider and Aleksander Gelfond independently proved that if x is an algebraic number not equal to 0 or 1 and y is an irrational algebraic number, then x^y is transcendental.

So far we have noted that real numbers are either rational or irrational. They are also either algebraic or transcendental; their decimals are either finite, infinite and repeating, or infinite and nonrepeating. Figure 6 shows how these three partitions of the real numbers are related.

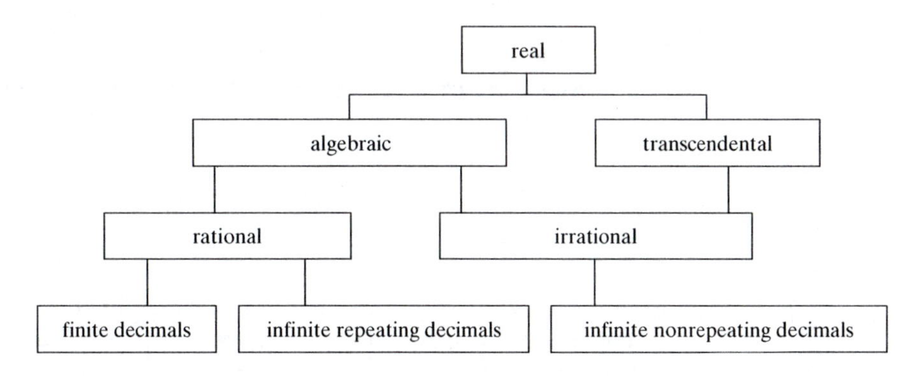

Figure 6 Partitions of the real numbers.

Cardinality

The sets **N** of natural numbers, **Z** of integers, **Q** of rational numbers, **R** of real numbers, and **C** of complex numbers form a nested chain of infinite sets:

$$\mathbf{N} \subset \mathbf{Z} \subset \mathbf{Q} \subset \mathbf{R} \subset \mathbf{C}.$$

Each set in this chain of sets contains the preceding set as a proper subset. Thus, if we use set inclusion as the measure of size, it follows that the *size* of each set in the chain is larger than and not equal to its predecessor.

But set inclusion is not the only way to compare the sizes of sets. In 1895, Georg Cantor introduced the following definition which compares two sets in terms of their "counts" or *cardinality*.

Definitions Suppose that A and B are sets. Then A and B **have the same cardinality** if and only if there is a **one-to-one correspondence** between A and B; that is,

i. For each $a \in A$, there is exactly one corresponding $b \in B$.

ii. For each $b \in B$, there is exactly one corresponding $a \in A$.

A one-to-one correspondence can be defined by a formula, a diagram, a rule, or a table. The essential point is that the definition should clearly specify the unique element of B that corresponds to each given element of A and, vice versa, the unique element of A that corresponds to each given element of B.

To determine the cardinality of a nonempty finite set, we form a correspondence between its elements and the natural numbers beginning with 1. For instance, we usually use the set $\mathbf{N}_7$ of natural numbers from 1 to 7 to "count" the elements of the set W of days of the week, because $\mathbf{N}_7$ has the same cardinality as W. More generally, a nonempty set F is **finite** if and only if F is in one-to-one correspondence with the set $N_k = \{n \in \mathbf{N}: 1 \leq n \leq k\}$ for some positive integer k. In this case, we say that the cardinality of F is k.

Countably infinite sets

The set **N** of natural numbers and the set **Z** of integers have the same cardinality. A simple one-to-one correspondence between **N** and **Z** is specified by the zig-zag pattern for counting **Z** with **N** shown in Table 5.

Table 5

N	.	.	.	5	3	1	2	4	6	.	.	.
↓												
Z	.	.	.	−2	−1	0	1	2	3	.	.	.

This correspondence from **N** to **Z** can be specified also by the formula

$$n \text{ corresponds to } \begin{cases} \dfrac{n}{2} & \text{if } n \text{ is even} \\ -\dfrac{n-1}{2} & \text{if } n \text{ is odd.} \end{cases}$$

Any set that has the same cardinality as either the set **N** or some finite set is a **countable** or **denumerable** set. Thus, $\mathbf{N}_7$, W, **N**, and **Z** are all countable sets. Any set that has the same cardinality as the set **N** of natural numbers is said to be **countably infinite**. Of the sets named in this paragraph, only **N** and **Z** are countably infinite.

The set $\mathbf{Q}^+$ of all positive rational numbers is certainly larger than the set **N** of natural numbers in terms of set inclusion. However, $\mathbf{Q}^+$ can be shown to have the same cardinality as **N** by the following scheme due to Cantor.

Step 1: Imagine a triangular array whose nth row contains the $n - 1$ fractions $\frac{a}{b}$, where a and b are positive integers and $a + b = n$, written in order from $b = 1$ to $b = n - 1$.

Step 2: Delete the fractions that are not in lowest terms (because they appear elsewhere in the array) (Figure 7).

Figure 7

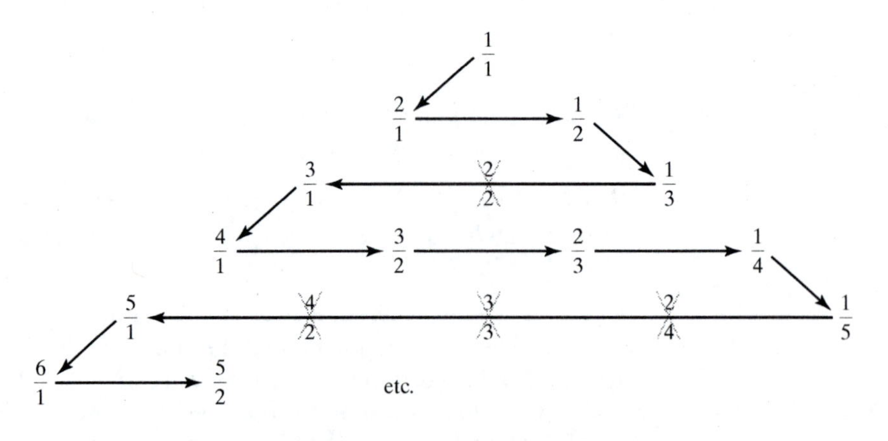

Step 3: Correspond the remaining array entries with the elements of **N** by counting across in the array, row by row. That is, $\frac{1}{1} \to 1, \frac{2}{1} \to 2, \frac{1}{2} \to 3, \frac{1}{3} \to 4, \frac{3}{1} \to 5, \frac{4}{1} \to 6, \frac{3}{2} \to 7$, etc.

This scheme establishes a one-to-one correspondence between the set $\mathbf{Q}^+$ of positive rational numbers and the set **N** of natural numbers. Consequently, $\mathbf{Q}^+$ and **N** have the same cardinality and so $\mathbf{Q}^+$ is countably infinite.

In Problem 9, you are asked to prove the following slightly stronger result.

Theorem 2.16 The set **Q** of all rational numbers is countably infinite.

There is a more surprising result.

Theorem 2.17 The set A of all algebraic numbers is countable.

You are guided through a proof of this result in Project 2 for this chapter.

Thus, the sets of natural numbers, integers, rational numbers, and algebraic numbers are all countably infinite; that is, all of these sets have the same size if size is measured by cardinality.

Sets that are not countable

Not all infinite sets are countably infinite. The following famous result, also due to Cantor, identifies one infinite set that is not countably infinite. It is proved by an indirect proof using a procedure called the *Cantor diagonalization process*.

Theorem 2.18 The open interval $(0, 1)$ of real numbers between 0 and 1 is not countably infinite.

Proof: Suppose to the contrary that there is a one-to-one correspondence between the set **N** of natural numbers and the interval $(0, 1)$. Imagine an infinite tabular description of this correspondence (Table 6) in which all the successive integers are listed at the left with their corresponding decimal representations on the right. [If a real number in $(0, 1)$ listed in Table 6 has two decimal representations, we show the infinite decimal.]

Table 6

N	$(0, 1)$
1	$0.\mathbf{d_{11}}d_{12}d_{13}d_{14}d_{15}d_{16}\ldots$
2	$0.d_{21}\mathbf{d_{22}}d_{23}d_{24}d_{25}d_{26}\ldots$
3	$0.d_{31}d_{32}\mathbf{d_{33}}d_{34}d_{35}d_{36}\ldots$
4	$0.d_{41}d_{42}d_{43}\mathbf{d_{44}}d_{45}d_{46}\ldots$

We assert that there is a real number between 0 and 1 that is not listed. Let $x = 0.d_1d_2d_3d_4\ldots$, with $d_i = 5$ unless $d_{ii} = 5$, in which case make $d_i = 4$. By this construction, each decimal place d_i differs from the ith number in the list of Table 6 in the ith decimal place (in the place we have identified in bold). Hence this real number x is not included in the list of real numbers put into the one-to-one correspondence we assumed existed. (Of course, there are many possibilities for each d_i, but we need only one at each stage of the process.) Thus, any list like that in Table 6 cannot contain all the real numbers between 0 and 1. So there cannot be a 1-1 correspondence between this interval and **N**. ┘

Since all the real numbers in the open interval $(0, 1)$ cannot be put in a list, neither can any set that contains $(0, 1)$. Among those sets is **R** itself.

Corollary 1: The set **R** of real numbers is not countably infinite.

If two sets A and B are countably infinite, then so is their union, because a list of the union can be made by alternating members of lists for A and B. **R** is the union of the set **Q** of all rational numbers and the set **I** of all irrational numbers. Since the rationals can be listed yet **R** cannot, it must be impossible to list all the irrationals.

Corollary 2: The set **I** of irrational numbers is not countably infinite.

Corollary 2 indicates that most real numbers are irrational. In fact, if a real number from **R** is chosen at random, then the probability is 0 that it is rational. How is this possible? Think about choosing the random real number $D.d_1d_2d_3\ldots$ by selecting the decimal part of the real number one digit at a time. In order for the decimal to represent a rational number, it must ultimately repeat a set of digits forever. The probability that the next digit will continue the pattern of the repetition is 0.1, and so the probability that the next n digits will continue the pattern is $(0.1)^n$. As n gets larger and larger, which has to be the case if the repetition is to be forever, the limit of this probability is 0.

There is also a geometric way of seeing that most numbers are irrational. Consider the list of rational numbers created as a part of the proof of Theorem 2.16. Cover the nth rational number by an interval of length 2^{-n}. That is, cover $\frac{1}{1}$ by an interval of length $\frac{1}{2}$; $\frac{2}{1}$ by an interval of length $\frac{1}{4}$; $\frac{1}{2}$ by an interval of length $\frac{1}{8}$; $\frac{3}{1}$ by an interval of length $\frac{1}{16}$; and so on, as shown in Figure 8.

Figure 8

The total length of these intervals is the sum of the geometric series

$$\frac{1}{2}+\frac{1}{4}+\frac{1}{8}+\frac{1}{16}+\cdots+\frac{1}{2^n}+\cdots,$$

which is 1. So all the rational numbers can fit into a set of intervals whose total length is at most 1 (the intervals could overlap). In fact, there is a stronger result (see Problem 11). A similar kind of squeezing cannot be done with the irrationals, because all the irrationals cannot be put into a single list.

2.1.4 Problems

1. Identify a rational number and an irrational number between the two given real numbers.

a. -86 and -87 b. $-.00004$ and $-.00005$

c. π and $\pi - \frac{1}{2^7}$ d. $.01$ and $\sin(.01)$

2. Show that, for any real numbers x and y with $0 < x < y$, there exist positive integers p and q such that the irrational number $s = \frac{p\sqrt{2}}{q}$ is in the interval (x, y).

3. Prove Theorem 2.13(a).

4. Prove Theorem 2.13(b).

5. Show that each number is algebraic.

a. $-\sqrt{12}$ b. $1+\sqrt{3}$ *c. $\sqrt{2}+\sqrt{3}$ *d. $1-\sqrt[3]{5}$

6. Prove that the set of transcendental numbers is not countably infinite.

7. Three partitions of the real numbers are shown in Figure 6. Name a fourth partition of the real numbers.

8. Show that the following pairs of sets have the same cardinality:

a. The set E of all even integers and the set **Z** of all integers

b. For any real numbers a, b, c, and d with $a < b$ and $c < d$, the open intervals (a, b) and (c, d) of real numbers

9. Prove that the set $\mathbf{Q}$ of all rational numbers is countably infinite.

10. Prove that a countable union of countably infinite sets is itself countably infinite. (*Hint*: Use the result of Problem 9.)

11. Show that the set of rationals can be squeezed into a set of intervals whose total length is less than ε, where ε is any fixed positive number.

ANSWERS TO QUESTIONS

1. Samples: 40.111001 and 40.1110010100100010000 1..., where the number of 0s between 1s increases by 1 each time.

2. Sample: $17x^5 - 11 = 0$.

3. The definitions are equivalent because if $p(x) = \sum_{i=0}^{n} \frac{a_i}{b_i} x^i = 0$, where the a_i and b_i are all integers, then multiplying both sides by the product of all the b_i yields a polynomial with integer coefficients and the same solutions.

Unit 2.2 The Complex Numbers

Complex numbers first arose in connection with the solution of equations. The familiar quadratic formula

$$x = \frac{-b \pm \sqrt{b^2 - 4ac}}{2a}$$

for the solutions of the quadratic equation $ax^2 + bx + c = 0$ was essentially known to the Babylonians, but negative solutions and solutions corresponding to negative values of the discriminant $b^2 - 4ac$ were rejected as meaningless. The next great step in the solution of algebraic equations was made in the sixteenth century with the solution of the general cubic and quartic equations. These exact solutions were expressed in terms of the coefficients by radical expressions similar to but more complicated than the quadratic formula. They became well known through the publication in 1545 of the book *Ars magna* by Girolamo Cardano (1501–1576). In that book, Cardano credits Niccolo Tartaglia[6] (c. 1499–1557) with the idea for the solution of the cubic, and Ludovico Ferrari (1522–1565) with the solution of the quartic.

Like the quadratic formula, the solution of cubic and quartic equations often led directly to expressions involving square roots of negative numbers. More specifically, Rafael Bombelli (c. 1526–1573) observed that the Tartaglia–Cardano solution method applied to the cubic equation $x^3 - 15x - 4 = 0$ led to intermediate expressions involving square roots of negative numbers ($\sqrt{-1}$, etc.) even though the solutions of this equation,

$$x = 4, \quad x = -2 + \sqrt{3}, \quad \text{and} \quad x = -2 - \sqrt{3},$$

are all real numbers. He and others began to use the usual rules of arithmetic with these expressions involving $\sqrt{-1}$ to solve other problems. For instance, in the 1700s, Euler used complex numbers to derive some new results in number theory. In the early 1800s, Cauchy developed an extensive theory of functions of complex numbers. However, general acceptance of these expressions as numbers was slow in coming. A measure of this reluctance is indicated by the fact that the term "real" was first coined to distinguish the numbers in general use from these new "imaginary" numbers.

[6]Tartaglia's birth name was Fontana. His nickname, Tartaglia, which means "stutterer," derives from a speech impediment that he had as a result of a saber cut he received as a child during an attack on his city by the French.

Complex numbers continued to be viewed with suspicion until their geometric representation as points in the complex plane was developed and utilized. The complex plane representation of the complex number system was discovered in 1797 by Casper Wessel (1745–1818) and independently by Jean Robert Argand (1768–1822) in 1806. This representation was not well known until the 1830s, when Gauss used it extensively in proofs of the Fundamental Theorem of Algebra (that every polynomial equation of degree ≥ 1 with complex coefficients has at least one solution in the set of complex numbers).

The geometric aspects of complex numbers were exploited by A. F. Möbius (1790–1868), of *Möbius strip* fame, a student of Gauss, who used the complex plane and complex functions to classify geometric transformations of the plane. Bernhard Riemann (1826–1866) developed the calculus of complex functions from a geometric standpoint and applied these functions to geometry and number theory.

Today, complex numbers permeate mathematics, engineering, and science. Many of their applications exploit the remarkable connections that exist between the algebraic and geometric structures of the complex numbers, some of which we examine in this unit and in other chapters of this book, especially Chapters 7, 8, and 9.

2.2.1 The complex numbers and the complex plane

As we described above, complex numbers were invented and gradually gained acceptance as a means to solve polynomial equations. The solution of such equations led to binomial expressions of the form $x + y\sqrt{-1}$, where x and y are real numbers. The nature of the factor $\sqrt{-1}$ in these "complex" numbers remained rather mysterious for some time, since no real number could possibly equal $\sqrt{-1}$ because $r^2 \geq 0$ for any real number r. However, if the question of the meaning of $\sqrt{-1}$ was set aside, mathematicians found that by treating expressions of the form $x + y\sqrt{-1}$ as if they were binomials with the added property that $(\sqrt{-1})^2 = -1$, they would obtain sensible results.

Calculations based on this binomial model such as

$$(3 + 2\sqrt{-1}) + (-5 - \sqrt{-1}) = (3 - 5) + (2 - 1)\sqrt{-1} = -2 + \sqrt{-1}$$

and

$$\begin{aligned}(3 + 2\sqrt{-1})\cdot(-5 - \sqrt{-1}) &= [(3)(-5) + (2)(-1)(-1)] + [(3)(-1) + (2)(-5)]\sqrt{-1} \\ &= -13 - 13\sqrt{-1}\end{aligned}$$

were carried out routinely from the sixteenth through the eighteenth centuries. Some discomfort with these calculations remained throughout this period because the nature of the quantity $\sqrt{-1}$ had not been satisfactorily resolved and because no geometric representation for the quantities $x + y\sqrt{-1}$ was available.

The complex plane

We can identify each complex number $x + y\sqrt{-1}$, where x and y are real numbers, with the point (x, y) in the coordinate plane $\mathbf{R} \times \mathbf{R}$, or $\mathbf{R}^2$. The resulting coordinate plane, called the **complex plane**, provides a geometric model for the set $\mathbf{C}$ of all complex numbers (Figure 9).

In this model, the point $(0, 1)$ in the complex plane represents $\sqrt{-1}$. The standard notation for this number in mathematics books is $\boldsymbol{i}$ (for imaginary), while engineering texts typically use the symbol $\mathbf{j}$ (because $\boldsymbol{i}$ or $\mathbf{I}$ stands for electric current).

The geometric placement of i as the point $(0, 1)$ is natural for the following reason. The rotation of 180° about the origin takes the point $(1, 0)$ onto the point $(-1, 0)$, and more generally maps the point (x, y) onto the point $(-x, -y)$. Thus a 180° rotation can be viewed as a geometric picture of multiplication by -1. But

Figure 9

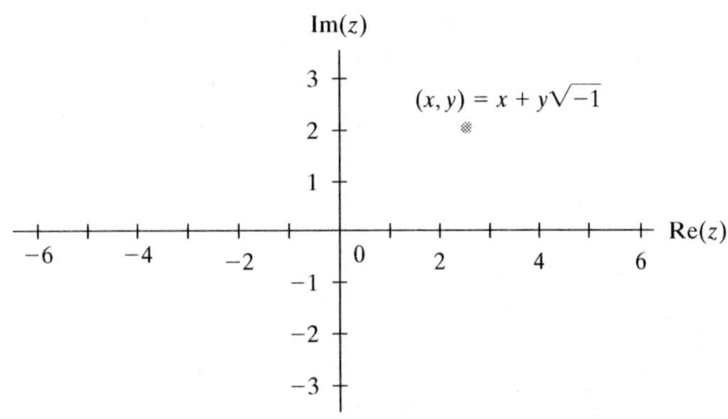

The complex plane

Figure 10

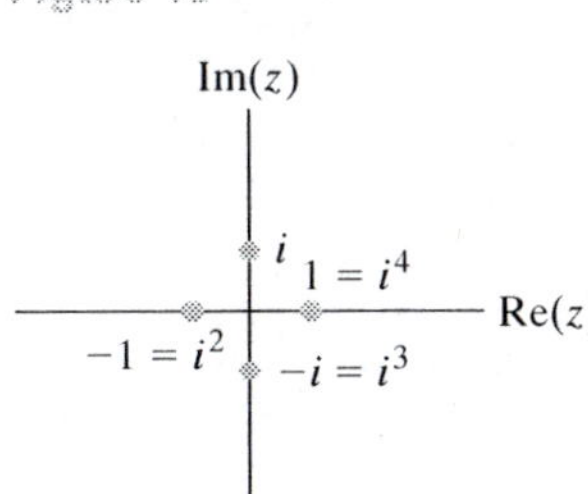

a rotation of 180° is equivalent to two counterclockwise rotations of 90°. So a rotation of 90° can be viewed as a geometric picture of multiplication by a square root of -1, that is, as multiplication by $\sqrt{-1}$ or i.

Then $i \cdot i$ or i^2 becomes pictured by two rotations of 90°, or the rotation of 180° about the origin (see Figure 10). Three 90° rotations of $(0, 1)$ about the origin map it onto $(0, -1)$, which corresponds to $i^3 = -1 \cdot \sqrt{-1} = -i$. Four 90° rotations of $(0, 1)$ about the origin map it back onto itself, which corresponds to multiplication by 1 and the identity $i^4 = 1$.

With $i = (0, 1)$, we can also view the complex number $x + y\sqrt{-1}$ as a linear combination of the unit vectors $(1, 0)$ and $(0, 1)$. That is, $x + y\sqrt{-1} = x \cdot 1 + y \cdot \sqrt{-1} = x \cdot (1, 0) + y \cdot (0, 1)$, where x and y are real numbers and the multiplication is scalar multiplication of a vector by a real number.

Having identified $x + y\sqrt{-1}$ (or $x + iy$) with the ordered pair (x, y), we must have operations on these ordered pairs in order to define the *complex number system*. Now the **binomial notation** $x + iy$ comes in handy. We define addition and multiplication on the ordered pairs as the binomial notation together with $i^2 = -1$ suggests. That is, we convert each ordered pair into binomial notation and then operate on the ordered pairs as binomials before converting back. For addition, we obtain

$$(x_1, y_1) + (x_2, y_2) = (x_1 + iy_1) + (x_2 + iy_2) = (x_1 + x_2) + (y_1 + y_2)i = (x_1 + x_2, y_1 + y_2).$$

Definition A **complex number** is an ordered pair (x, y) of real numbers x and y with addition and multiplication defined as follows:

$$(x_1, y_1) + (x_2, y_2) = (x_1 + x_2, y_1 + y_2)$$
$$(x_1, y_1) \cdot (x_2, y_2) = (x_1x_2 - y_1y_2, x_1y_2 + x_2y_1).$$

Question: Show the intermediate steps on binomials that yield the ordered pair definition of multiplication of complex numbers.

The complex number $(x, 0)$ can be identified with the real number x because addition and multiplication of numbers of this form yield normal addition and multiplication of real numbers. That is,

$$(x_1, 0) + (x_2, 0) = (x_1 + 0, y_1 + 0) = (x_1 + x_2, 0)$$
$$(x_1, 0) \cdot (x_2, 0) = (x_1 \cdot x_2 - 0 \cdot 0, x_1 \cdot 0 + x_2 \cdot 0) = (x_1x_2, 0).$$

By identifying each real number x as the complex number $(x, 0)$ and defining complex addition and multiplication of complex numbers as the binomial notation suggests, we find that $z = x + iy$ is indeed the complex number sum of the real number x and the number iy. If $z = (x, y)$, the real number x is called the **real part** of z and is denoted by **Re(z)**, while the real number y is called the **imaginary part** of z and is denoted by **Im(z)**.[7] Notice that both the real and imaginary parts of z are real numbers.

Since each real number x is identified with the ordered pair $(x, 0)$, the real number line corresponds to the horizontal axis in the complex plane. For this reason, the horizontal axis is usually called the **real axis**. The vertical axis is called the **imaginary axis** for a similar reason: Multiples yi of $i = (0, 1)$ are identified with the ordered pairs $(0, y)$ in the complex plane. Complex numbers on the imaginary axis are called **imaginary numbers**. The number i is often called the **imaginary unit**, and the point $(1, 0)$ representing the real number 1 is called the **real unit**, because they mark unit distances on the imaginary and real axes.

When a and b are rational numbers, and c is a positive rational number but not the square of a rational number, then $a + b\sqrt{c}$ and $a - b\sqrt{c}$ are called **irrational conjugates** (from the Latin "conjugatus" for "joined together"). This name arose because these numbers appear together as solutions to the same quadratic equation with rational coefficients. Similarly, for any complex number $z = x + iy$, the complex number $\bar{z} = x - iy$ is called the **complex conjugate** of $z = x + iy$. Geometrically, the complex conjugate $\bar{z} = x - iy$ is the reflection image of $z = x + iy$ over the real axis (see Section 7.2.3).

Just as $|x|$ is the distance from x to 0 on the real number line, $|z|$ is the distance of z from the origin in the complex plane. So $|z| = \sqrt{x^2 + y^2}$ is the **absolute value** or **modulus** of $z = x + iy$. For example, if $z = -2 + 3i$, then $\bar{z} = -2 - 3i$, $\text{Re}(z) = -2$, $\text{Im}(z) = 3$, and $|z| = \sqrt{(-2)^2 + (3)^2} = \sqrt{13}$. The set U of all complex numbers z such that $|z| = 1$ is the **unit circle** in the complex plane (see Figure 11).

Figure 11

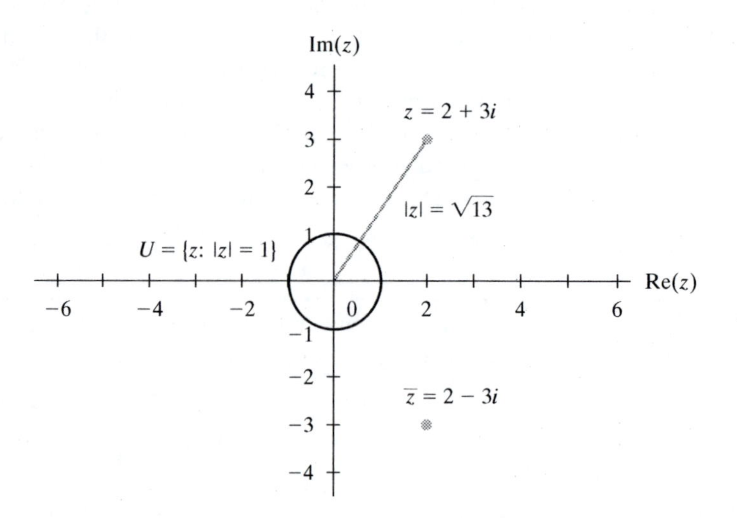

The migration of polynomial zeros in the complex plane

Before we proceed further with the discussion of complex numbers, we present the following example to show that the complex plane representation of complex numbers fits well with their historical origins in the solution of polynomial equations. It illustrates the important fact that the locations in the complex plane of the solutions of a polynomial equation $p(x) = 0$ vary continuously with the coefficients of $p(x)$; that is, continuous changes in the coefficients of the polynomial $p(x)$ result in continuous changes in the locations of the solutions in the complex plane.

---

[7] Other notations in use include $\text{Re}(z)$ and $\text{Im}(z)$, $\text{R}(z)$ and $\text{I}(z)$, and Rz and Iz.

EXAMPLE 1 Graph the solutions to the following three quadratic equations in the complex plane, and discuss what has happened.

a. $x^2 + 2x - 8 = 0$

b. $x^2 + 2x + 1 = 0$

c. $x^2 + 2x + 2 = 0$

Solution The given equations differ only by their constant coefficients. The quadratic formula yields the following pairs of solutions:

a. $x = -1 \pm 3 = -4$ or 2

b. $x = -1$ (double root)

c. $x = -1 \pm i$

The graphs of the three equations, $y = x^2 + 2x - 8$, $y = x^2 + 2x + 1$, and $y = x^2 + 2x + 2$, are displayed in Figure 12a. The corresponding solutions are displayed in Figure 12b in the complex plane.

Figure 12

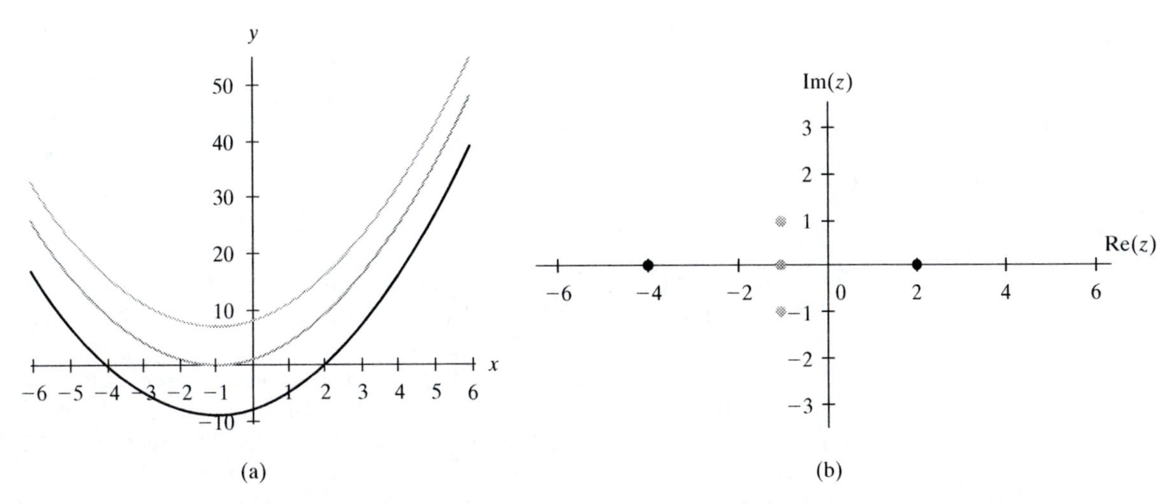

(a) (b)

As the constant term increases continuously from -8 (for the lowest graph in Figure 12a) to 1 (for the middle graph in Figure 12a), the two solutions $x = -4$ and $x = 2$ move continuously inward along the real axis (in Figure 12b) toward the double root $x = -1$ for equation b. As the constant term continues to increase from 1 (for the middle graph in Figure 12a) to 2 (for the upper graph in Figure 12a), the double root splits (in Figure 12b) into a pair of complex conjugate solutions that migrate continuously toward $x = -1 + i$ and $x = -1 - i$.

The continuous migration of the solution set in Example 1 becomes clearer if we compute the solution set S_c of $x^2 + 2x + c = 0$ using the quadratic formula.

$$S_c = \left\{ \frac{-2 \pm \sqrt{4 - 4c}}{2} \right\} = \{-1 \pm \sqrt{1 - c}\}$$

When $c < 1$, S_c consists of two points on the real axis that are symmetric to -1, and the distance between these points increases as c decreases. At $c = 1$, S_c consists of a single number -1, and as c increases from 1, the set S_c consists of two points symmetric to the real axis on the vertical line $x = -1$. The distance between these points steadily increases as the value of c increases.

Polar forms of complex numbers

As we have seen, a complex number $z = x + iy$ is represented in the complex plane by the point with rectangular coordinates (x, y). For this reason, $x + iy$ is also called the **rectangular form** of the complex number z. Complex numbers also have a very useful and descriptive polar representation based on polar coordinates $[r, \theta]$ of points in the complex plane.

Figure 13

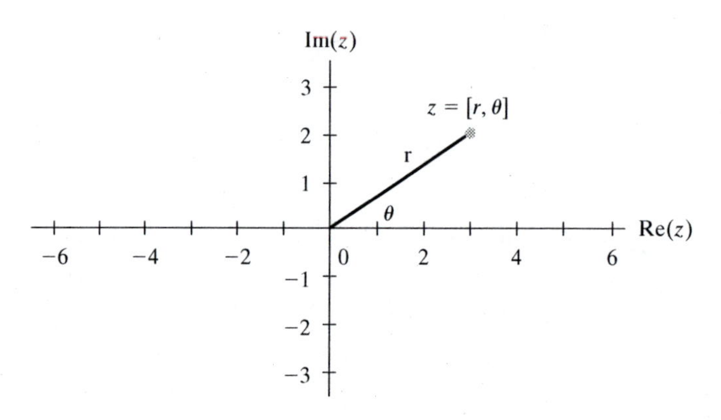

Recall that if (x, y) is the point in the complex plane representing the nonzero complex number $z = x + iy$, then **polar coordinates** $[r, \theta]$ of that point (see Figure 13) are given by

(1) $$r = \sqrt{x^2 + y^2}$$

(2) $$\cos\theta = \frac{x}{r},\ \sin\theta = \frac{y}{r},\ \text{and } -\pi < \theta \leq \pi.$$

We use brackets [,] to distinguish polar coordinates from rectangular coordinates (,). Many books use parentheses for both types of coordinates, relying on the context to inform the reader.

Every point has infinitely many polar coordinates. If $[r, \theta]$ are polar coordinates of a point z in the complex plane, then other polar coordinates of z are $[r, \theta + 2k\pi]$ for any integer k, and $[-r, \theta + (2k + 1)\pi]$ for any integer k.

The r-coordinate given by (1) is just the absolute value $|z|$ or modulus of the complex number z. The θ-coordinate determined by (2) is called the **principal angle** or **principal argument** of z and is denoted by **Arg(z)**.

Conversely, if $[r, \theta]$ is any pair of polar coordinates of a point z in the complex plane, then $x = r\cos\theta$ and $y = r\sin\theta$ determine the rectangular coordinates (x, y) and the rectangular representation $x + iy$ of the complex number z. In particular, the rectangular form of a complex number $z = x + iy$ can be expressed in terms of any polar representation $[r, \theta]$ of z using the conversion

$$z = r(\cos\theta + i\sin\theta).$$

Any such representation is called a **polar representation** or **polar form** of z.

For example, if $z = -3 + 4i$, then $r = \sqrt{(-3)^2 + 4^2} = 5 = |z|$, $\sin\theta = \frac{4}{5}$, and $\cos\theta = -\frac{3}{5}$. From these values of $\sin\theta$ and $\cos\theta$, $\text{Arg}(z) \approx 2.214$ radians. So $[5, 2.214]$ is an approximate pair of polar coordinates of z. Other polar coordinates of z are given by $[5, 2.214 + 2k\pi]$ and by $[-5, 2.214 + (2k + 1)\pi]$ for any integer k. You can check this by overlaying a rectangular coordinate system on the polar coordinate system. The point $(-3, 4)$ should coincide with $[5, 2.214\ldots]$.

The polar form of complex numbers is used extensively for the construction and geometric interpretation of products, quotients, powers and roots of complex numbers, as we show in Section 2.2.2.

The Riemann sphere

Although the complex plane is the geometric model of the complex number system that is used most commonly, another geometric model is very interesting and useful for understanding the complex number system. It is called the *Riemann sphere.*

The Riemann sphere is depicted in the three-dimensional diagram in Figure 14, which a sphere of diameter 1 is placed on the complex plane so that the south pole of the sphere rests on the origin.

Figure 14

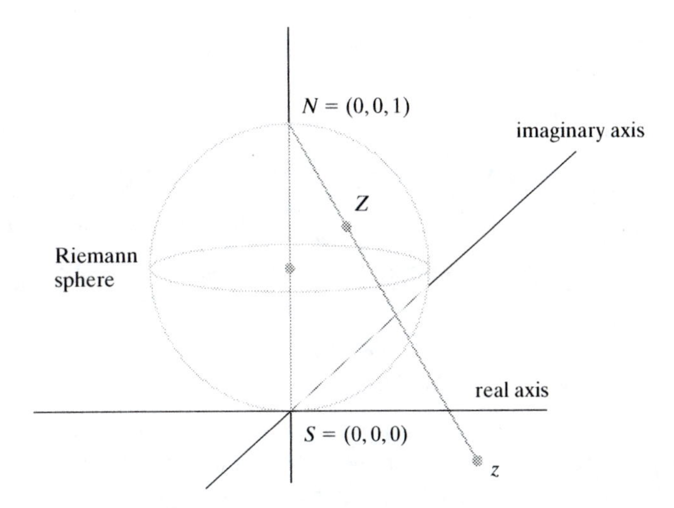

The line joining the north pole N of the Riemann sphere and a given complex number z in the complex plane intersects the Riemann sphere at a point Z different from N. This point on the sphere serves as a representation of the complex number z on the Riemann sphere. In this way, every complex number corresponds to a unique point on the Riemann sphere, and every point on the Riemann sphere except the north pole corresponds to a point in the complex plane. This correspondence is called **stereographic projection**.

Although no complex number corresponds to the north pole of the Riemann sphere under stereographic projection, as a point z in the complex plane moves further and further from the origin in the complex plane, the corresponding point z on the Riemann sphere moves closer and closer to the north pole N on the Riemann sphere. For this reason, the North Pole on the Riemann sphere is often called the **point at infinity** and denoted by ∞. The Riemann sphere (with the point at infinity removed) is a geometric model of the complex number system that is an alternative to the complex plane model.

2.2.1 Problems

1. Rewrite each complex number in all the forms not given: binomial form, rectangular coordinates, polar form, polar coordinates. Then give its real part, imaginary part, and absolute value.

a. $-5 + 2i$ b. $[3, 235°]$ c. $0.5\cos(\frac{\pi}{5}) + 0.5i\sin(\frac{\pi}{5})$

d. $(12, -2)$ e. $\frac{4+i}{4-i}$ f. i^{430}

2. Prove the following properties of any complex number z.

a. $\text{Re}[z] = \frac{z+\bar{z}}{2}$ b. $\text{Im}[z] = \frac{z-\bar{z}}{2i}$

3. Suppose $[r, \theta]$ is a point z in the complex plane, $x = r\cos\theta$, and $y = r\sin\theta$. If $r' = -r$ and $\theta' = \theta + \pi(2n+1)$, where n is an integer, prove that $[r', \theta']$ determines the same rectangular representation (x, y) of z.

4. Track the solution set in the complex plane of the quadratic equation $x^2 + bx + 2 = 0$ as the value of the real coefficient b varies.

5. Use a calculator or computer graphing and computer algebra utility to track the solution sets in the complex plane of the quartic polynomial equation $4x^4 + 8x^3 - 3x^2 - 9x + c = 0$ as the constant term c ranges from -4 to 6 in increments of 2.

6. Describe precisely the set on the Riemann sphere that corresponds to the indicated set in the complex plane.

a. the unit circle $\{z \in C: |z| = 1\}$

b. the real axis $\{z \in C: \text{Im}(z) = 0\}$

c. the imaginary axis $\{z \in C: \text{Re}(z) = 0\}$

d. a straight line in the complex plane that passes through the origin

e. a straight line in the complex plane that does not pass through the origin

7. In the complex plane, the point representing the complex conjugate $\bar{z}$ of z is the reflection image of z over the real axis. Give a similar description of the relationship between z and $\bar{z}$ on the Riemann sphere.

ANSWER TO QUESTION

1. $(x_1, y_1) \cdot (x_2, y_2) = (x_1 + iy_1)(x_2 + iy_2) = x_1x_2 + ix_1y_2 + iy_1x_2 + i^2y_1y_2 = (x_1x_2 - y_1y_2) + (x_1y_2 + x_2y_1)i$
$= (x_1x_2 - y_1y_2, x_1y_2 + x_2y_1)$

2.2.2 The geometry of complex number arithmetic

In Section 2.2.1, we introduced four different notations for complex numbers:

binomial form $a + bi$ or $a + b\sqrt{-1}$
rectangular form (a, b)
polar coordinates $[r, \theta]$
polar form $r(\cos\theta + i\sin\theta)$

A fifth common form, involving the same variables as polar coordinates and polar form, was first derived by Leonhard Euler (1707–1783) by manipulating the infinite series for the sine, cosine, and the base e of natural logarithms.

exponential form $re^{i\theta}$

As with most mathematical ideas, the existence of a variety of notations often signals the importance of the idea. With complex numbers, this variety provides additional benefits because certain properties can be more simply stated in one notation than another, and perhaps better understood in one notation than another.

Addition and subtraction in C

As we mentioned in the previous section, complex numbers were first treated as if they were binomials with the additional property that $i^2 = -1$.

When $z_1 = x_1 + iy_1$ and $z_2 = x_2 + iy_2$, their sum is

$$z_1 + z_2 = (x_1 + x_2) + i(y_1 + y_2),$$

and their difference is given by

$$z_1 - z_2 = (x_1 - x_2) + i(y_1 - y_2).$$

Addition and subtraction of complex numbers corresponds exactly to vector addition and subtraction if we regard complex numbers as vectors in the complex plane with their real and imaginary parts as components. This is seen by using rectangular rather than binomial form in the definitions of addition and subtraction. If $z_1 = (x_1, y_1)$ and $z_2 = (x_2, y_2)$, then

$$z_1 + z_2 = (x_1 + x_2, y_1 + y_2)$$

and

$$z_1 - z_2 = (x_1 - x_2, y_1 - y_2).$$

When $0, z_1$, and z_2 are not collinear, the sum $z_1 + z_2$ is the fourth vertex of a parallelogram with consecutive vertices z_1, 0, and z_2. This is called the **Parallelogram Law** in both vector and complex number addition.

Figure 15

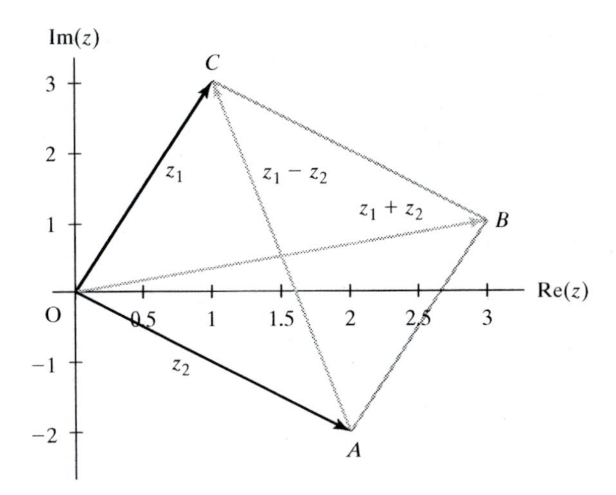

We have pictured addition and subtraction in Figure 15. There the vector $\overrightarrow{OB} = z_1 + z_2$ is the sum of the vectors $\overrightarrow{OC} = z_1$ and $\overrightarrow{OA} = z_2$. Notice also that $\overrightarrow{OA} + \overrightarrow{AC} = \overrightarrow{OC}$, or in complex number terms, $z_2 + \overrightarrow{AC} = z_1$. Consequently, $\overrightarrow{AC} = z_1 - z_2$. You may recall that the **norm** or **length** $\|\mathrm{v}\|$ **of a 2-dimensional vector** $\vec{v} = (x, y)$ is $\sqrt{x^2 + y^2}$, which is identical to the formula for the absolute value of the complex number (x, y). Consequently, $\|\overrightarrow{AC}\| = |z_1 - z_2|$. We have given a vector proof of the following theorem.

This theorem can also be proved algebraically using the components of z_1, z_2, and $z_1 - z_2$.

Theorem 2.19 **(Distance Formula):** The distance between z_1 and z_2 in the complex plane is $|z_1 - z_2|$.

Recall that the distance between the real numbers x and y on the number line is given by $|x - y|$. Theorem 2.19 shows that the same property holds for complex numbers in the complex plane. With real numbers, when $b > 0$ the equation $|x - a| = b$ yields the points x whose distance from a is b. Similarly, in **C**, if r is a positive real number, the equation $|z - z_0| = r$ yields those complex numbers whose distance from z_0 is r. For example, an equation for the circle K centered at $3 + i$ with radius 2 is $|z - (3 + i)| = 2$.

Question:

a. Identify 4 points on the circle $\boldsymbol{K}$.

b. Find an equation in rectangular coordinates for this circle.

These properties demonstrate that the complex numbers can provide an alternate way of describing some figures in Euclidean geometry. (See Problems 6 and 7.)

Recall one of the most important propositions in Euclidean geometry—the *Triangle Inequality*: that for any points A, B, and C, $AC \leq AB + BC$, with equality if and only if the points are collinear with B between A and C. If the points are not collinear, this inequality tells us that the sum of the lengths of any two sides of a triangle is greater than the length of the third side.

The triangle inequality can be stated in terms of complex numbers as

$$|z_1| + |z_2| > |z_1 + z_2|$$

Figure 16

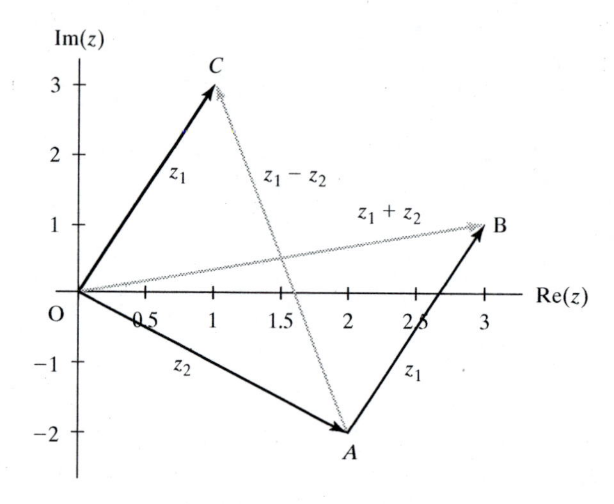

provided that 0, z_1, and z_2 are vertices of a triangle, as in ΔOAB of Figure 16. When 0, z_1, and z_2 are collinear, then

$$|z_1| + |z_2| > |z_1 + z_2| \text{ if 0 is between } z_1 \text{ and } z_2,$$

and

$$|z_1| + |z_2| = |z_1 + z_2| \text{ if 0 is not between } z_1 \text{ and } z_2.$$

As a result, we have proved the following theorem, which is often the way the triangle inequality is stated for complex numbers. The absolute value of the sum of two complex numbers does not exceed the sum of their absolute values.

Theorem 2.20 **(Triangle Inequality):** For all complex numbers z_1 and z_2,

$$|z_1 + z_2| \leq |z_1| + |z_2|.$$

From the Triangle Inequality, the following related inequality can be deduced. It is seen from ΔOAC in Figure 16. Since $OA + AC \geq OC$, $|z_2| + |z_1 - z_2| \geq |z_1|$. So $|z_1 - z_2| \geq |z_1| - |z_2|$. Similarly, since $OC + AC \geq OA$, $|z_1| + |z_1 - z_2| \geq |z_2|$. So $|z_1 - z_2| \geq |z_2| - |z_1|$. Putting these conclusions together, the absolute value of the difference of two complex numbers is greater than the absolute value of the difference of the absolute values of the numbers.

Corollary: For all complex numbers z_1 and z_2,

$$|z_1 - z_2| \geq \big||z_1| - |z_2|\big|.$$

Multiplication and division in C

The binomial form provides a way to determine the product of two complex numbers. If $z_1 = x_1 + iy_1$ and $z_2 = x_2 + iy_2$, then $z_1 z_2 = (x_1 + iy_1)(x_2 + iy_2) = (x_1x_2 - y_1y_2)$

$+ i(x_1y_2 + x_2y_1)$. In particular, the product of a complex number $z = x + iy$ and its complex conjugate $\bar{z} = x - iy$ is $x^2 + y^2 = |z|^2$. That is, the product of a complex number and its conjugate is the square of its absolute value:

$$z\bar{z} = |z|^2.$$

This identity is useful for obtaining the **quotient** of two complex numbers $z_1 = x_1 + iy_1$ and $z_2 = x_2 + iy_2$ with $z_2 \neq 0$. We note that

$$\frac{z_1}{z_2} = \frac{z_1\overline{z_2}}{z_2\overline{z_2}} = \frac{z_1\overline{z_2}}{|z_2|^2} = \frac{(x_1x_2 + y_1y_2) + i(y_1x_2 - x_1y_2)}{x_2^2 + y_2^2}.$$

That is, to find the quotient $\frac{z_1}{z_2}$, multiply the numerator z_1 by the complex conjugate $\overline{z_2}$ of the denominator and divide the result by the square of the absolute value of the denominator.

The binomial form expressions for z_1z_2 and $\frac{z_1}{z_2}$ do not have the immediate obviousness of the corresponding expressions for $z_1 + z_2$ and $z_1 - z_2$. In rectangular coordinates, the expressions are as complicated. If $z_1 = (x_1, y_1)$ and $z_2 = (x_2, y_2)$, then

$$z_1z_2 = (x_1x_2 - y_1y_2, x_1y_2 + x_2y_1)$$

and

$$\frac{z_1}{z_2} = \left(\frac{x_1x_2 + y_1y_2}{x_2^2 + y_2^2}, \frac{y_1x_2 - x_1y_2}{x_2^2 + y_2^2}\right).$$

However, the geometric nature of multiplication and division of complex numbers is revealed by their polar form. If

$$z_1 = r_1(\cos\theta_1 + i\sin\theta_1) \quad \text{and} \quad z_2 = r_2(\cos\theta_2 + i\sin\theta_2),$$

then the definitions of multiplication and division of complex numbers given above result in the following polar forms for z_1z_2 and $\frac{z_1}{z_2}$:

$$\begin{aligned}
z_1z_2 &= (x_1x_2 - y_1y_2) + i(x_1y_2 + x_2y_1) \\
&= r_1r_2[(\cos\theta_1\cos\theta_2 - \sin\theta_1\sin\theta_2) + i(\cos\theta_1\sin\theta_2 + \sin\theta_1\cos\theta_2)] \\
\frac{z_1}{z_2} &= \frac{z_1\overline{z_2}}{z_2\overline{z_2}} = \frac{z_1\overline{z_2}}{|z_2|^2} = \frac{(x_1x_2 + y_1y_2) + i(y_1x_2 - x_1y_2)}{x_2^2 + y_2^2} \\
&= \frac{r_1}{r_2}[(\cos\theta_1\cos\theta_2 + \sin\theta_1\sin\theta_2) + i(\sin\theta_1\cos\theta_2 - \cos\theta_1\sin\theta_2)].
\end{aligned}$$

We can then apply the trigonometric identities

$$\begin{aligned}
\cos(\alpha)\cos(\beta) \pm \sin(\alpha)\sin(\beta) &= \cos(\alpha \pm \beta) \\
\sin(\alpha)\cos(\beta) \pm \cos(\alpha)\sin(\beta) &= \sin(\alpha \pm \beta)
\end{aligned}$$

to obtain the following simple and very revealing polar forms, first discovered by Abraham DeMoivre (pronounced di mwav') (1667–1754).

$$z_1z_2 = r_1r_2[\cos(\theta_1 + \theta_2) + i\sin(\theta_1 + \theta_2)]; \qquad \frac{z_1}{z_2} = \frac{r_1}{r_2}[\cos(\theta_1 - \theta_2) + i\sin(\theta_1 - \theta_2)]$$

Thus, to multiply two complex numbers in polar form, multiply their absolute values and add their arguments, and to divide two such numbers, divide the corresponding absolute values and subtract the corresponding arguments. Geometric pictures of these operations are shown in Figures 17 and 18.

Figure 17

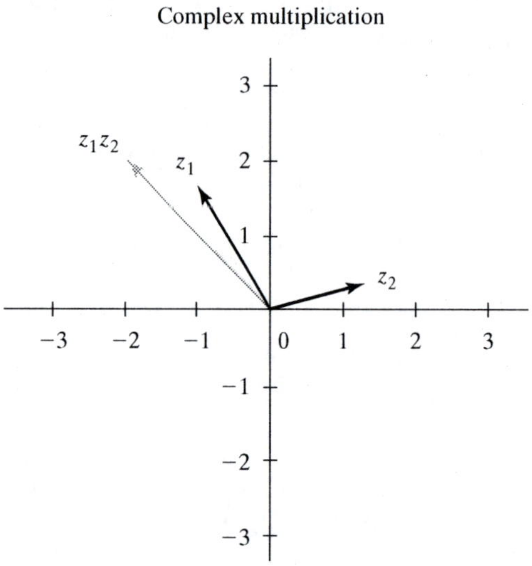

Complex multiplication

Figure 18

Complex division

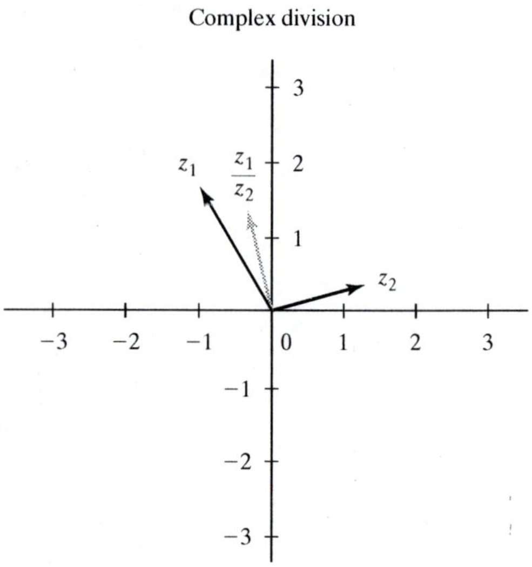

The result is even more revealing if polar coordinates are used instead of polar form. We then obtain the following result.

Theorem 2.21 Let $z_1 = [r_1, \theta_1]$ and $z_2 = [r_2, \theta_2]$. Then

$$z_1 z_2 = [r_1 r_2, \theta_1 + \theta_2]$$

and

$$\frac{z_1}{z_2} = \left[\frac{r_1}{r_2}, \theta_1 - \theta_2\right].$$

EXAMPLE 1 Let $z_1 = 2(\cos(\frac{2\pi}{3}) + \mathrm{i}\sin(\frac{2\pi}{3}))$ and $z_2 = \sqrt{3}(\cos(\frac{\pi}{12}) + \mathrm{i}\sin(\frac{\pi}{12}))$. Find $z_1 z_2$ and $\frac{z_1}{z_2}$.

Solution Think in polar coordinates: $z_1 = [2, \frac{2\pi}{3}]$ and $z_2 = [\sqrt{3}, \frac{\pi}{12}]$. Then to find $z_1 z_2$, multiply the r-coordinates and add the θ-coordinates.

$$z_1 z_2 = 2\sqrt{3}\left(\cos\left(\frac{3\pi}{4}\right) + i\sin\left(\frac{3\pi}{4}\right)\right)$$

To find the quotient, divide the r-coordinates and subtract the θ-coordinates.

$$\frac{z_1}{z_2} = \frac{2}{\sqrt{3}}\left(\cos\left(\frac{7\pi}{12}\right) + i\sin\left(\frac{7\pi}{12}\right)\right).$$

The geometry of multiplication is perhaps clearer to understand if multiplication by a specific complex number $z = [r, \theta]$ is viewed as a unary operation. This operation maps a complex number k onto another whose distance from the origin is r times as far as k and rotated θ counterclockwise around the origin. Specifically, because $i = [1, 90°]$ in polar coordinates, multiplying k by the number i does not change the distance of k from the origin, but it rotates it 90°, just as we saw in the previous section.

Powers in C

Perhaps the most visual of all basic complex number operations is the one that is the most difficult to describe geometrically with real numbers: the calculation of powers.

In polar coordinates, $[r,\theta]^2 = [r,\theta][r,\theta] = [r\cdot r, \theta+\theta] = [r^2, 2\theta]$. It is not difficult to prove by mathematical induction that $[r,\theta]^n = [r^n, n\theta]$. The theorem is usually given in polar form.

Theorem 2.22 **(Calculation of Powers in C):** Let $z = r(\cos\theta + i\sin\theta)$ be any complex number and n be any positive integer. Then $z^n = r^n(\cos(n\theta) + i\sin(n\theta))$ or, equivalently, $[r,\theta]^n = [r^n, n\theta]$.

Proof: A proof is left to you. ┘

In particular, note that $i^2 = -1$, $i^3 = -i$, $i^4 = 1$, $i^5 = i$, and, in general, $i^{4k} = 1$; $i^{4k+1} = i$; $i^{4k+2} = -1$; and $i^{4k+3} = -i$ for each nonnegative integer k. That is, the set of powers of i contains exactly 4 points. This set is called the *orbit* of i.

Definition Suppose that z is a complex number. Then the set $O(z) = \{z^n : n \in N\}$ of all positive integer powers of z is called the **orbit** of z.

For example, $O(i) = \{i, -1, -i, 1\}$.
The orbit of a complex number z with $|z| > 1$, such as

$$z = 1.1\left(\cos\left(\frac{\pi}{8}\right) + i\sin\left(\frac{\pi}{8}\right)\right),$$

is a sequence of infinitely many different points, spiraling outward from the origin, each 1.1 times as far from the origin as the previous, with the arguments of successive points in the sequence increasing by $\frac{\pi}{8}$. (See Figure 19.)

Figure 19

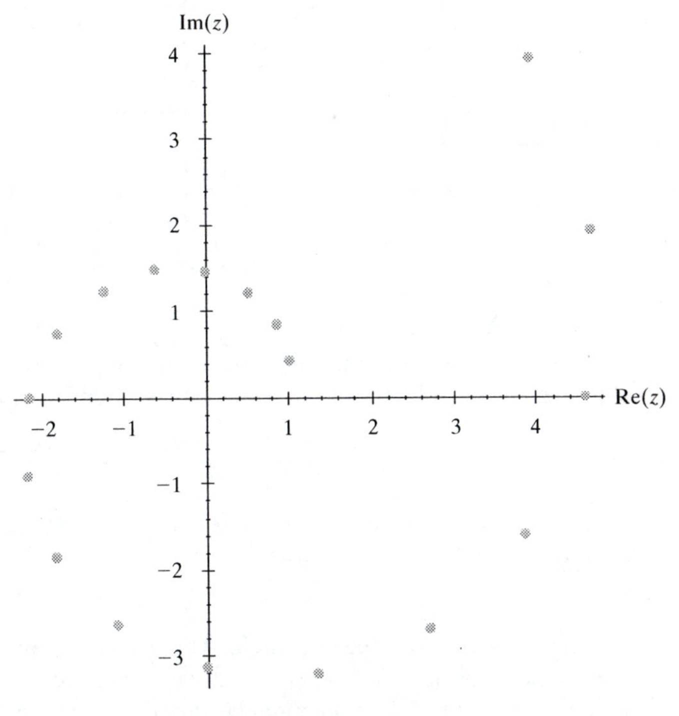

First 18 powers of $[1.1, \frac{\pi}{8}]$

Figure 20

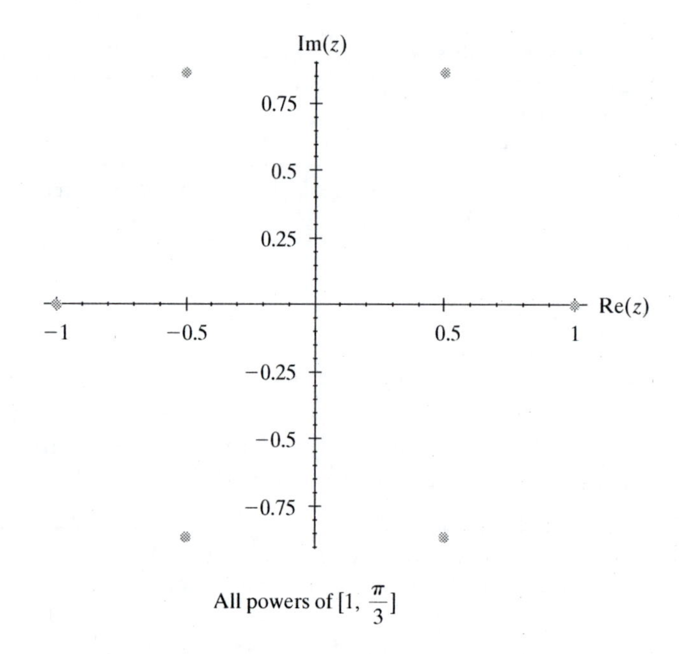

All powers of $[1, \frac{\pi}{3}]$

Orbits of complex numbers on the unit circle have a number of interesting properties (see Problem 10). In particular, very small changes in the complex number z can result in very large changes in the corresponding orbit. To illustrate the chaotic dependence of the orbit $O(z)$ of a complex number z on the unit circle, compare the orbits of the two nearby complex numbers $z_1 = [1, \frac{\pi}{3}]$ and $z_2 = [1, 1.05]$. The graph in Figure 20 displays the orbit of z_1. Figure 21 displays the first 50, 200, and 400 terms of the orbit of z_2.

Figure 21

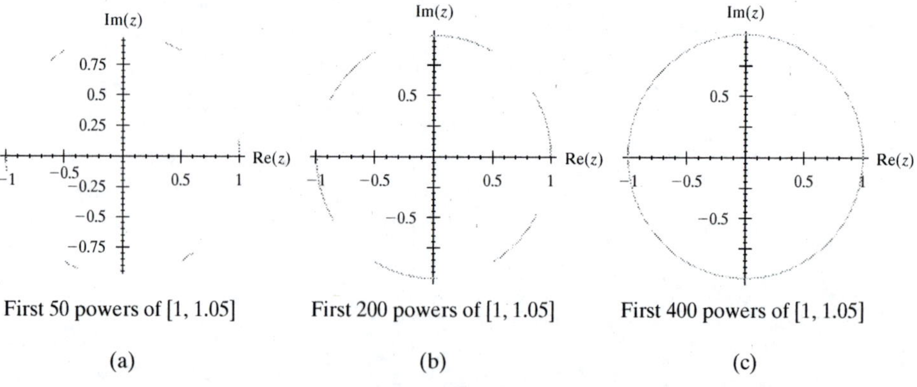

First 50 powers of [1, 1.05] First 200 powers of [1, 1.05] First 400 powers of [1, 1.05]

(a) (b) (c)

Notice the difference in the set of integer powers of $[1, \frac{\pi}{3}]$ and $[1, 1.05]$. Although the number $\frac{\pi}{3}$ is very close to 1.05, the orbit of the complex number

$$z_1 = \cos\left(\frac{\pi}{3}\right) + i \sin\left(\frac{\pi}{3}\right)$$

consists of the six vertices of a regular hexagon inscribed in the unit circle with z_1 as one vertex, while the orbit of the complex number

$$z_2 = \cos(1.05) + i \sin(1.05)$$

is everywhere dense in the unit circle. This behavior is in sharp contrast to the migration of the zeros of $x^2 + 2x + c = 0$ discussed in Section 2.2.1. There, small changes in the coefficients of the polynomial resulted in correspondingly small changes in the set of zeros of the polynomial in the complex plane.

Roots of complex numbers

If n is a natural number, any solution x to the equation $x^n = a$ is called an ***n*th root** of a. That is, if a is the nth power of x, then x is an nth root of a. When $a = 0$, the equation $x^n = 0$ has the solution 0 for all natural numbers n, so the nth root of 0 is 0.

If a is a nonzero real number, the existence of its real nth roots is somewhat complicated:

1. If n is an even integer and a is positive, then a has two real nth roots, one that is positive and is denoted by $\sqrt[n]{\boldsymbol{a}}$ or $\mathbf{a}^{1/n}$ and the other negative and denoted by $-\sqrt[n]{\boldsymbol{a}}$ or $-\boldsymbol{a}^{1/n}$.
2. If n is an even integer and a is negative, then a has no real nth roots.
3. If n is an odd integer, then a has a unique real nth root, denoted by $\sqrt[n]{a}$.

For complex numbers, the structure of nth roots is much simpler than it is for real numbers, as is suggested by the following example. This example also shows the elegance of polar coordinate notation.

EXAMPLE 2 Compute all of the 5th roots of the complex number $a = -1 - i$.

Solution First, we rewrite a in polar coordinates.

$$|a| = \sqrt{(-1)^2 + (-1)^2} = \sqrt{2} \quad \text{and} \quad \operatorname{Arg}(a) = -\frac{3\pi}{4}.$$

Consequently, we want to determine all complex numbers $z = [r, \theta]$ such that $z^5 = a = [\sqrt{2}, -\frac{3\pi}{4}]$. By Theorem 2.22,

$$z^5 = [r^5, 5\theta]$$

So we solve $r^5 = \sqrt{2}$ and $5\theta = -\frac{3\pi}{4}$ for r and θ to obtain one fifth root.

$$[r, \theta] = \left[\sqrt[10]{2}, -\frac{3\pi}{20}\right]$$

Now we use other polar coordinates $[r, \theta]$ of the point $[\sqrt{2}, -\frac{3\pi}{4}]$. Those with $r > 0$ are given by $[\sqrt{2}, -\frac{3\pi}{4} + 2\pi k]$, where k is any integer. Because for any integer k the equation

$$5\theta = -\frac{3\pi}{4} + 2\pi k$$

has the solutions $\theta = \dfrac{-\frac{3\pi}{4} + 2k\pi}{5} = -\frac{3\pi}{20} + \frac{2\pi k}{5}$, it follows that $z_k = [\sqrt[10]{2}, -\frac{3\pi}{20} + \frac{2\pi k}{5}]$ are also 5th roots of $a = -1 - i$ for each integer k. However, the sine and cosine functions are periodic with period 2π, so all distinct 5th roots of $a = -1 + i$ are given by $z_k = [\sqrt[10]{2}, -\frac{3\pi}{20} + \frac{2\pi k}{5}]$ for $k = 0, 1, 2, 3,$ and 4. The graph in Figure 22 displays the five distinct 5th roots of $a = -1 - i$. Notice that they are located at the vertices of a regular pentagon.

Figure 22

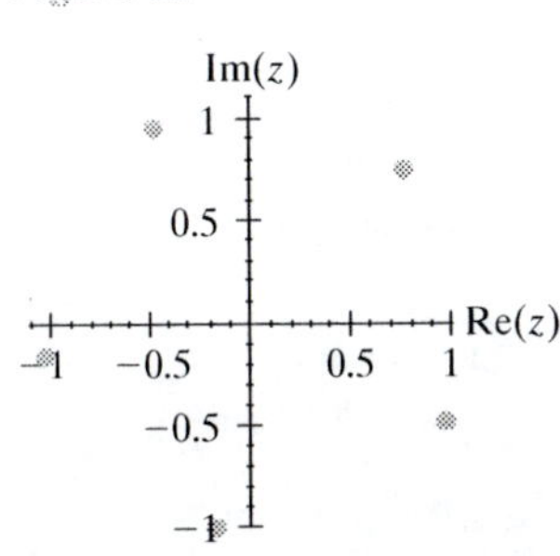

The procedure of Example 2 generalizes to prove the following important result also due to DeMoivre. Few results in all of mathematics so elegantly display the interrelationship of algebra and geometry.

Theorem 2.23 **(DeMoivre's Theorem):** For every natural number $n > 1$, every nonzero complex number z has exactly n distinct complex nth roots:

$$z_k = \sqrt[n]{|z|}\left(\cos\left(\frac{\mathrm{Arg}[z] + 2k\pi}{n}\right) + i\sin\left(\frac{\mathrm{Arg}[z] + 2k\pi}{n}\right)\right),$$

for $k = 0, 1, \ldots, n - 1$.

These points in the complex plane are the vertices of a regular polygon inscribed in the circle of radius $\sqrt[n]{|z|}$ centered at the origin in the complex plane.

Proof: You are asked to carry out the details of this generalization in Problem 11. ∎

The nth roots of 1

For any positive integer n, DeMoivre's Theorem shows the real number 1 has n complex nth roots

$$\left[1, \frac{2k\pi}{n}\right], \quad \text{for } k = 0, 1, \ldots, n - 1.$$

These numbers are customarily called the ***n*th roots of unity**. These solutions to $z^n = 1$ are the vertices of a regular n-gon inscribed in the unit circle with 1 at one vertex. In polar form, the first vertex counterclockwise from 1 is the point

$$\omega_n = \cos\left(\frac{2\pi}{n}\right) + i\sin\left(\frac{2\pi}{n}\right)$$

Notice that $\omega_n^2 = \cos(\frac{4\pi}{n}) + \mathrm{i}\sin(\frac{4\pi}{n})$, $\omega_n^3 = \cos(\frac{6\pi}{n}) + \mathrm{i}\sin(\frac{6\pi}{n})$, and, in general, that

$$\omega_n^k = \cos\left(\frac{2k\pi}{n}\right) + i\sin\left(\frac{2k\pi}{n}\right)$$

for any positive integer k. Thus, the set of nth roots of unity can be written as

$$\{1, \omega_n, \omega_n^2, \ldots, \omega_n^{n-1}\}$$

The nth roots of any complex number z can be expressed in terms of the nth roots of unity as follows: If z_0 is any particular nth root of z, then the set of all nth roots of z is given by

$$\{z_0, z_0\omega_n, z_0\omega_n^2, \ldots, z_0\omega_n^{n-1}\}.$$

2.2.2 Problems

1. Find $z_1 + z_2$, $z_1 - z_2$, z_1z_2, and $\frac{z_1}{z_2}$.
   a. $z_1 = -2 + 8i$, $z_2 = -2 - 8i$
   b. $z_1 = \left(\frac{1}{2}, \frac{\sqrt{3}}{2}\right)$, $z_2 = (1, 0)$
   c. $z_1 = [3, 225°]$, $z_2 = 7\left(\cos\frac{5\pi}{3} + i\sin\frac{5\pi}{3}\right)$
   d. z_1 and z_2 are the solutions to $x^2 + x + 1 = 0$.

2. Let $z_1 = 3 - 6i$ and $z_2 = 2 + 5i$. Graph z_1, z_2, $z_1 + z_2$, and $z_1 - z_2$ and explain how the absolute values of these numbers are related geometrically.

3. Suppose that z_1, z_2, and z_3 are complex numbers such that $|z_1| = |z_2| = |z_3|$. Explain why $\mathrm{Arg}\left(\frac{z_3 - z_2}{z_3 - z_1}\right) = \frac{1}{2}\mathrm{Arg}\left(\frac{z_2}{z_1}\right)$.

4. Describe the set of points z satisfying $|z - 2i| = 6$.

5. State the result symbolically and prove.

a. In the complex numbers, the conjugate of a sum equals the sum of the conjugates.

b. In the complex numbers, the conjugate of a product equals the product of the conjugates.

c. In the complex numbers, the conjugate of a quotient equals the quotient of the conjugates.

d. In the complex numbers, the conjugate of the nth power of a number equals the nth power of the conjugates.

6. Prove that if z_1, z_2, and z_3 are complex numbers on the unit circle such that $z_1 + z_2 + z_3 = 0$, then z_1, z_2, and z_3 are vertices of an equilateral triangle.

7. Suppose that z_1 and z_2 are distinct points in the complex plane. Explain why the equation $|z - z_1| + |z - z_2| = c$ describes an ellipse in the complex plane with foci z_1 and z_2, provided $c > |z_1 - z_2|$.

8. Write and prove Theorem 2.21 for z_1 and z_2 written in exponential form $re^{i\theta}$.

9. Use mathematical induction to prove Theorem 2.22.

10. Let $z = \cos\theta + i\sin\theta$.

a. Compute the orbit of z when $\theta = \frac{3\pi}{5}$.

b. Compute the orbit of z when $\theta = 1$ radian.

c. If θ is a rational multiple of π, explain why the orbit of z is a finite subset of the unit circle.

d. If $\theta = 1$ radian, explain why

i. the orbit of z is not a finite set;

ii. the point $(1, 0)$ is not in the orbit of z;

iii. there are points z^n in the orbit of z that are very near $(1, 0)$.

11. Prove DeMoivre's Theorem by generalizing the procedure of Example 2.

12. Compute all the 4th roots of $a = 1 + \sqrt{3}$ and describe where they are located in the complex plane.

13. Determine the seven 7th roots of 1 and describe their graph.

14. Notice that for all θ,

$$\begin{aligned}\cos(2\theta) + i\sin(2\theta) &= (\cos\theta + i\sin\theta)^2 \\ &= \cos^2\theta - \sin^2\theta + i(2\cos\theta\sin\theta).\end{aligned}$$

Equating the real and imaginary parts of the complex numbers on each side provides a proof that $\cos(2\theta) = \cos^2\theta - \sin^2\theta$ and $\sin(2\theta) = 2\sin\theta\cos\theta$.

a. Extend this idea to find formulas for $\cos(3\theta)$ and $\sin(3\theta)$ in terms of $\sin\theta$ and $\cos\theta$.

b. Modify the given information and your result from part a to find formulas for $\cos(2\theta)$ and $\cos(3\theta)$ in terms of $\cos\theta$.

*c. Prove that there exists a formula for $\cos(n\theta)$ in terms of $\cos\theta$ for all natural numbers n.

15. Suppose $z = re^{i\theta}$. Describe the n complex nth roots of z in terms of r and θ.

ANSWER TO QUESTION

a. Possible points $(1, 1)$, $(5, 1)$, $(3, 3)$, $(3, -1)$

b. $(x - 3)^2 + (y - 1)^2 = 4$

Chapter Projects

1. An alternate definition of decimal representation. In Section 2.1.2, we defined $d = \{D, d_1, d_2, d_3, \ldots, d_k, \ldots\}$ to be a **decimal representation** of a real number x if and only if x and d are related by the following inequalities:

$$(*)\ D + \frac{d_1}{10} + \cdots + \frac{d_k}{10^k} \le x \le D + \frac{d_1}{10} + \cdots + \frac{d_k}{10^k} + \frac{1}{10^k},$$

for all $k \in \mathbf{N}$.

Suppose that the following alternate definition of decimal representation were adopted instead: $d = \{D, d_1, d_2, d_3, \ldots, d_k, \ldots\}$ is a **decimal representation** of a real number x if and only if x and d are related by the following inequalities:

$$(**)\ D + \frac{d_1}{10} + \cdots + \frac{d_k}{10^k} \le x < D + \frac{d_1}{10} + \cdots + \frac{d_k}{10^k} + \frac{1}{10^k},$$

for all $k \in \mathbf{N}$.

[Note that the second inequality in (**) is strict.]

a. Explain why, according to definition (**) the decimal 1.000...000... (all 0s) is a decimal representing the real number 1, but the decimal .999999...9999... (all 9s) does not represent 1.

b. Retrace the development of the decimal representation in Section 2.1.2 with definition (*) replaced by (**). Show that every real number has a unique decimal representation in this development.

2. Countability of algebraic numbers. The following sequence of questions proves Theorem 2.17, that the set of all algebraic numbers is countably infinite.

a. Explain why the set S of all solutions x to the equation

$$ax^2 + bx + c = 0,$$

where a, b, and c are integers between -5 and $+5$, is a finite set. Provide a reasonable estimate for the cardinality of S.

b. If n is a positive integer and if $p(x) = a_n x^n + a_{n-1}x^{n-1} + \cdots + a_1 x + a_0$ is a polynomial of degree $\leq n$ with integer coefficients $a_n, a_{n-1}, \ldots, a_1, a_0$, then the **height h(p)** of p is defined as

$$h(p) = n + |a_n| + |a_{n-1}| + \cdots + |a_1| + |a_0|.$$

For any positive integer k, define $S(k)$ to be the set of all real solutions x to $p(x) = 0$, where p is a polynomial with integer coefficients that has height k. Find the maximum height of the polynomials described in part a. Compute the cardinality of the sets $S(1)$, $S(2)$, and $S(3)$.

c. Explain why $S(k)$ is a finite set for every positive integer k. Show by example that a real number x may belong to $S(k)$ for more than one value of k. Explain why the set A of all algebraic numbers is the union of all of the sets $S(k)$; that is,

$$A = \bigcup_{k=1}^{\infty} S(k).$$

d. Describe a scheme for "counting" A based on the result of part **c**. In this way, prove that the set A of all algebraic numbers is countable.

3. Dedekind cuts. Write an essay on the definition of real numbers using Dedekind cuts. Include in your essay how operations on real numbers are defined and how some of the basic properties of addition and multiplication are deduced.

4. Incommensurability in the pentagram. A **pentagram** is the figure formed by the diagonals of a regular polygon (see Figure 23). It was known to ancient Greeks that a pentagram has segments that are incommensurable; that is, that there are ratios of lengths of segments in a pentagram that are irrational. James Choike gives a modern proof in *The College Mathematics Journal* 11 (1980), pp. 312–316.

Figure 23

a. Find and describe his proof.

b. Apply the idea in the proof to show that the leg and hypotenuse of an isosceles right triangle are incommensurable.

(This provides yet another proof that $\sqrt{2}$ is irrational.)

5. The Cardano-Tartaglia method for solving cubic equations.

a. Given a cubic polynomial equation

$$(*) \qquad x^3 + ax^2 + bx + c = 0,$$

where a, b, c are real numbers, show that the change of variables $x = y - \frac{a}{3}$ reduces (*) to a cubic equation

$$(**) \qquad y^3 + py + q = 0$$

without a quadratic term.

b. Show that the change of variables $y = z - \frac{p}{3z}$ reduces (**) to a quadratic equation in z^3 of the form

$$(z^3)^2 + qz^3 - \frac{p^3}{27} = 0.$$

If the latter equation is now solved for z, the six solutions lead back to solutions of the given cubic in x, not more than three of which can be distinct.

c. Apply the Cardano–Tartaglia method to find the solutions of the Bombelli cubic equation $x^3 - 15x - 4 = 0$. (Rafael Bombelli [1526–1573] was among the first to operate with complex numbers as if they had properties like other numbers have.) Observe that the solution of the quadratic equation in z^3 that results from the substitution $x = z - \frac{p}{3z} = z + \frac{5}{z}$ involves complex numbers even though the solutions of this cubic, $x = 4$ and $x = -2 \pm \sqrt{3}$, are all real numbers.

d. Use a calculator (such as the TI-92) or a mathematical software package (such as *Mathematica*) with symbolic algebra capabilities to solve the Bombelli cubic. Explore the solution of $p(x) = 0$ for cubic and quartic polynomials with integer coefficients to see if exact solutions are given.

6. The Complex Exponential and Trigonometric Functions In calculus the following power series representations for e^x, $\sin(x)$, and $\cos(x)$ are derived.

$$e^x = \sum_{n=0}^{\infty} \frac{x^n}{n!} = 1 + x + \frac{x^2}{2!} + \frac{x^3}{3!} + \frac{x^4}{4!} + \cdots$$

$$\sin x = \sum_{n=0}^{\infty} (-1)^n \frac{x^{2n+1}}{(2n+1)!} = x - \frac{x^3}{3!} + \frac{x^5}{5!} - \cdots$$

$$\cos x = \sum_{n=0}^{\infty} (-1)^n \frac{x^{2n}}{(2n)!} = 1 - \frac{x^2}{2!} + \frac{x^4}{4!} - \cdots$$

*a. Show that all of these series converge absolutely for all real x. Use this fact to explain why these series converge for all complex numbers z.

b. Use these complex series to define the complex functions e^z, $\sin(z)$, and $\cos(z)$ for all complex z. Prove that for all complex numbers $z = x + iy$,

$$e^z = e^x(\cos y + i \sin y).$$

c. Apply this identity to verify Euler's formula: For any real number θ,

$$e^{i\theta} = \cos\theta + i\sin\theta.$$

d. The real exponential function $y = e^x$ is positive for all real x, and it assumes each positive value once and only once. Use Euler's formula to show that there are infinitely many complex numbers z such that $e^z = -1$. Then use mathematical software or hardware with symbolic algebra capabilities to solve the equation $e^x = -1$. Describe and explain the output.

e. Solve the equation $e^{2x} - e^x - 2 = 0$ for its real solutions by hand and then graph $f(x) = e^{2x} - e^x - 2$ on your graphing calculator to check your work. Then use mathematical software or hardware with symbolic algebra capabilities to solve this equation. Explain the output. Describe the set of all complex solutions of these equations.

f. Show that the complex exponential function $w = e^z$ takes on *all* complex values except $w = 0$ infinitely many times. [*Hint*: For any nonzero complex number $w = u + iv$, show that $z = \ln(|w|) + i \operatorname{Arg}(w)$ is one value of z such that $e^z = w$.]

g. Use hand calculation to find all real solutions of the trigonometric equation

$$2 \sin^2 x + 3 \sin x - 2 = 0.$$

Then use mathematical software or hardware with symbolic algebra capabilities to solve this equation. Explain the output.

h. What conclusions can you draw from parts **d–g** about using a symbolic algebra facility to solve equations?

7. Transformations of the Complex Plane If x and y are real number variables related by a function $y = f(x)$, we can graph this equation in the xy-plane to understand its behavior. If $z = x + iy$ and $w = u + iv$ are complex number variables related by a function $w = f(z)$, we can understand the behavior of this function by regarding it as a *transformation* of the complex plane; that is, we can observe the geometric relationship between a point z in the complex plane and its image $w = f(z)$. For example, the function $w = z + (1 + i)$ is often called the *translation by* $1 + i$ because the point (u, v) in the complex plane representing the complex number $w = u + iv$ is the translation image of the point (x, y) representing $z = x + iy$ by the vector joining the origin to $1 + i$. Similarly, the function $w = iz$ is called a 90° (or $\frac{\pi}{2}$) rotation about the origin because $|i| = 1$ and $\operatorname{Arg}(i) = \frac{\pi}{2}$, so $|w| = |z|$ and $\operatorname{Arg}(w) = \operatorname{Arg}(z) + \frac{\pi}{2}$ [or $\operatorname{Arg}(w) = \operatorname{Arg}(z) - \frac{3\pi}{2}$].

a. Provide a similar geometric description of the following complex functions.

i. $f(z) = \bar{z} + 5$ ii. $f(z) = (1 + i)z + (1 - i)$

b. For a given nonzero complex number in polar form

$$z = r(\cos\theta + i \sin\theta),$$

compute the polar form of $\frac{1}{z}$. Let t be a point of tangency from z to the unit circle. Show that if $|z| > 1$, then $w = \frac{1}{z}$ is the intersection of the line through t perpendicular to the segment connecting the origin 0 to z, as shown in Figure 24.

Figure 24

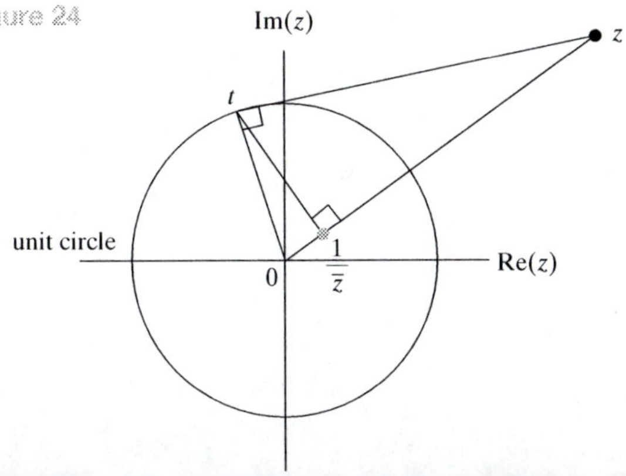

(The point at $\frac{1}{z}$ satisfies $|z||\frac{1}{z}| = 1$ and is called the **inverse of the point z for the unit circle**. If $0 < |z| < 1$, the same diagram with the points z and $\frac{1}{z}$ interchanged shows that z has an inverse point for the unit circle that is outside the unit circle. Points on the unit circle are their own inverse points. For these reasons, the function that maps z onto $\frac{1}{z}$ is called **inversion in the unit circle**.)

8. Coordinatization of the Riemann Sphere We can establish a coordinate system for the Riemann sphere and complex plane as indicated in Figure 25.

Figure 25

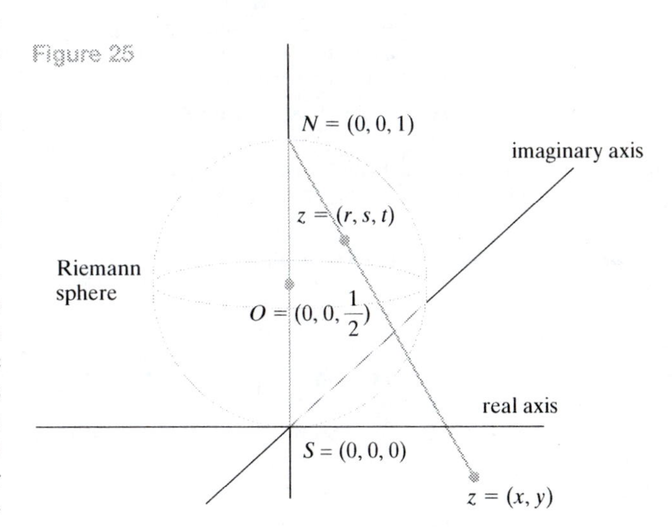

a. Derive the following relationships between the coordinates (x, y) of the point representing a complex number z in the complex plane and the coordinates (r, s, t) of the point representing z on the Riemann sphere:

$$r = \frac{x}{1 + |z|^2}, \quad s = \frac{y}{1 + |z|^2}, \quad t = \frac{|z|^2}{1 + |z|^2},$$

$$x = \frac{r}{1 - t}, \quad \text{and} \quad y = \frac{s}{1 + t}.$$

b. Use the relationships in part **a** to prove the following results:

i. The image under stereographic projection of any circle in the complex plane is a circle on the Riemann sphere that does not pass through the North Pole N.

ii. The image under stereographic projection of any line in the complex plane is a circle on the Riemann sphere that passes through the North Pole N with N removed.

iii. Inversion in the unit circle $w = \frac{1}{z}$ (see Project 7b) corresponds to reflection in the plane of the equator on the Riemann sphere.

c. Use part **b** to conclude that inversion in the unit circle transforms objects in the set LC of lines and circles in the complex plane into other objects in this set (with the understanding the points S or N may be excluded from objects in LC when necessary).

Bibliography

Unit 2.1 References

Ifrah, Georges. *The Universal History of Numbers*. New York: John Wiley, 2000.
A detailed discussion of the history of numbers and counting from prehistory to the present.

Kalman, Dan, Robert Mena, and Shahriar Shariari. "Variations on an Irrational Theme—Geometry, Dynamics, Algebra." *Mathematics Magazine* 70(2), April 1997.
This article discusses the irrationality of $\sqrt{2}$ and the Pythagorean discomfort with the idea, and then relates irrationals to dynamical systems, eigenvalues, and complex numbers.

Maor, Eli. *e: The Story of a Number*. Princeton, NJ: Princeton University, 1994.
The entire book is devoted to the base of natural logarithms, with discussions of its irrationality and other properties.

Niven, Ivan. *Irrational Numbers.* Carus Mathematical Monographs, No. 11. Washington, DC: Mathematical Association of America, 1956.
This classic short book summaries much of the known information on irrational numbers.

Rademacher, Hans, and Otto Toeplitz. *The Enjoyment of Mathematics*. Princeton, NJ: Princeton University Press, 1957.
Chapter 17 treats the topic of approximating irrational numbers by rationals. Chapter 23 discusses periodic decimals.

Ribenboim, Paulo. *My Numbers, My Friends*. New York: Springer-Verlag, 2000.
A collection of light, accessible, and sometimes funny essays on many topics, including irrationals.

Wilder, Raymond L. *Evolution of Mathematical Concepts*. New York: John Wiley, 1968.
Chapter 4 gives a detailed examination of the real numbers, discussing many of the topics found in this unit.

Unit 2.2 References

Hahn, Liang-shin. *Complex Numbers and Geometry*. Washington, DC: Mathematical Association of America, 1994.
This book assumes no background in complex numbers and demonstrates connections between geometry and complex numbers that result in proofs and natural generalizations of many theorems in plane geometry.

Nahim, Paul J. *An Imaginary Tale—The Story of* $\sqrt{-1}$. Princeton, NJ: Princeton University Press, 1998.
The number may be called imaginary but the tale is not. This book is devoted to the history, mathematical properties, and some applications of i and other complex numbers.

Dunham, William. *Euler—The Master of Us All*. Washington, DC: Mathematical Association of America, 1999.
The French mathematician Laplace wrote, "Read Euler, read Euler. He is the master of us all." This book, written by an eminent historian, surveys Euler's contributions in a variety of branches of mathematics. Chapter 5 deals with Euler's work with complex numbers and complex variables.

Chapter 3

FUNCTIONS

Functions provide a means of expressing relationships between variables. The values of these variables can be numbers or nonnumerical objects such as geometric figures, functions, or nonmathematical objects. Many of the functions that you have studied in mathematics are *real functions*; that is, functions relating two variables x and y whose values are real numbers. Among the familiar types of real functions are polynomial and rational functions as well as trigonometric, exponential, and logarithmic functions. Typically, real functions are prescribed by formulas of the form $y = f(x)$, such as $y = 2x^2 - x$, $y = \sin(x^2)$, or $y = 2^{-x}$, and much of what you have learned about analyzing and using real functions has depended on making use of formulas such as these.

Figure 1

Even functional relationships that are simple to describe may lead to functional formulas that are relatively complex. Consider, for example, the function that expresses the volume V of fuel in an underground cylindrical tank of length ℓ and radius r whose axis is horizontal, in terms of the depth d of the fuel (Figure 1).

Geometric analysis of this relationship leads to the rather complicated formula:

$$V = f(d) = \ell\left(\pi r^2 - r^2\cos^{-1}\left(\frac{d}{r} - 1\right) + (d - r)\sqrt{2rd - d^2}\right)$$

Functional relationships of this complexity are often difficult to analyze purely on the basis of formula manipulation. For example, it seems reasonable that the depth d of fuel in the tank is also a function of the volume V of fuel; that is, that the function $V = f(d)$ can be "inverted" to obtain a function $d = f^{-1}(V)$. From its definition, the inverse f^{-1} exists because, for any given volume V of fuel between 0 and the capacity of the tank, pouring that amount of fuel into the tank would fill the tank to one and only one level $d = f^{-1}(V)$. However, solving the equation

$$V = \ell\left(\pi r^2 - r^2\cos^{-1}\left(\frac{d}{r} - 1\right) + (d - r)\sqrt{2rd - d^2}\right)$$

for d in terms of V is not feasible. Consequently, we cannot find a formula for f^{-1}.

Without a formula, how can we show mathematically that this inverse function exists? The answers to this question and others like it often rest on qualitative features of the functional relationship that may not be apparent from the functional formula. Our look at some familiar real functions in this chapter focuses on features of these functions that may not have been emphasized in your previous courses.

In prior courses, you have also studied many other types of functions that are not real functions. Geometric transformations such as reflections, translations, and rotations are functions relating variables whose values are points or geometric figures. Operations such as addition and multiplication for numbers and functions, the dot product and length for vectors, and determinants of square matrices are functions whose independent and/or dependent variables have values that are not real numbers. In Chapter 2, we used the special type of function called a one-to-one correspondence to show that certain pairs of infinite sets have the same cardinality. Thus, the function concept arises in a wide variety of mathematical contexts. Part of the purpose of this chapter is to illustrate this diversity and to discuss the common ideas about function that are relevant in all or most of these contexts.

Unit 3.1 The Definitions, Historical Evolution, and Basic Machinery of Functions

We begin this chapter by asking "What is a function?" In mathematics, it is common to answer the question "What is a _____?" by giving a definition of _____. We use definitions in mathematics as we do in everyday speech, to help us understand the meaning of an idea and the context in which we use it. Definitions in mathematics also serve another purpose. They are powerful resources for facilitating a proof or solving a problem—providing specific information about the meaning of each condition in the proof or problem.

As important as definitions are, however, they can sometimes mask alternative ways in which an idea can be approached or described, and may not signal why the idea is important. A full answer to "What is a _____?" includes not only a formal definition, but also *alternate definitions and descriptions, why* the idea has been defined, *how* the idea is described, and *to what purposes* the idea is put. A full answer leads us to concept analysis.

3.1.1 What is a function?

The idea of a function f is to express a relationship between the elements of two sets. If A and B are sets, then a *function f from A to B* is often described as a *rule* or *process* that associates with each element a of the set A one and only one element b of the set B. We often think of each element in the first set as *determining* the corresponding element of the second set.

Definition A function is a rule that assigns to each element of a set A a unique element of a set B (where B may or may not equal A).

The set A is called the **domain** of the function f, the set B the **codomain**, and the subset of the set B consisting of those elements that are images under the function f of some element of its domain is called the **range** of the function f. When f associates a in A to b in B, the element b is called the **image of a under f** or **the value of f at a**, and a is called a **preimage of b under f**. Figure 2 depicts these sets and elements schematically.

Figure 2

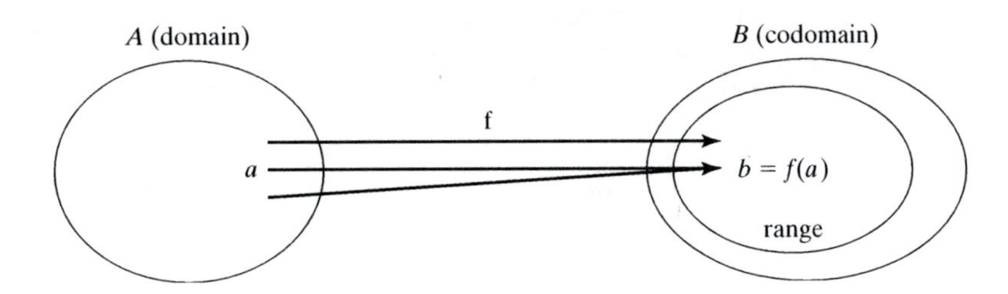

A number of different notations have evolved for functions in mathematics and, more recently, in computer science. Two notations are particularly common in mathematics. When the function f associates a with b, then we write

	f(a) = b,	called **f(x)** or **f() notation**,
or	**f: a → b,**	called the **arrow** or **mapping** notation.

The arrow notation conveys the idea of an action associating the elements from set A to their corresponding values in set B. Some writers use $f\colon A \to B$ only for indicating the domain and codomain, and use $f\colon a \mapsto b$ to identify corresponding elements. We use $\to$ for both. When arrow notation is used, we often say that the function f **maps** the element a onto b and we call f a **mapping** or **map** from A into B. We say that f maps the set A **onto** the codomain B if every element in B is in the range.

A value in the domain of a function is called an **argument** of the function. The variable that stands for the argument is called the **independent variable**. The variable that stands for the values of the function is called the **dependent variable**. In some applications these are called the **input variable** and **output variable**, respectively, reflecting the influence of computer science.

For the function $f\colon x \to y$, many books in precollege mathematics consistently use the single letter f to name the function and distinguish this from the symbol $f(x)$ used to identify the values of the function. In mathematics more broadly, and in computer science, this distinction is not always made, and the symbol $f(x)$ may stand for a function and also its values. Using the symbol $f(x)$ to stand for a function allows the independent variable to be explicitly identified.

A function relationship may be expressed by a *formula* such as $y = x^2$, which squares each real number x, or $y = \sin(x)$, which associates with each real number x the real number y that is the sine of an angle of x radians. A formula that represents a function equates the dependent variable with an algebraic expression written in terms of the independent variable.

A function relationship may also be expressed as a *verbal description* of the correspondence between two sets, such as the correspondence between the set **N** of natural numbers and the set P of prime numbers that associates with the natural number n the nth prime number p_n. In this case, there is no formula for computing the nth prime number p_n for a given natural number n, but the precise meaning of the relationship is nonetheless clear from the description or from the following correspondence diagram.

$$\begin{array}{ccccccccccc} \mathbf{N} & 1 & 2 & 3 & 4 & 5 & . & . & n & . & . \\ \downarrow & \downarrow & \downarrow & \downarrow & \downarrow & \downarrow & . & . & \downarrow & . & . \\ P & 2 & 3 & 5 & 7 & 11 & . & . & p_n & . & . \end{array}$$

A function relationship may also be expressed by a *table* listing all of its values. For example, the population given by the U.S. Bureau of the Census is a function from the set of years divisible by 10, from 1790 to the present, to the set of natural numbers. This function is partly described in Table 1.

Table 1 U.S. Population $P(t)$ (in Millions) for the Period 1780–1870

Year *t*	1780[a]	1790	1800	1810	1820	1830	1840	1850	1860	1870
***P*(*t*)**	2.8	3.9	5.3	7.2	9.6	12.9	17.1	23.2	31.4	39.8

[a] An official census was first taken in 1790, but the Census Bureau has estimated the population for 1780 and earlier.

To allow correspondences where the idea of a function "rule" is not particularly appropriate, mathematicians use a formal definition of function in terms of the language of sets. In this definition, explicitly stated below, we think of each pair of corresponding elements in the two sets as being the components a and b of an ordered pair (a, b) and we think of the function as the set of ordered pairs.

Recall that the **Cartesian product** of two sets A and B, denoted $A \times B$, is the set of *all* ordered pairs whose first components are from A and whose second components are from B. For instance, if $S = \{1, 2\}$ and $T = \{3, 4, 5\}$, then $S \times T = \{(1, 3), (1, 4), (1, 5), (2, 3), (2, 4), (2, 5)\}$. The word "Cartesian" in the name of this operation comes from the fact that ordered pairs were first introduced in coordinate graphs on the Cartesian plane. The word "product" and the symbol "$\times$" are quite appropriate for this operation because if A has m elements and B has n elements, then $A \times B$ has mn elements.

Definition For any sets A and B, a **function f from A to B, f: A $\rightarrow$ B**, is a subset f of the Cartesian product $A \times B$ such that every $a \in A$ appears once and only once as the first element of an ordered pair (a, b) in f.

For example, let $A =$ the set of all circles in a plane M, and $B =$ the set of all points in M. Then $A \times B$ is the set of all ordered pairs (C, P), where C is a circle and P is a point in M. The subset of $A \times B$,

$$f = \{(C, P) \in A \times B: P \text{ is the center of } C\},$$

is a function $f: A \rightarrow B$ because each circle in A has exactly one center. That is, no C is associated with two or more values of P.

On the other hand, the subset of $B \times A$,

$$g = \{(P, C) \in B \times A: P \text{ is the center of } C\},$$

is not a function $g: B \rightarrow A$ because any point P is the center of (infinitely) many different circles in the plane. So every point P appears with infinitely many different values of C. However, if we restrict g to the set

$$g_1 = \{(P, C) \in B \times A: P \text{ is the center of } C \text{ and } C \text{ has radius } 1\},$$

then $g_1: B \rightarrow A$ is a function.

The ordered pair characterization of function is particularly appropriate for real functions because we can picture the ordered pairs in a *graph*. For this reason, some authors prefer to define a function as a correspondence and define the graph of a function to be the set of ordered pairs created by the correspondence.

The ordered pair definition of function has the advantage that it is precise and unambiguous, but it presents a functional relationship in a rather passive, static way. The description of a function by a rule of correspondence has the advantage that it suggests that a function provides an active procedure for producing range elements from elements of the domain of the function. Together, these descriptions of a function present the concept in a very general yet very precise and useful manner.

Functions and equations

We have noted that some people distinguish $f(x)$ from f, while others do not. It is also the case that some people identify a function with its equation, as in "the function $f(x) = 2x + 5$," while others distinguish a function from a description of its formula or rule. Here we provide a short discussion of notation used in writing functions and equations to clarify the conceptual relationship between functions and equations.

Consider the problem of Jane's average from Chapter 1. Jane has an average of 87 out of 100 after four tests, and we wish to know what score is needed on the 5th test for her average on all 5 tests to be y. (We use y here instead of A.) If Jane scores x points on the 5th test, her average after 5 tests can be written as

$$\frac{4 \cdot 87 + x}{5}. \tag{1}$$

(1) is called *an expression in* x. It is common to use the term "expression" for forms such as (1) that have no equality or inequality signs. Expressions such as (1) are central in that they can be used both in defining a function and in stating an equation, as we now illustrate.

Jane's average after 5 tests is a *function of* the point score on the fifth test. Let us use the symbol f to refer to the function relating x and y. In talking about such a function it is convenient to be able to refer to three different things, the input (independent) variable, the output (dependent) variable, and the function itself. Here is an explicit description of the function f in function notation:

$$f(x) = \frac{4 \cdot 87 + x}{5}. \tag{2}$$

This establishes x as the input variable and f as the name of the function, and links them through the expression in x given in (1).

What we have written in (2) is called a *defining formula* or a *defining equation* for the function f. Perhaps *formula* is the more appropriate word, since *equation* is used universally in a rather different role [see, for example, equation (7), below]. Moreover, an alternate way (3) of giving an explicit definition of the function f uses *mapping notation* and no equation.

$$f: x \rightarrow \frac{4 \cdot 87 + x}{5} \tag{3}$$

This gives the name f to the mapping that sends x to the defining expression (1). Mapping notation also makes clear that there is no notion of equality involved in a function, but only the notion of a relationship between input and output.

Another way of giving an explicit definition of the function f is by relating the expression in x given in (1) to the output variable y:

$$y = \frac{4 \cdot 87 + x}{5}. \tag{4}$$

This establishes y as the output variable and x as the input variable, and links them through the defining expression given in (1). But notice that attempting to link all three of x, y, and f at once in a single formula leads to

$$y = f(x). \tag{5}$$

This is the generic form of function notation and does not define a particular function. To link this to a specific defining expression such as (1) requires a double equality such as

(6) $$y = f(x) = \frac{4 \cdot 87 + x}{5}.$$

Experienced users of mathematics get accustomed to idiosyncrasies of notations for defining functions, but they can create confusion for beginners.

Recall from Chapter 1 that the original question asked what score x Jane needed to average 90. This leads to an *equation in the unknown* x.

(7) $$\frac{4 \cdot 87 + x}{5} = 90$$

There is a definite connection between the function defined in (6) and the equation stated in (7). The equation amounts to a condition on the input x, namely that the function (6) have the specific numerical output 90. Writing the condition as $f(x) = 90$ would leave out the specific content of the equation. To include this content and also make specific the fact that this is a condition on inputs of the function f, a double equality such as (8) could be used.

(8) $$f(x) = \frac{4 \cdot 87 + x}{5} = 90$$

The variable y is commonly used when we discuss the graph of the function f. We can regard (4) above as an alternative way of defining a function, one that makes explicit the name of the output variable y rather than the name of the function. Alternatively, (4) could be regarded as an equation relating the variables x and y. Then the equation (4) also describes the graph of the function as a geometrical object.

Functions of two variables

Equations with two variables (usually x and y) are often used to define lines or curves in the plane, especially when we are interested in them as geometric objects rather than as graphs of functions. In their general form they are not restricted to a $y = \ldots$ formulation. For instance, (9) is an equation that defines the ellipse of Figure 3.

Figure 3

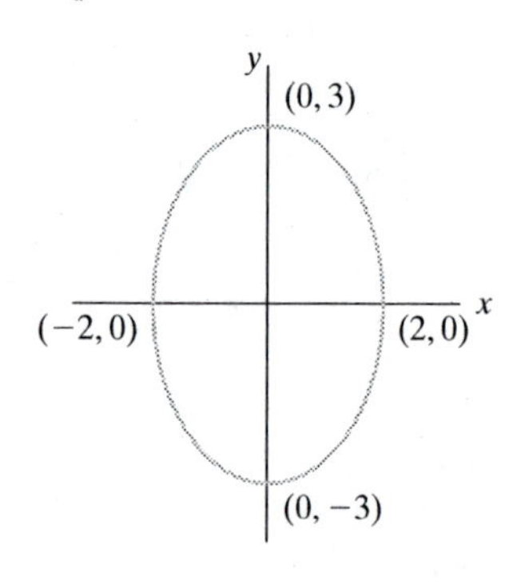

(9) $$\frac{x^2}{4} + \frac{y^2}{9} = 1$$

Study of equations such as (9), and in general of equations of the form $F(x, y) = k$ (where k is a fixed number), are the subject of the analytic geometry of the plane. They do not necessarily describe a function in the single variable x, because [as is the case in (9)], for some values of x there may exist two values of y. In such cases, the equation $F(x, y) = k$ may identify more than one function of a single variable. For example, equation (9) can be viewed as identifying the union of the two functions f_1 and f_2 defined by

$$f_1(x) = 3\sqrt{1 - \frac{x^2}{4}} \quad \text{and} \quad f_2(x) = -3\sqrt{1 - \frac{x^2}{4}}.$$

Figure 4

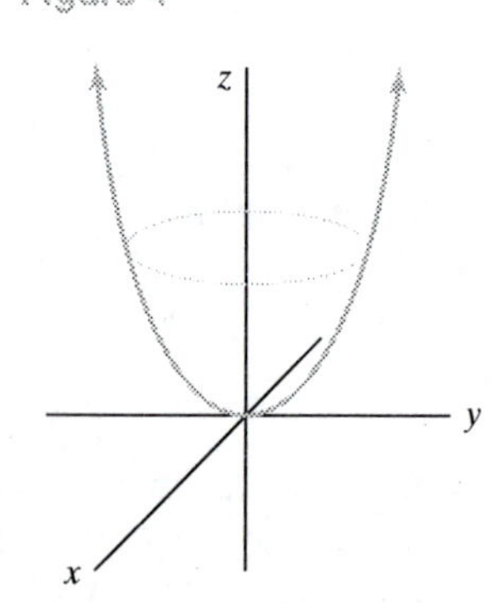

Equations of the form $F(x, y) = k$ also serve to identify functions of *two* variables x and y. Specifically, equation (9) is related to the function F of two variables,

(10) $$z = F(x, y) = \frac{x^2}{4} + \frac{y^2}{9},$$

or, in mapping notation, $F\colon (x, y) \to \frac{x^2}{4} + \frac{y^2}{9}$. The graph of (10) is an elliptic paraboloid above the xy-plane (Figure 4). Equation (9) amounts to a requirement on the output of the function (10), and the pairs (x, y) that meet this requirement are the solutions of (10). To express (9) graphically we can add the plane $z = 1$ to the graph of (10).

These two surfaces intersect in an ellipse, and the graph of all the solutions, projected down to the xy-plane, is the ellipse of Figure 3.

Sequences

A **sequence** can be defined formally as a function whose domain is the set of integers greater than or equal to a fixed integer k, where k is usually 0 or 1. The image of an integer n in a sequence S is usually denoted by $\mathbf{S_n}$ [rather than $S(n)$] and called the **nth term** of the sequence. The sequence itself is often denoted $\{\mathbf{S_n}\}$ or **S**. Although a sequence is formally a set of ordered pairs, we often list only the terms in order of n. For instance, the sequence $\{(1, 2), (2, 4), (3, 8), (4, 16), \ldots\}$ with rule $S_n = 2^n$ is usually written as $2, 4, 8, 16, \ldots 2^n, \ldots$.

Since sequences are functions, they may be described in the same ways as functions are, by formulas, tables, or correspondences (the correspondence earlier in this section between **N** and the set of primes is a sequence). However, sequences possess a fundamental property that distinguishes them from other types of functions—the possibility of being defined *recursively*. For example, the geometric sequence

$$g = \{g_n\} = 2, 6, 18, 54, \ldots, 2(3^n), \ldots$$

has initial term $g_0 = 2$ and common ratio $r = 3$, so that $g_{n+1} = 3g_n$ for all integers $n \geq 1$. These conditions define the sequence *recursively*, that is, after some given terms, later terms are found from earlier terms. The same sequence can also be *explicitly* described by the formula

$$g_n = 2(3^n) \qquad \text{for all integers } n \geq 0.$$

The famous Fibonacci sequence $F = \{F_n\}$ has as its first few terms

$$F = \{F_n\} = 1, 1, 2, 3, 5, 8, 13, 21, 34, \ldots$$

and can be described recursively by

$$F_1 = 1, \quad F_2 = 1, \quad F_{n+2} = F_{n+1} + F_n \qquad \text{for any integer } n \geq 1.$$

The possibility of defining sequences recursively greatly enhances the utility of sequences for modeling and analyzing problems. Recursive definitions are rooted in the most basic property of the natural number system **N**—mathematical induction. With mathematical induction, recursively defined sequences become a powerful tool for mathematical modeling and analysis. We examine mathematical induction in some detail in Chapter 5.

A brief history of the concept of function

The function concept was not discovered or conceived by a single individual or at a particular time. Rather, it evolved over a period of several centuries and continues to evolve today in response to important problems in a number of different fields both within and outside of mathematics.

It is interesting that, although the concept and notation for functions were not introduced until the 18th century, graphs of functions were used to analyze their properties as early as the 14th century. An early instance of this use of graphs was by Nicole Oresme (1323–1382), who used velocity-time graphs to study the motions of bodies under uniform acceleration. Although he did not state the law of falling bodies [distance $=$ constant(time)^2] later attributed to Galileo, his results essentially yielded that conclusion.

The historical evolution of the definition of function began with a much less general and less precise description than either the rule of correspondence or the ordered pair definition for a function given earlier in this section. Gottfried Leibniz used the term "function" for the first time in 1694 to describe six very specific line segments associated with a variable point on a given plane curve. (See Project 1 for details about these "Leibniz segments".)

In 1718, Jean Bernoulli (1667–1748) significantly broadened the meaning of function by stating "a quantity composed in any manner whatever of a variable and any constants" is a function of the variable magnitude.[1] He also began experimenting with various notations for functions, with his symbol fx being the closest to the modern $f(x)$ notation.

The notation $f(x)$ for a function of a variable quantity x was introduced in 1748 by Leonhard Euler in his text *Algebra*, which was the forerunner of today's algebra texts. Many other mathematical symbols in use today, including e for the base of the natural logarithm and π for the ratio of the circumference of a circle to its diameter, were introduced by Euler in his writings.

The evolution of the function concept during the eighteenth century took place completely within the context of real functions and centered around a lively interaction between Jean Bernoulli, his sons Daniel (1700–1782) and Nicolaus (1695–1726), Euler, and Jean Le Rond d'Alembert (1717–1783). This interaction was prompted by their common efforts to analyze and describe the motions of a vibrating, tightly stretched string such as a guitar or violin string. As their work progressed, the concept of a function evolved from a rule expressed by a single formula with a finite number of terms, to formulas that would allow infinite series and limits, and finally to piecewise-defined functions that required more than one formula to describe the function. This extension of the meaning of function was used independently in the work of J. B. J. Fourier (1768–1830), whose treatise on heat conduction published in 1822 used infinite series composed of sine and cosine functions (later called *Fourier series*) to represent functions. He observed that his functions included the piecewise-defined functions of earlier mathematicians.

In 1837, Lejeune Dirichlet (pronounced Direesh'lay) (1805–1859) gave the following definition of function: "If a variable y is so related to a variable x that when a numerical value is assigned to x, there is a rule according to which a unique value of y is determined, then y is said to be a function of the independent variable x."[2] Perhaps to emphasize the generality of his definition, Dirichlet introduced the following "badly behaved" function $f: \mathbf{R} \to \mathbf{R}$:

$$\begin{aligned} f(x) &= 1 \qquad \text{if } x \text{ is a rational number;} \\ f(x) &= 0 \qquad \text{if } x \text{ is an irrational number.} \end{aligned}$$

This function, which is everywhere discontinuous, now is called the **Dirichlet function**.

Functions commonly studied in calculus may have rather complicated formulas, but they are almost all continuous and also differentiable (except possibly at a finite number of points in their domains). The Dirichlet function is an example of what is sometimes called a *pathological function*. The term "pathological" is used in mathematics for examples that illustrate unexpected or unusual behavior. The Dirichlet function, which is discontinuous at every point, is pathological with respect to continuity. It is perhaps more surprising that there are functions $f: \mathbf{R} \to \mathbf{R}$ that are continuous at every point but not differentiable at any point! Of the many such functions, one of the most intuitively appealing is described in Project 5.

---

[1]Cited in Carl B. Boyer, *A History of Mathematics*. Second edition, revised by Uta C. Merzbach. New York: John Wiley, 1991, p. 422.

[2]Boyer, *op. cit*, p. 510.

Until the last half of the 20th century, the concept of function included **multivalued functions**, those in which a domain value may be associated with more than one range value. Examples of rules for multivalued functions $x \to y$ are $y = \pm\sqrt{x}$, $y =$ a factor of x, and $y =$ an angle whose sine is x. The term "multivalued function" is still common in the study of complex variables, but in most other parts of mathematics these formulas are said to define *relations*, not functions.

The extension of the use of the function concept to contexts in which the domain or range is not necessarily a set of numbers occurred during the latter part of the 19th century and the first half of the 20th century as a result of the developments in set theory, abstract algebra, and analysis. The ordered pair definition of a function $f: A \to B$ was part of that final evolution to the modern concept. It provided the added generality necessary to make the function concept a central organizational tool and unifying thread for virtually all fields of mathematics.

3.1.1 Problems

1. Show that the formula

$$V = V(d) = \ell\left(\pi r^2 - r^2\cos^{-1}\left(\frac{d-r}{r}\right) + (d-r)\sqrt{2rd - d^2}\right)$$

describes the real function that relates the volume V to the depth d of fuel in an underground cylindrical fuel tank of length ℓ and radius r whose axis is horizontal (Figure 5). [*Hint*: Let $d = BE$, $r = AO = OC$, and $V(d) = \ell$(Area of region $AECBA$), using Figure 6. Notice that Area of region $AECBA = \pi r^2 - 2$(Area of sector ODC) + Area of triangle OAC, and that $\cos(\angle DOC) = \frac{BO}{r}$.]

Figure 5

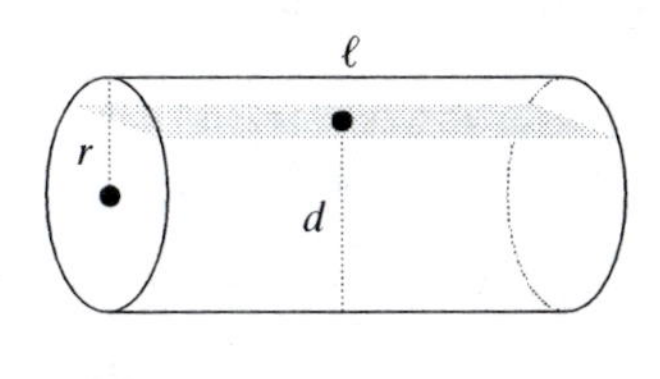

Figure 6

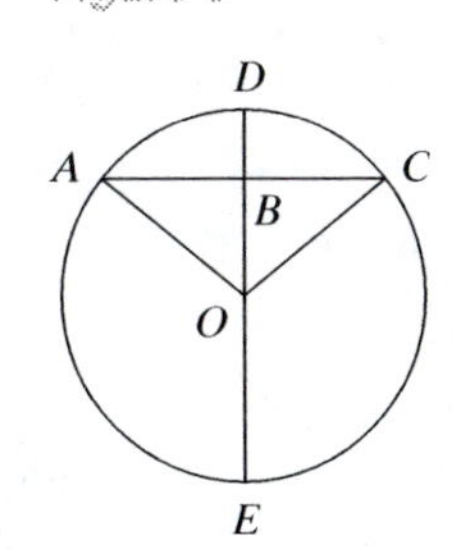

2. The following statements each describe a relationship between two sets A and B in which the given elements are in the set A. Identify the sets A and B precisely in each case. Determine whether or not the relationship defines a function $f: A \to B$.

Example: The point P is the center of a given circle C in the plane.

Answer: A is the set of circles in the plane. B is the set of points in the plane. The relationship defines a function (because each circle has exactly one center).

a. The triangle T is circumscribed by the given circle C.

b. The circle C circumscribes the given triangle T.

c. The area of a given triangle T is A.

d. The pair $\{p, q\}$ of points in the plane are the foci of a given ellipse E.

3. Give a formula for a function with the indicated domain and range.

a. domain **R**, range **R**

b. domain **Z**, range the set of integers ≥ 2

c. domain **R**, range the set of reals $\geq k$, where k is a given constant

d. domain **R**, range $\{y: a \leq y \leq b\}$

e. domain $\{x \in \mathbf{R}: x > 2\}$, range $\{y \in \mathbf{R}: y > 1\}$

4. a. Suppose A is the empty set. Using either the rule or ordered pair definition of function, are there any functions from A to a set B? If so, characterize them.

b. Suppose B is the empty set. Using either the rule or ordered pair definition of function, are there any functions from set A to set B? If so, characterize them.

5. a. Let A and B be finite sets. Suppose A contains x elements and B contains y elements, with $x \geq y$. How many different functions are there from A to B?

b. Does the answer to part **a** change if $x < y$? Why or why not?

6. Give a precise description of the Euclidean distance d in the coordinate plane as a function of points (x_1, y_1) and (x_2, y_2) in $\mathbf{R}^2$; that is, identify the domain A and range B and a rule for the distance function $d: A \to B$.

7. a. Give the ordered pairs of the correspondence that maps the letters of the alphabet other than Q and Z onto the telephone digits 2 through 9.

b. Is this correspondence a function? Why or why not?

8. Consider the greatest integer function (or floor function) defined by $\lfloor x \rfloor$ = the largest integer $\leq x$. For example, $\lfloor 1.7 \rfloor = 1$, $\lfloor -\pi \rfloor = -4$. Plot this function with domain the interval $-3 < x < 3$, first by hand and then with a calculator. Explain the difference between the hand plot and the calculator plot.

*9. The Dirichlet function $f: \mathbf{R} \to \mathbf{R}$, which is defined to have the value 1 at all rational numbers and the value 0 at all irrational numbers, can be expressed as a double limit as follows:

$$(*) \quad f(x) = \lim_{m\to\infty} \left(\lim_{n\to\infty} (\cos(m!\pi x))^n \right).$$

Although this expression looks a bit mysterious at first glance, you can unravel the mystery rather quickly as follows:

a. Explain why $(\cos(m!\pi x))^n = 1$ for any integer x and any positive integers m and n with $m > 1$.

b. For any rational number $x = \frac{p}{q}$ with $q > 0$, explain why $(\cos(m!\pi x))^n = 1$ for any positive integers m and n with $m \geq 2q$.

c. For any irrational number x, explain why $-1 < \cos(m!\pi x) < 1$. Then explain why this implies that $\lim_{n\to\infty} (\cos(m!\pi x))^n = 0$ for any positive integer m.

d. Finally, use the results of parts **a–c** together to explain the double limit (*)

3.1.2 Problem analysis: from equations to functions

The purpose of this section is to exemplify once again how pursuing the idea of generalization can raise questions from the level of exercises to the level of mathematical analysis. An earlier example was the problem of matching an average, discussed in Chapter 1. Problems of this sort can be generalized in a systematic way: If the numbers given in the problem statement are replaced by general parameters, the result of the analysis gives the answer as a *function* of these parameters.

A "numbers-in / number-out" problem

We use the following school-level problem to illustrate this idea.

> Person A sets out in a car going at 50 mph. Starting 3 hours later, person B tries to catch up. If person B goes at 75 mph, how long does it take to catch up?

Question 1: Before reading on, solve this problem.

A solution to this problem as stated can be based on the fact that the distances traveled by the two people are the same. Let t be the time it takes person B to catch up. Equating expressions for the two distances leads to a linear equation.

$$(1) \qquad 50 \cdot (t + 3) = 75 \cdot t$$

Solving the equation gives the answer: $t = 6$ hours.

Treated in this way, this is a "numbers-in / number-out" problem. The numbers 50 mph, 3 hours, and 75 mph are the input, and a simple analysis using algebraic techniques produces an output, 6 hours. A particular answer has been given to a particular question.

This type of problem can be useful in giving students practice in setting up and solving equations. Yet if we take the problem one step further to give a general answer to a general problem, much richer mathematics can be illustrated.

A "parameter-in / function-out" problem

As a start on the process of generalizing, suppose we replace person B's speed of 75 mph with a parameter: the speed w.

Question 2: Before reading on, solve the problem in terms of the speed w of person B.

Solving the problem in this form leads to the same equation as before, except with 75 mph replaced by w:

$$50 \cdot (t + 3) = w \cdot t \tag{2}$$

Question 3: Before reading on, solve equation (2) for t.

When we solve equation (2) for t, we do not get a numerical solution. Instead, we obtain $t = \dfrac{3}{\frac{w}{50} - 1}$, which shows the catch-up time t as a function of any catch-up speed w. The variable w has become the argument of the function

$$t = f(w) = \frac{3}{\dfrac{w}{50} - 1}. \tag{3}$$

A graph of f can exhibit all solutions. The special case of the original situation (75 mph catch-up speed) and its solution (6 hours) is represented by the single point (75, 6) (see Figure 7).

Figure 7

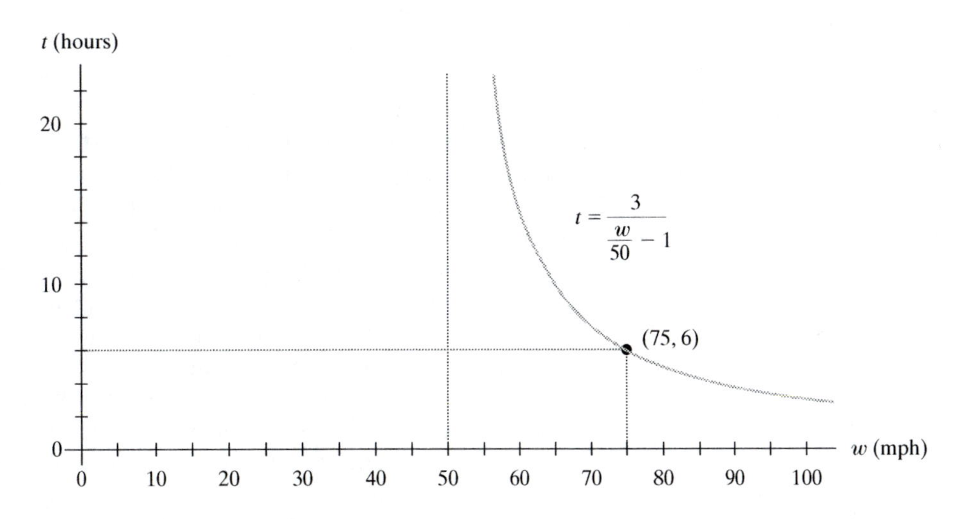

Having a function as a solution to the problem is much more useful and powerful than having a number as a solution. Formula (3) tells us exactly how the catch-up time depends on the catch-up speed. For example, we can find the slope at $w = 75$ mph (by calculus or estimating from the graph). This slope, about $-\frac{1}{4}\frac{\text{hours}}{\text{mph}}$, tells us that for every one mile per hour person B's catch-up speed increases from 75 mph, about $\frac{1}{4}$ hour is subtracted from the catch up time. We can see also what happens when w is very fast (the time t approaches zero) and what happens when w is just a little over 50 mph (as the catch-up speed w approaches 50 mph, the catch-up time t goes to ∞). Both of these results make sense in the original problem situation.

An "intersection of functions" approach

Another approach to this problem is to model the speed of each car with a linear function. The distance $d_B(t)$ traveled by car B in time t defines a function d_B described by the formula

$$d_B(t) = 75t. \tag{4}$$

Similarly, the distance $d_A(t)$ traveled by car A in time t defines a function d_A with

$$d_A(t) = 50(t + 3). \tag{5}$$

Figure 8 shows a graph of each of these functions. Notice that negative values of t from -3 to 0 are meaningful, indicating how far car A had traveled t hours *before* car B started.

Figure 8

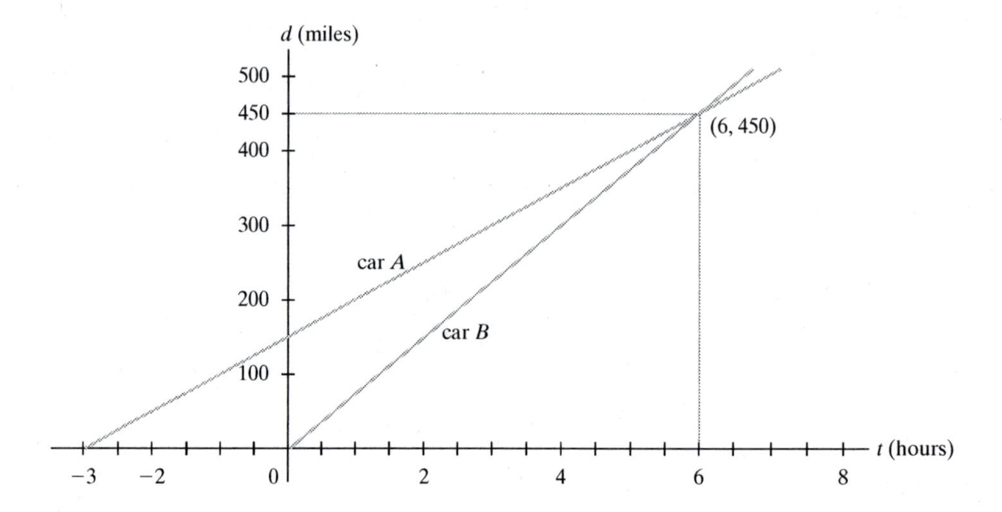

With this representation, the intersection of the graphs indicates a time when the cars have traveled the same distance. So it is a solution to the original problem. The catch-up time t at the intersection is found by setting $d_B(t) = d_A(t)$, which is equation (1), and which we found to be 6 hours. When $t = 6$, $d = 450$, indicating that both cars have traveled 450 miles.

The graph in Figure 8 generalizes the original problem in a different way than the graph of the function in Figure 7. The vertical distance between the two lines in Figure 8 is the distance between cars A and B at any time t, even when $t > 6$ (assuming the speeds of the cars remain constant).

We have here therefore two different uses of functions and their graphs. A relationship between these two different graphs can be illustrated in a powerful way. Using dynamic geometry software, the graph of Figure 7 can be produced from an animated version of the graph of Figure 8 in which the slope of the function for car B is varied while the t-axis intercept is kept fixed.

Other ways of generalizing the problem

This is not the only way to generalize the problem. To get a fuller picture, note the relevant variables in the problem, as shown in Table 2.

Table 2

Quantity	Symbol	Original Value
speed of person A	v	50 mph
speed of person B (catch-up speed)	w	75 mph
speed increase (how much faster B is than A)	Δv	25 mph
head start time (head start A has over B)	h	3 hours
catch-up time (time for B to catch up with A)	t	6 hours

The function (3) we have derived above gives t as a function of w, while keeping the other variables with their original numerical values (3 hours and 50 mph). Some other ways to generalize are presented in the Problems.

3.1.2 Problems

1. Consider a general "catch-up" situation such as the one analyzed in this section.

a. Show that the time required for a person to catch up is directly proportional to the delay (the elapsed time between the time the first person starts and the time the second person starts).

b. Show that the constant of proportionality in part **a** is dependent only on the ratio of the two speeds.

2. The text claims that, roughly, for every one mile per hour B increases his catch-up speed, the catch-up time is decreased by about $\frac{1}{4}$ hour. What is the basis of this claim? How close is "roughly" if the speed is increased from 75 mph to 80 mph? How close is "roughly" if the speed is decreased from 75 mph to 70 mph?

3. The function f graphed in Figure 7 expresses the catch-up time t as a function of person B's speed w.

a. Find a function that expresses the catch-up *distance* d as a function of person B's speed w. Graph this function, assuming the values $v = 50$ mph for person A's speed and $b = 3$ hours for the delay of person B.

b. Express the catch-up distance d solely in terms of general parameters: person B's speed w, person A's speed v, and the time delay b of person B.

4. The text mentions a way the two graphs in the section could be related using a dynamic geometry program. Carry this out, producing the graph of Figure 7 from a dynamic version of the graph of Figure 8.

5. A general "meeting" problem concerns two people starting off at the same time and heading toward each other.

a. Express the amount of time it will take to meet as a function of the speeds of the people and the initial distance between them.

b. Express the location of the meeting place as a function of the speeds of the people and the initial distance between them.

6. **Round trips with and against a wind.** Here is a problem of a type most students encounter in their study of elementary algebra.

> An airplane makes a round trip where the one-way distance is 1000 km. On the out-leg the plane faces a headwind of 50 km/h, while on the return there is a tailwind of 50 km/h. If the speed of the plane in still air is 400 km/h, what is the total time for the trip?

a. *A qualitative argument.* Before you solve the problem, think about it in a "qualitative" way: Sketch a rough graph of a function giving the total time for the round trip in terms of the wind speed as the wind speed varies from 0 to 400 kph. Compared with the total time for a round trip with *no* wind, do you think the time for the round trip *with* the wind is (i) less, (ii) the same, or (iii) more?

b. *A numerical answer.* Answer the question of the problem. Does your answer support your qualitative response in part **a**?

c. *A general answer.* The numerical answer does not reveal much about the structure of the situation. Solve the problem again, this time expressing the total time in terms of general parameters for the total distance, the air speed of the plane, and the wind speed. There are many different equivalent symbolic expressions that will express the total time. Try to "coax" the expression you arrive at into a simple form.

d. *The general answer refined.* Express the total trip time with no wind (call it t_0) in terms of the given parameters. Use t_0 to get a more revealing expression for the total time *with* wind. There is a connection of this problem to special relativity through a "Lorenz transformation." Look this up and show what the connection is.

e. *The motion functions.* Functions have not played a role so far in the analysis we have outlined. Give an alternative approach by modeling the situation with the motion functions of the plane's outbound and return trip. Graph these functions.

f. *The dimensionless factor.* A dimensionless factor $\frac{1}{1-r^2}$, where r is the ratio of the wind speed to the plane's speed, appears in the expression for the total time found in parts **c** and **d**. Analyze this factor as a function of r, and graph this function.

7. **A mixture problem.** Consider this question, also of a common type of problem from elementary algebra.

> How many ounces of a solution that is 90% alcohol need to be mixed with 5 ounces of a solution that is 50% alcohol in order to obtain a solution that is 80% alcohol?

a. Answer the question, letting y be the answer.

b. Generalize the question by replacing 80% by x%. Then graph the function that maps x onto y. Interpret the graph in terms of the original question.

c. Find a formula for the function that maps y onto x and graph that function. Interpret the graph in terms of the original question.

d. Generalize the problem, replacing 90% by A and letting y be the answer. Find a formula for the function that maps A onto y and graph that function. Interpret the graph in terms of the original question.

e. Generalize the problem in a different way, replacing 5 by G and letting y be the answer. Find a formula for the function that maps G onto y and graph that function. Interpret the graph in terms of the original question.

3.1.3 Some types of functions

Functions are important in a wide variety of contexts in which the domains or ranges are not sets of numbers. Functions are also given a variety of other names that are suggestive of the additional special properties that they have. In this section, we explore some of these instances of functions to demonstrate the utility and broad applicability of the concept in mathematics.

One-to-one correspondences as one-to-one functions

In Chapter 2, we used one-to-one correspondences to establish when sets have the same cardinality. For instance, the existence of a one-to-one correspondence between the set **N** of natural numbers and the set $\mathbf{Q}^+$ of positive rational numbers meant that **N** and $\mathbf{Q}^+$ have the same cardinality. We indicated that a one-to-one correspondence can be defined by a formula, a diagram, a rule, or a table. One-to-one correspondences are *one-to-one functions*.

Definition A function $f\colon A \to B$ is a **one-to-one function** or **1-1 function** if and only if every element b in B is the image of *at most* one element a in A. Symbolically, f is 1-1 if and only if for all x_1 and x_2 in A, $f(x_1) = f(x_2)$ implies $x_1 = x_2$.

For instance, with domain **R**, $y = f(x) = x^3$ defines a 1-1 real function because ${x_1}^3 = {x_2}^3$, $x_1 \in \mathbf{R}$, and $x_2 \in \mathbf{R}$ implies $x_1 = x_2$. On the other hand, with domain **R**, $y = \cos x$ does not define a 1-1 function because $\cos 0 = \cos(2\pi)$ yet $0 \neq 2\pi$.

Often 1-1 correspondences can be presented geometrically. For example, Figure 9 pictures a one-to-one correspondence between the set **R** of real numbers and the open interval $(-\frac{\pi}{2}, \frac{\pi}{2})$ in **R**.

Figure 9

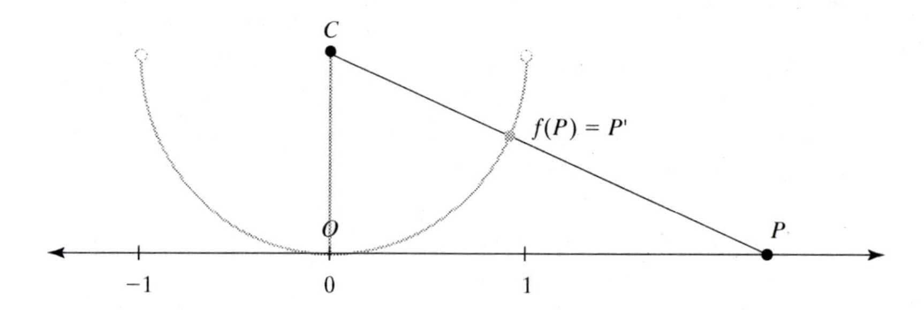

It is described geometrically as follows: Construct a semicircle of radius 1 that is tangent to the real number line at the point O with coordinate 0. Then the line joining a point P on the real line to the center of C of this semicircle determines a point P' on this semicircle. Now $\angle P'CO$ has radian measure between $-\frac{\pi}{2}$ and $\frac{\pi}{2}$, measuring positive and negative from ray CO. Let the coordinate of P correspond to the measure of $\angle P'CO$. For instance, the point with coordinate 1 corresponds to $\frac{\pi}{4}$. This one-to-one correspondence $f\colon \mathbf{R} \to (-\frac{\pi}{2}, \frac{\pi}{2})$ provides an alternative geometric model of the real number system to the number line—a semicircle—if we identify the point $P' = f(P)$ on the semicircle with the coordinate of P on the number line. In this model, it is natural to interpret the two endpoints of the semicircle as representations of $+\infty$ and $-\infty$. Of course, $+\infty$ and $-\infty$ are not real numbers, but they do represent limits of sets of real numbers. The model has the following property: For any sequence $\{P_n\}$ of real numbers, $\lim_{n\to\infty} P_n = L$ if and only if $\lim_{n\to\infty} f(P_n) = f(L)$. (See Problem 1.)

Inverses of functions

The importance of 1-1 functions lies in the fact that their action can be reversed; that is, given any element y in the range $f(A)$ of a 1-1 function $f: A \to B$, there is exactly one element x in A such that $f(x) = y$. The rule that assigns each y in $f(A)$ to the unique x in A for which $y = f(x)$ defines a function with domain $f(A)$ and range A called the *inverse of* f.

Definition If $f = \{(x, y): y = f(x)\}$ and f is one-to-one, then the function $\{(y, x): (x, y) \in f\}$ is called the **inverse** of f and denoted $\mathbf{f^{-1}}$.

More precisely, we have the following:

Theorem 3.1 If $f: A \to B$ is a one-to-one function with range $f(A)$, then

$$f^{-1} = \{(y, x) \in f(A) \times A: (x, y) \in f\}$$

is a one-to-one function with domain $f(A)$ and range A.

Proof: To show that f^{-1} is a function from $f(A)$ into A, we need only show that for each y in $f(A)$ there is only one x in A such that (y, x) is in f^{-1}. But

$$(y, x) \in f^{-1} \Leftrightarrow (x, y) \in f$$

and for each y in $f(A)$ there is exactly one x in A such that $(x, y) \in f$ because f is a one-to-one function. By similar reasoning, we see that f^{-1} is a one-to-one function because f is a function.

If $f(x_1) = f(x_2) = y$ and $x_1 \neq x_2$, then if g were the inverse of f, we would have $g(y) = x_1$, and $g(y) = x_2$, and g would not be a function. So a function that is not one-to-one does not have an inverse. ┘

Composition of functions

Simple functions are often combined to produce other functions. *Function composition* gives the result when a second function operates on the images of a first function. Here is a formal definition in the language of ordered pairs.

Definition If $f: A \to B$ and $g: B \to C$ are functions, then the **composite function** $(g \circ f): A \to C$ is the subset $g \circ f$ of $A \times C$ defined as follows:

$$g \circ f = \{(a, g(f(a))) \in A \times C: a \in A\}.$$

Read "$g \circ f$" as "the composite of f followed by g". We distinguish the operation "composition" from the result "composite" of performing the operation. Some authors use the word "composition" for both.

For example, if $f: x \to x^2$ (the squaring function) and $g: x \to x - 5$ (the "subtracting 5" function), then $g(f(x)) = g(x^2) = x^2 - 5$, so $g \circ f: x \to x^2 - 5$.

Inverse functions and composition of functions

The term "inverse function" comes from the fact that the composite of a function and its inverse is an identity function under the operation of function composition.

The function $f: x \to x^2$ with domain $\mathbf{R}$ does not have an inverse, but if we restrict its domain to $[0, \infty)$, then the restricted function does have an inverse. In general, if $f: A \to B$ is a function and if C is a subset of A, then the set

$$f_C = \{(a, b): a \in C\}$$

is a function from the set C into B called the **restriction of f to C**. The restriction $f_C: C \to B$ has the equation

$$f_C(x) = f(x) \qquad \text{for all } x \text{ in } C.$$

For any set C, the symbol I_C denotes the **identity function** on C; that is,

$$\mathbf{I_C} = \{(x, x): x \in C\}.$$

Equivalently, I_C is the function from C to C that satisfies $I_C(x) = x$ for all $x \in C$. $\mathbf{I_C}$ maps any element of C onto itself.

With this language, we can connect function composition, inverse functions, and identity functions.

Theorem 3.2 Suppose $f: A \to B$ is a given function. Then there is a function $g: f(A) \to A$ such that

$$g \circ f = I_A \quad \text{and} \quad f \circ g = I_{f(A)}$$

if and only if f is a one-to-one function and $g = f^{-1}$.

Proof:

($\Rightarrow$) Suppose f is 1-1 and $g = f^{-1}$. Then for all $a \in A$, if $f(a) = b$, then $g(b) = a$. So $g(f(a)) = g(b) = a$. Thus $g \circ f = I_A$. Similarly, $f(g(b)) = f(a) = b$ for any element b of $f(A)$, so $f \circ g = I_{f(A)}$.

($\Leftarrow$) This is left to you as Problem 3. ┘

By substituting f^{-1} for g in Theorem 3.2, we see that a 1-1 function f with domain A satisfies the following properties of composition:

$$f^{-1} \circ f = I_A \quad \text{and} \quad f \circ f^{-1} = I_{f(A)}.$$

Thus, just as with the operations of addition and multiplication of real or complex numbers, composing a function with its inverse in either order results in an identity for that operation.

Operations as functions

From the addition of whole numbers that you learned while very young through the operations of differentiation and integration in calculus, you have encountered a host of *operations*. Most of the common operations you have encountered are *unary* or *binary operations*, and both unary and binary operations are functions.

Binary operations are special types of functions of two variables. A **binary operation** takes as input an ordered pair of elements and from them yields a single element as output. If both components of the ordered pair and the output are from the same set S, we say the binary operation is *on the set* S. In function language, **a binary operation on a set S** is any function $S \times S \to S$. (Recall that $S \times S$ is the set of ordered pairs of elements S.) For instance, the operation A of addition can be described in function language as $A: (m, n) \to m + n$. Addition, subtraction, and multiplication are binary operations on the set of integers. Division is not a binary operation on the set of integers, because there are integers whose quotient is not an integer, but division is a binary operation on the set of nonzero rational numbers. (*Note*: Some authors do not require that the output be in S to have a binary operation on S. Instead, they specify that the operation is called **closed** if the output is in S.)

Binary operations need not be on numbers. For example, union and intersection are binary operations on the set of all subsets of a set. Composition of real functions is also a binary operation.

Unary operations are special types of functions of one variable. A **unary operation** takes as input a single element and from it yields another single element. If both the domain and codomain are the same set A, then the unary operation is said to be *on the set A*. That is, a **unary operation on a set A** is any function $u: A \to A$ that maps the elements of the set into elements of that set. For instance, the reciprocal operation r defined by $r(x) = \frac{1}{x}$ is a unary operation on the set of nonzero rational numbers. On the other hand, r is not a unary operation on the set of nonzero integers because there are nonzero integers whose reciprocals are not nonzero integers.

As with binary operations, unary operations need not be on numbers. For example, given a subset A of a set S, the **complement of A in S**, sometimes denoted by $\overline{\mathbf{A}}$, is the set of elements of S not in A. The function $c: A \to \overline{A}$, "taking the complement," is a unary operation on the subsets of a set.

Every function $f: \mathbf{R} \to \mathbf{R}$ is a unary operation on the set of real numbers. For instance, the function f with rule $f(x) = x + 5$ might be described as the unary operation "adding 5." More generally, for any set S, any function $f: S \to S$ is a unary operation on S.

Operations are often discussed not as special types of functions, but rather with a special terminology of their own. For instance, the idea of *closure* is common with operations, but not with functions. Let f be a unary or binary operation on a set A. If B is a subset of A, then B is said to be **closed with respect to the operation f** if and only if the restriction of f to B is an operation on B.

For example, for the binary operation of addition on the set $\mathbf{Z}$ of integers, the subset E of even integers is closed with respect to addition, but the subset O of odd integers is not. That is, the restriction of the binary operation of addition $A: \mathbf{Z} \times \mathbf{Z} \to \mathbf{Z}$ to $E \times E$ is a binary operation on E, while its restriction to $O \times O$ is a binary operation but it is not a binary operation on O. Similar conclusions hold for the binary operation of multiplication on the set $\mathbf{R}$ of real numbers on the subset $\mathbf{Q}$ of rational numbers (closed) and the subset $\overline{\mathbf{Q}}$ of irrational numbers (not closed).

Inverse operations are not necessarily inverse functions

Sometimes operations use the same terminology as functions, but with different meanings. Specifically, certain pairs of unary or binary operations are called *inverses* of one another even through they are not pairs consisting of a function and its inverse function. Examples of such pairs are the operations of addition and subtraction for numbers; and the operations of multiplication and division for numbers. In each of these examples, the term "inverse" is used differently for operations than for functions.

For example, the binary operation of addition on the set $\mathbf{Z}$ of integers, the set $\mathbf{Q}$ of rational numbers, and the set $\mathbf{R}$ of real numbers does not have an inverse as a function because it is not one-to-one. For instance,

$$7 = 4 + 3 = 8 + (-1) \Rightarrow 7 \text{ is the image of } (4, 3) \text{ and } (8, -1) \text{ under addition.}$$

However, there is a natural way in which addition does have an inverse in the context of functions: For a given set S (such as $\mathbf{Z}$, $\mathbf{Q}$, or $\mathbf{R}$ but not $\mathbf{N}$) on which addition is a binary operation, and a given k in S, define

$$A_k: S \to S \quad \text{and} \quad S_k: S \to S$$

by $A_k(x) = x + k$ and $S_k(x) = x - k$ for all x in S. We might call A_k the "adding k" function and S_k the "subtracting k" function. Then A_k and S_k are a pair of inverse functions for each k in S because if $A(x) = x + k = y$, then $S(y) = y - k = x$. You are asked to describe a similar interpretation of multiplication and division in Problem 6.

In calculus, the operations of differentiation and integration are often described as inverse operations as a means of summarizing the statement called the *Fundamental Theorem of Calculus*. We state that theorem here without proof.

Theorem 3.3 **(Fundamental Theorem of Calculus):** If f is a real function that is continuous on an interval $[a, b]$ and if F is the real function defined for each x in $[a, b]$ by

$$F(x) = \int_a^x f(t)\,dt,$$

then $F(x)$ is differentiable on $[a, b]$ and $\frac{d}{dx}(\int_a^x f(t)\,dt) = f(x)$ for all x in $[a, b]$.

It looks as if integration and differentiation are inverse operations on functions, since integration followed by differentiation yields the original function. That is, $\frac{d}{dx}(\int_a^x f(t)\,dt) = f(x)$ for any continuous function on $[a, b]$. But since $\int_a^x f'(t)\,dt = f(x) - f(a)$, the value of the composite function in the other order differs by a constant from the "original" function f. In fact, since $\frac{d}{dx}f(x) = \frac{d}{dx}(f(x) + c)$ for any constant c, the unary operation of differentiation does not have an inverse. Thus, to describe integration and differentiation as inverse operations is technically not valid.

Transformations

Some functions that suggest a change of a set, often geometric in nature, are called *transformations*. The most common transformations in school mathematics are *geometric transformations*. A **geometric transformation** is a function whose domain and range are sets of points. Most often the domain and range of a geometric transformation are both $\mathbf{R}^2$ or both $\mathbf{R}^3$. Often geometric transformations are required to be 1-1 functions, so that they have inverses. Examples of geometric transformations are reflections, rotations, size changes, scale changes, and shears. The name "transformation" for these functions comes from the fact that figures in the domain are viewed as having been transformed by the function into their corresponding image figures in the range, and the requirement that they be 1-1 ensures that there is a unique transformation "back" from the image to the original figure.

A rule for a geometric transformation of the plane requires that for each point there be a way to find its image. This rule may be given in geometric language. For example, part of the definition of the reflection r in the plane over line m is that the image of point P is the point P' such that m is the perpendicular bisector of $\overline{PP'}$. Function notation is commonly used, so if r is the reflection over m, we may write $r: P \rightarrow P'$, and $r(P) = P'$. A rule may also be given in terms of coordinates of points. For instance, the transformation T under which the point $(x + 2y, y)$ is the image of (x, y) can be described as $T: (x, y) \rightarrow (x + 2y, y)$, or $T(x, y) = (x + 2y, y)$. This transformation happens to be a shear.

Transformations may be composed. For example, we can compose

$$r_{x=y} = \text{reflection over the line } x = y$$

and $$r_x = \text{reflection over the } x\text{-axis}$$

to obtain $r_{x=y} \circ r_x$, the rotation of 90° counterclockwise about $(0, 0)$, or $r_x \circ r_{x=y}$, the rotation of 90° clockwise about $(0, 0)$. Composition of transformations is basic to the study of congruence and similarity (see Chapters 7 and 8).

Transformations may describe functions in areas of mathematics outside geometry, but almost always from a situation in which there is underlying geometry. For instance, linear transformations are particular functions that map vectors onto vectors, or matrices onto matrices.

Morphisms

Recall from Chapter 1 that $\langle \mathbf{R}^+, \cdot \rangle$ and $\langle \mathbf{R}, + \rangle$ are sets with operations on their elements satisfying certain general rules. The set $\mathbf{R}^+$ of positive real numbers is a group (it satisfies the group properties) under the binary operation of multiplication, while the set $\mathbf{R}$ of real numbers is a group for the binary operation of addition. The function $\log_b$: $\mathbf{R}^+ \rightarrow \mathbf{R}$ is a one-to-one correspondence between $\mathbf{R}^+$ and $\mathbf{R}$ under which results of the operations correspond. That is, if $\log_b(x) = m$ and $\log_b(y) = n$, then the log of the product in $\mathbf{R}^+$ corresponds to the sum of the logs in $\mathbf{R}$: $\log_b(xy) = m + n$. Such one-to-one correspondences between algebraic systems are called **isomorphisms**, and the two systems are said to be **isomorphic** algebraic systems. The term *isomorphism* (from the Greek "iso", meaning "same," and "morph", meaning "form or structure") indicates that the two systems $\langle \mathbf{R}^+, \cdot \rangle$ and $\langle \mathbf{R}, + \rangle$ are algebraically identical. See Problem 15 for another example of isomorphism.

The term **homomorphism** is used for functions from one algebraic system onto another under which results of operations correspond, but that are not necessarily one-to-one functions. For example, if m is an integer greater than 1, the function h: $N \rightarrow N_m$ maps any natural number n onto its congruence class n^* modulo m (see Section 6.1.2). Both the set N of all natural numbers and the set N_m of congruence classes modulo m are groups under addition. The function h is operation preserving: $h(a + b) = (a + b)^* = a^* + b^* = h(a) + h(b)$ for all a and b in N.

3.1.3 Problems

1. Consider the correspondence f: $\mathbf{R} \rightarrow (-\frac{\pi}{2}, \frac{\pi}{2})$ described in this section.

a. Find the values of $f(-1)$ and $f(10)$.

b. Generalize part **a** to derive a formula for $f(x)$.

c. Use part **b** or a geometric argument based on Figure 9 to explain why $\lim_{n\to\infty} P_n = L$ if and only if $\lim_{n\to\infty} f(P_n) = f(L)$.

2. a. Prove that any two open intervals (a, b) and (c, d) in the set $\mathbf{R}$ of real numbers have the same cardinality.

b. Find a formula for a 1-1 correspondence that demonstrates part **a**.

3. Prove the ($\Leftarrow$) direction of Theorem 3.2, that if f: $A \rightarrow B$ is a given function and there is a function g: $f(A) \rightarrow A$ such that $g \circ f = I_A$ and $f \circ g = I_{f(A)}$, then f is a one-to-one function and $g = f^{-1}$.

4. Decide if the following subsets of the set $\mathbf{Z}$ of integers are closed for the unary operation of squaring. Support your conclusion.

a. the set E of even integers

b. the set O of odd integers

c. the set P of prime (positive) integers

d. the set C of composite positive integers

5. Describe each of the following as a unary or binary operation on an appropriate set of numbers.

a. the greatest common divisor $\gcd(x, y)$ of x and y

b. the maximum $\max\{x, y\}$ of x and y

6. Provide an interpretation of the statement in the context of functions:

The binary operations multiplication and division of real numbers are inverse operations.

7. Explain in the context of functions the meaning of the following statement:

The reciprocal operation $a \rightarrow \frac{1}{a}$ *and the squaring operation* $a \rightarrow a^2$ *commute on the set of nonzero rational numbers, but the "taking the opposite" operation* $a \rightarrow -a$ *does not commute with the squaring operation.*

8. Let P be a binary operation on the set of positive real numbers such that $P(x, y) = 2x + 2y$.

a. Is P commutative?

b. Is P associative?

c. Give a geometric application of P and interpret the result of part **a** in terms of that application.

9. Repeat Problem 8 for the binary operation A on the set of positive real numbers such that $A(x, y) = xy$.

10. Let S be a finite set with n elements.

a. How many binary operations are there on S?

b. How many of these are commutative?

11. The number of permutations of n objects taken r at a time, written nPr or $P(n, r)$, can be considered as a function of two variables n and r described by the rule $P(n, r) = \frac{n!}{(n-r)!}$, where $n \geq r$. P also can be considered as a binary operation.

a. Is P commutative?

b. Is P closed on $\mathbf{N}$?

12. Answer the questions of Problem 11 for the number of combinations of n objects taken r at a time, written nCr, $C(n, r)$, or $\binom{n}{r}$, described by the rule $C(n, r) = \frac{n!}{r!(n-r)!}$, where $n \geq r$.

13. Consider the binary operation of powering (or exponentiation) defined by p: $(x, y) \rightarrow x^y$.

a. Is p a binary operation on $\mathbf{R}^+ \times \mathbf{R}^+$, where $\mathbf{R}^+$ is the set of positive real numbers?

b. Explain why p is not a binary operation on $\mathbf{R} \times \mathbf{R}$.

c. What value(s), if any, does your calculator give for $p\left(-8, \frac{1}{3}\right)$ and $p\left(-8, \frac{2}{6}\right)$?

d. Explain what your answer to part **c** means in the context of binary operations.

14. Let X be a set with 5 distinct elements a, b, c, d, e, and let Y be the set of 5 distinct prime numbers $2, 3, 5, 7, 11$. The set $P(X)$ of all subsets of the set X is an algebraic system for the binary operations of union $\cup$ and intersection $\cap$ of sets. The set $D(2310)$ of all positive divisors of 2310 is an algebraic system for the binary operations of least common multiple (lcm) and greatest common factor (gcf). (2310 is the product of 2, 3, 5, 7, and 11.)

a. Explain why the sets $P(X)$ and $D(2310)$ both have $2^5 = 32$ elements.

b. Show that the algebraic systems $\langle P(X), \cup, \cap \rangle$ and $\langle D(2310), \text{lcm}, \text{gcf} \rangle$ are isomorphic by defining a one-to-one correspondence f between $P(X)$ and $D(2310)$ such that

$$f(A \cup B) = \text{lcm}(f(A), f(B)) \quad \text{and}$$
$$f(A \cap B) = \text{gcf}(f(A), f(B))$$

for all subsets A and B of X.

15. Let i be the complex number such that $i = \sqrt{-1}$, and let S_4 be the set $\{1, i, -1, -i\}$.

a. Prove that the operation $\cdot$ of multiplication of complex numbers is a binary operation on S_4 and that $\langle S_4, \cdot \rangle$ is a group.

b. Prove that the function h: $Z \rightarrow S_4$, defined by $h(n) = i^n$ for each integer n, is a homomorphism of $\langle N, + \rangle$ onto $\langle S_4, \cdot \rangle$.

Unit 3.2 Properties of Real Functions

We use the term **real function** to refer to a function whose domain and range are subsets of the set $\mathbf{R}$ of real numbers. The function f that opens this chapter and all the functions of the problem analysis of Section 3.1.2 are real functions. A function may involve only real numbers but still not be a real function. For instance, a binary operation on real numbers cannot be a real function because it is of the form $\mathbf{R} \times \mathbf{R} \rightarrow \mathbf{R}$; that is, the elements of its domain are not single real numbers, but ordered pairs. The most important special feature of real functions as far as their analysis is concerned is that they can be graphed in the Cartesian coordinate plane.

A small number of categories of real functions dominate the study of precalculus and calculus.

i. linear, quadratic, and, more generally, the *polynomial functions*, that is, functions that can be expressed as

$$y = p_n(x) = a_n x^n + a_{n-1}x^{n-1} + \cdots + a_1 x + a_0 \qquad \text{for all } x \text{ in } R,$$

with real coefficients;

ii. *rational functions*, that is, the functions f that can be expressed as

$$f(x) = \frac{p(x)}{q(x)}, \qquad \text{for all } x \text{ in } R \text{ such that } q(x) \neq 0,$$

where $p(x)$ and $q(x)$ are polynomials with real coefficients;

iii. *exponential functions*, functions of the form

$$f(x) = cb^x, \qquad \text{for all } x \text{ in } \mathbf{R},$$

where c is any real number and the *base* b is a positive real number other than 1;

iv. *logarithmic functions*, functions of the form

$$f(x) = \log_b(x) \qquad \text{for all } x > 0,$$

for any positive base $b \neq 1$;

v. the six *trigonometric functions*

$$\sin, \quad \cos, \quad \tan, \quad \cot, \quad \sec, \quad \csc,$$

and the corresponding *inverse trigonometric functions*; and

vi. *sequences* of real numbers

$$a = \{a_n\} = a_0, a_1, a_2, \ldots, a_n, \ldots.$$

Typically, these functions are introduced one category at a time and, once introduced, the graphs, properties, and applications of the functions in that category are developed and analyzed. As a result of the category-by-category approach, it is not unusual to miss some of the general principles for analyzing *any* real function.

In this unit, we discuss important properties of real functions from perspectives that cross-function category lines. In Section 3.2.1, we outline our objectives for analyzing real functions, and we introduce two general types of real functions, *discrete* real functions and *interval-based* real functions, that together essentially include all of the functions just listed. Section 3.2.3 treats properties of monotone real functions and their inverses. Section 3.2.4 focuses on the growth properties and limiting behavior of real functions. In Sections 3.3.1 and 3.3.2 we discuss how to use real functions to model data. Section 3.3.3 is an extended analysis of a familiar optimization problem that illustrates the importance of generalization in the analysis of real functions.

3.2.1 Analyzing real functions

The domain D of a real function f can be any subset of the real numbers, but typically it is either

(Type 1) a finite set of real numbers or a set of integers greater than or equal to a fixed integer k, where k is often 0 or 1; or

(Type 2) **R** itself, or an interval in **R**, or a union of intervals in **R**.

Real functions of Type 1 are called **discrete real functions**. Discrete real functions for which the domain D is the set of all integers greater than or equal to some fixed integer are called **real sequences**.

Unfortunately, there is no standard term to identify real functions of Type 2. We use the term **interval-based real functions** for functions of this type. The important point about our choice of definition for functions of Type 2 is that it is general enough to include all of the real functions of categories (i) through (v) mentioned above. That is, Type 2 functions include polynomial, rational, exponential, logarithmic, trigonometric, and inverse trigonometric functions. Also, the basic concepts of calculus, such as limits at a point, derivatives, and continuity, are applicable to functions of Type 2.

Some common rules for functions apply to both discrete and interval-based real functions. The formula $f(x) = ax + b$ defines a **linear function** (that is, a polynomial function of degree 1) with **R** as a domain, as well as a **linear** or **arithmetic sequence** with **N** as its domain. In analogous fashion, the formula $g(x) = ab^x$ defines an **exponential function** with **R** as its domain as well as an **exponential** or **geometric**

sequence with **N** as its domain. Consequently, we can apply what we know about linear and exponential functions to arithmetic and geometric sequences.

Finding values of real functions

You are familiar with a wide variety of real functions. In fact, much of the traditional work in high school and early college mathematics courses is to provide algorithms and theorems for determining the behavior of real functions. Today, with graphing technology and computer algebra systems, it is possible to describe the behavior of a wide variety of functions that hitherto were very difficult to analyze.

It goes without saying that before you can analyze a function f in any detail, you need to know how to determine $f(x)$ for any value of x in the domain of f. However, what "determining" means depends on the situation. For instance, suppose you need to determine $\sin\frac{2\pi}{5}$, one of the values of the sine function.

1. You might use a calculator and obtain $\sin\frac{2\pi}{5} \approx 0.95105651$. This decimal value is not exact; it is the value of $\sin\frac{2\pi}{5}$ rounded to eight decimal places. However, for most calculations and for many other purposes the value of $\sin\frac{2\pi}{5}$ has been "determined." Yet this value would not help us know whether $\sin\frac{2\pi}{5}$ is rational or not.
2. You might draw a picture as shown in Figure 10, knowing that $\sin\frac{2\pi}{5}$ is the second coordinate of a particular point on the unit circle. You have "determined" the value of $\sin\frac{2\pi}{5}$ geometrically. This would be most helpful in finding the coordinates of a regular pentagon.
3. You might know (perhaps from properties of regular pentagons) that $\sin\frac{2\pi}{5}$ can be expressed in terms of radicals: $\sin\frac{2\pi}{5} = \sqrt{\frac{5+\sqrt{5}}{8}}$. This *exact* expression for $\sin\frac{2\pi}{5}$ suggests that $\sin\frac{2\pi}{5}$ is an irrational number. This can be proved. (See Problem 13.)
4. For some uses, for example, if you were going to apply properties of the sine function, it might be best to consider $\sin\frac{2\pi}{5}$ as already "determined," since $\sin\frac{2\pi}{5}$ is an exact way to state the value, too. In particular, if you ask a computer algebra program such as *Mathematica* to evaluate $\sin(\frac{2\pi}{5})$, the output is $\sin(\frac{2\pi}{5})$.

Figure 10

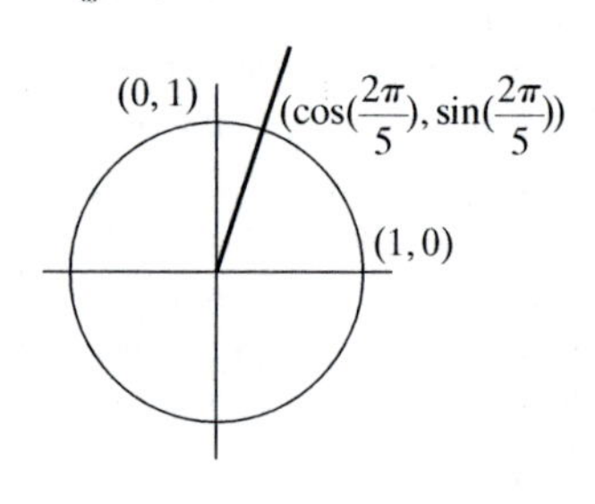

Graphs of real functions

Because a real function is a set of ordered pairs of real numbers, it has a graph in a coordinate plane. But there is more than one way to coordinatize a plane. While by far the most common coordinate system is the rectangular (Cartesian) coordinate system, the polar coordinate system is also quite useful, as we showed with complex numbers in Chapter 2.

Figures 11a and 11b show the graphs of the sets of ordered pairs of real numbers in which one component of the pair is constant. For rectangular coordinates, the graph of the set is a *vertical line* if the first coordinate is kept constant and a *horizontal line* if the second coordinate is constant. For polar coordinates, the graph of the set is a *circle centered at the origin* if the first of the pair is kept constant and a *line through the origin* if the second of the pair is constant. Thus the polar coordinate graph of a set of ordered pairs of real numbers is quite different from the graph of the same formula in rectangular coordinates.

For this reason, we need some way to distinguish the systems. Customarily this is done by using x and y as variables for rectangular coordinates, and r and θ as variables for polar coordinates, as we have done in Figure 11. In polar coordinates, the second component of the pair, θ, is usually the independent variable, and r is the dependent variable. How different are these systems? Figure 12 shows the graphs of the sine function in the two coordinate systems. Figure 13 presents the graph of a linear (!) function in the two systems.

Thus the names for many common functions have come from thinking of the Cartesian graph, and not from the graph in polar coordinates. So, too, is our thinking about functions from their graphs.

The requirement in the definition of a function f that every $a \in A$ appears once and only once as the first element of an ordered pair (a, b) has a simple graphical interpretation for Cartesian graphs of real functions. But the Vertical Line Test is not true if the graph is in polar coordinates, as any of Figures 11b, 12b, or 13b show.

Figure 11a

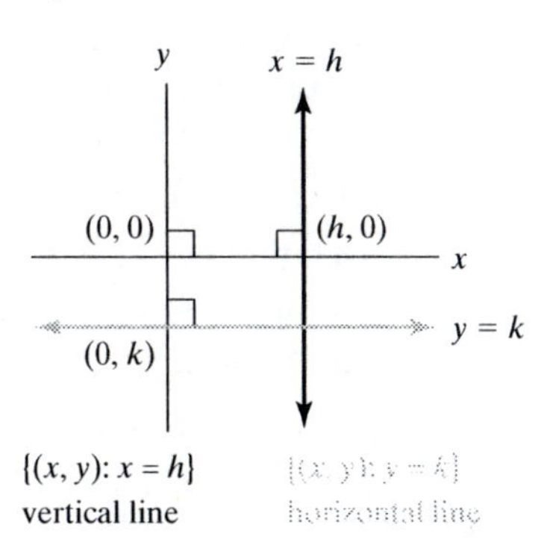

$\{(x, y): x = h\}$
vertical line

$\{(x, y): y = k\}$
horizontal line

Figure 11b

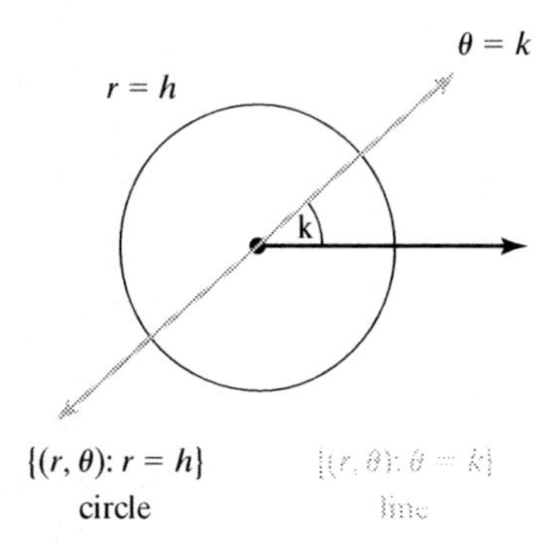

$\{(r, \theta): r = h\}$
circle

$\{(r, \theta): \theta = k\}$
line

Figure 12a

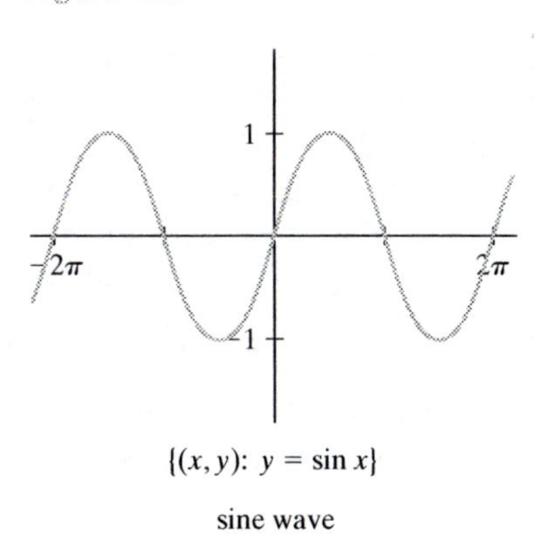

$\{(x, y):\ y = \sin x\}$

sine wave

Figure 12b

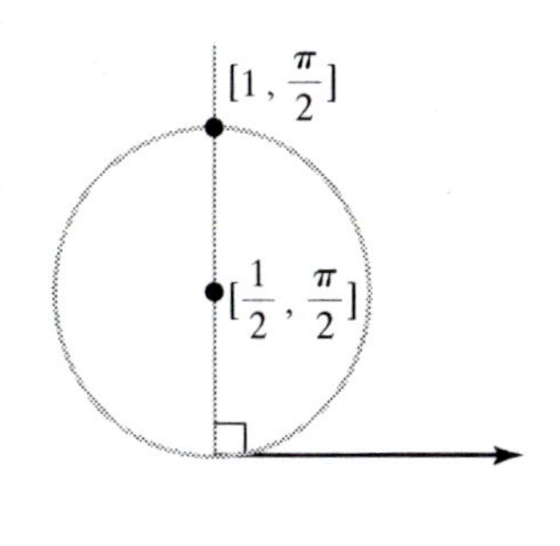

$\{(r, \theta):\ r = \sin \theta\}$

circle (center $[\frac{1}{2}, \frac{\pi}{2}]$, radius $\frac{1}{2}$)

Figure 13a

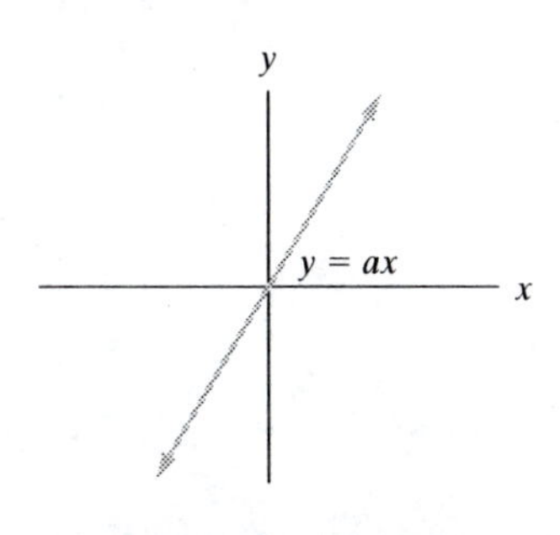

$\{(x, y): y = ax\}$
line through origin

Figure 13b

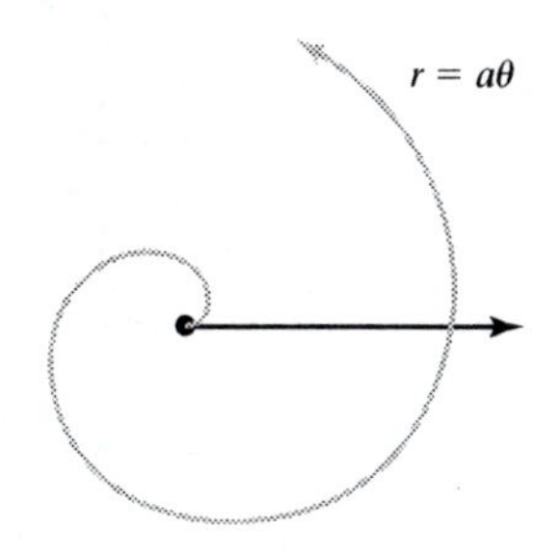

$\{(r, \theta): r = a\theta\}$
Archimedean spiral

Theorem 3.4 **(Vertical Line Test for Cartesian graphs):** A Cartesian graph G is the graph of a real function if and only if no vertical line intersects G in more than one point.

Figures 12 and 13 also demonstrate the ability of polar coordinates to describe curves that are difficult to describe in rectangular coordinates. Furthermore, these curves are graphs of functions. Consequently, those who study curves must be able to work in polar coordinates and be able to translate from one system to the other. Still, most people work with rectangular coordinates, and our discussion for the remainder of this chapter assumes we are dealing with Cartesian graphs and the functions that determine them.

Estimating values of real functions

When we have a rule or formula for a real function, we can usually calculate the values of the function for any domain value. This is not the case if we have only a table or graph for a function with an infinite domain. It would be ideal if formulas could be constructed for functions presented by tables or graphs that would

i. produce the exact values specified by the table or graph, and

ii. enable the computation of sensible intermediate values (interpolation) or subsequent function values (extrapolation) not specified in the table.

There are methods for obtaining formulas that accomplish objective (i), or that come close enough to accomplishing both (i) and (ii) to be adequate for most practical purposes.

For example, the U.S. population function $P: t \to P(t)$ presented in Table 1 of Section 3.1.1 is only defined at the 10-year census intervals between 1780 and 1870, so it has only 10 data points. These points are graphed in Figure 14. Of course, we know that the U.S. population existed for times between the census dates as well as for times before 1780 and after 1870. It would be useful to have a formula for $P(t)$ that we could use to interpolate and extrapolate the census information. The graph of the census information in Figure 14 suggests that an exponential function might provide a reasonable fit to these data. Indeed, the method of exponential regression to this data (either by hand or with a calculator or data analysis program) yields the exponential function $Q: x \to Q(x)$ with formula

$$Q(x) = 2.906(1.03)^x, \text{ where } x = \text{years after } 1780 = t - 1780,$$

as a reasonable approximation for the population function. The graph of this exponential function, together with the population data points it approximates, is shown in Figure 14.

Figure 14

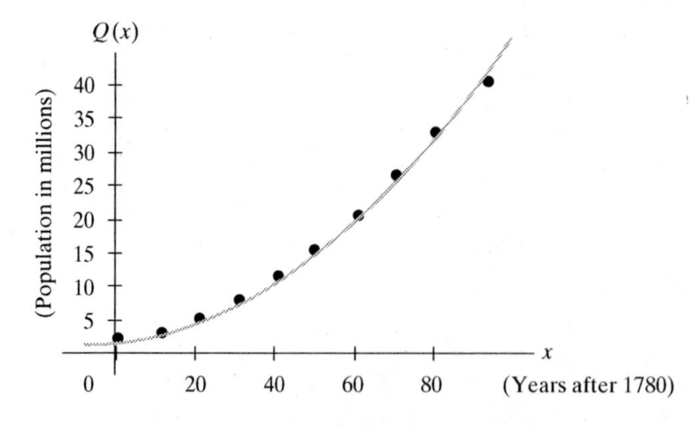

The formula for $Q(x)$ provides a means to interpolate and extrapolate the tabular data for the U.S. population function P.

Characteristics to examine in analyzing a real function

No hard and fast rules exist for "determining" or "analyzing" the behavior of a real function f. Sometimes approximations will do, sometimes a good picture is sufficient, and sometimes exact values are needed. But you should look for certain characteristics.

1. **Domain.** What is the domain of f in the given situation? Is f discrete? Interval-based?
2. **Singularities and asymptotes.** Determine the singular points of f (that is, the points at which f is undefined) and whether f has any vertical asymptotes at these points.
3. **Range.** What is the set of all possible values of f?
4. **Zeros.** At what points does the graph of f intersect the x-axis?
5. **Maxima (minima) or relative maxima (relative minima).** Find the greatest (or least) values of f, or find the greatest (or least) values on a given interval.
6. **Increasing or decreasing.** Determine those intervals on which f is increasing and those intervals on which f is decreasing.
7. **End behavior.** Describe what happens to $f(x)$ as $x \to \infty$ and as $x \to -\infty$.
8. **General properties.** Is f continuous? Is f differentiable and, if so, what is its derivative? Is f integrable and, if so, what is $\int f(x)\,dx$? What is a power series for f?
9. **Special properties.** Is f a composite of known functions? Does the graph of f have a particular known shape? Does the graph have symmetry? Is f periodic?
10. **Models and applications.** What types of situations are naturally modeled by f? How is f used in these situations?

If the given real function f is a sequence $\{f_n\}$, this analysis checklist is shortened because the domain of $\{f_n\}$ is the set of all integers n such that $n \geq n_0$ for some integer n_0. So characteristic 1 is automatic and claracteristics 2 and 8 do not have to be examined. On the other hand, for a recursively defined sequence $\{f_n\}$, a crucial part of the analysis is to determine the nth term f_n *explicitly* so that any given term can be computed without first computing all of its predecessors. Often, such an explicit description of f_n can be conjectured by carefully analyzing the first few terms of the sequence $\{f_n\}$, and then verified by mathematical induction. We shall do this for the famous Tower of Hanoi problem in Chapter 5. In other cases, the form of the recursive definition for $\{f_n\}$ allows us to compute f_n explicitly by special methods. Some of these methods are discussed later in this chapter.

Examples of analyses of real functions

Entire volumes have been devoted to analyses of real functions. In later sections of this unit, we discuss important properties of real functions and apply them in analyzing characteristics of real functions as described above. Here we present two examples to illustrate the analysis procedure. For the first example, recall that **linear functions** are those functions f with rules of the form $f\colon x \to mx + b$.

EXAMPLE 1 Analyze the linear functions $f\colon x \to mx + b$ with $m > 0$.

Solution

1. Any real number can be in the domain of f.
2. f has no singularities; it is defined for all real numbers.
3. The range of f is $\mathbf{R}$.

4. There is one zero, $\frac{-b}{m}$.
5. There are no maxima or minima.
6. f is increasing on the entire domain.
7. $f(x) \to \infty$ as $x \to \infty$, and $f(x) \to -\infty$ as $x \to -\infty$.
8. f is continuous and differentiable with derivative $f'(x) = m$. $\int f(x)\,dx = \frac{mx^2}{2} + bx + C$. Since f is a polynomial function, its power series in x is $f(x)$ itself, namely, $mx + b$.
9. The graph of f is a line; f is not periodic.
10. f is the model for situations of constant increase or decrease, and for situations involving linear combinations of two variables. Over short intervals, f is also a good approximation to many curves. f is often used to model data (via linear regression).

The many uses of linear functions account for their importance in mathematics.

Other classes of functions are not as easily analyzed as linear functions. Some classes have such varying characteristics that functions in the class are more easily analyzed individually.

EXAMPLE 2 Let P be the sequence defined by $P_0 = 2500$ and $P_{n+1} = 1.15P_n - \frac{0.15}{20000}P_n^2$, for all $n \geq 0$. Analyze the behavior of P, referring to characteristics (1) to (10) above.

Solution We see immediately from the definition that (1) P is a sequence defined for all nonnegative integers. A table of the first 100 values of P (Table 3) and the corresponding graph shown in Figure 15 suggest that (4) P has no zeros, (5) 2500 is the minimum value of P, (6) P is increasing on its entire domain, and (7) as $n \to \infty$, $P_n \to 20000$. So (3) its range is a subset of the interval $[2500, 20000]$.

Table 3

n	P_n	n	P_n	n	P_n	n	P_n	n	P_n
0	2500	1	2828	2	3192	3	3595	4	4037
5	4520	6	5045	7	5611	8	6217	9	6859
10	7535	11	8240	12	8966	13	9708	14	10458
15	11206	16	11945	17	12667	18	13364	19	14029
20	14657	21	15244	22	15788	23	16287	24	16740
25	17150	26	17516	27	17843	28	18131	29	18385
30	18608	31	18802	32	18971	33	19118	34	19244
35	19353	36	19447	37	19528	38	19597	39	19656
40	19707	41	19750	42	19787	43	19819	44	19846
45	19869	46	19888	47	19905	48	19919	49	19931
50	19941	51	19950	52	19958	53	19964	54	19969
55	19974	56	19978	57	19981	58	19984	59	19986
60	19988	61	19990	62	19992	63	19993	64	19994
65	19995	66	19996	67	19996	68	19997	69	19997
70	19998	71	19998	72	19998	73	19999	74	19999
75	19999	76	19999	77	19999	78	19999	79	19999
80	20000	81	20000	82	20000	83	20000	84	20000
85	20000	86	20000	87	20000	88	20000	89	20000
90	20000	91	20000	92	20000	93	20000	94	20000
95	20000	96	20000	97	20000	98	20000	99	20000

Figure 15

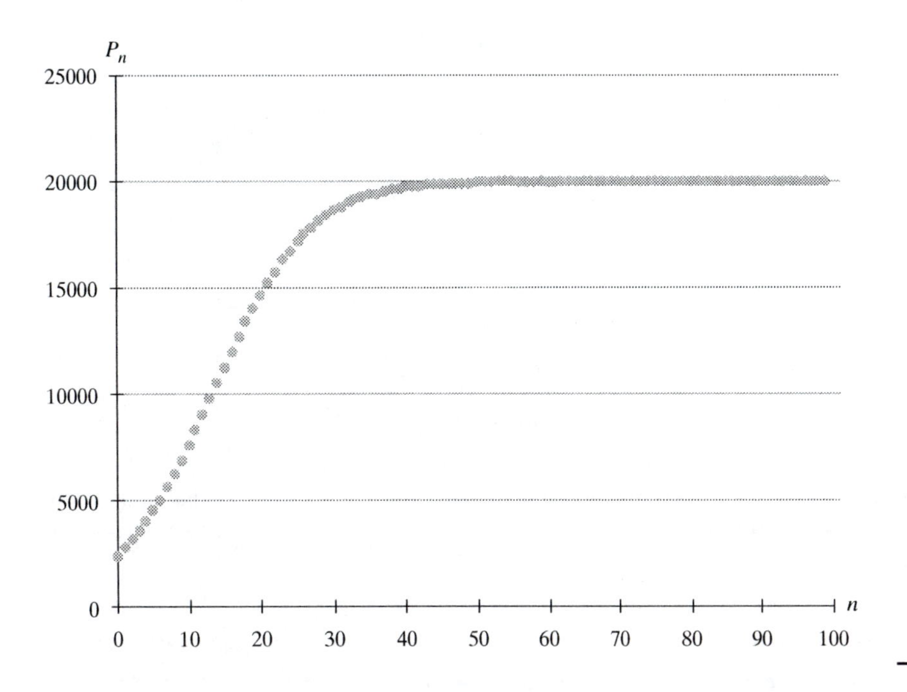

P is an example of a discrete *logistic function* and its graph lies on a *logistic curve*. Logistic functions are used to model situations of growth where there is a bound to the growth, such as occurs with populations of animals in finite regions. Suppose that an animal population in a game reserve is currently 2500 and growing at 15% a year, but that a population of at most 20,000 can be supported in the reserve. The function P provides one possible mathematical model for the population P_n n years from now. (See Problem 12.)

What today we call the logistic function was first studied by Pierre François Verhulst (1804–1849). In 1838, Verhulst proposed a population model describing how a population of animals living in a fixed region initially increases, but after some period approaches a limiting value called the *saturation point*. His model was the differential equation

$$(*) \quad \frac{du}{dt} = (A - ku)u,$$

where u = population at time t, and A and k are positive constants. He applied this model to Belgium, whose population at that time was about 3.8 million, and used models based on this function to predict a limiting population for Belgium of 9,400,000. In fact, Belgium's population stabilized in the late 20th century at about 10,100,000. In the 1920s, P. F. Raymond Pearl and Lowell J. Reed, not knowing of Verhulst's work, rediscovered this idea and applied it to the U.S. population. For this reason, some authors call (*) the **Verhulst-Pearl equation**. A solution of (*) is what is usually called a **logistic function**. As the graph in Example 2 shows (Figure 15), its graph is a monotone-increasing curve that tends to the saturation value $\frac{A}{k}$ as $t \to \infty$. As Verhulst's prediction suggests, this function does seem to model quite well the growth pattern of some real populations, and variants of it are used by biologists and demographers throughout the world to predict future populations.

In Example 2, we did not have an explicit formula for P_n. And, if you were not familiar with functions of this type, you would not know of its applications. As with all mathematical objects, the properties of functions are many, and a single analysis cannot be expected to include them all.

3.2.1 Problems

1. Indicate whether the described function is a real function.

a. the binary operation of addition of real numbers

b. the operation that corresponds each nonzero real number with its two square roots

c. the function that associates with each quadratic polynomial function p defined by

$$y = p(x) = ax^2 + bx + c \qquad \text{with } a \neq 0,$$

the coordinates (r, s) of the vertex of the graph of p in the xy-plane

d. the function that maps each positive integer onto the number of its different prime factors

e. the function f with $f(x) = 0$ for all real numbers x

f. the Dirichlet function D with $D(x) = 1$ if x is rational and $D(x) = 0$ if x is irrational

2. In the section is an analysis of the class of linear functions $f: x \to mx + b$, where $m > 0$.

a. Indicate where and how this analysis is different if $m < 0$.

b. Indicate where and how this analysis is different if $m = 0$.

3. A singularity is a value v at which a function is undefined. The singularity is **removable** if by assigning a value to the function at v, the function will be continuous on an interval containing v. For each real function described, give all values of v for which the function has singularities and tell whether the singularities are removable. If the singularity is removable, redefine the function to make it continuous over R.

a. $f(x) = \tan x$

b. $g(t) = \frac{t^2 - t}{t - 1}$

c. $h(a) = \frac{a}{3a^2 - 2a - 8}$

d. $j(y) = (y - 1)(y + 2)(y - 3)$

In Problems 4–11, analyze the function using the ideas of this section.

4. the quadratic function defined by $y = ax^2 + bx$, given $b \neq 0$ and $a > 0$

5. the exponential function defined by $y = 5000(1.06)^x$

6. the function f with the following piecewise definition:

$$f(x) = \begin{cases} x^2, & x \geq 0 \\ 4x + 5, & x < 0 \end{cases}$$

7. the function whose formula is $\frac{x}{|x|}$, for all $x \in R$, $x \neq 0$

*8. The function $y = e^{-kt}(A \cos ct + B \sin ct)$, when $A = -4$, $B = 3$, $c = 2$, and $k = 0.125$

9. the sequence $\{g_n\}$ defined recursively by

$$g_0 = 3,\ g_{n+1} = 0.5g_n \qquad \text{for } n \geq 0$$

10. the sequence $\{a_n\}$ defined recursively by

$$a_0 = 3,\ a_{n+1} = a_n + 4 \qquad \text{for } n \geq 0$$

11. the sequence $\{s_n\}$ that is defined recursively by

$$s_0 = 10,\ s_{n+1} = 2s_n - 9 \qquad \text{for } n \geq 0$$

12. a. Show that all terms of the logistic growth sequence $\{P_n\}$ in Example 2 satisfy the equation

$$P_{n+1} - P_n = .15P_n\left(1 - \frac{P_n}{20000}\right).$$

b. Use a calculator or computer to explore the growth for the first 50 years of an animal population that begins with an initial population of 2500 animals and grows at a rate of 15% per year with no limit on the number of animals that the environment can support. Specifically, let

$$P_0 = 2500 \quad \text{and} \quad P_{n+1} - P_n = .15P_n.$$

c. Explain why the difference equation in part **a** models an initial growth rate of nearly 15% but a decreasing annual growth rate as n increases, with a limiting population of 20,000 animals.

13. Since it can be proved that for any rational number r, $\cos(r\pi)$ is an algebraic number [Theorem 2.14(b)], it follows that $\sin(r\pi)$ is also an algebraic number. For some values of r, a specific proof can be given. Complete the following steps to prove that $\sin\frac{2\pi}{5} = \sqrt{\frac{5 + \sqrt{5}}{8}}$ is irrational.

a. Show that if $x = \sqrt{\frac{5 + \sqrt{5}}{8}}$, then x must be a root of the polynomial equation

$$16x^4 - 20x^2 + 5 = 0.$$

b. Apply the Rational Root Test (Footnote 3 in Section 2.1.1) to identify all possible rational roots of the polynomial equation in part **a**.

c. Verify that none of the potential rational roots of the polynomial equation in part **a** is $x = \sqrt{\frac{5 + \sqrt{5}}{8}}$.

d. Prove that $\sqrt{\frac{5 + \sqrt{5}}{8}}$ cannot be rewritten in the form $a + \sqrt{b}$ if a and b are rational.

3.2.2 Composition and inverse functions

Section 3.1.3 introduced function composition and inverse functions. In this section, we discuss some specific properties of these ideas that relate to real functions.

Reasons to consider function composition

When we encounter a new function, it is always very helpful if we can view it as a composite of simpler, more familiar functions. For example, to analyze the behavior of the real function f defined by

$$y = f(x) = e^{\sin x}$$

it is useful to view it as the composite of the functions with formulas $y = e^z$ and $z = \sin x$. That is, $f = h \circ g$, where $g(x) = \sin x$ and $h(x) = e^x$. In this way, the algebraic properties of f can be traced back to properties of g and h. Among the properties of f are the following.

1. The values of f are all positive because $e^z > 0$ for all z.
2. f is a periodic function with period 2π because the sine function has this period.
3. Because $y = e^z$ is a strictly increasing function of z, and the values of $z = \sin x$ vary from -1 at $x = -\frac{\pi}{2}$ to 1 at $x = \frac{\pi}{2}$, the range of f is the interval $[e^{-1}, e]$, or about $[.368, 2.718]$.

These properties can be checked with a graph (Figure 16).

Figure 16

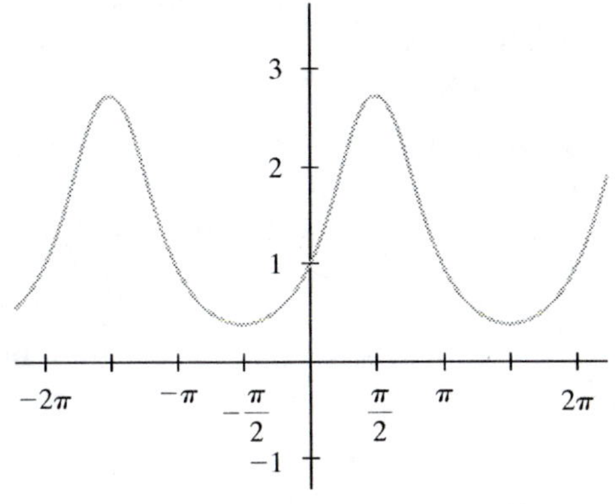

You can also use function composition to solve equations and inequalities. For example, solving the trigonometric equation $2\sin^2 x - \sin x - 1 = 0$ is equivalent to finding the zeros of the function $f(x) = 2\sin^2 x - \sin x - 1$, where f is the composite of $z = \sin x$ and the quadratic function $y = 2z^2 - z - 1$. From this viewpoint, we see that

$$2z^2 - z - 1 = (2z + 1)(z - 1) = 0$$

when $z = -\frac{1}{2}$ or $z = 1$. Consequently,

$$(\sin x = 1 \quad \text{or} \quad \sin x = -\tfrac{1}{2}) \Rightarrow x = (\tfrac{\pi}{2}) + 2k\pi \quad \text{or} \quad x = -(\tfrac{\pi}{6}) + 2k\pi \quad \text{or} \quad x = -(\tfrac{5\pi}{6}) + 2k\pi,$$

where k is any integer. These values check in the original equation.

Thus, composition of functions is an important tool for building or analyzing complicated functions with simpler parts. It is precisely for this reason that you used the chain rule for differentiation when you studied calculus.

Inverses of real functions

There is an important geometric relationship between the Cartesian graphs of a 1-1 real function f and its inverse f^{-1}: The graph of f^{-1} is the reflection image of the graph of f over the line $y = x$. For example, for a linear function and its inverse, this relationship is displayed in Figure 17.

Figure 17

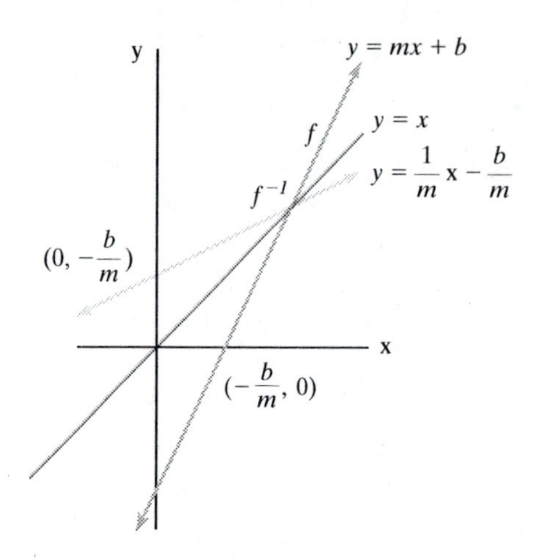

When a horizontal line is reflected over the line $y = x$, its image is a vertical line. Consequently, there is a *horizontal line test* for the Cartesian graph that indicates when a real function is 1-1.

Theorem 3.5 **(Horizontal Line Test for Cartesian graphs):** A real function is 1-1 if and only if no horizontal line intersects its Cartesian graph in more than one point.

When different scales are used on the two axes (as is often the case with trigonometric functions, or with the use of graphing technology), the graphs of f and f^{-1} may not look like reflection images of each other, but algebraically they remain reflection images since the reflection image of (x, y) over the line $y = x$ is (y, x). For example, for the natural logarithm function and its inverse, this relationship is displayed in Figure 18. (Notice that in Figure 18, due to the difference in scales on the axes, the line with equation $y = x$ does not make a 45° angle with the axes.)

Figure 18

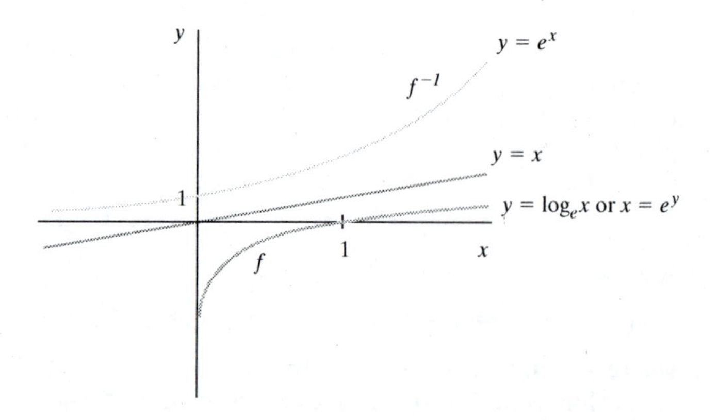

The relationship between a function and its inverse stated in Theorem 3.1 provides a method for finding the inverse of a given real function. If the real function f is described by an equation $y = f(x)$, then switching x and y and solving the resulting equation for y results in a formula for the inverse function $y = f^{-1}(x)$. This procedure is illustrated by the following two important examples of inverse functions with domains and ranges in the set **R**.

EXAMPLE 1 Find the inverse of the linear function f on $\mathbf{R}$, where

$$(*) \quad y = f(x) = mx + b, m \neq 0.$$

Solution Interchanging x and y, an equation for the inverse is $x = my + b$. Solving this equation for y, $y = \frac{x-b}{m}$. Thus, the inverse of the linear function (*) exists and is the linear function described by

$$f^{-1}(x) = \frac{x-b}{m} = \frac{1}{m}x - \frac{b}{m}.$$

A common instance of Example 1 lies in the formulas for converting Fahrenheit temperatures F to Celsius temperatures C and back. The formula for converting C to F is $F = \frac{9}{5}C + 32$. (This formula is easily derived by realizing that $0°\text{C} = 32°\text{F}$; i.e., when $C = 0$, $F = 32$, and then noting that the difference between freezing and boiling on the Fahrenheit scale is $\frac{9}{5}$ times the difference on the Celsius scale.) This is a linear function with slope $m = \frac{9}{5}$ and F-intercept $b = 32$. Its inverse will have slope $\frac{1}{m}$ and intercept $-\frac{b}{m}$, so will have equation $C = \frac{5}{9}F - \frac{160}{9} = \frac{5}{9}(F - 32)$.

If a real function $f: D \to f(D)$ is not one-to-one on its entire domain D, it is often one-to-one on certain subsets C in its domain for which $f(C) = f(D)$. In such cases, the restriction f_C of f to C has an inverse. The trigonometric functions provide importance instances of this situation.

EXAMPLE 2 Identify restrictions on the domain of the sine, cosine, and tangent functions on which these functions have inverses.

Solution Figure 19 pictures the graphs of $\{(x, y): y = \sin x, x \in \mathbf{R}\}$ and $\{(x, y): x = \sin y, y \in \mathbf{R}\}$.

Figure 19

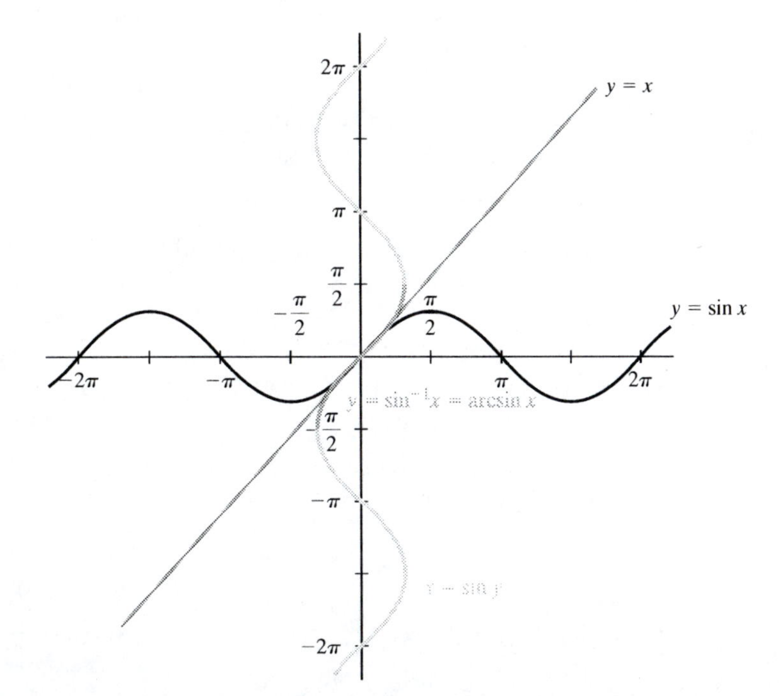

The graph of $x = \sin y$ shows that this equation does not represent a function. Infinitely many values of y correspond to each value of x from -1 to 1. However, by restricting the domain of $y = \sin x$ to $\{x: -\frac{\pi}{2} \leq x \leq \frac{\pi}{2}\}$, there is an inverse function (whose graph is shown in darker blue). This inverse is called the **arcsin** or $\mathbf{sin^{-1}}$ function.

Choosing $\{x: -\frac{\pi}{2} \leq x \leq \frac{\pi}{2}\}$ as a domain is not the only way to restrict the domain of $\{(x, y): y = \sin x, x \in \mathbf{R}\}$ so that it has an inverse. But this is the unique domain that satisfies four properties. First, it includes the basic acute angle measures $0 < x < \frac{\pi}{2}$. Second, the restriction has the same range as its parent, the sine function with domain $\mathbf{R}$. Third, each restricted function is a 1-1 function that is continuous on its selected domain. Fourth, on these domains the sine function and its inverse are increasing functions. One consequence of this latter fact is the following appealing (although not entirely necessary!) sign consistency with respect to differentiation:

$$\frac{d}{dx}(\sin x) = \cos x$$

$$\frac{d}{dx}(\arcsin x) = \frac{1}{\sqrt{1 - x^2}}$$

Thus, the choice of this restriction is attractive for a number of reasons.

In Example 2, we have used the prefix *arc* to signify the inverse of the sine function. The use of arcsin, along with arccos and arctan, arises historically from the definitions of these functions in terms of angles and the unit circle. It is also very common to use $\sin^{-1}$, $\cos^{-1}$, and $\tan^{-1}$ to signify these inverses.

Another form, more commonly found in high school texts, is the use of arcsin, arccos, ... to signify the multivalued relations $\{(x, y): x = \sin y\}$, $\{(x, y): x = \cos y\}, \ldots$, without the restriction on the domain of x, and then to use Arcsin, Arccos, ... (with capital "A") to signify the functions arcsin, arccos, ..., with the restriction on the domain.

Exponential and logarithmic functions as inverse functions

A third important set of real functions and their inverses are the exponential and logarithmic functions

$$y = b^x \qquad \text{and} \qquad y = \log_b x.$$

In each case, b is the base of the function, $b > 0$, and $b \neq 1$. That the functions are inverses is due to the relationship

(1) $\quad y = b^x \Leftrightarrow x = \log_b y$

discovered by Euler over a century after John Napier had introduced logarithms.

The relationship (1) among these functions leads to corresponding properties. Let $\exp_b$ be the exponential function defined by $\exp_b(x) = b^x$ for all real numbers x. Then $\log_b$ and $\exp_b$ satisfy the following basic function properties for all real x and y:

(2) $\quad \log_b(xy) = \log_b(x) + \log_b(y)$ (Logarithm of a Product) $\qquad \exp_b(x) \cdot \exp_b(y) = \exp_b(x + y)$ (Product of Powers)

From these properties, two other pairs of basic properties follow:

(3) $\quad \log_b\left(\frac{x}{y}\right) = \log_b(x) - \log_b(y)$ (Logarithm of a Quotient) $\qquad \frac{\exp_b(x)}{\exp_b(y)} = \exp_b(x - y)$ (Quotient of Power)

(4) $\quad \log_b(x^y) = y \log_b(x)$ (Logarithm of a Power) $\qquad (\exp_b(x))^y = \exp_b(xy)$ (Power of a Power)

Question: Rewrite properties (2), (3), and (4) of powers in more customary notation.

Historically, logarithmic functions were used to simplify calculations because they transform the operations of multiplication, division, and exponentiation in $\mathbf{R}^+$ into the simpler operations of addition, subtraction, and multiplication in $\mathbf{R}$. In Chapter 1, we detailed some of the ways properties are transformed by these functions. Calculators and computers have replaced logarithms as a means to carry out numerical calculations, but logarithm functions and these functional equations continue to be important for simplifying algebraic and analytic calculations and for modeling real-world problems.

Although a different logarithm function $\log_b(x)$ exists for each positive number $b \neq 1$, these functions are all multiples of one another because of the **change of base formula**: If b and c are positive numbers both unequal to 1, then

$$\log_c(x) = \log_c(b)\log_b(x) \qquad \text{for all } x > 0.$$

Because $\log_c(b) = \frac{1}{\log_b(c)}$ by this change of base formula, we only need to evaluate one of these logarithm functions to know them all. The choice of the irrational base e, which is approximately equal to 2.71828, is the simplest from the point of view of calculus because

$$\frac{d}{dx}\log_e(x) = \frac{1}{x} \quad \text{and} \quad \frac{d}{dx}\log_b(x) = \frac{1}{x\log_e(b)} \qquad \text{for } b \neq e.$$

The logarithm of x to the base e is called the **natural logarithm of x** and is usually denoted by **ln(x)** rather than $\log_e(x)$. Because $\frac{d}{dx}\log_b(x) = \frac{1}{x\log_e(b)} > 0$, if $b > 1$ the slope of the tangent line to the graph of $\log_b$ is always positive, so the function $\log_b$ is strictly increasing throughout its domain. If $0 < b < 1$, the slope of the tangent line to the graph of $\log_b$ is always negative, and so $\log_b$ is strictly decreasing throughout its domain.

The exponential function $\exp_e(x) = e^x$ with the natural logarithm base e is the simplest choice of base for computations in calculus and differential equations because

$$\frac{d}{dx}(e^x) = e^x \quad \text{whereas} \quad \frac{d}{dx}(b^x) = (\log_e b)e^x \qquad \text{for } b \neq e.$$

Composition and inverses of functions that are not real functions

All of the examples of composition and inverses of functions presented in this section have been of real functions. But the definitions of composition of functions and of the inverse of a function allow these ideas to be applied to all types of functions.

For instance, consider the following geometric example. Let A be the set of all triples $\{p, q, r\}$ of three noncollinear points in a plane P, let C be the set of all circles in P, and let T be the set of all triangles in P. Define the functions $f\colon A \to C$ and $g\colon A \to T$ as follows:

$$f(\{p, q, r\}) = \text{the circle passing through } p, q, \text{and } r,$$
$$g(\{p, q, r\}) = \text{the triangle with vertices } p, q, \text{and } r.$$

Then f is not a 1-1 function because any circle passes through (infinitely) many triples of points in P. However, for any triangle in P, there is only one triple of its vertices, so g is a 1-1 function, and the inverse of g is defined by $g^{-1}(\Delta ABC) = \{A, B, C\}$. That is, g^{-1} maps a triangle onto the triple of points that are its vertices.

As is the case with real functions, a function that is not a real function that has no inverse may have a restriction with an inverse. Suppose that C is the set of all circles in a plane P, and that U (for unit circle) is the set of all circles of radius 1 in P. Then the function $f: C \to P$ defined by

$$f(K) = \text{center of circle } K, \text{ for all circles } K \text{ in } C$$

is not 1-1 because infinitely many circles have the same center. However, the restriction of f to U is 1-1 because there is only one circle of radius 1 with a given center. That is, there is a function that maps each circle onto its center because we can determine the center of a circle if we know the circle. However, we cannot in general determine the circle if we are only given its center; we need more information, such as its radius.

3.2.2 Problems

1. Let f and g be real functions with $f(x) = x^2$ and $g(x) = \sin x$.
   a. Give a rule and describe the domain and range of $f \circ g$.
   b. Give a rule and describe the domain and range of $g \circ f$.

2. Repeat Problem 1 for the real functions $f: x \to \sqrt{x-3}$ and $g: x \to x^2 - 4$.

3. Use function composition and what you know about the behavior of $y = g(x) = \frac{1}{x}$ and $y = h(x) = e^x$ to analyze the behavior near $x = 0$ of the function f defined by

$$f(x) = \frac{1}{1 + e^{-1/x}} \qquad \text{for } x \neq 0.$$

4. Consider the function $g \circ f$, where $f: A \to B$ and $g: B \to C$.
   a. What can be said about $g \circ f$ if both g and f are 1-1 and onto?
   b. What can be said about $g \circ f$ if only f is 1-1 and onto?
   c. What can be said about $g \circ f$ if only g is 1-1 and onto?

5. Find all solutions of the equation

$$x^2 - 3 - \frac{1}{x^2 - 3} = 0.$$

6. a. Explain why the function (first discussed at the opening of this chapter) that expresses the volume V in terms of the depth d of fuel in an underground cylindrical fuel tank of length ℓ and radius r whose axis is horizontal has an inverse even though it is not possible to solve the formula (*) for $V(d)$ for d in terms of V.

$$(*) \quad V = V(d) = \ell\left(\pi r^2 - r^2\cos^{-1}\left(\frac{d}{r} - 1\right) + (d - r)\sqrt{2rd - d^2}\right).$$

   b. Without using a calculator or computer, sketch a graph of V that includes information on the domain and range of V as well as how V varies on this domain. (For example, on what part of the domain will V increase most rapidly? least rapidly? What symmetries will the graph exhibit?)

7. For any real numbers a, b, c, and d such that c and d are not both 0, the formula $f(x) = \frac{ax+b}{cx+d}$ defines a real function.
   a. Find the domain and range of f.
   b. Show that f is a one-to-one function if and only if $ad - bc \neq 0$.
   c. Show that if f is a one-to-one function, then its inverse function f^{-1} is given by $f^{-1}(x) = \frac{dx-b}{-cx+a}$, where c and a are not both 0.
   d. Show that if f is a 1-1 function and $a = -d$, then $f^{-1} = f$.

8. Consider each of the following functions restricted to the indicated domain.
   a. Tell why this restriction is not used to determine the usual inverse of the function with domain **R**.
      i. sine, domain $\{x: 0 \leq x \leq \pi\}$
      ii. cosine, domain $\{x: -\frac{\pi}{2} \leq x \leq \frac{\pi}{2}\}$
      iii. cosine, domain $\{x: -\pi < x \leq \pi\}$
      iv. tangent, domain $\{x: 0 \leq x < \frac{\pi}{2}\}$
   b. Give the restriction on the cosine function that satisfies the four desired properties for its inverse.
   c. Give the restriction on the tangent function that satisfies the four desired properties for its inverse.

9. Determine an exact value and an approximation to four decimal places of each of the following.
   a. $\cos^{-1}(\cos(6\pi))$ b. $\sin(\sin^{-1}(0.5))$ c. $\tan(\cot^{-1}(\sqrt{3}))$

10. Give the domain, range, and a formula for each of the following functions.
   a. $\sin^{-1} \circ \sin$ b. $\cos \circ \cos^{-1}$ c. $\sin \circ \cos^{-1}$
   d. $\tan \circ \sin^{-1}$ e. $\sec \circ \cos^{-1}$ f. $\cot \circ \csc^{-1}$

11. Look up (if necessary) the definitions of the arcsecant, arccosecant, and arccotangent functions and develop a rationale for the choices that are made for the restrictions of the secant, cosecant and cotangent functions in these definitions.

12. From the inverse relationship (1) between logarithmic and exponential functions and the three properties of powers (2), (3), and (4), deduce the corresponding properties of logarithms (2), (3), and (4).

13. Prove: For all x, a, and b for which the logarithms are defined,

$$\log_b(x) + \log_c(x) = \frac{\log_x(bc)}{\log_x(b) \cdot \log_x(c)}.$$

14. Consider the **power functions** $f(x) = x^b$, where $b \in Z$ and $b > 1$, and $x \in \mathbf{R}$. For what values of b does an inverse function f^{-1} exist? Justify your answer.

15. In the coordinate plane $\mathbf{R}^2$, consider the square S centered at $(0, 0)$ with sides of length 2 parallel to the axes, and the circle C centered at the origin of radius 2.

a. Describe geometrically a one-to-one function $f: S \to C$ with range C, and its inverse function $f^{-1}: C \to S$.

b. Find a formula that represents the function $f: S \to C$.

ANSWER TO QUESTION

1. $b^x \cdot b^y = b^{x+y}$, $\frac{b^x}{b^y} = b^{x-y}$, and $(b^x)^y = b^{xy}$.

3.2.3 Monotone real functions

We have seen in the previous section that a function has an inverse only if it is one-to-one. We now seek a sufficient condition for a real function to be one-to-one. For this, it is helpful to have some precise language. A real function f is **monotone** (or **monotonic) increasing** on a subset S of its domain D if and only if for all x and y in S, $x < y \Rightarrow f(x) \leq f(y)$. That is, as the values of x increase, the values of $f(x)$ either stay the same or increase. It is **monotone** (or **monotonic) decreasing** on S if and only if as x increases, the values of $f(x)$ either stay the same or decrease. That is, f is monotone decreasing if and only if for all x and y in S, $x < y \Rightarrow f(x) \geq f(y)$.

Important to us are those situations in which no two values of a monotone function are the same. A function f is **strictly increasing** on S if for all x and y in S, $x < y \Rightarrow f(x) < f(y)$. Likewise, f is **strictly decreasing** on S if for all x and y in S, $x < y \Rightarrow f(x) > f(y)$. And f is **strictly monotone** on S if it is either strictly increasing or strictly decreasing on S.

The greatest integer function defined by $g(x) = \lfloor x \rfloor$ is an example of a monotone increasing function on the set of real numbers that is not strictly monotone. The function $f: x \to \frac{1}{x^2}$ is strictly decreasing on the set of positive real numbers and strictly increasing on the set of negative real numbers.

It is usually easy to recognize that a real function is strictly monotone on S by its graph. The graph of a function that is strictly increasing on S increases steadily from left to right across S, while the graph of a function that is strictly decreasing on S steadily decreases from left to right across S. However, sometimes a rough graph is not enough. For instance, though its table suggests that after the 80th term its values are equal, the sequence analyzed in Example 2 of Section 3.2.1 is actually strictly increasing.

Sometimes it is not necessary to graph a real function to be certain that it is strictly monotone on its domain. Think once again of the function V that expresses the volume $V(d)$ of fuel in terms of the depth d of fuel in an underground cylindrical fuel tank of length ℓ and radius r whose axis is horizontal. The domain of V is the closed interval $[0, 2r]$, and it is clear from the context that if $d_1 < d_2$ in this domain, then $V(d_1) < V(d_2)$. (Greater depth of fuel means more fuel.)

Theorem 3.6 If $f: D \to \mathbf{R}$ is a real function that is strictly monotone on a subset S of its domain D, then the restriction f_S of f to S is a one-to-one function.

Corollary: If f is a real function that is strictly monotone on its domain, then f is a one-to-one function.

You are asked to prove Theorem 3.6 and its corollary in the Problems.

The following theorem from calculus provides a sufficient condition for a differentiable function to be strictly monotone on an interval.

Theorem 3.7 Suppose that a real function f is differentiable at each point of an open interval (a, b). Then

i. f is strictly increasing on (a, b) if $f'(x) > 0$ for all $x \in (a, b)$;

ii. f is strictly decreasing on (a, b) if $f'(x) < 0$ for all $x \in (a, b)$.

Thus if f is a differentiable real function on an open interval (a, b), and if f' is continuous and $f'(x) = 0$ has no solutions on (a, b), then f is either strictly increasing or f is strictly decreasing on that interval. This means f is 1-1 on that interval. So it has an inverse when restricted to that interval.

As an application of Theorem 3.7, we note that if k and b are real numbers such that k is positive and $b > 1$, then the exponential function $f: \mathbf{R} \to \mathbf{R}$ defined by

$$f(x) = b^{kx} \qquad x \in \mathbf{R}$$

is strictly increasing on $\mathbf{R}$ because $f'(x) = kb^{kx} > 0$ for all real numbers x. This is another way of showing that the function $\exp_b$ has an inverse.

Before giving the situation any thought, some people think that if a function is strictly increasing, then its inverse will be strictly decreasing. (They are probably confusing "inverse" with "reciprocal".) Theorem 3.8 shows that this is never the case.

Theorem 3.8 If a real function f is strictly increasing (strictly decreasing) on an interval (a, b), then its inverse f^{-1} is also strictly increasing (strictly decreasing) on (a, b).

Proof: The proof follows from the definitions of the terms in the theorem and is left to you as a problem. ┘

Theorems 3.7 and 3.8 explain a number of observations you may have had about individual categories of functions. Here are two examples.

The functions $f: x \to b^x$ and its inverse $g: x \to \log_b x$ are strictly increasing on their domains if $b > 1$, and strictly decreasing functions on their domains if $0 < b < 1$.

The restriction of the function $f: x \to \sin x$ to the interval $(-\frac{\pi}{2}, \frac{\pi}{2})$ and its inverse function $g: x \to \arcsin(x)$ are both strictly increasing functions on their domains, while the restriction of the function $f: x \to \cos x$ to the interval $(0, \pi)$ and its inverse function $g: x \to \arccos(x)$ are both strictly decreasing functions on their domains.

If f is a strictly monotone real function on an open interval (a, b), then the restriction to (a, b) of f is a one-to-one function by Theorem 3.6, and this restriction has an inverse f^{-1} defined on the interval $(f(a), f(b))$ with range (a, b). The graphs of f and f^{-1} are reflection images of each other over the line $y = x$, and the lines that are tangents to these graphs at the points (x_0, y_0) and (y_0, x_0) are also inverse functions. We have just given almost all that is needed to prove the following theorem. The rest is left to you as Problem 9.

Theorem 3.9 Suppose that f is a real function with equation $y = f(x)$, that f is differentiable and strictly monotone on an open interval (a, b), and that $f'(x) \neq 0$ for all x in (a, b). Then, for all x in (a, b), the value of the derivative of f^{-1} at $f(x)$ equals the reciprocal of the value of the derivative of f at x. That is,

$$(f^{-1})'(f(x)) = \frac{1}{f'(x)} \quad \text{for all } x \text{ in } (a, b).$$

Monotone sequences

Suppose that $\{s_n\}$ is a real sequence whose domain D_k is the set of all integers greater than or equal to an integer k (where k is usually 0 or 1). Then, from the definitions at the beginning of this section, $\{s_n\}$ is strictly increasing if and only if $s_n < s_{n+1}$ for all $n \geq k$, and it is strictly decreasing if and only if $s_n > s_{n+1}$ for all $n \geq k$.

If we allow equality in the preceding inequalities, that is, if we replace $>$ and $<$ by $\leq$ and $\geq$ in these inequalities, we obtain the definitions of an **increasing sequence** and a **decreasing sequence**. A **monotone sequence** is a real sequence that is either an increasing sequence or a decreasing sequence.

A sequence is not a differentiable function, so Theorem 3.7 cannot be used to determine if a sequence is monotone (unless the formula for the nth term is such that the function with that formula and domain **R** is differentiable). For instance, although we know that the factorial sequence s with $s(n) = n!$ is monotone, we could not use Theorem 3.7 to prove that. So how can we determine if a sequence $\{s_n\}$ is a monotone or strictly monotone sequence?

Listing a large number of terms of the sequence by hand or with your calculator certainly can confirm that a sequence is *not* monotone, but it cannot prove that the sequence is monotone or strictly monotone. For instance, consider the sequence s defined by

$$(*) \quad s_n = n^3 2^{-n/4} \quad \text{for all } n \in \mathbf{N}.$$

A table of the first 15 terms of the sequence suggests that this sequence is strictly increasing, but a continuation of the table to 30 terms confirms that the sequence is not monotone. (See Problem 10.)

When a sequence is defined by an explicit formula for the nth term, it is often possible to use that formula to determine whether or not it is monotone. When a sequence is defined recursively, then the process usually has to involve both parts of the recursive definition.

EXAMPLE 1 Determine whether the sequence $s = \{s_n\}$, defined recursively by

$$s_0 = 1$$

$$s_{n+1} = \frac{1}{8}(3s_n + 4) \qquad \text{for } n \geq 0,$$

is monotone or strictly monotone.

Solution To obtain an idea about how the sequence behaves, we calculate the first few terms of s:

$$s_0 = 1, \quad s_1 = \frac{1}{8}(3s_n + 4) = \frac{7}{8}, \quad s_2 = \frac{1}{8}\left(3 \cdot \frac{7}{8} + 4\right) = \frac{53}{64}.$$

Thus, if this sequence is monotone, it is decreasing or strictly decreasing. The sequence is strictly decreasing if for all positive integers n,

$$(*) \quad s_{n+1} < s_n.$$

That is, we wish to know when

$$\frac{1}{8}(3s_n + 4) < s_n.$$

Solving this inequality, we find that $\frac{1}{8}(3s_n + 4) < s_n$ exactly when $s_n > \frac{4}{5}$. Now we use mathematical induction to show that for all n, $s_n > \frac{4}{5}$. (For a detailed discussion of mathematical induction, see Unit 5.1.) We have already verified that the first three terms are greater than $\frac{4}{5}$. Suppose $s_k > \frac{4}{5}$. Then $s_{k+1} = \frac{1}{8}(3s_k + 4) > \frac{1}{8}\left(3 \cdot \frac{4}{5} + 4\right) = \frac{4}{5}$. So by the principle of mathematical induction (Theorem 5.1), for all n, $s_n > \frac{4}{5}$. Consequently, the sequence is strictly decreasing.

Examining sequences that define e

The number e, the base of the natural logarithm and natural exponential functions, has the following two equivalent limit definitions, both of which are significant for understanding the nature of this number as well as its applications:

$$e = \lim_{n \to \infty}\left(1 + \frac{1}{n}\right)^n, \qquad e = \sum_{n=0}^{\infty} \frac{1}{n!} = 1 + \frac{1}{1!} + \frac{1}{2!} + \cdots + \frac{1}{n!} + \cdots.$$

The definition at the left, as the limit of the sequence $\{b_n\}$ (b is for binomial) defined by

$$b_n = \left(1 + \frac{1}{n}\right)^n \quad \text{for } n \geq 1,$$

is crucial to the applications of e to modeling growth and decay problems. However, $\{b_n\}$ converges rather slowly to e in comparison with the sequence $s = \{s_n\}$ of partial sums of the infinite series representation of e defined by

$$s_n = \sum_{k=0}^{n} \frac{1}{k!} = 1 + \frac{1}{1!} + \frac{1}{2!} + \cdots + \frac{1}{n!} \qquad \text{for } n \geq 0.$$

The infinite series representation of e is important for analyzing the properties of the number e and the natural exponential function exp: $x \to e^x$. For example, in Chapter 2, we used the infinite series for e to prove that e is irrational. This is not obvious since the first nine decimal places have a rather nice pattern: $e \approx 2.718281828$.

It is obvious from its definition that s is monotone increasing. It is not so obvious that the terms of the sequence $b = \{b_n\}$ are always less than the terms of the sequence $s = \{s_n\}$, but the two sequences have the same limit (see Project 6). This is one way of seeing that $\{s_n\}$ converges to e faster than $\{b_n\}$.

3.2.3 Problems

1. Prove Theorem 3.6 and its corollary.

2. Explain why the function $f: x \to \frac{1}{x}$ is not strictly monotone on the set of nonzero real numbers.

3. Show that a converse to Theorem 3.7 is false. That is, exhibit a function f that is strictly monotone on an interval (a, b) but for which there is at least one value $x \in (a, b)$ such that $f'(x) = 0$.

4. Prove that if $a > 0$, the function f defined by $f(x) = a(x - h)^2 + k$ is strictly increasing on the interval $[h, \infty)$ and strictly decreasing on the interval $(-\infty, h]$.

5. Give a restriction on the function $y = f(x) = ax^2 + bx + c$ so that its inverse f^{-1} exists, is either strictly increasing throughout its domain or strictly decreasing throughout, and so that no larger domain possesses these properties.

6. The graph of the natural logarithm function $\ln(x)$ on a calculator or computer makes it seem obvious that this function is strictly increasing on its domain $D = \{x \in \mathbf{R}: x > 0\}$. Prove this fact algebraically by using the properties of logarithms.

7. a. Prove that if f is strictly increasing on a domain D, then its inverse f^{-1} is strictly increasing on $f(D)$.

   b. What has to be changed in your answer to part **a** if "increasing" is replaced by "decreasing" in both places?

8. Show that Theorem 3.9 holds for the function $y = f(x) = x^3$ when $x > 0$.

9. Prove Theorem 3.9.

10. a. Use a graphing calculator or computer to tabulate and graph the first 30 terms of the sequence defined by $s_n = n^3 2^{-n/4}$, for all $n \in \mathbf{N}$.

    b. On what interval(s) is this sequence monotone increasing? On what interval(s) is this sequence monotone decreasing?

11. Determine whether the sequence s defined recursively below is strictly monotonic.

    a. $s_0 = 1,\ s_{n+1} = \frac{1}{8}(3s_n + 6)$ for $n \geq 0$

    b. $s_0 = 1,\ s_{n+1} = \frac{1}{8}(3s_n + 5)$ for $n \geq 0$

12. Suppose that a and b are positive numbers. Determine the conditions on a and b that assure that the sequence s defined recursively by

$$s_0 = 1,\ s_{n+1} = as_n + b \qquad \text{for } n \geq 0$$

is a strictly increasing or strictly decreasing sequence.

13. Explain why $\frac{1}{p!} < \frac{1}{2^{p-1}}$ holds for all integers $p > 2$.

3.2.4 Limit behavior of real functions

The function $f: x \to \frac{1}{x}$, whose graph is the familiar hyperbola pictured in Figure 20, exhibits the two types of limiting behavior that we discuss in this section.

Figure 20

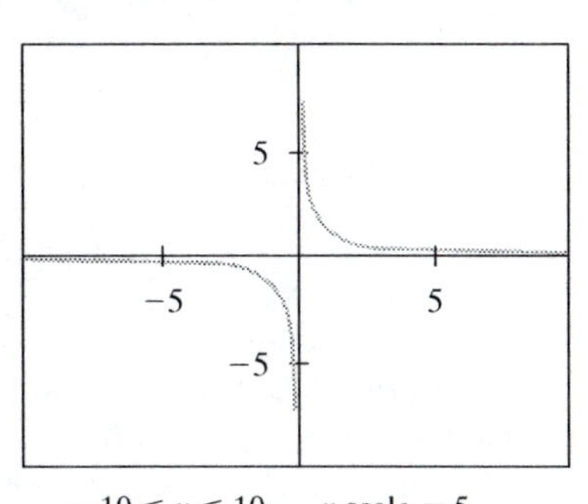

$-10 \leq x \leq 10$, x-scale $= 5$
$-10 \leq y \leq 10$, y-scale $= 5$

The first type of limiting behavior concerns what happens near $x = 0$, where the function f is not defined. Recall from calculus the following definitions of right- and left-hand limits. Suppose that f is an interval-based real function, that c is a real number, and that an interval (c, k) just to the right of c belongs to the domain of f. If, for each positive number M, there corresponds a real number k_M in the interval (c, k) such that $f(x) > M$ for all x in the interval (c, k_M), then **f diverges to ∞ as x approaches c from the right**. We write

$$\lim_{x \to c^+} f(x) = \infty$$

and say "The limit of $f(x)$ as x approaches c from the right is positive infinity."

Question 1: Sketch the situation $\lim_{x \to c^+} f(x) = \infty$ including the interval (c, k), a particular M, and a particular k_M.

Divergence of f to $-\infty$ as x approaches c from the left or from the right, as well as divergence to ∞ as x approaches c from the left,

$$\lim_{x \to c^-} f(x) = -\infty \quad \lim_{x \to c^+} f(x) = -\infty \quad \text{and} \quad \lim_{x \to c^-} f(x) = \infty,$$

are defined in a similar way. When $\lim_{x \to c^+} f(x) = \lim_{x \to c^-} f(x)$, then we avoid the + and − and speak and write $\lim_{x \to c} f(x)$.

In the case of the function $f: x \to \frac{1}{x}$, when $c = 0$, values of $f(x)$ can be made greater than any positive number M you might choose by taking small enough values of x, which indicates that $\lim_{x \to 0^+} f(x) = \infty$. Similarly, values of $f(x)$ can be smaller than any negative number M you might choose by taking negative values of x that are close enough to zero, which indicates that $\lim_{x \to 0^-} f(x) = -\infty$.

Vertical asymptotes

Graphically, if f diverges to ∞ or $-\infty$ as x approaches c from the left or from the right, then the graph of f approaches the vertical line $x = c$ accordingly, and this vertical line is called a **vertical asymptote of f**.

For example, the function $f: x \to \frac{1}{x}$ is a simple example of a rational function of the form $f(x) = \frac{p(x)}{q(x)}$, where $p(x)$ and $q(x)$ are polynomials and $p(c) \neq 0$ whenever $q(c) = 0$. For all such functions, $x = c$ is a vertical asymptote of f.

The function g in Question 2 has been picked to have almost all the various types of factors of the numerator and denominator of a rational function. You should attempt to answer the question before reading beyond it.

Question 2: Determine the zeros and vertical asymptotes of the function g defined by

$$g(x) = \frac{(x + 4)(x - 7)^2(x + 10)}{(x + 1)(x - 7)(x + 10)^2}.$$

Let us analyze the function g. Let $p(x)$ be the numerator of the fraction for $g(x)$ and let $q(x)$ be the denominator. The possible zeros of g can only be those values of x for which $p(x) = 0$, that is, -4, 7, or -10. However, when $x = 7$ or $x = -10$, the denominator $q(x) = 0$ and g is undefined. Consequently, -4 is the only zero of g.

For the asymptotes, we examine the values of x for which $q(x) = 0$, that is, when $x = -1, 7$, or -10. We simplify the fraction for $g(x)$ and obtain $g^*(x) = \frac{(x+4)(x-7)}{(x+1)(x+10)}$. The functions g^* and g are identical *except when* $x = 7$ *or* $x = -10$, where g is undefined. For the behavior of g near $x = 7$, we examine g^* and see that $\lim_{x \to 7} g(x) = \lim_{x \to 7} g^*(x) = 0$. So there is no asymptote at $x = 7$. For the behavior of g near $x = -10$, we again examine g^*. Here g^* is also undefined, and we determine that $\lim_{x \to -10^+} g^*(x) = -\infty$ and $\lim_{x \to -10^-} g^*(x) = \infty$. Either one of these limits indicates that g has a vertical asymptote at $x = -10$.

Figure 21 shows a calculator plot of the function g defined in Question 2. It looks as if g has two zeros, at -4 and 7, but the function is not defined at 7. You should graph g^* and compare it to g.

Figure 21

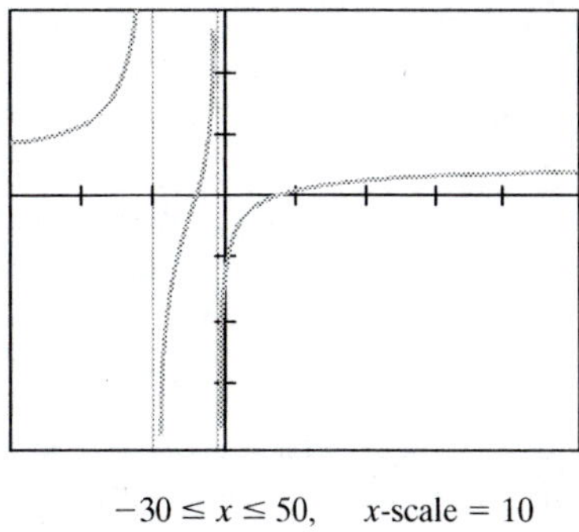

$-30 \le x \le 50$, x-scale = 10
$-8 \le y \le 6$, y-scale = 2

There was nothing in the preceding argument that required that $p(x)$ and $q(x)$ be polynomials. Thus there is a more general theorem.

Theorem 3.10 **(Vertical Asymptote Theorem):** Let p and q be continuous real functions of x defined in an open interval with c as an endpoint. Then a function f with a rule of the form $f(x) = \frac{p(x)}{q(x)}$ will have a vertical asymptote $x = c$ if $q(c) = 0$ and $p(c) \neq 0$.

For another example of Theorem 3.10, consider the tangent function. $\tan x = \frac{\sin x}{\cos x}$, so letting $p(x) = \sin x$ and $q(x) = \cos x$ in Theorem 3.10 explains why the tangent function has vertical asymptotes $x = \frac{\pi}{2} + k\pi$ for any integer k.

End behavior

The second type of limiting behavior of functions that we discuss concerns what happens to $f(x)$ as x becomes larger and larger, or smaller and smaller. For this end behavior, we want not only to know $\lim_{x\to\infty} f(x)$ and $\lim_{x\to-\infty} f(x)$ but also whether there is some other simpler function $g(x)$ as $|x|$ becomes larger and larger such that $\lim_{|x|\to\infty} \frac{f(x)}{g(x)} = 1$.

Definition Two real functions f and g have the same end behavior if and only if $\lim_{x\to\infty} \frac{f(x)}{g(x)} = 1$ and $\lim_{x\to-\infty} \frac{f(x)}{g(x)} = 1$.

We begin with those discrete real functions that are real sequences, so that we need only examine $\lim_{n\to\infty} s_n$.

Limits of real sequences

When a real sequence $s = \{s_n\}$ models a real-world situation, the existence of the limit of the sequence $\lim_{n\to\infty} s_n$ usually tells us something interesting about the situation. For example, for the sequential model $\{P_n\}$ of logistic growth of a fish population in Example 2 of Section 3.2.1,

$$\lim_{n\to\infty} \{P_n\} = 20{,}000$$

tells us that the "carrying capacity" of the lake is 20,000 fish. That model would predict, for any given initial population P_0 of fish, the population P_n after n years will approach 20,000. We refer to this as the *end behavior* of the sequence $\{P_n\}$.

Recall from your study of limits that a real sequence $\{s_n\}$ converges to a real number L if and only if, for each positive number ε there is an integer k_ε such that

$$L - \varepsilon < s_n < L + \varepsilon$$

for all $n > k_\varepsilon$. That is, the limit exists if for any ε there is a certain point beyond which all terms of the sequence are within ε of the limit. Each term does not have to be closer to the limit than the preceding term, but after a while all terms have to be closer to the limit than any preceding term.

If a real sequence is monotone and bounded, then it is certain to have a limit. Intuitively, the monotonicity means that the terms are points in order on a number line. The boundedness means the terms cannot go beyond the limit. So they have nowhere to go but toward the limit.

Theorem 3.11 If $\{s_n\}$ is a monotone real sequence contained in an interval $a < x < b$, then $\{s_n\}$ converges to a real number L.

Theorem 3.11 follows from the Nested Interval Property of real numbers. We omit its proof.

If the sequence $\{s_n\}$ is monotone increasing, then the limit L is the *least upper bound* of the range of the sequence; that is, L is the smallest real number r such that $r \geq s_n$ for all n. Similarly, if $\{s_n\}$ is a monotone decreasing sequence, then the limit L is the *greatest lower bound* of the range of the sequence; that is, L is the largest real number r such that $r \leq s_n$ for all n.

For instance, in Section 3.2.1, we showed that the sequence $\{s_n\}$ defined recursively by $s_0 = 1$ and $s_{n+1} = \frac{1}{8}(3s_n + 4)$ for $n \geq 0$ is strictly decreasing. Also, because the initial value of the sequence is positive and the coefficients in the recurrence formula are positive, all terms of the sequence are in the interval between 0 and 1. Therefore, by Theorem 3.11, the sequence $\{s_n\}$ converges to a number L. Here is how to find that number.

$$L = \lim_{n\to\infty} s_{n+1} = \lim_{n\to\infty} \frac{1}{8}(3s_n + 4) = \frac{3}{8} \lim_{n\to\infty} s_n + \frac{1}{2} = \frac{3}{8}L + \frac{1}{2}$$

From solving the equation, $L = \frac{4}{5}$. After the analysis of the last section, this value should not come as a surprise.

Horizontal asymptotes

When f is a real function and there is a real number L such that

$$\lim_{x\to\infty} f(x) = L \quad \text{or} \quad \lim_{x\to-\infty} f(x) = L,$$

then the graph of f in the coordinate plane approaches the horizontal line $y = L$ as $x \to \infty$ or as $x \to -\infty$, respectively. In either case, the line $y = L$ is a **horizontal asymptote of f**. Notice that if g is the constant function $g(x) = L$ for all x, then when L is a horizontal asymptote of f, $\lim_{x\to\infty} \frac{f(x)}{g(x)} = \lim_{x\to\infty} \frac{f(x)}{L} = \frac{1}{L} \lim_{x\to\infty} f(x) = 1$. So f has the same end behavior as the constant function. For instance, the graph of the sequence $\{P_n\}$ of logistic growth of a fish population in Example 2 of Section 3.2.1 has the horizontal asymptote $P_n = 20{,}000$ as $n \to \infty$. The graph of $f: x \to \frac{1}{x}$ has the horizontal asymptote $y = 0$ both as $x \to \infty$ and as $x \to -\infty$.

Suppose f is an interval-based real function with domain D that has an inverse f^{-1} defined on its range $f(D)$. Suppose also that the graph of $y = f(x)$ has a vertical asymptote at an endpoint $x = c$ of an interval in the domain of f. Then, because the graph of f^{-1} is the reflection image of the graph of f over the line $y = x$, the graph of $y = f^{-1}(x)$ has a corresponding horizontal asymptote $y = c$.

For example, we have mentioned that the tangent function has the vertical asymptotes $x = \frac{\pi}{2} + k\pi$ for any integer k. If f is the restriction of the tangent function to the interval $-\frac{\pi}{2} < x < \frac{\pi}{2}$, then f has an inverse, the arctangent function $\tan^{-1}$ or arctan. The horizontal lines $y = -\frac{\pi}{2}$ and $y = \frac{\pi}{2}$ are horizontal asymptotes of $y = \arctan(x)$.

Also, the horizontal asymptote $y = 0$ of the exponential function defined by $y = f(x) = b^x$ corresponds to the vertical asymptote $x = 0$ of the inverse function $y = f^{-1}(x) = \log_b(x)$.

End behavior of rational functions

Before reading on, work on the following question.

Question 3: Graph the functions f and g defined by $f(x) = \frac{3x^3-5x^2+12x-100}{6x+1}$ and $g(x) = 0.5x^2$ on the window $-10 \le x \le 20$ and $-50 \le y \le 100$.

You should notice that despite g being a polynomial function and f being a rational function with a vertical asymptote, the graphs of the two functions are quite near each other for most values of x. In fact, they get relatively closer together the larger $|x|$ becomes. That is, $\lim\limits_{x\to\infty} \frac{f(x)}{g(x)} = 1$. This can be seen by dividing both the numerator and denominator in the formula for $f(x)$ by $6x$. The resulting function $f^*(x) = \dfrac{\frac{1}{2}x^2 - \frac{5}{6}x + 2 - \frac{100}{6x}}{1 + \frac{1}{6x}} =$ has values identical to those of f except when $x = 0$. As $|x|$ gets larger and larger, the denominator gets closer and closer to 1, while the value of the term $\frac{1}{2}x^2$ in the numerator dominates all other values. As a result $\lim\limits_{x\to\infty} \frac{f^*(x)}{g(x)} = \lim\limits_{x\to\infty} \frac{f(x)}{g(x)} = 1$.

Another way to see this end behavior is to perform the polynomial division of $3x^3 - 5x^2 + 11x - 100$ by $6x + 1$. (See Unit 5.3 if you have forgotten the procedure.) The quotient polynomial is $\frac{1}{2}x^2 - \frac{11}{12}x + 2\frac{11}{72}$, and the remainder is $-102\frac{11}{72}$. As $|x|$ gets larger, the quadratic term in the quotient dominates all others and so $\lim\limits_{x\to\infty} \dfrac{f(x)}{\frac{1}{2}x^2} = 1$.

Can we generalize this result? Yes. g is a power function. It turns out that the end behavior of any rational function is like that of a power function.

Theorem 3.12 **(End Behavior of Rational Functions):**
Suppose $f(x) = \frac{a_m x^m + a_{m-1}x^{m-1} + \cdots + a_1 x + a_0}{b_n x^n + b_{n-1}x^{n-1} + \cdots + b_1 x + b_0}$ for all real numbers x for which the denominator is not zero. Then the end behavior of f is the same as the end behavior of the real function g defined for all x by $g(x) = \frac{a_m}{b_n}x^{m-n}$.

Proof: Rewrite the formula for $f(x)$ by factoring x^m from the numerator and factoring x^n from the denominator. The resulting function f^* is identical to f except when $x = 0$, which does not affect the end behavior.

$$f^*(x) = \frac{\left(a_m + \dfrac{a_{m-1}}{x} + \dfrac{a_{m-2}}{x^2} + \cdots + \dfrac{a_1}{x^{m-1}} + \dfrac{a_0}{x^m}\right)x^m}{\left(b_n + \dfrac{b_{n-1}}{x} + \dfrac{b_{n-2}}{x^2} + \cdots + \dfrac{b_1}{x^{n-1}} + \dfrac{b_0}{x^n}\right)x^n}$$

As $|x| \to \infty$, all the terms within the parentheses approach zero except a_m and b_n, which remain constant. Thus $\lim\limits_{|x|\to\infty} f(x) = \lim\limits_{|x|\to\infty} f^*(x) = \frac{a_m x^m}{b_n x^n}$. Dividing the powers of x yields the theorem. ┘

To fully understand Theorem 3.12, it is helpful to partition all rational functions into four subsets based on the degrees m and n. When $m \geq n + 2$, then the end behavior of f is a power function, as we have seen with the function g in Question 3. When $m = n + 1$, then the end behavior approaches that of a nonconstant linear function. Then f has an **oblique asymptote**. When $m = n$, the rational function approaches a constant limit and so f has a horizontal asymptote. And when $m < n$, as $|x| \to \infty$ the denominator ultimately dwarfs the numerator, so the values of the function approach 0 and the x-axis becomes an asymptote to the function.

Orders of growth of real functions

One of the most interesting and useful applications of the idea of end behavior comes when we restrict our attention to functions whose values increase without bound as $x \to \infty$, that is, functions f with $\lim_{x\to\infty} f(x) = \infty$. There are many such functions, including the following:

1. the linear functions $f(x) = ax + b$ with $a > 0$
2. the quadratic functions $f(x) = ax^2 + bx + c$ with $a > 0$
3. (more generally) the polynomial functions $f(x) = a_nx^n + a_{n-1}x^{n-1} + \cdots + a_1x + a_0$ with $a_n > 0$
4. logarithm functions $f(x) = \log_b x$ with base $b > 1$
5. exponential functions $f(x) = cb^x$ with $c > 0$ and base $b > 1$
6. nth root functions $f(x) = ax^{1/n}$ with $a > 0$
7. the factorial function $f(x) = x!$ with domain the set of nonnegative integers

Given two functions of different types from the preceding list, the relative sizes of the two functions for a given value of x will depend on the coefficients of the functions. However, as x increases without bound, the values of one of the functions will eventually exceed and continue to grow much larger than the values of the other function. For example, the values of the linear function g with $g(x) = 100x$ are larger than those of the quadratic function h with $h(x) = x^2$ on the interval $(0, 100)$ but $g(x) < h(x)$ for all $x > 100$ and the values of h become much larger than those of g as x continues to increase. If we were to change the leading coefficients of g and h to other positive values, the specific details of the comparative growth of g and h might change but the general conclusion would be the same: At some point, values of h would become larger than values of g.

This idea is formalized in the idea of *order of growth*. Suppose that two real functions f and g have the property that $\lim_{x\to\infty} f(x) = \lim_{x\to\infty} g(x) = \infty$. Then we say that f and g have the **same order of growth** if there is a nonzero real number L such that $L = \lim_{x\to\infty} \frac{f(x)}{g(x)}$. So two functions with the same end behavior have the same order of growth, but not necessarily conversely. We say that ***f* has a higher order of growth than *g*** if $\lim_{x\to\infty} \frac{f(x)}{g(x)} = +\infty$ or, equivalently, that $\lim_{x\to\infty} \frac{g(x)}{f(x)} = 0$.

For example, because

$$\lim_{x\to\infty} \frac{100x}{x^2} = \lim_{x\to\infty} \frac{100}{x} = 0,$$

it follows that $g(x) = x^2$ has a higher order of growth than $f(x) = 100x$. On the other hand, the function g has the same order of growth as $k(x) = 22x^2 + 50x + 1000$ because

$$\lim_{x\to\infty} \frac{22x^2 + 50x + 1000}{x^2} = \lim_{x\to\infty} \left(22 + \frac{50}{x} + \frac{1000}{x^2}\right) = 22.$$

We can also compare the order of growth of functions of entirely different types. For instance, consider the orders of growth of the functions f and g, where $f(x) = \log x$ and $g(x) = x^{1/10}$.

Question 4: Graph $f(x) = \log x$ and $g(x) = x^{1/10}$ with a computer or graphing calculator to see if either function seems to have a higher order of growth than the other.

We must emphasize the importance of the limit here. It is likely that your graphs in Question 4 showed the values of f to be larger than the values of g. The values of f may be larger than the values of g even for large values of x, but the important aspect is what happens for even larger values of x. With the functions of Question 4, the values of f are larger than the values of g when $0 < x < 10{,}000{,}000{,}000$, but after that the values of g are larger, and ultimately g dominates. We can prove this applying L'Hôpital's Rule, whose proof can be found in most calculus texts. (The rule is named after the French mathematician Guillaume François Antoine de L'Hôpital [1661–1704].) There are several forms of L'Hôpital's Rule; the form that we use is for analyzing the orders of growth of real functions.

Theorem 3.13 **(L'Hôpital's Rule):** Suppose that f and g are differentiable real functions and $\lim_{x\to\infty} f(x) = \lim_{x\to\infty} g(x) = \infty$. Then $\lim_{x\to\infty} \frac{f(x)}{g(x)} = \lim_{x\to\infty} \frac{f'(x)}{g'(x)}$, provided $\lim_{x\to\infty} \frac{f'(x)}{g'(x)}$ exists.

For example, consider the functions of Question 4. First we apply L'Hôpital's Rule.

$$\lim_{x\to\infty} \frac{\log x}{x^{1/10}} = \lim_{x\to\infty} \frac{(\ln 10)x^{-1}}{\frac{1}{10}x^{-9/10}}$$

$$= \lim_{x\to\infty} \frac{10 \ln 10}{x^{1/10}}$$

Since the denominator increases (although very slowly) without bound, the limit is 0. So the tenth root function g dominates and has a higher order of growth than the log function f. This is true even though the values of f are larger until x reaches 10 billion.

Orders of growth have become important in recent decades for subjects as diverse as the interpretation of mathematical models and the comparison of the speed of computer algorithms. For instance, if models based on data for competing businesses A and B indicate that A's revenues are well represented by a quadratic function while those of B are modeled by an exponential function, then even if A's revenue is now larger than B's, ultimately B's revenue will overtake A's. In comparing the speed of computer algorithms f and g, the variable (call it x) is a measure of the size of the input (for example, the number of digits in an integer) and $f(x)$ and $g(x)$ are the times it takes the algorithms to accomplish the task (for example, to find the prime factorization of the integer). Faster algorithms are those that grow more slowly as x gets larger.

The result for the functions of Question 4 may have surprised you. In fact, the orders of growth within and among the seven function types mentioned in this section are not always obvious. In Problems 4–12, you are asked to prove the various parts of the following theorem:

Theorem 3.14 The orders of growth of functions compare as follows, from higher to lower:

the factorial function $f(x) = x!$

all exponential functions $f(x) = cb^x$ with $c > 0$ and $b > 1$, with larger bases having higher orders of growth

all polynomial functions $f(x) = a_n x^n + \cdots + a_1 x + a_0$ with $a_n > 0$, and $n \geq 1$, with larger degrees having higher orders of growth

all nth root functions $f(x) = ax^{1/n}$ with $a > 0$, with larger values of $\frac{1}{n}$ having higher orders of growth

all logarithm functions $f(x) = \log_b x$ with $b > 1$.

3.2.4 Problems

1. Determine the zeros, vertical asymptotes, and end behavior of the function defined by the given rule.

a. $f(x) = x^{-7}$ b. $g(x) = \frac{4}{x^8}$

c. $h(x) = 2 \cdot 3^{x+2}$ d. $k(t) = \frac{2t - 5}{7t + 1}$

e. $g(u) = \frac{(u-1)(u-2)^2(u-3)^3}{(u-1)^3(u-2)^2(u-3)}$ f. $v(x) = 6\ln(x + 3)$

g. $w(t) = \frac{9t^2 - 4}{-3t - 2}$ h. $r(x) = \frac{x^3 - 8x^2 + 4x - 32}{x^5 - 1}$

2. Give all asymptotes of the cotangent function and its inverse.

3. A sequence s is defined recursively as follows:

$$\begin{cases} s_1 = 100 \\ s_{n+1} = \frac{1}{2}\left(s_n + \frac{1}{3}\right) \end{cases}. \text{ Find } \lim_{n\to\infty} s_n.$$

4. Explain why the factorial function has a higher growth rate than any exponential function.

5. Prove: Two exponential functions g and h with positive coefficients and bases b_g and b_h have the same order of growth if $b_g = b_h > 1$, and g has a higher order of growth than h if $b_g > b_h > 1$.

6. Use L'Hôpital's Rule to explain why any exponential function $g(x) = cb^x$ with $c > 0$ and $b > 1$ has a higher order of growth than any polynomial function $h(x)$ with a positive leading coefficient.

7. Show that any two polynomial functions g and h with positive leading coefficients and of degrees m and n, respectively, have the same order of growth if $m = n$, and g has a higher order of growth than h if $m > n$.

8. a. Show that the polynomial function $g(x) = 3x^3 + 2x + 1$ has a higher growth rate than $h(x) = \log_{10}(x)$.

b. Explain why any polynomial function g with degree ≥ 1 and a positive leading coefficient has a higher growth rate than any logarithm function h with base $b > 1$.

9. Prove: Any polynomial function with positive leading coefficient has a higher growth rate than any nth root function.

10. Compare the growth rates of the real functions f and g when $f(x) = \sqrt[m]{x}$ and $g(x) = \sqrt[n]{x}$.

11. Prove: Any nth root function has a higher growth rate than any logarithm function.

12. Prove: Two logarithmic functions g and h with bases b_g and b_h not only have the same order of growth, but the two are just positive multiples of one another.

13. a. Consider the function $f: x \rightarrow x^{5/2}$. How does the order of growth of this function compare with the orders of growth of the functions in Theorem 3.14?

b. Generalize part **a**.

14. Compare the growth rate of the function $f: x \rightarrow \log(\log x)$ with that of the functions in Theorem 3.14. (The time taken by some algorithms is of the same growth rate as this f.)

ANSWERS TO QUESTIONS:

1. See Figure 22.

Figure 22

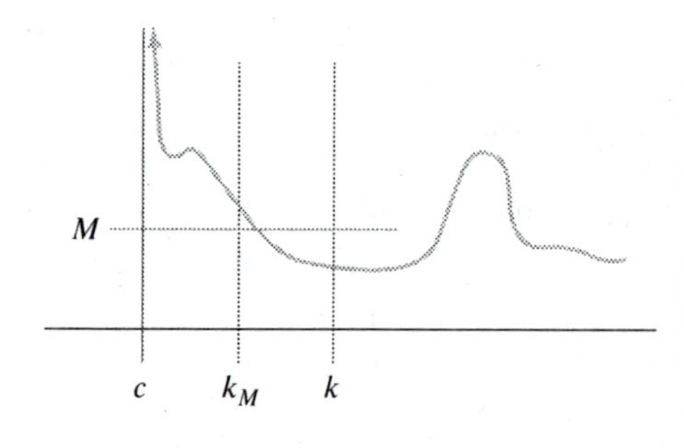

3. See Figure 23.

Figure 23

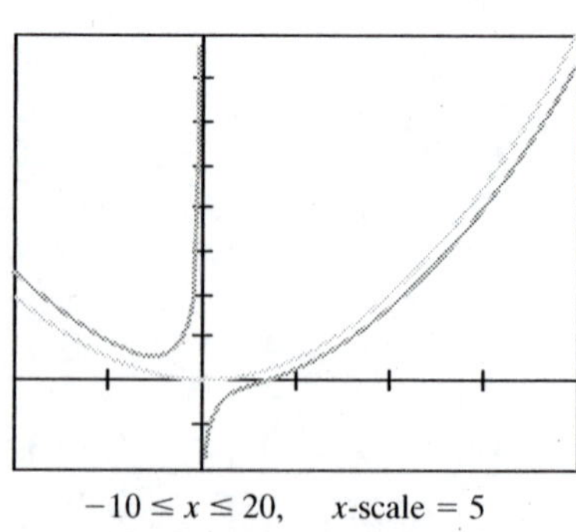

Unit 3.3 Problems Involving Real Functions

Real functions arise in many ways. In Section 3.1.2, we discussed one of these ways: as a generalization of a "numbers-in/number-out" problem. In this unit, we discuss two other important ways in which functions arise. One of these is from the attempt to find a formula that fits bivariate (two-variable) data. In this kind of situation, there may be no cause and effect between the two variables, but merely an association of simultaneity (at a particular time t, variable x had the following value). So we may have no theory from which a formula for the function can be derived and thus we must depend entirely on the data that are offered. Sections 3.3.1 and 3.3.2 cover such situations.

At times, we do know that a change in one variable affects another. Then, perhaps from physical laws or from mathematical theory, we try to find a formula relating the variables and use that formula to determine how the variables are related mathematically. In Section 3.3.3 we engage in an extended analysis of a standard problem that results in some particularly interesting functional relationships.

3.3.1 Fitting linear and exponential functions to data

If a function is specified by a formula, it is easy to recognize whether it is a certain type of function such as a polynomial function, an exponential function, or a rational function. However, in many applications, we may not be given a formula for a real function. We may only have information about its behavior, or a partial table of its values, or its graph. This is typically the case when we are trying to fit a real function to data from an experiment or other data records. In such cases, is it possible to recognize that a certain type of function can adequately describe the given information while another cannot? Also, if it is possible to determine that a certain type of function is most appropriate for the given information, can a formula for that function be determined?

For example, some values to two decimal places of three functions f, g, and h, known to be monotone for $x > 0$, are given in Table 4.

Table 4

x	0	1	2	3	4	5	6
$f(x)$	1.9	2.51	3.31	4.37	5.76	7.6	10.03
$g(x)$	2.05	2.30	3.05	4.30	6.05	8.30	11.05
$h(x)$	2	2.67	3.39	4.19	5.15	6.40	8.23

On the basis of these data, it is possible to conclude that one of these functions can be well described by an exponential function, that one can be described well by a polynomial function, and that the third cannot be well described by either. This is possible because certain types of elementary functions leave evidence of their type by the way in which their values change with changes in the independent variable. Knowledge of this evidence is not only very helpful for answering the question raised above but also for understanding the behavior of these functions and for using them to model real-world data. We will revisit these data later.

Fitting data with a linear function

EXAMPLE 1 Table 5 is a partial list of values for a different function $f: \mathbf{R} \to \mathbf{R}$.

Table 5

x	2	3	4	5	6	7	8
$f(x)$	5.53	7.68	9.83	11.98	14.13	16.28	18.43

a. Find a function that represents this data well.

b. Can we be certain that no other function could have generated this data?

Solution **a.** We compute the differences $f(x+1) - f(x)$ for the given data and obtain Table 6.

Table 6

x	2	3	4	5	6	7	8
$f(x+1) - f(x)$	2.15	2.15	2.15	2.15	2.15	2.15	

The values of a linear function given by $L(x) = mx + b$ change in a very predictable way with incremental changes in the independent variable: For any x, the difference between the values $L(x+1)$ and $L(x)$ is m because $L(x+1) - L(x) = m(x+1) + b - mx - b = m$.

This shows that the given data can be fit exactly by a linear function of the form

$$L(x) = 2.15x + b.$$

Also, because we are given that the value of the function at $x = 2$ is 5.53, we can conclude that $b = f(2) - 2.15(2) = 5.53 - 4.30 = 1.23$. Thus, the linear function

$$L(x) = 2.15x + 1.23$$

exactly fits the given data.

b. Does that mean that the given function f is necessarily equal to L for all values of x? The answer is no. Even if we assume that the pattern of differences displayed in the data persists for all real x; that is, if we assume that $f(x+1) - f(x) = 2.15$ for all real x, the answer is no. The function h: $\mathbf{R} \to \mathbf{R}$ defined by

$$h(x) = L(x) + \sin(2\pi x) \qquad \text{for all real } x$$

is also a function for which $h(x+1) - h(x) = m$ for all real x, yet h is not a linear function.

Question 1: Refer to the last line of Example 1. Explain why $h(x+1) - h(x) = m$ for all real x, and why h is not a linear function.

From Example 1, we can draw general conclusions about fitting a linear function to the $n+1$ data points (x_0, y_0), $(x_0 + 1, y_1)$, $(x_0 + 2, y_2)$, $(x_0 + 3, y_3), \ldots,$ $(x_0 + n, y_n)$. We call this the *Linear Data Fitting Principle.*

Linear Data Fitting Principle: Suppose $n+1$ data points $(x_0 + k, y_k)$, $k = 0, 1, 2, \ldots, n$, are known. Consider the successive differences $y_{k+1} - y_k$ for $k = 0, 1, 2, \ldots, n-1$.

a. If all of these differences are equal to (or nearly equal to) the same constant m, then the linear function L defined by

$$L(x) = mx + b, \text{ where } b = y_0 - mx_0$$

exactly fits (or nearly fits) the given data.

b. If these differences are not all the same, then no linear function exactly fits the given data.

The meaning of "nearly equal to" used in the statement of the Linear Data Fitting Principle depends on the context in which the data fit is taking place. A measure of "nearly equal to" is given by the linear correlation coefficient of the data. If the correlation is high enough, the *line of best fit*, found by linear regression, will fit the data well enough to be useful.

If the differences are equal, then there is an exact fit, as described in Theorem 3.15.

Theorem 3.15 Suppose that $f\colon \mathbf{R} \to \mathbf{R}$.

a. If f is a linear function, then f has the following property (*): There is a real number m such that $f(x + 1) - f(x) = m$ for all real numbers x.

b. If f has the property (*), then the linear function L defined by $L(x) = mx + f(0)$ for all real numbers x agrees with f at all nonnegative integers n.

Proof:

a. If f is a linear function, there are real numbers m and b such that $f(x) = mx + b$ for all real x. We have already shown that $f(x + 1) - f(x) = m$ for all real numbers x. This proves (a).

b. A proof uses mathematical induction. Note that $L(0) = f(0)$ by definition of L, so the conclusion holds for $n = 0$. Assume that for some nonnegative integer k, $L(k) = f(k)$. Then, because $f(k + 1) - f(k) = m$, it follows that

$$\begin{aligned} L(k+1) &= m(k+1) + f(0) \\ &= (mk + f(0)) + m \\ &= L(k) + m \\ &= f(k) + (f(k+1) - f(k)) \\ &= f(k+1). \end{aligned}$$

That is, the conclusion must also hold for $k + 1$. Therefore, by the Principle of Mathematical Induction (see Section 5.1.2), $L(n) = f(n)$ for all nonnegative integers n. ∎

The property (*) in Theorem 3.15b is called the **constant difference property** for a function f.

Example 1b showed that there are functions $f\colon \mathbf{R} \to \mathbf{R}$ with the constant difference property that are not linear functions. However, for real functions that are sequences, the constant difference property does characterize the sequences that are linear. If in Theorem 3.15, we replace x by n and m by d, and use sequence notation for f, we obtain the following corollary.

Corollary: A real sequence $\{f_n\}$ has a constant difference d, that is,

$$f_{n+1} - f_n = d \qquad \text{for all } n \geq 0$$

if and only if

$$f_n = dn + f_0 \qquad \text{for all } n \geq 0.$$

The sequence $f = \{f_n\}$ defined by

$$f_n = dn + f_0 \qquad \text{for all } n \geq 0$$

is an arithmetic sequence or linear sequence with first term f_0 and common difference d. By using the corollary, arithmetic sequences can model sequential data with a constant increase or decrease.

An application to two-dimensional sequences

Most of the sequences you have seen have one variable. But some sequences are themselves elements of sequences. Consider a sequence of connected squares made from toothpicks. We call the sequence a *train* and the squares the *cars* (Figure 24).

Figure 24

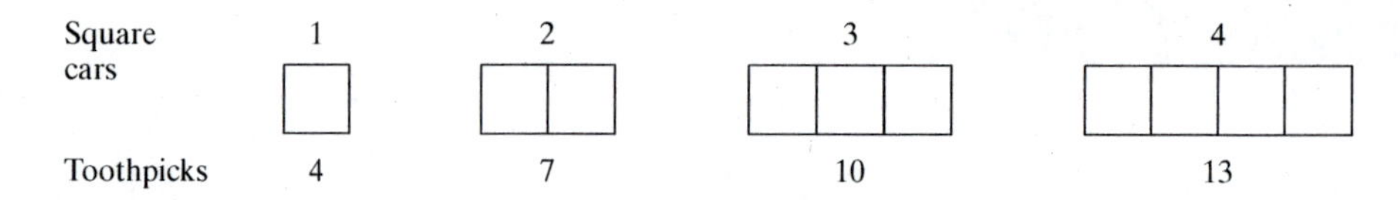

How many toothpicks does it take to make a train with c cars? The constant difference here is 3, since 3 more toothpicks are needed for each successive car. This implies that it takes $3c + b$ toothpicks to make c squares, where b is unknown. But from the situation with 1 square, we can deduce that $3 \cdot 1 + b = 4$, and so $b = 1$. Thus it takes $3c + 1$ toothpicks to make a train with c squares.

Now consider two analogous patterns. First is a train of equilateral triangles (Figure 25).

Figure 25

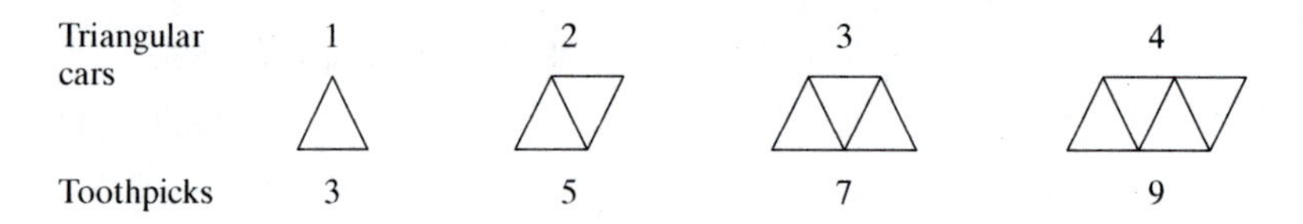

The second is a train of equilateral hexagons (Figure 26).

Figure 26

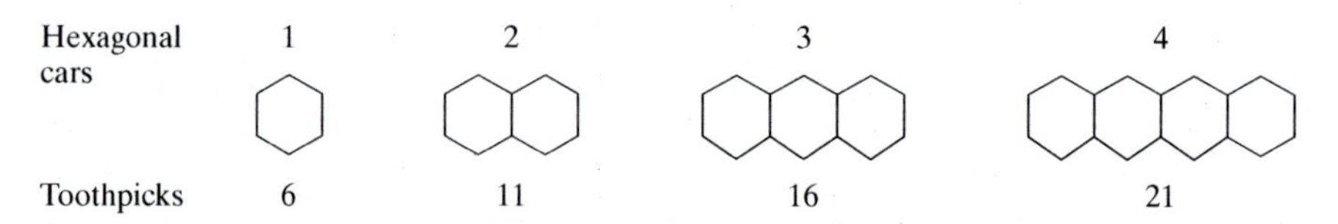

Our curiosity leads us to want a formula for the number of toothpicks $T(n, c)$ needed for a train of c cars made by a polygon with n sides. We have, in essence, a *two-dimensional sequence* in the variables n and c. The work with squares tells us that $T(4, c) = 3c + 1$. Similarly, the triangles show us that $T(3, c) = 2c + 1$. And the hexagons indicate that $T(6, c) = 5c + 1$. You can build a train like this with polygons of any number of sides. These instances suggest that a train of n-gons that is c cars long has $(n - 1)c + 1$ sides. That is, we conjecture that $T(n, c) = (n - 1)c + 1$.

We can prove this result using the corollary to Theorem 3.15. Think of n as fixed. We now want a formula for the number of sides $T(n, c)$ in terms of c. Thus c (not n) plays the role of n in the corollary. For any value of c, each additional car adds $n - 1$ sides. So this sequence has a common difference $n - 1$. By the corollary, $T(n, c) = dc + T(0, c)$, where d is the common difference. So $T(n, c) = (n - 1)c + T(0, c)$. $T(0, c)$ does not have meaning in this situation. But $T(n, 1) = n$ (a train with 1 car has n sides). So $n = T(n, 1) = n - 1 + T(0, c)$, from which $T(0, c) = 1$. Consequently, $T(n, c) = (n - 1)c + 1$, proving the conjecture.

The result can also be derived using a combinatorial (counting) argument. Consider a train c cars long consisting of n-sided polygons. If we separate the cars, nc toothpicks are needed. When we put the cars back together, there are $c - 1$ cases where two sides are in common, so $c - 1$ toothpicks are not needed. So

$$T(n, c) = nc - (c - 1) = (nc - c) + 1 = (n - 1)c + 1.$$

Measures of rates of change

Data fit by an exponential function $h(x) = cb^x$ cannot have the constant difference property. In fact, for this function h,

$$h(x + 1) - h(x) = cb^{x+1} - cb^x = cb^x(b - 1) = h(x)(b - 1).$$

Notice that although the difference $h(x + 1) - h(x)$ is not constant, the quotient $\frac{h(x + 1) - h(x)}{h(x)}$ is equal to the constant $b - 1$. This quantity measures the increase or decrease in the value of h between x and $x + 1$ relative to the value of h at x. This ratio, often called the *growth rate*, usually provides a better means of comparing the increase or decrease in value of two real functions f and g between x and $x + 1$ when the relative sizes of f and g at x are very different, as the following example indicates.

EXAMPLE 2 Compare the growth in assets of two companies, Company A and Company B, between the years 2000 and 2001, using the data in the following table.

Company	Assets on Jan. 1, 2000	Assets on Jan. 1, 2001
A	37.2 million	41.7 million
B	1.7 million	2.6 million

Solution During the year 2000, the assets of Company A grew by 4.5 million while the assets of Company B grew by 0.9 million, one-fifth of the increase for Company A. However, the growth rate of Company A during that period was $\frac{4.5}{37.2} \approx .12$, while that of Company B was larger: $\frac{.9}{1.7} \approx .53$. This comparison of growth rates is typically expressed in percentages; the assets of Company A grew by about 12% during that year while those of Company B grew by about 53%.

While the president of Company A can correctly claim that the increase in assets of his company was 5 times that of Company B, many people would be more impressed with the observation by the president of Company B that the percentage growth rate of her company was more than 4 times that of Company A.

Example 2 suggests the following measures of *rates of change* of a real function. Suppose that f is a real function and that x and $x + 1$ are in the domain of f.

Definitions The **rate of change of f between x and $x + 1$** [or the **first difference $\Delta f(x)$ of f at x**] is

$$\Delta f(x) = f(x + 1) - f(x).$$

The **relative rate of change of f** or **growth rate between x and $x + 1$** is

$$\frac{f(x + 1) - f(x)}{f(x)}.$$

The **percentage rate of change of f between x and $x + 1$** is

$$100\left(\frac{f(x + 1) - f(x)}{f(x)}\right).$$

Recall from calculus that, for a function f defined on $[a, b]$, the rate of change of f on $[a, b]$ is defined to be $\frac{f(b)-f(a)}{b-a}$. If $b = a + 1$, then $\frac{f(b)-f(a)}{b-a} = f(a + 1) - f(a) = \Delta f(a)$. So our definition of rate of change of f between x and $x + 1$ is consistent with the meaning of rate of change in calculus.

Let us summarize what we have shown so far in terms of these various rates of change. For any real number a, the rate of change between a and $a + 1$ [or first difference $\Delta f(a)$] of the linear function $f(x) = mx + b$ is equal to m. The relative rate of change between a and $a + 1$ of the exponential function $f(x) = cb^x$ is $b - 1$. The percentage rate of change of the exponential function f between a and $a + 1$ is $100(b - 1)$.

Because $\frac{f(x+1)-f(x)}{f(x)} = \frac{f(x+1)}{f(x)} - 1$, the relative rate of change of f between x and $x + 1$ can be expressed in terms of the **first quotient $\Theta f(x) = \frac{f(x+1)}{f(x)}$ of f at x** as follows:

$$\frac{f(x+1) - f(x)}{f(x)} = \Theta f(x) - 1.$$

Consequently, the first quotient of the exponential function f is constant and equal to its base b. This number is sometimes called the **growth factor**.

Fitting data with exponential functions

Analogous to the linear data fitting principle is one for exponential functions.

Exponential Function Data Fitting Principle: Suppose the $n + 1$ data points $(x_0 + k, y_k)$, $k = 0, 1, 2, \ldots, n$, are known. Consider the successive quotients $\frac{y_{k+1}}{y_k}$ for $k = 0, 1, 2, \ldots, n - 1$.

a. If all these quotients are equal to (or nearly equal to) the same constant b, then the exponential function E defined by

$$E(x) = y_0 b^x$$

exactly fits (or nearly fits) the given data.

b. If these quotients are not all the same, then no exponential function exactly fits the given data.

The quality of the fit of the function depends on how near the "nearly equal" numbers are to b, and the context of the situation. Some calculators and computer programs will show an *exponential function of best fit* and produce a *correlation coefficient* from the data. This correlation coefficient is typically the *linear correlation coefficient* between x and $\log y$.

Now consider situation (a) described in the Exponential Function Data Fitting Principle. Suppose that $f: \mathbf{R} \to \mathbf{R}$ is a function with the property that there is a constant b such that $\Theta f(a) = b$ for all a in $\mathbf{R}$. Then is f necessarily an exponential function? As in the case of constant difference for linear functions, the answer is no (see Problem 8). However, the following result is adequate for many applications of exponential functions.

Theorem 3.16 Suppose that $f: \mathbf{R} \to \mathbf{R}$.

a. If f is an exponential function, then f has the following property (**): There is a real number b such that

$$\frac{f(x+1)}{f(x)} = b \qquad \text{for all real numbers } x.$$

b. If f has the property (**), then the exponential function G defined by

$$G(x) = f(0)b^x \qquad \text{for all real numbers } x,$$

agrees with f at all nonnegative integers n.

Proof: Theorem 3.16 can be proved by mathematical induction in a way analogous to the proof of Theorem 3.15. (See Problem 7.)

Corollary: A real sequence $\{g_n\}$ has a constant ratio r, that is,

$$\frac{g_{n+1}}{g_n} = r \qquad \text{for all } n \geq 0$$

if and only if

$$g_n = g_0 r^n \qquad \text{for all } n \geq 0.$$

Proof: You are asked to prove this result in Problem 10.

Exponential sequences

Traditionally, the sequence $f = \{f_n\}$ defined by

$$f_n = f_0 r^n \qquad \text{for all } n \geq 0$$

is called the **geometric sequence** or **exponential sequence** with first term f_0 and common ratio r. Because of the property of Theorem 3.16, geometric sequences can be used to model sequential data with a constant percentage rate of increase or decrease.

We can apply Theorem 3.16 to the following data given at the beginning of this section on the three "mystery functions" f, g, and h.

x	0	1	2	3	4	5	6
$f(x)$	1.9	2.51	3.31	4.37	5.76	7.6	10.03
$g(x)$	2.05	2.30	3.05	4.30	6.05	8.30	11.05
$h(x)$	2	2.67	3.39	4.19	5.15	6.40	8.23

It is evident that no linear function can fit these tabular data because the first differences of the tabular values are far from constant. We check the first quotients of the data to see if we can observe an exponential function pattern.

x	0	1	2	3	4	5	6
$f(x)$	1.9	2.51	3.31	4.37	5.76	7.6	10.03
$\Theta f(x)$	1.321	1.318	1.320	1.318	1.319	1.319	
$g(x)$	2.05	2.30	3.05	4.30	6.05	8.30	11.05
$\Theta g(x)$	1.122	1.326	1.410	1.407	1.372	1.331	
$h(x)$	2	2.67	3.39	4.19	5.15	6.40	8.23
$\Theta h(x)$	1.335	1.270	1.236	1.229	1.243	1.286	

This first quotient data show that the function f can be well fitted at nonnegative integer values by the exponential function $k(x) = 1.9(1.32)^x$, but that the values of the functions g and h are not well suited to an exponential fit.

Question 2: The percent of error of a from b is $\frac{a-b}{b}$ written as a percent.

a. Compare the data for f with the values of the fitting exponential $k(x) = 1.9(1.32)^x$. What is the maximum percent of error?

b. Use this fitting exponential to project a value of $f(10)$ and to interpolate a value of $f(3.5)$.

Instantaneous rates of change

Any linear function f has a constant rate of change

$$f(x + 1) - f(x)$$

between x and $x + 1$ for any $x \in \mathbf{R}$, yet there are elementary functions whose rates of change on any such interval are constant that are not linear functions. Similarly, any exponential function g has constant relative rate of change

$$\frac{g(x + 1) - g(x)}{g(x)}$$

between x and $x + 1$ for any $x \in \mathbf{R}$, but there are elementary functions with constant relative rates of change that are not exponential functions. However, it can be proved that:

(i) If a function f: $\mathbf{R} \to \mathbf{R}$ has a constant *instantaneous rate of change*

$$f'(x) = \lim_{h \to 0} \frac{f(x + h) - f(x)}{h} = m$$

at each $x \in \mathbf{R}$, then f must be the linear function defined by

$$f(x) = mx + f(0) \qquad \text{for all } x \in \mathbf{R}.$$

(ii) If a function g: $\mathbf{R} \to \mathbf{R}$ has a constant *instantaneous relative rate of change*

$$g_r(x) = \lim_{h \to 0} \left(\frac{g(x + h) - g(x)}{hg(x)} \right) = m$$

at each $x \in \mathbf{R}$, and if $b = e^m$, then g must be the exponential function defined by

$$g(x) = g(0)b^x \qquad \text{for all } x \in \mathbf{R}.$$

Project 3 at the end of this chapter discusses these characterizations.

3.3.1 Problems

1. The following table gives U.S. census populations for California, Oregon, and Washington from 1850 (the first census for which they were included) to 1990.

Year	California	Oregon	Washington
1850	92,597	12,093	1,201
1860	379,994	52,465	11,594
1870	560,247	90,923	23,955
1880	864,694	174,768	75,116
1890	1,213,398	317,704	357,232
1900	1,485,053	413,536	518,103
1910	2,377,549	672,765	1,141,990
1920	3,426,861	783,389	1,356,621
1930	5,677,251	953,786	1,563,396
1940	6,907,387	1,089,684	1,736,191
1950	10,586,223	1,521,341	2,378,963
1960	15,717,204	1,768,687	2,853,214
1970	19,971,069	2,091,533	3,413,244
1980	23,667,764	2,633,156	4,132,353
1990	29,785,857	2,842,337	4,866,669

Compute differences and percentage rates of change between each successive census for each state. (You may find it helpful to enter these data into a spreadsheet.) Consider the 40-year intervals (with five censuses) 1850–1890, 1900–1940, and 1950–1990.

a. For each state, does any linear function fit the data well for one or more of these intervals?

b. Does any exponential function fit the data well for one or more of these intervals?

c. From your analysis, discuss the suitability of using linear or exponential functions to describe these data.

2. Consider the data points and function f of Example 1.

a. Suppose the values of the function were rounded to the nearest tenth. Can the data be fit exactly by a linear function? If so, find an equation for the function. If not, use a calculator to find the line of best fit, give the correlation coefficient, and tell what is the farthest this line in vertical distance is from the actual data points.

b. Repeat part **a** if the values are rounded to the nearest integer.

c. Comment on the effects of rounding in this situation.

3. The following table gives years of life expected at birth in the United States, for census years from 1900 to 1990, as estimated by the National Center for Health Statistics. Consider the intervals 1900–1940, 1950–1990, and 1900–1990.

Year	All	Male	Female
1900	47.3	46.3	48.3
1910	50.0	48.4	51.8
1920	54.1	53.6	54.6
1930	59.7	58.1	61.6
1940	62.9	60.8	65.2
1950	68.2	65.6	71.1
1960	69.7	66.6	73.1
1970	70.8	67.1	74.7
1980	73.7	70.0	77.5
1990	75.4	71.8	78.8

a. Does any linear function fit the data well for one or more of these intervals?

b. Does any exponential function fit the data well for one or more of these intervals?

c. From your analysis, discuss the suitability of using linear or exponential functions to describe these data.

4. Draw a sequence of trains of equilateral pentagons like the trains of other polygons shown in this section. Determine how many toothpicks are needed for a train of c cars.

5. Given in this section is a combinatorial argument for a formula for $T(n, c)$, the number of toothpicks needed for a train of c cars made by a polygon with n sides. Other combinatorial arguments are possible.

a. Show that the same formula results from counting an initial side and adding "defective" polygons, each with one side open.

b. Show that the same formula results from adding the numbers of noncommon sides and common sides.

c. Give a table of values for $T(n, c)$ for $3 \le n \le 7$ and $1 \le c \le 5$.

6. Company A had a net income of 5 million dollars in 2000, while a small competing company, Company B, had a net income of 2 million dollars. The management of Company A develops a business plan for future growth that projects an increase in net income of .5 million per year, while the management of Company B develops a plan aimed at increasing its net income by 15% each year.

a. Express these business plans in terms of first differences and first quotients for their respective sequences of projected annual net incomes.

b. Explain why if both companies are able to meet their net income growth goals, the net income of Company B will eventually be larger than that of Company A. In what year will the net income of Company B be larger than that of Company A?

7. Prove parts (a) and (b) of Theorem 3.16.

8. Consider the function g defined by $g(x) = 2^{x+\sin(2\pi x)}$, $x \in \mathbf{R}$.

a. Show that $\Theta g(a) = 2$ for all real numbers a but that g is not an exponential function.

b. What property of the sine function is crucial to the conclusion of part **a**?

9. Prove: If $\{a_n\}$ is an exponential sequence with constant ratio r, then $\{\log a_n\}$ is a linear sequence with constant difference $\log r$.

10. Write a proof of the corollary to Theorem 3.16.

ANSWERS TO QUESTIONS

1. $h(x + 1) - h(x) = (L(x + 1) + \sin(2\pi(x + 1))) - (L(x) + \sin(2\pi x)) = (m(x + 1) + b + \sin(2\pi x + 2\pi)) - (mx + b + \sin(2\pi x)) = m$ [since for all x, $\sin(x + 2\pi) = \sin x$].

2. a. The percents of error are all under 0.21%, a very fine fit. b. $f(10) \approx 30.51$, $f(3.5) \approx 5.02$.

3.3.2 Fitting polynomial functions to data

Theorem 3.15 states that for any linear function $f(x) = mx + b$, the first difference at each real number x is equal to m. Is there a corresponding difference property for quadratic functions? To answer this question, consider the quadratic function

$$f(x) = ax^2 + bx + c.$$

At any real number x, the first difference $\Delta f(x)$ is given by

$$\begin{aligned}\Delta f(x) &= f(x + 1) - f(x) = a(x + 1)^2 + b(x + 1) + c - ax^2 - bx - c \\ &= 2ax + (a + b).\end{aligned}$$

Notice that the first difference function Δf is a linear function of x with slope $2a$. Consequently, the **second difference** $\Delta^2 f$, that is, the first difference at each real number x of the first difference function Δf, equals $2a$, a constant. Consequently, every quadratic function $f(x) = ax^2 + bx + c$ has the property that its second difference at any real number x is the constant $2a$.

For the general polynomial of degree 3,

$$p(x) = ax^3 + bx^2 + cx + d,$$

the first difference function Δf is given by

$$\begin{aligned}\Delta f(x) &= a(x+1)^3 + b(x+1)^2 + c(x+1) + d - ax^3 - bx^2 - cx - d \\ &= 3ax^2 + (3a + 2b)x + (b + c),\end{aligned}$$

a quadratic function of x. Therefore, the second difference function $\Delta^2 f$ is linear and the third difference function $\Delta^3 f$ is constant.

Question: For the general polynomial $p(x)$ of degree 3 given above, show that $\Delta^2 f(x) = 6ax + (6a + 2b)$ and $\Delta^3 f(x) = 6a$.

Consequently, a telltale mark of a polynomial of degree 3 is that its third differences are the nonzero constant $6a$, where a is the leading coefficient of p.

Similarly, we can show that polynomials of degree 4 have nonzero constant fourth differences, polynomials of degree 5 have nonzero constant fifth differences, and, in general, we can prove by mathematical induction that nth-degree polynomials have nonzero constant nth differences.

The converse statement is not true. There are functions $f: \mathbf{R} \to \mathbf{R}$ with a nonzero constant nth-difference function $\Delta^n f(x)$ that are not nth-degree polynomials. However, any function $f: \mathbf{R} \to \mathbf{R}$ with a nonzero constant nth difference function can be well-fitted by an appropriate nth-degree polynomial.

Theorem 3.17 Suppose that $f: \mathbf{R} \to \mathbf{R}$. If the nth-difference function $\Delta^n f$ of f is a nonzero constant, then there is an nth-degree polynomial that agrees with the function f at every nonnegative integer.

Proof: A proof of this result is outlined in Project 5 for this chapter. ┘

You may have recognized the analogy between nth differences and nth derivatives when it comes to polynomials. Both the 1st derivative and the 1st difference of a polynomial of degree k are polynomials of degree $k - 1$; the 2nd derivatives and 2nd differences are polynomials of degree $k - 2$, and so on. If the polynomial is of degree n, then its nth derivative and its nth difference will be constant functions. The results for nth differences are the *discrete* analogs of the *continuous* results studied in calculus.

The method of finite differences

As an application of Theorem 3.17, consider again the question of identifying the function types from the function value data given in the preceding section (Table 7).

Table 7

x	0	1	2	3	4	5	6
$f(x)$	1.9	2.51	3.31	4.37	5.76	7.6	10.03
$g(x)$	2.05	2.30	3.05	4.30	6.05	8.30	11.05
$h(x)$	2	2.67	3.39	4.19	5.15	6.40	8.23

EXAMPLE 1 We have already shown that the given data for f are fit well by the exponential function $f(x) = 1.9(1.32)^x$, but that the data for g and h are not well suited to exponential fits. Do the data for either g or h display a polynomial pattern?

Solution To answer this question, we compute a table of successive differences for both g and h (Table 8).

Table 8

x	0	1	2	3	4	5	6
$g(x)$	**2.05**	**2.30**	**3.05**	**4.30**	**6.05**	**8.30**	**11.05**
$\Delta g(x)$	.25	.75	1.25	1.75	2.25	2.75	
$\Delta^2 g(x)$	.50	.50	.50	.50	.50		
$h(x)$	**2**	**2.67**	**3.39**	**4.19**	**5.15**	**6.40**	**8.23**
$\Delta h(x)$	.67	.72	.8	.96	1.25	1.83	
$\Delta^2 h(x)$	.05	.08	.16	.29	.58		
$\Delta^3 h(x)$	.03	.08	.13	.29			
$\Delta^4 h(x)$	.05	.05	.16				

These calculations indicate that the data for g can be fitted with a polynomial of degree 2, while the data for h do not appear to follow a polynomial pattern. For a polynomial $p(x) = ax^2 + bx = c$ of degree 2, we showed previously that

$$\Delta p(x) = 2ax + (a + b), \qquad \Delta^2 p(x) = 2a, \qquad p(0) = c.$$

Because $\Delta^2 p(x) = .5 = 2a$, and $f(0) = 2.05 = p(0)$, it follows that $a = .25$ and $c = 2.05$. We can compute b by equating $g(1) = 2.30$ with $p(1) = .25 + b + 2.05$. Thus, $b = 0$, and the polynomial $p(x) = .25x^2 + 2.05$.

Example 1 illustrates the **Method of Finite Differences**. In general, for any set of $n + 1$ data points

$$(x_0, y_0), (x_1, y_1), (x_2, y_2), (x_3, y_3), \ldots, (x_n, y_n)$$

in the plane with distinct x-coordinates, there is a polynomial

$$y = p_n(x) = a_n x^n + a_{n-1}x^{n-1} + \cdots + a_1 x + a_0$$

whose graph passes through all of these points. One way to construct such a polynomial is to solve the system of $n + 1$ linear equations

$$y_k = p_n(x_k) \qquad k = 0, 1, \ldots, n,$$

for the $n + 1$ unknown coefficients $a_0, a_1, a_2, \ldots, a_n$.

For example, for the four points $(-1, 3), (0, -1), (1, 2), (2, -2)$, the coefficients a_0, a_1, a_2, a_3 of the cubic polynomial,

$$y = p_3(x) = a_3 x^3 + a_2 x^2 + a_1 x + a_0$$

are obtained by solving the corresponding system of equations:

$$\begin{aligned} 3 &= p_3(-1) = -a_3 + a_2 - a_1 + a_0 \\ -1 &= p_3(0) \quad = \qquad\qquad\qquad\quad + a_0 \\ 2 &= p_3(1) \quad = a_3 + a_2 + a_1 + a_0 \\ -2 &= p_3(2) \quad = 8a_3 + 4a_2 + 2a_1 + a_0 \end{aligned}$$

The solution to this system is easily obtained by elimination as $a_0 = -1, a_1 = \frac{11}{6}, a_2 = \frac{7}{2}$, and $a_3 = -\frac{7}{3}$. So the cubic polynomial that exactly fits these four points is

$$p_3(x) = -\frac{7}{3}x^3 + \frac{7}{2}x^2 + \frac{11}{6}x - 1.$$

The Lagrange Interpolation Formula developed in Project 4 for this chapter provides another way to construct a polynomial of degree n that exactly fits $n + 1$ data points. That construction is done directly from the data without the need to solve the corresponding system of linear equations for the unknown coefficients as we did above. As you will see if you complete that project, it is straightforward to write a computer program that accepts the given data as input and then uses the Lagrange Interpolation Formula to construct the appropriate fitting polynomial as output. Some calculators and computers are programmed to use this formula.

While it is possible to fit any data set of $n + 1$ points exactly with a polynomial of degree n, it may not be very helpful to do so unless there is some reason to believe that the data represent the values of a polynomial of degree n. This is because as the number of points in the data set increases, the oscillation of the graph of the corresponding polynomial between the data points increases more and more. As a result, the resulting fit may not be very meaningful for interpolating or extrapolating information from the data. In numerical analysis, this phenomenon is sometimes referred to as the "polynomial wiggle problem."

3.3.2 Problems

1. Let $d(n)$ stand for the number of diagonals of a polygon of n sides. Here is a table of values of $d(n)$.

n	3	4	5	6	7	8	9	10	11
$d(n)$	0	2	5	9	14	20	27	35	44

a. Use the methods of this section to find a polynomial formula for $d(n)$ in terms of n.

b. Give a geometric argument to show that your formula of part **a** is true for all n.

2. Let $S(n)$ = the sum of the 4th powers of the integers from 1 to n. Use the methods of this section to find a polynomial formula for $S(n)$ in terms of n.

3. a. Find a polynomial function P that contains these data points for the population (in thousands) of Seattle in certain years: (1950, 468), (1960, 557), (1970, 531), (1980, 494), (1990, 516), (2000, 537).

b. Does the polynomial wiggle problem appear?

c. Add the data point (1900, 81) to the set of part **a**. Discuss what happens.

4. For each positive integer n and each real number x, define

$$x^{(n)} = x(x-1)\cdot\cdots\cdot(x-n+1).$$

For example, $x^{(1)} = x$, $x^{(2)} = x(x-1)$, $x^{(3)} = x(x-1)(x-2)$, and so on. We also define $x^{(0)} = 1$. If a polynomial $p(x)$ is expressed in the form

$$p(x) = b_n x^{(n)} + b_{n-1}x^{(n-1)} + \cdots + b_1 x^{(1)} + b_0,$$

we say that $p(x)$ is in **factorial form**. For example, the factorial form of the polynomial $p(x) = x^2 - 2x + 3$ is $p(x) = x^{(2)} - x^{(1)} + 3$, because $x^2 - 2x + 3 = x(x-1) - x + 3$.

a. Show that $\Delta x^{(n)} = nx^{(n-1)}$ for any positive integer n.

b. If $p(x) = b_n x^{(n)} + b_{n-1}x^{(n-1)} + \cdots + b_1 x^{(1)} + b_0$ is a polynomial in factorial form, explain why the factorial form of the first difference $\Delta p(x)$ is

$$\Delta p(x) = nb_n x^{(n-1)} + (n-1)b_{n-1}x^{(n-2)} + \cdots + 2b_2 x^{(1)} + b_1.$$

The results in parts **a** and **b** indicate that the factorial form of polynomials is very helpful in forming successive differences.

c. There is a convenient way to find the factorial form of a polynomial that is written in standard form. Show that if $p(x)$ is a polynomial with standard form

$$p(x) = a_n x^n + a_{n-1}x^{n-1} + \cdots + a_1 x + a_0,$$

then the coefficients of the factorial form of $p(x)$,

$$p(x) = b_n x^{(n)} + b_{n-1}x^{(n-1)} + \cdots + b_1 x^{(1)} + b_0,$$

are, in reverse order, the remainders on successive divisions of $p(x)$ by $x, x-1, \ldots, x-n+1$; that is,

$$p(x) = xq_1(x) + b_0 \text{ (Note that } b_0 = a_0.)$$
$$q_1(x) = (x-1)q_2(x) + b_1, \text{ and so on.}$$

d. Use the result of part **c** to describe a scheme employing synthetic division for computing the coefficients of the factorial form of a polynomial from the coefficients of its standard form.

e. Write $p(x) = 5x^4 - 2x^3 + 12x^2 + 16x - 31$ in factorial form.

ANSWER TO QUESTION

1. $\Delta^2 f(x) = (3a(x+1)^2 + (3a+2b)(x+1) + (b+c)) - (3ax^2 + (3a+2b)x + b + c)$
$= 3a(x+1)^2 - 3ax^2 + (3a+2b)(x+1) - (3a+2b)x$
$= 3a(2x+1) + (3a+2b) = 6ax + 6a + 2b$
$\Delta^3 f(x) = (6a(x+1) + 6a + 2b) - (6ax + 6a + 2b) = 6a$

3.3.3 An extended analysis of the box problem

The following problem is typical of max-min problems solved in calculus and pre-calculus. We call it the **box problem.**

> Find the dimensions of the open box of maximum volume that can be made from an 8″ by 11″ rectangular sheet of cardboard by cutting small squares out of the corners, folding up the sides, and taping the corners.

The solution of this and many other max-min problems follows a single pattern: The problem presents a functional relationship f between two variables, a dependent variable (here, the volume V) and an independent variable x (here the side of the square cutout), and the solution of the problem is obtained by determining the value or values of x in the domain of f that maximize or minimize the value of the dependent variable.

Once V has been expressed in terms of a single independent variable x by formula $y = f(x)$, the exact solution of the problem can be found by identifying and testing the critical points of f, or approximate solutions can be obtained from the graph of $y = f(x)$ by using a graphing calculator.

Although this solution pattern is straightforward and usually results in a correct solution to the given max-min problem, the solution often does not disclose important features of the problem that explain why the solution turned out as it did. There is often rich and surprising mathematics lurking in these problems that is completely missed in their standard solution. Of course, these max-min problems are typically given for a particular purpose, namely to give students practice in using algebraic or graphical techniques or calculus to find the maximum and minimum values of functions, and they do achieve that purpose. Still, it is useful to subject the problem to extended analysis. The mathematics needed for this sort of deeper analysis is not technically difficult, but it requires a different attitude toward what it means to work on a problem. This attitude is to identify and pursue interesting possibilities and in general to be "mathematically curious" when approaching a problem.

Solving the box problem

Figure 27 depicts the given information in the preceding box problem. As Figure 27 suggests, the volume V of the finished box is a function of the size x of the square cutouts at the corners of the rectangle given by

$$V(x) = x(8-2x)(11-2x) = 4x^3 - 38x^2 + 88x. \quad (1)$$

The value of x at which the volume V achieves it maximum can be found either by the techniques of calculus or by graphing the function carefully. For example, in the calculus approach, we find the critical points of $V(x)$ by solving the equation $0 = V'(x) = 12x^2 - 76x + 88$. One of the solutions to this equation lies outside the

Figure 27

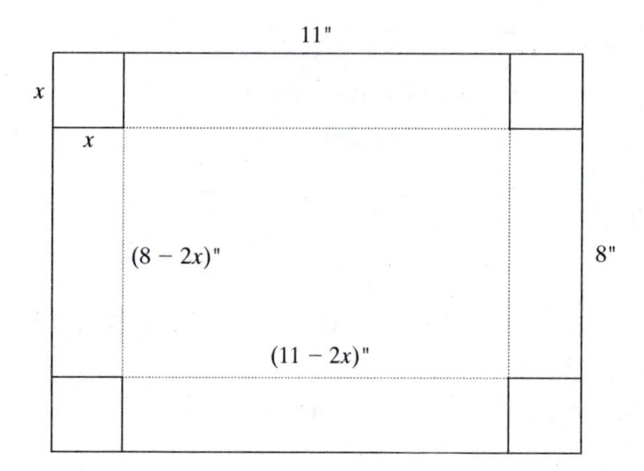

range $0 < x < 4$ of possible values of x; and the other is approximately 1.525 and is a local maximum. The corresponding approximate dimensions of the box of maximum volume are length $L = 11 - 2(1.525) = 7.95''$; width $W = 8 - 2(1.525) = 4.95''$; height $H = 1.525''$; and maximum volume $V_{max} = 60.01$ in.^{3}. In many texts, this is the end of the story. That is unfortunate because there is considerable mathematical richness in this situation that has not been tapped in the limited analysis done so far.

Generalizing the problem

To obtain more interesting and insightful results, we begin by asking what happens if we replace the 8″ by 11″ sheet of cardboard with an arbitrary rectangular sheet. Of course, if the dimensions of a rectangular sheet are just a multiple of 8″ by 11″ such as 16″ by 22″ or 10″ by 13.75″, nothing much changes—the dimensions that maximize the volume are simply the corresponding multiples of the dimensions obtained for the given problem. Thus, the only essential feature of the rectangle in this problem is its *shape*. The shape can be expressed with a single parameter L representing the length of the rectangle, letting the width W of the rectangle be 1. Figure 28 shows the situation in terms of L.

Let $V(x, L)$ represent the volume as a function of the parameter L and the variable x. Then

$$V(x, L) = x(1 - 2x)(L - 2x) = 4x^3 - 2(L + 1)x^2 + Lx. \tag{2}$$

Figure 28

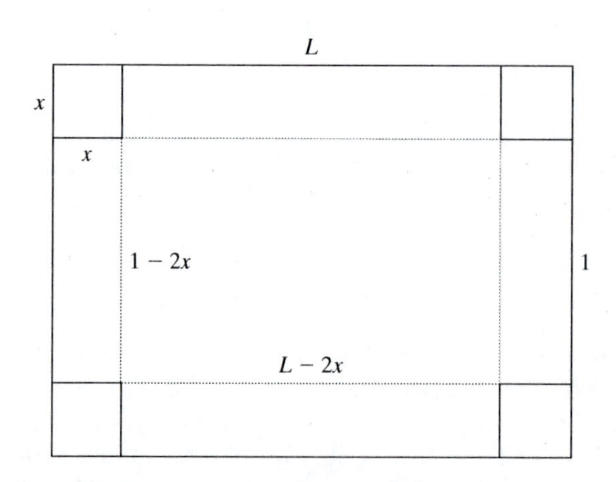

In order to find where the function V in (2) achieves a maximum for a given value of L, we differentiate V with respect to x and set the derivative equal to zero.

$$\frac{d}{dx}V(x, L) = 12x^2 - 4(L+1)x + L = 0 \tag{3}$$

Solving (3), we find that

$$x = \frac{4(L+1)}{24} \pm \frac{1}{24}\sqrt{16(L+1)^2 - 48L}. \tag{4}$$

Since (2) shows that $V(x, L)$ is a cubic function of x with positive leading coefficient, the local maximum of V will be at the smaller of these two values of x, so it is the negative sign of the $\pm$ that we want in (4). Also note that the right side of (4) can be simplified algebraically somewhat. The result is that the size x of the optimal cutout of a rectangle with unit width and length L is given by

$$x = f(L) = \frac{L + 1 - \sqrt{(L+1)^2 - 3L}}{6}. \tag{5}$$

The function f in (5) represents the height of the open box of maximum volume that can be made from an 1″ by L″ rectangular sheet of cardboard with $L \geq 1$ by cutting small squares out of the corners and folding up the sides. Figure 29 depicts a graph of $x = f(L)$.

Figure 29

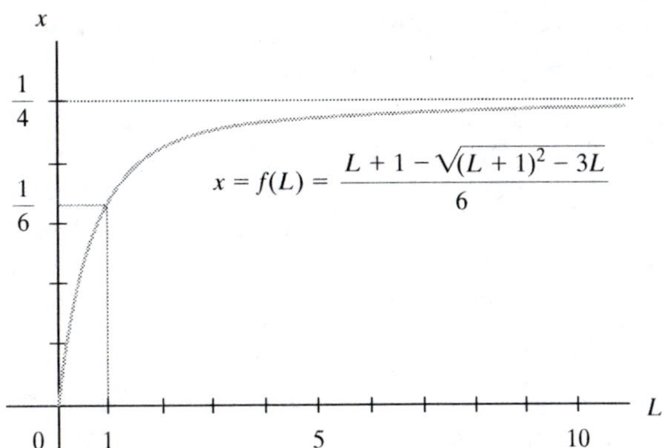

Figure 30

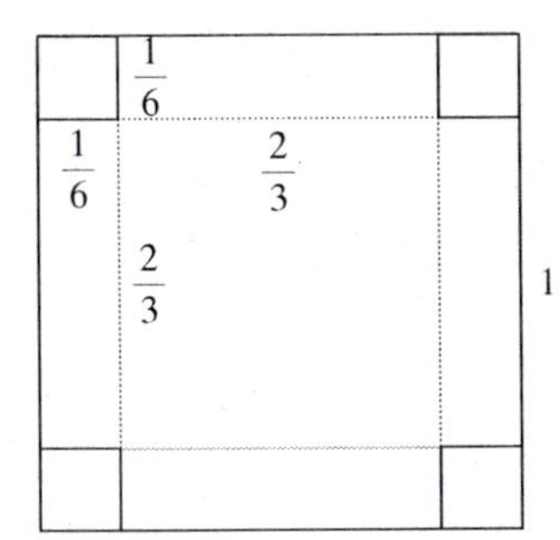

Figure 31

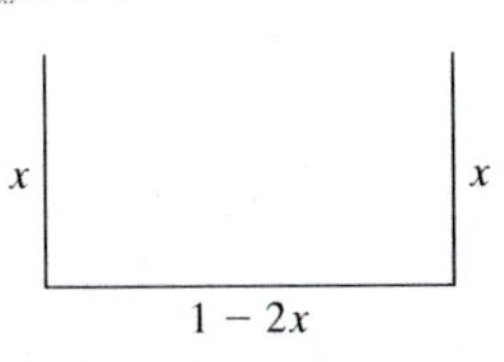

Note that if the given rectangle is a square, then $L = 1$ and $x = f(1) = \frac{1}{6}$. That is, the maximum volume box, starting with a given square, is obtained by cutting out square corners with side length $\frac{1}{6}$ of the side length of the given square. This is shown in Figure 30.

On the other hand, as the given rectangle elongates more and more, that is, as $L \to +\infty$, the graph suggests that $f(L) \to \frac{1}{4}$. This means that for a long thin rectangle the optimal cutout size is larger than the $\frac{1}{6}$ of the side length for a square, and increases toward $\frac{1}{4}$ of the side length. (See this book's front and back covers.)

Why is the fraction $\frac{1}{4}$ in this limiting case for a given rectangular shape? Notice that if L is large, the length $L - 2x$ of the bottom of the box is approximately L, a constant. If the length of the bottom of the box is constant, the maximum volume of the box is obtained by maximizing the rectangular area of the end of the box. This area, in turn, is formed from a "perimeter" of length 1, since the width of the starting rectangle is 1 (Figure 31).

The problem of choosing x to maximize the area of the rectangle is equivalent to the standard problem of forming a rectangular pen against a wall using a fixed length of fence. The familiar answer to this problem tells us that the maximum area occurs when the pen has the shape of half a square. That is, the maximum area occurs when $x = \frac{1}{2}(1 - 2x)$, which is when $x = \frac{1}{4}$. This argument gives a deeper way of understanding *why* the limit is $\frac{1}{4}$ and not some other number.

Figure 32a

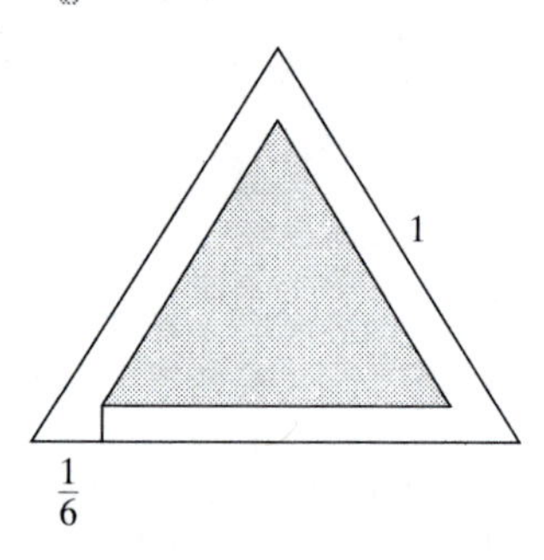

Specifying the problem to regular polygons

Let us return to the special case of a square for which we determined that the side length of the optimal cutout is simply $\frac{1}{6}$ of the side length of the square. What if we had started with a piece of cardboard in the shape of an equilateral triangle, or a regular pentagon or regular hexagon? How would this optimal cut-out fraction change? The fact is that the same $\frac{1}{6}$ proportion also holds if we start with any regular polygon (Figure 32a). You are asked to check this in Problem 3.

Another way of stating this is that, for regular polygons, the maximum volume is obtained when the inner figure is $\frac{2}{3}$ of the (linear) size of the outer figure. What is special about this proportion? *It is precisely the proportion that makes the base area of the container formed exactly equal to the "wall" area of the container.* This result is easy to check for the figures we have illustrated so far: square and equilateral triangle.

Figure 32b

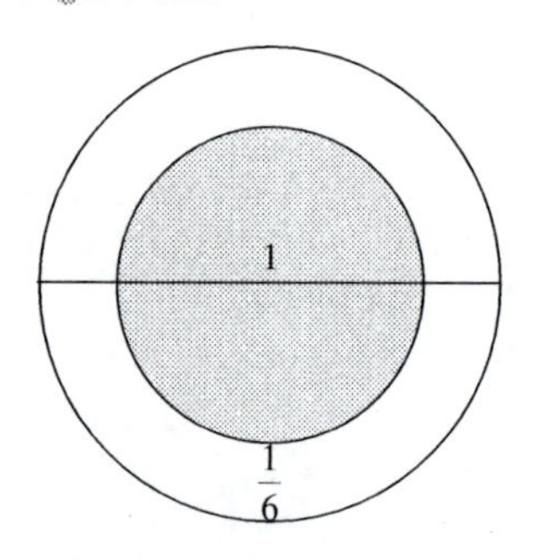

Generalizing once again

This relationship between base area and wall area in containers of maximum volume holds for a wide variety of container shapes, including some that cannot be obtained by literally cutting out pieces and folding up sides. An example is circular cupcake papers that are creased rather than cut before being folded up (see Figure 32b).

All of these examples are special cases of the following more general result. Recall that a **simple closed curve** is a curve that encloses a region and does not intersect itself. A region is **convex** if and only if for any two points A and B in the region, the segment $\overline{AB}$ is in the region.

Theorem 3.18 Let C be any simple closed convex curve in the plane for which a new simple closed curve C' can be created inside C at a constant distance d from C. Form the open cylindrical container whose base is the interior of C' and whose height is d. Of all the containers formed from the curve C in this way, the one with the largest volume is the container whose base area equals its wall area.

Proof: We represent the situation in Figure 33. Let the area of the region enclosed by curve C' be A and the perimeter of C' be p. It follows that the wall area is pd.

Figure 33

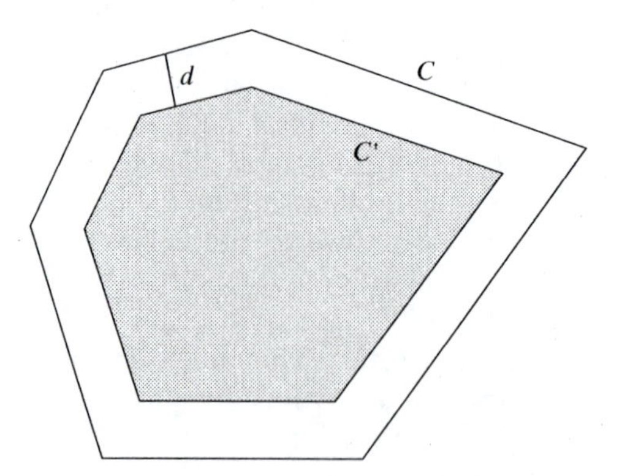

Suppose that for a particular d, the base area equals the wall area. Then

$$A = pd.$$

Since the container is a cylinder, its volume is given by

$$V = Ad.$$

Suppose now d is increased by a small amount x. The new container thus formed has a greater height, namely $d + x$, and a base area that has been decreased by a border strip of approximate area px. Hence the new base area is approximately $A' = A - px$. So the new volume is

$$V' = (A - px)(d + x) = Ad - pdx + Ax - px^2.$$

Using the facts that $V = Ad$ and $A = pd$, we see that

$$V' = V - px^2.$$

That is, $V' < V$.

A similar calculation shows that if d is decreased by a small amount x, the new volume is $V' = V - px^2$ and hence again $V' < V$. The conclusion is that *any* change in d from the container for which the base area equals wall area decreases the volume. Hence the maximum volume is obtained when the base area equals the wall area. This completes the proof. ┘

What we have shown here is a general condition that characterizes the proportions of the wall and base that give open containers of maximum volume. This condition explains the particular results we found for particular containers such as those with square or triangular bases. Thus, hidden within this innocent box problem, there is some simple, surprising, general, and elegant mathematics.

3.3.3 Problems

1. Prove the following results mentioned in this subsection.
   a. Verify that the area of a rectangular pen against the wall of a barn (along a river, etc.) with a fixed length of fencing material is maximized when the pen is in the shape of half a square.
   b. Explain why the solution to the box problem for a square sheet of cardboard is not half of a cube.

2. Verify algebraically that $\lim_{L\to\infty} \frac{L+1-\sqrt{(L+1)^2-3L}}{6} = \frac{1}{4}$, as is suggested by the graph. (*Hint*: Rationalize the numerator of the left-hand side.)

3. Prove the following results mentioned in this section without using Theorem 3.18.
   a. To make a cylindrical box with maximal volume from a given circle, the height of the box will be one-third the radius of the original circle.
   b. To make a triangular box of maximal volume from an equilateral triangle, the height of the box will be $\frac{\sqrt{3}}{12}$ the length of a side.

4. Knowing that the area of the bottom of the box equals the area of the sides in the case of the maximal volume, use the diagram in Figure 34 to give a simple proof of the fact that the height of a box folded from a square should be $\frac{1}{6}$ the side of the original square.

Figure 34

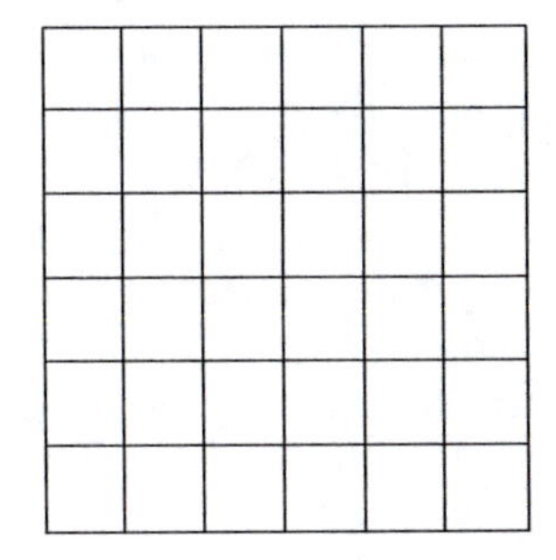

5. A rectangle (the term includes a square) is drawn on graph paper, as shown in Figure 35, and its border cells are shaded. In this case, the number of shaded cells in the border is not equal to the number of unshaded cells in the interior. Is it possible to draw a rectangle such that the border—one cell wide—contains the same number of cells as the interior?

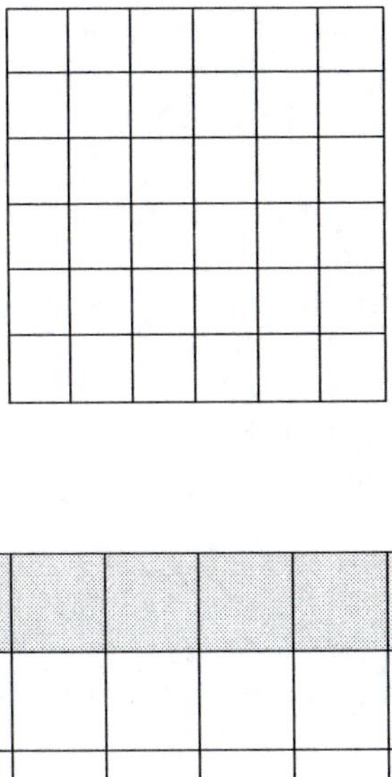

Figure 35

6. a. Find the point on the line $y = 4x - 1$ that is closest to the point $(-1, 3)$.
   b. Generalize the problem and its solution in some way.

Chapter Projects

1. The Leibniz segments. Given a curve C in a coordinate plane and a point P on C, let T be the point of intersection of the tangent line to C at P with the horizontal axis, let N be the point of intersection of the normal line to C at P with the horizontal axis, and let Q be the foot of the perpendicular from the horizontal axis through P (Figure 36).

Figure 36

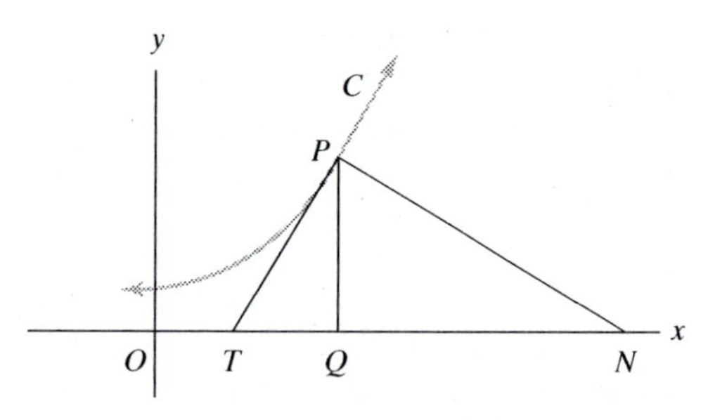

Leibniz introduced the following six "functions" of the point P on C.

i. The **abscissa** at P is the line segment $\overline{OQ}$.
ii. The **ordinate** at P is the line segment $\overline{PQ}$.
iii. The **tangent** at P is the line segment $\overline{PT}$.
iv. The **normal** at P is the line segment $\overline{PN}$.
v. The **subtangent** at P is the line segment $\overline{TQ}$.
vi. The **subnormal** at P is the line segment $\overline{NQ}$.

a. Leibniz was interested in the notion that the slope of $\overline{PT}$ is equal to three different ratios of these six functions. Identify these ratios.

b. Prove that the length PQ of the ordinate at P is the geometric mean of the length QT of the subtangent and the length QN of the subnormal at P.

c. Find the lengths of all six of the line segments defined above for the point $P = (6, 3)$ on the parabola defined by the equation

$$2x = y^2 + 3.$$

d. Recall that a parabola is the set of points equidistant from a given point (its focus) and a given line (its directrix). For the parabola described by $y^2 = 2px$, explain why the focus is at $(\frac{p}{2}, 0)$ and the directrix is the line $x = -\frac{p}{2}$. For a given point $P = (x_1, y_1)$ on this parabola, find the x-coordinates of the points T, Q, and N in terms of y_1.

2. Leibniz segments in a dynamic setting. Use a dynamic geometric construction program (such as Geometer's Sketchpad) to create a dynamic sketch that displays the six line segments in Project 1 for the point P on the parabola determined by a fixed vertical directrix d, a fixed vertex at the origin, and a fixed focus F on the horizontal axis as in Figure 37. The point S on the directrix should be a free point in the sketch so that you can drag it and observe how P and the six Leibniz segments vary. (To create this sketch, it is helpful to use the fact that the point P on the parabola is the point of intersection of the perpendicular bisector of the line segment $\overline{FS}$ and the line perpendicular to the directrix at S. [See the light gray lines in Figure 37.])

a. Use your sketch to trace the path of P along the parabola as you drag the point S along the directrix d.

b. Keep track of the length QN of the subnormal and the length QT of the subtangent in your sketch as you drag S along the directrix d. What theorem about the Leibniz segments of this parabola is suggested by this experiment?

c. Use part **d** of Problem 1 to prove the theorem that you conjectured in part **b**.

3. Characterizing linear and exponential functions with calculus. It is possible to characterize linear and exponential functions in terms of their *instantaneous* rates of change. Recall from calculus that if $f\colon \mathbf{R} \to \mathbf{R}$ is a differentible function and if x is a real number, then the **instantaneous rate of change of f at x**, which is usually called the **derivative $f'(x)$ of f with respect to x**, is defined by the limit

$$f'(x) = \lim_{h \to 0} \frac{f(x+h) - f(x)}{h}.$$

a. Suppose that $f\colon \mathbf{R} \to \mathbf{R}$ is a differentiable function. Prove that there are real constants m and b such that f is the linear function $f(x) = mx + b$ for all real numbers x if and only if $b = f(0)$ and $f'(x) = m$ for all real x.

Figure 37

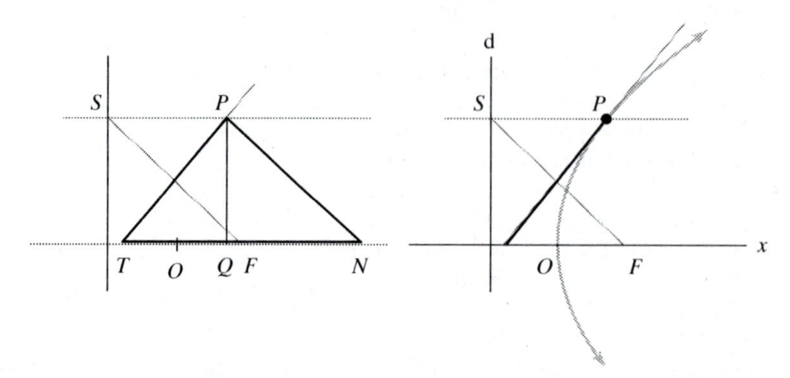

The **instantaneous relative rate of change of f at x, $f'_r(x)$**, is defined as the limit of the relative rate of change of f between x and $x + h$ as h approaches 0:

$$f'_r(x) = \lim_{h\to 0}\left(\frac{f(x+h) - f(x)}{hf(x)}\right).$$

The **instantaneous percentage rate of change of f at x, $f'_{\%}(x)$**, is defined by the limit of the percentage rate of change of f between x and $x + h$ as h approaches 0:

$$f'_{\%}(x) = \lim_{h\to 0}\left(100\frac{f(x+h) - f(x)}{hf(x)}\right).$$

Note that the instantaneous percentage rate of change of f is just 100 times its instantaneous relative rate of change, so only one of these rates is really necessary in any context.

b. Suppose that $B(t)$ is the balance in a bank account t years after an initial deposit of $1000 is left to accumulate interest at an 8% annual interest rate compounded continuously. Explain why the function $B(t)$ can be simply described by the statement $B'_{\%}(t) = 8$ *for all* $t \geq 0$ *and* $B(0) = 1000$.

c. Explain why $f'_{\%}(x) = 100\frac{f'(x)}{f(x)}$ for any differentiable function $f\colon \mathbf{R} \to \mathbf{R}$.

d. Use part **c** and some calculus to conclude that $B(t) = 1000e^{.08t}$ for all $t > 0$.

e. Suppose that $f\colon \mathbf{R} \to \mathbf{R}$ is a differentiable function. Prove that there are real constants c and b with $c \neq 0, 0 < b$, $b \neq 1$ such that f is the exponential function $f(x) = cb^x$, for all real numbers x, if and only if $c = f(0)$ and $f'_r(x) = \ln(b)$ for all real numbers x.

4. The Lagrange Interpolation Formula for polynomial fits. In Section 3.2.2, we showed that if

(*) $\quad (x_0, y_0), (x_1, y_1), (x_2, y_2), (x_3, y_3), \ldots, (x_n, y_n)$

are $n + 1$ points in the plane with distinct x-coordinates, then there is a polynomial

(**) $\quad y = p_n(x) = a_nx^n + a_{n-1}x^{n-1} + \cdots + a_1x + a_0$

whose graph passes through all of these points. We constructed such a polynomial for specific sets of points by solving the system of $n + 1$ linear equations in the $n + 1$ unknowns $a_n, a_{n-1}, a_{n-2}, \ldots, a_0$ obtained by substituting the points (*) into (**). Although the method in Section 3.2.2 is quite practical for determining a polynomial fit for a single set of points, it is not efficient as a computational method because it requires the solution of a new linear system of equations each time it is implemented. It is more efficient to have a formula for $p_n(x)$ expressed in terms of the coordinates of the points (*). Such a formula was developed by Joseph-Louis Lagrange (1736–1813). This project develops and applies his formula, which is referred to as the **Lagrange Interpolation Formula**.

For any integer k with $0 \leq k \leq n$, let $r_k(x)$ be the polynomial of degree n that is the product of the n factors: $(x - x_j)$ for $0 \leq j \leq n$ and $j \neq k$; that is,

$$r_k(x) = \prod_{j=0;\, j\neq k}^{n} (x - x_j).$$

For example, for the set S of four points $S = (-1, 3), (0, -1), (1, 2), (3, -2)$,

$$r_2(x) = (x + 1)(x)(x - 3) = x^3 - 2x^2 - 3x.$$

a. Compute the remaining polynomials $r_0(x)$, $r_1(x)$, and $r_3(x)$ for the set S.

b. Explain why, for each integer k with $0 \leq k \leq n$, the function $q_k(x) = \frac{r_k(x)}{r_k(x_k)}$ is a polynomial of degree n such that $q_k(x_j) = 0$ for $j \neq k$, and $q_k(x_k) = 1$.

c. Let $p_n(x)$ be the polynomial defined by

$$p_n(x) = y_0q_0(x) + y_1q_1(x) + \cdots + y_nq_n(x).$$

Contruct the polynomial $p_3(x)$ for the set S.

d. Explain why $p_n(x)$ is a polynomial with the required fit properties for the given set of $n + 1$ points. That is, explain why $p_n(x)$ is a polynomial of degree at most n such that

$$p_n(x_k) = y_k \qquad \text{for } k = 0, 1, \ldots, n.$$

e. Write a computer program that accepts as input a set (*) of $n + 1$ points in the plane with distinct x-coordinates, and outputs the Lagrange interpolation polynomial. Test your program by using it to construct the Lagrange interpolation polynomial for the set S. Plot these points and the graph of the Lagrange interpolation polynomial on a window that includes all of the points.

f. Complete the following table of first, second, and third differences for $p(x)$.

***x*-value**	−1	0	1	3	4
***p*(*x*)-value**	3	−1	2	−2	?
$\Delta p(x)$	?	?	?	?	
$\Delta^2 p(x)$	?	?	?		
$\Delta^3 p(x)$	?	?			

Use this difference table to reconstruct $p(x)$.

5. nth-differences and polynomial functions.

a. Prove that for each positive integer n, the nth-difference operator Δ^n is linear; that is, $\Delta^n[\alpha f + \beta g](x) = \alpha\Delta^n f(x) + \beta\Delta^n g(x)$ for all functions $f\colon \mathbf{R} \to \mathbf{R}$ and $g\colon \mathbf{R} \to \mathbf{R}$ and for all real numbers α, β, and x. (*Hint*: Use mathematical induction.)

b. Conjecture and prove a formula for the first difference function $\Delta f_k(x)$ for the functions $f_k\colon \mathbf{R} \to \mathbf{R}$ defined for each nonnegative integer k by $f_k(x) = x^k$. [Recall that if p is a positive integer and q is an integer such that $0 \leq q \leq p$, then $\binom{p}{q} = \frac{p!}{q!(p-q)}$ = coefficient of x^q in the binomial expansion of $(x + 1)^p$.]

c. Prove the following result: Suppose that $f\colon \mathbf{R} \to \mathbf{R}$ is a function. If the nth-difference function $\Delta^n f$ of f is a nonzero constant, then there is an nth-degree polynomial that agrees with the function f at every nonnegative integer n.

6. Limit definitions for e. In this text, we have discussed two limit descriptions of the base e of the natural logarithm function:

$$(*) \qquad e = \lim_{n\to\infty}\left(1+\frac{1}{n}\right)^n \quad \text{and} \quad \sum_{n=0}^{\infty}\frac{1}{n!}.$$

This project is to show that the corresponding sequences $b = \{b_n\}$ and $s = \{s_n\}$ defined by

$$b_n = \left(1+\frac{1}{n}\right)^n \qquad s_n = \sum_{k=0}^{n}\frac{1}{k!} = 1 + \frac{1}{1!} + \frac{1}{2!} + \cdots + \frac{1}{n!}$$

for n ≥ 1

converge to the same limit, so that either of the expressions in (*) can be used to describe e.

a. Follow steps i–iv below to prove that the following inequality holds for all positive integers n:

$$2 \le b_n = \left(1+\frac{1}{n}\right)^n \le s_n = 1 + \frac{1}{1!} + \frac{1}{2!} + \cdots + \frac{1}{n!} < 3.$$

Hint: First expand the expression for b_p using the binomial theorem to obtain

$$b_p = \left(1+\frac{1}{p}\right)^p = 1 + p\frac{1}{p} + \frac{1}{2!}\frac{p(p-1)}{p^2} + \cdots$$
$$+ \frac{1}{k!}\frac{p(p-1)\ldots(p-k)}{p^k} + \cdots + \frac{1}{p!}\frac{p!}{p^n}$$
$$= 1 + 1 + \frac{1}{2!}\left(1-\frac{1}{p}\right) + \frac{1}{3!}\left(1-\frac{1}{p}\right)\left(1-\frac{2}{p}\right) + \cdots$$
$$+ \frac{1}{p!}\left(1-\frac{1}{p}\right)\left(1-\frac{2}{p}\right)\ldots\left(1-\frac{p-1}{p}\right),$$

and find a similar expression for b_{p+1}.

Careful observation of these expressions will help you to prove the following:

i. For each integer $p > 1$, $b_p < s_p$.

ii. For each integer $p > 1$, $b_p < b_{p+1}$. (That is, the sequence b is strictly increasing.)

iii. For each integer $p > 1$, $s_p < s_{p+1}$. (The sequence s is strictly increasing.)

iv. For each integer $p > 1$, $2 \le s_p < 3$. (Use mathematical induction to prove for each integer $p > 2$, $\frac{1}{p!} < \frac{1}{2^{p-1}}$. Then use the sum of a finite geometric series $1 + r + r^2 + \cdots + r^{p-1} = \frac{1-r^p}{1-r}$.)

b. Explain why the sequences $\{b_n\}$ and $\{s_n\}$ are both strictly increasing convergent sequences.

c. Suppose that $b = \lim_{n\to\infty}(1+\frac{1}{n})^n$ and $s = \sum_{n=0}^{\infty}\frac{1}{n!}$. Explain why for any fixed integer $q > 2$ and any integer $p > q$,

$$b_p = 1 + 1 + \frac{1}{2!}\left(1-\frac{1}{p}\right) + \frac{1}{3!}\left(1-\frac{1}{p}\right)\left(1-\frac{2}{p}\right) + \cdots$$
$$+ \frac{1}{q!}\left(1-\frac{1}{p}\right)\left(1-\frac{2}{p}\right)\ldots\left(1-\frac{q-1}{p}\right).$$

d. For each fixed $q > 2$, use part **c** to conclude that $b \ge s_q$.

e. Conclude from parts **a** and **d** that $b = s$.

7. The gamma function. The gamma function generalizes a well-known sequence to include noninteger real values in its domain. Explain how the gamma function is defined and give some of its properties and uses.

Bibliography

Unit 3.1 References

Hershkowitz, Rina, Abraham Arcavi, and Theodore Eisenberg. "Geometric Adventures in Functionland." *The Mathematics Teacher* 80 (5), 1987, 346–352.
This article interweaves the algebraic concept of function and the geometric concept of family of quadrilaterals by studying the change in area produced by making a change in some property of a particular figure.

Kronfellner, Manfred. "The History of the Concept of Function and Some Implications for Classroom Teaching." In *Vita Mathematica* (Ronald Calinger, Ed.). Washington, DC: Mathematical Association of America, 1996, 317–320.
This chapter starts with a brief review of the historical evolution of the function concept and then outlines a method of teaching functions that follows the main ideas of its historical development: formulas as preparation for the function concept, working with provisional definitions, generalization from real functions to arbitrary functions, and functions as special relations.

Sui, Man-Keung. "Concept of Function—Its History and Teaching." Chapter 9 in *Learn from the Masters* (Frank Swetz, John Fauvel, Otto Bekken, Bengt Johansson, Victor Katz, Eds.). Washington, DC: Mathematical Association of America, 1995, 105–122.
This chapter traces the development of the concept of function and attempts to incorporate this mathematical-historical vein into the teaching of mathematics at various levels, from secondary school to university.

Unit 3.2 References

Gel'fand, I. M., E. G. Glagoleva, and E. E. Shnol. *Functions and Graphs*. Boston, MA: Birkhäuser, 1990.
This book, originally published in Russian in 1966, is an introduction to the graphing of functions from their equations. It was written for the Mathematical School by Correspondence in the Soviet Union, which was organized and directed by Gel'fand.

Markovitz, Zvia, Bat-sheva Eylon, and Maxim Bruckheimer. "Functions Today and Yesterday." *For the Learning of Mathematics* 6(2), 1986, 18–24, 28.
Through a research study, the authors analyze student understanding of the notion of function as a set of ordered pairs. All functions considered are real functions.

Unit 3.3 References

Dossey, John A., Sharon McCrone, Frank R. Giordano, and Maurice D. Weir. *Mathematics Methods and Modeling for Today's Mathematics Classroom*. Pacific Grove, CA: Brooks/Cole, 2001, Chapter 9.
This chapter, entitled "Model Fitting and Empirical Model Construction," elaborates on the ideas found in Sections 3.3.1 and 3.3.2.

Projects References

Dennis, David, and Jere Confrey. "Functions of a Curve: Leibniz's Original Notion of Functions and Its Meaning for the Parabola." *College Mathematics Journal* 26(2), March 1995, 124–131.

Chapter

4

EQUATIONS

Equations are such an important part of algebra and other mathematics because they have many uses. Among these uses are the following, in no particular order.

An equation may express a relationship among numbers or variables. For instance, we may use the relationship $(n - 1)(n + 1) = n^2 - 1$ to multiply 39 by 41 quickly in our heads, or the equation $a^2 + b^2 = c^2$ of the Pythagorean Theorem to find the length of a side of a right triangle.

An equation may express one variable in terms of another, as in the formula $C = \frac{5}{9}(F - 32)$ to convert Fahrenheit to Celsius temperatures, or $C = 34 + 23\lceil w - 1 \rceil$ for the cost (in the year 2001) of mailing inside the U.S. a first-class letter weighing w ounces.

An equation may define a function, as when we write $y = \sin x$ to obtain the values of the sine function. We may also think of $y = \sin x$ as describing a curve. Likewise, $|x| + |y| = 1$ describes a square, while $y = 5x^2 + 2x - 6$ describes a parabola. With three variables, an equation may describe a surface, such as the equation $x^2 + y^2 + z^2 = 1$ for the unit sphere.

An equation may provide information about when a particular quantity is maximized or minimized, a property we apply when we set the derivative of a function equal to zero to find an extreme point.

Each of these uses of equations and many others involve variables. Other equations in the same variables may arise while working with the given equations. We often need to solve these equations, that is, to determine all values of the variables for which the two sides of the equation have equal values.

In this chapter we discuss general ideas relating to equations and their solutions. You may be surprised that some of the methods for solving equations that you learned in different courses are special cases of beautiful and powerful theorems. Several of these methods can be adapted to solve inequalities.

Unit 4.1 The Concept of *Equation*

In mathematical discourse, an **equation** or **equality** is a sentence of the form $A = B$, where A and B may be numbers, algebraic expressions, sets, or other objects.

4.1.1 Equality, equivalence, and isomorphism

Defining equality

Equality, in a given mathematical context, is usually defined in terms of equality in a more primitive context where its meaning has already been defined or is taken as understood. For example, two *complex numbers* $z_1 = x_1 + iy_1$ and $z_2 = x_2 + iy_2$ (or $z_1 = (x_1, y_1)$ and $z_2 = (x_2, y_2)$) are defined to be equal if and only if their real and imaginary parts are equal; that is, if and only if $x_1 = x_2$ and $y_1 = y_2$. This definition assumes that the meaning of equality for real numbers is understood. As another example, we say that two *functions* $f\colon A \to B$ and $g\colon A \to B$ are equal if and only if $f(x) = g(x)$ for all x in A or, in other words, f and g are equal as subsets of $A \times B$.

Ultimately, the equality of most objects in mathematics can be traced back to equality of sets. In this respect, sets provide a most fundamental notion of equality in mathematics. *Sets* A and B are equal if and only if they contain the same elements; that is, every element of A is an element of B, and every element of B is an element of A. From the set definition and equality of sets, we can prove that two *ordered pairs* are equal if and only if their respective components are equal. That is, $(a, b) = (c, d)$ if and only if $a = c$ and $b = d$. (See Problem 3.)

Thus the meaning of equality of complex numbers and equality of functions rests on the meaning assigned to equality of ordered pairs. But what do we mean by an ordered pair (a, b)? It is not the same thing as the set $\{a, b\}$ whose elements are a and b because we want (a, b) to be a different ordered pair than (b, a) if a and b are different elements. To provide this distinction, it is customary to define the **ordered pair (a, b)** as the set $\{\{a\}, \{a, b\}\}$; that is, the ordered pair (a, b) is the set whose elements are the singleton set $\{a\}$ and the set $\{a, b\}$. This may look a little peculiar but it accomplishes the desired result. It defines the concept of ordered pair solely in terms of sets, and the ordered pair (a, b) is different from the ordered pair (b, a) whenever a and b are different elements.

We also use equality of sets to express equality in a geometric context. It is customary to define a *figure* in geometry as any set of points. This means that two *figures* in the plane are equal if and only if they are identical subsets of that plane. For instance, the segments $\overline{AB}$ and $\overline{BA}$ are equal, and we may write $\overline{AB} = \overline{BA}$, because they consist of the same points.

Sets, elements, and set membership provide a basis from which all of mathematics can be constructed. However, when we teach and learn mathematics, we almost always begin "somewhere in the middle" with a context that we regard as well understood, and from that basis we develop new contexts or ideas. For example, in our definition of functions in Chapter 3, or complex numbers in Chapter 2, we treated ordered pairs as well-understood objects without tracing the concept back to set theory. Because we proceed in this way, we usually define the objects and equality of objects in a new context in terms of those in the context that we are taking as "given". The definition of complex numbers as ordered pairs of real numbers and equality of complex numbers as equality of their respective real and imaginary parts is another instance of this practice.

The equal sign

We use the equality sign "=" to connect equal numbers or objects on its two sides. But there are subtle variants of this use. Children usually first see the equal sign used as *a signal to compute* in exercises such as $3 + 4 = ?$ or $3 + 4 = \_\_\_\_$. In this use, the equality sign has an implicit direction from left to right. This signal and direction become so strong that, when asked what number makes $7 + \_\_\_\_ = 10 + 5$ a true statement, many children who know addition well will answer 3, seeing 10 as the answer to an addition problem and ignoring the 5 on the right, or they answer 22, viewing the equal sign as a signal to add all the numbers that are present.

The notion that equality has a direction from left to right and is a signal to compute is even stronger in extended computations, where the result from one computation is used in a succeeding computation. For instance, if two items are bought for $3.50 and $4.29 and there is a 7.3% tax, some people write "$3.50 + 4.29 = 7.79 + .57 = 8.36$". In mathematical parlance, this statement is considered incorrect, because it implies that $3.50 + 4.29 = 8.36$. One way to correct the written work is to write 7.79 twice, as in

$$3.50 + 4.29 = 7.79$$
$$7.79 + .57 = 8.36.$$

Another way to avoid this problem is to write the original numbers in a column rather than a row. Then there are no equal signs. Instead, horizontal underscores are often the signals to compute, and no mathematical properties are associated with those underscores.

We often employ the equal sign *to assign meaning to a variable*, as when we write

$$\text{Let } t = \text{the time the ball was in the air}$$
$$\text{or} \quad \text{let } f(t) = \text{the height of the ball at time } t.$$

This use is different enough that some mathematicians and computer languages employ a modified equality symbol:

$$t: = \text{the time the ball was in the air.}$$
$$f(t): = \text{the height of the ball at time } t.$$

In some older computer languages, replacement of the value of an integer variable by a value one greater was signified by

$$x = x + 1,$$

a statement that would never be considered true in mathematical discourse. The statement $x: = x + 1$ avoids the confusion.

Still another variant of the equal sign is $\equiv$, used to denote *identities*, statements such as $\sin^2 x + \cos^2 x \equiv 1$ that are true for all values of the variable x.

In almost all these uses, the mathematical idea that is conveyed by the equality sign is that two objects in a set S are *equal* when each can be substituted for each other in virtually any of their applications.

Equality and equivalence relations

To distinguish valid uses of equality from invalid ones, we require equality in a set S to satisfy certain properties. Certainly an object should be able to be substituted for itself. That is, for any a in S,

$a = a.$ Reflexive Property of Equality

Since each must be able to be substituted for the other, equality is symmetric. That is, for any a and b in S,

If $a = b$, then $b = a$. Symmetric Property of Equality

To provide for the equality of more than two objects, and to avoid the unrestricted use of the equal sign as a signal to compute, we require that equality satisfy the transitive property. For any a, b, and c in S,

If $a = b$ and $b = c$, then $a = c$. Transitive Property of Equality

A relation that satisfies the reflexive, symmetric, and transitive properties is called an *equivalence relation*. So equality $(=)$ is an equivalence relation. There are many other examples of equivalence relations in mathematics. One that does not involve numbers is the **iff** (if and only if) relation between statements, denoted by $\Leftrightarrow$. It has the following properties, for any statements p, q, and r:

$p \Leftrightarrow p$ Reflexive Property of $\Leftrightarrow$
If $p \Leftrightarrow q$, then $q \Leftrightarrow p$. Symmetric Property of $\Leftrightarrow$
If $p \Leftrightarrow q$ and $q \Leftrightarrow r$, then $p \Leftrightarrow r$. Transitive Property of $\Leftrightarrow$

But when $p \Leftrightarrow q$, we do not call p and q equal. Rather we call them *logically equivalent*.

The congruence relation in geometry is another equivalence relation. In Chapter 7, we study the congruence relation in some detail. Here we merely note that the statement $\angle ABC \cong \angle DEF$ about angles is logically equivalent to the statement $m\angle ABC = m\angle DEF$ about numbers. In words, two angles are congruent if and only if their measures are equal. This means that in proofs, one may be substituted for the other. The substitution is so natural and obvious that mathematicians often ignore stating the measures explicitly, and will write $\angle ABC = \angle DEF$ even though these angles may not refer to the same set of points.

The difference between equivalence and equality is that we cannot always substitute equivalent elements for each other. For instance, when $\Delta ABC \cong \Delta DEF$, a particular circle might be inscribed in ΔABC, but it might not be inscribed in ΔDEF.

Equality and isomorphism

Equal objects are identical in the sense that they may be substituted for each other throughout a system, but equivalent objects do not necessarily have this property. *Equivalence* is the name given to a relation that satisfies the reflexive, symmetric, and transitive properties. *Isomorphism* takes this idea one step further.

Recall from Section 3.1.3 that when there is a 1-1 correspondence between two structures so that operations in one structure give answers that correspond to operations in the other structure, then the structures are called *isomorphic*.

As an example, consider the set of real numbers and the set of complex numbers. One point of view is that every real number is a complex number. In particular, the real number x is viewed as being identical (i.e., equal) to the complex number $x + 0i$. With this point of view, the set of real numbers is a subset of the set of complex numbers. A second point of view emerges from the complex numbers being ordered pairs of real numbers. If the real number x were identical to the complex number $(x, 0)$, then we would have in theory $x = \{\{x\}, \{x, 0\}\}$. With this point of view, a real number x corresponds to the complex number $(x, 0)$ and the system of real numbers is isomorphic to the system of complex numbers whose second component is 0. The first point of view might be thought of as top-down, beginning with complex numbers and working down to real numbers. The second point of view is bottom-up. Either point of view is defensible.

A similar situation exists among whole numbers and fractions. Does the simple fraction $\frac{4}{1}$ equal the whole number 4? In general, are the simple fractions with denominator 1 equal to whole numbers? Most books call them *equal*, but some mathematics educators believe they are more appropriately called *equivalent* than *equal*. For instance, in the usual algorithm for adding two fractions, we must use numerators and denominators, and since the denominators of $\frac{4}{1}$ and $\frac{8}{2}$ are different, one fraction cannot always be substituted for the other. This conception treats simple fractions as ordered pairs of whole numbers specifying *indicated quotients*, so they cannot equal whole numbers.

The reason books call $\frac{4}{1}$ and $\frac{8}{2}$ *equal* is that, as *performed quotients*, these equal the same real number. Furthermore, there is a one-one correspondence between the set of fractions with denominator 1 and the set of whole numbers

$$\frac{n}{1} \leftrightarrow n$$

in which addition and multiplication give corresponding answers.

$$\frac{m}{1} + \frac{n}{1} = \frac{m+n}{1} \leftrightarrow m + n$$

$$\frac{m}{1} \cdot \frac{n}{1} = \frac{mn}{1} \leftrightarrow mn$$

So there is a sense in which we cannot tell simple fractions with denominator 1 apart from the whole numbers.

In general, elements in isomorphic structures are sometimes called equal, sometimes equivalent. Both words convey important aspects of isomorphism. The elements are equal in the sense that they differ only in how they look. The elements are equivalent in that we can distinguish them even if only by notation.

Here is a related example of isomorphism. Consider the set S of ordered pairs of integers (a, b), where b is not allowed to be zero, and define two operations $\oplus$ and $\otimes$ as follows:

$$(a, b) \oplus (c, d) = (ad + bc, bd)$$
$$(a, b) \otimes (c, d) = (ac, bd).$$

Question 1: Before reading on, calculate $(2, 3) \oplus (6, 5)$ and $(-4, 1) \otimes (7, 1)$.

Question 2: Write each as a single fraction: $\frac{a}{b} + \frac{c}{d}$; $\frac{a}{b} \cdot \frac{c}{d}$.

The answers to Questions 1 and 2 illustrate that there is nothing to distinguish S from the set of simple fractions with normal addition and multiplication. Thus the operations $\oplus$ and $\otimes$ provide an alternative way to describe addition and multiplication of simple fractions, by representing the fractions as ordered pairs.

Would you then say that in this case (a, b) *equals* $\frac{a}{b}$, or would you say (a, b) *is equivalent to* $\frac{a}{b}$? There is no one correct answer to this question. Either interpretation is defensible.

4.1.1 Problems

1. According to the set definition of ordered pair, what is (b, a)?
2. Extend the definition of ordered pair to create a definition of *ordered triple* or *3-tuple* in terms of sets.
3. Use the definitions of equality of sets and equality of ordered pairs to prove that $(a, b) = (c, d)$ if and only if $a = c$ and $b = d$.

4. In this section, we noted that the relations "is equal to" and "is congruent to" are reflexive, symmetric, and transitive. The relation "is a cousin of" is symmetric but neither reflexive nor transitive. Give an example of a relation (mathematical or nonmathematical) that is
   a. reflexive and symmetric but not transitive
   *b. symmetric and transitive but not reflexive
   c. reflexive and transitive but not symmetric

5. Call two positive integers *close* if and only if their difference is less than 5% of the smaller integer. For example, 96 is close to 100, but 4 is not close to 8. Is closeness an equivalence relation?

6. Which of the three properties of an equivalence relation are satisfied by implication ($\Rightarrow$)? Explain why you have picked the properties you identify.

7. a. Give the results of the operations of complex number addition and multiplication on ordered pairs (a, b) and (c, d).
   b. In your opinion, is the complex number $a + bi$ *equal to* or is it merely *equivalent to* the ordered pair (a, b)? Explain the reason behind your choice.

8. Let S be the set of ordered pairs of positive integers. Let equivalence, addition, and multiplication be defined on the elements of S as follows: For all (a, b) and (c, d) in S: (a, b) is equivalent to $(c, d) \Leftrightarrow a + d = b + c$; $(a, b) + (c, d) = (a + c, b + d)$, and $(a, b) \cdot (c, d) = (ac + bd, ad + bc)$.
   a. Prove that if addends in S are equivalent, then their sums in S are also equivalent.
   b. Prove that if factors in S are equivalent, then their products in S are also equivalent.
   c. Allowing equivalence, show that multiplication is distributive over addition.
   d. Show that $\langle S, +, \cdot \rangle$ is isomorphic to $\langle \mathbf{Z}, +, \cdot \rangle$ if equivalent elements in S are considered identical and (a, b) in S corresponds to $a - b$ in $\mathbf{Z}$.

9. Identified are 4 systems, each consisting of a set with four elements and a single operation defined on the set in the usual way. Which system is not isomorphic to the others, and why?
   (a) $\{1, -1, i, -i\}$, multiplication
   (b) $\{f, g, h, j: f(x, y) = (-x, -y), g(x, y) = (-y, x), h(x, y) = (x, y), j(x, y) = (y, -x)\}$, composition
   (c) set of rotations with center $(0, 0)$ and magnitudes $0°$, $90°$, $180°$, and $270°$, composition
   (d) $\{r, s, t, u: r(x, y) = (x, y); s(x, y) = (x, -y), t(x, y) = (-x, -y), u(x, y) = (-x, y)\}$, composition

ANSWERS TO QUESTIONS

1. $(28, 15)$ and $(-28, 1)$

2. $\frac{ad+bc}{bd}$ and $\frac{ac}{bd}$

4.1.2 Solving equations

A **statement** is a sentence that is either true or false, and not both. The equation $3 + 5 = 8$ is a true statement, whereas the equation $3 + 5 = 9$ is a false statement. In the logical context of truth and falsity, the equation $3 + x = 9$ is neither true nor false. Sometimes the truth value of the equation is called **open** to signify that, until substitutions are made for x, its truth value is open to question.

When we translate an equation into English or another spoken language, then we realize that an equation is a sentence with "=" as its verb. An inequality is a sentence with "<", "≤", ">", or "≥" as its verb. When an equation or inequality involves a variable, then **solving** it means looking for all allowable values of the variable(s) that make the sentence a true statement. We call the variable an **unknown** and the values that make the equation true are the **solutions**. The variable can be represented by a letter, blank, or word. In this context, solving $x + 5 = 8$, filling in the blank to make ____ + 5 = 8 true, and finding a number that, when added to 5, equals 8 represent the same problem. We prefer letters for variables because with complicated equations, the process of solution is more succinctly described and generalizations are more easily proved. Also, letters can be abbreviations for quantities, such as c for cost. We can also use subscripts $c_1, c_2, \ldots$ for different (but usually related in some way) variables.

The **domain** or **replacement set** or **universe** for a variable is the set of possible values of a variable.

Existential and universal statements

When the truth value of an equation is open, the equation can often be turned into a true or false statement by the use of a quantifier such as $\forall$ (for all) or $\exists$ (there exists). For instance, consider the equation $6 + x = 6x$. It is open. However, since 1.2 is a solution to the equation, it is true that

$$\exists \textit{ a real number } x \textit{ such that } 6 + x = 6x.$$

$\exists$ is called an **existential quantifier** and the statement in italics is called an **existential statement** because it asserts the existence of an object satisfying the condition that follows it.

The domain of a variable is critical to the existence of solutions. If the domain of the variable x above is understood to be **R**, then $\exists$ *x such that* $6 + x = 6x$ is a true statement. On the other hand, if the domain of x is the set **Z** of integers, then it is false that $\exists$ *x such that* $6 + x = 6x$.

A **universal statement** is one that asserts the truth about all objects in a set. The universal statement

$$\forall \textit{ numbers } x, 6 + x = 6x$$

is false in either **R** or **Z** because there are values of x in each set for which the equation is false. If an equation is true for all values of the variable in its domain, then the equation is called an **identity**. When an equation is an identity, the corresponding universal and existential statements are both true. For instance, consider the equation $x^2 - 1 = (x + 1)(x - 1)$. The associated universal statement

$$(*) \qquad \forall \textit{ real numbers } x, x^2 - 1 = (x + 1)(x - 1),$$

is true, so $x^2 - 1 = (x + 1)(x - 1)$ is an identity on **R**. For this reason, $\forall$ is called a **universal quantifier**. For the identity (*), it is also true that

$$\exists \textit{ a real number } x \textit{ such that } x^2 - 1 = (x + 1)(x - 1).$$

On the other hand, if an equation is false for all values of the variable in its universe, then the corresponding universal and existential statements are both not true. We tend to call the equation *false* without any quantification. For instance, when y is a complex number, then $y = y + 5$ is false even though no quantifier is exhibited.

When quantifiers are not written, confusion can arise. For instance, when simplifying expressions, we think of two expressions as equal only if they are equal for all values of the variable. Thus, for instance, we would not substitute $6 + x$ for $6x$. But when solving a problem, we might well set $6 + x$ equal to $6x$, and in fact we might need to find all values of x for which $6 + x = 6x$ is true. So although $\forall x, 6 + x = 6x$ is false, it is true that $\exists x$ with $6 + x = 6x$. Some mathematics educators believe that overt statement of the quantifiers in solving equations can help students distinguish between solving equations (in which we try to determine the [usually] few values of the unknown for which a statement is true) and simplifying expressions (which are true for all values of the unknown).

Confusion between existential and universal statements appears when solving linear equations that do not have a unique solution. For instance, consider solving the equation

$$3(2t - 4) = 2 + 2(3t - 7).$$

In solving, we are asking whether $\exists t$ that satisfies the equation, and, if so, what is that value? Solving this equation in typical fashion, by removing parentheses, we obtain

$$6t - 12 = 2 + 6t - 14$$
$$6t - 12 = 6t - 12.$$

We began by looking for *any* value of t that solves the first sentence and wound up with an identity, signifying that *all* values of t are solutions, not only to the last equation but also to the first equation. The confusion of students can be described as having been caused by the move from an existential statement to a universal statement.

The same confusion appears in the study of trigonometry when students first meet the term *identity* and are asked to prove whether statements are identities. Is

$$\cos(2x) - 2\cos^2 x + 1 = 0$$

an identity, or is it an equation to be solved? One way of resolving this question is to work through the process of solving the equation using reversible steps. Then, if we reach a statement that is true for all values in the domain, we can assert that the statement is an identity. We discuss reversible steps in some detail in Unit 4.3.

Indicating solutions to equations

The general method in solving equations is to write a sequence of *equivalent equations*, in which each equation is either simpler or more along the pathway toward solutions than the previous. **Equivalent equations** are those that have exactly the same solutions. In solving, we search for equations equivalent to given equations whose solutions can be read immediately. For instance,

$$x^2 + 4 = 4x,$$

$$x^2 - 4x + 4 = 0,$$

and

$$(x - 2)^2 = 0$$

are all equivalent because they have the same solution, 2. The equation with obvious solutions equivalent to these is $x = 2$. Sometimes we explicitly identify the *set* of solutions, the **solution set** $\{2\}$.

When there are many solutions to an equation, we may list them, or graph them, or sometimes we just give a simple equation or equations equivalent to the given sentence. For instance, solutions to the equation $|y - 5| = 3$ are described in any of the ways shown in Figure 1.

Figure 1

Listing: 8, 2

Graphing:

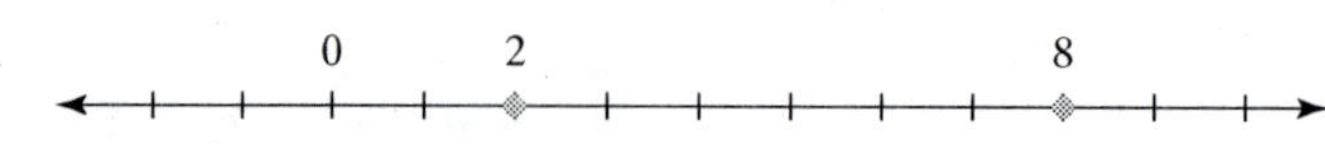

"Simpler" equations: $y = 8$ or $y = 2$.

Solution set: $\{8, 2\}$

An equation may have two variables, in which case each solution is an ordered pair. For instance, a solution to $3x + 5y = 9$ is an ordered pair. The pair is usually written with x as the first component and y as the second component, and there may be infinitely many solutions. For $3x + 5y = 9$, there are infinitely many solutions, one of which is $(-2, 3)$. A list is often impossible, as there are infinitely many solutions, and the set of solutions may be pictured by a graph. Individual elements of the solution set cannot all be listed, but *set-builder notation* $\{(x, y): 3x + 5y = 9\}$ or $\{(x, y)|3x + 5y = 9\}$, read "the set of all ordered pairs x, y such that $3x$ plus $5y$ equals 9," is sometimes found, particularly in schoolbooks.

Sentences with three, four, or more variables, have their solutions accordingly as 3-tuples (x, y, z), 4-tuples (x, y, z, w), or n-tuples $(x_1 x_2, \ldots, x_n)$.

Compound sentences and systems of equations

In English grammar and in mathematics, a **compound sentence** is a sentence containing two or more clauses, usually connected by a conjunction such as *and* or *or*. The sentence $y = 8$ *or* $y = 2$ is a compound sentence.

In mathematics, a **system of equations** or **system** is a compound sentence in which the individual equations are connected by the conjunction *and*. These equations are sometimes called **simultaneous equations**. We also may have a **system of inequalities**. Systems arise from situations in which there is more than one condition on the variables.

Think of p and q as the equations or inequalities in a system. Then, because p *and* q is true if and only if p is true and q is true, the **solution set to a system** is the intersection of the solution sets of the individual equations or inequalities. These solutions may be listed if there is a finite number of them. They may be graphed.

We often desire to find an equivalent simpler system. For instance, suppose we wish to find all points that are (1) five units from the origin *and* (2) ten units from the point $(10, 0)$. Geometrically this problem is known as a *locus* problem. The set of points satisfying each of conditions (1) and (2) is a circle.

Figure 2

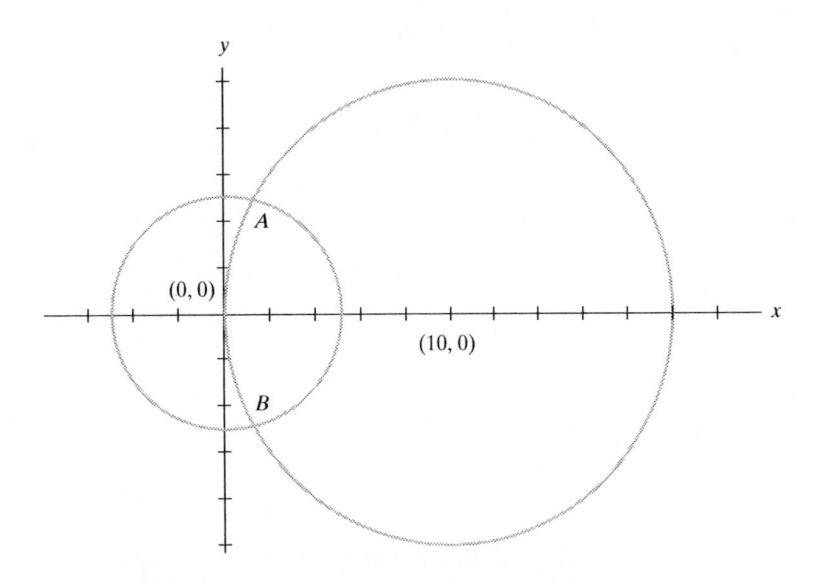

Figure 2 pictures all points on either of these circles. (Ironically, we are viewing the "or" situation in order to determine the "and" situation.) Each point of intersection of the circles is a solution to the system. So there are two solutions, A and B. To identify these points algebraically, we can solve the system whose equations are the equations of the two circles in Figure 2. Let (x, y) be the coordinates of a point of intersection. Then

$$x^2 + y^2 = 25 \quad \text{and} \quad (x - 10)^2 + y^2 = 100.$$

Solving this system, we are led eventually to an equivalent (compound) sentence:

$$x = \frac{5}{4} \quad \text{and} \quad y = \pm\frac{5}{4}\sqrt{15}.$$

The second part of this compound sentence is itself a compound sentence.

$$x = \frac{5}{4} \quad \text{and} \quad \left(y = \frac{5}{4}\sqrt{15} \quad \text{or} \quad y = -\frac{5}{4}\sqrt{15}\right)$$

From either of these, we can list the solutions: $\left(\frac{5}{4}, \frac{5}{4}\sqrt{15}\right)$ and $\left(\frac{5}{4}, -\frac{5}{4}\sqrt{15}\right)$. Or we may list the solutions as $(x, y) = \left(\frac{5}{4}, \frac{5}{4}\sqrt{15}\right)$ or $(x, y) = \left(\frac{5}{4}, -\frac{5}{4}\sqrt{15}\right)$.[1]

Writing compound sentences as a single sentence

Above, we noted that the equation $|y - 5| = 3$ is equivalent to the compound sentence $y = 8$ *or* $y = 2$. For efficiency, we sometimes wish to reverse this process and find a single equation that is equivalent to a compound sentence.

If only real number solutions are sought, two theorems from algebra enable any equations separated by *and* or *or* to be written as a single equation. To change an *or* statement to a single equation, use the **Zero Product Property**: $\forall a, b \in \mathbf{R}$, $ab = 0$ if and only if $a = 0$ or $b = 0$.

By this means, to convert

$$y = 8 \quad \text{or} \quad y = 2$$

to a single sentence, first work with the equations so that 0 is on one side.

$$y - 8 = 0 \quad \text{or} \quad y - 2 = 0.$$

Then apply the Zero Product Property.

$$(y - 8)(y - 2) = 0$$

You could expand the multiplication to obtain another equation equivalent to $y = 8$ or $y = 2$.

$$y^2 - 10y + 16 = 0$$

To change two equations separated by *and* to a single equation, use another property: For all $a, b \in \mathbf{R}$, $a^2 + b^2 = 0$ if and only if $a = 0$ and $b = 0$.

For example, for the system solved above,

$$x^2 + y^2 = 25 \quad \text{and} \quad (x - 10)^2 + y^2 = 100,$$

let $a = x^2 + y^2 - 25$ and $b = (x - 10)^2 + y^2 - 100$. Then an equivalent single equation is

$$(x^2 + y^2 - 25)^2 + ((x - 10)^2 + y^2 - 100)^2 = 0.$$

It now may be quite obvious why systems are not usually solved by using these theorems. The equivalent single equation is a 4th-degree equation in x and y. Still, for testing solutions, perhaps by computer, it may be more efficient to have one equation rather than a system of two equations.

Writing single equations as a compound sentence

Consider the equation $x^3 = 5x^2 - 2x - 8$. We can write this as the compound sentence $y = x^3$ *and* $y = 5x^2 - 2x - 8$. Although the rewriting introduces a second variable, the resulting system of equations can now be graphed in $\mathbf{R}^2$ and any solutions can be found or estimated by looking for points of intersection on the graphs.

---

[1]Notice that we list the solutions first as $\left(\frac{5}{4}, \frac{5}{4}\sqrt{15}\right)$ **and** $\left(\frac{5}{4}, -\frac{5}{4}\sqrt{15}\right)$, but we write $(x, y) = \left(\frac{5}{4}, \frac{5}{4}\sqrt{15}\right)$ **or** $(x, y) = \left(\frac{5}{4}, -\frac{5}{4}\sqrt{15}\right)$. The *and* is required because there is more than one solution. The *or* is necessary because (x, y) cannot take on more than one pair of values at a time. Semantically, a similar situation arises in English: A family has two children, Robert *and* Laura. So a child in this family has the name Laura *or* the name Robert.

Figure 3 shows the graphs of each equation and also shows that there are at least two points of intersection. Calculation shows that two solutions are exactly when $x = -1$ and $x = 2$. A different window of the graph shows a third solution, which we ask you to find. In Unit 4.3, we view the original equation as being an equality of the values of two functions f and g, where $f(x) = x^3$ and $g(x) = 5x^2 - 2x - 8$.

Figure 3

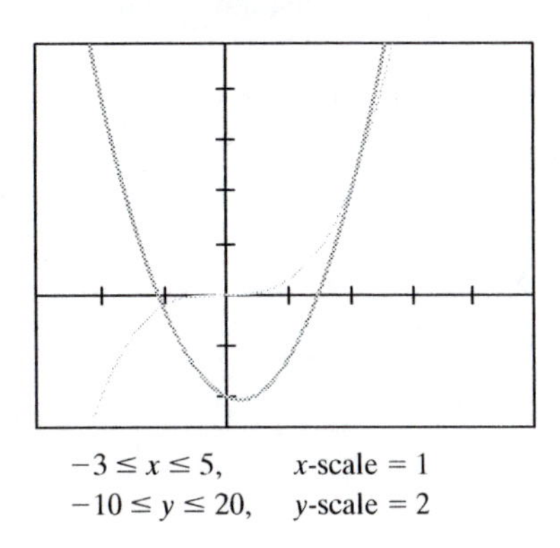

$-3 \le x \le 5$, x-scale $= 1$
$-10 \le y \le 20$, y-scale $= 2$

4.1.2 Problems

In Problems 1–3, suppose that the replacement set for the variable is the set of real numbers.

a. *Tell whether the given equation is true when considered as an existential statement.*

b. *Tell whether the given equation is true when considered as a universal statement.*

c. *Justify your answers to parts* **a** *and* **b**.

1. $(x + 10)^2 = x^2 + 100$

2. $\sin y = \cos y \cdot \tan y$

3. $\frac{10+2z}{6+z} = 2$

4. Explain how the domain of x affects the number of solutions to the equation

$$(x^4 - 1)(4x^4 - 9x^2 + 2) = 0.$$

5. Suppose that the domain of r is the set of real numbers. Is $\frac{r^6}{r^2} = r^3$ (a) an identity, (b) a false statement, or (c) neither an identity or false? Explain the reason for your answer.

6. If \$5000 is presently in an account that is yielding 4% annual interest compounded semiannually, then t years from now there will be $A = 5000(1.02)^{2t}$ dollars in the account.

a. In this situation, what is an appropriate domain for t?

b. If the word "semiannually" is changed to "continuously", what is an appropriate domain for t and formula for A?

c. Interpret the meaning of $5000(1.02)^{2t}$ when $t = -3.5$.

7. Which two equations are equivalent? Explain why the other two are not.

a. $|x| = -5$ b. $x = 5$ c. $|x| = 5$ d. $x = |5|$

8. Which two equations are equivalent? Explain why the other two are not.

a. $3A = 5B$ b. $3B = 5A$ c. $\frac{A}{B} = \frac{3}{5}$ d. $5A = 3B$

9. Which equation is not equivalent to the others? Why is it not equivalent?

a. $x^2 = 9$ b. $x = \sqrt{9}$ c. $|x| = 3$ d. $x = \pm 3$

10. In July 2001, the cost C of mailing a first-class letter weighing w ounces, $0 < w \le 13$, became $C = 34 + 23\lceil w - 1 \rceil$, where $\lceil x \rceil$ is the result when x is rounded up to the nearest integer. (The function $f: x \to \lceil x \rceil$ is called the **ceiling** or **rounding-up function**.)

a. List five solutions to the postage-cost equation $C = 34 + 23\lceil w - 1 \rceil$. Include some where w is not an integer.

b. Graph all solutions to $C = 34 + 23\lceil w - 1 \rceil$

c. Compare the graph of $C = 34 + 23\lceil w - 1 \rceil$ with the graph of $C = 34 + 23(w - 1)$ for the allowable values of w.

d. The equation $C = 34 + 23(w - 1)$ is equivalent to $C = 23w + 11$. Is $C = 34 + 23\lceil w - 1 \rceil$ equivalent to $C = 23\lceil w \rceil + 11$? Why or why not?

11. a. Write a single equation that describes the coordinates (x, y) of all points that are 4 units from $(-3, 9)$ and 6 units from $(2, 5)$.

b. Describe the graph of this equation.

12. a. Write a single equation that describes the set of all points (x, y) that are 5 units from the origin or less than 10 units from the point $(10, 0)$.

b. Describe the graph of all solutions to this equation.

13. a. Add the equations for the circles in Figure 2. Graph the result.

b. Subtract the equation of one circle in Figure 2 from an equation for the other, and graph the result.

c. Generalize parts **a** and **b** and explain why your generalization is true.

14. Graph all solutions to $(x^2 + y^2)^2 - 13(x^2 + y^2) + 36 = 0$.

15. Consider the two circles with centers (h, k) and (j, m) and radii r and s, respectively. Deduce answers to each part algebraically and geometrically.

a. Under what conditions will these circles intersect in two points?

b. Under what conditions will these circles intersect in exactly one point?

c. Under what conditions will these circles have no points in common?

16. Find the solution to the equation $x^3 = 5x^2 - 2x - 8$ that is not shown in this section.

17. Graph two equations to determine every real number whose square equals its cube.

18. a. Show that $3x^2 + 5xy - 2y^2 - 9x + 10y - 12 = 0$ is an equation for a pair of lines.

b. Under what conditions is $Ax^2 + Bxy + Cy^2 + Dx + Ey + F = 0$ an equation for a pair of lines?

c. Give a single equation for a pair of parallel oblique lines of your own choosing.

d. Give a single equation whose graph consists of the union of the x-axis, the y-axis, and the two lines that bisect the right angles formed by the axes.

19. List equations and inequalities, connected by appropriate *ands* and *ors*, whose graph resembles the face shown in Figure 4.

Figure 4

Unit 4.2 Algebraic Structures and Solving Equations

The equation $712 + x = 48$ exemplifies the simplest kind of equation to solve, for it involves exactly one unknown and one operation. And yet such equations do not always have unique solutions. For instance, in the set of real numbers $0y = 3$ has no solution, while $0y = 0$ allows any real number to be a solution. The equation $z^3 = 2$ has no solution in the set of rational numbers, one solution in the set of real numbers, and three solutions in the set of complex numbers. So the domains of the constants and the unknown are of crucial importance.

In Section 4.1.1 we mentioned the difficulties caused when an equation to be solved is found to be an identity. A similar difficulty occurs with an equation to be solved that has no solution, such as $15(3 - x) = 5(8 - 3x)$.

Why do some equations have unique solutions and others not? Can this be predicted in advance? An answer can be found by examining the relationship between operations and domains of variables.

4.2.1 Solving equations of the form $a * x = b$

In Chapter 1, we pointed out that addition and multiplication have corresponding properties. In this section we exploit that correspondence by considering both operations simultaneously. However, unlike the presentation in Chapter 1, in order to reinforce correspondence, we use the same variables for both operations.

Consider the equations $a + x = b$ and $ax = b$. They are both of the form $a * x = b$, where the $*$ can stand for any binary operation. Because $*$ is a binary operation, $*$ is well defined in the sense that equal inputs give equal results. That is, for all a, b, c, and d, if $a = b$ and $c = d$, then $a * c = b * d$.

Solving $a + x = b$ and $ax = b$

Since addition and multiplication have many of the same properties, we can analyze the solving of $a + x = b$ and $ax = b$ together.

Consider first equations of the form $a + x = b$, where x is the unknown. An equivalent equation is $x = b - a$. But is $b - a$ always a solution to the original equation? The answer to this question depends on the replacement set for the variables. For example, if we restrict a, b, and x to be positive integers, then there is no solution when $a > b$. If we restrict a, b, and x to be even integers, then there is always a unique solution. But if we restrict a, b, and x to be odd integers, then there is never a solution. Can we predict these results in advance? And what about the equation $ax = b$; when will it have a unique solution? To answer these questions, we examine the solution process for these equations in some detail.

Let us assume that the constants a and b and the unknown x are in some set S. A rigorous solution of these two equations goes through the same steps as solving the more general equation $a * x = b$.

Step 0: Given

$a + x = b$	$ax = b$	$a * x = b$

Step 1: Select the inverse of a under the operation, and apply the operation with this inverse to both sides of the equation. Thus, for this process, S must contain the inverse of a.

$-a + (a + x) = -a + b$	$\frac{1}{a}\cdot(ax) = \frac{1}{a}\cdot b$	$a^{-1} * (a * x) = a^{-1} * b$

Step 2: Use the associative property of the operation to regroup the left side.

$(-a + a) + x = -a + b$	$\left(\frac{1}{a}\cdot a\right)x = \frac{1}{a}\cdot b$	$(a^{-1} * a) * x = a^{-1} * b$

Step 3: The result of combining a and its inverse under this operation is the identity under the operation. (We name I as the identity under the operation $*$.) So for this procedure, S must contain the identity.

$0 + x = -a + b$	$1\cdot x = \frac{1}{a}\cdot b$	$I * x = a^{-1} * b$

Step 4: The result of combining the identity with x is x itself. The right side of each equation is in S if the set is closed under the operation.

$x = -a + b$	$x = \frac{1}{a}\cdot b$	$x = a^{-1} * b$

Check: We check the middle equation. (Problem 1 asks for checks of the other two.) Substitute $\frac{1}{a}\cdot b$ for x in the original equation.

$a\left(\frac{1}{a}\cdot b\right)$

$= \left(a\cdot\frac{1}{a}\right)\cdot b$ Associative property of multiplication

$= 1\cdot b$ The product of a number and its reciprocal is 1.

$= b$ The product of 1 and any number b is b.

To solve $y * a = b$, $y + a = b$ and $ya = b$ for y, then we need $I * a = a$, $a * a^{-1} = I$, and their equivalents. You are asked to solve these equations in Problem 2.

Examination of the steps and the check shows that four properties are sufficient to solve the equations $a * x = b$ and $y * a = b$. A set S and an operation $*$ that satisfy all these properties constitute the **group** $\langle \mathbf{S}, * \rangle$ and the four properties are consequently called the **group properties**. Specifically, $\langle S, * \rangle$ is a group if and only if the following four conditions are satisfied.

S is closed under the operation $*$. (closure)

$*$ is associative in S. (associativity)

There is an element I in S such that, for all a, $I * a = a * I = a$. (identity)

For all a in S, there is an element a^{-1} such that $a^{-1} * a = a * a^{-1} = I$. (inverse)

We have proved:

Theorem 4.1 Suppose $\langle S, *\rangle$ is a group. Then for all a and b in S, the equations $a * x = b$ and $y * a = b$ have unique solutions $a^{-1} * b$ and $b * a^{-1}$, respectively, in S.

For instance, $\langle$set of integers, $+\rangle$ is a group. The theorem we have proved explains why any equation of the form $a + x = b$, where a and b are integers, has exactly one solution, and that solution is an integer. For instance, $712 + x = 48$, has exactly one solution, -664. The existence of the group $\langle \mathbf{R} - \{0\}, \cdot\rangle$ also explains why there is exactly one solution to $ax = b$ when a and b are not zero.

In both these cases, the operation is commutative. So the equations $a + x = b$ and $y + a = b$ have the same solutions, as do the equations $ax = b$ and $ya = b$. If $\langle S, *\rangle$ was a group and $*$ was not commutative, then the equations $a * x = b$ and $y * a = b$ would have unique solutions but it is possible that x would not equal y.

Question 1: Explain why $\langle\{0\}$, multiplication$\rangle$ is a group.

Notice that if 0 is in a set S of complex numbers that contains an element other than 0, then the set is not a group under multiplication, because 0 has no inverse. So there are difficulties with $ax = b$ when either a or b is 0.

Question 2: Suppose a and b are real numbers. Identify all solutions to each equation.

a. $0x = 0$

b. $0x = b$ and $b \neq 0$

c. $ax = 0$ and $a \neq 0$

d. $ax = b, a \neq 0$ and $b \neq 0$

In a group, the solution to the equation $a * x = a$ or $y * a = a$ is the identity I. The solution to $a * x = I$ or $y * a = I$ is the inverse of a. So, if S is closed and $*$ is associative, the existence of unique solutions in a set S for all equations of the form $a * x = b$ also determines whether S is a group under $*$. This result, a converse to Theorem 4.1, demonstrates the intimate connection between groups and the solving of equations.

Theorem 4.2 Let S be a nonempty set that is closed under a binary operation $*$ that is associative and for which, for all a and b in S, the equations $a * x = b$ and $y * a = b$ have unique solutions in S. Then $\langle S, *\rangle$ is a group.

4.2.1 Problems

1. a. Check that $-a + b$ is a solution to $a + x = b$.

b. Check that $a^{-1} * b$ is a solution to $a * x = b$.

2. Solve $y * a = b$, indicating the properties needed at each step.

3. Suppose a and b are chosen from the given set and $*$ is the indicated operation. Does the equation $a * x = b$ always have a unique solution in the set? If not, give an example of an equation with the operation that does not have a unique solution in S.

a. set of even integers, addition

b. set of even integers, multiplication

c. set of odd integers, addition

d. set of odd integers, multiplication

4. Repeat the directions of Problem 3 for these sets and operations.

a. set of integers divisible by 7, addition

b. set of integers divisible by 7, multiplication

c. set of integers, subtraction

d. set of 2×2 matrices with rational elements, multiplication

e. set of irrational numbers, addition

ANSWERS TO QUESTIONS

1. $0 \cdot 0 = 0$, so $\{0\}$ is closed under multiplication. $0 \cdot (0 \cdot 0) = (0 \cdot 0) \cdot 0$, so the operation is associative in S. $0 \cdot 0 = 0$, so 0 is an identity for the operation and also is its own inverse.
2. a. Any real number is a solution to $0x = 0$.
b. No real number is a solution to $0x = b \neq 0$.
c. If $ax = 0$ and $a \neq 0$, then $x = 0$ (a special case of importance).
d. $x = b/a$, a unique solution

4.2.2 Solving equations of the form $ax + b = cx + d$

A **linear equation** is one that is equivalent to an equation of the form $ax + b = cx + d$, where either $a \neq 0$ or $c \neq 0$. We now examine the solving of this equation in detail. As we did in Section 4.2.1, we assume that the coefficients and x are to be from a particular set S, and we wish to know when the solution will be unique. (There are several choices for S, as you will see.)

Step 0: We begin with

$$ax + b = cx + d.$$

The purpose of Steps 1–4 is to obtain an equivalent equation with the variable x on only one side.

Step 1: Add $-cx$, the additive inverse of cx, to both sides.

$$-cx + (ax + b) = -cx + (cx + d)$$

Step 2: Apply the associative property of addition to the right side.

$$-cx + (ax + b) = (-cx + cx) + d$$

Step 3: We chose to add $-cx$ so that we could add it to its additive inverse cx to obtain the additive identity.

$$-cx + (ax + b) = 0 + d$$

Step 4: The additive identity has the property that $0 + d = d$ for all d.

$$-cx + (ax + b) = d.$$

Step 5: On the left side, use the associative property of addition.

$$(-cx + ax) + b = d$$

Step 6: Use the distributive property of multiplication over addition.

$$(-c + a)x + b = d$$

Steps 7–10 are analogous to Steps 1–4. Their purpose is to obtain an equivalent equation of the form $Ax = B$.

Step 7: Add $-b$ to both sides.

$$((-c + a)x + b) + -b = d + -b$$

Step 8: Use the associative property of addition on the left side.

$$(-c + a)x + (b + -b) = d + -b$$

Step 9: Now use the property of inverses to add b and $-b$.

$$(-c + a)x + 0 = d + -b$$

Step 10: Again we apply the additive identity property.

$$(-c + a)x = d + -b$$

Now the solution process branches.

Case 1: If $-c + a = 0$, then the problems described with $0x = \text{k}$ mentioned in Section 4.2.1 arise. There may be no solution (if $d + -b \neq 0$) or many solutions (if $d + -b = 0$).

Case 2: If $-c + a \neq 0$, then this is an equation of the form $Ax = B$ with $A \neq 0$. From Theorem 4.1, it has a unique solution, $\frac{1}{A} \cdot B$. So in this case, applying the group properties of a set with multiplication, $x = \frac{1}{-c+a}(d + -b) = \frac{d+-b}{-c+a}$. Translating into subtraction, $x = \frac{d-b}{a-c}$. Thus the only possible solution is $\frac{d-b}{a-c}$. Substituting $\frac{d-b}{a-c}$ for x in the original equation shows that this value works, so that it is the only solution.

The properties sufficient to obtain a unique general solution to $ax + b = cx + d$ in a set S are as follows:

$\langle S, +\rangle$ is a group.	(used in Steps 1–5 and 7–10)
$\cdot$ is distributive over $+$.	(used in Step 6)
$\langle S - \{0\}, \cdot\rangle$ is a group.	(used in Steps Case 2)

(Remember that the symbol $S - \{0\}$ signifies the set S without the element 0.) These 9 properties (4 for each group and 1 for distributivity) are precisely the defining properties of a **field** $\langle \mathbf{S}, +, \cdot\rangle$. The properties of the additive and multiplicative groups, the commutative properties of addition and multiplication, and the distributive property of multiplication over addition are consequently called **field properties**. The following theorem is a consequence of the definition of field and the analysis we gave to solving the equation $ax + b = cx + d$.

Theorem 4.3 If a, b, c, and d are in a set S, $\langle S, +, \cdot\rangle$ is a field, and $a \neq c$, then the equation $ax + b = cx + d$ has exactly one solution, $\frac{d-b}{a-c}$, in S. If $a = c$, then $ax + b = cx + d$ has no solution if $d \neq b$ and any element of S as a solution if $d = b$.

The idea of the theorem is simple: In a field, all linear equations can always be solved without going outside the field. The systems $\langle \mathbf{R}, +, \cdot\rangle$ of real numbers and $\langle \mathbf{C}, +, \cdot\rangle$ of complex numbers are fields. To determine whether a subset S of $\mathbf{R}$ or of $\mathbf{C}$ with these operations is a field, since $+$ and $\cdot$ of complex numbers are associative, and the distributive property holds, all we need check are that S is closed under $+$ and $\cdot$ and that S contains 0, 1, opposites, and inverses of its elements.

Thus, for the set $\mathbf{Q}$ of rational numbers, $\langle \mathbf{Q}, +, \cdot\rangle$ is a field, because 0 and 1 are rational, and the sums, products, opposites, and inverses of rationals are rational. Consequently, even without solving, Theorem 4.3 tells us that there is exactly one solution to $\frac{35}{92}x + 2.46 = -8x + \frac{3}{7}$, and that solution is a rational number.

Many common subsets of the reals do not form fields with addition and multiplication, including the set of integers, the set of even integers, the set of irrationals, and the set of primes. For instance, the equation $4x + 6 = 12x + 14$, in which all the coefficients and constants are all even, has a solution that is not even.

Matrix multiplication and systems

The equations treated above are linear equations, but Theorems 4.1–4.3 have broader applicability. The example we now examine involves matrix multiplication, which is not a commutative operation. We noted in Section 4.1.2 that a system of equations can be thought of as a single compound sentence. If the equations are linear, then the system can be represented as a single matrix equation. For instance, the system (1) of two linear equations in two unknowns is equivalent to the matrix equation (2).

$$(1)\ \begin{cases} ax + by = e \\ cx + dy = f \end{cases} \qquad (2)\ \begin{bmatrix} a & b \\ c & d \end{bmatrix}\begin{bmatrix} x \\ y \end{bmatrix} = \begin{bmatrix} e \\ f \end{bmatrix}$$

The solution of such systems or their matrix counterparts is studied in courses on linear algebra, and we shall not detail them here. However, an analogy to the solving of $ax = b$ is worth mentioning. For this, we examine multiplication of 2×2 matrices.

Recall that a **matrix** is a rectangular array. A matrix with m rows and n columns is said to have **dimension** $m \times n$. (Rows go across; columns go down.) Two matrices are **equal** if and only if each of the mn pairs of corresponding elements are equal. Multiplication of matrices is a row-by-column multiplication, by which we mean that to multiply two matrices A and B each row in A is combined with a column in B to obtain a single number in the product matrix AB. Specifically, the element in the ith row and jth column of AB is found by combining the ith row of A with the jth column of B. This combining is done by multiplying corresponding elements in the row and column and adding the products. Consequently, in order to be combined, each row of A and column of B must have the same number of elements. So A must have as many *columns* as B has *rows*. Another way to put this is that AB exists only if the dimensions of A are $m \times n$ and the dimensions of B are $n \times r$, and then the dimensions of AB are $m \times r$. In system (2) above a 2×2 matrix is multiplied by a 2×1 matrix to obtain another 2×1 matrix. The two elements of the product matrix are found by the multiplications and addition identified in the two equations on the left.

The definition of matrix multiplication means that 2×2 matrices can be multiplied in either order (but do not necessarily give the same result). Specifically, for all a, b, c, d, e, f, g, and h,

$$\begin{bmatrix} a & b \\ c & d \end{bmatrix}\begin{bmatrix} e & f \\ g & h \end{bmatrix} = \begin{bmatrix} ae + bg & af + bh \\ ce + dg & cf + dh \end{bmatrix}.$$

So the set of 2×2 matrices is closed under multiplication. Also, matrix multiplcation is associative (see Problem 6), because matrices act like functions, as can be seen in Section 7.2.2. To find an identity, we solve

$$\begin{bmatrix} a & b \\ c & d \end{bmatrix}\begin{bmatrix} w & x \\ y & z \end{bmatrix} = \begin{bmatrix} a & b \\ c & d \end{bmatrix}$$

for w, x, y, and z. We obtain

$$\begin{cases} aw + by = a \\ cw + dy = c \end{cases} \quad \text{and} \quad \begin{cases} ax + bz = b \\ cx + dz = d \end{cases}.$$

Solving these systems, $w = 1$, $x = 0$, $y = 0$, $z = 1$. So the only possible identity I is

$$I = \begin{bmatrix} 1 & 0 \\ 0 & 1 \end{bmatrix}.$$

You should convince yourself that for any 2×2 matrix A, $A \cdot I = I \cdot A = A$.

To find the (multiplicative) inverse of $A = \begin{bmatrix} a & b \\ c & d \end{bmatrix}$, solve

$$\begin{bmatrix} w & x \\ y & z \end{bmatrix}\begin{bmatrix} a & b \\ c & d \end{bmatrix} = \begin{bmatrix} 1 & 0 \\ 0 & 1 \end{bmatrix}.$$

Provided that $ad - bc \neq 0$, you should find (see Problem 7) that

$$\begin{bmatrix} w & x \\ y & z \end{bmatrix} = \begin{bmatrix} \dfrac{d}{ad - bc} & \dfrac{-b}{ad - bc} \\ \dfrac{-c}{ad - bc} & \dfrac{a}{ad - bc} \end{bmatrix}.$$

So the multiplicative inverse of A, written $\mathbf{A^{-1}}$, exists only if $ad - bc \neq 0$, and then

$$\underset{A^{-1}}{\begin{bmatrix} \dfrac{d}{ad - bc} & \dfrac{-b}{ad - bc} \\ \dfrac{-c}{ad - bc} & \dfrac{a}{ad - bc} \end{bmatrix}} \cdot \underset{A}{\begin{bmatrix} a & b \\ c & d \end{bmatrix}} = \underset{I.}{\begin{bmatrix} 1 & 0 \\ 0 & 1 \end{bmatrix}}$$

This yields a theorem that is analogous to Theorems 4.1 and 4.2.

Theorem 4.4 With multiplication, the set of 2×2 matrices $\begin{bmatrix} a & b \\ c & d \end{bmatrix}$ with $ad - bc \neq 0$ is a group, and every equation of the form $\begin{bmatrix} a & b \\ c & d \end{bmatrix} \cdot X = \begin{bmatrix} e & f \\ g & h \end{bmatrix}$ has a unique solution.

Now let us return to the matrix equation

$$\begin{bmatrix} a & b \\ c & d \end{bmatrix}\begin{bmatrix} x \\ y \end{bmatrix} = \begin{bmatrix} e \\ f \end{bmatrix}$$

which we can write as

$$A \cdot X = B.$$

If $ad - bc \neq 0$, there exists A^{-1} and we can multiply both sides by A^{-1}. We now follow Steps 1–4 of Section 4.2.1.

$$\begin{aligned} A^{-1} \cdot (A \cdot X) &= A^{-1} \cdot B \\ (A^{-1} \cdot A) \cdot X &= A^{-1} \cdot B \\ I \cdot X &= A^{-1} \cdot B \\ X &= A^{-1} \cdot B \end{aligned}$$

This tells us that there will always be a unique solution to the system $\begin{cases} ax + by = e \\ cx + dy = f \end{cases}$ when $ad - bc \neq 0$. (Geometrically, the lines $ax + by = e$ and $cx + dy = f$ are not parallel, so they intersect in exactly one point.) When $ad - bc = 0$, there is not a unique solution. Just as in $0x = b$, either there is no solution or there are infinitely many solutions.

Question: What is the geometry of the situation when $ad - bc = 0$?

These results generalize to higher-dimensional $n \times n$ systems.

4.2.2 Problems

1. None of these sets, with the usual operations of + and •, forms a field. Create an equation of the form $ax + b = cx + d$, where a, b, c, and d are in the set but whose solution does not lie in the set.

a. set of integers

b. set of square roots of nonnegative integers

c. set of complex numbers of the form ki

d. set of integer powers of 2

2. Prove: The set of numbers of the form $m + n\sqrt{2}$, where m and n are rational numbers, with the usual operations of + and •, is a field.

3. Solve in the field of Problem 2: $(3 + 4\sqrt{2})x + (5 - 6\sqrt{2}) = (2 + \sqrt{2})x + \sqrt{32}$.

4. Explain why the set of numbers of the form $a + b\sqrt{2} + c\sqrt{3}$, where a, b, c, and d are rational numbers, with the usual operations of + and •, is not a field.

5. The salary scales in three school districts are as follows, for a teacher with a master's degree:

District P: \$30,000 plus \$1500 for each year of experience

District Q: \$30,000 plus \$1750 for each year of experience

District R: \$28,000 plus \$1750 for each year of experience

a. Give a formula for the salary in each district for a teacher with n years of experience.

b. Use your formulas to indicate the number of years experience teachers have in Districts P and R when they earn the same salary.

c. Use your formulas to indicate the number of years experience teachers have in Districts P and Q when they earn the same salary.

d. Use your formulas to indicate the number of years experience teachers have in Districts Q and R when they earn the same salary.

e. If in District T, teachers earn a salary of S_T dollars plus E_T dollars for each year of experience, and in District U, teachers earn a salary of S_U dollars plus E_U dollars for each year of experience, $S_T > S_U$, and $E_U > E_T$, how many years will it take District U teachers to catch up to District T?

6. Let $A = \begin{bmatrix} a & b \\ c & d \end{bmatrix}$, $M = \begin{bmatrix} e & f \\ g & h \end{bmatrix}$, and $X = \begin{bmatrix} w & x \\ y & z \end{bmatrix}$. Show that $(A \cdot M) \cdot X = A \cdot (M \cdot X)$, and thus prove that multiplication of 2×2 matrices is associative.

7. a. Show steps in solving the systems

$$\begin{cases} wa + xc = 1 \\ ya + zc = 0 \end{cases} \text{ and } \begin{cases} wb + xd = 0 \\ yb + zd = 1 \end{cases}$$

for w, x, y, and z in terms of a, b, c, and d, without using matrices.

b. Explain why part **a** determines the multiplicative inverse of a 2×2 matrix.

8. Use 2×2 matrices to solve each system for (x, y).

a. $\begin{cases} 2x - 5y = 11 \\ 4x - 15y = -4 \end{cases}$

b. $\begin{cases} ax + by = e \\ cx + dy = f \end{cases}$, assuming $ad - bc \neq 0$.

9. Suppose $ad - bc = 0$, $c \neq 0$, and $d \neq 0$. What can be deduced about $\frac{a}{c}$ and $\frac{b}{d}$?

ANSWER TO QUESTION

The lines $ax + by = e$ and $cx + dy = f$ are parallel and either have no points in common or they coincide.

4.2.3 Quadratic and other polynomial equations

The simplest quadratic equation, $x^2 = k$, describes the relationship between the area k of a square and a side x of the square. Other area formulas, such as $A = \pi r^2$, lead to equations of the form $ax^2 = b$.

Problems of counting can lead to quadratics. The number of games needed by n teams to play each other exactly once is $\frac{n^2-n}{2}$. This is also the sum of the integers from 1 to $n - 1$, since the nth team plays the other $n - 1$ teams, the $(n - 1)$st team plays the $n - 2$ teams that are left, and so on until the 2nd team plays the 1 team that is left. This is also the $(n - 1)$st triangular number, i.e., the $(n - 1)$st term in the sequence 1, 3, 6, 10, ... in which differences increase by 1.

More generally, if successive differences in a sequence increase by a constant amount, we saw in Section 3.3.2 that a formula for the nth term will be of the 2nd degree. This is what happens to the velocity of a heavier-than-air object dropped straight down—its velocity increases each second by approximately $32\frac{\text{ft}}{\text{sec}}$ or $9.8\frac{\text{m}}{\text{sec}}$. That is why we call $\frac{32\frac{\text{ft}}{\text{sec}}}{\text{sec}}$ or $\frac{9.8\frac{\text{m}}{\text{sec}}}{\text{sec}}$ acceleration due to gravity. Thus a formula for the height of an object that is dropped (or shot up) into the air is a quadratic expression.

Quadratics also arise as special cases of discrete exponential growth (or decay) when the exponent is 2. For instance, P dollars invested at an annual yield of r will grow to $A = P(1 + r)^2$ after 2 years.

Quadratics are also intimately related to distance through the Pythagorean Theorem $a^2 + b^2 = c^2$ and the Euclidean distance formula $d(P, Q) = \sqrt{(x_1 - x_2)^2 + (y_1 - y_2)^2}$. Consequently, some curves defined in terms of distance, such as the circle, ellipse, parabola, and hyperbola (the conic sections) are described by quadratic equations. In 3 dimensions, the distance formula $d(P, Q) = \sqrt{(x_1 - x_2)^2 + (y_1 - y_2)^2 + (z_1 - z_2)^2}$ leads to surfaces such as spheres, paraboloids, etc., being described by quadratic equations.

The variety of applications of quadratics, together with their interesting mathematical properties, indicate why quadratic equations are an important topic of study.

An ancient problem leading to a quadratic equation

There is evidence that the Egyptians solved quadratic equations as early as 2000 B.C. The ancient Babylonians used quadratic equations in calculating prices for their wares, and we can trace the origins of the quadratic formula to their calculations. They also attempted to find a process of working with figures in such a way that predetermined solutions were found. For example, 650 systems of Babylonian equations have been found, all of which have the exact solution $x = 30$, $y = 20$.

Certain systems led to quadratic equations. The following problem is posed and solved on a Babylonian tablet dated around 1760 B.C.

> Find two numbers whose sum is 10 and whose product is 18.

The solution given on the tablet can be translated and generalized to yield a solution to any quadratic equation in one variable.

First, let us see how the problem leads to a quadratic equation. Suppose the numbers are m and n. Then the problem is to solve the system

$$\text{(1)} \qquad \begin{cases} m + n = 10 \\ mn = 18. \end{cases}$$

Solving the second equation for m, and substituting back in the first equation, $\frac{18}{n} + n = 10$. Since $n \neq 0$, we can multiply both sides by n to obtain $18 + n^2 = 10n$. This is a quadratic equation in n. Put in the **standard form** $ax^2 + bx + c = 0$, the problem above is equivalent to solving the quadratic equation

$$\text{(2)} \qquad n^2 - 10n + 18 = 0.$$

The visible similarity of the numerical coefficients in the system (1) and the quadratic equation (2) is not coincidence. The two solutions to (2) are the numbers whose sum is 10 and product is 18. The general result is named after François Viète (or Vieta), the French mathematician who in the 1590s was the first to use variables and constants as we do today in algebra and, in so doing, discovered this theorem.[2]

Theorem 4.5 **(Viète's Theorem):** If m and n are solutions to $x^2 + bx + c = 0$, then

a. $m + n = -b$ and **b.** $mn = c$.

Proof: We prove the theorem for the case when m and n are distinct solutions to $x^2 + bx + c = 0$.

Then $\quad m^2 + bm + c = 0 \quad$ and $\quad n^2 + bn + c = 0.$

Thus $\quad m^2 + bm + c = n^2 + bn + c,$

from which $\quad m^2 - n^2 + bm - bn = 0.$

The left side can be factored.

$$(m - n)(m + n + b) = 0$$

So either $m - n = 0$ or $m + n + b = 0$. Since $m \neq n$, $m + n = -b$. This proves part (a). To show (b), since $m^2 + bm + c = 0$, then $c = -m^2 - bm = m(-m - b)$. But $n = -m - b$. So $mn = c$. ∎

Viète stated his theorem for positive m and n because he did not consider negative solutions, but his proof holds for any distinct solutions. In fact, from the Quadratic Formula, it is easy to show that the relationships in Theorem 4.5 hold for any solutions to a given quadratic of the form

$$x^2 + bx + c = 0. \tag{3}$$

To apply Viète's Theorem to the familiar quadratic equation $ax^2 + bx + c = 0$, first divide both sides of the equation by a. This can be done because $a \neq 0$ when the equation is quadratic. You should prove the following corollary (see Problem 2).

Corollary: m and n are two solutions to the quadratic equation $ax^2 + bx + c = 0$ if and only if $m + n = \frac{-b}{a}$ and $mn = \frac{c}{a}$.

Neither Theorem 4.5 nor its corollary exhibits the solutions to a quadratic equation. Still, they are powerful, because they can lead to the solutions to any quadratic equation.

Solving the original problem

First we apply Theorem 4.5 to the original problem or, equivalently, the system (1). Since the two numbers m and n add to 10, their average is 5. So we can let $m = 5 + x$ and $n = 5 - x$. Since $mn = 18$,

$$(5 + x)(5 - x) = 18$$

from which

$$25 - x^2 = 18.$$

So $x^2 = 7$, and $x = \pm\sqrt{7}$. Thus $m = 5 + \sqrt{7}$ and $n = 5 - \sqrt{7}$, or vice versa.

[2]Victor Katz, *A History of Mathematics*, p. 345.

Solving the quadratic $x^2 + bx + c = 0$

Now we apply Theorem 4.5 to a more general problem. Let m and n be the two solutions of the equation (3). From the theorem, $m + n = -b$. So their average is $\frac{-b}{2}$. Consequently,

$$\text{let } m = \frac{-b}{2} + x \quad \text{and} \quad n = \frac{-b}{2} - x. \tag{4}$$

Since (again from Theorem 4.5) $mn = c$,

$$\left(\frac{-b}{2} + x\right)\left(\frac{-b}{2} - x\right) = c.$$

$$\frac{b^2}{4} - x^2 = c \tag{5}$$

$$x^2 = \frac{b^2}{4} - c$$

$$x = \sqrt{\frac{b^2}{4} - c} \quad \text{or} \quad x = -\sqrt{\frac{b^2}{4} - c}.$$

Now we substitute back into the two equations in (4). Which solution should be substituted in which equation? There are four possible substitutions but the same two different solutions arise for each

$$m = \frac{-b}{2} \pm \sqrt{\frac{b^2}{4} - c} \quad \text{and} \quad n = \frac{-b}{2} \mp \sqrt{\frac{b^2}{4} - c}. \tag{6}$$

(The symbol $\mp$ is only used together with the $\pm$ symbol. It signifies that when the $+$ aspect of the $\pm$ symbol is chosen for m, the $-$ aspect of the $\mp$ symbol applies to n, and when the $-$ aspect of the $\pm$ symbol is chosen for m, the $+$ aspect of the $\mp$ symbol applies to n. That is, the upper aspects and lower aspects of these symbols go together.) The equations (6) provide a formula [Theorem 4.6a] that enables any quadratic equation to be solved. The formula in Theorem 4.6b is a variant taught in many countries. The formula in Theorem 4.6c is the familiar variant taught in the United States.

Theorem 4.6 **(Quadratic Formula):**

a. For all real numbers b and c,

$$x^2 + bx + c = 0 \Leftrightarrow x = \frac{-b}{2} \pm \sqrt{\frac{b^2}{4} - c}.$$

b. For all real numbers b and c,

$$x^2 + 2bx + c = 0 \Leftrightarrow x = -b \pm \sqrt{b^2 - 4c}.$$

c. For all real numbers a, b, and c, with $a \neq 0$,

$$ax^2 + bx + c = 0 \Leftrightarrow x = \frac{-b \pm \sqrt{b^2 - 4ac}}{2a}.$$

In Theorem 4.6, any of the expressions under the radical sign $\sqrt{\ }$ can be negative. When this occurs, the quadratic has nonreal solutions. For example, for the quadratic $3x^2 - 4x + 2 = 0$, $a = 3$, $b = -4$, and $c = 2$, and by Theorem 4.6(c), $x = \frac{-(-4) \pm \sqrt{(-4)^2 - 4(3)(2)}}{2(3)} = \frac{4 \pm \sqrt{-8}}{6}$. At this point, a person who knows very little about complex numbers can still see that the solutions are not real, and may even realize that the solutions can be rewritten as $\frac{2\pm\sqrt{-2}}{3}$. With the additional knowledge that $\sqrt{-2} = \sqrt{-1}\cdot\sqrt{-2} = i\sqrt{2}$, the solutions can be written in $a + ib$ form as $\frac{2}{3} \pm \frac{i\sqrt{2}}{3}$. By arithmetic on the two nonreal solutions, we see that their sum is $\frac{4}{3}$ and their product is $\frac{2}{3}$, as predicted by the corollary to Theorem 4.5.

At the beginning of Unit 2.2, we noted that some mathematicians of the 16th and 17th centuries encountered and dealt with complex numbers in exactly this way. Their general methods for solving certain equations would lead them to numbers that were not real. They performed arithmetic operations on these numbers using as many of the properties of real numbers as they could (the only basic property they could not use is that $\sqrt{rs} = \sqrt{r}\cdot\sqrt{s}$, since it is not true when both r and s are negative), and they found that properties of the solutions such as Viète's Theorem held for these nonreal numbers.

The number inside the radical sign $\sqrt{\ }$ in any of the variants of the Quadratic Formula in Theorem 4.6 is called the **discriminant** because its value determines how many real solutions the equation has. It is easy to see that, if the coefficients of the quadratic equation are real numbers, then there are 2, 1, or 0 real solutions accordingly as the discriminant is greater than, equal to, or less than 0.

Later mathematicians began to consider quadratic equations whose *coefficients* were not real. They found that the Quadratic Formula, with a suitable interpretation of square roots of complex numbers, $\sqrt{a + bi}$, worked equally well when coefficients were complex. However, when the coefficients are not real, the value of the discriminant no longer determines the number of real solutions.

A geometric picture

Graphing the system (1) is quite instructive. We show a graph in Figure 5. The two points of intersection of the hyperbola and the line are the two solutions to the system. The graph displays the symmetry of the solutions. Either $5 + \sqrt{7}$ or $5 - \sqrt{7}$ can be chosen for m, and then the other of the pair is the value of n. The graph also suggests that if values other than 10 and 18 had been chosen for the sum and product, then it is possible that there would be no real solution.

Question: Consider the system $\begin{cases} m + n = k \\ mn = 18 \end{cases}$, where m and n are real numbers.

1. Find all real numbers k for which there is exactly one real solution to the system.
2. Describe the real numbers k for which there are no real solutions to the system.

Figure 5

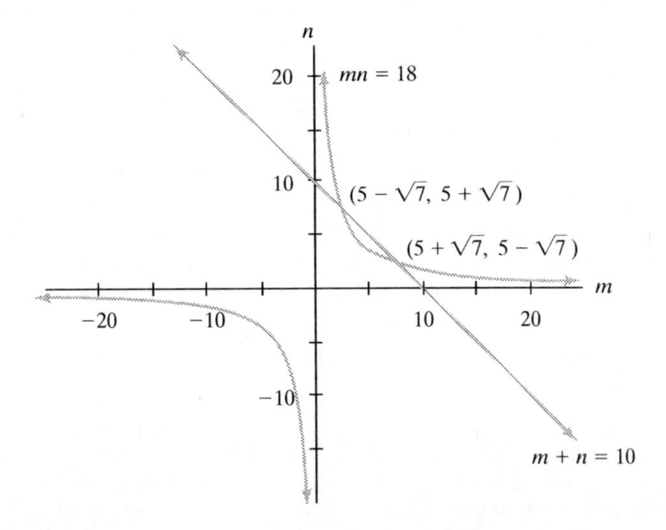

The limiting values of k in the Question are those for which the line $m + n = k$ is tangent to the hyperbola $mn = 18$. That occurs when $m = n$, so it occurs when $m = n = \pm\sqrt{18}$. So the values of k for which there is exactly one real solution to the system are $\pm 2\sqrt{18}$. Between these values, the line does not intersect the hyperbola and there is no solution. That is, for $-2\sqrt{18} < k < 2\sqrt{18}$, there is no real solution. You should verify these results algebraically (see Problem 4).

Cubic and quartic equations

A **cubic equation** is an equation of the form $P(x) = Q(x)$, where $P(x)$ and $Q(x)$ are polynomials of degree ≤ 3, and one or the other has degree 3. Every cubic equation in one variable x is equivalent to an equation of the form

$$ax^3 + bx^2 + cx + d = 0. \tag{7}$$

Attempts to solve cubic equations date back thousands of years. The Greek mathematicians, including Menaechmus (c. 350 B.C.) and Archimedes (287–212 B.C.), were able to solve specific cubic equations using geometric reasoning about volumes of cubes. The Persian astronomer and mathematician Omar bin Ibrahim al-Khayyami (c.1048–1122) demonstrated processes for solving thirteen different types of cubic equations.[3] Omar needed this many types because his geometric solutions did not allow the coefficients of the original equation to be negative. So he would not consider an equation such as $2x^3 - 5x^2 - x + 6 = 0$, but he would solve what today we view as an equivalent equation, $2x^3 + 6 = 5x^2 + x$.

We noted at the beginning of Unit 2.2 that Girolamo Cardano, building on the work of his predecessors, first published (in 1545) a solution to the general cubic equation (7). Cardano began by providing a rule for solving a specific example of a cubic equation (like the Greeks, using a geometric approach). Having found one solution to the cubic, he was then in the position to find others. Cardano's formula for cubics is an analogue of the quadratic formula. It provides a general formula in terms of coefficients, which is applicable to all polynomials of degree 3.

The solution of the cubic was a decisive step in the development of the theory of polynomial equations, and one which led to other important discoveries. Very shortly after the solution to the cubic, Cardano's student Ludovico Ferrari (1522–1565) then derived a general formula for **quartic equations** (i.e., polynomial equations of degree 4) from Cardano's result for the cubic.

Higher-degree polynomial equations

The speed with which Ferrari discovered a general solution for 4th-degree polynomial equations led mathematicians to search for formulas that would describe solutions for polynomial equations of higher degrees allowing a finite number of arithmetic operations and rational powers of their coefficients. Within the next hundred years, Marin Mersenne (1588–1648), René Descartes (1596–1650), and Pierre de Fermat (1601–1665) modernized the theory, promoting the standardization of algebraic notation. Descartes discovered how to lower the degree of an equation if a solution for it were known, and how to remove a term of degree $n - 1$ in an equation of degree n. He also invented the famous rule of signs (see Chapter Project 3) which tells us where to find the solutions of an equation. But for over two centuries no one was able to obtain a general formula even for solutions to 5th-degree polynomial equations. It was not even known whether all polynomial equations of higher degrees had solutions in the complex numbers.

---

[3]Omar is more widely known today as the poet Omar Khayyam than as an astronomer or mathematician. His popularity arose from the publication of the *Rubaiyat of Omar Khayyam*, a collection and translation of about a thousand 4-line poems (rubai) attributed to Omar by Edward Fitzgerald in 1859. It is now thought that Omar wrote perhaps 200 to 250 of these poems himself and collected the others.

The question of whether or not every polynomial equation has a solution was answered by Gauss in 1799 when he proved that any polynomial equation with real coefficients has at least one solution in the complex numbers, a theorem we now call the Fundamental Theorem of Algebra (see Section 6.2.3). In 1848, Gauss published his fourth proof of this theorem and allowed the coefficients to be complex.

A major step toward resolving the difficulty of solving higher-degree polynomial equations was taken by the French mathematician Joseph Louis Lagrange (1736–1813). In work published in 1770, Lagrange was able to express solutions to cubic and quartic equations in terms of permutations of solutions to reduced equations. His work, which today can be described in the language of groups of permutations, provided a foundation for the later efforts on the problem.

In 1798, Paolo Ruffini, an Italian mathematician and physician, thought he had proved that a general formula for solutions of polynomial equations of degree higher than 4 does not exist. Ruffini's proof was flawed, and his discovery (although correct) was largely ignored by his contemporaries. It wasn't until 1824 that the Norwegian mathematician Niels Abel (1802–1829) independently produced and correctly proved that no algebraic formula exists for the general fifth-degree polynomial equation in terms of its coefficients. This showed that there does not exist such a formula for the general polynomial equation of any degree greater than 4.[4]

In 1831, the French mathematician Evariste Galois (1811–1832) used group theory to establish necessary and sufficient conditions under which a polynomial equation is solvable allowing a finite number of arithmetic operations and rational powers of its coefficients. This work showed why third- and fourth-degree polynomial equations could be solved and confirmed Abel's results for higher-degree polynomial equations.

Abel's result excludes only the possibility of a *general* formula for higher-degree polynomial equations. We can still find solutions for specific ones. We can also approximate solutions using computers, or use successive approximations to solutions applying methods such as the method of tangents discovered by Isaac Newton (1642–1727) in 1669.

[4]It is possible to read Abel's proof in David E. Smith's *A Source Book in Mathematics* (Dover reprint, New York, 1959), or to see a computational proof of insolvability, using the ideas of Abel, in "On the Nonsolvability of the General Polynomial," by R.G. Ayoub, in *The American Mathematical Monthly* 89, 1982, pages 397–402.

4.2.3 Problems

1. a. Find two numbers whose sum is -10 and whose product is 5.
   b. Find two numbers whose sum is 1 and whose product is 1.
   c. Find the dimensions of a rectangle whose area is 60 cm^2 and whose perimeter is 200 cm.

2. Prove the Corollary to Theorem 4.5.

3. a. Show that Theorem 4.6b follows from Theorem 4.6a.
   b. Show that Theorem 4.6c follows from Theorem 4.6a.

4. Show algebraically that when $-2\sqrt{18} < k < 2\sqrt{18}$, the line with equation $x + y = k$ does not intersect the hyperbola $xy = 18$.

5. Determine the positive difference of the solutions to the quadratic equation $ax^2 + bx + c = 0$.

6. a. If m and n are the solutions to $ax^2 + bx + c = 0$, find a formula for $m^2 + n^2$ in terms of a, b, and c.
   b. If m and n are the solutions to $ax^2 + bx + c = 0$, find a formula for $m^3 + n^3$ in terms of a, b, and c.

7. Let r_1, r_2, and r_3 be the three solutions to $ax^3 + bx^2 + cx + d = 0$.
   a. Prove that $r_1 + r_2 + r_3 = -\frac{b}{a}$.
   b. Prove that $r_1r_2r_3 = -\frac{d}{a}$.
   c. Find an expression for $r_1r_2 + r_1r_3 + r_2r_3$ in terms of a, b, c, and/or d.
   d. Extend the results of parts **a**, **b**, and **c** in some way to equations of higher degree.

Unit 4.3 The Solving Process

In Unit 4.2, a major focus of discussion was on determining the number of solutions of linear and quadratic equations from information about the coefficients of the variables in those equations. In this unit, we expand the scope of the discussion to include a wide variety of equations—including those involving powers and roots, exponents and logarithms, and trigonometric functions. We also examine some inequalities. Our focus is on process and, unless otherwise stated, we are interested only in those solutions that are real numbers.

Some processes are so general that they can be employed in almost all cases. For instance, if an equation has a single unknown, then the equation has the form $f(x) = g(x)$. If f and g are continuous functions or otherwise reasonably well-behaved, we can graph them as discussed in Section 4.1.2 and read solutions off the graph. We can also generate tables of values of f and g and close in on solutions to a desired degree of accuracy.

Although these general processes do not tend to give exact values except if we know in advance that solutions are integers or other familiar numbers, these processes are nonetheless quite powerful because of their scope. Methods for finding exact solutions do not exist even for equations as simple as $\log x = x^2 - 2$ or $e^t = 5 + t$ or $\sin\theta = \frac{\theta}{10}$ (with θ in radians). For such equations approximations must be used.

In fact, there is a sense in which school mathematics misleads students into believing that there exist exact methods to solve all equations because only those equations for which exact methods exist are discussed. It is more accurate to say that there exist general methods that provide approximate solutions to a wide variety of equations (such as the three mentioned above) and special methods that provide exact solutions to a smaller but a still large and important set of equations.

4.3.1 Generalized addition and multiplication properties of equality

Procedures involving the solving of what we now call linear equations have been found in Sumerian tablets, Egyptian papyri, and Chinese writing whose origins date back 3000 to 4000 years. Around 825 A.D., the Arab mathematician Ja'far Mohammed Ben Musa, more popularly known today as al-Khowarizmi (the man from Khwarazm), wrote a treatise entitled "al-jabr w'al muqabalah", which is Arabic for "the consolidation and the balancing", or "the reduction and the setting equal". The title refers to the processes of solving a problem by combining like terms and transposing a term to the opposite side of an equation by changing its sign. When European mathematicians learned of this work several hundred years later, they referred to it by the first word of the title, "al-jabr", which became written as "algeber" by the English mathematician Robert Recorde in the 16th century but has come down to us as our word *algebra*.

The Arab mathematicians used a mechanical procedure still found in some algebra classes. To solve

$$8D - 41 = 6D + 12,$$

the adding of $6D$ on the right side would be changed to subtracting $6D$ from the left side,

$$8D - 41 - 6D = 12,$$

the subtraction of 41 on the left side would be replaced by adding 41 on the right side,

$$8D - 6D = 41 + 12,$$

the terms would be collected,

$$2D = 53,$$

and lastly, the 2 multiplied on the left would be changed to dividing by 2 on the right.

$$D = \frac{53}{2}$$

This procedure, when generalized, gives an algorithm for the solution of linear equations (and the word "algorithm" is named after al-Khowarizmi). But it does not indicate *why* the procedures work. With the discovery of number systems in which not all properties of real numbers hold, it became important to identify the properties that enabled equations to be dealt with in a systematic manner. Thus, over the centuries, the word "algebra" has become applied also to the structures of the systems in which equations are solved. This is one reason why the groups and fields discussed in Unit 4.2 are considered a part of the branch of mathematics called *algebra* and are studied in courses often titled *higher algebra* or *abstract algebra*.

Until 1960, school mathematics tended not to use the language of these structures.[5] With the new math texts of the 1960s, using the field properties became a unifying idea in the solving of equations, and more careful attention was paid to the logic behind each step. In this section, we examine the reasoning in some detail.

Linear equations can be solved by adding the same quantities to both sides or multiplying both sides by the same nonzero quantity. So students often think that in any equation they can always add the same quantity to both sides, or multiply both sides by the same quantity without affecting the solutions. We shall see that this is not always the case.

Adding the same quantity to both sides

Nothing seems more harmless than adding the same quantity to both sides of an equation. There is, after all, the **Addition Property of Equality**: If $a = b$, then $a + c = b + c$.

But consider finding all real solutions to

$$x + \log(1 - x) = 4 + \log(1 - x). \tag{1}$$

It seems natural to add $-\log(1 - x)$ to each side. The result is

$$x = 4. \tag{2}$$

Yet substitute 4 for x in equation (1) and you will see that 4 is not a solution. When $x = 4, 1 - x = -3$, and there is no real log of -3.

This example points out the importance of the domains of the functions involved in an equation. Think of equation (1) as

$$f(x) = g(x),$$

[5] Examination of first-year algebra texts from five different publishers published over the years 1936–1956 showed that two texts mentioned the commutative property of addition, one text mentioned the commutative property of multiplication and the associative properties of multiplication and addition, and a different text identified the "rule of distribution" when multiplying a monomial by a polynomial. No text indicated that 0 and 1 were additive or multiplicative identities and, consequently, no text could discuss inverses under these operations. Four of the texts had basic rules allowing the same number to be added to or subtracted from both sides of an equation, and allowing an equation to be multiplied by the same number or divided by the same nonzero number. The fifth text dealt with these rules by asserting that equals could be added to equals, subtracted from equals, etc.

with $f(x) = x + \log(1-x)$ and $g(x) = 4 + \log(1-x)$. Let $h(x) = -\log(1-x)$. In the intersection of the domains of the functions f, g, and h, we can add $h(x)$ to both sides and obtain the equivalent equation

$$f(x) + h(x) = g(x) + h(x).$$

If the domains of f, g, and h are not the same, then solutions may be lost [as can be seen by adding $h(x) = \log(1-x)$ to both sides of the equation $x = 4$] or gained [as can be seen by adding $h(x) = -\log(1-x)$ to both sides of the equation $x^2 + \log(1-x) = 4 + \log(1-x)$]. The intersection of the domains of f and g in the above example is $\{x: x < 1\}$, and on this domain there is no solution either to the original equation (1) or the resulting equation (2).

Theorem 4.7 For all values of x in the intersection of the domains of f, g, and h,

$$f(x) = g(x) \Leftrightarrow f(x) + h(x) = g(x) + h(x).$$

Multiplying both sides by the same quantity

The above example with addition suggests being careful also with multiplication. Here are two examples in which we multiply both sides by the same quantity. Consider the equation

$$x^2 = 5x, \tag{3}$$

which has solutions 0 and 5. Multiply both sides by $\frac{1}{x}$ and the result is the equation

$$x = 5, \tag{4}$$

which clearly has only one solution.

Or consider the equation

$$y = 2. \tag{5}$$

Since y equals 2, you might think that you could multiply the left side of (5) by y and the right side by 2 and get an equivalent equation. But the resulting equation,

$$y \cdot y = 2 \cdot 2,$$

$$y^2 = 4, \tag{6}$$

has two solutions, 2 and -2, whereas the original equation (5) had only the single solution 2. Clearly we must be careful in what we do, or we will gain or lose solutions.

In general, consider the equation

$$f(x) = g(x).$$

If we multiply by any number k other than 0, then the resulting equation is equivalent.

$$f(x) = g(x) \Leftrightarrow kf(x) = kg(x)$$

Consequently, for those values in the intersection of the domains of f, g, and h,

$$f(x) = g(x) \Leftrightarrow h(x)f(x) = h(x)g(x),$$

except for those values of x for which $h(x) = 0$.

Theorem 4.8 For all values of x in the intersection of the domains of f, g, and h, if $h(x) \neq 0$, then

$$f(x) = g(x) \Leftrightarrow f(x) \cdot h(x) = g(x) \cdot h(x).$$

For instance, begin with

$$(7) \qquad \frac{t-30}{t^2-9} - \frac{3-t}{3+t} = \frac{3}{t^2-9}.$$

Multiply both sides by $t^2 - 9$.

$$(8) \qquad (t-30) - (3-t)(t-3) = 3$$

$$t - 30 + t^2 - 6t + 9 = 3$$

$$t^2 - 5t - 24 = 0$$

$$(t-8)(t+3) = 0$$

$$(9) \qquad t = 8 \quad \text{or} \quad t = -3.$$

But one of these values does not work in equation (7).

Question: Before reading on, determine which solution in (9) works in equation (7) and which does not. Also determine the first step that results in a nonequivalent equation.

What has happened? In this case, multiplying both sides by $t^2 - 9$ has introduced a new equation (8) with a solution -3 that is not a solution to the original equation. (-3 is sometimes called an *extraneous solution* to the original equation. But it is not a solution to that equation, so we try to avoid the term.) In function language, we have here

$$f(t) = \frac{t-30}{t^2-9} - \frac{3-t}{3+t}, \qquad g(t) = \frac{3}{t^2-9}, \quad \text{and} \quad h(t) = t^2 - 9.$$

The intersection of the domains of f, g, and h is $\{t: t \neq 3 \text{ or } t \neq -3\}$. In that intersection, the multiplication can take place.

There is another explanation for the nonequivalence of equations (7) and (8). When $t = -3$ and we multiply both sides of equation (7) by $t^2 - 9$, then we have multiplied both sides by zero. If we multiply by zero to get from equation (7) to (8), then we would need to divide by zero to show the equivalence. And we cannot divide by 0.

4.3.1 Problems

1. Give the number of real solutions to each equation and find one of these solutions to the nearest hundredth.

a. $\log x = x^2 - 2$

b. $e^t = 1 + t$

c. $\sin\theta = \frac{\theta}{10}$ (with θ in radians)

2. Find all real solutions to each equation.

a. $x + \sqrt{2-x} = 4 + \sqrt{2-x}$

b. $y^2 + 6\log(3-2y) = 25 + 6\log(3-2y)$

c. $(z-15)(z+1)^2 = (z+1)(z-15)\sqrt{(z-20)}$

d. $\frac{1+w}{1+w} = 0$

3. Consider this "proof" that $1 = 2$. In which step is the error?

Step 1: Let $x = 1$ and $y = 1$.

Step 2: By substitution, $xy = y^2$.

Step 3: Multiplying both sides by -1, $-xy = -y^2$.

Step 4: Adding x^2 to both sides, $x^2 - xy = x^2 - y^2$.

Step 5: Factoring, $x(x-y) = (x+y)(x-y)$.

Step 6: Dividing both sides by $x - y$, $x = x + y$.

Step 7: By substitution, $1 = 1 + 1$.

Step 8: $1 = 2$

4. Let a and b be any real numbers. Prove: If $1 = 2$, then $a = b$. (*Note*: This question exemplifies that a proof can be valid, but if the hypothesis is false, you cannot trust the conclusion.)

5. Show that, of four consecutive integers, the product of the middle two is never equal to the product of the first and last.

6. a. Show that, of four consecutive integers, the sum of the middle two is always equal to the sum of the first and last.

b. Generalize part **a** and prove your generalization.

4.3.2 Applying the same function to both sides of an equation

Consider the equation

$$\sqrt{5v+6} = -\sqrt{2v+15}$$

Squaring both sides,

$$5v + 6 = 2v + 15.$$

Continuing to solve the equation, we are led to $v = 3$. But 3 does not satisfy the original sentence; substitution gives $\sqrt{21} = -\sqrt{21}$. You may have realized from the start that the left side of this equation is never negative and the right side is never positive, so they could be equal only if the radicands were each 0.

Does the function have an inverse?

Let us analyze algebraically what went wrong. When we take powers or roots of both sides of an equation, or take the log of both sides, or the absolute value of both sides, we are applying a function to both sides. For instance, squaring both sides of

$$\sqrt{5v+6} = -\sqrt{2v+15}$$

to yield

$$5v + 6 = 2v + 15$$

can be interpreted as beginning with $f(v) = g(v)$, where $f(v) = \sqrt{5v+6}$ and $g(v) = -\sqrt{2v+15}$. Then, applying the squaring function $h(v) = v^2$ to both sides. $h(f(v)) = 5v + 6$, and $h(g(v)) = 2v + 15$.

The difficulty is that the squaring function h is not 1-1, so the squaring step is not reversible. That is, on the set of all real numbers, the squaring function has no inverse function. And, as a result, we cannot assume that equivalent equations are generated by squaring. We may still square both sides, but we must be very careful to check any solutions to the new equation back in the original problem.

Question: Change the square roots in $\sqrt{5v+6} = -\sqrt{2v+15}$ to cube roots and solve by the method of the previous paragraph.

a. What are the functions f, g, and h in this situation?

b. Solve the equation by applying h to both sides. Does the method work?

The squaring example and the result of the question are special cases of a very general and revealing theorem. The theorem indicates why cubing both sides of an equation is a reversible step, but squaring is not. Specifically, since $h: x \rightarrow x^3$ has an inverse $h^{-1}: x \rightarrow \sqrt[3]{x}$ on the reals, you can cube both sides of an equation without gaining or losing real solutions. On the other hand, since $h: x \rightarrow x^2$ has no inverse, you cannot square both sides of an equation without potentially affecting the solutions.

Theorem 4.9 Let h be a 1-1 function. Then for all x in the domains of f and g for which $f(x)$ and $g(x)$ are in the domain of h,

$$f(x) = g(x) \Leftrightarrow h(f(x)) = h(g(x)).$$

Proof:

($\Rightarrow$) Since h is a function, $a = b \Rightarrow h(a) = h(b)$. Thus for all x, $f(x) = g(x)$ implies $h(f(x)) = h(g(x))$ as long as $f(x)$ and $g(x)$ are in the domain of h.

($\Leftarrow$) Because h is a 1-1 function, h^{-1} exists. So $h(f(x)) = h(g(x))$ implies $h^{-1}(h(f(x))) = h^{-1}(h(g(x)))$. But $h^{-1} \circ h$ is the identity function, so $f(x) = h^{-1}(h(f(x))) = h^{-1}(h(g(x))) = g(x)$. ┘

Logarithm and exponential functions are 1-1 functions, so they can be applied to both sides of an equation without affecting its solutions. Since they are inverses of each other, one can be used to help solve an equation involving the other. For instance, consider solving the equation

$$2^{n^2-n+3} = 2^{2n^2-3n}, \quad (1)$$

which involves values of exponential functions with base 2. We think "the exponents must be equal". But why? The function applied to both sides of (1) that leads to

$$n^2 - n + 3 = 2n^2 - 3n \quad (2)$$

is the $\log_2$ function. We have begun with

$$f(n) = g(n),$$

where $f(n) = 2^{n^2-n+3}$ and $g(n) = 2^{2n^2-3n}$, and ended with

$$\log_2(f(n)) = \log_2(g(n)).$$

The function $\log_2$ can be applied to both sides of equation (1) without fear of modifying the roots to the equation (1) because $\log_2$ has an inverse.

Solving $f(x) = k$

In some cases, it is rather easy to determine what function should be applied to both sides. Examine these equations. What do they have in common?

$$a + x = b$$
$$ax = b$$
$$\sin\theta = k$$
$$\log x = c$$
$$2y = b$$
$$x^5 = n$$

They are all of the form $f(\text{unknown}) = \text{constant}$; that is, $f(x) = k$. Each is typically solved by applying the inverse of f to both sides. This can be done because of the following theorem, which follows directly from Theorem 4.9.

Theorem 4.10 If f is a 1-1 function, then $f(x) = k \Leftrightarrow x = f^{-1}(k)$.

Proof: The proof is left for you.

EXAMPLE 1 Find all real solutions to $x^5 = 12$.

Solution 1 Take the 5th root of each side.

$$(x^5)^{1/5} = 12^{1/5}$$
$$x = 12^{1/5}$$

Solution 2 Think: $f(x) = x^5$. Apply the inverse of f, $f^{-1}(x) = x^{1/5}$ to each side.

$$f^{-1}(x^5) = f^{-1}(12)$$
$$x = 12^{1/5}$$

EXAMPLE 2 Find all values of θ between 0 and 2π such that $\sin\theta = \frac{\sqrt{3}}{2}$.

Solution We would like to apply the inverse of the sine function to both sides, but on the interval $(0, 2\pi)$, there is no inverse function. So we first restrict the interval to one on which the sine has an inverse. We find all values of θ between 0 and $\frac{\pi}{2}$ such that $\sin\theta = \frac{\sqrt{3}}{2}$.

$$\sin^{-1}(\sin\theta) = \sin^{-1}\left(\frac{\sqrt{3}}{2}\right)$$

This yields one solution: $\theta = \frac{\pi}{3}$. To obtain all other solutions between 0 and 2π, we use properties of the sine function, not Theorem 4.10. (See Problem 3.)

Solving an exponential equation points out an advantage of applying the inverse function to both sides. Example 3 shows an equation of the form $f(x) = k$, in which f is the exponential function with base 2. Its inverse is the log function with base 2. In Solution 1, we apply a log function, but not with that base, and we wind up with a linear equation to solve. In Solution 2, we apply the inverse and obtain a solution in one step.

EXAMPLE 3 Solve $2^x = 14$.

Solution 1 Take the log base 10 of both sides.

$$\log(2^x) = \log 14$$

Use the Power Property of Logarithms:

$$x \log 2 = \log 14$$

Divide both sides by log 2.

$$x = \frac{\log 14}{\log 2}$$

Solution 2 Because the base of the power is 2, apply the log base 2 to both sides.

$$\log_2(2^x) = \log_2 14$$

$$x = \log_2 14 = \frac{\log 14}{\log 2}$$

The two solutions are equal because of the Change of Base Theorem (Section 3.2.2).

4.3.2 Problems

1. Solve each equation in the set of real numbers. Indicate the function(s) you have applied to each side in the solution.

a. $e^{(t^2)} = e^{3t}$

b. $\log_5(m^2) = 4$

c. $\log_5(p^2) = \log_5 4$

d. $6^z = 3$

e. $n^{12} = 12n$

f. $\log_x 3 = \log_{2x}(6)$

2. Prove Theorem 4.10.

3. Describe all real solutions to the equation of Example 2.

4. Solve each equation over the interval $[0, 2\pi)$. Give solutions to the nearest thousandth.

a. $\cos v = 0.5$

b. $\tan \frac{w}{3} = 1$

5. Consider the set L of linear functions with domain $\mathbf{R}$, $L = \{f: f(x) = ax + b, a, b \in \mathbf{R}, a \neq 0\}$. It can be shown that $\langle L, \circ\rangle$ is a group, where $\circ$ is function composition.

a. What is the identity for this group?

b. What is the inverse of $f: x \rightarrow 5x - 4$?

c. Use your answer to part **b** to solve $5x - 4 = 12$ in one step by using an inverse function.

6. Begin with $x = 1$, which clearly has the unique solution 1. Cube both sides to obtain $x^3 = 1$. This equation has three solutions: 1, $\frac{-1+i\sqrt{3}}{2}$, and $\frac{-1-i\sqrt{3}}{2}$. Has this violated Theorem 4.9? Why or why not?

7. **Average yearly inflation.** If an item costs C at one time and D n years later, and $Cx^n = D$, then we call x the average annual inflation factor (for example, $x = 1.04$ refers to an inflation rate of 4%).

a. If a house cost \$112,000 in 1990 and sold for \$155,000 in 1998, what was the average annual inflation factor?

b. At a 6% average annual inflation factor, how many years will it take for the cost of a house to double?

c. Answer the question of part **b** when 6% is replaced by 2%, 4%, and 8%.

d. If 6% in part **b** is replaced by r%, bankers use $\frac{72}{r}$ to estimate the number of years. This is called the *rule of* 72. Graph this estimate and the actual answer over the interval $1 \leq r \leq 12$. Comment on the accuracy of the rule of 72.

ANSWER TO QUESTION

1. a. $f(x) = \sqrt[3]{5x + 6}$, $g(x) = -\sqrt[3]{2x + 15}$, and $h(x) = x^3$.
   b. $h(f(x)) = h(g(x))$ is $5x + 6 = -(2x + 15)$, from which $x = -3$. This value does work in the original equation. The method works.

4.3.3 Solving inequalities

Inequality is, as its name implies, the counterpart of equality. For any numbers a and b, either $a = b$ or $a \neq b$, and in the latter case we say that a and b are not equal. When $a \neq b$, there are two possibilities: Either $a < b$ or $a > b$. From this, it is not obvious that inequality should share any properties with equality. But solving inequalities does turn out to have some similarities with solving equations. In this section, we compare the two ideas of equality and inequality.

How do inequalities arise?

Inequalities arise when we wish to know when one quantity is larger or smaller than another in time, in cost, in rate, or in any other attribute that a variable might measure. Here is a typical situation leading to inequalities.

Suppose cell-phone company A charges \$15 a month plus 21¢ a minute for calls. Then, if t minutes are used in a month, the cost is $15 + .21t$ dollars. If we let $A(t)$ be the cost for t minutes with company A, then $A(t) = 15 + .21t$. This is a nicely behaved linear function with intercept 15 and slope .21. Now suppose a second cell-phone company B charges \$20 a month but only 18¢ for each minute of calls. Proceeding as before, we obtain $B(t) = 20 + .18t$, another linear function.

If we are trying to compare costs, then we are not particularly interested in when the costs are equal, but when one *is less than* the other. Consequently, this cost-comparison situation leads more naturally to inequalities than to equations.

In particular, company A costs less when $A(t) < B(t)$, that is, when t is a solution to the inequality (1).

$$15 + .21t < 20 + .18t \tag{1}$$

This inequality (1) can be solved by doing the same things to both sides as if it were an equation. Add -15 to both sides, then add $-.18t$ to both sides, then divide both sides by .03.

$$\begin{aligned} .21t &< 5 + .18t \\ .03t &< 5 \\ t &< 166.\overline{6} \end{aligned} \tag{2}$$

So if 166 or fewer minutes are used, the first company (with the lower monthly base rate) will cost less.

Starting by adding negative quantities on both sides of (1) does not change the solutions. Suppose $-.21t$ is first added to both sides.

$$15 + .21t < 20 + .18t$$
$$15 < 20 - .03t$$
$$-5 < -.03t$$

However, now the multiplication of both sides by the negative number $-\frac{1}{.03}$ changes the direction or sense of the inequality.

$$\frac{-5}{-.03} > t \quad (3)$$
$$166.\overline{6} > t$$

This inequality is equivalent to the inequality (2), as it must be.

A second major source of inequalities arises in situations in which a quantity is known to be largest or smallest. If M is the maximum value of a function f, then for all x, $f(x) \leq M$. Similarly, if m is the minimum value of f, then for all x, $f(x) \geq m$. From this, for all x, $m \leq f(x) \leq M$.

Why is special attention needed for inequalities?

Because the steps in solving

$$15 + .21t < 20 + .18t \quad (1)$$

are so much like those for solving the corresponding equation

$$15 + .21t = 20 + .18t, \quad (4)$$

many students wonder why they need to learn any special procedures for solving inequalities. Why not just solve the equation (4) to obtain $t = 166.\overline{6}$ and proceed from there to answer the question of when $A(t)$ is less than $B(t)$? One reason is that $t = 166.\overline{6}$ does not tell us the direction of the inequality, while the solutions (2) and (3) to inequality (1) both indicate the direction. A second reason is that the solutions (2) and (3) help to indicate the *interval* of solutions. Specifically, because t is positive, t can be any integer in the open interval $(0, 166.\overline{6})$. Still, when f and g are continuous functions, then solving the equation $f(t) = g(t)$ is typically of assistance in solving $f(t) < g(t)$.

A third reason for special attention to inequalities is that in certain situations, changing inequalities to equations can be unwise. For instance, solving a system of linear inequalities in two variables may yield an entire region of points as its solution. Replacing the $<$ or $>$ signs with $=$ and solving the corresponding system of linear equations may yield part of the boundary of the region but little else.

For instance, the Triangle Inequality states that the sum of the lengths of two sides of a triangle is greater than the third side. So, if the three sides of a triangle have lengths a, b, and c, these lengths must satisfy the system of inequalities $a + b > c$, $a + c > b$, and $b + c > a$. Thus if two sides of a triangle have lengths 23 and 35, the third side (call it c) must satisfy $23 + 35 > c$, $23 + c > 35$, and $35 + c > 23$. The solution to the system of inequalities is $12 < c < 58$, and there is no solution to the corresponding system of equations.

Consider the following extension of this problem.

> One side of a triangle has length 10. What are the lengths of the other two sides?

Let x and y be the lengths of the other two sides. From the Triangle Inequality, possible pairs of lengths (x, y) are the solutions to the system

$$\begin{cases} x + y > 10 \\ x + 10 > y \\ y + 10 > x. \end{cases}$$

The result is a region (which you are asked to graph in Problem 5). Although the lines $x + y = 10$, $x + 10 = y$, and $y + 10 = x$ are boundaries to that region, knowing only the equations does not tell you any solutions.

Solving $f(x) < g(x)$

Even the simple inequality $ax < b$, where $a \neq 0$, requires two cases for its solution. If $a > 0$, then $x < \frac{b}{a}$. If $a < 0$, then $x > \frac{b}{a}$. If such a simple inequality has two cases, we might expect more complicated inequalities to require more cases. However, there is a sense in which many inequalities are no more complicated than $ax < b$.

When we multiply both sides of $f(x) < g(x)$ by a number a, we can think of this as applying the function $h(x) = ax$ to both sides. h is an increasing function when a is positive, and this is why the sense of the inequality remains the same. So if $a > 0$, $af(x) < ag(x)$. h is decreasing when a is negative, and this is why the sense of the inequality is switched. That is, if $a < 0$, $af(x) > ag(x)$.

Multiplying both sides of an inequality by the same number is an instance of the following theorem.

Theorem 4.11 For any continuous real functions f and g, with domain D:

a. If h is strictly increasing on the intersection of $f(\text{D})$ and $g(\text{D})$, then

$$f(x) < g(x) \Leftrightarrow h(f(x)) < h(g(x)).$$

b. If h is strictly decreasing on the intersection of $f(\text{D})$ and $g(\text{D})$, then

$$f(x) < g(x) \Leftrightarrow h(f(x)) > h(g(x)).$$

Proof:

a. Recall the definition of a strictly increasing function. h is strictly increasing $\Leftrightarrow$ for all a, b on an interval, $a < b \Rightarrow h(a) < h(b)$. The $(\Rightarrow)$ direction of the theorem follows by substituting $f(x)$ for a, and $g(x)$ for b. For the $(\Leftarrow)$ direction because h is strictly increasing, it has an inverse h^{-1} that is strictly increasing. Think of $h(f(x))$ and $h(g(x))$ as being a and b, and h^{-1} as the strictly increasing function. Then apply h^{-1} to both sides.

b. This proof is similar to (1) and is left for you as Problem 7. ┘

To illustrate Theorem 4.11, we ask what condition on two positive numbers a and b holds when $a^x > b^x$. We consider this as the inequality $f(x) > g(x)$, where the functions f and g with $f(x) = a^x$ and $g(x) = b^x$. Let $h(x) = \log x$. Now h is increasing throughout its domain, so

$$f(x) > g(x) \Leftrightarrow h(f(x)) > h(g(x)).$$

That is,

$$\begin{aligned} a^x > b^x &\Leftrightarrow \log a^x > \log b^x \\ &\Leftrightarrow x \log a > x \log b. \qquad \text{(Log of a Power)} \end{aligned}$$

If $x > 0$, we can divide by x without changing the sense of the inequality. If $x < 0$, then division by x changes the sense of the inequality. This answers the question at the beginning of the paragraph. When $a^x > b^x$, then $a > b$ if x is positive, and $a < b$ if x is negative.

Solving $f(x) < k$

As a further example of Theorem 4.11, consider the two inequalities, $\sin\theta < .5$ and $\cos\phi < .5$, where $0 < \theta < \frac{\pi}{2}$, and $0 < \phi < \frac{\pi}{2}$. These are both of the form $f(\theta) < k$. Graphs of the sine and cosine functions (Figures 6a and 6b) show the solutions geometrically.

Figure 6a

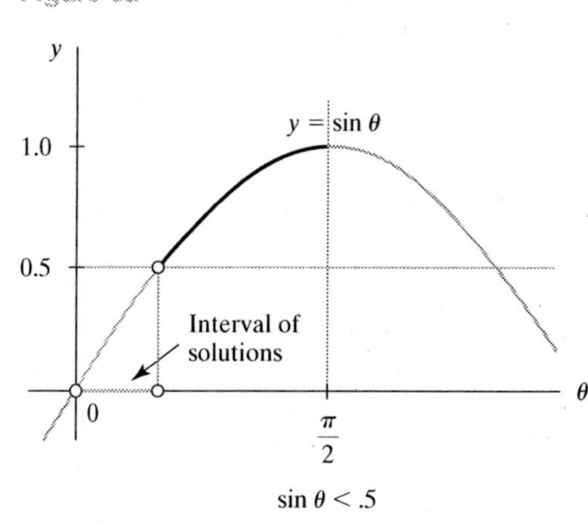

sin θ < .5

Figure 6b

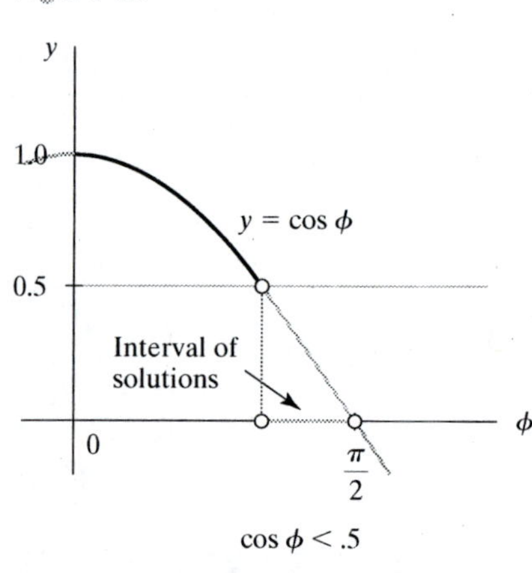

cos ϕ < .5

By applying Theorem 4.11, the inequalities can be solved algebraically. Begin with the inequalities to be solved over the interval $\left(0, \frac{\pi}{2}\right)$. We show the solutions in two columns.

$$\sin\theta < .5 \quad \Big| \quad \cos\phi < .5$$

Apply the inverse functions to both sides. Because the sine function is increasing on $\left(0, \frac{\pi}{2}\right)$, so is $\sin^{-1}$, and so the sense of the inequality remains the same. Since the cosine function is decreasing on $\left(0, \frac{\pi}{2}\right)$, so is $\cos^{-1}$, and the sense of its inequality reverses.

$\sin^{-1}(\sin\theta) < \sin^{-1}(.5)$	$\cos^{-1}(\cos\phi) < \cos^{-1}(.5)$
$0 < \theta < \sin^{-1}(.5)$	$\frac{\pi}{2} > \phi > \cos^{-1}(.5)$
$0 < \theta < \frac{\pi}{6}$	$\frac{\pi}{2} > \phi > \frac{\pi}{3}$

As another example, consider all positive solutions to the equation $x^{-3} \geq 125$. The function $f: x \rightarrow x^{-3}$ is decreasing over the positive reals, so the sense of the inequality will change when its inverse is applied to both sides.

$$x^{-3} \geq 125$$
$$(x^{-3})^{-1/3} \leq 125^{-1/3}$$
$$0 \leq x \leq \tfrac{1}{5}$$

Inequalities, Continuity, and Discontinuity

Continuous functions are sometimes intuitively described as those functions whose graphs can be drawn without taking the pencil off of the paper. This aspect of continuous functions is embodied in the *Intermediate Value Theorem*, stated and pictured below (Figure 7). A proof of this property can be found in many calculus and real variable texts and we omit it here.

Theorem 4.12 **(Intermediate Value Theorem):** Suppose f is a continuous function on the interval $[a, b]$. Then for every real number y_0 between $f(a)$ and $f(b)$, there is at least one real number x_0 between a and b such that $f(x_0) = y_0$.

Figure 7

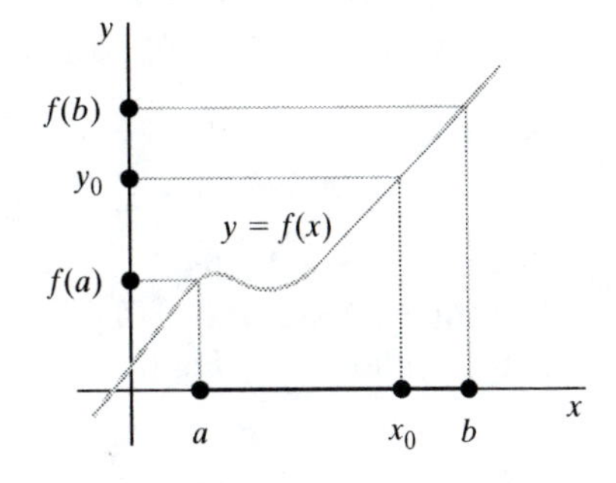

For instance, because the speed a car is traveling can be considered as a continuous function of time, if $f(t)$ is the speed you are driving at time t, and $f(2{:}15) = 50$ mph and $f(2{:}30) = 35$ mph, then at some time between 2:15 and 2:30 you must have been driving at exactly 40 mph.

The impact of the Intermediate Value Theorem is that on any interval for which a function of one variable is continuous, a solution to an inequality will itself be a union of intervals. This you have seen for all the functions discussed so far in this section, because all polynomial, exponential, logarithmic, and the sine and cosine functions are continuous.

Figure 8

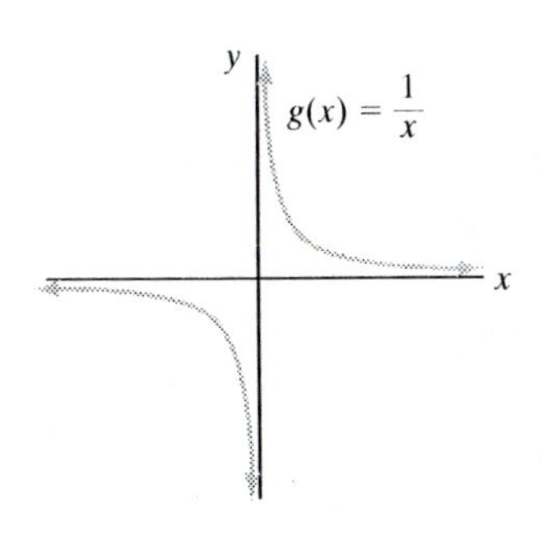

But the ideas of continuous functions can be extended to certain noncontinuous functions by examining intervals of their domain on which the function is continuous. This is a useful strategy for solving inequalities involving rational functions.

The function g with $g(x) = \frac{1}{x}$ is a simple example of a rational function, a quotient of polynomial functions. It is not defined when $x = 0$, so it is not continuous on the set $\mathbf{R}$ of real numbers. But g is continuous on the set $\mathbf{R}^+$ of positive real numbers, and it is continuous on the set $\mathbf{R}^-$ of negative real numbers. Its graph is a hyperbola (Figure 8).

Figure 9

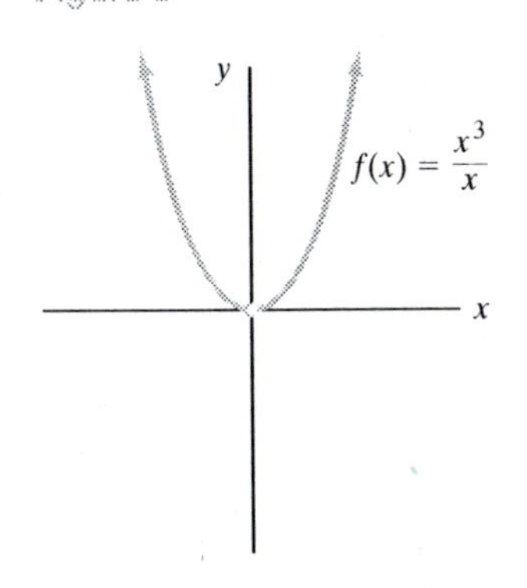

Likewise, the rational function f, with $f(x) = \frac{x^3}{x}$, which is not defined when $x = 0$, is not continuous on $\mathbf{R}$, but is continuous on $\mathbf{R}^+$ and $\mathbf{R}^-$. The graph of f (Figure 9) is a parabola with a point (its vertex) missing.

But there is a fundamental difference between f and g when it comes to the discontinuity at $x = 0$. The discontinuity at f can be fixed by setting $f(0) = 0$. When the discontinuity is removed by setting $f(0) = 0$, the new function is continuous throughout $\mathbf{R}$. For this reason, we call the discontinuity *removable.* A discontinuity of a function f is **removable** at x_0 if and only if the limit of the values of the function approach the same finite limit L as x approaches x_0 from the left or from the right. That is, the discontinuity is removable at x_0 if and only if there is a real number L with $L = \lim_{x \to x_0^+} f(x) = \lim_{x \to x_0^-} f(x)$. The discontinuity is removed by setting $f(x_0) = L$.

Because the values of the function g in Figure 8 do not have the same finite limit on either side of the discontinuity (they don't even have a finite limit), the discontinuity is not removable. A discontinuity that is not removable is called an **essential** discontinuity.

Solving $f(x) < 0$ when the zeros of f are known

The locations of removable and essential discontinuities of a rational function f are of great assistance in solving $f(x) < 0$. This is because the function f is continuous between these discontinuities. Now, if in addition to the discontinuities, we also know neighboring zeros a and b of f, then the Intermediate Value Theorem implies that either $f(x) > 0$ for all x between a and b, or $f(x) < 0$ for all x between a and b. (Otherwise there would be a zero between a and b and they would not be neighboring.) Thus if you separate the domain of a function f into the nonoverlapping open intervals with all of its zeros and discontinuities as endpoints, you can determine whether $f(x) > 0$ or $f(x) < 0$ on the interval by finding the value of the function for a point in the interval. This procedure for solving inequalities is called the **test-point method.**

EXAMPLE 1 Solve $\frac{(x+1)(x-3)^2}{5x-6} > 0$.

Solution 1 Let $f(x) = \frac{(x+1)(x-3)^2}{5x-6}$. Then f has zeros at -1 and 3 and an essential discontinuity when $5x - 6 = 0$, that is, when $x = 1.2$. Because f is continuous other than at the discontinuity, the solutions to $f(x) > 0$ are within the intervals determined by the points $-1, 1.2$, and 3 (See Figure 10).

Figure 10

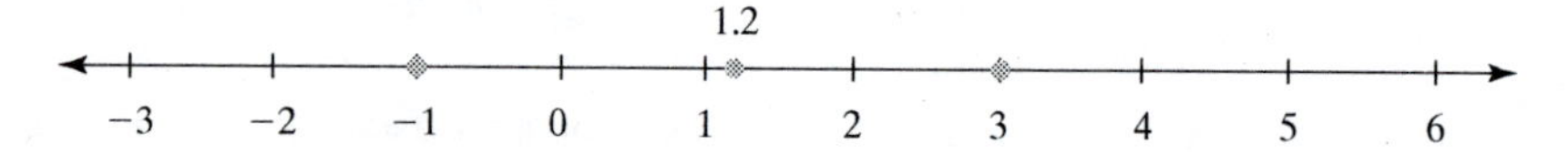

Now find a representative value of f for x in each interval.

Interval $(-\infty, -1)$:	We let $x = -2$.	$f(-2) = \frac{-1\cdot(-5)^2}{-16} = \frac{25}{16} > 0$
Interval $(-1, 1.2)$:	We let $x = 0$.	$f(0) = \frac{1\cdot(-3)^2}{-6} = -1.5 < 0$
Interval $(1.2, 3)$:	We let $x = 2$.	$f(2) = \frac{3\cdot(-1)^2}{4} = \frac{3}{4} > 0$
Interval $(3, \infty)$:	We let $x = 5$.	$f(5) = \frac{6\cdot(2)^2}{19} = \frac{24}{19} > 0$

Thus the solution set to $\frac{(x+1)(x-3)^2}{5x-6} > 0$ is $(-\infty, -1) \cup (1.2, 3) \cup (3, \infty)$.

Solution 2 Having found the zeros as in Solution 1, graph f (Figure 11).

Figure 11

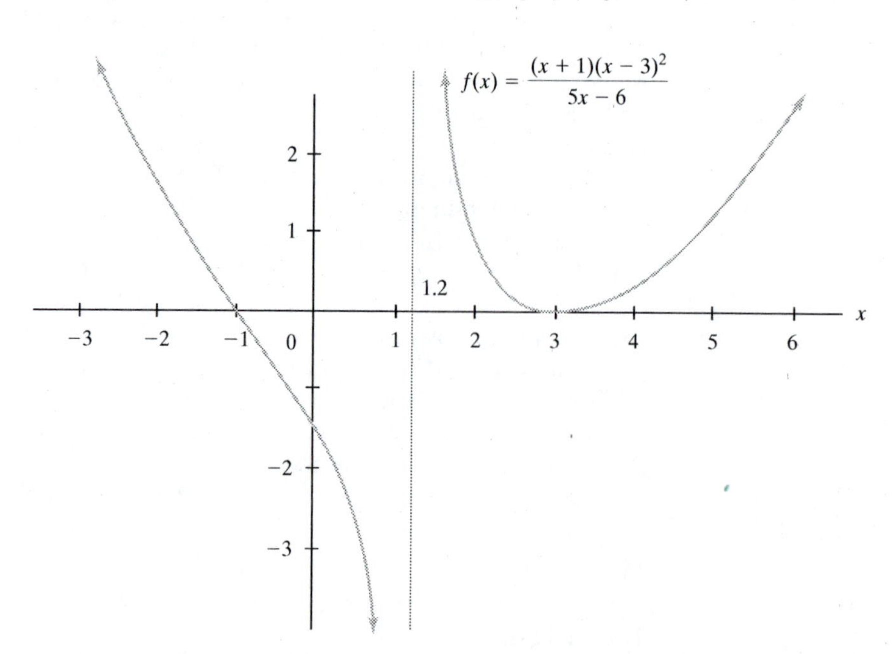

The graph in Figure 12 displays the solutions on a number line.

Figure 12

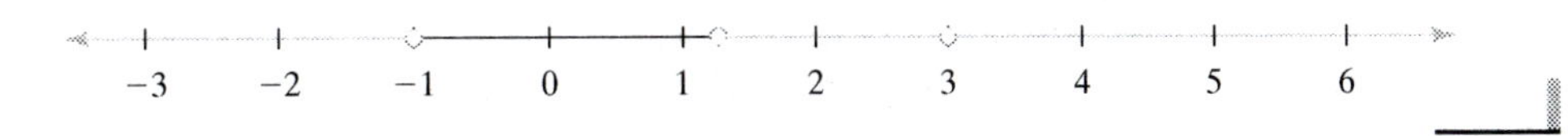

4.3.3 Problems

1. Graph the functions A and B defined at the beginning of this section and interpret the graph to give a solution to the inequality $A(t) < B(t)$.

2. a. Consider cell-phone companies with the following rates.

   Company C: \$19.99 per month plus 21¢ per minute.

   Company D: \$19.99 per month plus 19¢ per minute.

   Set up an inequality to be solved to determine the amount of usage for which Company C is preferred over Company D. Solve the inequality and interpret the mathematical solution.

   b. Repeat part **a** if Company D charges \$18.99 per month plus 19¢ per minute.

   c. Repeat part **a** if Company D charges \$5 per month plus 35¢ per minute.

3. If two sides of a triangle have sides a and b, what are the possible lengths c of the third side?

4. Two circles have radii r and s, with $r > s$ and a distance d between their centers. Give the conditions on $r, s,$ and d under which each situation arises.

   a. One circle lies entirely in the interior of the other circle.

   b. The circles are internally tangent.

   c. The circles intersect in two distinct points.

   d. The circles are externally tangent.

   e. Each circle lies entirely in the exterior of the other.

5. Let 10, x, and y be the lengths of the three sides of a triangle. Graph the possible pairs of values (x, y).

6. Consider Theorem 4.11. Describe h when both sides of on inequality $f(x) < g(x)$ are multiplied by -1, and give the resulting inequality.

7. Write a proof of part (2) of Theorem 4.11.

8. In the situation of Theorem 4.11, what happens if h is the constant function?

9. Graph $y = x^{-3}$ with a window appropriate to the inequality $x^{-3} \geq 125$. Does the result found in this lesson check?

10. Give an algebraic solution to $\frac{1}{x} < 2$ over the set of all real numbers. Justify each algebraic step.

11. The half-life of carbon-14, used to date ancient artifacts, is about 5730 years. If between 15% and 20% of the carbon-14 originally in an artifact remains, about how old is the artifact?

12. a. Let θ be an acute angle. For what values of θ is $\sin\theta > \sin 2\theta$?

    b. Suppose $0 \leq \theta \leq 2\pi$. For what values of θ is $\sin\theta > \sin 2\theta$?

13. Solve each inequality in **R**.

    a. $\frac{(3y+6)(2y-3)}{(1-y)^2} > 0$

    b. $t^3(4t + 1) \leq 4t + 1$

    c. $(x - 1)^3 < 8$

    d. $d + 11 \geq d^2 + 3d - 50$

    e. $3^{2^x-1} < \frac{1}{9}$

    f. $\log(y + 1)^2 > -2$

    g. $\frac{1}{(x+a)(x+b)} \leq 0$

    h. $\frac{x^2+10x+25}{x+5} > 6$

14. If a, b, and c are real, $a > 0$ and $b^2 - 4ac > 0$, describe all solutions to $ax^2 + bx + c < 0$.

4.3.4 Extended analysis: averages of speeds

Problems similar to Problem 1 are found in some high school mathematics textbooks.

> **Problem 1:** Suppose that on a round trip you average 30 mph on the way out and average 60 mph on the way back. What is your average speed for the entire trip?

Question 1: Before reading further, solve this problem, and write down a few difficulties students might experience with it.

An initial analysis

The arithmetic average of 30 and 60 is 45, but the average speed for the entire trip is *not* 45 mph, but rather 40 mph. Below we analyze this simple problem and derive the correct solution. But first, let us acknowledge that there is a little bit of a mystery here. Why isn't the answer the simple arithmetic average? Is there something special about averaging rates?

To answer this last question, we pose a similar problem involving rates.

Problem 2: On a trip I average 30 mph for a certain time, and then you take over and average 60 mph for the same amount of time. What is the average speed for the total time we have driven?

The answer here is that the average speed is 45 mph, and this *is* the arithmetic average. (See the analysis below.) This means that average rates are sometimes arithmetic averages, sometimes not. What is the big picture here?

We proceed with an analysis. As a start, we need to be clear that average speed means $\frac{\text{total distance}}{\text{total time}}$. That is, total time $= \frac{\text{total distance}}{\text{average speed}}$. In Problem 1, we do not know the one-way distance, but we can represent it as d. Since time is $\frac{\text{total distance}}{\text{average speed}}$, the time for the out leg of the round trip is $\frac{d}{30}$ while the time for the return leg is $\frac{d}{60}$. We then have

$$\text{(1)}\quad \text{average speed} = \frac{\text{total distance}}{\text{total time}} = \frac{2d \text{ miles}}{\left(\frac{d}{30}+\frac{d}{60}\right)\text{hr}} = \frac{2 \text{ miles}}{\left(\frac{1}{30}+\frac{1}{60}\right)\text{hr}} = 40\frac{\text{mi}}{\text{hr}}.$$

Notice that the d in the numerator is divided by the d in the denominator, so the answer does not involve d.

Notice that the average speed, 40 mph, is less than the arithmetic average of 30 and 60 (which is 45). There is actually an easy way to see this intuitively. Since the distances out and back are equal, you must travel at the lower rate 30 mph for a longer time; it has more of an effect on the average rate than the higher rate 60 mph. Specifically, you must travel twice as long at 30 mph as at 60 mph, and 40 is the weighted average $\frac{2\cdot 30+1\cdot 60}{3}$.

In Problem 2 we don't know the time each of us drove, but we can represent it as t, since both of us drove for the same amount of time. Then

$$\text{(2)}\qquad \text{average speed} = \frac{\text{total distance}}{\text{total time}} = \frac{30t+60t}{t+t} = \frac{30+60}{2} = 45.$$

In this case the t in the numerator is divided by the t in the denominator, and the average speed, 45 mph, does not involve t. This average speed is the arithmetic average of the two speeds. Thus averaging rates over equal times gives an arithmetic average (Problem 2), while averaging rates over equal distances does not (Problem 1).

Harmonic means

Question 2: Before reading further, think of a way to generalize Problem 1.

One possible generalization is from the specific speeds 30 and 60 to arbitrary speeds v and w. When we do this in formula (1), a definite structure appears:

$$\text{(3)}\qquad \text{average speed} = \frac{2d}{\frac{d}{v}+\frac{d}{w}} = \frac{2}{\frac{1}{v}+\frac{1}{w}}$$

The structure is even clearer if this is written in a slightly different form:

$$\text{(4)}\qquad \text{average speed} = \frac{1}{\frac{1}{2}\left(\frac{1}{v}+\frac{1}{w}\right)}.$$

This shows that *the average speed is the reciprocal of the arithmetic mean of the reciprocals of the individual speeds v and w.* This "average" of positive quantities v and w is called the **harmonic mean** of v and w. Let $H(v, w)$ be the harmonic mean of v and w. From (3), $H(v, w) = \frac{2}{\frac{1}{v} + \frac{1}{w}}$. Thus $\frac{1}{H(v, w)} = \frac{\frac{1}{v} + \frac{1}{w}}{2}$. So the reciprocal of the harmonic mean of v and w is the arithmetic mean of the reciprocals of v and w.

The name "harmonic mean" comes from the study of music by the ancient Greeks. Similar strings with lengths in the ratios 1:2 yield tones an octave apart; strings with lengths in the ratio 1:4 yield tones two octaves apart. In between these, strings with lengths in the ratio 1:3 yield tones an octave plus a musical fourth—for example, from middle C to the F above high C. In corresponding manner, $\frac{1}{3}$ is the harmonic mean of $\frac{1}{2}$ and $\frac{1}{4}$. The harmonic mean, along with the arithmetic mean and the geometric mean appears frequently in mathematics and its applications. (See Problems 12–15 and also Sections 8.3.1 and 10.1.4.)

It is interesting to see how the harmonic mean compares with the arithmetic mean. To give an analysis in terms of functions of one variable, suppose we fix one speed at 30 mph, and look at the two kinds of averages of a variable speed w with 30:

(5) arithmetic mean of w and 30: $$A(w) = \frac{1}{2}(30 + w) = 15 + \frac{w}{2}$$

(6) harmonic mean of w and 30: $$H(w) = \frac{1}{\frac{1}{2}\left(\frac{1}{30} + \frac{1}{w}\right)} = \frac{1}{\left(\frac{1}{60} + \frac{1}{2w}\right)}.$$

We see that $A(w)$ is a *linear* function of w, while $H(w)$ is a *rational* function of w. Graphs of these two functions (Figure 13) help us picture further the structure of the round trip problem.

Figure 13

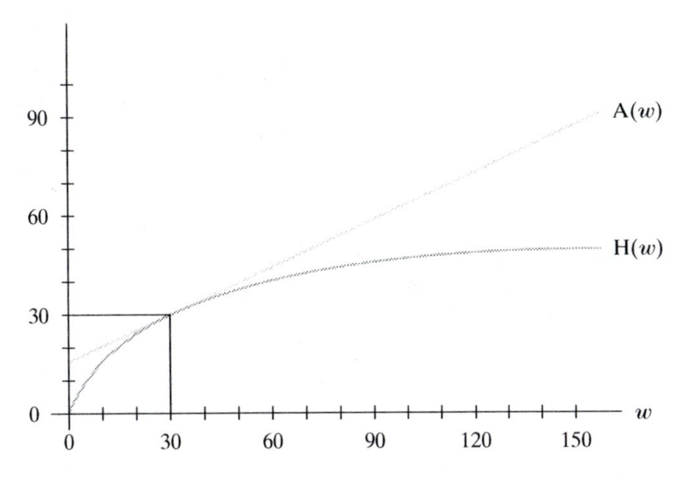

From the graph, it appears that the two means $A(w)$ and $H(w)$ yield the same result at one particular value of the speed w, namely $w = 30$ mph (which is the speed we are holding constant). This result is easily verified by substitution into the formulas for $A(w)$ and $H(w)$. At $w = 30$ both averages are 30, but at any speed w other than 30 the harmonic mean is *strictly less than* the arithmetical mean.

Further, from the graph it appears that $H(w)$ approaches a maximum value. This is easy to find analytically: $\lim_{w \to \infty} H(w) = 60$. (See Problem 3.) There is a natural interpretation of this limit. Suppose the trip is 120 miles each way and we average 30 mph on the out leg. This takes 4 hours. To average 60 mph for the entire trip of 240 miles would also take 4 hours. So no matter how fast we go on the return leg, we cannot possibly average 60 mph. In general, the harmonic mean of two numbers must be less than twice the smaller number.

We can verify this analytically. Notice the applications of Theorem 4.11 in this analysis. Suppose $x > y > 0$ and $H(x) \geq 2y$. Then

$$\frac{1}{\frac{1}{2}\left(\frac{1}{x}+\frac{1}{y}\right)} \geq 2y \qquad \text{Definition of } H(x)$$

$$\Leftrightarrow \frac{1}{\frac{1}{x}+\frac{1}{y}} \geq y \qquad \text{Multiply both sides by } \frac{1}{2}.$$

(Apply $f_1(x) = \frac{1}{2}x$, an increasing function, to both sides.)

$$\Leftrightarrow \frac{1}{x}+\frac{1}{y} \leq \frac{1}{y} \qquad \text{Take the reciprocal of both sides.}$$

(Apply $f_2(x) = \frac{1}{x}$, a decreasing function, to both sides. This changes the sense of the inequality.)

$$\Leftrightarrow \frac{1}{x} \leq 0 \qquad \text{Add } -\frac{1}{y} \text{ to each side.}$$

(Apply $f_3(x) = x - \frac{1}{y}$, an increasing function, to each sides.)

The inequality $\frac{1}{x} \leq 0$ has no solution since $x > 0$, so the original inequality has no solution. So we can conclude that the harmonic mean of two numbers must be less than twice the smaller number.

4.3.4 Problems

1. Prove that the reciprocal of the harmonic mean of x and y is the arithmetic mean of the reciprocals of x and y.
2. Write the harmonic mean of x and y as the quotient of two polynomials in x and y.
3. Let $H(x)$ be the harmonic mean of w and 30. Show why $\lim_{w\to\infty} H(w) = 60$.
4. a. In Figure 14 are graphed three line segments, labeled 1, 2, and 3. What is the relationship of the *slopes* of these segments?

Figure 14

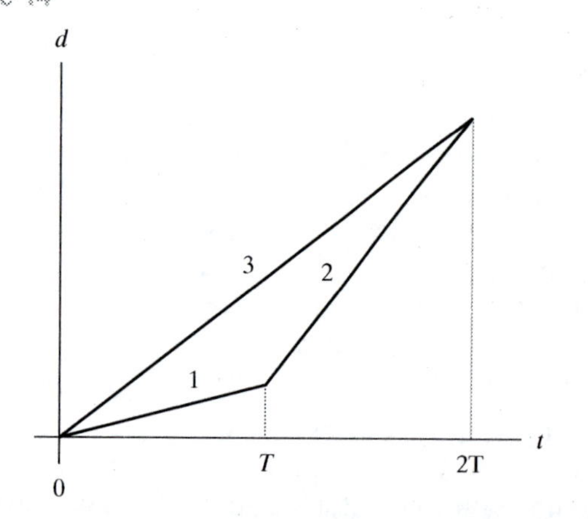

b. In Figure 15 are graphed three different line segments, also labeled 1, 2, and 3. What is the relationship of the slopes of *these* lines?

c. Discuss the relevance of this problem to the question of averages of rates.

Figure 15

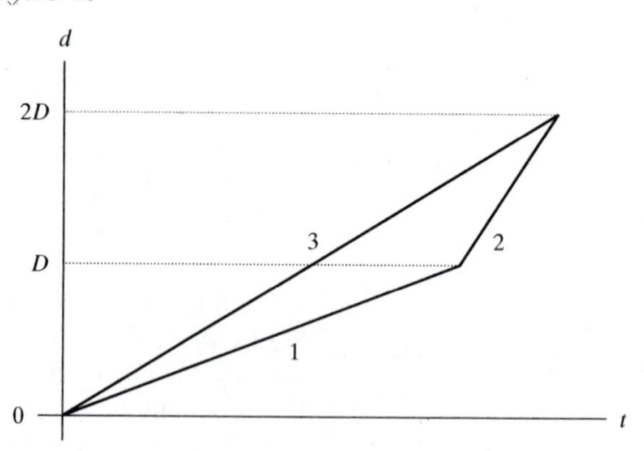

5. On a round trip you run at a pace of 6 minutes per mile on the way out and at a pace of 12 minutes per mile on the way back.
   a. What is your average pace for the entire trip, in minutes per mile?
   b. Generalize part **a**.

6. On a round trip you run at a pace of 6 minutes per mile for a certain time, and then at a pace of 12 minutes per mile for the same time.

a. What is your average pace for the entire trip, in minutes per mile?

b. Generalize part **a**.

7. Prove that the arithmetic mean of two different numbers lies between the numbers.

8. Prove that the harmonic mean of two different positive numbers lies between the numbers.

9. Recall that the geometric mean of x and y is $\sqrt{xy}$. Prove that the geometric mean of two different positive numbers lies between the numbers.

10. a. Prove that the geometric mean of two positive numbers is the geometric mean of the harmonic mean and the arithmetic mean of the two numbers.

b. Establish a general ordering relationship among the arithmetic mean, the geometric mean, and the harmonic mean of two positive numbers.

11. Find the harmonic mean of the two solutions to the quadratic equation $ax^2 + bx + c = 0$.

12. The **harmonic sequence** is the sequence $\frac{1}{1}, \frac{1}{2}, \frac{1}{3}, \frac{1}{4}, \frac{1}{5}, \frac{1}{6}, \ldots, \frac{1}{n}, \ldots$. Find a specific way in which this sequence illustrates harmonic means.

13. In Figure 16, segment $\overline{AB}$ of length x is parallel to segment $\overline{CD}$ of length y. Segment $\overline{EF}$ is parallel to the other two.

a. Find the length z of this segment in terms of x and y.

b. Does the length of $\overline{EF}$ depend in any way on d, the length of $\overline{AC}$?

c. Verify your answer to part **a** using a dynamic geometry drawing program in which A, B, and C are fixed and D moves along the parallel to $\overleftrightarrow{AB}$ through C.

14. Two cities, Bear Lake and Moose City, are 100 miles apart. Going at a constant speed, suppose car A makes the 100-mile trip from Bear Lake to Moose City in time t_A. Starting at the same time, and also going at a constant speed, car B makes the 100-mile trip from Moose City to Bear Lake in time t_B.

a. Find the elapsed time t when they passed each other as a function of t_A and t_B, and explain why the distance does not affect the answer.

b. A typical "work" problem has the following general form: Person A can complete a job in time t_A, while person B can complete the same job in time t_B. How long will it take A and B to complete the job working together (assuming they work at the same rates together as separately)?

c. Use your answer to part **b** to interpret part **a** as a "work" problem.

15. **Earthquake waves.** Here is a motion problem of a type not usually seen in high school texts.

> In an earthquake, longitudinal "S-waves" and transverse "P-waves" travel outward from the epicenter. The two types of waves start at the same time, at the moment the earthquake begins. It is known that S-waves travel at $3.4\,\frac{\text{km}}{\text{sec}}$, and P-waves travel at $5.6\frac{\text{km}}{\text{sec}}$. Suppose that at a seismograph station, the P-waves arrive 15 sec before the S-waves for a particular earthquake. What is the distance from the station to the epicenter?

a. *An incorrect solution.* Here is a "solution" that has been proposed occasionally even by experienced problem solvers: "The distance asked for in the problem will be the product of a velocity and a time. The difference between the two velocities, $5.6\frac{\text{km}}{\text{sec}}$ and $3.4\frac{\text{km}}{\text{sec}}$, gives us a velocity, $2.2\,\frac{\text{km}}{\text{sec}}$, and the difference between the arrival times, 15 seconds, gives us a time. Therefore, the required distance is $2.2\frac{\text{km}}{\text{sec}}$ times 15 sec, which is 33 km."

i. Find the flaw in the above "solution" to the earthquake problem.

ii. The 33 km distance in this "solution" does have meaning in the situation. What is that meaning, and how does it differ from what is asked for?

b. *A graphical solution.* On the same set of axes, graph the distance vs. time motion functions for each of the two waves. Without calculating anything, solve the problem through an analysis of (and measurement on) these graphs.

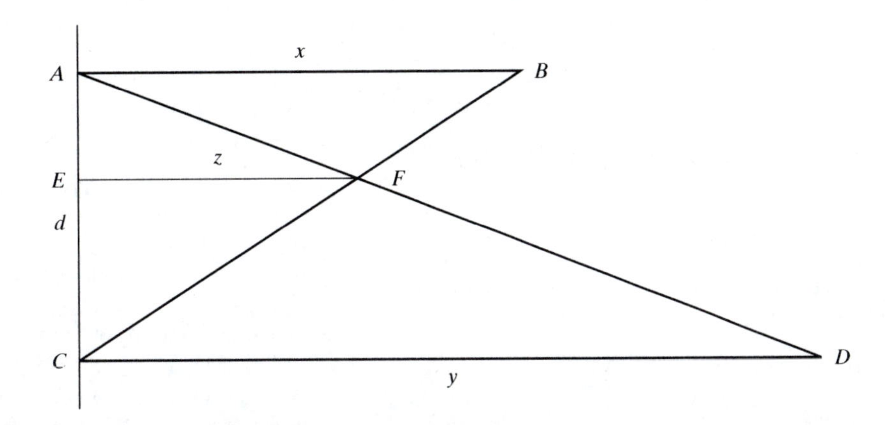

c. *An algebraic solution.* Solve the problem algebraically. Relate the equations in your algebraic solution to the graphical solution in part **b**.

d. *A general solution.* Solve the problem one more time, this time for any speeds s and p of waves and any difference of times d. "Coax" your expression for the required distance into a form that reveals the structure of the situation in a clear way.

e. Problem 2 in this section led to the harmonic mean of two quantities. The general formula for d found in part **d** has an expression involving two speeds that is similar to, but not identical to, the formula for the harmonic mean of two speeds. We call the expression found in part **d** the *harmonic difference* of p and s. How is the harmonic difference of p and s similar to, and how does it differ from, the harmonic mean of p and s?

f. *Other applications.* Several formulas in physics have the same general form as the harmonic mean or the harmonic difference.

i. Look in a physics book for a formula for the resistances of two resistances in parallel, and analyze how it is derived.

ii. Look in a physics book for a formula for the focal length of a lens, and analyze how it is derived.

16. Use the situations and properties of the harmonic mean of two numbers to develop a definition for the harmonic mean of n numbers. Defend your definition.

Chapter Projects

1. Interest in solving algebraic equations dates back to the pre-Hellenic civilizations of Sumeria, Babylon, and Egypt. Write an essay describing the results found by these ancient civilizations. Use at least three sources and identify them.

2. a. Investigate the probability that $ax^2 + bx + c = 0$ has real solutions when $a, b,$ and c are random nonzero integers each with absolute value less than or equal to some positive integer k. Does the probability increase or decrease as k increases?

b. Investigate the probability that $ax^2 + bx + c = 0$ has rational solutions when a, b, and c are random nonzero integers each with absolute value less than or equal to some positive integer k. Does the probability increase or decrease as k increases?

3. Find out what *Descartes' Rule of Signs* is, give some examples of its use, and provide a general proof.

4. a. Find Cardano's derivation of the solutions of the general cubic equation $ax^3 + bx^2 + cx + d = 0$. Explain how each step follows from the previous.

b. Pick three different cubic equations whose solutions you know. At least one should have nonreal solutions. Show that Cardano's solution process works for these equations.

5. Ferrari discovered that the solution of the general quartic equation $ax^4 + bx^3 + cx^2 + dx + e = 0$ can be obtained from the solution of a cubic equation. Examine the work of Ferrari to show how to solve any quartic equation and apply his method to two examples of your choosing.

6. Newton discovered a method for approximating solutions to equations of the form $f(x) = 0$, now usually called *Newton's method of approximation*. Describe this method, give some examples of its use, and carefully indicate the functions f to which it can be applied.

7. We mentioned in Chapter 2 that Gauss gave the first rigorous proof of the fundamental theorem of algebra, and that Gauss ultimately provided five different proofs of this result. Give the details of one of Gauss's proofs, or of a more recent proof of this theorem.

Bibliography

Unit 4.1 Reference

Katz, Victor J. *A History of Mathematics*. New York: HarperCollins, 1993, pp. 13–17.
Here is a discussion of early Egyptian and Chinese methods for solving linear equations.

Unit 4.2 Reference

Urpensky, J. V. *Theory of Equations*. New York: McGraw-Hill, 1948.
This book contains an in-depth treatment of the solution of polynomial equations, including details about the solution of cubic and quartic equations and one of the proofs of Gauss of the Fundamental Theorem of Algebra. It was written as a text for a course that now is rarely taught.

Unit 4.3 References

Korovkin, P. *Inequalities*. Translated from the Russian by Halina Moss. New York: Blaisdell, 1961.
This book discusses a large number of inequalities accessible to high school students.

Rademacher, Hans, and Otto Toeplitz. *The Enjoyment of Mathematics*. Princeton, NJ: Princeton University Press, 1957, Chapter 15.
A discussion of the inequality between the arithmetic and geometric means for n numbers.

Schoenberg, Isaac J. *Mathematical Time Exposures*. Washington, DC: Mathematical Association of America, 1982, Chapter 15.
This chapter develops a number of relationships between arithmetic, geometric, and harmonic means. It also discusses the Euclidean mean $\sqrt{\frac{x^2+y^2}{2}}$ of two positive numbers x and y, and some generalizations.

Chapter

5

INTEGERS AND POLYNOMIALS

The three units of this chapter study three familiar mathematical objects: natural numbers, integers, and polynomials. In Unit 5.1, we provide an *intrinsic* description of the system of natural numbers. By an intrinsic description of the system of natural numbers, we mean one that describes $\langle \mathbf{N}, +, \cdot \rangle$ in terms of its elements and relationships between these elements rather than in terms of larger number systems such as the system $\langle \mathbf{R}, +, \cdot \rangle$ of real numbers. In trying to develop such a description, mathematicians of the late 19th and early 20th centuries discovered that these numbers are intimately connected with the process of mathematical induction. In turn, mathematical induction is closely connected with recursion and recursive processes, ideas that are critical in the study of algorithms and computer programming.

The set of *integers* comprises the natural numbers, their additive inverses (opposites), and zero. Addition, subtraction, and multiplication are possible within the set of integers, but division is not (for example, $2 \div 5$ is not an integer). Yet important properties of division underlie the integers, such as the existence of quotients and remainders, and the representations of integers as products of primes. Proof by mathematical induction is an important tool for the establishment of these properties and this method is used in many other parts of mathematics. In Unit 5.1, we discuss mathematical induction and its roots in the natural number system.

Two different, though related, connections are made in this chapter between integers and polynomials. First is the representation of integers in various bases, base 10 being so familiar that a numeral such as

$$825$$

gives us the usual name for the number ("eight hundred twenty-five"), rather than the literal translation

$$800 + 20 + 5,$$

or the representation utilizing the base 10,

$$8 \cdot 10^2 + 2 \cdot 10 + 5.$$

The second connection may be a surprise. The divisibility properties of integers, and their consequences, have counterparts for polynomials. The division properties underlying the integers are discussed in Unit 5.2, and the similarities between these numbers and polynomials are explored in Unit 5.3.

Unit 5.1 Natural Numbers, Induction, and Recursion

The set of natural numbers is considered here and by most sources to consist of the numbers 1, 2, 3, ... The choice of 1 as the first natural number is supported by the historical fact that the use of zero came much later than the other natural numbers. On the other hand, some mathematicians consider 0 to be a natural number because it is a digit used in writing decimals, because it is the possible result of a count, and because many processes are more easily described if they are viewed as beginning with a step labeled 0. These different definitions do not arise out of ignorance or out of flaws in the mathematics, but because the same mathematics can be put to different uses and may have originated from different considerations. In this particular case, whether 0 is or is not a natural number has little significance. If necessary, statements that are true about natural numbers excluding 0 can easily be modified so that they are true including 0, and vice versa.

In writing "1, 2, 3, ...," the three dots, called an *ellipsis*, have a particular meaning above. We understand from the ellipsis in this case that each natural number is followed by another that is one greater, and that the numbers go on forever. This very simple idea has wide applicability and deep consequences. It is the first pattern that children learn about mathematics and its structure underlies many other patterns.

5.1.1 Recursion and proof by mathematical induction

n	f_n
1	2
2	4
3	7
4	11
5	...

Suppose that $f = \{f_n\}$ is a sequence with 2 and 4 as the first two terms and with the property that the difference between successive terms grows by 1. The first four terms are shown in the table at the left.

We can determine that $f_5 = 16$ because $f_5 - f_4$ must be 5. Similarly, $f_6 = 22$ because $f_6 - f_5$ must be 6. We could go on in this way to compute successive terms of the sequence $f = \{f_n\}$. Many different formulas give the same values for the first few terms of this sequence. N. A. Sloane's *Handbook of Integer Sequences*, accessible on the Internet, lists several. Here are two:

$$(1) \qquad f_n = \frac{n^2}{2} + \frac{n}{2} + 1$$

$$(2) \qquad f_n = n^6 - 21n^5 + 175n^4 - 735n^3 + 1624.5n^2 - 1763.5n + 721.$$

Question 1: Show that formulas (1) and (2) give the same values for f_n when n is 1, 2, 3, 4, 5, and 6.

How can we determine whether either formula for f_n is correct for all natural numbers n? The answer lies in checking these explicit formulas against the given information describing the sequence. The given information already supplies a definition that determines the value of f_n for any value of n. That definition can be written as follows:

$$(3) \qquad \begin{cases} f_1 = 2, \\ \text{for all integers } n > 1, f_n = f_{n-1} + n. \end{cases}$$

Such a definition is called a *recursive definition* for the sequence f.

Definition A **recursive definition for a sequence** consists of two statements:

1. A specification of one or more initial terms of the sequence, called **initial values**;
2. An equation, called a **recurrence relation**, that relates each subsequent term to one or more of the previous terms.

Given a sequence $f = \{f_n\}$ defined recursively, as in (3), we often want to find an explicit formula for f that allows us to compute, for any natural number n, the nth term of f directly without computing all preceding terms first. Such a formula is usually conjectured on the basis of observation of a pattern in the first few terms. To prove that a conjectured particular explicit formula is valid, we apply a very general and very powerful method of proof called *mathematical induction*. (We will show in the next section how the validity of this method of proof rests on a fundamental property of the natural number system.)

> ***Proof by Mathematical Induction:*** For each natural number n, let $S(n)$ be a statement that is either true or false. To prove that $S(n)$ is true for all n, it is sufficient to
>
> 1. Prove $S(1)$ is true.
> 2. Prove that for all natural numbers k, the truth of $S(k)$ implies the truth of $S(k + 1)$.

In a proof by mathematical induction the verification of (1) is often called the **basis step**, while the verification of (2) is called the **inductive step**. The supposition of the truth of $S(k)$ in (2) for the proof is often called the **induction hypothesis**.

EXAMPLE For the sequence $f = \{f_n\}$ defined recursively by (3), use mathematical induction to prove that

$$S(n): f_n = \frac{n^2}{2} + \frac{n}{2} + 1$$

is true for every natural number n. (This means that $f_n = \frac{n^2}{2} + \frac{n}{2} + 1$ is an explicit formula for the sequence f.)

Solution **Basis step:** $S(1)$: Does this formula for f_1 yield $f_1 = 2$? We calculate.

$$f_1 = \frac{1^2}{2} + \frac{1}{2} + 1 = 2 \quad \text{Yes it does.}$$

Inductive step: Suppose that $S(k)$ is true for the natural number k; that is, suppose that

$$f_k = \frac{k^2}{2} + \frac{k}{2} + 1 \qquad \text{for some natural number } k.$$

Then, by the recurrence relation for f,

$$f_{k+1} = f_k + (k + 1) = \frac{k^2}{2} + \frac{k}{2} + 1 + (k + 1).$$

The preceding equation can be reduced to the form

$$f_{k+1} = \frac{(k + 1)^2}{2} + \frac{k + 1}{2} + 1.$$

This shows that $S(k + 1)$ is true based on the supposition that $S(k)$ is true.

Therefore, by mathematical induction, $S(n)$ is true for all natural numbers n. That is, for all natural numbers n,

$$f_n = \frac{n^2}{2} + \frac{n}{2} + 1$$

is an explicit formula for this sequence.

Question 2: Show that the formula (2) for f_n (immediately preceding Question 1) does not give the desired value when $x = 7$.

The method of proof by mathematical induction is very useful in a much broader context than that of verifying explicit formulas for recursively defined sequences, as you will see in this chapter and throughout the book. As another example, we consider a problem based on a puzzle toy invented by the French mathematician Édouard Lucas (1842–1891) in 1883 known today as the *Tower of Hanoi.*

Figure 1

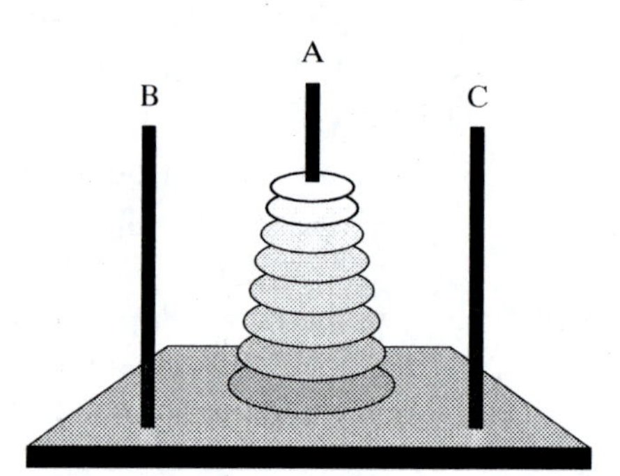

The Tower of Hanoi problem

In the Tower of Hanoi toy, there are three posts. On one post (A) there is a stack of disks with different diameters, increasing in size downward like those pictured in Figure 1. The problem is to transfer all the disks to one of the other posts (B or C) in the fewest possible moves, *moving one disk at a time in such a way that no disk ever is placed on top of a smaller disk.*

Lucas romanticized his problem by calling it a simplified version of a "Tower of Brahma" in a temple in the city of Benares, India. (In Hinduism, Brahma is the creator god, and Brahmins represent the highest caste. Benares is a real city, but there is no evidence that such a tower has ever existed.) He further went on to describe the disks as being made of gold and 64 in number. According to Lucas's story, the task of transferring the disks was given to the temple priests, and when the priests finish their task, the tower, temple, and Brahmins would crumble into dust and with a thunderclap the world would vanish.

We do not know how or why the location of the legend has switched from Benares, India, to Hanoi, Vietnam, while maintaining the priests and vanishing world, but Lucas's problem now has a name and a mathematical life of its own. The **Tower of Hanoi problem** is: What is the smallest number of moves required to transfer the 64 disks from one post to another following the instructions?

Finding a recursive formula

A counting problem with small numbers can often be solved by *enumeration*, that is, by listing all possibilities. But if a problem involves larger numbers, enumeration is generally impossible and often a general formula needs to be found. Some process is needed by which a potential solution is conjectured, and another process is needed to show that this conjecture is a solution.

Our first goal is to find a recursive formula for f_k, the number of moves required to transfer k disks. To find this formula, we need the initial conditions. So we solve the problem for the smallest numbers of disks not only to determine the initial conditions but also to see if there is some way in which a solution for k disks helps in finding a solution for $k + 1$ disks. So we solve the given problem with a smaller number of disks than 64, specifically, with 1 disk, then 2 disks, then 3 disks. That is, we find f_1, f_2, and f_3 and look for patterns.

In the work shown here, A, B, and C represent the three posts. One disk on A and no disks on B and C is represented as 1 0 0. A transfer from A to B is represented by an arrow connecting the previous step to the next step, in this case 1 0 0 → 0 1 0.

	A B C		A B C	(disk from A to B)
1-disk problem:	┃ │ │	→	│ ┃ │	

We write this as

$$1 \quad 0 \quad 0 \quad \rightarrow \quad 0 \quad 1 \quad 0.$$

So the 1-disk problem takes one move. Clearly, no smaller number of moves is possible. So $f_1 = 1$.

2-disk problem:	2 0 0	→	1 1 0	(top disk from A to B)
	1 1 0	→	0 1 1	(bottom disk from A to C)
	0 1 1	→	0 0 2	(top disk from B to C)

The 2-disk problem takes 3 moves if we move the disks according to the procedure prescribed above. No smaller number of moves is possible by any other procedure because any procedure must use one move to uncover the bottom disk, another move to transfer the bottom disk to another needle, and a third move to transfer the top disk on top of the bottom disk. Thus, $f_2 = 3$.

3-disk problem:	3 0 0	→	2 1 0	
	2 1 0	→	1 1 1	(top 2 disks from A to C)
	1 1 1	→	1 0 2	
	1 0 2	→	0 1 2	(Bottom disk from A to B)
	0 1 2	→	1 1 1	
	1 1 1	→	1 2 0	(2 disks from C to B)
	1 2 0	→	0 3 0	

To solve the 3-disk problem, we first transfer the top 2 disks from post A to post C. This is a 2-disk problem, so it can be done in 3 but no fewer moves. Then we transfer the bottom disk from A to B. Finally, we transfer the two top disks from post C on top of the disk on post B, as in the 2-disk problem. That again can be done in 3 and no fewer moves. So the total number of moves using this procedure is 7. How do we know that there is not another procedure that can achieve the desired result in fewer than 7 moves? Any successful procedure must eventually uncover the bottom disk. That part of the procedure will require at least 3 moves. The procedure would require another move to transfer the bottom disk to another peg and then at least 3 more moves to place the other two disks on top of the bottom disk. Thus, any successful procedure would require at least 7 moves, and so $f_3 = 7$.

This process is easily generalized. It shows that to transfer k disks, first transfer the top $k - 1$ disks from post A to post C (Figures 2a and b). This takes at least f_{k-1} moves. Then transfer the bottom disk in the original stack to post B (Figures 2b and 2c). That requires one move. Finally, transfer the $k - 1$ disks from post A on top of the disk on post B (Figures 2c and 2d). This again takes at least f_{k-1} moves. Thus, the total number of moves for transferring k disks by this procedure is at least $f_{k-1} + 1 + f_{k-1}$. Also, for any other procedure for transferring k disks successfully, there is a point at which the bottom disk on the original peg must be uncovered, and the other $k - 1$ disks must be properly stacked on the third peg. At that point at least f_{k-1} moves would have been used. Another move is necessary to transfer the bottom disk to the empty peg, and then at least another f_{k-1} moves to transfer the $k - 1$ remaining disks on top of the bottom disk. Consequently, any successful procedure uses at least $f_{k-1} + 1 + f_{k-1}$ moves and so $f_k = 2f_{k-1} + 1$. We now have both the initial values and a recurrence relation for the Tower of Hanoi problem.

Figure 2

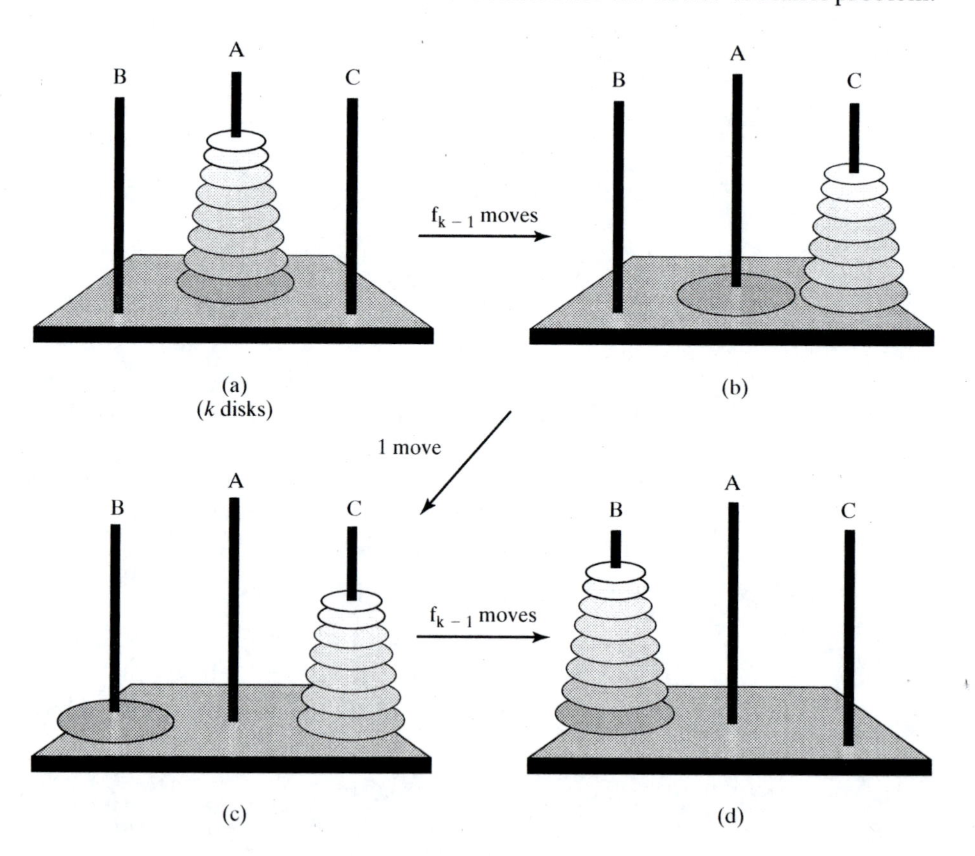

Question 3: By following the analysis in the 3-disk case show that f_4, the number of moves needed in the 4-disk problem, is 15.

Conjecturing an explicit formula

We summarize the known values of f_k in the following table.

Number of disks	k	1	2	3	4
Required moves	f_k	1	3	7	15

The table shows that the required moves form a sequence that begins 1, 3, 7, 15, ... Without some further analysis, we cannot tell what the ellipsis signifies. However, because $2^1 - 1 = 1, 2^2 - 1 = 3, 2^3 - 1 = 7$, and $2^4 - 1 = 15$, it would seem reasonable to conjecture that $f_n = 2^n - 1$ for all n, and so $2^{64} - 1$ moves would be required for the original 64 disks.

To distinguish what we wish to prove from what we know, let us call the conjectured sequence C. C is defined by $C_n = 2^n - 1$ for all n. Notice the difference between C and f. C is a sequence we have defined *explicitly* on the basis of a conjecture, whereas f is the Tower of Hanoi sequence for which we have a known *recursive* definition. If we can show that for all n, $C_n = f_n$, then our formula for C_n is an explicit formula for the Tower of Hanoi sequence.

Proving that the explicit and recursive formulas determine the same sequence

C_n is the sequence defined by $C_n = 2^n - 1$ for all positive integers n. f_n is the sequence defined by $f_1 = 1$ and $f_{n+1} = 2f_n + 1$ for all positive integers n. To prove that $C_n = f_n$ for all n, we use mathematical induction.

Let $S(k)$ be the statement that $C_k = f_k$.

Basis step: We need to show that $S(1)$ is true. $C_1 = 2^1 - 1 = 1$ and $f_1 = 1$, so $S(1)$ is true.

Inductive step: Suppose $S(k)$ is true. Then $C_k = f_k$ for some positive integer k. That is, $f_k = 2^k - 1$ for some integer k. We need to show that $C_{k+1} = f_{k+1}$. Because C_k is defined explicitly, $C_{k+1} = 2^{k+1} - 1$. Because f_{k+1} is defined recursively, $f_{k+1} = 2f_k + 1$, which by the induction hypothesis equals $2(2^k - 1) + 1 = 2^{k+1} - 2 + 1 = 2^{k+1} - 1$. Thus $C_{k+1} = f_{k+1}$ and so $S(k + 1)$ is true.

Thus, $f_n = C_n = 2^n - 1$ for all natural numbers n. In particular, the number of moves needed to transfer 64 disks is $C_{64} = 2^{64} - 1$. This solves the Tower of Hanoi problem.

Reviewing the process

In general, the method of proof by mathematical induction enables a proof of the validity of an explicit formula for a sequence from a recursive definition for that sequence. In the Tower of Hanoi problem, we began not knowing the initial conditions or the recursive definition. Finding f_1, f_2, and f_3 gave us the initial condition $f_1 = 1$ (as well as f_2 and f_3) and also showed the process by which f_{n+1} is obtained from f_n. We proved that our description of the process (shown in Figure 2) was accurate. (For many other sequences, we begin already knowing a recursive description.) The values of f_1, f_2, f_3, and f_4 also suggested an explicit formula. We let C_n be the nth term of the sequence with that explicit formula. Then we showed that $C_n = f_n$ for all n.

We have distinguished C and f here to clarify what happens in the inductive proof. In practice, the sequence C is called f too! A proof using mathematical induction will look like this.

Let $S(n)$ be the statement $f_n = 2^n - 1$. (On this line, f is what we have called C.)

Basis step: We know (from actual testing) that $f_1 = 1$. Since $2^1 - 1 = 1$, $S(1)$ is true.

Inductive step: Suppose $S(k)$ is true for some k. Then $f_k = 2^k - 1$. By the recursive definition, $f_{k+1} = 2f_k + 1$. Substituting, $f_{k+1} = 2(2^k - 1) + 1 = 2^{k+1} - 2 + 1 = 2^{k+1} - 1$. So $S(k + 1)$ is true for that k.

By mathematical induction, $S(n)$ is true for all integers $n \geq 1$.

In the next section, we show how the validity of the method of proof by mathematical induction rests on the structure of the natural number system.

5.1.1 Problems

1. Each sequence contains an ellipsis but no rule for next terms. Provide two rules that result in different next terms of the sequence. Give the next three terms of the sequence under each rule.

a. $2, 3, 5, \ldots$ b. $2, 4, 8, \ldots$ c. $5, 3, 6, \ldots$ d. $\frac{1}{2}, \frac{2}{3}, \frac{3}{5}, \ldots$

2. Show steps for solving the Tower of Hanoi problem when $k = 5$.

3. The original toy of Édouard Lucas had 8 disks. Suppose you were to carry out the moves to transfer the disks from one post (A) to another post (B) on this toy.

a. At least how many moves are required?

b. If the disks are moved in the manner described in this section, what would be the number of disks on each post A, B, and C after 10 moves?

c. If the disks are moved in the manner described in this section, what would be the number of disks on each post A, B, and C after 100 moves?

4. a. How many digits are there in the decimal (base 10) representation of $2^{64} - 1$?

b. If the priests in the Tower of Hanoi problem move one disk per second, approximately how many years are required to transfer the 64 disks from one post to another according to the rules? (An answer to the nearest thousand years is accurate enough.)

5. A *balance scale* is a scale in which weights are placed on two pans on opposite sides of a fulcrum. The scale balances if the weights are equal. Use a strategy similar to that employed with the Tower of Hanoi problem to consider the following balance-scale problems.

a. Suppose 9 coins are identical in appearance. Eight of these are identical in weight but the ninth is lighter than the others because it is counterfeit. Explain how it is possible to identify the counterfeit coin by using a balance scale just twice.

b. More generally, prove that if 3^n coins are identical in appearance, and if $3^n - 1$ of these are identical in weight but one is a lighter counterfeit coin, it is possible to identify the counterfeit coin by using a balance scale just n times.

*6. Explore and conjecture a solution of the variation of the Tower of Hanoi problem in which a fourth post is added.

ANSWER TO QUESTION

3. For 4 disks, first transfer the top 3 disks to post C as in the 3-disk problem (requires 7 moves), then move the bottom disk in the original stack to post B (1 move), and finally move the three disks from post C to post B (requires 7 moves). Total number of required moves for 4 disks $= 7 + 1 + 7 = 15$ for this transfer procedure. Because any successful transfer procedure must eventually uncover the bottom disk by moving the remaining 3 disks to another peg and then move them back after the bottom disk is moved, any such procedure uses at least $7 + 1 + 7 = 15$ moves. Thus, $f(4) = 15$.

5.1.2 Mathematical induction

In the preceding section, we used the method of proof by mathematical induction to prove that the minimum number of moves required to transfer a stack of n disks from one post to another, subject to the rules of the Tower of Hanoi problem, is $2^n - 1$. Now we examine this method in more detail.

The principle of proof by mathematical induction is sometimes described using a metaphor of infinitely many dominoes placed on their edges, one next to the other, so that if one is tipped over, the next also falls (if the kth domino is tipped over, then the $(k + 1)$st domino falls). Then if the first is tipped over (1 is tipped), all the dominoes fall.

This metaphor has one important difference from the mathematics. In the domino situation, time elapses between the falling of one domino and the falling of the next. In the mathematical situation, all the statements are proved true simultaneously.

It is important to realize that the method of proof by mathematical induction is quite different from inductive reasoning in mathematics or science. Roughly speaking, inductive reasoning is used to *conjecture* general results on the basis of the verification of these results for a variety of special cases, as we did to conjecture the explicit formula for f_n in the Tower of Hanoi problem in Section 5.1.1. On the other hand, mathematical induction is a form of deductive reasoning that can be used to *prove* results.

A classic proof using mathematical induction

Proofs involving mathematical induction are sometimes considered difficult. We find it helpful to identify, from the start, the statement $S(n)$ that we wish to prove for all natural numbers n. Even in obvious cases, just writing $S(n)$ may help in seeing how to get started.

Example 1 exhibits a classic type of problem proved using mathematical induction. Before beginning the induction proof, it is helpful to analyze the problem using the ideas of Section 5.1.1.

EXAMPLE 1 Prove: For every positive integer n, $1 + 2 + 3 + \cdots + n = \sum_{i=1}^{n} i = \frac{n(n+1)}{2}$.

Solution Let $S(n)$ be the statement $\sum_{i=1}^{n} i = \frac{n(n+1)}{2}$.

Basis step: $S(1)$: Does $\sum_{i=1}^{1} i = \frac{1(1+1)}{2}$? Yes, they both equal 1.

It is useful to write $S(k)$ and $S(k+1)$ before beginning the inductive step.

$$S(k): \quad \sum_{i=1}^{k} i = \frac{k(k+1)}{2}$$

$$S(k+1): \quad \sum_{i=1}^{k+1} i = \frac{(k+1)((k+1)+1)}{2} = \frac{(k+1)(k+2)}{2}$$

Inductive step: Does the truth of $S(k)$ imply the truth of $S(k+1)$?

Suppose $S(k)$ is true. Then $\sum_{i=1}^{k} i = \frac{k(k+1)}{2}$.

Because $\sum_{i=1}^{k+1} i = \sum_{i=1}^{k} i + (k+1)$,

our supposition implies
$$\sum_{i=1}^{k+1} i = \frac{k(k+1)}{2} + (k+1) = \frac{2k+2+k^2+k}{2} = \frac{k^2+3k+2}{2} = \frac{(k+1)(k+2)}{2}.$$

So $S(k+1)$ is true.

Therefore, by mathematical induction, $S(n)$ is true for all natural numbers n. ∎

Mathematical induction with basis step not at S(1)

It is sometimes convenient for the basis step not to refer to $n = 1$, but rather to $n = 0$, or some other integer n_0. Then, if you verify the basis step for n_0, and the inductive step for any integer $k \geq n_0$, the conclusion of the method of mathematical induction is that $S(n)$ is true for all $n \geq n_0$. In Question 1, you are asked to complete a proof of a statement about an n-gon (a polygon with n sides). For an n-gon, $n \geq 3$. Thus, in Question 1, the induction cannot begin with a number smaller than $n_0 = 3$.

Here, as in Example 1, the move from $S(k)$ to $S(k+1)$ is critical in the proof. Notice how the inductive step in the proof relies on knowing how the $(k+1)$-gon is related to the k-gon.

Question 1: Fill in the blanks to complete the proof that the sum of the degree measures of the interior angles of a convex n-gon is $180(n-2)$. (You may use the theorem that the sum of the degree measures of the interior angles of a triangle is 180°.)

Proof (by mathematical induction): Let $S(n)$: The sum of the degree measures of the interior angles of a convex n-gon is $180(n-2)$.

Basis step: $S(3)$: ________
$S(3)$ is true because ________.

Write $S(k)$: The sum of the degree measures of the interior angles of a convex k-gon is $180(k-2)$.

$S(k+1)$: ________

Inductive step: If $S(k)$ is true, then the sum of the degree measures of the interior angles of the convex k-gon $A_1A_2\ldots A_k$ is $180(k-2)$. Let $A_1A_2\ldots A_{k+1}$ be the convex $(k+1)$-gon formed by replacing $\overline{A_kA_1}$ with $\overline{A_kA_{k+1}}$ and $\overline{A_{k+1}A_1}$ (see Figure 3). The point A_{k+1} has to be in the exterior of the convex k-gon $A_1A_2\ldots A_k$ in order for the $(k+1)$-gon to be convex. Then the sum of the measures of the interior angles of $A_1A_2\ldots A_kA_{k+1}$ is ________. (You should fill in the blanks to finish the proof.) ┘

Figure 3

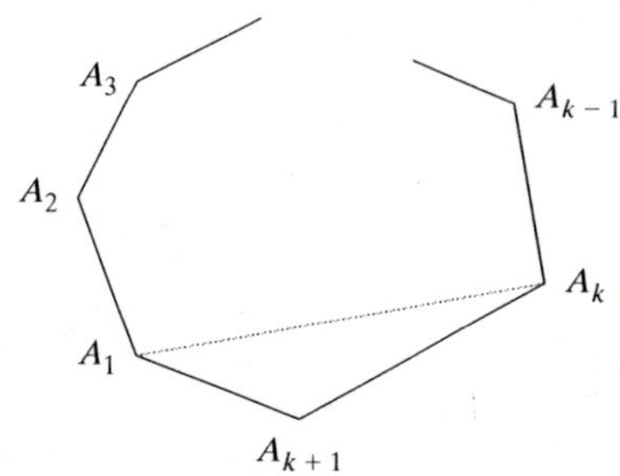

Origins of mathematical induction

Proof by mathematical induction has origins over 1600 years ago. In the fourth century, Pappus of Alexandria used arguments similar to the reasoning employed in mathematical induction to prove geometric theorems. The Arab mathematician and engineer al Karaji (c. 1000 A.D.) employed a kind of mathematical induction "in reverse", reasoning from a specific case of the result *backward* to the initial case. The French mathematician, astronomer, and philosopher Levi ben Gerson (1288–1344) used a method of proof that he called "rising step-by-step without end", in which he carried out what was essentially the inductive step first and then noted the truth of the result for some initial case.

The first explicit statements and applications of mathematical induction as a logical procedure close to its modern form were by the Italian priest Francisco Maurolico in 1575, in his book on arithmetic, and by the French mathematician and scientist Blaise Pascal (1623–1662), in 1654. Pascal used mathematical induction to prove relationships among the binomial coefficients in the triangular array of numbers called Pascal's Triangle[1] (Figure 4).

Figure 4

$$
\begin{array}{ccccccccccc}
 & & & & & 1 & & & & & \\
 & & & & 1 & & 1 & & & & \\
 & & & 1 & & 2 & & 1 & & & \\
 & & 1 & & 3 & & 3 & & 1 & & \\
 & 1 & & 4 & & 6 & & 4 & & 1 & \\
\cdot & & \cdot & & \cdot & & \cdot & & \cdot & & \cdot
\end{array}
$$

Proof by mathematical induction as a property of natural numbers

Why is proof by mathematical induction a valid method of reasoning? On what mathematical principles is it based? Proof by mathematical induction is based on *properties of the natural number system.* Many of these properties are familiar to us. For example, the set **N** of natural numbers is closed under the operations of addition and multiplication, these operations are commutative and associative, and multiplication distributes over addition [that is, $n(p + q) = np + nq$ for all natural numbers n, p, and q]. Also, for any pair m, n of natural numbers, one and only one of the relations

$$m < n, m = n, n < m$$

holds. Other properties are less familiar to us. For example, the property that any set of natural numbers that contains the smallest natural number 1 and that is closed under adding 1 must contain all natural numbers, or the property that each natural number n has a unique successor in the ordering of natural numbers.

Mathematicians in the nineteenth century discovered that these and all other known properties of natural numbers could be derived from a small set of assumptions. Two such mathematicians were Richard Dedekind[2] (whom we mentioned in Chapter 2) and Giuseppe Peano (1858–1932). Peano was a professor at the University of Turin in Italy who made important contributions to the foundations of mathematics. These mathematicians independently produced different, but logically equivalent, axiomatic descriptions of the natural number system in 1888 and 1889, respectively. The following set of axioms is a version of their systems written in modern notation and that, in most textbooks, is credited to Peano.

The **Peano axioms** (or **Peano postulates**): The natural number system is a set **N**, an element 1 of **N**, and a function s: **N** → **N** called the successor operation with the following properties:

P1: $1 \neq s(n)$ for any number n in **N**.

P2: For any numbers m and n in **N**, if $s(m) = s(n)$, then $m = n$.

P3: If M is a subset of **N** that contains the number 1 and if $s(n)$ is in M for each number n of M, then M = **N**.

[1] Although this triangular array bears his name in the West, he was not its discoverer. The array was known and used by the Arab mathematician al-Karaji in the 11th century and by the Chinese mathematician Jia Xian in the 13th century. See Katz, op. cit., pp. 191, 240–242.

[2] See Katz, op. cit., pp. 664–665, or *A History of Mathematics*, Second Edition, by Carl B. Boyer (John Wiley & Sons, 1989).

The Peano axioms can be paraphrased as follows: (P1) There is a natural number that is not the successor of any other. (P2) No number is the successor of more than one number. So the natural numbers do not circle back on themselves. (P3) The set of natural numbers **N** is the smallest set with properties P1 and P2.

The general idea of the derivation of the basic properties of addition and multiplication of natural numbers from the Peano postulates is not difficult. First, we need to obtain the natural numbers themselves. We define $s(1)$, the successor of 1, to be 2. Then we define $3 = s(2), 4 = s(3)$, and so on. Notice that this is exactly the way very young children learn their numbers, with a particular order but without any notion of operation. Also notice that the Peano postulates define the natural number system recursively.

Addition can be introduced first by defining "adding 1", $a + 1$, as the same as finding the successor of a. So $a + 1 = s(a)$. This, too, is a way that addition is often introduced in elementary school. Then $a + 2 = s(a + 1) = s(s(a)), a + 3 = s(a + 2) = s(s(s(a)))$, and so on. The derivation of these properties is outlined in greater detail in Project 1 of this chapter.

Then, with the symbol + for addition, P3 becomes the following theorem.

Theorem 5.1 Suppose that M is a subset of the set **N** natural numbers with the following two properties:

i. The natural number 1 belongs to M;

ii. If a natural number k belongs to M, then the next natural number $k + 1$ also belongs to M.

Then all natural numbers belong to M; that is, $M = \mathbf{N}$.

Now we prove that Theorem 5.1 implies the validity of the method of proof by mathematical induction. Suppose that for each natural number n, that $S(n)$ is a statement such that

a. $S(1)$ is true.

b. The assumption that $S(k)$ is true for some natural number k implies that $S(k + 1)$ is also true.

We must show that these suppositions (a) and (b) imply that $S(n)$ is true for all natural numbers n. To this end, let M be the set of all natural numbers n for which the statement $S(n)$ is true. Then supposition (a) says that the natural number 1 belongs to M; that is, M satisfies condition (i) of Theorem 5.1. Next, suppose that the natural number k belongs to M; that is, suppose that $S(k)$ is true for the natural number k. Then by supposition (b), it follows that $S(k + 1)$ is also true; that is, $k + 1$ also belongs to M, and so M satisfies condition (ii) of Theorem 5.1. Because M satisfies hypotheses (i) and (ii) of Theorem 5.1, it follows that every natural number n belongs to M; that is, the statement $S(n)$ is true for all natural numbers n.

The well-ordering property

Peano and Dedekind were not the only mathematicians to develop a set of axioms to describe the natural numbers. Mario Pieri (1860–1913), a contemporary of Peano's, developed an axiom system (see Problem 21) that replaces Axiom P3 with a principle equivalent to the *well-ordering property*, based on the existence of a smallest (or least) element in an ordered set.

Question 2:

a. Identify a subset of the positive rational numbers that does not have a least element.

b. Identify a nonempty subset of the integers that does not have a least element.

c. Try to explain why it is impossible to identify a nonempty subset of the natural numbers that does not have a least element.

We can derive the well-ordering property by using mathematical induction.

Theorem 5.2 **(Well-Ordering Property):** Any nonempty set of natural numbers contains a smallest number.

Proof: We use an indirect proof. Suppose T is a nonempty set of natural numbers that has no smallest number. We want to show that T cannot exist.

Let $S(n)$ be the statement: The natural numbers from 1 to n are not in T.

Basis step: $S(1)$: 1 is not in T. $S(1)$ is true, for if 1 were in T, then T would have a smallest number, for 1 is the smallest natural number.

Inductive step: Suppose $S(k)$ is true. Then 1 through k are not in T. But then $k + 1$ cannot be in T either, for it would be the smallest element of T. So $S(k + 1)$ is true.

Consequently, $S(n)$ is true for all n. But this means T has no elements. This contradicts the supposition that T is nonempty. So any set of natural numbers that has some elements must have a smallest number. ∎

Written as an if-then statement, the well-ordering property is

> If S is a nonempty set of natural numbers, then S has a smallest element.

The logically equivalent contrapositive is

> If S is a set of natural numbers without a smallest element, then S is empty.

The contrapositive is often the more convenient form in proofs. You will encounter some nonobvious applications of the well-ordering property in Unit 5.2.

5.1.2 Problems

1. Let $S(n)$ be the statement: The sum of the first n natural numbers is $\frac{1}{2}n^2 + \frac{1}{2}n + 1000$.

a. Show that if $S(k)$ is true, so is $S(k + 1)$.

b. Explain why $S(n)$ is not true for all natural numbers n.

In Problems 2–10, prove using the method of mathematical induction.

2. $1^2 + 2^2 + \cdots + n^2 = \frac{n(n+1)(2n+1)}{6}$ for all natural numbers n.

3. $1 \cdot 3 + 3 \cdot 5 + 5 \cdot 7 + \cdots + (2n - 1)(2n + 1) =$

$$\sum_{i=1}^{n}[(2i - 1)(2i + 1)] = \tfrac{1}{3}(4n^3 + 6n^2 - n)$$

4. $\frac{1}{1 \cdot 2} + \frac{1}{2 \cdot 3} + \cdots + \frac{1}{n \cdot (n+1)} = \frac{n}{n+1}$

5. $\sum_{i=1}^{n} i(i + 1) = \frac{n(n+1)(n+2)}{3}$

6. Every convex n-gon has $\frac{n(n-3)}{2}$ diagonals.

7. If $0 < x < 1$, then $0 < x^{(n+1)} < x^n < 1$ for each natural number n.

8. $2^n < n!$ for all integers $n > 3$.

9. There is an integer k such that $100^n < n!$ for all integers $n > k$.

10. a. If n people are standing in line and if the line starts with a woman and ends with a man, then somewhere in the line there is a man standing immediately behind a woman.

b. If a sequence of integers begins with an odd integer and ends with an even integer, then somewhere in the sequence an odd integer is followed immediately by an even integer.

11. From Example 1 in this section and the result of Problem 5, find an explicit formula for $\sum_{i=1}^{n} i(i+1)(i+2)$ and prove the result using mathematical induction.

12. Are there formulas like $1 + 2 + 3 + \cdots + n = \frac{n(n+1)}{2}$ that are true for the first million cases and then false after that? If so, can you find one? If not, explain why not.

13. Give an example of a statement $S(n)$ different from the one in the section:

a. that is false for all natural numbers n, but for which the assumption that $S(k)$ is true implies $S(k+1)$ is true.

b. that is true for all natural numbers n but for which the assumption that $S(k)$ is true for some k is not needed to show that $S(k+1)$ is true.

14. Consider the statement:

> *Any set of coins chosen from a box will all have the same denomination.*

This is obviously not a true statement, since in general a set of coins will contain more than one denomination. Yet we present an argument that seems to prove that the statement is true by mathematical induction. Find the flaw in the "Proof".

"Proof": Let $P(n)$ be the statement "Any set of n coins chosen from a box all have the same denomination."

$P(1)$ is clearly true.

Suppose that $P(k)$ is true. Consider any set of $k+1$ coins: Remove one coin. Since $P(k)$ is true, the remaining k coins all have the same denomination: Now put that coin back, and remove a different coin. Again, since $P(k)$ is true, the remaining k coins all have the same denomination. Therefore, all $k+1$ coins have the same denomination, that is, $P(k+1)$ is true. Thus, by the principle of mathematical induction, $P(n)$ is true for all n.

15. Use the well-ordering property of **N** to prove the formula

$$1 + 2 + \cdots + n = \frac{n(n+1)}{2} \quad \text{for all } n \in \mathbf{N}$$

that was proved by mathematical induction in this section.

16. Prove the following statement using the well-ordering property: Every natural number >1 is divisible by a prime number.

17. a. Conjecture an explicit formula for s_n if $s_n = 1 \cdot 1! + 2 \cdot 2! + 3 \cdot 3! + \cdots + n \cdot n!$.

b. Use mathematical induction to prove that your conjecture is correct.

18. Conjecture and prove a statement describing the parity (even or odd character) of the terms of the sequence u defined by $u_1 = 1, u_2 = 2, u_{n+2} = 2u_n + u_{n+1}$.

19. Prove each statement from the Peano axioms.

a. For all $n \in \mathbf{N}$, $s(n) \neq n$.

b. For each $n \in \mathbf{N}$ such that $n \neq 1$, there is an $m \in \mathbf{N}$ such that $s(m) = n$.

20. Show that if any one of the Peano axioms is eliminated, the remaining two axioms are satisfied by sets very different from the set **N** of natural numbers by verifying the following statements:

a. The set $\mathbf{Q}^+$ of rational numbers with the function $s: \mathbf{Q}^+ \to \mathbf{Q}^+$ defined by $s(r) = r + 1$ for all r in $\mathbf{Q}^+$ satisfies axioms P1 and P2 but not P3.

b. The singleton set $S = \{1\}$ consisting of the natural number 1 and the identity function on S satisfies P2 and P3 but not P1.

c. The set $S = \{1, 2\}$ consisting of the two natural numbers 1 and 2, and the function s on S defined by $s(1) = 2$ and $s(2) = 2$ satisfies P1 and P3 but not P2.

21. Here is the alternative set of axioms for the system **N** of natural numbers based on the primitive notions of number and the successor operation s, as developed by Pieri in 1908:

*M*1: There exists at least one number.

*M*2: The successor of a number is a number.

*M*3: Two numbers, neither of which is the successor of a number, are equal.

*M*4: In any nonempty set of numbers there exists at least one number that is not the successor of any number in the set. [This is the well-ordering property, written in terms of the successor operation.]

a. From these axioms show that there is exactly one element in **N** that is the not the successor of any number.

b. Show that the principle of mathematical induction, Peano's axiom P3, can be deduced from Pieri's postulates.

22. Properties of multiplication of natural numbers can be derived from the Peano axioms and addition as follows:

> For all natural numbers n, define $n \cdot 1$ to be n. Then define $n \cdot (k+1)$ to be $n \cdot k + n$ and $(k+1) \cdot n$ to be $k \cdot n + n$.

Now use mathematical induction to prove the following properties:

a. $1 \cdot n = n$ for all natural numbers n.

b. $k \cdot n = n \cdot k$ for all natural numbers n and k.

c. $n \cdot (k + \ell) = n \cdot k + n \cdot \ell$ for all natural numbers n, k, and ℓ.

23. Al Karaji's proof of the formula for the sum of the first n cubes,

$$(*) \quad 1^3 + 2^3 + 3^3 + \cdots + n^3 = (1 + 2 + 3 \cdots + n)^2$$

for all positive integers n, was based on the following diagram for a "backward" induction from the case $n = 10$ (Figure 5).

Figure 5

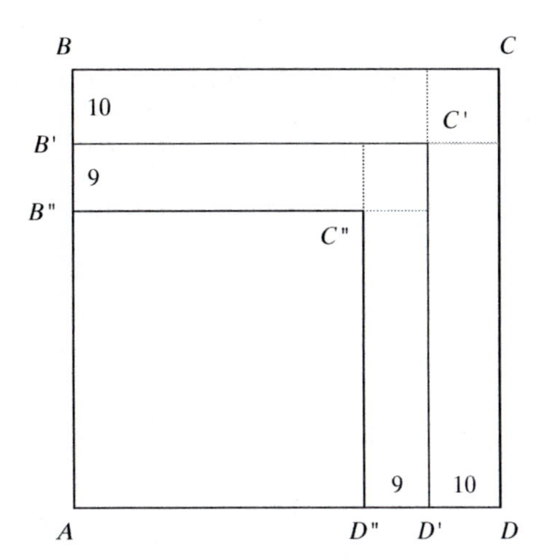

Side length of square $ABCD$

$= 1 + 2 + \ldots + 10 = \dfrac{(10)(11)}{2}$.

Side length of square $AB'C'D'$

$= 1 + 2 + \ldots + 9 = \dfrac{(9)(10)}{2}$.

a. Conclude from this diagram that

i. The "inverted L" polygon $B'C'D'DCBB'$ has area 10^3.

ii. The "inverted L" polygon $B''C''D''D'C'B'B''$ has area 9^3.

b. Explain how the conclusions in part **a** can be extended to an explanation of the formula

$$1^3 + 2^3 + 3^3 + \cdots + 9^3 + 10^3 = (1 + 2 + 3 \cdots + 9 + 10)^2.$$

Use mathematical induction to prove the result ($*$).

24. a. Prove: If $\frac{a}{b} = \sqrt{2}$, then $\frac{a}{b} = \frac{a+2b}{a+b}$.

b. In part **a**, we know that a and b cannot both be integers, but part **a** tells us that if $\frac{a}{b}$ is close to $\sqrt{2}$, so is $\frac{a+2b}{a+b}$. Let $S_n = \frac{a_n}{b_n}$, where $a_1 = 7, b_1 = 5, a_{n+1} = a_n + 2b_n$, and $b_{n+1} = a_n + b_n$ for all $n \geq 1$. Write the first five terms of $\{S_n\}$ and verify that S_{n+1} is closer to $\sqrt{2}$ than S_n.

*c. Prove that if $a > 0$ and $b > 0$, then $\frac{a+2b}{a+b}$ is closer to $\sqrt{2}$ than $\frac{a}{b}$.

ANSWERS TO QUESTIONS

1. The sum of the degree measures of a triangle is $180(3 - 2)$. Thus the sum of the degree measures of the angles of a triangle is 180°. The sum of the degree measures of the interior angles of a convex $(k + 1)$-gon is $180(k + 1 - 2)$, or $180(k - 1)$. Thus the sum of the measures for $A_1A_2 \ldots A_k + m\angle A_kA_1A_{k+1} + m\angle A_{k+1} + m\angle A_1A_kA_{k+1} = 180(k - 2) + 180 = 180(k - 1)$. So $S(k + 1)$ is true. Thus, by mathematical induction, $S(n)$ is true for all integers $n \geq 3$.

2. a. For example, the set of all rational numbers between 2 and 3.

b. For example, the set of all integers less than 10.

5.1.3 More applications of mathematical induction

The examples and problems in this section illustrate some aspects of mathematical induction that did not arise in the preceding section.

An application to a recursively defined sequence

Consider the famous Fibonacci sequence $1, 1, 2, 3, 5, 8, 13, 21, 34, 55, 89, 144, 233, 377, 610, \ldots$, defined by

$$\begin{cases} F_1 &= 1 \\ F_2 &= 1 \\ F_{n+1} &= F_n + F_{n-1}, \qquad \text{for all } n \geq 2. \end{cases}$$

Unlike our earlier examples, the recurrence relation specifies the value of F_{n+1} in terms of the *two* immediately preceding terms, F_n and F_{n-1}. For this reason, the recurrence relation for the Fibonacci sequence is said to be of *second order*, while those in the earlier examples were of *first order*. For second-order recurrence relations, two initial values, F_1 and F_2, must be specified to "start" the recursion.

There are so many relationships involving Fibonacci numbers that a journal (*The Fibonacci Quarterly*) and numerous books (see the references at the end of the chapter) have been devoted to them. You can sometimes discover a relationship by playing around with a function based on the numbers. For instance, let $A(n)$ be the sum of the first n Fibonacci numbers. That is, $A(n) = \sum_{i=1}^{n} F_i$. Here are the first six values of $A(n)$.

$$\begin{aligned} A(1) &= 1 \\ A(2) &= 1 + 1 = 2 \\ A(3) &= 1 + 1 + 2 = 4 \\ A(4) &= 1 + 1 + 2 + 3 = 7 \\ A(5) &= 1 + 1 + 2 + 3 + 5 = 12 \\ A(6) &= 1 + 1 + 2 + 3 + 5 + 8 = 20 \end{aligned}$$

Question: Cover the page below this question. How are the values of $A(n)$ related to the Fibonacci sequence? [*Hint:* Calculate more values of $A(n)$ if you need to.]

For these six values of n, $A(n) = F_{n+2} - 1$. Is this true for all natural numbers n? Because we know the recurrence relation is $A(k+1) = A(k) + F_{k+1}$, a mathematical induction proof is a reasonable strategy. Let $S(n)$ be the statement: $A(n) = F_{n+2} - 1$.

Basis step: We need two statements in this step because each term in the Fibonacci sequence is defined using the two previous terms.

$$A(1) = 1 \quad \text{and} \quad F_{1+2} - 1 = F_3 - 1 = 2 - 1 = 1, \text{ so } S(1) \text{ is true.}$$
$$A(2) = 2 \quad \text{and} \quad F_{2+2} - 1 = F_4 - 1 = 3 - 1 = 2, \text{ so } S(2) \text{ is true.}$$

Inductive step: Suppose $S(k)$ is true. That is, suppose $A(k) = F_{k+2} - 1$.

$$\begin{aligned} \text{Now } A(k+1) &= \sum_{i=1}^{k+1} F_i = F_{k+1} + \sum_{i=1}^{k} F_i = F_{k+1} + A(k) \\ &= F_{k+1} + F_{k+2} - 1 \qquad \text{by the inductive hypothesis} \\ &= F_{k+3} - 1 \\ &= F_{(k+1)+2} - 1 \end{aligned}$$

Thus $S(k+1)$ is true.

Thus, by mathematical induction, $S(n)$ is true for all n, and so

$$A(n) = F_{n+2} - 1 \text{ is true for all natural numbers } n.$$

The following example illustrates that it is sometimes necessary to modify the straightforward choice of the statement $S(n)$ in proofs by mathematical induction.

Consider the sequence $U = \{u_n\}$ defined recursively as follows:

$$u_1 = 1, u_2 = 1, \qquad u_{n+2} = 2u_{n+1} + u_n \qquad \text{for } n \geq 1$$

The first ten terms of u are

$$1, 1, 3, 7, 17, 41, 99, 239, 577, \text{ and } 1393.$$

They are all odd numbers. Is it true that, for all natural numbers n, u_n is an odd natural number?

Let $S(n)$: u_n is an odd natural number.

We try to prove $S(n)$ is true for all natural numbers n by mathematical induction.

Basis step: $S(1)$ is true because $u_1 = 1$ is an odd number.

Inductive step: Suppose that $S(k)$ is true for some natural number k. That is, suppose that u_k is an odd number. Now $u_{k+1} = 2u_k + u_{k-1}$ and $2u_k$ is even. But it would follow that u_{k+1} is odd only if u_{k-1} were odd, and we do not know that.

Thus, the supposition that u_k is an odd number is not sufficient to conclude that u_{k+1} is odd. However, if we replace the statement $S(n)$ by

$$S'(n): \quad u_n \text{ and } u_{n-1} \text{ are odd natural numbers,}$$

then mathematical induction can be used to prove that u_n is odd for all n. (See Project 2 for a general approach to the use of mathematical induction for problems of this sort.)

An application to a basic property of exponents

We noted earlier that recursive definitions provide a way to deal with definitions in which there is an ellipsis.... These dots can occur in the middle of a mathematical expression as well as at an end. Consider the usual definition of x^m when m is a natural number.

$$x^m = \underbrace{x \cdot x \cdot \cdots \cdot x}_{m \text{ factors}}$$

The explicit definition of x^m does not fit well when $m = 1$. The recursive definition of x^m that follows avoids the ellipsis and also explicitly defines x^1:

Initial value: $x^1 = x$

Recurrence relation: For all $m \geq 1$, $x^{m+1} = x \cdot x^m$.

With a recursive definition, the product of powers property $x^m \cdot x^n = x^{m+n}$ (the most basic of the laws of exponents) can be deduced using mathematical induction.

Let $S(k)$ be the proposition that $x^m \cdot x^k = x^{m+k}$ for all positive integers m.

Basis step: We need to show $S(1)$, namely that $x^m \cdot x^1 = x^{m+1}$ for all positive integers m. Begin with the left side. From the initial value in the recursive definition of x^m, $x^m \cdot x^1 = x^m \cdot x$. The recurrence relation, together with the commutative property of multiplication, shows that $x^m \cdot x = x^{m+1}$.

Inductive step: Suppose $S(k)$ is true. Then $x^m \cdot x^k = x^{m+k}$ for all positive integers m. Then $x^m \cdot x^{k+1}$

$= x^m \cdot (x^k \cdot x)$ for all positive integers m, from the recursive definition

$= (x^m \cdot x^k) \cdot x$ from the associative property of multiplication

$= x^{m+k} \cdot x$, since $S(k)$ is true by the inductive hypothesis

$= x \cdot x^{m+k}$ by the commutative property of multiplication

$= x^{m+(k+1)}$ from the recursive definition applied once again.

Thus $S(k + 1)$ is true.

And so the proposition $x^m \cdot x^n = x^{m+n}$ for all positive integers m *and* n.

The other two basic laws of exponents can be deduced in a similar way. (See Problems 9 and 10.)

5.1.3 Problems

1. a. Give a recursive description of the sequence of increasing even numbers 2, 4, 6, ...

   b. Give a recursive description of the sequence of increasing even numbers 0, 2, 4, ...

2. Let F_n be the nth Fibonacci number. Deduce these properties without using the method of mathematical induction.

   a. $F_n = 2F_{n-2} + F_{n-3}$

   b. $F_n = 5F_{n-4} + 3F_{n-5}$

   c. $F_n^2 - F_{n-1}^2 = F_{n+1} \cdot F_{n-2}, \quad \text{for all } n > 2$

3. Suppose that $\{F_n\}$ is the Fibonacci sequence.

   a. Evaluate the expression
   $$F_n^2 - F_{n+1} \cdot F_{n-1}, \quad \text{where } n \text{ is an integer} > 1$$
   for the first few values of n. Use the results to formulate a conjecture concerning the values of this expression.

   b. Prove or find a counterexample for your conjecture in part **a**.

   c. Repeat parts **a** and **b** for the expression
   $$F_n^2 - F_{n+2} \cdot F_{n-2} \quad \text{where } n \text{ is an integer} \geq 2.$$

   d. Repeat parts **a** and **b** for the expression
   $$F_n^2 - F_{n+3} \cdot F_{n-3} \quad \text{where } n \text{ is an integer} \geq 3.$$

4. Find a simpler expression for $F_{n+3} \cdot F_{n+2} - F_{n+1} \cdot F_n$ in terms of Fibonacci numbers and prove that your expression is correct.

5. Let $Q(n)$ be the sum of the squares of the first n Fibonacci numbers.

   a. Find $Q(1), Q(2), \ldots, Q(6)$.

   b. Conjecture a formula for $Q(n)$ in terms of F_n.

   c. Prove your formula.

6. The **Lucas numbers** (named after Édouard Lucas, who studied them) are determined by the following recursive definition:
$$\begin{cases} L_1 = 1 \\ L_2 = 3 \\ L_{n+1} = L_n + L_{n-1}, & \text{for all } n \geq 2. \end{cases}$$

   a. Write the smallest 10 Lucas numbers.

   b. The recursion in the Lucas sequence is the same as in the Fibonacci sequence. Which of the properties of the Fibonacci numbers mentioned in this section and in Problems 2–5 are also properties of the Lucas numbers?

7. Prove that, if a fair coin is tossed n times, the probability that somewhere two consecutive tosses will be heads is $1 - \frac{F_{n+2}}{2^n}$. (*Hint*: Let P_n be the probability that two consecutive heads do not appear in the n tosses. Use mathematical induction.)

8. Consider the sequence U defined in this section. Use mathematical induction to prove that the terms of u are all odd numbers.

9. Use the recursive definition of x^m and mathematical induction to prove: For all natural numbers m and n, $(x^m)^n = x^{mn}$. (*Hint*: Use the recursive definition of this section.)

10. Use the recursive definition of x^m and mathematical induction to prove: For all natural numbers m, $(xy)^m = x^m y^m$.

11. Conjecture a formula for the nth term of the sequence that is defined recursively by
$$x_1 = 3;\, x_2 = 5, \text{ and } x_{n+2} = 3x_{n+1} - 2x_n \quad \text{for all } k \geq 1.$$
Use mathematical induction to prove your conjecture.

12. a. Conjecture a simple formula for the product
$$P(n) = \left(1 - \frac{1}{2^2}\right)\left(1 - \frac{1}{3^2}\right)\cdots\left(1 - \frac{1}{n^2}\right) \quad \text{for } n > 1$$
   and prove your conjecture with mathematical induction.

   b. Use the factoring formula $a^2 - b^2 = (a - b)(a + b)$ to simplify the product to the same formula you conjectured in part **a**. Does this simplification prove the formula in part **a** without using mathematical induction?

13. The coefficients in the expansions of the $(a + b)^n$ are often presented in the form of **Pascal's Triangle**:

$$\begin{array}{lccccccccc}
n = 0 & & & & & 1 & & & & \\
n = 1 & & & & 1 & & 1 & & & \\
n = 2 & & & 1 & & 2 & & 1 & & \\
n = 3 & & 1 & & 3 & & 3 & & 1 & \\
n = 4 & 1 & & 4 & & 6 & & 4 & & 1
\end{array}$$

in which the numbers on the edges of the array are all equal to one and each interior number is the sum of the two numbers diagonally above it. For each nonnegative integer n, there are $n + 1$ entries in row n. Number these from left to right with $k = 0, 1, 2, \ldots, n$. The entry in position k of the nth row is denoted by $\binom{n}{k}$.

   a. Explain why the construction of Pascal's Triangle implies that the following relationship holds for the coefficients in the binomial expansion of $(a + b)^n$:
   $$\binom{n+1}{k} = \binom{n}{k-1} + \binom{n}{k} \quad \text{for } k = 1, 2, \ldots, n.$$

   b. Use mathematical induction to prove the Binomial Theorem:
   $$(a + b)^n = \sum_{k=0}^{n} \binom{n}{k} a^{n-k} b^k.$$
   [*Hint*: For the inductive step, expand $(a + b)^{n+1}$ as $(a + b)(a + b)^n$ and collect like terms. Then apply the formula in part **a**.]

5.1.4 An extended analysis of an induction situation

In this section, we examine a related set of problems whose solutions involve mathematical induction.

The problem and its initial analysis: Regions of the plane

A point P on a line ℓ divides that line into two regions. Two distinct points P_1 and P_2 on a line ℓ divide ℓ into 3 regions. Three distinct points P_1, P_2, and P_3 on a line ℓ divide ℓ into 4 regions. These situations are shown in Figure 6.

Figure 6

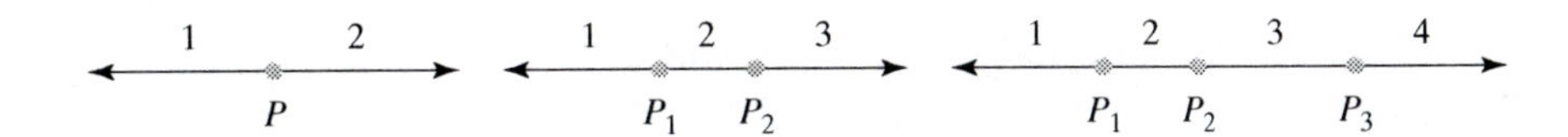

Stepping up a dimension, observe that a line ℓ in a plane M divides the plane into two regions. Two distinct nonparallel lines ℓ_1 and ℓ_2 in a plane M divide M into 4 regions.

Question 1: How many regions are formed if the two lines are parallel?

Three distinct lines ℓ_1, ℓ_2, and ℓ_3 in a plane M such that no two are parallel and they are not concurrent divide M into 7 regions (Figure 7).

Figure 7

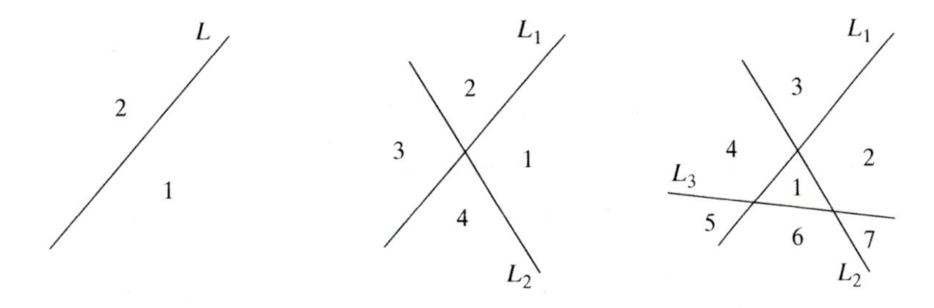

Question 2: If the three lines are concurrent, how many regions are formed?

Going on to three dimensions, note that a plane M divides three-dimensional space S into two regions. Two planes M_1 and M_2 in general position (distinct and nonparallel) divide S into 4 regions (Figure 8).

Figure 8

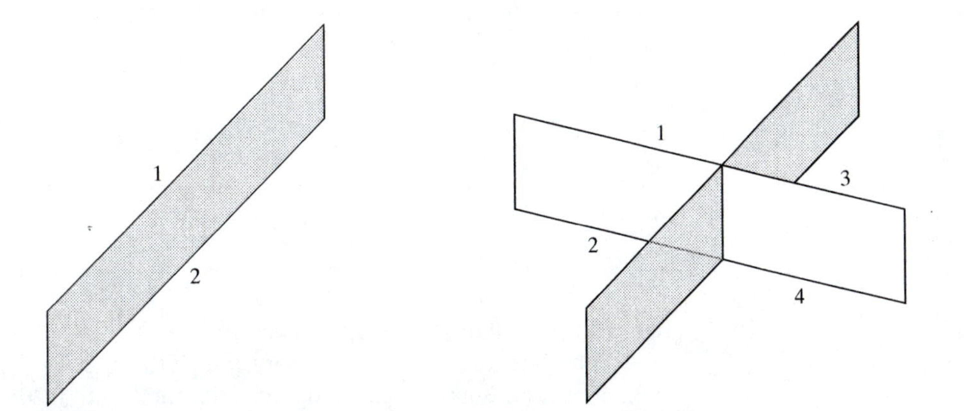

Question 3: If the two planes are parallel, how many regions are formed?

With a little more effort, you may be able to visualize that the maximum number of regions determined by three planes M_1, M_2, and M_3 in space where no two are parallel and no line is common to all three planes is 8. However, if we increase the number of planes beyond three, the problem becomes quite difficult to visualize. Still we would like to explore this problem, as well as the lower dimension problems, and find patterns.

These examples suggest the following problems:

Problem 1: At most how many regions on a line are determined by n points on the line?

Problem 2: At most how many regions in a plane are determined by n lines in the plane?

Problem 3: At most how many regions in space are determined by n planes in space?

The information that we have obtained for the cases $n = 1, 2$, and 3, with the given conditions on points, lines, and planes, is summarized in Table 1.

Table 1

Problem	Description	$n = 1$	$n = 2$	$n = 3$
1	n points on line	2 regions	3 regions	4 regions
2	n lines in plane	2 regions	4 regions	7 regions
3	n planes in space	2 regions	4 regions	8 regions

Let us create some notation that will facilitate a solution. Let

f_n = the maximum number of regions on a line determined by n points on the line,

g_n = the maximum number of regions in a plane determined by n lines in the plane,

and

h_n = the maximum number of regions in space determined by n planes.

Can we find explicit formulas or recurrence relations that would allow us to compute any desired term of the sequences $\{f_n\}$, $\{g_n\}$, and $\{h_n\}$? Are these three sequences related?

Further exploration of Problem 1 as in Figure 9 leads us to Table 2.

Figure 9

Table 2

n	1	2	3	4	5	6	7	8
f_n	2	3	4	5	6	7	8	9

Based on these data, we might conjecture that $f_n = n + 1$ for all natural numbers n.

The conjecture $f_n = n + 1$ is an explicit formula for f_n in term of n. But to use mathematical induction, we need a formula for f_{n+1} in terms of f_n and n. So we ask ourselves what happens when an additional point is added to the line, and observe the following: Each time a new point P is added to the line, it must be inserted in an existing region R determined by the previous points. As a result, that region R is divided into two regions, R' and R'', while all other previous regions remain unaltered. Thus, exactly one new region is added when a new point is added to the line. That is,

$$f_{n+1} = f_n + 1 \quad \text{for each natural number } n.$$

This equation for f_{n+1} in terms of f_n is a recurrence relation for f_n. It allows us to deduce the explicit formula

$$f_n = n + 1 \quad \text{for all natural numbers } n$$

easily, using mathematical induction. Let $S(n)$ be the statement $f_n = n + 1$.

Basis step: $f_1 = 1 + 1 = 2$. So $S(1)$ is true.

Inductive step: Suppose $S(k)$ is true; that is, that $f_k = k + 1$ for some positive integer k. Then by the recurrence relation,

$$f_{k+1} = f_k + 1 = (k + 1) + 1.$$

Thus $S(k + 1)$ is true.

Therefore, by mathematical induction, $f_n = n + 1$ for all natural numbers n.

Next, we proceed to an analysis of Problem 2. We have already observed that if g_n is the maximum number of regions in a plane determined by n lines in the plane, then $g_1 = 2$, $g_2 = 4$, and $g_3 = 7$. Figure 10 shows that $g_4 = 11$.

Figure 10

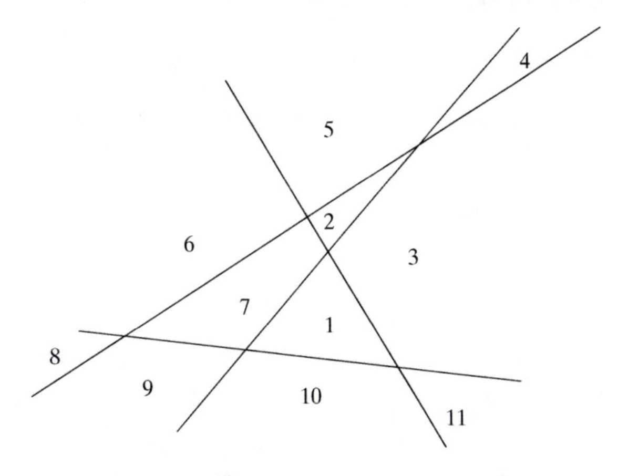

We have recorded the results from Figures 7 and 10 in Table 3.

Table 3

Lines $n =$	1	2	3	4	5	6	7
Regions $g_n =$	2	4	7	11			

We could go on to find g_5, g_6, etc., in the same way, but the diagrams get harder and harder to construct. To find an explicit formula for this sequence, we need a strategy.

Conjecturing an explicit formula

The Method of Finite Differences (Section 3.3.2) can help construct explicit formulas for sequences for which partial data lists like that in Table 3 are available. The first differences are 1, 2, 3, 4, and the second differences are 1, 1, 1. Since the second differences are constant, we expect that g_n might be a quadratic function of n. To follow up this idea, we let $g_n = An^2 + Bn + C$, where A, B, and C are constants to be determined by the partial data in Table 3. Then, using the values of g from the table, we have

$$\begin{aligned} 2 &= A + B + C &\quad (\text{since } g_1 = 2)\\ 4 &= 4A + 2B + C &\quad (\text{since } g_2 = 4)\\ 7 &= 9A + 3B + C &\quad (\text{since } g_3 = 7). \end{aligned}$$

We solve this system of three equations for A, B, and C.

$$A = \frac{1}{2}, \quad B = \frac{1}{2}, \quad \text{and } C = 1.$$

Hence a conjecture for an explicit formula for g_n is

$$g_n = \frac{1}{2}n^2 + \frac{1}{2}n + 1. \tag{1}$$

Note that formula (1) has been derived from just the first three values in Table 3. If we use formula (1) to compute the remaining entry g_4, we get the correct value of 11. This gives us some confidence that we are on the right track with formula (1). But we have not proved that formula (1) is correct. All we know is that it gives the right number of regions for $n = 1$ to $n = 4$.

Since it is hard to construct diagrams for the next two or three higher values of n, and impossible to construct diagrams for all values of n, we now search for a recursive formula for the sequence $\{g_n\}$.

Finding a recursive formula

Figure 11 displays the four lines in Figure 10 that were used to evaluate g_4 together with a fifth dashed line ℓ.

Figure 11

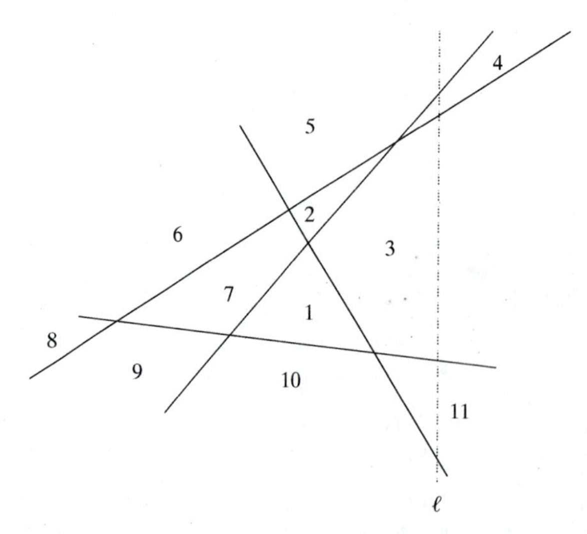

Because ℓ is not parallel to any other line and goes through no earlier point of intersection of lines, each time ℓ intersects an existing line, it divides the region it has just passed through into two regions. After ℓ passes through the 4th line, it also divides the last region it passes through into two regions. In short, the *5th* line creates *5* new regions.

The same argument shows that adding the $(n + 1)st$ line to a configuration of n lines in general position creates $n + 1$ new regions. We can express it as a recurrence relation.

$$g_{n+1} = g_n + (n + 1) \qquad \text{for all natural numbers } n \tag{2}$$

Equation (2) says that the number of regions created with $n + 1$ lines is $n + 1$ more than the number of regions created with n lines. Together with the initial value $g_1 = 2$, equation (2) provides a recursive definition for the sequence $\{g_n\}$.

Equation (2) does *not* give us a way of directly computing g_n for some prescribed n without first computing the previous term. However, we do know that (2) holds for all positive integers n. On the other hand, equation (1) is an explicit formula for the sequence $\{g_n\}$ that allows us to calculate any desired term g_n without first calculating all preceding terms. However, based on our analysis to this point we cannot be sure that formula (1) is correct beyond $n = 4$. Mathematical induction allows us to connect the two formulas and to complete the solution of Problem 2.

Finishing the solution

We now prove that the formula (1) for the sequence $\{g_n\}$ is correct for all natural numbers n. We base the proof on mathematical induction and the facts that $g_1 = 2$ and that the recurrence relation (2) holds for all natural numbers n.

Basis step: For $n = 1$, the formula

$$g_n = \frac{1}{2}n^2 + \frac{1}{2}n + 1 \tag{1}$$

gives $g_1 = 2$, which we know is correct.

Inductive step: Suppose that (1) is correct for a natural number k. That is, suppose

$$g_k = \frac{1}{2}k^2 + \frac{1}{2}k + 1$$

for some natural number k. Use the recurrence relation (2) to find the value of g_{k+1}.

$$g_{k+1} = g_k + (k + 1) = \left(\frac{1}{2}k^2 + \frac{1}{2}k + 1\right) + (k + 1)$$

The right-hand side of the preceding equation can be expressed in the form

$$g_{k+1} = \frac{1}{2}(k + 1)^2 + \frac{1}{2}(k + 1) + 1,$$

which is exactly what formula (1) gives for g_{k+1}.

Therefore, by mathematical induction, formula (1) is a correct explicit formula for the maximum number of regions in a plane formed by n lines in the plane.

Extending the ideas to solve Problem 3

To solve Problem 2, first we drew some diagrams and used them to find directly a few values of the sequence $\{g_n\}$. Then we used a data analysis tool, the method of finite differences, to conjecture an explicit formula, formula (1), for this sequence that fit the data from our diagrams but that we could not be sure solved Problem 2. Then we searched for a recursive relationship between the diagrams for successive values of n. This resulted in a difference equation (2) that described this relationship for all n. Finally, we used equation (2) and mathematical induction to verify that formula (1) was an explicit formula for the sequence $\{g_n\}$.

You should now be able apply a similar analysis to solve Problem 3 with perhaps just a few guiding suggestions and observations.

Problem 3: At most how many regions in space are determined by n planes in space?

Conjecturing an explicit formula: Recall that we have set h_n to be the maximum number of regions in space determined by n planes on the line, and that we showed earlier by direct counting that $h_1 = 2$, $h_2 = 4$, and $h_3 = 8$.

> Notice that these three data suggest the conjecture $h_n = 2^n$ for all natural numbers n. Check that conjecture for $n = 4$ to show that it is false; actually, $h_4 = 15$, not 16.

You will probably find counting the regions for $n = 4$ to be difficult with a diagram because the diagram is very complicated. You can do it with a three-dimensional model made of cardboard and a good visual imagination. Or think of trying to slice a cheese ball into the largest number of pieces with n cuts and no rearranging of the pieces.

Finding a recursive formula: Imagine the result of adding a fourth plane P to a configuration of three planes. The intersection of each of the three given planes with the new plane P will be a line in P. Also, P can be chosen so as to divide each region (determined by the original three planes) that P passes through into two new regions.

> Explain why adding a new plane P to a configuration of three planes can add 7 more regions to the configuration so that $h_4 = h_3 + 7 = 15$.

Conjecturing a second explicit formula: Use the method of finite differences in a manner similar to its use in Problem 2 on the data for $n = 1, 2, 3$, and 4 to conjecture the following explicit formula for the sequence $\{h_n\}$:

(3) $$h_n = \frac{n^3 + 5n + 6}{6}, \qquad \text{for each natural number } n.$$

Relating Problems 1, 2, and 3: Notice that that the recursive definition (2) of the sequence $\{g_n\}$ in Problem 2 can be written in the form

(2) $$g_{n+1} = g_n + f_n, \qquad \text{for all natural numbers } n,$$

where $\{f_n\}$ is the sequence that solved Problem 1. Explain why the sequence $\{h_n\}$ can be described by the difference equation

(4) $$h_{n+1} = h_n + g_n, \qquad \text{for all natural numbers } n,$$

where $\{g_n\}$ is the sequence that solved Problem 2.

Finishing the solution: Use mathematical induction and equation (4) to prove that the explicit formula (3) is correct and provides a solution to Problem 3.

A related problem and its initial analysis: Regions of the circle

Now we show an example where reasoning without using recurrence leads one astray. Problem 4 is an interesting (and famous) problem that demonstrates the pitfalls that may arise from generalizing from initial data too quickly.

Problem 4: If n distinct points are on a circle, and each pair of points is connected by a line segment, at most how many separate regions are created within the circle?

The problem asks for an explicit formula for the sequence $\{c_n\}$ defined by

$c_n =$ the maximum number of regions formed by all the chords connecting n distinct points on the circle

It is natural to begin by drawing diagrams for each case $n = 1, 2, 3, \ldots$, and counting the number c_n of regions for each n.

Figure 12

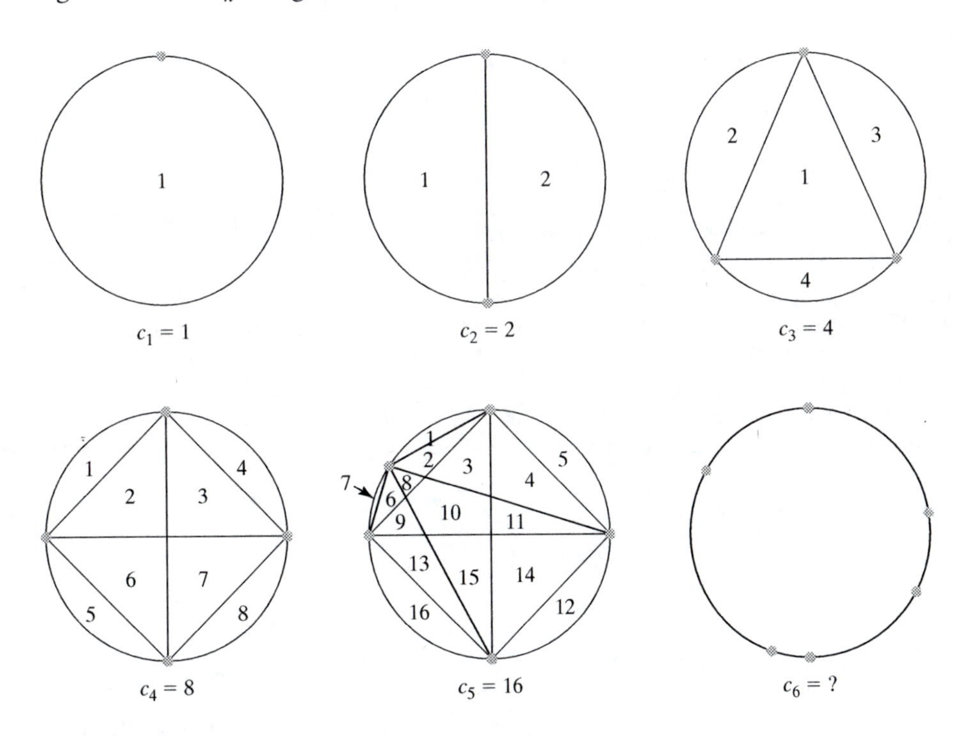

We record the results from Figure 12 in Table 4.

Table 4

Points n =	1	2	3	4	5	6	7
Regions c_n =	1	2	4	8	16		

The numerical pattern here is a familiar one that suggests the following conjecture:

$$c_n = 2^{n-1}$$

Is this conjecture correct? The formula states that when each new point is added to the circumference of the circle the number of regions is doubled. If this is true, then we should be able to show directly that adding a point doubles the number of regions.

However, if we look closely at, say, the relationship between the case of 4 points and the case of 5 points in Figure 12, we see that although c_5 is in fact double c_4, the *mechanism* is not one of doubling at all. Specifically, region 1 is split into 5 regions, region 2 into 3 regions, and each of regions 3 and 6 into 2 regions, while the other regions are not affected at all. The fact that this has led to doubling might just be a coincidence. In Problem 1, you are asked to make a direct count from a careful diagram to show that $c_6 = 31$, not 32, as the above conjecture would predict.

5.1.4 Problems

1. Consider the sequence $\{c_n\}$ used in this section to count the maximum number of regions inside of a circle that can be created by the chords connecting n points on the circle.
   a. Analyze the case $n = 6$ carefully, and show that in fact $c_6 = 31$.
   b. Use the method of finite differences on the data for the cases $n = 2$ through $n = 6$ to show that a polynomial of degree 4 that fits these data points is given by

$$d_n = \frac{n^4 - 6n^3 + 23n^2 - 18n + 24}{24}$$

   (See Conway and Guy for a proof of this formula.) Use this formula to compute d_7 and d_8.
   c. Use a graphing calculator or computer to compare the graphs on the interval $1 \le n \le 10$ of the formula for d_n in part **b** with the formula 2^{n-1} that we had conjectured earlier for c_n.
2. What is the maximum number of points of intersection of n distinct lines? Prove your result using mathematical induction.
3. What is the maximum number of regions formed by n parallel lines? Prove your result.
4. What is the maximum number of regions formed by n concurrent lines, that is, by n lines that all contain a particular point? Prove your result.
5. Prove: The regions formed in the plane by n lines can be colored with just two colors in such a way that any two regions whose borders share a line segment are colored differently.
6. What is the maximum number of regions formed by n planes that all pass through a common point?
7. A point P is connected to m points on a line not containing P. How many triangles are formed? Prove your result.
8. *Let A, B, and C be three noncollinear points. Choose m points between points A and B and connect each point to C. Then choose n points between points B and C and connect them to A. At most how many regions are formed inside the triangle? Prove your result.

ANSWERS TO QUESTIONS

1. 3 2. 6 3. 3

Unit 5.2 Divisibility Properties of the Integers

The system $\langle \mathbf{Z}, +, \cdot \rangle$ of integers appears to be a straightforward extension of the natural number system $\langle \mathbf{N}, +, \cdot \rangle$; that is, **Z** consists of the natural numbers, their negatives and zero. Yet, although the natural numbers were created and used by humankind in prehistoric times, the negative numbers and zero came much later. Even as late as the 1400s, when Italian mathematicians were solving cubic equations, they took pains to avoid negative numbers. The word "negative" itself comes from the Latin verb "negare," meaning "to deny" and retains that meaning in the English words *negate* and *abnegate*.

It may have been the applications of arithmetic to business—profit and loss, assets and liabilities—that first showed the utility of having negative numbers. A look at a daily newspaper shows how far we have come. Negative numbers are seen in temperatures, in gains and losses of stocks, in golf scores relative to par, in graphs such as those depicting the national debt, and so on. The set of integers—positive, negative, and zero—is among the most important number systems in all of mathematics.

As we have already observed, the system $\langle \mathbf{Z}, +, \cdot \rangle$ of integers is not closed for the binary operation of division by nonzero integers. Nevertheless, division is a fundamental mathematical idea for integers. Basic concepts and tools in school mathematics, such as prime factorization, greatest common factor and least common multiple, and the base representation of integers, rest ultimately on divisibility properties of the integers.

5.2.1 The Division Algorithm

Recall that an **algorithm** is a step-by step procedure for carrying out a prescribed type of computation that terminates with the answer in a finite number of steps. The procedures that are learned in grade school for the addition, subtraction, and multiplication of multidigit numbers are algorithms. We now examine some mathematics behind the long division algorithm, the most complicated algorithm of elementary school arithmetic.

Suppose we wish to divide 3742 (the **dividend**) by 23 (the **divisor**). In the United States, the long division algorithm applied to $3742 \div 23$ is usually written as shown here, but in some other countries of the world it looks different.

$$\begin{array}{r} 162 \\ 23\overline{)3742} \\ \underline{23} \\ 144 \\ \underline{138} \\ 62 \\ \underline{46} \\ 16 \end{array}$$

We conclude from this algorithm that the integer division $3742 \div 23$ yields a *quotient* of 162 and a *remainder* of 16 or, in symbols,

$$3742 = 23 \cdot 162 + 16.$$

Integer division

In general, the long division algorithm that is learned in grade school for computing the quotient $a \div b$ for two positive integers a and b yields a unique **integer quotient** q and a unique **integer remainder** r such that

$$a = bq + r \quad \text{where } 0 \le r < b.$$

Because the quotient and remainder are both integers, this procedure is often referred to as **integer division** to distinguish it from **rational number division**, in which the quotient is the single number $\frac{a}{b}$ and there is no remainder. The rational number division $3742 \div 23$ yields $\frac{3742}{23} \approx 162.69565\ldots.$

Integer division of a by b is also defined if a is a negative integer. For example, if $a = -7$ and $b = 3$, then

$$-7 = 3(-3) + 2;$$

that is, integer division of -7 by 3 yields the integer quotient of -3 and the integer remainder of 2.

If $a > 0$, the integer division of $a \div b$ has the following geometric interpretation on the number line. For a given positive integer b, locate the given integer a between two successive multiples of b, qb, and $(q + 1)b$, as in Figure 13.

Figure 13

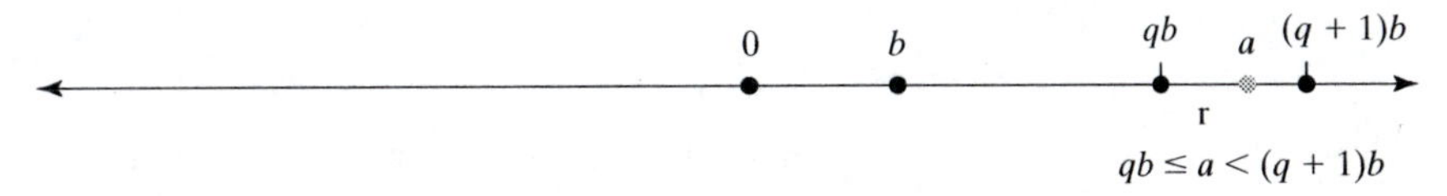

Then q is the integer quotient while the integer remainder is $r = a - qb$ because

$$0 \le a - qb = r < [(q + 1)b - qb] = b.$$

The existence of an integer quotient q and a nonnegative integer remainder r smaller than the divisor b in an integer division $a \div b$ is a consequence of an algorithmic procedure of repeated subtraction that is illustrated by the following example. For the integer division $3742 \div 23$, construct the sequence of integers beginning with 3742 and successively subtract positive integer multiples of the divisor 23 as long as the difference is nonnegative.

$$3742, 3742 - 23, 3742 - 2(23), 3742 - 3(23), \ldots, 3742 - 162(23) = 16$$

The algorithm stops because the next subtraction of 23 would result in a negative difference (-7). Like long division, this algorithm shows that the integer division $3742 \div 23$ results in an integer quotient $q = 162$ and an integer remainder $r = 16$. The algorithmic construction used in this example can be generalized to prove the following important theorem, which derives its name from that construction.

Theorem 5.3 **(The Division Algorithm):** If a and b are integers and if $b > 0$, then there are unique integers q and r such that $a = bq + r$ and $0 \le r < b$.

Proof: The division problem in the Division Algorithm is $a \div b$. That is, a is the dividend and b is the divisor. Here we prove the existence of the integer quotient q and the integer remainder r for given integers a and b with $b > 0$. You are guided through a proof that q and r are unique in Problem 9.

If $a = 0$, then we can choose $q = 0$ and $r = 0$ because $0 = b \cdot 0 + 0$. If $a > 0$ and a is a positive multiple of b, say $a = kb$ with k in **N**, choose $q = k$ and $r = 0$. Otherwise, use repeated subtraction to generate a set S of integers: a, $a - b$, $a - 2b$, $a - 3b$, and so on, as long as these integers are positive. That is,

$$S = \{a - kb: a - kb > 0 \text{ and } k \text{ is a nonnegative integer}\}.$$

Then S is not empty because it contains at least the single natural number a. Consequently, by the Well-Ordering Property (Theorem 5.2), the set S has a smallest member $a - sb$. Because $a - sb$ is the smallest member of S and a is not a multiple of b, the integer $a - (s + 1)b$ must be negative and so $a - sb < b$. Therefore, we can take $q = s$ and $r = a - sb$ to obtain $a = bq + r$ and $0 \le r < b$ as required.

If the given integer a is negative, then $-a > 0$. Consequently, by the preceding part of the proof, there are integers q^* and r^* such that

$$-a = bq^* + r^* \quad 0 \le r^* < b.$$

But then $a = b(-q^*) - r^* = b(-q^* - 1) + (b - r^*)$ and $0 \le (b - r^*) < b$. Therefore, if we take $q = -q^* - 1$ and $r = b - r^*$, we obtain $a = bq + r$ with $0 \le r < b$ in this case also. This completes the proof of the existence of an integer quotient q and an integer remainder r for any integer a and any positive integer b. ∎

Integer division $a \div b$ with $a < 0$ has a geometric interpretation on the number line (Figure 14).

Figure 14

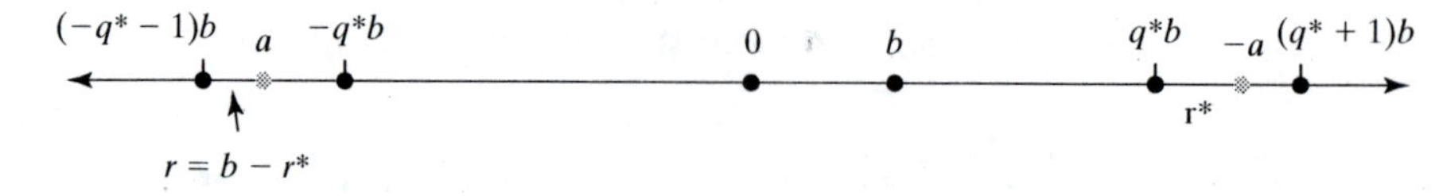

The traditional long division algorithm applied to $3742 \div 23$ is a shortcut of the algorithm found in the proof of Theorem 5.3.

$$\begin{array}{rl}
162 & \\
23\overline{)3742} & \\
\underline{23} & \\
144 & \qquad 1442 = 3742 - 1 \cdot 100 \cdot 23 \\
\underline{138} & \\
62 & \qquad 62 = 1442 - 6 \cdot 10 \cdot 23 \\
\underline{46} & \\
16 & \qquad 16 = 62 - 2 \cdot 1 \cdot 23
\end{array}$$

Thus $3742 = 23(1 \cdot 100 + 6 \cdot 10 + 2 \cdot 1) + 16 = 23 \cdot 162 + 16$.

First we subtract $1 \cdot 100 \cdot 23$ to obtain 1442. (To save time, some 0 digits and other digits [such as the 2 in 1442] are usually not written.) Then we subtract $6 \cdot 10 \cdot 23$ from 1442 to obtain 62. Then we subtract $2 \cdot 1 \cdot 23$ from 62 to obtain 16. Having found a number less than the divisor 23, we stop. We had to subtract 23 a total of 162 times; 162 becomes the quotient and 16 becomes the remainder.

When the dividend is negative, a slightly different procedure is needed to ensure a positive remainder. For instance, consider the division $-3742 \div 23$. As rational number division, $\frac{-3742}{23} = -162.69565\ldots = -162 - 0.69565\ldots = -162 - \frac{16}{23}$. Each of these representations points to a negative remainder. Yet for the Division Algorithm, a positive remainder is needed. The solution is to subtract 23 one more time than is needed for the positive division $3742 \div 23$. $3742 = 163 \cdot 23 - 7$, so $-3742 = -163 \cdot 23 + 7$. (This is where $q = -q^* - 1$ in the proof of Theorem 5.3 and in Figure 14.) So $-3742 \div 23$ results in an integer quotient of -163 with a remainder of 7.

Why are negative remainders not allowed in integer division?

Why do we *not* allow negative remainders in integer division even though they appear in rational number division? The reason is that the remainders in integer division cycle. For instance, the remainder from $3742 \div 23$ is one less than the remainder from $3743 \div 23$, and one greater than the remainder from $3741 \div 23$. Increasing the dividend by 1 increases the remainder by 1 until 23 is reached, and then the remainder starts over at 0. By requiring nonnegative remainders, this process is true for all integer divisions. For instance, the remainder from $-3742 \div 23$ is 7. Increasing the dividend by 1, we obtain $-3741 \div 23$, where the remainder is 8. We take advantage of this cycling in the applications of modular arithmetic introduced in Section 6.1.

5.2.1 Problems

1. Give the quotient and remainder from the Division Algorithm when each of these integers is divided by 79.

a. 461,533 b. 461,534 c. 461,535 d. −461,533

e. −461,534 f. −461,535

2. Find q and r from the Division Algorithm for the given values of a and b.

a. $a = 0, b = 14$ b. $a = -42, b = 14$ c. $a = 9, b = 14$

3. If the integer formed by the last 3 digits of a larger integer is divisible by 8, then prove that the larger integer is also divisible by 8.

4. **Even and Odd Numbers.** Use these definitions: An integer n is *even* if and only if there exists an integer k with $n = 2k$. An integer n is *odd* if and only if there exists an integer k with $n = 2k + 1$.

a. Use the Division Algorithm to prove that every integer is either odd or even, and not both.

b. Prove that the sum, difference, and product of any two even numbers is even.

c. State and prove statements like those in part **b** for two odd numbers.

5. Prove the following statement: If n is an integer, then either $n = 3k$ for some integer k, $n = 3k + 1$ for some integer k, or $n = 3k + 2$ for some integer k, and no two of these hold.

6. Use the Division Algorithm to prove the following theorem: Suppose that m and n are positive integers such that $m > n$. Then exactly one of the following statements must be true:

(a) There is a positive integer k such that $m = kn$.

(b) There is a positive integer k such that $kn < m < (k + 1)n$.

7. From its definition, the greatest integer function $\lfloor x \rfloor$ satisfies the double inequality $x - 1 < \lfloor x \rfloor \leq x$, for each real number x. Use this property to show that the quotient q and the remainder r for given integers a and b with $b > 0$ in the Division Algorithm can be written as

$$q = \left\lfloor \frac{a}{b} \right\rfloor \text{ and } r = a - b\left\lfloor \frac{a}{b} \right\rfloor.$$

That is, prove that with these definitions of q and r, it follows that q and r are unique integers satisfying

$$a = qb + r \quad \text{and} \quad 0 \leq r < b.$$

8. The following procedure uses two variables, R and Q, to construct two nonnegative integer outputs, q and r, from two positive integer inputs a and b, such that $a = bq + r$.

Step 1: Begin the procedure by setting $R = a$ and $Q = 0$.

Step 2: If $R < b$, set $q = Q$ and $r = R$ and STOP; otherwise, if $R \geq b$, go to Step 3.

Step 3: Set $R = R - b$ and $Q = Q + 1$ and then repeat Step 2.

a. Explain why this procedure must STOP before Step 2 is repeated a times.

b. Explain why $a = bq + r$, where $0 \leq r < \mathrm{b}$ when this procedure reaches STOP.

c. Write a computer program that implements this procedure with inputs a and b and outputs q and r.

9. Prove that for given integers a and b with $b > 0$, there is at most one pair of integers q and r such that

$$(*) \quad a = bq + r \quad \text{with } 0 \leq r < \mathrm{b}$$

by completing the following steps.

a. For the specific example $a = 7$ and $b = 3$, suppose that

$$(**) \quad 7 = 3q_1 + r_1 = 3q_2 + r_2 \quad \text{with } 0 \leq r_1 < r_2 < b.$$

Show that, with this supposition, then $r_2 - r_1$ must be a positive integer less than 3, and yet it must also be an integer multiple of 3. Explain why these conclusions are contradictory. Conclude that (**) cannot hold.

b. Suppose that $a = bq_1 + r_1 = bq_2 + r_2$ for integers q_1, q_2, r_1, r_2. Prove that if either of the relations $r_1 \neq r_2$ or $q_1 \neq q_2$ holds, then both must hold.

c. Suppose that there are integers q_1, q_2, r_1, r_2 with $0 \leq r_1 < b$, $0 \leq r_2 < b$, and $q_1 \neq q_2$ such that

$$a = bq_1 + r_1 = bq_2 + r_2.$$

Then $r_1 \neq r_2$ by part **b** so assume that $r_1 < r_2$. Show that

i. $r_2 - r_1$ is a positive integer less than b.

ii. $r_2 - r_1$ is an integer multiple of b.

Explain why the conclusions (i) and (ii) are contradictory, and conclude that there can be at most representation (*).

5.2.2 Divisibility of integers

A variety of terms are associated with divisibility of integers. When a, b, and c are integers and $ab = c$, we may say any of the following equivalent statements:

c **is a multiple of** *b*

b **divides** *c* (or *b* divides *c* "evenly," or *b* divides *c* without remainder)

b **is a divisor of** *c*

b **is a factor of** *c*.

The same is true with respect to a: c is a multiple of a; a is a divisor of c; a divides c; and a is a factor of c. We sometimes write $b|c$ and $a|c$ to express these relationships. (Do not confuse the sentence $b|c$ with the fraction b/c; the former is a statement asserting divisibility while the latter is a quotient.)

From basic properties of addition and multiplication, the following can be proved: If a, b, c, and d are any integers, and if $d|a$ and $d|b$, then $d|(a + b)$, $d|(a - b)$, and $d|(ca)$. That is, if one integer is a divisor of two integers, then it is also a divisor of their sum, their difference, or any multiple of either of them. Consequently, if $d|a$ and $d|b$, then $d|(ax + by)$ for any integers x and y. In words, if an integer is a divisor of two integers, then it is also a divisor of any integer linear combination of them.

The division $a \div b$ is made easier if a and b have a common divisor k, because $\frac{a}{k} \div \frac{b}{k}$ has the same real number quotient as $a \div b$ and involves numbers with smaller absolute value. For instance, 2 is a common divisor of -24 and 30, and $-24 \div 30 = \left(\frac{-24}{2}\right) \div \left(\frac{30}{2}\right) = -12 \div 15$. But then, 3 is a common divisor of -12 and 15, and $-12 \div 15 = \left(\frac{-12}{3}\right) \div \left(\frac{15}{3}\right) = -4 \div 5$. Had we divided the original numbers by 6, the greatest common factor of -24 and 30, all this could have been done in one step. Recall that if a and b are integers, not both zero, then the **greatest common factor** of a and b, **gcf(a, b)** is the unique natural number such that

1. gcf(a, b) is a factor of both a and b;
2. If d is any integer that is a factor of both a and b, then d is a factor of gcf(a, b).

For example, gcf$(-24, 30) = 6$ and gcf$(15, -8) = 1$.

Question 1: Explain why, for any integers a and b,

$$\text{gcf}(a, b) = \text{gcf}(|a|, |b|)$$

Question 2: Explain why gcf$(0, b) = |b|$ for any nonzero integer b.

An integer >1 is **prime** if and only if its only positive integer divisors are itself and 1; otherwise it is **composite**. Recall from Chapter 2 that two nonzero integers a and b are *relatively prime* if and only if they have no common integer factor >1. This can be stated in terms of greatest common factors as follows: Two integers a and b are relatively prime if and only if gcf$(a, b) = 1$. If $|a|$ and $|b|$ are distinct prime numbers, then a and b are relatively prime. (Why?) However, two integers a and b may be relatively prime, without $|a|$ or $|b|$ being prime. For example, gcf$(-6, 25) = 1$, yet neither 6 nor 25 are prime numbers.

For relatively small integers a and b, gcf(a, b) can be easily computed mentally or by factoring a and b into products of primes. However, for large a and b such as

$$a = 1062347 \quad b = 775489,$$

when prime factorizations of a and b are not readily available, the following theorem is helpful.

Theorem 5.4 If a, b, q, r are integers and if

$$a = bq + r,$$

then gcf$(a, b) = $ gcf(b, r).

Proof: Any factor of a and b is also a factor of r. (Why?) So the greatest common factor of a and b will also be a factor of r. So, from the definition of gcf, gcf$(a, b) \leq$ gcf(b, r). You can finish the proof by showing that gcf$(b, r) \leq$ gcf(a, b). ┘

Question 3: Complete the proof of Theorem 5.4.

EXAMPLE 1 Find gcf(1062347, 775489).

Solution To shorten writing, let $a = 1062347$ and $b = 775489$. For this pair, gcf(a, b) is not obvious by inspection. However, by applying the Division Algorithm repeatedly to a and b and its successive remainders, Theorem 5.4 shows that we eventually reach gcf(a, b). Eight applications of the Division Algorithm are needed.

$$\begin{aligned}
1062347 &= (775489)(1) + 286858 \\
\Rightarrow \quad \text{gcf}(a, b) &= \text{gcf}(775489, 286858) \\
775489 &= (286858)(2) + 201773 \\
\Rightarrow \quad \text{gcf}(a, b) &= \text{gcf}(286858, 201773) \\
286858 &= (201773)(1) + 85085 \\
\Rightarrow \quad \text{gcf}(a, b) &= \text{gcf}(201773, 85085) \\
201773 &= (85085)(2) + 31603
\end{aligned}$$

$$
\begin{aligned}
\Rightarrow \quad \text{gcf}(a,b) &= \text{gcf}(85085, 31603) \\
85085 &= (31603)(2) + 21879 \\
\Rightarrow \quad \text{gcf}(a,b) &= \text{gcf}(31603, 21879) \\
31603 &= (21879)(1) + 9724 \\
\Rightarrow \quad \text{gcf}(a,b) &= \text{gcf}(21879, 9724) \\
21879 &= (9724)(2) + 2431 \\
\Rightarrow \quad \text{gcf}(a,b) &= \text{gcf}(9724, 2431) \\
9724 &= (2431)(4) + 0
\end{aligned}
$$

Therefore, $\text{gcf}(1062347, 775489) = 2431$.

The Euclidean Algorithm

The algorithmic procedure used in the preceding example to compute the gcf(1062347, 775489) is called the *Euclidean Algorithm* and is formally described by the following theorem.

Theorem 5.5 **(Euclidean Algorithm):** Given two positive integers a and b with $a > b$, the greatest common factor of a and b can be found as follows:

Step 1: Apply the Division Algorithm to find integers q and r such that $a = bq + r$ and $0 \leq r < b$.

Step 2: If $r = 0$, then b is the required greatest common factor. STOP. If $r > 0$, replace (a, b) by (b, r) and repeat Step 1 and Step 2 until the algorithm stops.

Proof: The Euclidean Algorithm works because by Theorem 5.4, $\text{gcf}(a, b) = \text{gcf}(b, r)$ for each repetition of Step 1.

If $r = 0$ on any repetition of Step 1, then $\text{gcf}(a, b) = \text{gcf}(b, r) = b$.

If $r > 0$ on any repetition of Step 1, then $r < b$.

Thus, the nonnegative integer remainders for the successive divisions in the algorithm get smaller and smaller. Consequently, a remainder of 0 must occur after a finite number of steps, and at that point we have found the greatest common factor.

The Euclidean Algorithm has the following geometric interpretation: For two positive integers a and b with $b < a$, draw a rectangle with side lengths a and b. Then the first step of the algorithm, $a = bq_1 + r_1$, partitions the rectangle into q_1 squares with side length b, with a rectangle with side lengths b and r_1 remaining if $r_1 \neq 0$. If $r_1 \neq 0$, the second step of the algorithm, $b = r_1q_2 + r_2$, partitions that remaining rectangle into q_2 squares with side length r_1, with a rectangle with side lengths r_1 and r_2 remaining if $r_2 \neq 0$. The algorithm continues until the partition no longer has a remaining rectangle. Then $\text{gcf}(a, b)$ is the side length of squares in the final partition.

For example, to find gcf(90, 21), the Euclidean Algorithm proceeds algebraically as follows:

$$
\begin{aligned}
90 &= 21 \cdot 4 + 6 \\
21 &= 6 \cdot 3 + 3 \\
6 &= 3 \cdot 2 \quad \text{Stop: gcf}(90, 21) = 3.
\end{aligned}
$$

A geometric interpretation of this calculation is displayed in Figure 15.

Figure 15

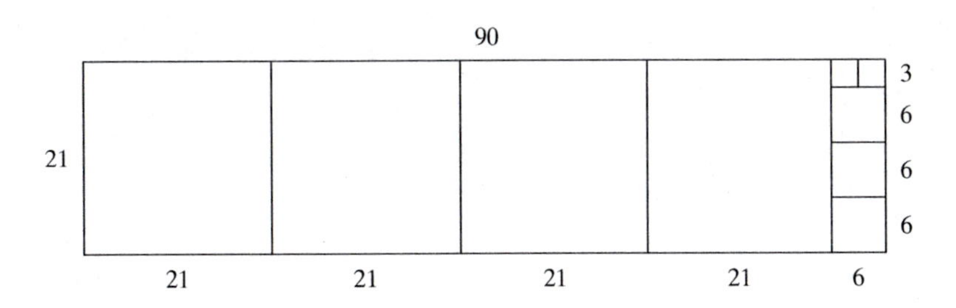

5.2.2 Problems

1. Use the Euclidean Algorithm to find the greatest common factor of 470141 and 4841.

2. Given two integers a and b, the **least common multiple** of a and b, written **lcm(a, b)**, is the smallest natural number m such that both a and b are factors of m.

a. Use the prime factorizations $180 = 2^2 \cdot 3^2 \cdot 5$ and $504 = 2^3 \cdot 3^2 \cdot 7$ to compute gcf(180, 504) and lcm(180, 504).

b. If p and q are distinct primes, and m and n are positive integers with $m > n$, find

i. $\text{gcf}(p^n, p^m)$

ii. $\text{gcf}((pq)^n, (pq)^m)$

iii. $\text{gcf}(p^n q^m, p^m q^n)$

iv. $\text{lcm}(p^n, p^m)$

v. $\text{lcm}((pq)^n, (pq)^m)$

vi. $\text{lcm}(p^n q^m, p^m q^n)$.

3. Give an example of two composite numbers that are relatively prime.

4. Let p_1, p_2, q_1, and q_2 be four distinct primes with $p_1 + q_1 = p_2 + q_2$. Explain why $\frac{p_1 q_1}{p_2 q_2}$ is in lowest terms.

5. a. Identify at least one important use of the least common multiple in arithmetic.

b. Verify that

$$\text{gcf}(a, b) \cdot \text{lcm}(a, b) = |ab|$$

for each of the pairs $\{a, b\}$ of integers in parts **a** and **b** of Problem 2.

*c. Prove that for all integers a and b, $\text{gcf}(a, b) \cdot \text{lcm}(a, b) = |ab|$.

d. If integers a and b are relatively prime, how do $\text{lcm}(a, b)$ and ab compare?

6. The Maya of Central America had many calendars. The Haab calendar was based on the Sun and had a length of 365 days. The Tzolkin calendar had a sacred year of 260 days. When the first days of these calendars coincided, to appease the gods there were elaborate rites including human sacrifices. The period from one day of coincidence to the next was called a Calendar Round. How long was a Calendar Round?

a. in days

b. in Haab years

c. in Tzolkin years

7. The "flow diagram" in Figure 16 shows a programmable procedure using the Euclidean Algorithm for finding the greatest common factor of two positive integers a and b.

a. Trace the flow of the positive integers $a = 15, b = 24$.

b. Write a computer program that implements the Euclidean Algorithm for any two positive integer inputs a and b and outputs $\text{gcf}(a, b)$.

Figure 16 Variables a, b, gcf, x are positive integers and r is a nonnegative integer.

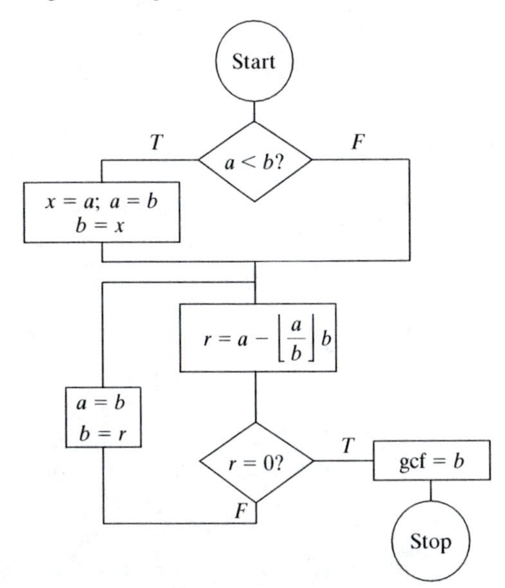

8. Let F_n be the nth Fibonacci number.

a. Prove: If $3|n$, then $2|F_n$.

b. Prove: If $4|n$, then $3|F_n$.

*9. Let F_n be the nth Fibonacci number. Prove that for all n the Euclidean Algorithm requires n divisions to conclude that $\gcd(F_{n+2}, F_{n+1}) = 1$.

(*Hint*: Use the facts that $F_{n+2} = F_{n+1} + F_n$ and $0 \leq F_n \leq F_{n+1}$.)

10. Suppose that a, b, and n are positive integers with $a > b$, and that a and b are the smallest relatively prime positive integers for which the Euclidean Algorithm stops in n steps.

a. Explain why, if $r_0 = a$ and $r_1 = b$, then all of the quotients q_k in the successive steps

$$r_{k-2} = r_{k-1}q_{k-1} + r_k, \quad \text{for } k > 2,$$

of the Euclidean Algorithm must be equal to 1.

b. Use part **a** to prove that $a = F_{n+1}$ and $b = F_n$, where F_n is the nth Fibonacci number.

ANSWERS TO QUESTIONS

1. For integers a, b, and c, if $ab = c$, then $-a(b) = -c$. Thus if a is a factor of c, then $-a$ is a factor of $-c$. Since $|a| = |-a|$, c and $-c$ have the same natural number factors. Thus $\gcd(a, b) = \gcd(-a, b) = \gcd(a, -b) = \gcd(-a, -b)$, from which $\gcd(a, b) = \gcd(a, b) = \gcd(|a|, |b|)$.

2. $|b|$ is a factor of b because $|b| = 1 \cdot b$ or $|b| = -b = -1 \cdot b$. $|b|$ is a factor of 0 because $|b| \cdot 0 = 0$. If c is any natural number that is a factor of both 0 and b, then $b = k \cdot c$ for some nonzero integer k, and so $|b| = |kc| \geq |c| = c$.

3. If d is a factor of b and r, then d is also a factor of a and b because $a = bq + r$. Therefore, every factor of b and r is a factor of a and b, so $\gcd(b, r) \leq \gcd(a, b)$. Since we have already proved $\gcd(a, b) \leq \gcd(b, r)$, the two must be equal.

5.2.3 Solving linear Diophantine equations

In many situations, the answer to a problem doesn't make sense unless the numbers are integers, and in some cases only positive integers make sense. Equations with integer coefficients for which we are only interested in integer solutions or nonnegative integer solutions are called **Diophantine equations**, after the Greek mathematician Diophantus, who lived in Alexandria during the third century A.D. In his long treatise *Arithmetika*, Diophantus gave the first systematic treatment of the solution of this type of equation.

Examples of Diophantine equations

Here are three examples of problems involving Diophantine equations.

1. *Stamp problems.* Can the exact 87-cent postage for a package be put on the package using only 15-cent stamps and 6-cent stamps? If insurance on the package is $1, can the total postage of $1.87 be paid using a combination of such stamps? If the answer to either question is yes, is there more than one combination that works? In general, what postage amounts less than $5 can be paid with a combination of 15-cent stamps and 6-cent stamps? (We explore stamp problems below.)

2. *Pythagorean triples.* Positive integer solutions (x, y, z) of the equation $x^2 + y^2 = z^2$ are called **Pythagorean triples** because they can be the lengths of sides of a right triangle. Examples are $(3, 4, 5)$ and $(5, 12, 13)$. Find a formula for x, y, and z in terms of the same variable n that generates all (infinitely many) Pythagorean triples. (A formula is derived in Section 5.2.4.)

3. *Fermat's Last Theorem.* In 1647, Fermat claimed to have a proof that the equation $x^n + y^n = z^n$ has *no* positive integer solutions (x, y, z) for any integer $n > 2$. After his death, many mathematicians tried to prove what became known as **Fermat's Last Theorem**. It became one of the most famous conjectures in all of mathematics. After efforts by many mathematicians who managed to prove partial results, the theorem was finally proved in general by Andrew Wiles in 1997. (See Project 6 at the end of this chapter for more details.)

Question 1: Answers for most of the stamp problems can be obtained by exploring various possibilities informally. Discover as many of these answers to the last of the stamp problem questions as you can before reading further.

Exploring the stamp problems

For the stamp problems, we seek nonnegative integer solutions (x, y) of the linear equation $15x + 6y = c$, where c is the total postage in cents. To determine whether 87¢ postage can be put on an envelope using only x 15¢ stamps and y 6¢ stamps, we can solve the Diophantine equation

$$15x + 6y = 87.$$

One way to find these solutions is with a table (Table 5).

Table 5

x	1	2	3	4	5
Amount left	72	57	42	27	12

The amounts left for 1, 3, and 5 15-cent stamps are divisible by 6. These choices indicate that the set of all solutions is $\{(1, 12), (5, 2), (3, 7)\}$.

If the 6¢ and 15¢ stamps are also to cover the $1 insurance, then we are led to solve the Diophantine equation

$$15x + 6y = 187.$$

The problem with finding solutions in this case is that regardless of the number x of 15-cent stamps that we use, the remaining balance is not divisible by 6 (Table 6).

Table 6

x	1	2	3	4	5	6	7	8	9	10	11	12
Amount left	172	157	142	127	112	97	82	67	52	37	22	7

Even if we allow x or y to be negative integers (which does not make sense in the context of the problem), we would still have no integer solutions because of the same divisibility problem.

These examples show that Diophantine equations may have no nonnegative solutions, finitely many nonnegative solutions, or infinitely many nonnegative solutions. The Diophantine equations corresponding to the stamp problems are of the form

$15x + 6y = 87$, which has 3 nonnegative integer solutions and infinitely many integer solutions;

and $15x + 6y = 187$, which has no integer solutions.

The Diophantine equation for Pythagorean triples (x, y, z),

$x^2 + y^2 = z^2$, has infinitely many solutions,

but when the exponent 2 is replaced by a larger number n, we obtain the Diophantine equation of Fermat's Last Theorem,

$x^n + y^n = z^n, n > 2$, which has no solutions.

Some Diophantine problems are quite easy to solve, while some are extremely difficult.

The general linear Diophantine equation $ax + by = c$

In the rest of this section, we consider the general linear Diophantine equation

$$ax + by = c.$$

That is, given integers a, b, and c, we seek all pairs of integers (x, y) satisfying this equation. As with the specific stamp problems, the solutions depend on divisibility relationships between the integer coefficients a, b, and c. The key element in this discussion is the Euclidean Algorithm.

The following result establishes a connection between greatest common factors, the Euclidean Algorithm, and linear Diophantine equations.

Theorem 5.6 For any integers a and b not both zero, the Diophantine equation

$$ax + by = \text{gcf}(a, b)$$

has solutions.

Partial proof: If $a = 0$, then $ax + by = \text{gcf}(a, b)$ reduces to $by = |b|$. Then $(x, y) = (0, 1)$ or $(x, y) = (0, -1)$ is a solution. Similarly, if $b = 0$, the equation has solutions. Since $\text{gcf}(a, b) = \text{gcf}(|a|, |b|)$, the equation $ax + by = \text{gcf}(a, b)$ has integer solutions if and only if the equation $|a|x + |b|y = \text{gcf}(a, b)$ has integer solutions. So we can assume $a > 0$ and $b > 0$. Since $\text{gcf}(a, b) = \text{gcf}(b, a)$, we can assume $a > b$. The idea of the proof is to reverse the steps in the Euclidean Algorithm for computing $\text{gcf}(a, b)$. For example, suppose that for two positive integers a and b with $a > b$, the Euclidean Algorithm stops after four applications of the Division Algorithm. That is, suppose

$$\begin{aligned} a &= bq_1 + r_1 \\ b &= r_1q_2 + r_2 \\ r_1 &= r_2q_3 + r_3 \\ \text{and} \quad r_2 &= r_3q_4 + 0, \end{aligned}$$

where all the q_i and r_i are integers, and $0 < r_{i+1} < r_i$ for all i. Then

$$\begin{aligned} \text{gcf}(a, b) &= r_3 = r_1 - r_2q_3 \\ &= r_1 - (b - r_1q_2)q_3 \\ &= a - bq_1 - bq_3 + (a - bq_1)q_2q_3 \\ &= a(1 + q_2q_3) + b(-q_1 - q_3 - q_1q_2q_3). \end{aligned}$$

Thus, $\text{gcf}(a,b) = ax + by$, where $x = 1 + q_2q_3$ and $y = -(q_1 + q_3 + q_1q_2q_3)$. Since all the q_i are integers, we have found a solution to $ax + by = \text{gcf}(a, b)$. Notice that the last three steps of the Euclidean Algorithm scheme for $\text{gcf}(a, b)$ are precisely the Euclidean Algorithm scheme for $\text{gcf}(b, r_1)$. Thus our calculation shows that

$$(*) \qquad \text{gcf}(b, r_1) = v = bx' + r_1y' \text{ where } x' = -q_3 \text{ and } y' = 1 + q_2q_3.$$

Now, by using (*) and the first step of the scheme, we obtain

$$\text{gcf}(a, b) = \text{gcf}(b, r_1) = v = bx' + (a - bq_1)y' = ay' + b(x' - q_1y') = ax + by.$$

Thus, we have not only proved the theorem for the case in which the Euclidean Algorithm terminates after four applications of the Division Algorithm, but also we have shown how to prove the theorem for this case from the case in which it terminates in three applications of the Division Algorithm. This sets the stage for a general proof by mathematical induction. You are asked to supply the details of the proof in Problem 7.

We now use Theorem 5.6 to derive a very basic divisibility result.

Theorem 5.7 Suppose that a, b, and c are integers, that a is a factor of the product bc, and that $\text{gcf}(a, b) = 1$. Then a is a factor of c.

Proof: By Theorem 5.6, there are integers x and y such that

$$1 = \text{gcf}(a, b) = ax + by.$$

Multiply this equation by c to obtain

$$c = acx + bcy.$$

Because a is clearly a factor of acx, and because a is given to be a factor of bc, it follows that a is a factor of c.

Corollary: If a prime number p divides the product $a_1 a_2 \ldots a_n$ of n integers, then p divides at least one of the a_i.

Question 2: Use mathematical induction to prove the corollary to Theorem 5.7.

Question 3:

a. Prove that two positive integers a and b are relatively prime if and only if there are integers x and y such that

$$ax + by = 1.$$

b. Use part **a** to explain why two consecutive positive integers are always relatively prime.

The solution of linear Diophantine equations

We now use Theorem 5.6 to derive the following complete description of the solution structure for linear Diophantine equations.

Theorem 5.8 Suppose that a, b, and c are integers and that $d = \text{gcf}(a, b)$. Then the linear Diophantine equation

$$ax + by = c$$

has solutions if and only if d is a factor of c. If (x_0, y_0) is one solution of this equation, then the set of all integer solutions is given by

$$(*) \qquad \left(x_0 - \frac{mb}{d}, y_0 + \frac{ma}{d}\right) \quad \text{for all integers } m.$$

Proof: Geometrically, we wish to prove that if (x_0, y_0) is a **lattice point** (a point with integer coordinates) through which the line $ax + by = c$ passes, then all other lattice point solutions also lie on the line, as shown in Figure 17.

Figure 17

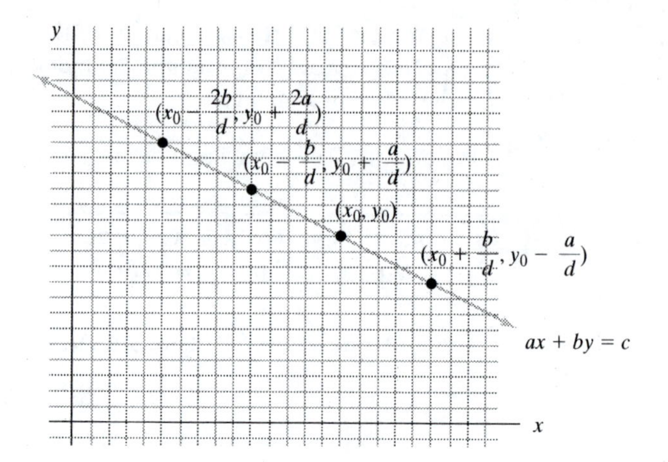

($\Rightarrow$) Because d is a factor of both a and b, it must also be a factor of $ax + by$ for any integers x and y. So d must be a factor of c.

($\Leftarrow$) Now suppose that $d = \text{gcf}(a, b)$ divides c. By Theorem 5.6, there exist integers x' and y' such that $|a|x' + |b|y' = \text{gcf}(a, b)$ because $\text{gcf}(|a|, |b|) = \text{gcf}(a, b)$. If necessary we can replace the integers x' or y' (or both) by their opposites to obtain integers x'' and y'' such that $ax'' + by'' = d$. Because d divides c, there is an integer r such that $c = rd$. Then the integers $x = rx''$ and $y = ry''$ satisfy the equation $ax + by = c$. So, in this case, the Diophantine equation has solutions. Let (x_0, y_0) be one solution of $ax + by = c$ and let m be any integer. Then $\frac{ma}{d}$ and $\frac{mb}{d}$ are integers because d divides both a and b. Therefore,

$$x = x_0 - \frac{mb}{d} \quad \text{and} \quad y = y_0 + \frac{ma}{d} \tag{1}$$

are integers for each integer m. We verify that (1) are solutions of $ax + by = c$ by direct substitution:

$$a\left(x_0 - \frac{mb}{d}\right) + b\left(y_0 + \frac{ma}{d}\right) = ax_0 + by_0 - \frac{mab}{d} + \frac{mab}{d} = c.$$

On the other hand, let (x_1, y_1) be any other integer solution of $ax + by = c$. Then both (x_0, y_0) and (x_1, y_1) satisfy the Diophantine equation

$$a^*x + b^*y = c^* \qquad \text{where } a^* = \frac{a}{d},\ b^* = \frac{b}{d}, \text{ and } c^* = \frac{c}{d} \tag{2}$$

and $\text{gcf}(a^*, b^*) = 1$; that is, a^* and b^* are relatively prime. But then

$$a^*(x_0 - x_1) + b^*(y_0 - y_1) = c^* - c^* = 0;$$

that is,

$$a^*(x_0 - x_1) = -b^*(y_0 - y_1).$$

Because a^* and b^* have no factors in common, a^* divides $(y_0 - y_1)$ and b^* divides $(x_0 - x_1)$. This means that there are integers p and q such that $pa^* = y_0 - y_1$ and $qb^* = x_0 - x_1$. So

$$x_1 = x_0 - qb^* \qquad y_1 = y_0 - pa^*.$$

Because (x_1, y_1) and (x_0, y_0) are solutions of (2), it follows that $p = -q$. Consequently,

$$x_1 = x_0 - qb^* = x_0 - \frac{qb}{d} \qquad y_1 = y_0 + \frac{qa}{d}$$

and so (x_1, y_1) is among the solutions in (*).

Finally, let us see how Theorem 5.8 relates to the stamp problems that opened this section. Because $d = \text{gcf}(15, 6) = 3$, and 3 is a factor of 87, the equation $15x + 6y = 87$ has infinitely many integer solutions including the solution (5, 2) found earlier. The set of all integer solutions of this equation is given by (*) in Theorem 5.8 as

$$\{(5 - 2m, 2 + 5m), \text{ where } m \text{ is any integer}\}.$$

Because we are only interested in nonnegative integer solutions, we only want those integers m such that $5 - 2m \geq 0$ and $2 + 5m \geq 0$, that is, those integers m such that

$$\frac{5}{2} \geq m \geq -\frac{2}{5}.$$

Thus, the only possible values of m are $m = 0, 1$, and 2, and the corresponding solutions are $\{5, 2\}$, $\{3, 7\}$, and $\{1, 12\}$. These are precisely the solutions we found in our exploration.

Since $d = 3$ and 3 is not a factor of 187, by Theorem 5.8 the equation $15x + 6y = 187$ has no solution. This also agrees with our earlier analysis.

5.2.3 Problems

1. Identify all postage amounts less than \$5 that can be paid with a combination of 15-cent stamps and 6-cent stamps.

2. If the postage for mailing a package is \$3.75, find all possible combinations of 34-cent and 23-cent stamps that can be used to pay that postage exactly.

3. Find a lattice point on the graph of $77x + 115y = 39$.

4. Let a, b, and c be integers with $ab \neq 0$.
   a. Prove: If (x_1, y_1) is a lattice point on the graph of $ax + by = c$, then so is $(x_1 - b, y_1 + a)$.
   b. Find a lattice point on the graph of $17x - 3y = 4$.
   c. Use part **a** to determine a second lattice point on the graph of $17x - 3y = 4$.

5. Consider the set S of all integers z such that $z = 1470x + 119y$, where x and y are integers. Describe S in terms of a single variable k.

6. Suppose $ax + by = c$ is an equation for a line that is neither horizontal nor vertical, and a, b, and c are integers. Prove or disprove: The graph of $ax + by = c$ either contains a lattice point or is arbitrarily close to but does not contain a lattice point.

7. Complete the proof of Theorem 5.6 by following the plan suggested by the partial proof in this section.

In Problems 8–12, use the theorems in this section about the solutions of linear Diophantine equations.

8. **Track problem.** Imagine the number line as a "track" along which you can move back and forth. You start at 0 and are allowed two kinds of "moves," each of which can be repeated any number of times: (i) Move 15 units right or left.; (ii) Move 6 units right or left.
   a. Show how you can end up at -27.
   b. Show that you can never end up at 88.
   c. Describe the locations you can never end up at, and why.

9. **Two-stack problem.** Type A coins are 13 mm thick, while Type B coins are 8 mm thick.
   a. Find a stack of Type A coins that is exactly 1 mm taller than some stack of Type B coins.
   b. Find a stack of Type B coins that is exactly 1 mm taller than some stack of Type A coins.

10. **Another two-stack problem.**
   a. Suppose you have two side-by-side stacks of ingots, one of silver ingots 101 mm thick, and one of gold ingots 23 mm thick. Can the two stacks ever be exactly the same height? If not, why not? If so, what is the shortest such height? Draw a diagram.

b. Generalize part **a** to involve side-by-side stacks, one stack with books p mm thick, the other stack with books q mm thick, where p and q are integers. [*Hint*: You may find $\gcd(p, q)$ useful in your answer.]

c. In part **b**, if p and q are relatively prime, what is the height of the shortest stacks that have the same height?

11. Consider stacks of two kinds of things, one 7 mm thick and one 12 mm thick. Given an arbitrary whole number d, can two such stacks ever differ in height by exactly d mm? If not, why not? If so, what are their heights in terms of d?

12. **Balance Scale Problem.** A certain object weighs c ounces, where c is a positive integer. Show that the weight of the object can be determined by using only 3-ounce and 5-ounce scale weights on two balance pans.

13. Suppose m and n are positive integers such that the third remainder in the application of the Euclidean Algorithm to find their gcf is 1. That is, suppose m and n are related as follows, where the q_i and r_i are constrained as in the Euclidean Algorithm.

$$m = nq_1 + r_1$$
$$n = r_1q_2 + r_2$$
$$r_1 = r_2q_3 + 1$$

a. Find integers x and y such that $mx + ny = 1$, where x and y are expressed in terms of m, n, and the quotients q_1, q_2, and q_3.

b. Apply the Euclidean Algorithm to the relatively prime integers 210 and 17. You should end up with a calculation in which the third remainder is 1. Test your formula from part **a** to find integers x and y such that $210x + 17y = 1$.

ANSWERS TO QUESTIONS

2. *Proof of the corollary to Theorem 5.7:* Because p is prime, it is relatively prime to any integer that is not a multiple of p. Consequently, if p divides the product a_1a_2 of $k = 2$ integers, it must divide a_1 or a_2. Thus the conclusion of the corollary is true for $k = 2$.

Suppose that if p divides a product $a_1a_2 \ldots a_k$ of k integers, then p divides at least one factor in the product. If p divides a product $a_1a_2 \ldots a_{k+1}$ of $k + 1$ factors, then it divides the product ab of the two integers

$$a = a_1a_2 \ldots a_k \quad \text{and} \quad b = a_{k+1}.$$

But then Theorem 5.7 asserts that p divides a or p divides b. If p divides a, then our inductive supposition implies that p divides a_p for some p with $1 \le p \le k$. If p divides b, then p divides a_{k+1}. Thus, in either case, p must divide one of the factors in the product $a_1a_2 \ldots a_{k+1}$ so the conclusion is true for $k + 1$ factors if we suppose it to be true for k factors. Therefore, the statement of the corollary is true for all $k \ge 2$.

3. Proof of (a): If the positive integers a and b are relatively prime, then $\gcd(a, b) = 1$. Therefore, by Theorem 5.6, there are integers x and y such that $ax + by = \gcd(a, b) = 1$. Conversely, suppose that there are integers x and y such that $ax + by = 1$. If an integer d is a divisor of both a and b, then it is a divisor of 1 because $1 = ax + by$. The only integer divisors of 1 are 1 or -1, so $\gcd(a, b) = 1$.

Proof of (b): Suppose that k and $k + 1$ are consecutive positive integers. Then $1 \cdot (k + 1) + (-1) \cdot (k) = 1$, so k and $k + 1$ are relatively prime by part **a**.

5.2.4 The Fundamental Theorem of Arithmetic

The natural numbers can be generated by beginning with 1 and repeatedly adding 1, the process that is embodied in the Peano axioms. We can think of 1 as the unit and all the other natural numbers as its successors. If we begin instead with 2 and repeatedly add 2, the numbers 4, 6, 8, ... and other multiples of 2 are its successors. Similarly, we may begin with any natural number and repeatedly add it to find its successors. The Greek mathematician Eratosthenes (c.276–c.194 B.C.) crossed out the successors of 2 and the successors of all higher natural numbers in this way, in a process we today call the **Sieve of Eratosthenes** and was left with the prime numbers.

It is not as obvious that the natural numbers greater than 1 can be generated by *multiplying* these primes.

Theorem 5.9 Any natural number $n \ge 2$ is either prime or is the product of prime numbers.

Proof: Let M be the set of all natural numbers ≥ 2 that are composite and cannot be written as the product of primes. By the well-ordering property of **N**, if $M \ne \emptyset$, then M has a smallest element. Let this element be x. Then x has a factor $y < x$ that is also not prime and cannot be written as the product of primes. Thus y is an element of M, which contradicts the assumption that x is the smallest element of M. Therefore, $M = \emptyset$. ∎

It follows immediately from Theorem 5.9 that any natural number $n > 1$ has a prime number factor. This has the following interesting consequence.

Theorem 5.10 For any natural number $n > 2$, there is a prime number p between n and $n!$.

Proof: Consider the number $q = n! - 1$. If q is prime, take $p = q$, and clearly $n < p < n!$. If q is not prime, then q has a prime number factor p. Now, we need only show that $n < p$. Suppose to the contrary that $p \leq n$. Then p would be a factor of $n!$ as well as $n! - 1$. Consequently, p would be a factor of the difference $n! - (n! - 1) = 1$, which is impossible. Therefore, since the supposition that $p \leq n$ led to a contradiction, it follows that $n < p$. ∎

The gap between n and $n!$ is very large even for relatively small natural numbers n. For example, for $n = 10$, the value of 10! is 3,628,800. So Theorem 5.10 is not of much help in finding primes. However, because Theorem 5.10 guarantees a prime on the closed interval $[n + 1, n! - 1]$, and there are infinitely many intervals of this kind, Theorem 5.10 implies the following famous result first proved by Euclid.

Theorem 5.11 There are infinitely many prime numbers.

Proof: The proof by Euclid is very much like the proof of Theorem 5.10. Suppose that there are only finitely many prime numbers $p_1, p_2, \ldots, p_n$, listed in order of their size ($p_1 = 2$, $p_2 = 3$, etc.). Then consider the number $q = p_1 p_2 \ldots p_n + 1$ that is one more than the product of all the primes. By Theorem 5.9, either q is prime or q is a product of primes. If q is prime, then because q is greater than p_n, the list does not contain all primes. This is a contradiction. If q is not prime, then it has a prime factor that must be one of the p_i on the list. But then this p_i must divide both q and $q - 1$. And if so, this p_i must divide 1, which is again a contradiction. So the number of primes is not finite. ∎

We now show that every integer greater than 1 can be factored into a product of primes in only one way. Euclid proved this theorem by repeated use of what we have called the Euclidean Algorithm. We use the well-ordering property of **N**, which is quite natural for this proof.

Theorem 5.12 **(Fundamental Theorem of Arithmetic):** Every integer $n \geq 2$ is either prime or can be represented as a product of prime numbers, and, except for the order of the factors, there is only one such representation of n.

Proof: Suppose, contrary to the conclusion of the theorem, that there are integers ≥ 2 that have more than one prime factorization, apart from the ordering of the factors. By the well-ordering property, there is a smallest number s with two different prime factorizations:

$$s = p_1 p_2 \ldots p_u = q_1 q_2 \ldots q_v$$

where p_k and q_j are prime numbers for all k and j. Because p_1 is a divisor of the product $q_1 q_2 \ldots q_v$, it must be a divisor of at least one factor q_p. But q_p is a prime number so $p_1 = q_p$. Now divide both sides by the factors p_1 and q_p to obtain two factorizations $p_2 \ldots p_u$ and the product of the remaining q_j (without q_p) of an integer $s^* < s$. These two factorization of s^* must be the same except for the order of the factors because s is the smallest natural number that has different prime factorizations. But then the two factorizations of s must also be the same except for the order of the factors, which contradicts the supposition that there are integers ≥ 2 with more than one prime factorization. ┘

The number 1 is not considered to be a prime number even though it seems to satisfy the definition of prime number (i.e., its only positive divisors are itself and 1). Notice that if 1 were considered prime, the Fundamental Theorem of Arithmetic would not be true as stated. Of course, it could be restated correctly by excluding the number 1 from prime factorizations, but the statement of the new theorem would be clumsy. The same would happen with other theorems. It is just easier to agree that 1 is not a prime number.

The standard prime factorization

Suppose that n is a natural number greater than 2 and that $p_1, p_2, \ldots, p_k$ are the distinct prime factors of n written in increasing order; that is, $p_1 < p_2 < \cdots < p_k$. For each prime factor p_i of n, let m_i be the number of times that p_i occurs as a factor of n in its prime factorization. Then a prime factorization of n can be expressed in the following form:

$$n = p_1^{m_1} \cdot p_2^{m_2} \cdots \cdot p_k^{m_k}.$$

This is called the **standard** or **canonical prime factorization** of n. For example, the standard prime factorization of 393,250 is

$$393250 = 2 \cdot 5^3 \cdot 11^2 \cdot 13.$$

The Fundamental Theorem of Arithmetic can be reformulated using this idea: *Every natural number $n > 2$ has a unique standard prime factorization.*

The form of the standard prime factorization of a natural numbers follows one systematic (and easily programmable) procedure for factoring a natural number n for which the prime factors are not obvious: Divide n repeatedly by the successive primes 2, 3, 5, 7, 11, ... until the factorization is achieved. For example, let $n = 393250$.

$$\begin{aligned}
393250 \div 2 &= 196625 \\
196625 \div 2 &\neq \text{an integer} \\
196625 \div 3 &\neq \text{an integer} \\
196625 \div 5 &= 39325 \\
39325 \div 5 &= 7865 \\
7865 \div 5 &= 1573 \\
1573 \div 5 &\neq \text{an integer} \\
1573 \div 7 &\neq \text{an integer} \\
1573 \div 11 &= 143 \\
143 \div 11 &= 13, \text{ which is prime.}
\end{aligned}$$

Therefore, $393250 = 2 \cdot 5^3 \cdot 11^2 \cdot 13$.

Applications of the Fundamental Theorem of Arithmetic

Why is the Fundamental Theorem of Arithmetic so fundamental? The main reason is that it shows that the prime numbers are the "building blocks" for all natural numbers, and that a given natural number can be constructed from these prime building blocks in essentially only one way. But what is so important about that? Part of the answer to this question rests in the way that the theory of numbers unfolds. It turns out that the Fundamental Theorem of Arithmetic is applied frequently as a technical tool for developing the theorems and special functions of number theory. For instance, in Section 2.1.1, we proved (as Theorem 2.2) that the square root of every integer that is not a perfect square is irrational. The statement that required the Fundamental Theorem of Arithmetic is the following theorem.

Theorem 5.13 If the product of two relatively prime integers u and v is a perfect square of an integer, then u and v are also perfect squares.

Proof: Let $uv = r^2$, where r is an integer. Since $r^2 = r \cdot r$ and r has a unique prime factorization, if a prime p is a factor of r^2, so is p^2. Yet, because u and v have no factors in common, each squared prime factor in r^2 must either come from the prime factorization of u or the prime factorization of v. Therefore, both u and v are the products of the squares of their prime factors. ∎

We used Theorem 5.13 in the proof of Theorem 2.2.

Pythagorean triples

Theorem 5.13 is also at the heart of a result that identifies all possible Pythagorean triples, that is, all triples (a, b, c) of natural numbers such that

$$a^2 + b^2 = c^2.$$

First note that, because any positive integral multiple of a Pythagorean triple is also a Pythagorean triple [e.g., since (3, 4, 5) is a Pythagorean triple, so is (6, 8, 10); see Problem 13], we only need to identify the Pythagorean triples that are **primitive**; that is, Pythagorean triples (a, b, c) for which a, b, and c have no integer factors in common.

You can obtain a complete description of all primitive Pythagorean triples (a, b, c) by proving the following (see Problem 14).

1. One of the numbers a or b must be even and the other odd.
2. c must be odd.

Together, (1) and (2) imply that when (a, b, c) is a primitive Pythagorean triple and $a^2 + b^2 = c^2$, we can always have b even while a and c are odd.

3. In the factorization $a^2 = c^2 - b^2 = (c - b)(c + b)$, $c - b$ and $c + b$ must be odd and have no integer factors in common other than 1 and -1.

Then you can conclude from (3) by Theorem 5.13, since their product is a perfect square, that $c - b$ and $c + b$ are perfect squares. Because $c + b > c - b$, this means that there are natural numbers u and v such that $v > u$,

$$a = uv, \quad c + b = v^2, \quad \text{and} \quad c - b = u^2.$$

Solving this system of equations for b and c, we have

$$a = uv; \quad b = \frac{v^2 - u^2}{2}; \quad c = \frac{v^2 + u^2}{2}.$$

This argument proves that every primitive Pythagorean triple can be generated by these formulas. Furthermore, direct computation shows that all triples generated by the formulas are Pythagorean triples.

Theorem 5.14 **(Pythagorean Triples Theorem):** Any primitive Pythagorean triple (a, b, c) of natural numbers such that a is odd and b is even is given by

$$(*) \quad a = uv; \qquad b = \frac{v^2 - u^2}{2}; \qquad c = \frac{v^2 + u^2}{2}$$

for suitable odd numbers u and v with $v > u$. Conversely, for any choice of odd numbers u and v with $v > u$, the formulas (*) produce a primitive Pythagorean triple (a, b, c) with a odd and b even.

Public key network security systems

There are also important practical applications of the Fundamental Theorem of Arithmetic outside of the theory of numbers. *Public key encoding* is a widely used security system that is designed to assure communication privacy in computer networks. The system relies on pairs of *keys* (or *passwords*) that determine key-specific schemes that are used to encode and decode messages.

Each pair of keys consists of a **public key**, which is usually made available to everyone using the network, and a **private key**, which is known only to the person who has been assigned that pair. Anyone who knows the public key can encode (encrypt) a message and send it to another person on the network, but only that other person knows the private key that can decode (decrypt) the message. This makes it possible to exchange messages securely over an insecure network.

Public and private keys are produced in pairs by computer programs. The most well-known of these is the RSA public key encryption algorithm, which is based on prime factorization and modular arithmetic (see Unit 6.1). The RSA algorithm derives its name from the first letters of the last names of its inventors, Ronald L. Rivest, Adi Shamir, and Leonard M. Adleman, who jointly published a six-page paper in 1978 that described the system.

The RSA algorithm takes two very large prime numbers, P and Q, and forms their product to obtain a very large composite number $N = PQ$. (N may have 150 digits or so.) The RSA algorithm then generates an "exponent" E, and an "inverse exponent" D, so that E and the number $(P - 1)(Q - 1)$ are relatively prime. The public key consists of E and N while D and N comprise the corresponding private key.

If someone knows the public key and also knows the two prime factors P and Q of N, it would not be difficult for them to compute the corresponding private key. However, factoring the number N into its large prime factors P and Q is virtually impossible because there are no known computer factorization algorithms that are efficient enough to find these factorizations in any reasonable length of time. Moreover, the Fundamental Theorem of Arithmetic assures that no simpler factorization of N is possible. Thus, the task of computing the private key from a public key is virtually impossible without knowing the prime factorization $N = PQ$.

5.2.4 Problems

1. In Section 5.2.2, gcf(1062347, 775489) was found using the Euclidean Algorithm. Find gcf(1062347, 775489) by factoring these numbers into primes.

2. Prove Theorem 5.9 by the method of mathematical induction, letting $S(n)$ be: *Any integer* q *with* $2 \le q \le n$ *is either prime or is the product of prime numbers.*

3. Explain why Theorem 5.11 follows from Theorem 5.10.

4. There are number systems in which prime factorization is not unique. Consider the set M of integers of the form $4k + 1$. That is,

$$M = \{4k + 1 : k \in N\} = \{1, 5, 9, 13, \ldots, 4k + 1, \ldots\}.$$

a. Show that M is closed under ordinary multiplication. That is, show that if p and q belong to M, then so does pq.

b. A number p in M is called an *M-prime* if its only factors in M are 1 and p. Show that $9, 21, 33, 77$ are M-prime, yet $9 \times 77 = 21 \times 33$. (It can be shown that every number in M can be written as a product of M-primes.)

5. **Bertrand's Postulate.** The following result (which is usually referred to as Bertrand's *Postulate* even though it is a *theorem*) shows that much smaller intervals than those between n and $n!$ must contain at least one prime.

> *For any integer $n > 1$, there is a prime p such that $n < p < 2n$.*

Bertrand's Postulate essentially tells us that the gap to the next prime number cannot be larger than the number from which the gap begins. This result was conjectured by a Parisian professor, Joseph Bertrand (1822–1900), in 1845 on the basis of experimentation. Seven years later, the Russian mathematician Pafnuty L. Tchebycheff (1821–1894) supplied a proof. In 1932, the Hungarian mathematician Paul Erdos (1913–1996) found a different proof.

a. Verify Bertrand's Postulate for $n = 10, n = 100$, and $n = 500$.

b. Consider the following variant of Bertrand's Postulate. For any integer $n > 1$, there is a prime p such that $n < p < kn$, where $k < 2$. By experimenting, how small do you think k can be made?

c. Explore the following unsolved problem: Is there always a prime between n^2 and $(n + 1)^2$?

6. **Gaps between prime numbers.** Show that for any integer $n > 1$, there exist strings of n consecutive composite numbers; that is, there are arbitrarily large gaps between successive prime numbers.

7. **An upper bound for the nth prime.** Suppose that $P(n)$ is the nth prime number; that is, $P(1) = 2, P(2) = 3, P(3) = 5, \ldots$.

a. Prove that, for all positive integers n, $P(n + 1) \leq [P(1)P(2) \ldots P(n)] + 1$.

*b. Use part **a** and mathematical induction to prove that, for all positive integers n, $P(n) \leq 2^{(2^n)}$.

8. **Yet another proof of the infinitude of primes.** Prove that infinitely many prime numbers p can be written in the form $p = 4k + 3$ for some positive integer k, by completing the following steps.

a. Explain why every positive integer n can be expressed in exactly one of the following forms for a suitable nonnegative integer k:

i. $n = 4k$

ii. $n = 4k + 1$

iii. $n = 4k + 2$

iv. $n = 4k + 3$

b. Explain why only forms (ii) and (iv) are possible if n is an odd prime.

c. Prove that the product mn of two positive integers of the form (ii) is of this same form.

d. More generally, prove that the product of any number of positive integers of the form (ii) can also be expressed in this form.

e. Suppose that, contrary to the statement of the desired result, there are only a finite number of primes

$$p_1, p_2, \ldots, p_m$$

of the form (iv). Let

$$q = 4p_1p_2 \ldots p_m + 3.$$

Explain why q must have a prime factor of the form (iv). (*Hint*: Use parts **b** and **d**.)

9. **And still another proof.** Consider the sequence q that is defined recursively as follows:

$$q_1 = 2, \quad \text{and} \quad q_{n+1} = q_n^2 - q_n + 1 \quad \text{for } n \geq 1.$$

a. Show that the first few terms of this sequence are prime numbers.

b. Find the first positive integer n such that q_n is not prime.

c. Show that $q_{n+1} = q_1q_2 \ldots q_n + 1$ and that q_{n+1} has a prime factor that is not a prime factor of any q_k when $1 \leq k \leq n$.

d. Use part **c** to obtain another proof of the fact that there are infinitely many prime numbers.

10. a. Find the standard prime factorizations of 360, 360^2, 360^3, and 360^n.

b. Generalize part **a**.

11. a. If $m = 2940$ and $n = 3150$, find the standard prime factorizations of m and n, $\gcd(m, n)$, and $\operatorname{lcm}(m, n)$.

b. Explain how $\gcd(m, n)$ and $\operatorname{lcm}(m, n)$ can be determined from the standard prime factorizations of m and n.

12. Find all primitive Pythagorean triples (x, y, z) with $z \leq 65$.

13. Prove: If (a, b, c) is a Pythagorean triple, then so is (ka, kb, kc), where k is any natural number.

14. Assume (a, b, c) is a primitive Pythagorean triple with $a^2 + b^2 = c^2$. Fill in these missing steps in the proof of Theorem 5.14.

a. Prove that one of a or b must be even and the other must be odd.

b. Prove that c is odd.

c. Solve the system $c + b = v^2$ and $c - b = u^2$ for b and c.

15. Write a computer or calculator program that accepts the input of two odd numbers u and v with $v > u$, and outputs a primitive Pythagorean triple (a, b, c) with a odd and b even.

16. Pythagoras[3] tried to generate what we today call Pythagorean triples using the following relation:

$$\left(\frac{m^2 - 1}{2}\right)^2 + m^2 = \left(\frac{m^2 + 1}{2}\right)^2.$$

---

[3]See Ronald Calinger, *A Contextual History of Mathematics* (Upper Saddle River, NJ: Prentice Hall, 1999), p. 72.

a. Does this relation generate any Pythagorean triples?
b. Does this relation generate all the Pythagorean triples?

17. If (x, y, z) is a primitive Pythagorean triple, prove that x, y, and z are relatively prime in pairs; that is,

$$\text{gcf}(x, y) = \text{gcf}(x, z) = \text{gcf}(y, z) = 1.$$

18. Prove: If (a, b, c) and (d, e, f) are Pythagorean triples, then so is $(|ad - be|, ae + bd, cf)$.

19. Define three sequences x, y, and z recursively and simultaneously as follows:

$$\begin{aligned} x_1 &= 3; & x_{n+1} &= 3x_n + 2z_n + 1 \\ y_1 &= 4; & y_{n+1} &= 3x_n + 2z_n + 2 \\ z_1 &= 5; & z_{n+1} &= 4x_n + 3z_n + 2 \end{aligned}$$

a. Show that (x_n, y_n, z_n) is a Pythagorean triple for $n = 1, 2$, 3, and 4.
b. Use mathematical induction to prove that (x_n, y_n, z_n) is a Pythagorean triple for all natural numbers n.

20. a. Show that the polynomial $p(x) = x^2 - x + 41$ has prime number values for all positive integers <41.
b. Without calculating $p(41)$, prove that $p(41)$ is not prime. [It has been proved that no polynomial $p(x)$ with integer coefficients exists such that $p(n)$ is prime for every natural number n.]

21. Suppose that a and n are integers >1. Prove that if $a^n - 1$ is a prime number, then a must be equal to 2. (*Hint*: What is the sum of the finite geometric sequence $1, a, a^2, \ldots, a^{n-1}$?)

22. Numbers of the form $M_m = 2^m - 1$, where m is a natural number, are called **Mersenne numbers**.

a. Make a list of the 12 smallest Mersenne numbers.
b. Prove that if M_m is prime, then m is prime.
c. Based on your list in part **a**, show that the converse of part **b** is not true.

23. A natural number n is **perfect** if $2n$ is the sum of all the positive divisors of n. The first perfect number is 6 because $(1 + 2 + 3 + 6 = 12 = 2 \cdot 6)$. It can be shown that n is an even perfect number if and only if

$$n = 2^{m-1}M_m$$

where M_m is the mth Mersenne number and M_m is prime. The "if" part was proved by Euclid in his *Elements* (Book IX, Proposition 36). The "only if" part was proved by Leonhard Euler about two thousand years later. (Are there any odd perfect numbers? Nobody knows for sure.)

a. Prove the "if" part. (*Hint*: List all of the positive divisors of $2^{m-1}M_m$ when M_m is prime.)
b. Use this result and the list in Problem 22**a** to generate some large even perfect numbers.

5.2.5 Base representation of positive integers

The Hindu-Arabic numeration system that we use today to represent integers is just one of many that have been used through history. The **place-value** feature of this system has resulted in its almost universal use today. That is, the numerals in the Hindu-Arabic system are strings of digits and the value of a digit in the string depends on its position in the string. For example, the Hindu-Arabic numeral for the number of days in the year, 365, represents

$$3 \text{ hundreds} + 6 \text{ tens} + 5 \text{ ones}.$$

More specifically, this place-value system of numeration is said to be a **base 10** system because each positive integer has a unique representation as a sum of multiples of powers of 10, that is, as a *polynomial in 10*, where the multiples the coefficients of the polynomial are integers from 0 to 9. These multiples are the **digits** of the system, and we list only the digits in representing the number. For example, the number that results from performing the operations in the expression $3 \cdot 10^2 + 6 \cdot 10^1 + 5 \cdot 10^0$ is written as 365 in base 10. The representation as a polynomial allows for reasonably efficient algorithms for adding, subtracting, multiplying, and dividing.

The choice of 10 as the base is not critical to efficient algorithms. Rather, as you know, the choice of the base 10 is a result of the fact that we have 10 fingers, and fingers have been used for counting throughout history. (The term "digit" that we use for the symbols 0, 1, 2, 3, 4, 5, 6, 7, 8, 9 that compose our numerals is derived from the Latin "digitus", meaning "finger.") Other bases are both possible and practical. The most important alternate base is 2, for which the "digits" are 0 and 1, because this base is easy to implement in electronic circuitry. However, the numbers 3, 8, 12, and 16 are also convenient bases for certain purposes.

To write an integer n in base b, we need to write n as $a_k b^k + a_{k-1} b^{k-1} + \cdots + a_1 b + a_0$, where the coefficients a_i are integers with $a_k \neq 0$ and $0 \leq a_i < b$ for all i. This is **expanded base b notation**. It is customary to show that the base is b by a small subscript to the right of the digits. For instance, $365_{10} = 1031_7$, because

$$3 \cdot 10^2 + 6 \cdot 10^1 + 5 \cdot 10^0 = 1 \cdot 7^3 + 0 \cdot 7^2 + 3 \cdot 7^1 + 1 \cdot 7^0.$$

The algorithms for the arithmetic operations on integers in base b are consequences of their polynomial representations. For addition, we add like powers.

$$\begin{array}{ccccccccc} a_k b^k & + & a_{k-1} b^{k-1} & + \cdots + & a_1 b & + & a_0 \\ c_k b^k & + & c_{k-1} b^{k-1} & + \cdots + & c_1 b & + & c_0 \\ \hline (a_k + c_k) b^k & + & (a_{k-1} + c_{k-1}) b^{k-1} & + \cdots + & (a_1 + c_1) b & + & (a_0 + c_0) \end{array}$$

The only provision that must be made in the algorithm is due to the restriction that $a_i + c_i < b$. When $a_i + c_i > b$, then b must be subtracted in the ith place and 1 added to the digit in the $(i + 1)$st place. For instance,

$$\begin{array}{rccccc} & 3 & 0 & 2 & 1 & 5_7 \\ + & 6 & 4 & 6 & 5 & 2_7 \\ \hline & (9) & 4 & (8) & 6 & (7)_7 \end{array}$$

but the 9s, 8s, and 7s are not allowed in base 7 representation, so the sum is rewritten as

$$= 1\,2\,5\,1(7)0_7,$$

and then rewritten again as

$$= 1\,2\,5\,2\,0\,0_7.$$

Subtraction is done similarly.

Likewise, multiplication of integers in any base b follows multiplication of polynomials. Here we show an example of the process in expanded base 10 notation. Here, as above we place numbers in parentheses when they are too large to be digits. Then we rewrite the expressions so that all digits are less than 10.

$$\begin{aligned} 302_{10} \cdot 47_{10} &= \\ &= (3 \cdot 10^2 + 0 \cdot 10 + 2)(4 \cdot 10 + 7) \\ &= (12) \cdot 10^3 + (21) \cdot 10^2 + 0 \cdot 10^2 + 8 \cdot 10 + 0 \cdot 10 + (14) \\ &= (1 \cdot 10^4 + 2 \cdot 10^3) + (2 \cdot 10^3 + 1 \cdot 10^2) + 8 \cdot 10 + (1 \cdot 10 + 4) \\ &= 1 \cdot 10^4 + 4 \cdot 10^3 + 1 \cdot 10^2 + 9 \cdot 10 + 4 \\ &= 14194_{10} \end{aligned}$$

Which integers can be bases? Clearly, 0 is not a suitable value for b because $0^p = 0$ for any nonnegative integer p, and this means that no nonzero integer can be represented by a base 0 numeral. The integer 1 is also not a suitable base because it only allows the single "digit" of 0 for base representation. However, the following result, which is a consequence of the Division Algorithm, shows that any positive integer b larger than 1 can be a base.

Theorem 5.15 For any integer $b > 1$, every positive integer n has a unique representation in base b.

The Main Idea of the Proof: Here is the essence of the proof. To produce the successive coefficients

$$a_0, a_1, \ldots, a_{k-1}, a_k$$

apply the Division Algorithm to divide n and the successive quotients by b until a zero quotient is obtained. The successive remainders are the coefficients in the order given above.

For example, to represent 365_{10} in base 7, we proceed as follows:

$$\begin{aligned} 365 &= 52 \cdot 7 + 1 \\ 52 &= 7 \cdot 7 + 3 \\ 7 &= 1 \cdot 7 + 0 \\ 1 &= 0 \cdot 7 + 1. \end{aligned}$$

Then

$365 = 1 \cdot 7^3 + 0 \cdot 7^2 + 3 \cdot 7^1 + 1 \cdot 7^0$ because

$$\begin{aligned} 365 &= 52 \cdot 7 + 1 \\ &= (7 \cdot 7 + 3) \cdot 7 + 1 = 7 \cdot 7^2 + 3 \cdot 7 + 1 && \text{(substitute above for 52)} \\ &= (1 \cdot 7 + 0) \cdot 7^2 + 3 \cdot 7 + 1 = 1 \cdot 7^3 + 3 \cdot 7 + 1 && \text{(substitute above for the first 7)} \\ &= (0 \cdot 7 + 1) \cdot 7^3 + 0 \cdot 7^2 + 3 \cdot 7 + 1. && \text{(substitute above for the first 1)} \end{aligned}$$

Proof of Theorem 5.15: Existence: We show first that, for each positive integer n, there is at least one representation

$$(*) \qquad n = a_k b^k + a_{k-1} b^{k-1} + \cdots + a_1 b + a_0$$

where the coefficients a_i for $0 \le i \le k$ are integers with $a_k \ne 0$ and $0 \le a_i < b$. To show this, start with n and successively apply the Division Algorithm as follows to obtain a decreasing series of nonnegative quotients until a zero quotient is reached:

$$\begin{aligned} n &= q_0 b + a_0 && \text{where} && 0 \le a_0 < b && \text{and} && n > q_0 \\ q_0 &= q_1 b + a_1 && \text{where} && 0 \le a_1 < b && \text{and} && q_0 > q_1 \\ q_1 &= q_2 b + a_2 && \text{where} && 0 \le a_2 < b && \text{and} && q_1 > q_2 \\ &\vdots \\ q_{k-2} &= q_{k-1} b + a_{k-1} && \text{where} && 0 \le a_{k-1} < b && \text{and} && q_{k-2} > q_{k-1} \\ q_{k-1} &= 0b + a_k && \text{where} && 0 \le a_k < b \end{aligned}$$

The required base b representation of the integer n is now obtained by beginning with the first equation in the preceding list and successively substituting the subsequent members of the list as follows:

$$\begin{aligned} n &= q_0 b + a_0 = (q_1 b + a_1) b + a_0 \\ &= q_1 b^2 + a_1 b + a_0 = (q_2 b + a_2) b^2 + a_1 b + a_0 \\ &= q_2 b^3 + a_2 b^2 + a_1 b + a_0 \\ &\vdots \\ &= q_{k-2} b^{k-1} + a_{k-2} b^{k-2} + \cdots + a_1 b + a_0 \\ &= (q_{k-1} b + a_{k-1}) b^{k-1} + a_{k-2} b^{k-2} + \cdots + a_1 b + a_0 \\ &= (a_k b + a_{k-1}) b^{k-1} + a_{k-2} b^{k-2} + \cdots + a_1 b + a_0 \\ &= a_k b^k + a_{k-1} b^{k-1} + a_{k-2} b^{k-2} + \cdots + a_1 b + a_0 \end{aligned}$$

This completes the proof that there is at least one representation of n in base b.

Uniqueness: To prove the uniqueness of this representation, suppose to the contrary that for some positive integer n, there is a second representation

$$(**) \qquad n = c_m b^m + c_{m-1} b^{m-1} + \cdots + c_1 b + c_0$$

where the coefficients c_i are integers with $c_m \neq 0$ and $0 \leq c_i < b$. We can assume that $m = k$ if we relax the requirements that $a_k \neq 0$ and $b_m \neq 0$ and allow the addition of some zero coefficients. Then

$$(***) \qquad 0 = (a_k - c_k)b^k + (a_{k-1} - c_{k-1})b^{k-1} + \cdots + (a_1 - c_1)b + (a_0 - c_0)$$

where at least one of the coefficients $(a_i - c_i)$ is nonzero. Let j be the smallest integer such that $a_j - c_j \neq 0$. Then from (***) we can solve for $a_j - c_j$ as follows:

$$(a_k - c_k)b^k + (a_{k-1} - c_{k-1})b^{k-1} + \cdots + (a_{j+1} - c_{j+1})b^{j+1} = -(a_j - c_j)b^j$$

$$(a_k - c_k)b^{k-j} + (a_{k-1} - c_{k-1})b^{k-j-1} + \cdots + (a_{j+1} - c_{j+1})b = -(a_j - c_j).$$

Because b is a divisor of the left side of the last equation, it also is a divisor of the right side, $-(a_j - c_j)$. This implies that $b \leq |a_j - c_j|$. But this contradicts the facts that the nonnegative integers a_j and c_j are both less than b, so that $-b < a_j - c_j < b$. Consequently, the representation of any positive integer in base b is unique. ┘

Question 1: Find the base 4 representation of 365_{10}.

When the base $b = 2$, the coefficients must be either 0 or 1. This provides a special case of Theorem 5.15.

Corollary: Every positive integer can be uniquely written as a sum of non-negative powers of 2.

Question 2: Find the base 2 (binary) representation of 365_{10}.

If $b > 10$, new symbols are needed for the base b coefficients between 10 and $b - 1$. For example, for the base 16 (hexadecimal) system, which is frequently used in analyzing computer output, the additional symbols are typically taken to be $A = 10$, $B = 11$, $C = 12$, $D = 13$, $E = 14$, and $F = 15$. For example, the symbol $(D2B7)_{16}$ is the hexadecimal representation of the integer 53,943 because

$$(D2B7)_{16} = 13 \cdot 16^3 + 2 \cdot 16^2 + 11 \cdot 16 + 7 = 53{,}943.$$

5.2.5 Problems

1. Find the base 3 representation of the integer 421 by following the Division Algorithm construction used in the proof of Theorem 5.15.

2. Let $x = 647_8$ and $y = 251_8$. Find $x + y$, $x - y$, and xy in base 8.

3. Let $f(b) = 3804_b$.

a. Find a polynomial expression for $f(b)$.

b. Evaluate $f(10)$ and $f(16)$.

c. What positive integer values of b cannot be in the domain of f?

4. Every positive integer n has a unique representation of the form:

$$n = a_k 3^k + a_{k-1} 3^{k-1} + \cdots + a_1 3 + a_0$$

where the coefficients a_p are the integers 1, -1 or 0 with the leading coefficient $a_k \neq 0$. This is called the **balanced ternary representation** of n. A proof of this result can be obtained by following the construction used in the proof of the base b representation (Theorem 5.15), except that the repeated use of the Division Algorithm is replaced by repeated use of the following result:

> (*) *For each integer $n > 1$, exactly one of the following statements is true*:
>
> i. $n = 3k$ *for some positive integer* k.
>
> ii. $n = 3k + 1$ *for some positive integer* k.
>
> iii. $n = 3k - 1$ *for some positive integer* k.

a. Write the numbers from 1 to 30 in balanced ternary form.

b. Find the balanced ternary representation of the integer 421 by following the construction used in the proof of Theorem 5.15, but using the result (*) in place of the Division Algorithm.

c. Prove (*) by using the Division Algorithm.

d. By using the proof of Theorem 5.15 as a model, prove the balanced ternary representation theorem stated above.

5. How many digits are there in the representation of 2^{1000} in the indicated base?

a. base 10 b. base 2 c. base 8 d. base 16

6. If a number n has k digits when represented in base 2, how many digits will it have when represented in base 8, and why?

7. a. Prove: If the sum of the digits of a number in base 10 is divisible by 9, so is the number.

b. Prove: If the sum of the digits of a number in base b is divisible by $b - 1$, so is the number.

8. a. Prove: If the alternating sum and difference of the digits of a number in base 10 is divisible by 11, so is the number.

b. Prove: If the alternating sum and difference of the digits of a number in base b is divisible by $b + 1$, so is the number.

9. Prove: If the digits of a 3-digit number in base b are reversed, and the resulting number is subtracted from the original number, then the difference is divisible by both $b - 1$ and $b + 1$.

10. **Negative Bases.** Any integer b less than -1 can be a base for the representation of integers. The possible digits are $0, 1, 2, \ldots, |b + 1|$.

a. Represent 365 in base -10 by using the Division Algorithm.

b. Modify the statement and proof of Theorem 5.15 for any negative base $b < -1$.

11. a. Write 1000 (base 10) in binary, octal (base 8), and hexadecimal notation.

b. If a positive integer n has 100 digits in its base 10 representation, what is the maximum number of digits that are required for the hexadecimal representation of n?

c. If a positive integer n has 100 digits in its base 10 representation, what is the maximum number of digits that are required for the octal representation of n?

12. **Binary coded decimals.** Computer circuits deal with numbers in binary form, whereas we are accustomed to numbers in decimal form. One way to deal with this difference is to convert decimal numbers to binary form, perform all arithmetic operations in binary form, and then convert the results back to decimal form.

a. Carry out this conversion-operation procedure for the decimal sum

$$15 + 9 = 24$$

b. It is also possible to add decimals as follows: First, store the binary form of each decimal digit as a string of 0s and 1s of length 4 to obtain the *binary coded decimal* (BCD) form of that digit. For example, the BCD for 7 is 0111.

i. Write the decimal numbers 15, 9 and their sum 24 in BCD form.

ii. Which strings of 0s and 1s of length 4 are not the BCDs of any decimal digit?

c. We can then perform arithmetic operations directly with the decimal numbers in BCD form using the same procedure as in part **a**, modified by adding 6 = (0110) before each carry.

i. Compute the sum $15 + 9 = 24$ directly on the decimal digits by using this modified procedure for binary addition.

ii. Explain why this modified procedure works.

ANSWERS TO QUESTIONS

1. $365 = 91 \cdot 4 + 1 = (22 \cdot 4 + 3) \cdot 4 + 1 = ((5 \cdot 4 + 2) \cdot 4 + 3) \cdot 4 + 1 = (((1 \cdot 4 + 1) \cdot 4 + 2) \cdot 4 + 3) \cdot 4 + 1 = 1 \cdot 4^4 + 1 \cdot 4^3 + 2 \cdot 4^2 + 3 \cdot 4 + 1$, so $365_{10} = 11231_4$

2. $365_{10} = 1 \cdot 2^8 + 1 \cdot 2^6 + 1 \cdot 2^5 + 1 \cdot 2^3 + 1 \cdot 2^2 + 1 = 101101101_2$

Unit 5.3 Divisibility Properties of Polynomials

A **monomial** is a product of numbers and integer powers of a variable or variables. For instance,

$$\frac{4}{3}\pi r^3 \qquad 5x^2yz \qquad \text{and} \qquad \sqrt{2}n \qquad \text{are monomials}$$

$$\text{while} \qquad \sin r \qquad 5x^2 + y + z \qquad \text{and} \qquad \sqrt{2n} \qquad \text{are not.}$$

The numerical factor in a monomial is its **coefficient**. A nonzero number such as $3\sqrt{2}$ or 31 can also be regarded as a monomial in which the powers of all the variables are 0; such a monomial is called a **constant monomial**. The sum of the powers of the variables in a monomial is the **degree** of the monomial. For example, the degree of the monomial $-5x^2yz$ is 4, while the degree of a constant monomial is 0.

A **polynomial** is a finite sum of monomials and the coefficients of the polynomial are the set of coefficients of those monomials. Any monomial is considered to be a polynomial. A polynomial is said to be *in* its variables and *over* the set D of its allowable coefficients. We refer to D as the **coefficient domain** of the polynomial. The **degree** of a polynomial is the largest degree of any of the monomials that comprise it. For instance, the polynomial $5x^2z + xy + z$ is a polynomial of degree 3 in the variables x, y, and z and any nonzero constant monomial is a polynomial of degree 0. The polynomial whose coefficients are all 0 is the **zero polynomial**. The degree of the zero polynomial is undefined.

In this unit, we restrict our attention to polynomials in a single variable x, with a coefficient domain that is **Z**, **Q**, **R**, or **C**.

We use the symbol **D[x]** for the set of all polynomials in a variable x over a coefficient domain D. Thus, for example, **Z[x]** is the set of all polynomials with integer coefficients. The sets of polynomials are nested just as the coefficient domain sets are.

$$\mathbf{Z}[x] \subset \mathbf{Q}[x] \subset \mathbf{R}[x] \subset \mathbf{C}[x].$$

There are many analogies between division of integers and division of polynomials, including the following:

Integer division and polynomial division with corresponding quotient and remainder integers and polynomials

The long division algorithm for integers and for polynomials

Factorization of integers and polynomials

Greatest common factors and least common multiples of pairs of integers and pairs of polynomials

Prime factors of integers and polynomials

These analogies are not coincidental or superficial. They flow from the similarity of the algebraic structures of the systems of integers and polynomials. In this unit, we explore these analogies in some detail, and explain how they derive from this similarity.

5.3.1 The Division Algorithm for polynomials

To check whether 17 is a factor of 391 without a calculator, you could do the following long division.

$$\begin{array}{r} 23 \\ 17\overline{)391} \\ \underline{34} \\ 51 \\ \underline{51} \\ 0 \end{array}$$

The division shows that 17 is a factor of 391, and, because 17 and 23 are prime numbers, it also shows that the prime factorization of 391 is $17 \cdot 23$.

When one integer is not a factor of another, then the result can be indicated in two ways, which in Section 5.2.1 we identified as *integer division* and *rational number division*. For instance, if we divide 395 by 17, the long division

$$\begin{array}{r} 23 \\ 17\overline{)395} \\ \underline{34} \\ 55 \\ \underline{51} \\ 4 \end{array}$$

shows that $\frac{395}{17} = 23$ with a remainder of 4 (integer division) or, by carrying the division out to more decimal places, that $\frac{395}{17} = 23\frac{4}{17} \approx 23.235$ to three decimal places (rational number division).

Factors of polynomials

Now we begin our analogies. The definitions of the basic terms are essentially identical for integers and polynomials. Given polynomials $a(x)$ and $b(x)$ over a set D,

$b(x)$ **divides** $a(x)$ $b(x)$ is a **divisor** of $a(x)$ $b(x)$ is a **factor** of $a(x)$

if and only if there is a polynomial $c(x)$ over D such that $a(x) = b(x)c(x)$ for all x.

To show that $x^2 + 1$ is a factor of

$$x^4 - x^3 + 2x^2 - x + 1,$$

you can use polynomial long division.

$$\begin{array}{r} x^2 - x + 1 \\ x^2 + 1\overline{)x^4 - x^3 + 2x^2 - x + 1} \\ \underline{x^4 \qquad + x^2 \qquad\qquad} \\ -x^3 + x^2 - x \qquad \\ \underline{-x^3 \qquad - x \qquad} \\ +x^2 \qquad + 1 \\ \underline{+x^2 \qquad + 1} \\ 0 \end{array}$$

This shows that $x^2 + 1$ is a factor of $x^4 - x^3 + 2x^2 - x + 1$ and yields the factorization

$$x^4 - x^3 + 2x^2 - x + 1 = (x^2 + 1)(x^2 - x + 1).$$

Polynomial division vs. rational expression division

When one polynomial is not a factor of another, then there are two ways of expressing the result. These two ways are analogous to the two ways that exist for division of integers. For instance, to divide $x^4 - x^3 + 2x^2 - x + 4$ by $x^2 + 1$, we can again use long division.

$$\begin{array}{r} x^2 - x + 1 \\ x^2 + 1\overline{)x^4 - x^3 + 2x^2 - x + 4} \\ \underline{x^4 \qquad + x^2 \qquad\qquad} \\ -x^3 + x^2 - x \qquad \\ \underline{-x^3 \qquad - x \qquad} \\ x^2 \qquad + 4 \\ \underline{x^2 \qquad + 1} \\ 3 \end{array}$$

By analogy with integer division, we call $q(x) = x^2 - x + 1$ the **quotient polynomial** and $r(x) = 3$ the **remainder polynomial** of this **polynomial division** of $a(x) = x^4 - x^3 + 2x^2 - x + 4$ by $b(x) = x^2 + 1$. Notice that $a(x) = b(x)q(x) + r(x)$. This is analogous to integer division not only in the form $a = bq + r$, but also in the fact that a, b, q, and r are all polynomials. That is, in both integer division and polynomial division, the quotient and remainder are expressed within the system.

The result of polynomial division of $a(x)$ by $b(x)$ can also be written as **rational expression division**, in which the quotient is the rational expression $\frac{a(x)}{b(x)}$ and there is no remainder. Rational expression division is analogous to writing the quotient of integers $a \div b$ as the single rational number $\frac{a}{b}$. We can also rewrite the rational expression $\frac{a(x)}{b(x)}$ as

$$\frac{a(x)}{b(x)} = q(x) + \frac{r(x)}{b(x)}. \tag{1}$$

This is analogous to writing the rational number $\frac{a}{b}$ as the mixed fraction

$$\frac{a}{b} = q + \frac{r}{b}. \tag{2}$$

For example, the simplification of the fraction $\frac{395}{17}$ to the mixed fraction form

$$\frac{395}{17} = 23\frac{4}{17} = 23 + \frac{4}{17} \tag{3}$$

is analogous to the reduction of the rational expression $\frac{x^4 - x^3 + 2x^2 - x + 4}{x^2 + 1}$ to the mixed fraction form

$$\frac{x^4 - x^3 + 2x^2 - x + 4}{x^2 + 1} = (x^2 - x + 1) + \frac{3}{x^2 + 1}. \tag{4}$$

One advantage of the mixed fraction form (3) is that it clearly indicates that the value of the fraction $\frac{395}{17}$ is close to the integer 23; in fact, 23 is the integer part of $\frac{395}{17}$. The analogous advantage of the mixed fraction form (4) is that the values of the rational expression $f(x) = \frac{x^4 - x^3 + 2x^2 - x + 4}{x^2 + 1}$ are close to those of the polynomial $q(x) = x^2 - x + 1$ for large $|x|$, in the sense that $f(x) - q(x) = \frac{3}{x^2 + 1}$ approaches 0 as $x \to \infty$ or $x \to -\infty$. Figure 18 shows the rational function f in black and the graph of the quadratic function q with $q(x) = x^2 - x + 1$ in blue on the interval $-6 < x < +6$.

Figure 18

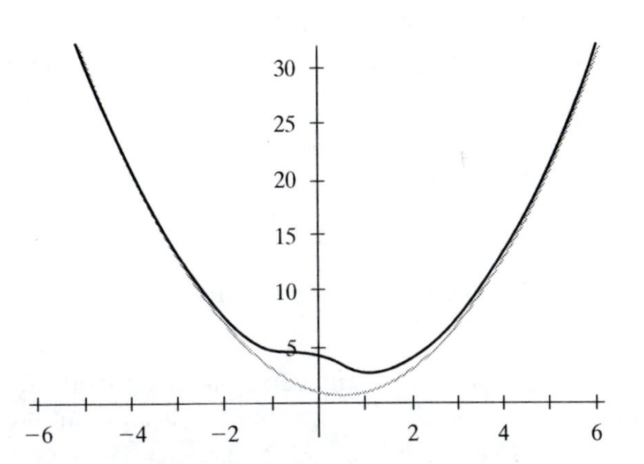

Thus the quotient polynomial in polynomial division approximates the rational expression in rational expression division just as the quotient 23 in the integer division of 394 by 17 approximates the rational number $23\frac{4}{17}$ in the rational number division of 394 by 17.

Integral domains

Why do these analogies between integer and polynomial division occur? Why does the long division procedure for polynomials work? Are there other analogies between integers and polynomials? The answers to these and other questions rest on the striking similarities that exist between the algebraic structures of the systems of integers and the system of polynomials.

Because the coefficient domains **Z**, **Q**, **R**, and **C** are closed under addition and multiplication, so too are addition and multiplication in **D[x]**. The set **D[x]** equipped with these binary operations is an algebraic system that we denote by $\langle \mathbf{D[x]}, +, \cdot \rangle$. This system has much in common with $\langle \mathbf{Z}, +, \cdot \rangle$.

These similarities are summarized in the language of abstract algebra by the statement that the system $\langle \mathbf{Z}, +, \cdot \rangle$ of integers and the systems $\langle \mathbf{D[x]}, +, \cdot \rangle$ of polynomials are examples of *integral domains.*

Definition An algebraic system $\langle \mathbf{A}, +, \cdot \rangle$ consisting of a set **A** together with two binary operations, $+$ and $\cdot$, on **A** is called an **integral domain** if it has the following properties. For all a, b, and c in **A**,

Addition:

(A-1) Addition is associative:

$$a + (b + c) = (a + b) + c.$$

(A-2) Addition is commutative:

$$a + b = b + a.$$

(A-3) Existence of an additive identity:

There is an element 0 in **A** such that $a + 0 = a$.

(A-4) Existence of an additive inverse for all elements:

There is an element $-a$ in **A** such that $a + (-a) = 0$.

Multiplication:

(M-1) Multiplication is associative:

$$a \cdot (b \cdot c) = (a \cdot b) \cdot c.$$

(M-2) Multiplication is commutative:

$$a \cdot b = b \cdot a.$$

(M-3) Existence of a multiplicative identity:

There is an element $1 \neq 0$ in **A** such that $a \cdot 1 = a$.

(C) Cancellation property:

If $a \cdot b = a \cdot c$ and $a \neq 0$, then $b = c$.

(D) Distributive property:

$$a \cdot (b + c) = a \cdot b + a \cdot c.$$

Question: What are the additive and multiplicative identities of the integral domain $\langle \mathbf{R[x]}, +, \cdot \rangle$?

The system $\langle \mathbf{Z}, +, \cdot \rangle$ of integers is an integral domain whose unit element is the integer 1 because all of the properties required of an integral domain are familiar properties of addition and multiplication for integers. The same is true for the system $\langle \mathbf{R}, +, \cdot \rangle$ of real numbers and the system $\langle \mathbf{C}, +, \cdot \rangle$ of complex numbers. The latter two systems have the following additional property for multiplication that does not hold for the system $\langle \mathbf{Z}, +, \cdot \rangle$ of integers.

(M-4) Existence of a multiplicative inverse for all elements not equal to the additive identity:

For each a in A such that $a \neq 0$, there is an element a^{-1} in A such that

$$a \cdot a^{-1} = 1.$$

In fact, the only nonzero integers n that have multiplicative inverses in $\mathbf{Z}$ are the integers 1 and -1, and these integers are their own multiplicative inverses. By comparing these properties with the field properties discussed in Section 4.2.2, we see that a *field* is an integral domain with the additional property M-4. Thus the systems $\langle \mathbf{R}, +, \cdot \rangle$ of real numbers and the systems $\langle \mathbf{C}, +, \cdot \rangle$ of complex numbers are fields as well as integral domains, while the system $\langle \mathbf{Z}, +, \cdot \rangle$ of integers is an integral domain but not a field.

Division in Z[x]

In $\langle \mathbf{Q[x]}, +, \cdot \rangle$, the integral domain of polynomials with rational coefficients, x^3 divided by $2x$ yields the result $\frac{1}{2}x^2$ with remainder 0. So, in $\langle \mathbf{Q[x]}, +, \cdot \rangle$, $2x$ is a factor of x^3. But in $\langle \mathbf{Z[x]}, +, \cdot \rangle$, the coefficients of all polynomials must be integers, so x^3 divided by $2x$ is impossible. That is, because $\mathbf{Z}$ is not a field, some simple divisions cannot be carried out in $\langle \mathbf{Z[x]}, +, \cdot \rangle$. (This is another analogy with integers.) So the Division Algorithm does not hold in $\langle \mathbf{Z[x]}, +, \cdot \rangle$. But it holds in any system $\mathbf{D[x]}$ where the set D of coefficients is a field.

The Division Algorithm for Polynomials

$\langle \mathbf{D[x]}, +, \cdot \rangle$ not only shares the algebraic properties of an integral domain with the system $\langle \mathbf{Z}, +, \cdot \rangle$ of integers, but also, as a consequence, versions of the Division Algorithm and the Euclidean Algorithm hold in $\langle \mathbf{D[x]}, +, \cdot \rangle$. We show the Division Algorithm here and the Euclidean Algorithm in Section 5.3.2.

Theorem 5.16 **(Division Algorithm for Polynomials):** Suppose that $a(x)$ and $b(x)$ are polynomials in $\mathbf{D[x]}$, where $\langle \mathbf{D}, +, \cdot \rangle$ is a field and that $b(x)$ is not the zero polynomial. Then there are unique polynomials $q(x)$ and $r(x)$ in $\mathbf{R[x]}$ such that

$$a(x) = b(x)q(x) + r(x)$$

and either $r(x)$ is the zero polynomial or the degree of $r(x)$ is less than the degree of $b(x)$.

This theorem is the basis of polynomial division. It also explains the familiar long division procedure for polynomials. In the example of long division given earlier,

$$\begin{array}{r}
x^2 - x + 1 \\
x^2 + 1 \overline{) x^4 - x^3 + 2x^2 - x + 4} \\
\underline{x^4 + x^2 } \\
- x^3 + x^2 - x \\
\underline{- x^3 - x } \\
x^2 + 4 \\
\underline{x^2 + 1} \\
3
\end{array}$$

we conclude from the first step that

$$\begin{aligned} x^4 - x^3 + 2x^2 - x + 4 &= (x^2 + 1)(x^2) + (-x^3 + x^2 - x + 4) \\ a(x) &= b(x)q_1(x) + r_1(x) \\ \text{where } \deg[r_1(x)] &= 3 > \deg[b(x)] = 2. \end{aligned}$$

The second step shows that

$$\begin{aligned} -x^3 + x^2 - x + 4 &= (x^2 + 1)(-x) + (x^2 + 4) \\ r_1(x) &= b(x)q_2(x) + r_2(x) \\ \text{where } \deg[r_2(x)] &= 2 = \deg[b(x)] = 2, \end{aligned}$$

and the last step shows that

$$\begin{aligned} x^2 + 4 &= (x^2 + 1)(1) + 3 \\ r_2(x) &= b(x)q_3(x) + r(x) \\ \text{where } \deg[r(x)] &= 0 < \deg[b(x)] = 2. \end{aligned}$$

The quotient $q(x)$ is obtained by substituting the expression for the remainder obtained from each step into the expression for $r_2(x)$ in the previous step, and then using the distributive property to collect the quotients. That is, $q(x) = q_1(x) + q_2(x) + q_3(x)$.

$x^4 - x^3 + 2x^2 - x + 4$	$a(x)$
$= (x^2 + 1)(x^2) + (-x^3 + x^2 - x + 4)$	$= b(x)q_1(x) + r_1(x)$
$= (x^2 + 1)(x^2) + (x^2 + 1)(-x) + (x^2 + 4)$	$= b(x)q_1(x) + b(x)q_2(x) + r_2(x)$
$= (x^2 + 1)(x^2) + (x^2 + 1)(-x) + (x^2 + 1)(1) + 3$	$= b(x)q_1(x) + b(x)q_2(x) + b(x)q_3(x) + r(x)$
$= (x^2 + 1)(x^2 - x + 1) + 4$	$= b(x)[q_1(x) + q_2(x) + q_3(x)] + r(x)$

The Division Algorithm for Polynomials can be proved in a way that is very similar to our proof of Theorem 5.3, the Division Algorithm for Integers. The key to the proof is that, at each step of the algorithm, the degree of the remainder polynomial is reduced by at least one. Thus the algorithm must end in a finite number of steps. A proof of Theorem 5.16 is outlined in Project 8 for this chapter.

The Remainder and Factor Theorems

If the divisor polynomial $b(x)$ in the Division Algorithm is a polynomial of the form $x - c$, for some real or complex number c, then the remainder polynomial $r(x)$ must be a constant k since its degree must be less than that of $b(x)$. If we replace x by c in the equation

$$a(x) = (x - c)q(x) + k,$$

we see that $a(c) = (c - c)q(c) + k = k$. That is, the constant k is the value of the dividend polynomial $a(x)$ at $x = c$. This proves the following important result known as the *Remainder Theorem.*

Theorem 5.17 **(Remainder Theorem):** Suppose that $a(x)$ is a polynomial of degree ≥ 1 and that c is a real or complex number. Then $a(c)$ is the remainder obtained by dividing $a(x)$ by $x - c$.

The Remainder Theorem has the following corollary that is usually called the *Factor Theorem.*

Theorem 5.18 **(Factor Theorem):** Suppose that $a(x)$ is a polynomial of degree ≥ 1 and that c is a real or complex number. Then $x - c$ is a factor of $a(x)$ if and only if $a(c) = 0$.

The Remainder and Factor Theorems are very useful tools for locating the roots of polynomial equations and for factoring polynomials. Some of these applications are explored in the problems.

The process of dividing a polynomial $a(x)$ of degree ≥ 1 by a polynomial $x - c$ of degree 1 is often facilitated by a scheme that is called *synthetic division*. For example, the scheme

$$\begin{array}{r|rrrr} -1 & 2 & -3 & 1 & -3 \\ & & -2 & 5 & -6 \\ \hline & 2 & -5 & 6 & (-9) \end{array}$$

is the synthetic division version for the long division of $2x^3 - 3x^2 + x - 3$ by $x + 1$.

$$\begin{array}{r} 2x^2 - 5x + 6 \\ x + 1 \overline{)2x^3 - 3x^2 + x - 3} \\ \underline{2x^3 + 2x^2} \qquad\qquad \\ -5x^2 + x \qquad \\ \underline{-5x^2 - 5x} \qquad \\ 6x - 3 \\ \underline{6x + 6} \\ -9 \end{array}$$

Notice that in synthetic division, only the coefficients of the variables are written. Also, note that changes in the sign of the constant in the divisor cause subtractions in the long division process to be replaced by corresponding additions in the synthetic division scheme. Although the synthetic division scheme is typically justified as a simple condensation of the long division algorithm, the scheme can be understood more precisely from the equation

$$a(x) = (x - c)q(x) + r, \tag{1}$$

that occurs in the division algorithm. For if

$$a(x) = a_n x^n + a_{n-1}x^{n-1} + \cdots + a_1 x + a_0, \tag{2}$$

then the quotient polynomial $q(x)$ is a polynomial of degree $n - 1$,

$$q(x) = q_{n-1}x^{n-1} + q_{n-2}x^{n-2} + \cdots q_1 x + q_0. \tag{3}$$

If we replace $a(x)$ and $q(x)$ in (1) by their equivalent expressions in (2) and (3),

$$a_n x^n + a_{n-1}x^{n-1} + \cdots + a_1 x + a_0 = (x - c)(q_{n-1}x^{n-1} + q_{n-2}x^{n-2} + \cdots q_1 x + q_0),$$

and equate coefficients of like powers of x, we obtain the coefficient relationships

$$q_{n-1} = a_n, \quad q_{n-2} = a_{n-1} + cq_{n-2}, \ldots, q_0 = a_1 + cq_2, \quad \text{and} \quad r = a_0 + cq_0.$$

These relationships fully justify and explain the general synthetic division scheme.

$$\begin{array}{r|rrrrr} c & a_n & a_{n-1} & \cdots & a_1 & a_0 \\ & & cq_{n-1} & \cdots & cq_1 & \\ \hline & q_{n-1} & q_{n-2} & \cdots & q_0 & r \end{array}$$

5.3.1 Problems

1. Consider the polynomials $a(x) = 2x^3 + 3x^2 - 4x - 7$ and $b(x) = x^2 + 1$.
   a. Perform polynomial division of $a(x)$ by $b(x)$. Call the quotient polynomial $q(x)$ and the remainder polynomial $r(x)$.
   b. Graph the rational function f with $f(x) = \frac{a(x)}{b(x)}$ and use part **a** to describe the behavior of $f(x)$ when $|x|$ is large.
2. Repeat Problem 1 if $a(x) = 5x^4 - 1$ and $b(x) = 10x^2 + 2$.
3. Factor $p(x) = x^4 - x^2 - 2$ over the given domain.
   a. **Z** b. **Q** c. **R** d. **C**
4. a. Show that $\frac{3}{2}$ is a solution to the polynomial equation
   $$p(x) = 2x^3 - 3x^2 - 4x + 6 = 0.$$
   b. Use the result of part **a** to write $p(x)$ as a product of polynomials of degree 1.
5. a. Prove: For any positive integer n, $x - c$ is a factor of $x^n - c^n$.
   b. Prove: For any odd positive integer n, $x + c$ is a factor of $x^n + c^n$.
6. Factors of some large numbers can be found by writing the numbers as polynomials. Use your knowledge of polynomial factors to find as many factors of each number as you can.
   a. 1000002000001
   b. $1\underbrace{000\ldots0}_{\text{50 zeros}}2\underbrace{000\ldots0}_{\text{50 zeros}}1$
   c. 11111111
   d. $\underbrace{111\ldots1}_{\text{63 ones}}$
   e. 827827
   f. 123123123123123123
7. a. Explain why the polynomial equation
   $$a_n x^n + a_{n-1}x^{n-1} + \cdots + a_1 x + a_0 = 0$$
   of degree $n \geq 1$ with nonnegative coefficients cannot have any positive real solutions.
   b. Find a similar coefficient criterion on the coefficients of polynomials that result in their having no negative real solutions.
8. Theorem 5.15 shows that every natural number n can be uniquely represented in the form $n = a_k b^k + a_{k-1}b^{k-1} + \cdots + a_1 b + a_0$, where b is a natural number ≥ 2 and the coefficients a_i are integers with $a_k \neq 0$ and $0 \leq a_i < b$ for all i. When $b = 10$ this is precisely the fact that every natural number can be uniquely represented in decimal (base 10) notation. The Division Algorithm for Polynomials can be generalized in an analogous way to achieve a similar result: If c is any element of a field $\langle F, +, \cdot \rangle$, then every nonzero polynomial $f(x)$ with coefficients in F can be uniquely represented in the form $f(x) = a_n(x - c)^n + a_{n-1}(x - c)^{n-1} + \cdots + a_1(x - c) + a_0$, where $a_n \neq 0$ and the a_i are elements of F for all i. Verify the theorem for the special case where $f(x) = x^9 - 1$, $c = 1$, and $F = \mathbf{R}$. That is, express $x^9 - 1$ as a sum of powers of $x - 1$

ANSWER TO QUESTION

The zero polynomial $z(x) = 0$ [the polynomial $z(x)$ that is equal to 0 for all real numbers x] is the additive identity of $\langle \mathbf{R[x]}, +, \cdot \rangle$, and the polynomial $p(x)$ defined for all real numbers x by $p(x) = 1$ is the multiplicative identity in $\langle \mathbf{R[x]}, +, \cdot \rangle$.

5.3.2 The Euclidean Algorithm and prime factorization for polynomials

The Division Algorithm in $\langle \mathbf{D[x]}, +, \cdot \rangle$ makes it possible to carry over to these systems of polynomials essentially the entire divisibility theory of the system $\langle \mathbf{Z}, +, \cdot \rangle$ of integers. However, as you might expect, polynomials are a little more complicated.

Specifically, when **D** is a field, there is a minor difference between divisors in **Z** and divisors in **D[x]**. If $a(x)$ is a factor of $b(x)$, then any nonzero multiple (in the field of coefficients) of $a(x)$ is also a factor of $b(x)$. For example, $2x + 1$ is a factor of $2x^2 - 9x - 5$ because

$$2x^2 - 9x - 5 = (2x + 1)(x - 5).$$

However, $\left(x + \frac{1}{2}\right) = \frac{1}{2}(2x + 1)$ is also a factor of $2x^2 - 9x - 5$ because

$$2x^2 - 9x - 5 = \frac{1}{2}(2x + 1)\cdot 2(x - 5)$$
$$= \left(x + \frac{1}{2}\right)(2x - 10).$$

Greatest common factors

Suppose $a(x)$ and $b(x)$ are in **D[x]**. If $a(x)$ and $b(x)$ are not both the zero polynomial, and if $d(x)$ is a polynomial in **D[x]**, then $d(x)$ is **a greatest common factor** of $a(x)$ and $b(x)$ if

i. $d(x)$ is a factor of $a(x)$ and $b(x)$, and

ii. if $p(x)$ is a polynomial in **D[x]** that is a factor of both $a(x)$ and $b(x)$, then $p(x)$ is a factor of $d(x)$.

From the preceding discussion, you can see that if $d(x)$ is a greatest common factor of $a(x)$ and $b(x)$, then any nonzero multiple of $d(x)$ is also a greatest common factor of $a(x)$ and $b(x)$.

However, for a given pair of polynomials $a(x)$ and $b(x)$ in **D[x]**, there is only one **monic polynomial** (a polynomial with leading coefficient 1) that is a greatest common factor of $a(x)$ and $b(x)$. For example, given the polynomials

$$a(x) = 4x^3 + 4x^2 + x = (2x + 1)^2 x$$
$$b(x) = 2x^3 - 9x^2 - 5x = (2x + 1)(x - 5)x$$

it follows from the factorizations shown that

$$d(x) = x^2 + \left(\frac{1}{2}\right)x = \left(x + \frac{1}{2}\right)x$$

is the unique monic greatest common factor of $a(x)$ and $b(x)$ in **D[x]**. We reserve the notation **gcf(a(x), b(x))** for the *monic* greatest common factor of two polynomials $a(x)$ and $b(x)$ in **D[x]**. With this notation, we showed above that

$$\text{gcf}(4x^3 + 4x^2 + x, 2x^3 - 9x^2 - 5x) = x^2 + \left(\frac{1}{2}\right)x.$$

The following divisibility result is the analog for **D[x]** to Theorem 5.4 for integers.

Theorem 5.19 Let $a(x)$ and $b(x)$ be polynomials in **D[x]** so that not both $a(x)$ and $b(x)$ are the zero polynomial. If $a(x) = b(x)q(x) + r(x)$, where $q(x)$ and $r(x)$ are polynomials in **D[x]**, then the set of greatest common factors of $a(x)$ and $b(x)$ is equal to the set of greatest common factors of $b(x)$ and $r(x)$, and in particular,

$$\text{gcf}(a(x), b(x)) = \text{gcf}(b(x), r(x)).$$

The Euclidean Algorithm

Just as Theorem 5.4 is the basis for the Euclidean Algorithm (Theorem 5.5) for calculating the greatest common factor for a given pair of integers, Theorem 5.19 yields the following corresponding result for polynomials.

Theorem 5.20 **(The Euclidean Algorithm for Polynomials):** Given two polynomials $a(x)$ and $b(x)$ in **D[x]**, where **D** is a field, the greatest common factor of $a(x)$ and $b(x)$ can be found as follows:

Step 1: Apply the Division Algorithm to find polynomials $q(x)$ and $r(x)$ such that $a(x) = b(x)q(x) + r(x)$, where $r(x)$ is the zero polynomial or $\deg[r(x)] < \deg[b(x)]$.

Step 2: If $r(x)$ is the zero polynomial, then $b(x)$ is the required greatest common factor. STOP. If $r(x)$ is not the zero polynomial and has leading coefficient c, replace $(a(x), b(x))$ by $(b(x), r(x))$ and repeat Step 1 and Step 2 until the algorithm stops. Then $d(x) = \frac{1}{c}r(x)$ is the required greatest common factor.

EXAMPLE Use the Euclidean Algorithm to compute $d(x) = \mathrm{gcf}(a(x), b(x))$ for the polynomials

$$a(x) = 2x^4 - 2x^3 + 3x^2 - x + 1 \quad \text{and} \quad b(x) = 2x^3 + 2x^2 + x + 1$$

Solution Apply the Division Algorithm as prescribed in Theorem 5.20 to obtain $q(x)$ and $r(x)$ when

$$a(x) = b(x)q(x) + r(x)$$
$$2x^4 - 2x^3 + 3x^2 - x + 1 = (2x^3 + 2x^2 + x + 1)(x - 2) + (6x^2 + 3).$$

Since $6x^2 + 3 \neq 0$, perform polynomial division of $b(x)$ by $r(x)$. That is, find $q_1(x)$ and $r_1(x)$ when $b(x) = r(x)q_1(x) + r_1(x)$.

$$2x^3 + 2x^2 + x + 1 = (6x^2 + 3)\left(\frac{x}{3}\right) + (2x^2 + 1).$$

Since $2x^2 + 1 \neq 0$, repeat with $r(x) = r_1(x)q_2(x) + r_2(x)$.

$$6x^2 + 3 = (2x^2 + 1)3 + 0.$$

Now $q_2(x) = 3$ and $r_2(x)$ is the zero polynomial. STOP.
Therefore, $r_1(x) = 2x^2 + 1$ is a greatest common factor of $a(x)$ and $b(x)$, and so $\mathrm{gcf}(a(x), b(x)) = x^2 + \left(\frac{1}{2}\right)$.

Prime factorization

In our discussion of the integers, we proved the Fundamental Theorem of Arithmetic, which states that every positive integer can be represented as the product of prime numbers, and that this representation is unique except for the order of the factors. There is an analogous result for polynomials. For this, we need to define what is meant by a *prime polynomial*.

Let $\langle \mathbf{D[x]}, +, \cdot \rangle$ be an integral domain and $\mathbf{D}$ be a field, and suppose $a(x)$ is in $\mathbf{D[x]}$ and $\deg[a(x)] \geq 1$. If the only divisors of $a(x)$ in $\mathbf{D[x]}$ are constant polynomials and constant multiples of $a(x)$, then $a(x)$ is a **prime polynomial** or **irreducible polynomial** in $\mathbf{D[x]}$.

From this definition, if $a(x)$ is a prime polynomial in $\mathbf{D[x]}$, then any nonzero multiple of $a(x)$ is also prime in $\mathbf{D[x]}$.

For example, let's analyze the prime factorization of the polynomial

$$p(x) = x^2 + 5x + 3$$

in $\mathbf{D[x]}$ for $\mathbf{D} = \mathbf{Q}, \mathbf{R}$, and $\mathbf{C}$.

The polynomial $p(x)$ is prime in $Q[x]$ because the corresponding equation

$$x^2 + 5x + 3 = 0$$

in a real variable x has two solutions $x = \frac{-5 \pm \sqrt{13}}{2}$, and so (by the Factor Theorem) the only factors of $p(x)$ of positive degree are $a(x) = \left(x - \frac{-5+\sqrt{13}}{2}\right)$ and $b(x) = \left(x - \frac{-5-\sqrt{13}}{2}\right)$ and their multiples, and neither of these polynomials are in $\mathbf{Q[x]}$. This also shows that a prime factorization of $p(x)$ in $\mathbf{R[x]}$ and $\mathbf{C[x]}$ is given by

$$p(x) = \left(x - \frac{-5 + \sqrt{13}}{2}\right)\left(x - \frac{-5 - \sqrt{13}}{2}\right),$$

and both factors are monic polynomials.

As another example, for the polynomial $a(x)$ we considered in the Example above,

$$a(x) = 2x^4 - 2x^3 + 3x^2 - x + 1,$$

a prime factorization of $a(x)$ in **Q[x]** or **R[x]** is

$$a(x) = 2\left[x^2 + \left(\frac{1}{2}\right)\right](x^2 - x + 1),$$

while a prime factorization of $a(x)$ in **C[x]** is

$$a(x) = 2\left[x + \left(\frac{\sqrt{2}}{2}\right)i\right]\left[x - \left(\frac{\sqrt{2}}{2}\right)i\right]\left[x - \left(\frac{1 + \sqrt{3}i}{2}\right)\right]\left[x - \left(\frac{1 - \sqrt{3}i}{2}\right)\right].$$

These examples are instances of the following theorem.

Theorem 5.21 **(The Unique Prime Factorization Theorem for Polynomials):** When **D** is a field, every polynomial $p(x)$ of positive degree in **D[x]** can be represented as a product

$$p(x) = ap_1(x)p_2(x)\dots p_k(x)$$

where a is in **D**, and $p_1(x), p_2(x), \dots, p_k(x)$ are monic prime polynomials in **D[x]**. This representation is unique except for the order of the polynomials $p_1(x)$, $p_2(x), \dots, p_k(x)$.

Theorem 5.21 can be proved by a method similar to that which we used in the proof of the Fundamental Theorem of Arithmetic. However, we do not present the details. Our primary purpose in stating the theorem here is as a final illustration of the striking analogies that exist between the divisibility structures of integers and of polynomials.

Differences among factorizations over Q, R, and C

Examples in this and the preceding section have shown that the prime factorization of a polynomial can vary if the set over which the coefficients of its factors varies over **Q**, **R**, and **C**. Several general theorems can help in determining when a factorization is prime.

Consider first polynomials over **C**. Let $p(x)$ be such a polynomial. Then, by the Fundamental Theorem of Algebra (Section 4.2.3), $p(x) = 0$ has a solution c_1 in **C**. Then, by the Factor Theorem, $x - c_1$ is a factor of $p(x)$, so

$$p(x) = (x - c_1)q_1(x).$$

Now $q_1(x)$ is of degree $n - 1$, and if $n - 1 \geq 1$, $q_1(x)$ has a solution also, call it c_2. Then $x - c_2$ is a factor of $q_1(x)$, and so $x - c_2$ is a factor of $p(x)$. Then there is a polynomial $q_2(x)$ with

$$p(x) = (x - c_1)(x - c_2)q_2(x).$$

This process will end after n steps, and so *over **C** every polynomial of degree n has a factorization with n factors $x - c_1$, $x - c_2, \dots x - c_n$.* Multiply the product of these factors by a and you have the prime factorization of the polynomial $p(x)$. The factorization of the 4th degree polynomial $a(x)$ shown preceding Theorem 5.21 is a prime example (!).

Now consider polynomials over **R**. If all the coefficients of $p(x) = 0$ are real numbers, then the nonreal solutions come in complex conjugate pairs, for if $p(a + bi) = 0$, then $p(a - bi) = 0$. As a result, using the Factor Theorem,

$$x - (a + bi) \quad \text{and} \quad x - (a - bi)$$

are factors of $p(x)$. Consequently, their product

$$x^2 - 2x + a^2 + b^2$$

is also a factor of $p(x)$. Thus, for polynomials over **R**, real solutions (and factors with real coefficients) come in pairs. Furthermore, this result implies that *the prime factorization of every polynomial with real coefficients has only linear and quadratic factors*. The linear factors come from its real solutions, while the quadratic factors arise from the pairs of complex conjugate solutions. This result is exemplified by the factorization of the polynomial $a(x)$ in **Q[x]** and **R[x]** preceding Theorem 5.21.

Now consider polynomials over **Q**, those with rational coefficients. If the coefficients are in lowest terms, we can multiply all of them by the least common multiple of their denominators to obtain a polynomial over **Z** with the same factors. For instance, to factor

$$p(x) = \frac{3}{2}x^2 + \frac{13}{4}x - 2,$$

multiply $p(x)$ by 4 to obtain

$$p^*(x) = 6x^2 + 13x - 8.$$

$p^*(x)$ is much easier to factor than $p(x)$.

$$p^*(x) = (3x + 8)(2x - 1),$$

so

$$p(x) = \frac{1}{4}(3x + 8)(2x - 1).$$

In this way, the problem of factoring a polynomial over **Q** can be converted into an equivalent problem of factoring a new polynomial over **Z**. When the polynomial is over **Z**, the Rational Root Theorem (Section 2.1.1) can be employed to search for all rational solutions. Specifically, *if a and b are in* **Z**, *and* $ax + b$ *is a factor of a polynomial* $p(x)$ *over* **Z**, *then a must be a factor in* **Z** *of the leading coefficient of* $p(x)$ *and b must be a factor of the constant coefficient of* $p(x)$. This explains why in schoolwork, students are asked to factor polynomials far more often with integer coefficients than noninteger rational coefficients.

5.3.2 Problems

1. a. Find a prime factorization of $8x^4 + 2x^2 - 1$ in **Z[x]**, **Q[x]**, **R[x]** and **C[x]**.

   b. For **Q[x]**, **R[x]**, and **C[x]**, find the unique prime representation with monic polynomials.

2. Find the monic greatest common divisor of the polynomials

$$a(x) = 4x^5 + 12x^4 + 20x^3 + 16x^2 + 16x + 4 \quad \text{and}$$

$$b(x) = 6x^5 + 12x^4 + 18x^3 + 12x^2 + 12x.$$

In problems 3 and 4, find the prime factorization of the polynomial in **R[x]** *and then in* **C[x]**.

3. $9x^4 + 440x^2 - 49$

4. $x^{106} - x^{100}$

5. Follow the directions of Problems 3 and 4. Assume a, b, and c are real.

   a. $ax^2 + bx + c$, given that $b^2 - 4ac \geq 0$

b. $ax^2 + bx + c$, given that $b^2 - 4ac < 0$

6. Factor the polynomial into prime factors over $\mathbf{R[x]}$, beginning with what is here.

a. $x^{12} - 1 = (x - 1)(\ldots) = \ldots$

b. $x^{12} - 1 = (x^6 - 1)(\ldots) = \ldots$

c. $x^{12} - 1 = (x^4 - 1)(\ldots) = \ldots$

d. $x^{12} - 1 = (x^2 - 1)(\ldots) = \ldots$

7. Just as we derived the representation in Theorem 5.6 of the greatest common divisor of two integers, we can obtain the following result by reversing the steps in the Euclidean Algorithm: If $a(x)$ and $b(x)$ are polynomials in $\mathbf{D[x]}$ that are not both the zero polynomial, then there are polynomials $p(x)$ and $q(x)$ in $\mathbf{D[x]}$ such that

$$\text{gcf}(a(x), b(x)) = a(x)p(x) + b(x)q(x).$$

Find this representation of $\text{gcf}(a(x), b(x))$ for the polynomials

$$a(x) = 2x^4 - 2x^3 + 3x^2 - x + 1 \quad \text{and}$$

$$b(x) = 2x^3 + 2x^2 + x + 1.$$

8. Use the proof of Theorem 5.4 in Section 5.2.2 as a guide to prove the corresponding Theorem 5.19 for polynomials.

Chapter Projects

1. **The Derivation of the Natural Number System from the Peano Axioms.** The Peano axioms are sufficient to give us the natural numbers and the operations of addition and multiplication on them. Do some research on this topic and answer the following questions regarding this powerful axiom set.

a. Show that 1 is unique. That is, prove that there is no number other than 1 that is not the successor of any number in **N**.

b. Show that addition can be completely described using two recursion equations in terms of 1 and s.

c. Show that multiplication $*$ can be completely described using two recursion equations in terms of 1, +, and s.

d. Use the well-ordering property to prove that addition and multiplication are closed in **N**.

e. Use mathematical induction to prove *commutativity* of addition and multiplication for all numbers in **N**.

f Use mathematical induction to prove *associativity* of addition and multiplication in **N**.

g. Prove that, for all x, y, and z in **N**, $x * (y + z) = x * y + x * z$. [*Hint*: Use mathematical induction letting $S(x)$ be the statement to be proved.]

2. **Strong Form of Proof by Mathematical Induction.** This method of proof can be stated as follows:

> Suppose that for each natural number n, $S(n)$ is a statement that is either true or false. To prove that $S(n)$ is true for all n, it is sufficient to
>
> 1. Prove that $S(1)$ is true, and
> 2. For any natural numbers k, prove that the supposition that $S(m)$ is true for $1 \leq m \leq k$ implies that $S(k+1)$ is true.

a. Explain why this method of proof is often called "strong" in comparison with the method of Proof by Mathematical Induction stated in Section 5.1.2.

b. Use the Strong Form of Proof by Mathematical Induction to prove that the following statement is true for all natural numbers $n > 1$:

> $S(n)$: n is either a prime number or a product of prime numbers.

and compare your proof with the proof of Theorem 5.9 in the text.

c. Explain why mathematical induction as stated in Section 5.1.2 is not enough to prove that the statement $S(n)$ in part **b** is true for all $n > 1$.

d. Modify the statement $S(n)$ in part **b**, and then prove it using the Principle of Mathematical Induction.

e. Generalize the method that you used in part **d** to prove that the Strong Form of Proof by Mathematical Induction is a consequence of mathematical induction as stated in Section 5.1.2.

3. **Paths along Lattices.** Consider the lattice points on and inside the rectangle with vertices $(0,0), (r,0), (r,s)$, and $(0,s)$, where r and s are positive integers. Consider all the broken line segment paths starting at $(0,0)$, moving from lattice point to lattice point, either horizontally to the right or vertically up, and ending at (r,s). How many such paths are there? (*Hint*: Let $P_{r,s}$ denote the number of paths through the given array where either r is greater than 0 or s is greater than 0. If $r = 0$ or $s = 0$, then there is only one path, i.e., $P_{1,r} = P_{s,1} = 1$ for all $r \geq 1, s \geq 1$. Obtain a relation between $P_{r,s}$, $P_{r-1,s}$, and $P_{r,s-1}$, and compare it with other relations you have seen.)

4. **Proving Irrationality by Infinite Descent.** The method of *infinite descent* was invented by Fermat. It states that an infinite sequence of decreasing natural numbers cannot exist. It follows from the well-ordering principle, which specifies that any such set must have a least element. Hippasus of Metapontum was a member of the Pythagorean school who discovered that there are some numbers that cannot be described by integers or ratios of integers. He is credited with the following geometric construction demonstrating that $\sqrt{2}$ is irrational. Start with a unit square $BACD$ and construct the circular arc $AD'D$ with center B cutting diagonal $\overline{BC}$ in D'.

Draw the tangent to the arc at D'. This will intersect side $\overline{AC}$ in a point. Call it B'. Construct the square $B'A'CD'$ Continue in this way generating a never-ending sequence of smaller and smaller squares. Then $\sqrt{2}$ is the ratio of the lengths of the sides of these smaller and smaller squares; namely, $\sqrt{2} = \frac{BC}{AB} = \frac{B'C}{A'B'} = \frac{B''C}{A''B''} = \ldots.$

a. Illustrate the geometrical construction of Hippasus.

b. Explain why the method of infinite descent applied to this construction shows that $\sqrt{2}$ is irrational.

c. Use the same idea to show that the golden mean $\frac{\sqrt{5}+1}{2}$ cannot be represented as the ratio of two integers.

5. Positive Solutions of Linear Diophantine Equations. Consider the Diophantine equation

$$(*) \quad px + qy = n$$

where p and q are positive integers and n is a nonnegative integer.

a. Prove that if $\gcd(p, q)$ is a factor of n, then

i. If $n > pq - (p + q)$, there is at least one solution in positive integers x and y. If (x_0, y_0) is a solution, then there are finitely many solutions all of the form $(x, y) = (x_0 + kq, y_0 - kp)$.

ii. If $n = pq - (p + q)$, there is no solution in positive integers x and y.

iii. If $0 \le n < pq - (p + q)$, there may or may not be a solution in positive integers x and y.

b. Verify part **a** by finding all integer solutions to $7x + 11y = 12$.

6. Fermat's Last Theorem. Complete parts **a–d** to prove the following special case of Fermat's Last Theorem: *The Diophantine equation $x^n + y^n = z^n$ has no positive integer solutions if n is any multiple of 4.*

a. Explain why proving that the equation $x^4 + y^4 = z^4$ has no positive integer solutions would imply the same conclusion for the equation

$$x^{4k} + y^{4k} = z^{4k}$$

for any positive integer k.

b. Explain why proving that the equation $x^4 + y^4 = z^2$ has no positive integer solutions would imply the same conclusion for the equation

$$x^4 + y^4 = z^4.$$

c. If (x, y, z) is an solution of $x^4 + y^4 = z^2$ consisting of positive integers x, y, and z with no common factors, then (x^2, y^2, z) is a primitive Pythagorean triple. Use the Pythagorean Triples Theorem to show that then the Diophantine equation $x^4 + y^4 = z^2$ has a solution (x', y', z'), where x', y', and z' are positive integers and $z' < z$.

d. Use the result of part **c** to prove that the equation $x^4 + y^4 = z^2$ has no positive integer solutions, and thus conclude from parts **a** and **b** that the Diophantine equation $x^n + y^n = z^n$ has no positive integer solutions if n is any multiple of 4.

7. Rational numbers in base 3. Extend the idea of decimals to base 3, as follows.

a. In base 10, $\frac{1}{2} = .5$. We can write $\left(\frac{1}{2}\right)_{10} = .5_{10}$. Suppose $\left(\frac{1}{2}\right)_{10} = x_3$. What is x?

b. Find base 3 equivalents to $\left(\frac{1}{3}\right)_{10}, \left(\frac{1}{4}\right)_{10}, \left(\frac{1}{5}\right)_{10}, \left(\frac{1}{6}\right)_{10}$, and $\left(\frac{1}{7}\right)_{10}$.

c. Which rational numbers have finite (terminating) representations in base 3?

8. Proving Theorem 5.16. Complete the following steps to construct a proof of the Division Algorithm for polynomials.

a. Suppose that $a(x)$ and $b(x)$ are polynomials in **R[x]** (or **C[x]**) and that $b(x)$ is not the zero polynomial. To show that there are polynomials $q(x)$ and $r(x)$ in **R[x]** (or **C[x]**) such that

$$a(x) = b(x)q(x) + r(x)$$

and such that either $r(x)$ is the zero polynomial or the degree of $r(x)$ is less than the degree of $b(x)$, consider three cases:

Case 1: $\deg[a(x)] \ge \deg[b(x)]$

Case 2: $0 \le \deg[a(x)] < \deg[b(x)]$

Case 3: $a(x)$ is the zero polynomial.

Prove the existence of the polynomials $q(x)$ and $r(x)$ for Cases 1 and 2.

b. Prove that if $a(x)$ and $b(x)$ are nonzero polynomials and if $\deg[b(x)] \le \deg[a(x)]$, then there is a polynomial $c(x)$ such that either $a(x) = b(x)c(x)$ or $\deg[a(x) - b(x)c(x)] < \deg[a(x)]$.

c. Prove the existence of $q(x)$ and $r(x)$ for Case 3 by considering two subcases:

Subcase (i): $a(x) = b(x)q(x)$ for some polynomial $q(x)$.

Subcase (ii): There is no polynomial $c(x)$ such that $a(x) = b(x)c(x)$ for all x.

[*Hint*: Use part **b** and the well-ordering property in Subcase (ii).]

d. Prove the uniqueness of the polynomials $q(x)$ and $r(x)$ by an argument similar to that used in Theorem 5.3 with the degree of a polynomial playing the same role as the order relation $<$ for integers.

Bibliography

Unit 5.1 References

Alfred, Brother U. *An Introduction to Fibonacci Discovery*. San Jose, CA: The Fibonacci Association, San Jose State College, 1965.
This 56-page softcover book contains many properties and problems about Fibonacci and Lucas numbers.

Ball, W. W. Rouse. *Mathematical Recreations and Essays*. New York: Macmillan, 1905.

Conway, John, and Richard Guy. *The Book of Numbers*. New York: Springer-Verlag, 1996, pp. 76–79.
An analysis of the maximum number of regions formed by connecting *n* points on a circle.

Dr. Math. Ask Dr. Math: FAQ Tower of Hanoi.
<http://mathforum.org/dr.math/faq/faq.tower.hanoi.html>

Gardner, Martin. "The Icosian Game and the Tower of Hanoi" in *The Scientific American Book of Mathematical Puzzles and Diversions*. New York: Simon and Schuster, 1959, pp. 57–59.

Hoggatt, Verner E., Jr. *Fibonacci and Lucas Numbers*. Boston: Houghton Mifflin, 1969.
A very nice introduction to the many properties of these numbers.

Lawrence Hall of Science, University of California at Berkeley. Tower of Hanoi.
<http://www.lhs.berkeley.edu/Java/Tower/towerhistory.html>

Thurston, H. A. *The Number System*. New York: Dover, 1967.
A fine source for the relationship between Peano Postulates and the natural numbers.

Vorob'ev, N. N. *Fibonacci Numbers*. New York: Blaisdell, 1961 (softcover); New York: New Classics Library, 1983 (hardcover).
Another introduction to the properties of these numbers.

Unit 5.2 References

Anderson, James. *Discrete Mathematics*. Upper Saddle River, NJ: Prentice Hall, 2001, pp. 794–799.
This contains a relatively brief elementary account of the RSA public key encryption algorithm.

Caldwell, Chris. *The Prime Pages*. Martin, TN: University of Tennessee at Martin.
<http://www.utm.edu/research/primes/>

Kaplan, Robert. *The Nothing That Is: A Natural History of Zero*. New York: Oxford, 2000, Chapter 8.

Unit 5.3 Reference

Beaumont, R. A., and R. E. Pierce. *The Algebraic Foundations of Mathematics*. Reading, MA: Addison-Wesley, 1963.
This book and its problem sets cover, illustrate, and extend many of the topics covered in Chapters 2, 5, and 6 of our book.

Chapter

6

NUMBER SYSTEM STRUCTURES

In Chapter 2, we discussed the real and complex number systems on the basis of their most familiar geometric models: the number line model of the real number system, $\langle \mathbf{R}, +, \cdot \rangle$ and the complex plane model of the complex number system $\langle \mathbf{C}, +, \cdot \rangle$. These models enabled us to investigate some important features of these systems, including decimal representation of real numbers, rational and irrational numbers and their distribution on the number line, and the geometric interpretation of the arithmetic operations for complex numbers.

In this chapter we investigate the real and complex number systems as *algebraic systems*. Recall from Section 4.2.1 that an algebraic system is a set of objects together with operations and relations on these objects that satisfy specified properties. Our objective is to identify sets of properties of $\langle \mathbf{R}, +, \cdot \rangle$ and $\langle \mathbf{C}, +, \cdot \rangle$ that completely describe them, in the same way that the Peano axioms describe the natural number system (Section 5.1.2).

One of the most important themes in all of mathematics is that some systems that look quite different on the surface share many properties. You have seen numerous examples of these commonalities in earlier chapters. In Chapter 1, we showed numerous properties shared by addition of the real numbers and multiplication of the positive real numbers. These correspondences were discussed in some detail in the solving of the equations $a + x = b$ and $ax = b$ in Chapter 4. In Chapter 2 we pointed out that addition of complex numbers can be viewed as addition of vectors. In Chapter 5, we showed the marvelous similarities between division of integers and division of polynomials.

Along the way you saw some of the names that are given to systems that possess specified properties: *group, field, integral domain*. These are **algebraic structures**. They are not the only algebraic structures but they are structures of systems most used in school mathematics.

Systems and the structures that underlie them are important for a variety of reasons. A theorem that is true in a structure of a particular kind is true in all systems that possess that structure. For instance, we showed in Chapter 4 that in any group $\langle S, * \rangle$, the equation $a * x = b$ has a unique solution in S. Since the systems $\langle \mathbf{R}^+, \cdot \rangle$

and $\langle \mathbf{R}, + \rangle$ each satisfy all the group properties, the general statement about groups implies that when a and b are positive real numbers, the equation $ax = b$ has a unique solution, and when a and b are any real numbers, the equation $a + x = b$ has a unique solution. The significance for teaching is that the group structure suggests a common method to obtain the solution: Operate on both sides with the inverse, obtain the identity, etc.

When systems possess the same structure, it may be possible to translate from one system to the other in order to solve a problem. In Unit 6.1 you will see this with the systems of modular arithmetic, which share many properties with the integers. This enables us to take some problems involving integers and translate them into modular arithmetic, solve them within modular arithmetic, and translate back.

When systems possess nearly the same structure, comparison of their structures can help in understanding them. In Unit 6.2 we examine the systems of real numbers and complex numbers with this goal in mind.

Unit 6.1 The Systems of Modular Arithmetic

In this unit we discuss a family of finite mathematical systems that have much in common with the infinite systems of natural numbers, integers, and polynomials. They are the systems of *modular arithmetic*, and they were first characterized by the German mathematician Karl Gauss (1777–1855) in his monumental work *Disquisitiones Arithmeticae* published in 1801. Modular arithmetic helps in understanding the structural properties of integers and other algebraic structures and presents an example of *equivalence* that is useful and important throughout mathematics. Furthermore, as you shall see, modular arithmetic has interesting and diverse applications outside mathematics.

6.1.1 Integer congruence

Figure 1 shows two clocks, one showing 8:00 and the other 1:00.

Figure 1

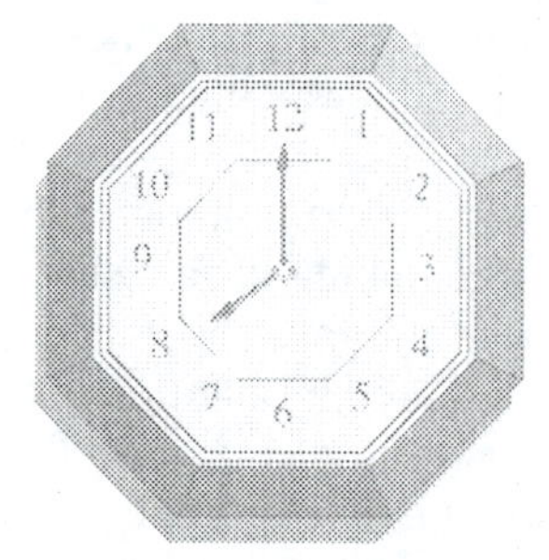

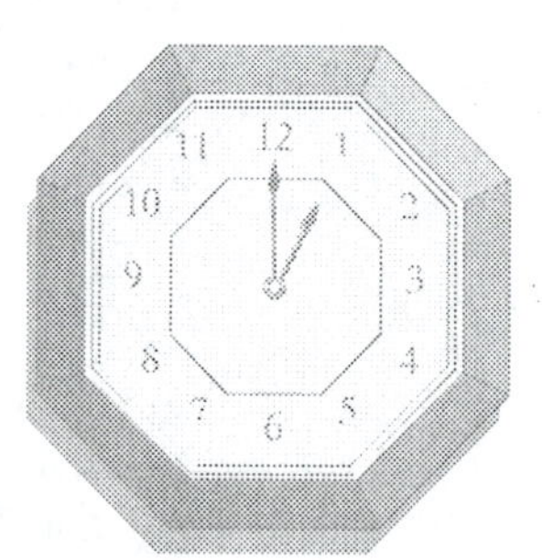

How are they related in time? The clock at the right may be 5 hours later than the clock at the left, or it may be 17 hours later, or 29 hours later, or 7 hours earlier, or 19 hours earlier, and so on. If we write $a \equiv b$ to denote the same clock time, we are led to the following statements in the arithmetic of 12-hour clocks:

$$8 + 5 \equiv 1$$

$$\text{but also } 5 \equiv 17 \equiv 29 \equiv -7 \equiv -19$$

$$\text{and, in general, } 5 \equiv 5 + 12k, \text{ where } k \text{ is any integer,}$$

$$\text{and even more generally, } n \equiv n + 12k, \text{ where } n \text{ and } k \text{ are any integers.}$$

Similar relationships could be based on a 24-hour clock. Then numbers of the form $n + 24k$ would be equivalent, and $5 \equiv 29$, but $5 \not\equiv 17$.

Equivalence of times in a 12-hour clock is an example of *congruence modulo 12*; for a 24-hour clock, equivalence exemplifies *congruence modulo 24*. In modular arithmetic, any positive integer greater than 1 can be the *modulus*, and congruence is defined as follows.

Definitions Suppose that a and b are integers, and that m is an integer (the **modulus**) greater than 1. Then **a and b are congruent modulo m**, written $\boldsymbol{a \equiv b}$ **Mod($\boldsymbol{m}$)**, if $a - b$ is an integer multiple of m; that is, if m is a factor of $a - b$. The set **a*** of all integers b such that $a \equiv b$ Mod(m) is called the **congruence class** (or **residue class**) **of a, modulo m.**

The definition of congruence modulo m states that two integers are in the same congruence class if their difference is divisible by m.

For example,

$$23 \equiv -5 \text{ Mod}(7) \quad \text{because } 23 - (-5) = 28 = 4 \cdot 7$$
$$-8 \equiv 0 \text{ Mod}(4) \quad \text{because } -8 - 0 = (-2) \cdot 4.$$

The clock arithmetic statement $5 - 9 = 8$ (5 o'clock $-$ 9 hours $=$ 8 o'clock) reflects the fact that $-4 \equiv 8$ Mod(12).

There are m congruence classes modulo m. For example, there are seven congruence classes modulo 7. Each congruence class is the set of those numbers that have the same remainder when divided by 7.

$$0^* = \{\ldots, -21, -14, -7, 0, 7, 14, 21, 28, \ldots\}$$
$$1^* = \{\ldots, -20, -13, -6, 1, 8, 15, 22, 29, \ldots\}$$
$$2^* = \{\ldots, -19, -12, -5, 2, 9, 16, 23, 30, \ldots\}$$
$$3^* = \{\ldots, -18, -11, -4, 3, 10, 17, 24, 31, \ldots\}$$
$$4^* = \{\ldots, -17, -10, -3, 4, 11, 18, 25, 32, \ldots\}$$
$$5^* = \{\ldots, -16, -9, -2, 5, 12, 19, 26, 33, \ldots\}$$
$$6^* = \{\ldots, -15, -8, -1, 6, 13, 20, 27, 34, \ldots\}$$

The Division Algorithm guarantees that each integer belongs to exactly one of these sets.

Notice that the congruence class **2*** modulo 7 consists of all integers of the form $2 + 7k$, where k is an integer. That is,

$$\mathbf{2}^* = (\ldots, -12, -5, 2, 9, 16, 23, \ldots\} = \{2 + 7k : k \in \mathbf{Z}\}.$$

Also notice that, modulo 7, $\mathbf{2^* = 9^* = (-5)^*}$, and in general $\mathbf{2^* = (2 + 7k)^*}$ for any integer k.

For any integer $m > 1$, and any integer n, the Division Algorithm tells us that

$$n = qm + r \qquad \text{where } 0 \leq r < m.$$

This means that $n \equiv r$ Mod(m) and that the remainder r is the smallest nonnegative integer that is congruent to n modulo m. For this reason, r is sometimes called the **least residue of $\boldsymbol{n}$ modulo $\boldsymbol{m}$**. Thus, for example, the least residue of 33 modulo 7 is 5, while the least residue of -23 modulo 9 is 4 because

$$-23 = (-3) \cdot 9 + 4.$$

The set of least residues modulo 7 is $\{0, 1, 2, 3, 4, 5, 6\}$. More generally, the set L_m of least residues modulo m is

$$L_m = \{0, 1, 2, \ldots, m - 1\}$$

and every integer is congruent to exactly one element of L_m.

For example, any integer k must be congruent modulo 8 to one and only one of the integers in the set L_8 of least residues modulo 8,

$$L_8 = \{0, 1, 2, 3, 4, 5, 6, 7\}.$$

However, it is also true that any integer k must be congruent modulo 8 to one and only one of the integers in the set

$$\{1, 2, 3, 4, 5, 6, 7, 8\},$$

[because $8 \equiv 0 \text{ Mod}(8)$]. The same is true of the set of integers

$$\{8, -7, 10, 11, 12, 13, 14, 39\}.$$

These three sets are examples of *complete residue sets* modulo 8. In general, a set of m integers is a **complete residue set modulo m** if every integer is congruent to exactly one of the integers in the set.

We are interested in operating on these residue sets. The set $3L_8 = \{0, 3, 6, 9, 12, 15, 18, 21\}$ of multiples by 3 of the least residue set L_8 modulo 8 is a complete residue set modulo 8 because

$$9 \equiv 1 \text{ Mod}(8);\ 12 \equiv 4 \text{ Mod}(8);\ 15 \equiv 7 \text{ Mod}(8);\ 18 \equiv 2 \text{ Mod}(8);\ 21 \equiv 5 \text{ Mod}(8).$$

On the other hand, the set $2L_8 = \{0, 2, 4, 6, 8, 10, 12, 14\}$ of multiples by 2 of the least residue set modulo 8 is not a complete residue set modulo 8 because, for example, the integer 5 is not congruent modulo 8 to any integer in this set. The key in these two examples is that 3 and 8 are relatively prime while 2 and 8 are not.

Theorem 6.1 Suppose that m is an integer > 1 and that b and m are relatively prime. Then the set $bL_m = \{0, b, 2b, 3b, \ldots, (m-1)b\}$ of multiples by b of the least residue set L_m modulo m is a complete residue set modulo m.

Proof: Because the sets L_m and bL_m both contain m integers, we need only show that each integer in L_m is congruent modulo m to one and only one integer in bL_m. Each integer in bL_m is congruent modulo m to at least one integer in L_m because L_m is a complete residue set modulo m. Suppose that two different elements of bL_m are congruent to the same integer in L_m. Then there are integers k_1 and k_2 such that $0 \le k_1 \le k_2 < m$, and that $bk_1 \equiv bk_2 \text{ Mod}(m)$. Then there is an integer q such that

$$bk_2 - bk_1 = b(k_2 - k_1) = qm.$$

Because b and m are relatively prime and m is a divisor of $b(k_2 - k_1)$, it follows that m is a divisor of $(k_2 - k_1)$. But $0 \le k_2 - k_1 < m$, so $k_2 - k_1 = 0$. So the two elements of bL_m are equal. This contradiction shows that each element of bL_m is congruent to a different element of L_m. So bL_m is a complete residue set modulo m. ┘

Operations on congruence classes

Look back at the seven congruence classes modulo 7 shown earlier in this section. Take any element of 2* and add it to any element of 3*. Because any element of 2* is of the form $7k + 2$, and any element of 3* is of the form $7k + 3$, their sum will be of the form $7k + 5$. That is, the sum will be an element of 5*. The same patterns hold for subtraction and multiplication. Consequently, for addition, subtraction, and multiplication, the congruence relation can be treated as we do equality.

Theorem 6.2 If m is an integer >1 and if $a \equiv b \text{ Mod}(m)$ and $c \equiv d \text{ Mod}(m)$, then

a. $a + c \equiv b + d \text{ Mod}(m)$
b. $a - c \equiv b - d \text{ Mod}(m)$
c. $ac \equiv bd \text{ Mod}(m)$

Proof:

a. If $a \equiv b \text{ Mod }(m)$ and $c \equiv d \text{ Mod}(m)$. Then m is a factor of $a - b$ and of $c - d$. So m is a factor of their sum $(a - b) + (c - d)$. Then, since $(a + c) - (b + d) = (a - b) + (c - d)$, $a + c \equiv b + d \text{ Mod}(m)$.

b. See the question immediately below.

c. This proof is left as a problem for you to do. ┘

Question: Use the proof of Theorem 6.2(a) as a model to prove Theorem 6.2(b).

In general, congruence of integers is an *equivalence relation*. That is, congruence for integers has the following properties:

Reflexive Property: For all integers a, $a \equiv a \text{ Mod}(m)$.
Symmetric Property: For all integers a and b, if $a \equiv b \text{ Mod}(m)$, then $b \equiv a \text{ Mod}(m)$.
Transitive Property: For all integers a, b, and c, if $a \equiv b \text{ Mod}(m)$ and $b \equiv c \text{ Mod}(m)$, then $a \equiv c \text{ Mod}(m)$.

You are asked to deduce these properties in Problem 9. (See Section 4.1.2 for a discussion of equivalence relations.)

The cancellation property in modular arithmetic

Since the system $\langle \mathbf{Z}, +, \cdot \rangle$ of integers satisfies all the properties of an integral domain (Section 5.3.1), it is natural to wonder which properties of an integral domain hold for the numbers and operations of modular arithmetic. The answer may surprise you: All properties hold but the cancellation property. In general, it is not possible to cancel common factors in congruence equations. For example,

$$5 \cdot 3 \equiv 1 \cdot 3 \text{ Mod}(6), \text{ but}$$
$$5 \not\equiv 1 \text{ Mod}(6)$$

because 6 is not a divisor of $5 - 1 = 4$. This means that you cannot in general divide both sides of a congruence by the same number and have the quotients be congruent.

On the other hand, sometimes there is a cancellation property. Note that

$$6 \cdot 5 \equiv 18 \cdot 5 \text{ Mod}(12)$$

because $30 - 90 = -60 = (-5)12$. If we cancel the common factor of 5 in the preceding congruence, we obtain

$$6 \equiv 18 \text{ Mod}(12),$$

which is true.

The key difference between these two examples is that the common factor 3 on the two sides of the congruence equation

$$5 \cdot 3 \equiv 1 \cdot 3 \text{ Mod}(6)$$

is not relatively prime to the modulus 6, while in the second case the common factor 5 in the congruence equation

$$6 \cdot 5 \equiv 18 \cdot 5 \text{ Mod}(12)$$

is relatively prime to the modulus 12. So there exists a modified cancellation property.

Theorem 6.3 Suppose that a, b, c, and m are integers and that $m > 1$. If $ac \equiv bc \text{ Mod}(m)$ and if c and m are relatively prime, then $a \equiv b \text{ Mod}(m)$.

Proof: If $a \cdot c \equiv b \cdot c \text{ Mod}(m)$, then m is a factor of $ac - bc = (a - b)c$. But if c and m are relatively prime and m divides the product $(a - b)c$, then m divides $(a - b)$ by Theorem 5.7. Therefore, $a \equiv b \text{ Mod}(m)$. ┘

Powers in modular arithmetic

The third part of Theorem 6.2 and mathematical induction help to prove the following corollary to Theorem 6.2.

Corollary: If $a \equiv b \text{ Mod}(m)$ and if p is any positive integer, then $a^p \equiv b^p \text{ Mod}(m)$.

For example, since $2 \equiv 17 \text{ Mod}(5)$, so $2^{43} \equiv 17^{43} \text{ Mod}(5)$. But the corollary does not tell us the congruence (residue) class to which 2^{43} and 17^{43} belong. A useful theorem for that purpose was discovered by Fermat.

Fermat stated the theorem in a letter to Marin Mersenne (1588–1648), a French theologian, in 1640. As with his famous Last Theorem, he did not provide a proof. In both cases, he excused the omission for a "lack of room". The name *Fermat's Little Theorem* has long been associated with this theorem because it was proved (by Euler) soon after he learned of its statement by Fermat.

Theorem 6.4 **(Fermat's Little Theorem):** Suppose that p is a prime number and that a is an integer such that p does not divide a. Then $a^{p-1} \equiv 1 \text{ Mod}(p)$.

Proof: Let L_p be the set of least residues modulo p,

$$L_p = \{0, 1, 2, 3, \ldots, p - 1\}.$$

Because p is a prime number and p does not divide a, it follows that a and p are relatively prime. Therefore, by Theorem 6.1, the set

$$aL_p = \{0, a, 2a, 3a, \ldots, (p - 1)a\}$$

of multiples of the least residues modulo p is a complete residue set modulo p. The set aL_p contains $p - 1$ positive integers, and each of the integers is congruent modulo p to one and only one positive integer in L_p. Therefore, by a generalization of Theorem 6.2c provable by mathematical induction (see Problem 6c in Section 6.1.3), the product of the positive integers in aL_p is congruent modulo p to the product of the positive integers in L_p.

$$(a)(2a)(3a)\ldots((p - 1)a) \equiv (1)(2)(3)\ldots(p - 1)\text{Mod}(p)$$
$$(p - 1)!\, a^{p-1} \equiv (p - 1)!\text{ Mod}(p).$$

But $(p - 1)!$ and p are relatively prime because p is prime. Consequently, by Theorem 6.3 we can cancel the $(p - 1)!$ in the preceding congruence to obtain the desired result. ┘

The following example shows how Fermat's Little Theorem and Theorem 6.2 can be used to simplify congruences with prime moduli.

EXAMPLE Find the remainder when 12^{158} is divided by 37.

Solution It is not possible to determine this remainder by direct division by hand or with most calculators or computers because 12^{158} is too large. However, the desired remainder is the least residue of 12^{158} Mod(37), so we can use modular arithmetic. By Fermat's Little Theorem, $12^{36} \equiv 1$ Mod(37). Consequently, by the corollary to Theorem 6.2, $12^{144} = (12^{36})^4 \equiv 1$ Mod(37).

Because $158 = 144 + 14$, and $(12, 37) = 1$, Theorem 6.3 implies that $12^{158} \equiv 12^{14}$ Mod(37).

By repeated squaring,

$$\begin{aligned} 12^1 &\equiv 12 \text{ Mod}(37) \\ 12^2 &\equiv 144 \text{ Mod}(37) \equiv 33 \text{ Mod}(37) \\ 12^4 &\equiv 33^2 \text{ Mod}(37) \equiv 1089 \text{ Mod}(37) \equiv 16 \text{ Mod}(37) \\ 12^8 &\equiv 16^2 \text{ Mod}(37) \equiv 256 \text{ Mod}(37) \equiv 34 \text{ Mod}(37). \end{aligned}$$

We multiply the powers whose exponents add to 14.

$$\begin{aligned} 12^{158} &\equiv 12^{14} \text{ Mod}(37) \\ &\equiv 12^{8+4+2} \text{ Mod}(37) \\ &\equiv (34)(16)(33) \text{ Mod}(37) \\ &\equiv (-3)\cdot 16\cdot(-4) \text{ Mod}(37) \\ &\equiv 192 \text{ Mod}(37) \\ &\equiv 7 \text{ Mod}(37) \end{aligned}$$

So, when 12^{158} is divided by 37, the remainder is 7.

6.1.1 Problems

1. What names are customarily given to the congruence classes Mod(2)?

2. Sort the multiples of 100, from 100 through 1000, into residue classes modulo 3.

3. Find the least residue modulo 13 of each integer.

a. 3783 b. -578 c. 16^{138}

4. Decide whether the set is a complete residue set modulo 10. Explain your decision in each case.

a. $\{0, 3, 6, 9, 12, 15, 18, 21, 24, 27\}$

b. $\{5, 6, 7, 8, 9, 10, 11, 12, 13, 14\}$

c. $\{0, 2, 4, 6, 8, 10, 12, 14, 16, 18\}$

d. $\{-7, 2, 19, 111, 0, 4, -5, -4, 87, 138\}$

5. Consider the complete set of least residues modulo 12.

a. What number is the additive identity?

b. What number is the additive inverse of 5?

c. What number is the multiplicative identity?

d. True or false? If $ab \equiv 0$ Mod(12), then $a \equiv 0$ Mod(12) or $b \equiv 0$ Mod(12). Explain your answer.

6. Repeat Problem 5 with modulus 11.

7. Prove Theorem 6.2c.

8. Prove the Corollary to Theorem 6.2.

9. Prove that congruence Mod(m) possesses the reflexive, symmetric, and transitive properties.

10. a. Find the remainder when 7^{315} is divided by 16.

b. Find the remainder when 17^{315} is divided by 16.

c. Find the remainder when 13^{315} is divided by 16.

d. Find the remainder when 14^{315} is divided by 16.

e. Arrange parts a, b, c, and d in order of difficulty and explain your ordering.

11. Use the congruence $5 = -1$ Mod(6) to determine the remainder when 5^{100} is divided by 6.

12. For what positive integers m is it true that $x^2 \equiv 0$ Mod(m) implies $x \equiv 0$ Mod(m)? Prove your result.

13. **Divisibility tests for 3, 9, 11:** Theorem 6.2 states that congruences are preserved by addition, subtraction, and multiplication. Use these properties of congruences together with the base 10 representation of a positive integer n,

$$n = a_k a_{k-1} \ldots a_1 a_0 = a_k 10^k + a_{k-1} 10^{k-1} \ldots + a_1 10 + a_0,$$

to provide a different proof than that of Section 5.2.5 of each of the following divisibility tests.

a. A positive integer n is divisible by 9 (or 3) if and only if the sum of the digits of n is divisible by 9 (or 3).

b. A positive integer n is divisible by 11 if and only if the alternating sum of the digits of n is divisible by 11.

[*Hint:* Note that $10 \equiv 1 \text{ Mod}(9)$, $10 \equiv 1 \text{ Mod}(3)$, and $10 \equiv -1 \text{ Mod}(11)$.]

14. a. Prove that the square of an integer is congruent to 0 or 1 modulo 4.

b. Use part **a** to prove that if z is a positive integer that is congruent to 3 modulo 4, then there are no integers x and y such that $x^2 + y^2 = z$.

15. Extend the definition of congruence modulo m to consider the set of real numbers Mod(1).

a. Identify three numbers congruent to 16.24.

b. Identify three numbers congruent to $\frac{3}{5}$.

c. Identify all the numbers that are congruent to π.

d. What is the additive inverse of 0.89?

e. True or false? If $ac \equiv bc \text{ Mod}(1)$, then $a \equiv b \text{ Mod}(1)$.

16. Congruence modulo m, where m is a positive real number, is defined as follows: $x \equiv y \text{ Mod}(m)$ if and only if $x - y$ is an integer multiple of m. When $m \in \mathbf{R}$, congruence modulo m can be used to describe periodic relationships. Find the smallest value of m for which each sentence is true.

a. For all x and y in $\mathbf{R}$, if $x \equiv y \text{ Mod}(m)$, then $\sin x = \sin y$.

b. For all x and y in $\mathbf{R}$, if $x \equiv y \text{ Mod}(m)$, then $\tan x = \tan y$.

c. For all x and y in $\mathbf{R}$, if $x \equiv y \text{ Mod}(m)$, then $x - \lfloor x \rfloor = y - \lfloor y \rfloor$.

ANSWER TO QUESTION

Proof of Theorem 6.2(b): Because $a \equiv b \text{ Mod}(m)$ and $c \equiv d \text{ Mod}(m)$, there are integers p and q such that $a - b = pm$ and $c - d = qm$. But then

$$(a - c) - (b - d) = (a - b) - (c - d) = pm - qm = (p - q)m,$$

which implies that $(a - c) \equiv (b - d)\text{Mod}(m)$.

6.1.2 Applications of integer congruence to calendars and cryptology

A phenomenon that cycles, as the hours of the day and days of the week do, provides a natural setting for use of modular arithmetic. In this section, we consider two applications. With calendars, the cycling is obvious. With cryptology, the study of codes, the use is due to a particular way of coding that is cyclic.

The Gregorian Calendar

An **Earth year**, that is, the length of time required for Earth to make one revolution around the Sun, is 365.2422 days, to four decimal places.

The calendar devised by the Egyptians in ancient times established a calendar year of exactly 365 days. Julius Caesar introduced leap years every four years to reduce the discrepancy between the lengths of the Egyptian calendar year and Earth year. This extra day every four years in the Julian calendar effectively increased the length of a calendar year to 365.25 days, a bit longer than an Earth year.

By the 16th century, the small difference of .0078 days per year between the Julian calendar year and the Earth year had caused the calendar to be 10 days different from the seasons. To correct this problem, in 1582, Pope Gregory refined the calendar further with the following change: Years divisible by 100 are not leap years unless the year is also divisible by 400. Thus, for example, the year 2000 was a leap year, but the year 1900 was not. In a given 400-year period, the Gregorian calendar has

$$365(400) + 100 - 4 + 1 = 146{,}097 \text{ days},$$

for an average calendar year of

$$\frac{146{,}097}{400} = 365.2425 \text{ days},$$

just a bit over the earth year of 365.2422 days.

Calculating the day of the week of a given date on the Gregorian calendar is an interesting exercise in modular arithmetic. The days of the week repeat in cycles of 7. Consequently, if we number the days of the week from 0 through 6,

$$1 = \text{Sun.}, 2 = \text{Mon.}, 3 = \text{Tue.}, 4 = \text{Wed.}, 5 = \text{Thu.}, 6 = \text{Fri.}, 0 = \text{Sat.},$$

then calculating the day of the week is doing arithmetic modulo 7. For example, if January 1 is on a Friday, which is congruent to 6 modulo 7, then the 50th day of the year (which is February 19th because January has 31 days) is also on Friday because $(50 - 1) + 6 = 55$ is congruent to 6 modulo 7. Or, thought of another way, the 1st day of the year is on the same day of the week as the 50th day of the year because $1 \equiv 50 \text{ Mod}(7)$.

In general, if January 1 is on day k of the week in a year, then the nth day of that year is on the pth day of the week, where p is the least residue modulo 7 of $(n - 1) + k$.

Question: If January 1 is on a Friday, on what day of the week is the 100th day of the year? If the year is not a leap year, what is the date on that day?

In a leap year, the extra day is added at the end of February as February 29. Because of this, it is simplest to think of March as the third month of year y and to think of January and February as the 13th and 14th months of the previous year $y - 1$. Let w = the day of the week, d = the day of the month, m = the number of the month, and y = the year. Then it turns out that

$$w \equiv d + 2m + \left\lfloor \frac{3(m+1)}{5} \right\rfloor + \lfloor y \rfloor + \left\lfloor \frac{y}{4} \right\rfloor - \left\lfloor \frac{y}{100} \right\rfloor + \left\lfloor \frac{y}{400} \right\rfloor + 2 \text{ Mod}(7).$$

We can say that

1. If the year y is not a leap year, then the day of the week for a given date in year $y + 1$ will be one day of the week later, because $365 \equiv 1 \text{ Mod}(7)$.
2. If year y is a leap year, then the day of the week for a given date in year $y + 1$ will be two days of the week later, because $366 \equiv 2 \text{ Mod}(7)$.

For example, the year 2000 (which we think of as beginning March 1) was a leap year while the years 1999 and 2001 were not. July 4, 1999 was a Sunday so July 4, 2000 was a Tuesday, and July 4, 1998 was a Saturday.

EXAMPLE On what day of the week was July 4, 1776, when the U.S. Declaration of Independence was signed?

Solution Here $m = 7$, $d = 4$, and $y = 1776$. So

$$\begin{aligned} w &\equiv 4 + 2 \cdot 7 + \left\lfloor \frac{3(7+1)}{5} \right\rfloor + 1776 + \left\lfloor \frac{1776}{4} \right\rfloor - \left\lfloor \frac{1776}{100} \right\rfloor + \left\lfloor \frac{1776}{400} \right\rfloor + 2 \text{ Mod}(7) \\ &\equiv 4 + 14 + 4 + 1776 + 444 - 17 + 4 + 2 \text{ Mod}(7) \\ &\equiv 2231 \text{ Mod}(7) \\ &\equiv 5 \text{ Mod}(7) \end{aligned}$$

So July 4, 1776 was a Thursday.

Cryptology

Julius Caesar, who on numerous occasions led Roman soldiers into battle, used a method for coding and decoding military messages that was based on shifting the letters of the alphabet cyclically by a fixed number of letters. For example, if the 26 letters of our alphabet are shifted cyclically by 4 letters, the code table shown in Table 1 results.

Table 1

Letter	A	B	C	D	E	F	G	H	I
Code	E	F	G	H	I	J	K	L	M
Letter	J	K	L	M	N	O	P	Q	R
Code	N	O	P	Q	R	S	T	U	V
Letter	S	T	U	V	W	X	Y	Z	
Code	W	X	Y	Z	A	B	C	D	

The message ATTACKATDAWN would be encoded as EXXEGOEXHEAR, and those who knew that the code is constructed by this shift could easily decode the message by the reverse shift using the same table.

The science of coding or **cryptology** is concerned with devising methods for encoding information so that the information can be transmitted securely and efficiently to specified individuals who know how to decode it, and so that other individuals outside that target group are not able to decode the information easily. Cryptology is also concerned with decoding messages where the code is initially unknown or partially known. In cryptology, the original message is called **plaintext** and the encoded message is called **ciphertext.**

Because ciphertext messages are transferred electronically in current applications, these messages are converted to numerical equivalents using ASCII[1] binary representation for letters, numbers, and symbols. However, for the sake of simplicity, our discussion of coding and decoding procedures will use base 10 representation and limit the plaintext symbols to the 26 letters of the English alphabet.

Suppose we correspond the letters in the English alphabet with two-digit blocks from 00 to 25, as in Table 2.

Table 2

Letter	A	B	C	D	E	F	G	H	I
Number	00	01	02	03	04	05	06	07	08
Letter	J	K	L	M	N	O	P	Q	R
Number	09	10	11	12	13	14	15	16	17
Letter	S	T	U	V	W	X	Y	Z	
Number	18	19	20	21	22	23	24	25	

Then plaintext and the ciphertext can both be expressed as a sequence of two-digit numbers. Then, for example, for Caesar's shift code a ciphertext number C corresponds to a plaintext number T according to the equation

$$C \equiv (T + 4) \text{ Mod}(26),$$

resulting in the code shown in Table 3.

Table 3

T	00	01	02	03	04	05	06	07	08
C	04	05	06	07	08	09	10	11	12
T	09	10	11	12	13	14	15	16	17
C	13	14	15	16	17	18	19	20	21
T	18	19	20	21	22	23	24	25	
C	22	23	24	25	00	01	02	03	

[1] ASCII stands for American Standard Code for Information Interchange, and was designed in 1963 for representing alphanumeric information.

The message ATTACKATDAWN would have the numerical plaintext form

00 19 19 00 02 10 00 19 03 00 22 13,

and would be encoded by the preceding Caesar shift code table as

04 23 23 04 06 14 04 23 07 04 00 17.

Decoding ciphertext to plaintext in this code is governed by the equation

$$T \equiv (C - 4) \text{ Mod}(26)$$

and decoding can be carried out with the same table.

More generally, any code in which C and T are related by an equation of the form

$$C \equiv (mT + b) \text{ Mod}(26) \qquad 0 \leq C \leq 25,$$

where m and b are integers and m is relatively prime to 26, is called a **linear code** or **affine code**. Theorems 6.1 to 6.3 guarantee that no two plaintext letters are coded by the same ciphertext. For example, if $m = 9$ and $b = 3$, then the corresponding affine code table is as given in Table 4. From it, we see that A (00) becomes D (03), B (01) becomes M (12), and so on.

Table 4

P	00	01	02	03	04	05	06	07	08
C	03	12	21	04	13	22	05	14	23
P	09	10	11	12	13	14	15	16	17
C	06	15	24	07	16	25	08	17	00
P	18	19	20	21	22	23	24	25	
C	09	18	01	10	19	02	11	20	

In any language, the relative frequency of the occurrence of a given letter varies considerably with the letter. For example, in English, the letters that occur most frequently are E, T, N, O, and A in that order. Together these five letters account for approximately 44% of typical text letters. The length and letter combinations found in ciphertext words also provide important clues to their plaintext meaning. Codes are broken by cryptologists through careful study of such patterns in ciphertext samples. On the other hand, by finding ways to hide such patterns, cryptologists also seek to devise codes that are more difficult to break.

One feature of modern codes that makes them more difficult to break is that the plaintext is broken into a series of blocks of a fixed length before it is converted to ciphertext blocks. These "block codes" hide the length of message words and make it more difficult to analyze letter combinations.

The individual letters in each plaintext block are used to determine the letters of the corresponding ciphertext block often by using some congruence scheme, such as the affine codes described above. The RSA algorithm mentioned in Section 5.2.4 uses another such scheme, more difficult to break than affine codes, called **exponential congruence encoding**. Fermat's Little Theorem (Theorem 6.4) is a very useful tool for this type of code.

6.1.2 Problems

1. Use the discussion of the Gregorian calendar to determine the day of the week of April 12, 1861, the date on which the American Civil War began with the Battle of Fort Sumter.

2. Use the formula for the day of the week given in this section.

a. Use today's date to verify that the formula works.

b. Determine the day of the week on which you were born.

c. Verify that January 1, 2001 and January 1, 2401 fall on the same day of the week.

d. Explain the parts of the formula that involve d and y.

*e. Explain the part of the formula that involves m, and explain why 2 is added at the end.

3. Write the word that combines your first and last names in plaintext, and then encode this word using the affine code

$$C \equiv (9T + 3) \text{ Mod}(26)$$

described in this section.

4. In the affine code $C \equiv (mT + b) \text{ Mod}(26)$, explain why m must be relatively prime to 26 in order for the code to work.

ANSWER TO QUESTION

The 100th day is on the pth day of the week, where $p \equiv (100 - 1) + 6 \equiv 105 \equiv 0 \text{ Mod}(7)$, so the 100th day is on Saturday. The date is April 10.

6.1.3 The Chinese Remainder Theorem

Chinese mathematicians of the third century A.D. were the first to pose divisibility problems of the following kind.

Problem 1: The Coconut Problem. A pile of coconuts of an unknown number is such that if the coconuts in the pile are divided into groups of 3, 2 coconuts remain; if divided into groups of 5, 3 coconuts remain; and if divided into groups of 7, 2 coconuts remain. How many coconuts can be in the pile and what is the smallest possible number of coconuts in the pile?[2]

Question 1: Restate this problem in terms of solving a system of congruences simultaneously.

Question 2: If x is one possible count of the pile of coconuts, explain why, for any positive integer k, $x + 105k$ is another possible count.

In the general version of this problem, we are seeking an integer x that can be written in several different ways:

$$x = q_1m_1 + a_1 = q_2m_2 + a_2 = \cdots = q_nm_n + a_n$$

where the m_i are integers that indicate the size of the groups (3,5, and 7 above), while the q_i are the number of groups of each size. The a_i are the remainders (2, 3, and 2 above).

The following theorem provides all solutions to such systems of equations and derives its name from its historical roots in problems such as the Coconut Problem.

Theorem 6.5 **(The Chinese Remainder Theorem):** Suppose that $m_1, m_2, \ldots, m_n$ are integers greater than 1 that are relatively prime in pairs, and that $a_1, a_2, \ldots, a_n$ are integers. Then the set of all solutions of the system of the congruences

(*) $$x \equiv a_1 \text{ Mod}(m_1), x \equiv a_2 \text{ Mod}(m_2), \ldots, x \equiv a_n \text{ Mod}(m_n)$$

is given by

(**) $$x \equiv a_1b_1\frac{m}{m_1} + a_2b_2\frac{m}{m_2} + \cdots + a_nb_n\frac{m}{m_n} \text{Mod}(m)$$

where $m = m_1m_2 \cdot \cdots \cdot m_n$ and each b_k, $k = 1, 2, \ldots, n$, is an integer satisfying the congruence equations

$$b_k\frac{m}{m_k} \equiv 1 \text{ Mod}(m_k).$$

---

[2] This problem is a paraphrase of a problem (Problem 26) found in Sunzi Suanjing (Master Sun's Mathematical Manual) written in the late third century A.D. In the fourth century, the Chinese mathematician Sun Zi provided an algorithmic solution of this problem equivalent to the computation in equation (**) in our solution, but gave no indication of how he arrived at his algorithm or how it might apply to other systems of congruences.

Before we prove the Chinese Remainder Theorem, let's use it to solve the Coconut Problem. If x is a possible number of coconuts in the pile, then the given information tells us that

(1) $$x \equiv 2 \text{ Mod}(3),\ x \equiv 3 \text{ Mod}(5), \text{ and } x \equiv 2 \text{ Mod}(7).$$

So $a_1 = 2$ and $m_1 = 3$, $a_2 = 3$ and $m_2 = 5$, and $a_3 = 2$ and $m_3 = 7$. Thus $m = m_1m_2m_3 = 105$. Now we need to find three integers b_1, b_2, and b_3 such that

$$35b_1 \equiv 1 \text{ Mod}(3),\ 21b_2 \equiv 1 \text{ Mod}(5), \text{ and } 15b_3 \equiv 1 \text{ Mod}(7).$$

Since (by trial and error), $35 \cdot 2 \equiv 1 \text{ Mod}(3)$, we can take $b_1 = 2$. Similarly, $21 \cdot 1 \equiv 1 \text{ Mod}(5)$ and $15 \cdot 1 \equiv 1 \text{ Mod}(7)$, so we can take $b_2 = 1$ and $b_3 = 1$. Therefore, by the Chinese Remainder Theorem, the set of all solutions of the system (1) is given by

(2) $$x \equiv 2 \cdot 2 \cdot 35 + 3 \cdot 1 \cdot 21 + 2 \cdot 1 \cdot 15 = 233 \text{ Mod}(105).$$

The smallest positive value of x is $233 - 2(105) = 23$, so the smallest possible number of coconuts in the pile is 23. However, the number of coconuts in the pile may also be $23 + 105 = 128$, or $23 + 2(105) = 243$, and so on.

Check: 23 coconuts can be divided into 7 groups of 3 with 2 left over, 4 groups of 5 with 3 left over, and 3 groups of 7 with 2 left over.

We now prove the Chinese Remainder Theorem.

Proof of Theorem 6.5: First, note that if x is a solution to the system

(*) $$x \equiv a_1 \text{ Mod}(m_1),\ x \equiv a_2 \text{ Mod}(m_2), \ldots, \text{ and } x \equiv a_n \text{ Mod}(m_n)$$

of congruences, and if m is the product of the n moduli $m_1, m_2, \ldots, m_n$, then $x + pm$ is also a solution of this system (*) for any integer p because, by Theorem 6.2,

$$x \equiv a_k \text{ Mod}(m_k) \quad \text{and} \quad pm \equiv 0 \text{ Mod}(m_k) \quad \text{imply} \quad x + pm \equiv a_k \text{ Mod}(m_k)$$

for each integer k with $1 \le k \le n$.

Now we show that for each integer k with $1 \le k \le n$, there is an integer b_k satisfying the congruence equation $b_k \frac{m}{m_k} \equiv 1 \text{ Mod}(m_k)$. For since m is the product of the moduli $m_1, m_2, \ldots, m_n$, and these moduli are relatively prime in pairs, m_k and $\frac{m}{m_k}$ are relatively prime. Therefore, by Theorem 5.6, there are integers b_k and c_k such that $b_k \frac{m}{m_k} + c_k m_k = 1$; that is, $b_k \frac{m}{m_k} - 1 = c_k m_k$. Consequently, $b_k \frac{m}{m_k} \equiv 1 \text{ Mod}(m_k)$ for $1 \le k \le n$.

If for each integer k, x_k is defined by $x_k = a_k b_k \frac{m}{m_k}$, then

$$x_k \equiv a_k \text{ Mod}(m_k), \text{ and } x_k \equiv 0 \text{ Mod}(m_p) \text{ for } p \ne k$$

because m_p is a factor of $\frac{m}{m_k}$ for each $p \ne k$ with $1 \le p \le n$. Therefore, if

(**) $$x = x_1 + x_2 + \cdots + x_n = a_1b_1\frac{m}{m_1} + a_2b_2\frac{m}{m_2} + \cdots + a_nb_n\frac{m}{m_n},$$

then

$$x \equiv a_1 \text{ Mod}(m_1),\ x \equiv a_2 \text{ Mod}(m_2), \ldots, x \equiv a_n \text{ Mod}(m_n)$$

by Theorem 6.2 and Problem 6. It follows that any integer x of the form (**) is a solution of the system (*) of congruences.

To complete the proof, we need to show that if y is any solution of the system (*) of congruences and if x_0 is any one of the solutions (**), then $y \equiv x_0 \text{ Mod}(m)$. But since both x_0 and y are congruent to a_k modulo m_k for $1 \leq k \leq n$, it follows that x_0 and y are congruent modulo m_k for $1 \leq k \leq n$. It follows that x_0 and y are congruent modulo the least common multiple of $m_1, m_2, \ldots, m_n$ (see Problem 6). This least common multiple is their product m of the moduli because these moduli are relatively prime in pairs. ⌟

There is an interesting connection between the Chinese Remainder Theorem method for solving systems of linear congruence and the method of solving linear Diophantine equations discussed in Section 5.2.3. This connection can be seen by comparing the two methods for solving a system of two congruences

$$x \equiv a_1 \text{ Mod}(m_1) \quad x \equiv a_2 \text{ Mod}(m_2), \tag{3}$$

where m_1 and m_2 are relatively prime integers >1, and a_1 and a_2 are any integers.

Chinese Remainder Theorem method

In this case,

$$m = m_1 m_2 \quad \text{and so} \quad \frac{m}{m_1} = m_2 \quad \text{and} \quad \frac{m}{m_2} = m_1.$$

If we can find integers b_1 and b_2 such that

$$b_1 m_2 \equiv 1 \text{ Mod}(m_1) \quad \text{and} \quad b_2 m_1 \equiv 1 \text{ Mod}(m_2), \tag{4}$$

then the Chinese Remainder Theorem states that the set of all integers x that satisfy both of the given congruences simultaneously is given by

$$x = a_1 b_1 m_2 + a_2 b_2 m_1 + p m_1 m_2 \quad \text{for any integer } p. \tag{5}$$

Linear Diophantine equation method

An integer x is a solution of the system (3) if and only if there are integers k_1 and k_2 such that

$$x - a_1 = k_1 m_1 \quad x - a_2 = k_2 m_2. \tag{6}$$

If we solve for x in the second equation and substitute the result for x in the first equation, we can conclude that the given system of congruences has solutions if and only if there are integer solutions (k_1, k_2) of the equation

$$m_1 k_1 - m_2 k_2 = a_2 - a_1. \tag{7}$$

By Theorem 5.8, this Diophantine equation has solutions because m_1 and m_2 are relatively prime. Moreover, if (k_1, k_2) is an integer solution of (7), then all other integer solutions are given by

$$(k_1 + q m_2, k_2 + q m_1) \quad \text{where } q \text{ is any integer.} \tag{8}$$

We can then generate all solutions of (3) by observing that $x = a_1 + k_1 m_1 = a_2 + k_2 m_2$ is a solution of (3) if and only if (k_1, k_2) is a solution of (7), where x and (k_1, k_2) are related by (6).

EXAMPLE 1 Find all solutions of the system of congruences

$$x \equiv 3 \text{ Mod}(4) \quad x \equiv 5 \text{ Mod}(7) \tag{3'}$$

by the Chinese Remainder Theorem and Diophantine equation methods.

Solution **Chinese Remainder Theorem method**

Here $m = m_1m_2 = 28$, and a solution of

$$(4') \qquad b_1 \cdot 7 \equiv 1 \text{ Mod}(4) \text{ and } b_2 \cdot 4 \equiv 1 \text{ Mod}(7)$$

is given by $b_1 = 3$ and $b_2 = 2$. Therefore, the set of all integer solutions x of this system is given by

$$(5') \qquad x = 3 \cdot 3 \cdot 7 + 5 \cdot 2 \cdot 4 + p \cdot 28 = 103 + 28p \quad \text{for any integer } p.$$

Diophantine equation method

We want to find all integer pairs (k_1, k_2) that satisfy

$$(7') \qquad 4k_1 - 7k_2 = 5 - 3 = 2.$$

Note that $(k_1, k_2) = (4, 2)$ is one such pair, and by equation (8) all integer pairs (k_1, k_2) that are solutions of $(7')$ are given by

$$(8') \qquad (4 + 7q, 2 + 4q) \quad \text{where } q \text{ is any integer.}$$

Therefore, all solutions x of $(3')$ are given by (6) as

$$(9') \quad x = 3 + 4(4 + 7q) = 19 + 28q \quad \text{or} \quad x = 5 + 7(2 + 4q) = 19 + 28q$$

where q is any integer. The solution sets to $(5')$ and $(9')$ are identical because

$$19 + 28q = 103 + 28p \text{ if and only if } q = p + 4.$$

Although the Diophantine equation method extends to systems of more than two congruences, it becomes more and more cumbersome to apply in comparison with the Chinese Remainder Theorem as the number of congruences increases.

There is also an interesting connection between the Chinese Remainder Theorem and polynomial interpolation that can be seen by considering the following problem.

Problem 2: Find a polynomial $p(x)$ that leaves a remainder of 3 when divided by x, a remainder of -1 when divided by $x - 1$, and a remainder of 2 when divided by $x - 3$.

The analogies between the divisibility properties of integers and polynomials prompt us to view this as a "polynomial congruence" problem:

Problem 2 (restated): Find a polynomial $p(x)$ such that $p(x) \equiv 3 \text{ Mod}(x)$, $p(x) \equiv -1 \text{ Mod}(x - 1)$, and $p(x) \equiv 2 \text{ Mod}(x - 3)$.

The meaning of polynomial congruence follows from the Division Algorithm for Polynomials (Theorem 5.16). Two **polynomials p(x) and q(x) are congruent** $\text{Mod}(r(x))$, where $\deg(r(x)) \geq 1$, if and only if $r(x)$ is a factor of $p(x) - q(x)$. We can use the analogy between integers and polynomials to construct $p(x)$ by the Chinese Remainder Theorem method. With $m_1(x) = x$, $m_2(x) = x - 1$, $m_3(x) = x - 3$ and $a_1 = 3$, $a_2 = -1$, and $a_3 = 2$, we find b_1, b_2, and b_3 such that $b_1m_2(x)m_3(x) \equiv 1 \text{ Mod}(m_1(x))$, $b_2m_1(x)m_3(x) \equiv 1 \text{ Mod}(m_2(x))$, and $b_3m_1(x)m_2(x) \equiv 1 \text{ Mod}(m_3(x))$. We solve these congruences for b_1, b_2, and b_3 by long division to obtain $b_1 = \frac{1}{3}$, $b_2 = -\frac{1}{2}$, $b_3 = \frac{1}{6}$, and then compute $p(x)$ by the Chinese Remainder Theorem formula (**).

$$\begin{aligned} p(x) &\equiv a_1b_1m_2(x)m_3(x) + a_2b_2m_1(x)m_3(x) + a_3b_3m_1(x)m_2(x) \text{ Mod}(m_1(x)m_2(x)m_3(x)) \\ &\equiv 3 \cdot \frac{1}{3} \cdot (x - 1)(x - 3) + (-1) \cdot \left(-\frac{1}{2}\right) \cdot x(x - 3) + 2 \cdot \left(\frac{1}{6}\right) \cdot x(x - 1) \text{Mod}(m_1(x)m_2(x)m_3(x)) \end{aligned}$$

so

$$p(x) = \frac{11}{6}x^2 - \frac{35}{6}x + 3 + k(x^3 - 4x^2 + 3x), \quad \text{where } k \text{ can be any integer.}$$

Question 3: Verify that this polynomial $p(x)$ has the remainders 3, −1, and 2 on division by x, $x - 1$, and $x - 3$.

According to the Remainder Theorem for polynomials (Theorem 5.17), the remainder on division of a polynomial $p(x)$ by $x - a$ is $p(a)$, the value of p at $x = a$. Using this fact, we can reformulate the above problem as a problem in polynomial "interpolation" as follows:

Problem 2 (second restatement): Find a polynomial $p(x)$ that has the value 3 when $x = 0$, the value −1 when $x = 1$, and the value 2 when $x = 3$.

The Lagrange Interpolation Formula (see Project 4 of Chapter 3) constructs a polynomial $q(x)$ of degree n with prescribed values at $n + 1$ distinct points. For this problem, the Lagrange Interpolation Formula polynomial is

$$\begin{aligned} q(x) &= p(0)\frac{m_2(x)\cdot m_3(x)}{m_2(0)\cdot m_3(0)} + p(1)\frac{m_1(x)\cdot m_3(x)}{m_1(1)\cdot m_3(1)} + p(3)\frac{m_1(x)\cdot m_2(x)}{m_1(3)\cdot m_2(3)} \\ &= 3\frac{(x-1)(x-3)}{(-1)(-3)} + (-1)\frac{x(x-3)}{(1)(-2)} + (2)\frac{x(x-1)}{(3)(2)} = \frac{11}{16}x^2 - \frac{35}{6}x + 3, \end{aligned}$$

which is exactly the Chinese Remainder Theorem polynomial with $k = 1$. This suggests that there is a Chinese Remainder Theorem for polynomial congruences and that this Chinese Remainder Theorem is essentially the classic Lagrange Interpolation Formula. This is in fact the case,[3] and this again illustrates the close connection that exists between integers and polynomials.

6.1.3 Problems

1. Use the Chinese Remainder Theorem to find all integer solutions and the smallest positive integer solution of the following systems of congruences.

a. $x \equiv 1 \text{ Mod}(5)$ $\quad x \equiv 5 \text{ Mod}(7)$ $\quad x \equiv 6 \text{ Mod}(9)$

b. $x \equiv 3 \text{ Mod}(4)$ $\quad 2x \equiv 1 \text{ Mod}(5)$ $\quad 3x \equiv 4 \text{ Mod}(7)$

(*Hint*: For part **b**, use properties of congruences to express the second and third congruences in the same form as the first.)

2. Ask a classmate or friend to think of a number N between 1 and 100 but not to tell you the value of N. Instead, ask them to tell you only the remainder R_3 when N is divided by 3, the remainder R_5 when N is divided by 5, and the remainder R_7 when N is divided by 7. Then do the following calculations in your head:

i. $K = 70R_3 + 21R_5 + 15R_7$

ii. If $K \le 100$, then K is N; otherwise, subtract from K the largest multiple of 105 that will yield a positive result to obtain N.

Use the Chinese Remainder Theorem solution of the Coconut Problem to explain why this game works. (Although this little game may appear to have amusement value only, it actually contains the central idea of a method used in computer science that enables computers to work with very large integers. For a description of this method, see Project 3.)

3. A troop of scouts has 17 members. After selling cookies they put all their dollar bills in a sack. When they tried to divide the bills evenly, they had three bills left over. So they went home and decided to try again the next day. One girl quit the troop so when they divided the money the next time, there were 10 bills remaining. They decided to sleep on it, and another girl quit in the morning. This worked out well because when the remaining members distributed the money there was none left over. What is the least amount of money that was in the sack?

4. **Another coconut problem.** A group of five sailors is stranded on an island and collects coconuts for food. They

[3] See "On the Chinese Remainder Theorem and Its Applications", by C. Tapia, D. Tello, and P. Kutzko at *http://www.math.uic.edu/~tello/papers/tapiatello7-27.pdf.*

decide to put the coconuts in a pile and split them evenly the next morning. But one sailor, not trusting the others, goes in the middle of the night to get his share. He splits the coconuts into 5 equal piles and, finding 1 coconut left over, throws it away. He then takes his share and puts the remaining coconuts back into a single pile. A little later a second sailor, also not trusting the others, does the same thing. He, too, splits the coconuts into 5 equal piles, finds 1 coconut left over, throws it away, takes his share, and puts the remaining coconuts back into a single pile. The same thing happens with the other three sailors. What is the smallest number of coconuts that could have been in the original pile?

5. Use the Chinese Remainder Theorem for polynomials to construct a polynomial $p(x)$ that leaves the remainder 1 upon division by $x + 1$, the remainder 0 when divided by x, and the remainder 1 upon division by $x - 1$.

6. a. Suppose that m_1 and m_2 are positive integers > 1 and that a, b, c, d are any integers. Prove the following:

If $a \equiv b \text{ Mod}(m_1)$ and $a \equiv b \text{ Mod}(m_2)$, then $a \equiv b \text{ Mod}(\text{lcm}(m_1, m_2))$.

b. Suppose that $m_1, m_2, \ldots, m_n$ are positive integers > 1 and that a, b are any integers. Use mathematical induction and part **a** to prove that

If $a \equiv b \text{ Mod}(m_1)$ and $a \equiv b \text{ Mod}(m_2), \ldots$ $a \equiv b \text{ Mod}(m_n)$, then

$$a \equiv b \text{ Mod}(\text{lcm}(m_1, m_2, \ldots, m_n)).$$

c. Suppose that m is an integer > 1, and that $a_1, a_2, \ldots, a_n$ and $b_1, b_2, \ldots, b_n$ are any integers. Use mathematical induction and Theorem 6.2 to prove the following:

If $a_1 \equiv b_1 \text{ Mod}(m)$, $a_2 \equiv b_2 \text{ Mod}(m), \ldots, a_n \equiv b_n \text{ Mod}(m)$ then

$$[a_1 + a_2 + \cdots + a_n] \equiv [b_1 + b_2 + \cdots + b_n] \text{ Mod}(m)$$
$$[a_1 \cdot a_2 \cdot \cdots \cdot a_n] \equiv [b_1 \cdot b_2 \cdot \cdots \cdot b_n] \text{ Mod}(m).$$

Unit 6.2 Number Fields

The essential algebraic features of addition and multiplication in the sets of rational numbers **Q** and real numbers **R** were listed in Section 5.3.1 with the statement that both $\langle \mathbf{Q}, +, \cdot \rangle$ and $\langle \mathbf{R}, +, \cdot \rangle$ are examples of *fields*. Recall that a set **F** equipped with two binary operations, called addition and multiplication, is a **field** if addition and multiplication are closed on **F** and have the following algebraic properties:

Addition and multiplication are commutative and associative operations in **F** (Properties A-1, A-2, M-1, M-2 in Section 5.3.1).

Multiplication is distributive over addition in **F** (Property D).

There is an element 0 in **F** that leaves any element in **F** unchanged under addition on **F** (Property A-3), and there is a nonzero element 1 that leaves any element unchanged under multiplication on **F** (Property M-3).

For each element a in **F**, there is an opposite (or additive inverse) element $-a$ in **F** such that $a + (-a) = 0$ (Property A-4), and for each nonzero element a in **F**, there is a reciprocal (or multiplicative inverse) element a^{-1} in **F** such that $aa^{-1} = 1$ (Property M-4).

Because the operations of addition and multiplication on **R** and **Q** have the same meaning, we say that $\langle \mathbf{Q}, +, \cdot \rangle$ is a *subfield* of $\langle \mathbf{R}, +, \cdot \rangle$. More generally, if $\langle \mathbf{F}, +, \cdot \rangle$ is a field and if **G** is a subset of **F** such that $\langle \mathbf{G}, +, \cdot \rangle$ is a field for the operations of addition and multiplication on **F**, then $\langle \mathbf{G}, +, \cdot \rangle$ is a **subfield** of $\langle \mathbf{F}, +, \cdot \rangle$.

The system $\langle \mathbf{C}, +, \cdot \rangle$ of all complex numbers $z = x + iy$, where x and y are real numbers and $i^2 = -1$, is also a field. The function $g\colon \mathbf{R} \to \mathbf{C}$ that maps each real number x onto the complex number $x + i0$ is an isomorphism of the field $\langle \mathbf{R}, +, \cdot \rangle$ onto the real subfield of the field of complex numbers.

In this unit, we discuss and compare the three fields $\langle \mathbf{Q}, +, \cdot \rangle$, $\langle \mathbf{R}, +, \cdot \rangle$, and $\langle \mathbf{C}, +, \cdot \rangle$. Our objective is to identify those properties that distinguish them from other fields.

6.2.1 Ordered fields

Our objective in this section is to find a set of axioms that *characterizes* the system of real numbers in the same way that the Peano axioms characterize the system of natural numbers. We will identify axioms for an order relation $<$ to augment the axioms for a field $\langle \mathbf{F}, +, \cdot \rangle$ to obtain a structure $\langle \mathbf{F}, +, \cdot, < \rangle$ called a *complete ordered field* that characterizes the real number system $\langle \mathbf{R}, +, \cdot, < \rangle$ in the sense that

i. All known properties of the arithmetic operations and natural ordering of the real number system $\langle \mathbf{R}, +, \cdot, < \rangle$ can be derived from the axioms for a complete ordered field.

ii. Any other algebraic system with the properties of a complete ordered field is isomorphic to the real number system $\langle \mathbf{R}, +, \cdot, < \rangle$.

The purpose of identifying such a defining list of properties is that it becomes an algebraic model of the real number system that can complement or even replace the geometric model provided by the number line.

As we construct this defining list, we give examples to show that there are other algebraic systems that are very different from $\langle \mathbf{R}, +, \cdot \rangle$ that have all of the properties of any partial list. However, as the list of properties expands, the list of examples of algebraic systems with these properties gets shorter. The list becomes a defining list for the real number system when any algebraic system with these properties is essentially the same as the system of real numbers.

Examples of fields

The fields $\langle \mathbf{Q}, +, \cdot \rangle$, $\langle \mathbf{R}, +, \cdot \rangle$, and $\langle \mathbf{C}, +, \cdot \rangle$ mentioned previously are not the only fields. Here are three other examples.

1. **$\langle \mathbf{Z_p}, +, \cdot \rangle$, the fields of arithmetic modulo a prime p**
   For a given prime number p, the set $\mathbf{Z_p}$ of all congruence classes modulo p is a field if addition and multiplication are defined by
   $$m^* + n^* = (m + n)^* \qquad m^* \cdot n^* = (m \cdot n)^*.$$
   This can be verified by using the results developed in Section 6.1.1. Theorems 6.2 and 6.3 are especially useful for this purpose. We do not present the details of this verification here.

2. **$\langle \mathbf{Q}(\sqrt{n}), +, \cdot \rangle$, the quadratic field over Q**
   For any positive integer n that is not a perfect square, the set $\mathbf{Q}(\sqrt{n})$ of all real numbers x that can be expressed in the form $x = r + s\sqrt{n}$ with r and s rational is a subfield of the field $\langle \mathbf{R}, +, \cdot \rangle$ of real numbers that in turn contains the field $\langle \mathbf{Q}, +, \cdot \rangle$ of rational numbers as a subfield. (You are asked to verify this fact in Problem 2.)

3. **The field $\langle \mathbf{Q(x)}, +, \cdot \rangle$ of rational expressions**
   Let $\mathbf{Q(x)}$ be the set of all rational expressions $r(x) = \frac{p(x)}{q(x)}$, where $p(x)$ and $q(x)$ are polynomials in x with rational number coefficients and $q(x)$ is not the zero polynomial.[4] Then, for any two rational expressions $r_1(x) = \frac{p_1(x)}{q_1(x)}$ and $r_2(x) = \frac{p_2(x)}{q_2(x)}$ in $\mathbf{Q(x)}$, define

[4] We used the symbol $\mathbf{Q[x]}$ in Section 5.3.1 to denote the set of all polynomials with rational number coefficients, while here we use the rather similar symbol $\mathbf{Q(x)}$ for the corresponding set of rational expressions. This is standard notation for these two sets in abstract algebra texts so we use it here. However, to avoid confusion, we will be careful that the context also makes it clear which of these sets is under consideration.

Equality: $r_1(x) = r_2(x)$ if and only if $p_1(x) \cdot q_2(x) - p_2(x) \cdot q_1(x)$ is the zero polynomial.

Addition: $(r_1 + r_2)(x) = \frac{p_1(x)q_2(x) + p_2(x)q_1(x)}{q_1(x)q_2(x)}$

Multiplication: $(r_1 \cdot r_2)(x) = \frac{p_1(x)p_2(x)}{q_1(x)q_2(x)}$

Notice that these definitions are the customary procedures for combining and simplifying rational expressions. In Problem 3, you are asked to verify that $\langle \mathbf{Q}(\mathbf{x}), +, \cdot \rangle$ is a field.

Thus, a rather diverse variety of algebraic systems are fields.

The real number system as an ordered field

In addition to the binary operations of addition and multiplication, the fields of rational numbers and real numbers are equipped with an inequality ($<$) relation. This relation was used in Chapter 2 in a crucial way in our discussion of decimal representation as well as in our discussion of the distribution of the rational and irrational numbers among the real numbers. The following properties of $<$ determine what is called an *ordered field*. For all a, b, and c,

O-1 (Trichotomy Property) Exactly one of the following holds:

$a < b, \quad a = b, \quad b < a.$

O-2 (Transitive Property) If $a < b$ and $b < c$, then $a < c$.

O-3 (Addition Property) If $a < b$, then $a + c < b + c$.

O-4 (Multiplication Property) If $a < b$ and $0 < c$, then $ac < bc$.

Definitions

A field $\langle \mathbf{F}, +, \cdot \rangle$ with a relation $<$ that satisfies properties O-1 through O-4 is called an **ordered field** $\langle \mathbf{F}, +, \cdot, < \rangle$. An element a of an ordered field $\langle \mathbf{F}, +, \cdot, < \rangle$ is **positive** if $0 < a$, and **negative** if $a < 0$.

In an ordered field, the symbols $>$, $\leq$, and $\geq$ are defined in their usual ways:

$$\mathbf{a} > \mathbf{b} \Leftrightarrow b < a$$
$$\mathbf{a} \leq \mathbf{b} \Leftrightarrow a < b \text{ or } a = b$$
$$\mathbf{a} \geq \mathbf{b} \Leftrightarrow a > b \text{ or } a = b.$$

From these properties, many other properties of order and inequality can be established. These include the properties of positive and negative elements. In Theorem 6.6 we identify the additive and multiplicative identities as 0 and 1, but if an ordered field is not a field of numbers, you could replace 0 and 1 by the additive and multiplicative identities in that field.

Theorem 6.6

Suppose that $\langle \mathbf{F}, +, \cdot, < \rangle$ is an ordered field with additive identity 0 and multiplicative identity 1. Then

a. An element x in $\mathbf{F}$ is positive if and only if $-x$ is negative.

b. If x and y in $\mathbf{F}$ are positive, then $x + y$ and xy are positive.

c. If $x \neq 0$, then x^2 is positive.

d. The multiplicative identity 1 is positive.

Proof: We show the steps of a proof of one direction of (a) and leave the other proofs for you as problems.

a. If x is positive, then $0 < x$ by the definition of positive. From this, using the Addition Property of Order O-3, $0 + -x < x + -x$. By A-3, $0 + -x = -x$, and by A-4, $x + -x = 0$, so by substitution $-x < 0$. This means that $-x$ is negative, by the definition of negative. ┘

Essentially, Theorem 6.6 states that, in an ordered field, if an element is positive, then its opposite is negative, and vice versa, that positive elements are closed under addition and multiplication, and that squares of all nonzero elements are positive.

The systems $\langle \mathbf{Q}, +, \cdot, < \rangle$ of rational numbers and $\langle \mathbf{R}, +, \cdot, < \rangle$ of real numbers are ordered fields. Any subfield $\mathbf{G}$ of an ordered field $\langle \mathbf{F}, +, \cdot \rangle$ is also an ordered field because properties O-1–O-4 are inherited by subsets. In particular, if n is any positive integer that is not a perfect square, then the field $\langle \mathbf{Q}(\sqrt{n}), +, \cdot, < \rangle$ introduced above and in Problem 2 is an ordered field with $+$, $\cdot$, and $<$ inherited from the ordered field $\langle \mathbf{R}, +, \cdot, < \rangle$ of real numbers.

The field $\langle \mathbf{Q}(\mathbf{x}), +, \cdot \rangle$ is an ordered field for the order relation $<$ defined as follows: If $r_1(x) = \frac{p_1(x)}{q_1(x)}$ and $r_2(x) = \frac{p_2(x)}{q_2(x)}$, and if the leading coefficients of $q_1(x)$ and $q_2(x)$ are positive, then $r_1 < r_2$ if and only if the leading coefficient of $p_2(x)q_1(x) - p_1(x)q_2(x)$ is positive. This is analogous to order in the set $\mathbf{Q}^+$ of positive rational numbers: $\frac{a}{b} < \frac{c}{d}$ if and only if $ad < cb$, which is the case if and only if $cb - ad > 0$. For example, $\frac{x^2+1}{2x-1} < \frac{x^3-1}{x^2}$ because the coefficient of x^4 in

$$(2x - 1)(x^3 - 1) - x^2(x^2 + 1) = x^4 - x^3 - x^2 - 2x + 1$$

is the positive number 1. This ordering can also be described as follows: A rational expression $r(x) = \frac{p(x)}{q(x)}$ in $\langle \mathbf{Q}(\mathbf{x}), +, \cdot, < \rangle$ is positive if and only if the leading coefficients of $p(x)$ and $q(x)$ have the same sign. If $r_1(x)$ and $r_2(x)$ are rational expressions in $\langle \mathbf{Q}(\mathbf{x}), +, \cdot, < \rangle$, then $r_1(x) < r_2(x)$ if and only if $r_2(x) - r_1(x)$ is a positive element of $\langle \mathbf{Q}(\mathbf{x}), +, \cdot, < \rangle$. This is analogous to order in all the rationals: $\frac{a}{b}$ is positive if and only if a and b have the same sign, and $\frac{a}{b} < \frac{c}{d}$ if and only if $\frac{c}{d} - \frac{a}{d} > 0$. We will not present the details of the verification that $\langle \mathbf{Q}(\mathbf{x}), +, \cdot, < \rangle$ is an ordered field.

Fields that are not ordered fields

Theorem 6.6 shows that there can be no order relation $<$ on the field of complex numbers $\langle \mathbf{C}, +, \cdot \rangle$ because, if there were, i^2 would be positive by Theorem 6.6(c), and yet $i^2 = -1$ is negative by Theorems 6.6(a) and 6.6(b).

In Problem 5, you are asked to show that if $\langle \mathbf{F}, +, \cdot \rangle$ is a finite field, then it is not possible to find an order relation $<$ on $\mathbf{F}$ that has properties O-1–O-4. In particular, if p is a prime number, then an order relation $<$ cannot be defined on the field $\langle \mathbf{Z}_p, +, \cdot \rangle$ of congruence classes modulo p that will have properties of an ordered field.

We see from the preceding examples that there is substantially less variety among ordered fields than among fields, but there are ordered fields that differ substantially in structure from the ordered field of real numbers.

Further properties of all ordered fields

You may have noticed that the multiplication property O-4 does not include multiplication by a negative. This and the other properties of positive and negative numbers can be deduced from the assumed properties of an ordered field and the properties deduced in Theorem 6.6. Theorems 6.7–6.9 begin by working from Theorem 6.6 to determine which elements are positive and end with the multiplication of both sides of an inequality by a negative.

Theorem 6.7 In an ordered field, if x is positive, then nx is positive for all natural numbers n.

Proof: A proof involves a very nice use of mathematical induction, applying Theorem 6.6. This is left to you. ┘

Although we think of Theorems 6.6 and 6.7 as applying to **R** and its subfields, they apply in any ordered field regardless of its elements. A corollary to Theorem 6.6 is that, in the field **R** and its subfields, the natural numbers themselves are positive.

Theorem 6.8 and its corollary enable us to deduce that the rational numbers we think of as positive, like $\frac{1}{15}$, $\frac{355}{113}$, and 8 million, are indeed positive. And, by Theorem 6.5(a), their opposites are negative.

Theorem 6.8 In an ordered field, if x is positive, then $\frac{1}{x}$ is positive.

Proof: The proof is left to you as Problem 8. (*Hint*: Begin with the supposition that x is positive yet $\frac{1}{x}$ is negative.)

Corollary: If x and y are positive, then $\frac{x}{y}$ and $\frac{y}{x}$ are positive.

Theorems 6.7 and 6.8 present another indication of the special nature of ordered fields: *Any set* **F** *in an ordered field* $\langle \mathbf{F}, +, \cdot, < \rangle$ *contains subsets* N^*, Z^*, *and* Q^* *such that* $\langle N^*, +, \cdot, < \rangle$ *has essentially the same structure as the system* $\langle \mathbf{N}, +, \cdot, < \rangle$ *of natural numbers,* $\langle Z^*, +, \cdot, < \rangle$ *has essentially the same structure as the system* $\langle \mathbf{Z}, +, \cdot, < \rangle$ *of integers, and* $\langle Q^*, +, \cdot, < \rangle$ *has essentially the same structure as the system* $\langle \mathbf{Q}, +, \cdot, < \rangle$ *of rational numbers.* The set N^* is the set of elements of **F** formed by repeatedly adding the multiplicative identity of F. The elements of the set N^* are called the **positive integral elements** of the ordered field $\langle \mathbf{F}, +, \cdot, < \rangle$. Of course, for the ordered fields **R** of real numbers, **Q** of rational numbers, and the intermediate field $\mathbf{Q}(\sqrt{n})$, the set N^* of positive integral elements is the set **N** of natural numbers itself.

For the ordered field $\langle \mathbf{Q(x)}, +, \cdot, < \rangle$ of rational expressions, the set N^* of positive integral elements of **Q(x)** is the set of rational expressions of the form $\frac{q_n(x)}{q_1(x)}$, where $q_n(x) = n$ for any positive integer n.

The next two properties of positive and negative elements require the distributive property for their proof.

Theorem 6.9 In an ordered field,

a. the product of a positive element and a negative element is negative;
b. the product of two negatives is positive.

Proof:

a. Let p be any positive element and n be any negative element. Then, by Theorem 6.6(a), $-n$ is positive. From Theorem 6.6(b), $p(-n)$, being the product of two positive elements, is positive. But, using the distributive property, $0 = p \cdot 0 = p(n + -n) = pn + p(-n)$, so $p(-n)$ and pn are opposites. Since $p(-n)$ is positive, pn must be negative.

b. This proof is left for you as Problem 9. ┘

Theorem 6.10 Let $\langle \mathbf{F}, +, \cdot, < \rangle$ be an ordered field. Then for all a, b, and $c \in \mathbf{F}$,

a. if $a < b$ and $c > 0$, then $ac < bc$;

b. if $a < b$ and $c < 0$, then $ac > bc$.

Proof: We prove part (b) and leave part (a) to you. Suppose $a < b$ and $c < 0$. We give here the steps and justifications.

$0 < -c$	Theorem 6.6(a)
$a(-c) < b(-c)$	Multiplication Property of Order O-4

By the distributive property as shown in the proof of Theorem 6.9, $a(-c) = -ac$ and $b(-c) = -bc$. We substitute.

$-ac < -bc$	
$ac + -ac < ac + -bc$	Addition Property of Order O-3
$0 < ac + -bc$	Property A-3 of a field
$bc + 0 < bc + (ac + -bc)$	Addition Property of Order O-3
$bc + 0 < ac + (bc + -bc)$	Properties A-1 and A-2 of a field
$bc < ac$	Properties A-3 and A-4 of a field

∎

6.2.1 Problems

1. If p is a prime number, prove that the set $\mathbf{Z_p}$ of all congruence classes modulo p is a field if addition and multiplication are defined by

$$m^* + n^* = (m + n)^* \quad \text{and} \quad m^* \cdot n^* = (m \cdot n)^*.$$

(*Hint*: Use Fermat's Little Theorem.)

2. Let n be a positive integer that is not a perfect square. Let $\mathbf{Q}(\sqrt{n})$ be the set of all real numbers x that can be expressed in the form $x = r + s\sqrt{n}$, where r and s are rational numbers.

a. Explain why 0 and 1 are in $\mathbf{Q}(\sqrt{n})$.

b. Prove that $+$ and $\cdot$ are closed in $\mathbf{Q}(\sqrt{n})$.

c. Prove that $\langle \mathbf{Q}(\sqrt{n}), +, \cdot \rangle$ is a subfield of the field $\langle \mathbf{R}, +, \cdot \rangle$ of real numbers that contains the field $\langle \mathbf{Q}, +, \cdot \rangle$ of rational numbers as a subfield.

d. Prove that $\sqrt{3}$ is not in $\mathbf{Q}(\sqrt{2})$.

3. Let $\mathbf{Q(x)}$ be the set of all rational functions of a real number variable x. Prove that $\langle \mathbf{Q(x)}, +, \cdot \rangle$ is a field if the operations $+$ and $\cdot$ are defined as

$$(r_1 + r_2)(x) = \frac{p_1(x)q_2(x) + p_2(x)q_1(x)}{q_1(x)q_2(x)}$$

$$(r_1 \cdot r_2)(x) = \frac{p_1(x)p_2(x)}{q_1(x)q_2(x)}$$

for all real numbers x for which both sides of these equations are defined.

4. a. Prove the part of Theorem 6.6(a) not proved in this section.

b. Prove Theorem 6.6(b).

c. Prove Theorem 6.6(c).

d. Prove Theorem 6.6(d).

5. Show that if $\langle \mathbf{F}, +, \cdot \rangle$ is a finite field, then it is not possible to find an order relation on $\mathbf{F}$ that has properties O-1 to O-4.

6. Suppose the complex number $a + ib$ were considered positive if and only if $a > 0$ and $b > 0$. What parts of Theorem 6.6 would be violated?

7. Prove Theorem 6.7.

8. Prove Theorem 6.8.

9. Prove Theorem 6.9(b).

6.2.2 Archimedean and complete ordered fields

In the development of the properties of the system **R** of real numbers in Chapter 2, we made use of the following feature of that system:

a. For any positive real number x there is one and only one integer n such that $n \le x < n + 1$.

We used property (a) to construct the decimal representation of positive real numbers, to define the greatest integer function, and to define the integer part of a nonnegative rational number.

Property (a) makes sense in an ordered field $\langle \mathbf{F}, +, \cdot, < \rangle$ if it is reformulated as follows:

a′. For any positive element x in F there is one and only one integral element n^* in F such that

$$n^* \le x < n^* + 1.$$

However, property (a′) does not hold in all ordered fields.

For example, in the ordered field $\langle \mathbf{Q(x)}, +, \cdot, < \rangle$ of rational expressions in a variable x, the integral element n^* corresponding to the integer n is the rational expression $n(x) = \frac{q_n(x)}{q_1(x)}$, where $q_k(x)$ is the polynomial of degree 0 with constant coefficient equal to k for $k \ne 0$, and $q_0(x)$ is the zero polynomial. That is, $n(x) = \frac{n}{1} = n$ for all x. Now let the rational expression $r(x) = \frac{p_n(x)}{q_1(x)}$, where $p_n(x)$ is a polynomial of degree $n \ge 1$ whose leading coefficient a_n is a positive (rational) number. Then $n^* < r(x)$ for each positive integer n because the leading coefficient of $p_n(x)q_1(x) - q_n(x)q_1(x)$ is a_n, which is positive. This shows that property (a′) cannot hold for the ordered field $\langle \mathbf{Q(x)}, +, \cdot, < \rangle$ of rational expressions.

On the other hand, the ordered field $\langle \mathbf{R}, +, \cdot, < \rangle$ of real numbers or any ordered subfield of this field has property (a). We call such an ordered field an Archimedean field.

Definition An ordered field $\langle \mathbf{F}, +, \cdot, < \rangle$ is **Archimedean** if and only if for each positive element x in F, there is an integral element k^* in **F** such that $x < k^*$.

The term "Archimedean" derives from a geometric axiom used by Archimedes that can be paraphrased as follows: Given three consecutive points A, B, and C on a line, the line segment $\overline{AB}$ can be extended by duplication a sufficient number n of times to produce a line segment of length greater than the length of $\overline{AC}$. That is, if AB and AC are positive real numbers, there is a positive integer n such that $n \cdot AB > AC$ or, equivalently, $\frac{AC}{AB} < n$. The latter form is the form we have used in the definition of an Archimedean field, but either form will do.

This axiom of Archimedes in geometry thus has an algebraic counterpart in the Archimedean property of the real number system. It is also reflected in the theory of limits in calculus. For example, the decreasing sequence of real numbers $\{a_n\}$ with $a_n = \frac{1}{n}$ converges to 0 on intuitive grounds, but a proof based on the definition of limit requires that we show that for each positive number ε, there is a positive integer n such that $\frac{1}{n} < \varepsilon$. The existence of n is a consequence of the Archimedean property of the real number system.

Any ordered field $\langle \mathbf{F}, +, \cdot, < \rangle$ with property (a′) is an Archimedean ordered field because, for a given $x > 0$, we can take k^* to be the integral element $n^* + 1$ in (a′). The following result shows that any Archimedean field has this property.

Theorem 6.11 For each positive element x in an Archimedean field $\langle \mathbf{F}, +, \cdot, < \rangle$ there is a unique integral element n^* such that $n^* \leq x < n^* + 1$.

Proof: Because $\langle \mathbf{F}, +, \cdot, < \rangle$ is an Archimedean field, the set S of all natural numbers k such that $x < k^*$ is nonempty. By the well-ordering property of $\mathbf{N}$ (Theorem 5.2), the set S contains a smallest natural number k_0. If $n = k_0 - 1$, then n is a natural number and n is not in S. Therefore, the corresponding integral element n^* of $\mathbf{F}$ satisfies $n^* \leq x < n^* + 1$. If m^* is any integral element of F that satisfies $m^* \leq x < m^* + 1$, then the integer $m + 1$ corresponding to $m^* + 1$ must be the smallest natural number in S. (Why?) Therefore, $m = n$ and so the integral element n^* is unique. ⌟

A property of ordered fields that is even stronger than the Archimedean property, called the *completeness property*, is the final entry in our defining list of properties for the algebraic system $\langle \mathbf{R}, +, \cdot, < \rangle$ of real numbers. To formulate this property, we introduce the following terminology.

Definitions Suppose that $\langle \mathbf{F}, +, \cdot, < \rangle$ is an ordered field and that A is a subset of $\mathbf{F}$.

a. An element b in $\mathbf{F}$ is an **upper bound** for A if $a \leq b$ for all $a \in A$. If there is an upper bound for A, then A is **bounded above.**

b. If b^* is an upper bound for A and if $b^* \leq b$ for any other upper bound b for A, then b^* is a **least upper bound** for A.

Lower bounds, bounded below and **greatest lower bounds** for A are defined by replacing $\leq$ by $\geq$ in (a) and (b).

Not all subsets of an ordered field have upper bounds or lower bounds. For example, the set A of all positive rational numbers r such that $r^2 > 2$ is a subset of the rational field $\langle \mathbf{Q}, +, \cdot, < \rangle$ and the real field $\langle \mathbf{R}, +, \cdot, < \rangle$ but has no upper bound in either field. [A is the open interval $(\sqrt{2}, \infty)$.] However, 0, 1, and 1.4 are lower bounds for A in either of these fields. $\sqrt{2}$ is a greatest lower bound for A in the field of real numbers, but A does not have a greatest lower bound in the field of rational numbers.

Question: Explain why a subset A of an ordered field can have at most one greatest lower bound (or least upper bound).

Definition An ordered field $\langle \mathbf{F}, +, \cdot, < \rangle$ is **complete** if and only if every subset of $\mathbf{F}$ that is bounded above has a least upper bound, and every subset that is bounded below has a greatest lower bound.

Being Archimedean does not make a field complete. The rational field $\langle \mathbf{Q}, +, \cdot, < \rangle$ is not complete because the set

$$A = \{r \in \mathbf{Q}: r^2 > 2\}$$

is bounded below but has no greatest lower bound in $\mathbf{Q}$. In Problem 3, you are asked to show that the field $\langle \mathbf{Q}(\sqrt{n}), +, \cdot, < \rangle$ is also Archimedean but not complete.

However, being complete is sufficient for a field to be Archimedean.

Theorem 6.12 Any complete ordered field $\langle \mathbf{F}, +, \cdot, < \rangle$ is Archimedean.

Proof: Suppose to the contrary that there is a positive element x in **F** but no positive integral element n^* such that $x < n^*$. But then by the Trichotomy Property O-1, $n^* \leq x$ for all n^*. So x is an upper bound for the set N^* of positive integral elements of **F**. Because $\boldsymbol{F}$ is complete, N^* has a least upper bound b^*. Because b^* is a least upper bound, the smaller element $b^* - 1^*$ cannot be an upper bound for N^*. Therefore, there is an integral element $n_0^* \in N^*$ such that $b^* - 1^* < n_0^*$. But then $b^* < n_0^* + 1^*$ and $n_0^* + 1^* \in N^*$. This contradicts the fact that b^* is an upper bound for N^*. Therefore, it must be true that for each positive element x in **F** there is a positive integral element n in **F** such that $x < n$. ┘

Because of the transitive property of $<$, whenever $a < b$ and $b < c$, we can write $a < b < c$ and say that b is between a and c. Thus, the terminology and notation used for intervals on the real number line is meaningful in any ordered field $\langle \mathbf{F}, +, \cdot, < \rangle$. The **length** of any of the intervals $[a, b]$, (a, b), $[a, b)$, and $(a, b]$ is the positive element $b - a$ of **F**.

Suppose that $\{c_n\}$ is a sequence of positive elements of **F**. Then $\{c_n\}$ **decreases to 0** if (i) $c_{n+1} \leq c_n$ for all n, and (ii) for each positive element $p \in \mathbf{F}$, there is a positive integer n such that $c_n < p$.

With this language we can state and prove a theorem that distinguishes complete ordered fields from other fields.

Theorem 6.13 **(The Nested Intervals Theorem):** Suppose that $\{I_n\}$ is a nested sequence of closed intervals in a complete ordered field $\langle \mathbf{F}, +, \cdot, < \rangle$. Then the intersection of all the I_n is a closed interval $[a, b]$ or a singleton set $\{a\}$. This intersection is a singleton $\{a\}$ if and only if the lengths of the intervals $\{I_n\}$ decreases to 0 in **F**.

Proof: Suppose that $I_n = [a_n, b_n]$ for elements a_n and b_n in **F** with $a_n < b_n$. Then because $\{I_n\}$ is a nested sequence,

$$a_n \leq a_{n+1} \leq b_{n+1} \leq b_n \qquad \text{for all } n.$$

It follows that the sequence $\{a_n\}$ has each b_k as an upper bound for all k, and that the sequence $\{b_n\}$ has each a_k as a lower bound for all k. Let a be the least upper bound of $\{a_n\}$ and let b be the greatest lower bound of $\{b_n\}$. We assert that the intersection of all the $I_n = [a, b]$. For if $x \in I_n$ for all n, then $a_n \leq x \leq b_n$ for all n; that is x is an upper bound for $\{a_n\}$ and a lower bound for $\{b_n\}$. By definition of least upper bound and greatest lower bound, it follows that $a \leq x \leq b$. Conversely, if $a \leq x \leq b$, then x is an upper bound for $\{a_n\}$ and a lower bound for $\{b_n\}$. It follows that $a_n \leq x \leq b_n$ for each positive integer n; that is, $x \in I_n$ for all positive integers n. Therefore, x is in all the I_n, and we conclude that $[a, b]$ is the intersection of the I_n. You are asked to prove the second statement in the Nested Intervals Theorem. ┘

We have already observed that the ordered field of rational expressions $\langle \mathbf{Q}(x), +, \cdot, < \rangle$ is not Archimedean. Consequently, by the contrapositive to Theorem 6.12, $\langle \mathbf{Q}(x), +, \cdot, < \rangle$ is not complete.

Let us now summarize our discussion of properties and examples of fields:

1. There are fields that are not ordered fields. Some examples are finite fields or the field $\langle \mathbf{C}, +, \cdot \rangle$ of complex numbers.
2. There are ordered fields that are not Archimedean fields. An example is the field $\langle \mathbf{Q}(x), +, \cdot, < \rangle$ of rational expressions in x with rational coefficients.
3. There are Archimedean ordered fields that are not complete ordered fields. An example is $\langle \mathbf{Q}, +, \cdot, < \rangle$ of rational numbers.
4. Every complete ordered field is an Archimedean field.

These examples show that the various types of fields form the following hierarchy of inclusions:

$$\text{fields} \supset \begin{matrix}\text{ordered}\\ \text{fields}\end{matrix} \supset \begin{matrix}\text{Archimedean}\\ \text{ordered fields}\end{matrix} \supset \begin{matrix}\text{complete}\\ \text{ordered fields}\end{matrix}$$

with each type properly including the following type.

The only example that we have given of the most restrictive of these types of fields, complete ordered fields, is the ordered field $\langle \mathbf{R}, +, \cdot, < \rangle$ of real numbers, and that example rested on the real number line model of the real number system.

Are there other examples of complete ordered fields different from the real number system $\langle \mathbf{R}, +, \cdot, < \rangle$? The answer is yes if we interpret "different" to mean a different set of objects, operations, and relations. For example, consider the ordered field $\langle \mathbf{R(x)}, +, \cdot, < \rangle$ of all rational expressions with *real* coefficients. Let the subset $\mathbf{K}(x)$ of $\mathbf{R(x)}$ consist of all rational expressions of the form $r_c(x) = \frac{q_c(x)}{q_1(x)}$, where c is any real number, and $q_c(x)$ is the polynomial of degree 0 with constant coefficient c if $c \neq 0$, and $q_0(x)$ is the zero polynomial. Then $\langle \mathbf{K(x)}, +, \cdot, < \rangle$ is a subfield of the ordered field $\langle \mathbf{R(x)}, +, \cdot, < \rangle$ that is different from the ordered field of real numbers. However in Problem 8, you are asked to show that the function g: $\mathbf{R} \rightarrow \mathbf{K(x)}$ defined by

$$g(c) = r_c(x) \qquad \text{for each real number } c$$

is an isomorphism of the ordered field $\langle \mathbf{R}, +, \cdot, < \rangle$ of real numbers onto the ordered field $\langle \mathbf{K(x)}, +, \cdot, < \rangle$, so these two fields are essentially the same.

Are there any complete ordered fields that are essentially different from the ordered field $\langle \mathbf{R}, +, \cdot, < \rangle$ of real numbers? The answer is no. That is, if $\langle \mathbf{F}, +, \cdot, < \rangle$ is *any* complete ordered field, then $\langle \mathbf{F}, +, \cdot, < \rangle$ must be isomorphic to the complete ordered field of real numbers! The proof of this is very nice, and here is an outline of how it proceeds: Suppose that $\langle \mathbf{F}, +, \cdot, < \rangle$ is a complete ordered field. Then, because the Nested Interval Theorem and the Archimedean property hold for $\mathbf{F}$, we can use basically the same construction that we used in Section 2.1.2 to construct a decimal representation for each element x of $\mathbf{F}$. This decimal determines a point x^* on the real line. Conversely, given the decimal representing a point x^* on the real line, we can construct a point x in $\mathbf{F}$ determined by that decimal. We can then use the correspondence

$$x \leftrightarrow x^*$$

to show that the number line model for the real numbers is isomorphic to the complete ordered field $\langle \mathbf{F}, +, \cdot, < \rangle$. This argument shows Theorem 6.14.

Theorem 6.14 Any complete ordered field is isomorphic to the ordered field $\langle \mathrm{R}, +, \cdot, < \rangle$ of real numbers.

6.2.2 Problems

1. Verify that the set N^* of positive integral elements of an ordered field $\langle \mathbf{F}, +, \cdot, < \rangle$ together with the successor function s defined by

$$s(x) = x + 1 \quad \text{for all } x \text{ in } N^*$$

satisfies the Peano axioms:

P1: There is a unique element e in N^* such that $e \neq s(a)$ for any a in N^*.

P2: The function s is one-to-one; that is, if $s(a) = s(b)$, then $a = b$.

P3: If M is a subset of N^* that contains e and the successor of all its elements, then $M = N^*$.

2. Consider the field $\langle \mathbf{Q}, +, \cdot, < \rangle$. Suppose $\frac{a}{b}$ and $\frac{c}{d}$ are positive elements of $\mathbf{Q}$, where a, b, c, and d are integers. Find an integer n in terms of a, b, c, and d such that $n \cdot \frac{a}{b} > \frac{c}{d}$.

3. If an integer n is not a perfect square, explain why the ordered field $\langle \mathbf{Q}(\sqrt{n}), +, \cdot, < \rangle$ is Archimedean but not complete.

4. Show that the Nested Intervals Theorem does not hold in the set **Q** of rational numbers, using the closed interval $[a, b]$ to mean the set of all rational numbers r such that $a \le r \le b$, where a and b are rational numbers and $a < b$.

5. Verify that the set **Q(x)** of all rational expressions $r(x) = \frac{p(x)}{q(x)}$, where $p(x)$ and $q(x)$ are polynomials with rational number coefficients, is a field whose additive identity is the zero polynomial and multiplicative identity is represented by $r(x) = \frac{1}{1}$ or any other $r(x) = \frac{p(x)}{p(x)}$, where $p(x)$ is in **Q[x]**.

6. Show that the isomorphism $h\colon \mathbf{Z} \to Z^*$ of the system $\langle \mathbf{Z}, +, \cdot, < \rangle$ onto the set Z^* of integral elements of an ordered field $\langle \mathbf{F}, +, \cdot, < \rangle$ can be extended in a similar way to an isomorphism $f\colon Q \to Q^*$ of the ordered field $\langle \mathbf{Q}, +, \cdot, < \rangle$ of rational numbers onto a subfield Q^* of $\langle \mathbf{F}, +, \cdot, < \rangle$. The elements of Q^* are called **rational elements** of F.

7. **Further Properties of Ordered Fields.** Suppose that $\langle \mathbf{F}, +, \cdot, < \rangle$ is an ordered field. Use the definition of an ordered field and the results in Theorem 6.11 to prove the following results:

a. Suppose a is an element of **F** and n is a natural number. Give recursive definitions of na, the sum of n terms all equal to a, and a^n. Prove that if $a > 0$, then $na > 0$ and $a^n > 0$ for all n.

b. If a and b are negative elements of **F**, then $\frac{a}{b}$ is positive.

8. Let **K(x)** consist of all rational expressions of the form $r_c(x) = \frac{q_c(x)}{q_1(x)}$, where c is any nonzero real number, and $q_c(x)$ is the polynomial of degree 0 with constant coefficient c and $q_0(x)$ is the zero polynomial. Show that the function $g\colon \mathbf{R} \to \mathbf{K(x)}$ defined by $g(c) = r_c(x)$ for each real number c is an isomorphism of the ordered field $\langle \mathbf{R}, +, \cdot, < \rangle$ of real numbers onto the ordered field $\langle \mathbf{K(x)}, +, \cdot, < \rangle$.

6.2.3 The structure of the complex number system

We have observed that the system $\langle \mathbf{C}, +, \cdot \rangle$ of complex numbers equipped with the operations of addition and multiplication is a field. However, in Section 6.2.1 we observed that there cannot exist an order relation $<$ on **C** that would enable **C** to be an ordered field. This also implies that there is no way to define positive nonreal complex numbers in such a way that they would satisfy the same properties as positive real numbers.

In Chapter 2 we noted that the real field **R** can be identified with the subfield of **C** by thinking of real numbers as complex numbers with imaginary parts equal to 0. The following theorem describes a relationship between the field **R** and **C** looking up from **R** rather than down from **C**.

Theorem 6.15 The complex number system $\langle \mathbf{C}, +, \cdot \rangle$ is the smallest field **F** that contains the system of real numbers $\langle \mathbf{R}, +, \cdot \rangle$ as a subfield and that contains an element e such that $e^2 = -1$.

Of course, the element e of **F** is just the imaginary unit i when **F** is taken to be the complex field **C**. The statement that **F** contains the real numbers **R** as a subfield means that **F** contains a subfield that can be ordered in such a way that it is a complete ordered field. Similarly, the statement in Theorem 6.15 that the complex number field is the smallest field means that the complex field **C** has this property (with $e = i$, of course), and that any other field **F** with the stated property contains a copy of **C** as a subfield.

We do not carry out the details of proving Theorem 6.15 but rather explore some of its consequences.

It is interesting to interpret Theorem 6.15 in the context of the solution of polynomial equations: The smallest field containing the real field **R** and the solution set of the single polynomial equation $p(x) = x^2 + 1 = 0$ is the field **C** of complex numbers. The remarkable fact is that the field **C** of complex numbers is large enough to contain the solution set of any nonconstant polynomial equation $q(x) = 0$ with arbitrary complex coefficients! This is a consequence of the Fundamental Theorem of Algebra.

Theorem 6.16 **(Fundamental Theorem of Algebra):** If $p(x)$ is a nonconstant polynomial with complex number coefficients, then there is a complex number c such that $p(c) = 0$.

The first statement of the Fundamental Theorem of Algebra was by Albert Girard (1595–1632), in his work *Invention nouvelle en l'algèbre (A New Invention in Algebra)* published in 1629. Girard asserted without proof that a polynomial equation of degree n had n solutions (though he did not use exactly that language). This required Girard to consider complex solutions, but he called them "impossible". Descartes knew of the work of Girard. In his analysis of solutions to equations, Descartes called positive solutions "true", negative solutions "false", and nonreal solutions "imaginary".

In the first proof of the Fundamental Theorem of Algebra, in 1799, Gauss does not explicitly consider the possibility that the coefficients of the polynomial might not be real. Nor is this possibility brought out in the different proofs which he gave in 1815 and 1816. However, in 1848, Gauss recast his original proof and allowed for nonreal coefficients. This is the generality with which the theorem is understood today.

Other fields containing complex numbers

An incorrect reading of Theorem 6.15 is that there are no fields that contain complex numbers other than the field $\langle \mathbf{C}, +, \cdot \rangle$, which contains them all. One subfield of the complex numbers involves the set of complex numbers $a + bi$, where a and b are rational. This is the smallest field that contains an element e with $e^2 = -1$ and the set $\mathbf{Q}$ of rational numbers and is akin to the field described in Example 2 of Section 6.2.1 in that its elements have the form $a + b\sqrt{c}$. Another subfield is the field of algebraic numbers. But neither of these subfields contains all the real numbers, so neither field violates Theorem 6.15.

6.2.3 Problems

1. Find the solution set of the equation $p(x) = 0$ and factor $p(x)$ into a product of linear factors.

a. $p(x) = (1 + i)x^2 + ix + (1 - i)$

b. $p(x) = x^8 - 1$

c. $p(x) = x^3 + 1$

2. Suppose that the product of two complex numbers $z_1 = x_1 + iy_1$ and $z_2 = x_2 + iy_2$ is defined as

$$z_1 \cdot z_2 = x_1x_2 + iy_1y_2.$$

If addition in $\mathbf{C}$ is left unaltered, show that $\mathbf{C}$ is not a field and identify the defining properties of a field that fail to hold.

3. Suppose that $p(x)$ is a polynomial with real number coefficients.

a. Prove that if c is a solution of the equation $p(x) = 0$, then so is its complex conjugate c^*.

b. If the degree of $p(x)$ is odd, explain why the equation $p(x) = 0$ has at least one real solution.

c. If the coefficients of the odd powers of x are zero and the remaining coefficients are positive, explain why the equation $p(x) = 0$ has no real solutions.

4. Suppose that $p(x) = a_4x^4 + a_3x^3 + a_2x^2 + a_1x + a_0$ is a polynomial with complex coefficients. Derive formulas that express each coefficient a_i in terms of the solutions r_1, r_2, r_3, and r_4 of the equation $p(x) = 0$.

5. Given that $x = 1 + i$ is a zero of the polynomial $p(x) = x^4 - 4x^3 + 5x^2 - 2x - 2$, find all other zeros of $p(x)$ and factor $p(x)$ completely over the complex numbers.

6. Suppose that r and s are zeros of quadratic polynomials (not necessarily the same polynomial) with rational coefficients. Prove that there is a polynomial with rational coefficients that has $r + s$ as a zero.

Chapter Projects

1. The Perfect Card Shuffle. We usually cut and shuffle a deck of cards several times after each hand of a card game to change the order in which the cards were collected into an order that we regard to be relatively random. But suppose that you are skillful enough to perform a **perfect shuffle** of the deck each time, that is, suppose that you are able to

i. cut a deck of 52 into two piles of 26 cards each, the top pile for your left hand and the bottom pile for your right hand;

ii. shuffle the two halves from the bottom of the two piles up so that the cards from each hand exactly alternate in the resulting pile with the bottom card coming from the left pile.

Thus, if the cards in the given deck were numbered 1 through 52, from top to bottom, then after a perfect cut and shuffle, the order of the cards would be

$$27, 1, 28, 2, 29, 3, 30, 4, 31, 5, \ldots, 52, 26.$$

Notice that the order of the cards after a perfect cut and shuffle is completely determined by the original order of the cards. Consequently, the order after n perfect shuffles is also completely determined by the given order.

a. Explain why the card in position k in the given deck is placed by one perfect shuffle in position $p(k)$, where $1 \leq p(k) < 53$ and $p(k) \equiv 2k \text{ Mod}(53)$.

b. A **not-so-perfect shuffle** is defined in the same way as a perfect shuffle except that (ii) is modified so that the bottom card comes from the right pile rather than the left pile. List the result of one not-so-perfect shuffle on a deck numbered from 1 to 52 from top to bottom. Comment on why the name "not-so-perfect" might be appropriate for this variation.

c. Write a calculator or computer program that will perform a perfect shuffle on a 52-card deck. Do the same for not-so-perfect shuffles. Then write programs that output the result of k perfect shuffles or k not-so-perfect shuffles for any given positive integer k.

d. By using the formulas in part **a** for $p(k)$ and $q(k)$, we can prove the following: The smallest number of perfect shuffles that will return the deck to its original order is 52, but only 8 not-so-perfect shuffles are needed to return the deck to the original order. Check that these statements are true by using the programs that you wrote in part **c**.

2. Congruence-based random number generators. Simulation of real-world problems with a computer requires the capacity to internally generate "random events" and "random data." This is accomplished with the use of internally generated real numbers that can be assumed to be randomly distributed on some interval such as the interval $0 < r < 1$. The programs used to generate these number distributions are called **random number generators**. Of course, the numbers produced as output of such programs are by no means random numbers. Rather, these are numbers completely determined by the input and the program itself. However, because they share certain characteristics with random number distributions, they are called **pseudo-random numbers**. These shared characteristics are:

i. Both random and pseudo-random numbers on the unit interval are uniformly distributed on that interval; that is, the probability that the generator produces a number in a subinterval $a < r < b$ should be equal to the length $b - a$ of that subinterval.

ii. At the level of the simulation, there is no pattern or regularity to the pseudo-random numbers that are generated.

One class of random number generators produces a sequence of numbers in the unit interval generated by multiplicative congruences. The basic idea is to use the least positive residues, $k = 1, 2, 3, \ldots, m - 1$ modulo a large integer m to generate a "scrambled" list of the rational numbers $r = \frac{k}{m}$ in the unit interval. This scrambling is accomplished by selecting an integer a in this range, then multiplying one of the numbers x_0 from among the integers between 1 and $m - 1$ by a and then finding the least positive residue of $ax_0 \text{ Mod}(m)$ to determine x_1. This procedure is iterated to produce a scrambled list of least residues $\text{Mod}(m)$. Explore this procedure as follows:

a. Suppose that $m = 13$, that $a = 3$, and that $x_0 = 4$. Generate a sequence as follows:

x_1 = least positive residue of ax_0 modulo m,
x_2 = least positive residue of ax_1 modulo m, and in general
x_{n+1} = least positive residue of ax_n modulo m for $a < 13$.

Use the resulting scrambled sequence of residues modulo 13 to construct the corresponding numbers in the unit interval.

b. Repeat part **a** for $m = 12$ and the same choices of a and x_0 and compare the results with the results in part **a**. Explain why it is desirable to choose m to be a prime number. What is the advantage of choosing m to be a relatively large prime?

3. Computer Arithmetic and the Chinese Remainder Theorem. Computers need to have the capacity to do arithmetic operations on very large integers. However, a computer processor is designed to work with "data words" representing input in binary or hexadecimal form, and these data words have a specified maximum length that is determined by the design of the computer. When integers are so large that the length of the data words representing them exceeds the maximum length allowed by the processor, some means must be found to carry out the required calculation with data words of lengths acceptable to the computer processor.

The Chinese Remainder Theorem provides one effective means for dealing with this problem. Suppose that we want to perform arithmetic operations on two integers p and q, but the data words needed to represent p and q are longer than the length L allowed by the processor. Choose positive integers

$m_1, m_2, \ldots, m_n$ greater than 1 that are relatively prime in pairs such that

1. The integers p and q are much smaller than the product m of these integers.
2. The integers $m_1, m_2, \ldots, m_n$ can be represented by data words of length L.

With these choices made, compute the least residues p_k and q_k modulo m_k of p and q for $k = 1, 2, \ldots, n$. Then p_k and q_k are smaller integers than m_k and their corresponding data words can be manipulated by the processor by (2). We know from Theorem 6.2 that arithmetic combinations of p and q such as the sum $p + q$ or product pq are congruent modulo m_k to the corresponding arithmetic combinations of p_k and q_k. Moreover, the Chinese Remainder Theorem allows us to recover the values of such combinations of p and q from the corresponding combinations of the p_k and q_k for $k = 1, 2, \ldots, n$.

To learn more about the Chinese Remainder Theorem and its applications in computer science, pursue some or all of the following suggestions:

a. Suppose that you have a primitive computer processor that can only deal with integers between 0 and 99. Take two integers p and q that lie between 10,000 and 99,999. Explain how you can implement the above procedure to find their sum $p + q$.

b. Suppose that m and n are relatively prime integers >1. Construct a table with m rows and n columns, with the successive columns labeled $0, 1, \ldots, n - 1$, and the successive rows labeled $0, 1, \ldots, m - 1$. Such a table is displayed here for $m = 3$ and $n = 5$.

	0	1	2	3	4
0					
1					
2					

Here the row labels are the successive least residues modulo m and the column labels are the successive least residues modulo n. Note also that the lower right portion of the table excluding the label row and label column has mn cells. Suppose that we agree to put the integers c in the range $0 \leq c \leq mn - 1$ so that c is in the row with label a and column with label b if and only if $c \equiv a \text{ Mod}(m)$ and $c \equiv b \text{ Mod}(n)$. Explain what the Chinese Remainder Theorem has to say about the entries in the table for $m = 3$ and $n = 5$, as well as for arbitrary relatively prime m and n.

Bibliography

Unit 6.1 References

Coutinho, S. C. *The Mathematics of Ciphers, Number Theory and RSA Cryptography*. Natick, MA: A. K. Peters, Ltd., 1999.
Section 7.6 details a "shared key" security system based on the Chinese Remainder Theorem. The book itself provides a broad introduction to number theory as well as its applications to cryptography.

Reichardt, Hans "Gauss." *Encyclopaedia Britannica*, 15th Edition. Chicago: Encyclopaedia Britannica.

Rothbart, Andrea. *The Theory of Remainders*. Dedham, MA: Janson Publications, 1995.
An elementary discussion of modular arithmetic.

Unit 6.2 References

Bartle, Robert G. *The Elements of Real Analysis*, Second Edition. New York: John Wiley, 1963.
Chapter 1 discusses the ordered field approach to the structure of the real number system.

Hamilton, N. T., and J. Landin. *Set Theory: The Structure of Arithmetic*. Boston: Allyn and Bacon, 1961.
This book constructs the natural numbers, integers, rational numbers, and real numbers using a set-theoretic approach different from the one we have used in our book.

Part II Geometry with Connections to Algebra and Analysis

Chapter

7 CONGRUENCE

Galileo wrote, "The universe ... is written in the language of mathematics, and its characters are triangles, circles, and other geometrical figures without which it is humanly impossible to understand a single word of it; without these, one wanders about in a dark labyrinth."[1]

Three hundred years later, Hilbert wrote, "Meaning is important in mathematics and geometry is an important source of that meaning."[2]

These quotes demonstrate the central position that geometry occupies in mathematics. Geometry connects mathematics to the physical and visual worlds. It uniquely represents other mathematics and has played a unique role in the evolution of the meaning of mathematical systems and mathematical truth.

The earliest recorded excursions into what we call geometry today were prompted by practical concerns: travel and navigation; measuring land for tax and inheritance purposes; astronomy and the prediction of celestial events, including the calendar; and the construction of buildings and monuments. But, beginning with Thales of Miletus (c. 624–c. 547 B.C.), Greek mathematicians took a more theoretical point of view and worked to systematize geometric knowledge by showing that certain propositions are logically related. The work of the Greek mathematicians most influential on today's school mathematics was done by Euclid.

Unit 7.1 Euclid and Congruence

7.1.1 Euclid's *Elements* (circa 300 B.C.)

The work of the Greek geometers reached a zenith with the appearance of the 13-book treatise by Euclid of Alexandria simply entitled *Elements*. Although we always speak of the "13 books" of Euclid, each "book" is closer to the length of a long chapter in today's books. The most popular English translation, that by Sir Thomas Heath (see the Bibliography at the end of this chapter) integrates the 13 books with an even greater amount of commentary into three volumes.

---

[1] Galileo, *The Assayer*, 1623.

[2] David Hilbert and S. Cohn-Vossen, *Geometry and the Imagination*, 1934.

Euclidean geometry can be studied from a variety of perspectives. The oldest of these, which we therefore call the *traditional perspective*, stems from the axioms and postulates of Euclid. It provides the core from which all other approaches emanate and to which they are compared. By denying or modifying certain postulates, we obtain finite geometries, non-Euclidean geometries, and other systems whose theorems can be compared to those deducible in Euclidean geometry. A second perspective, the *transformation perspective*, is due to the German mathematician Felix Klein (1849–1925) and the Norwegian mathematician Sophus Lie (1842–1899). Both Lie and Klein viewed geometry as studying properties of figures that are invariant under certain transformations. By modifying the properties selected to be invariant, we can obtain Euclidean geometry, affine geometry, projective geometries, or topology. A third perspective, not as easily traceable to a single person but also with origins in the 19th century, is the *vector perspective*. From the vector perspective, Euclidean geometry is the study of a vector space with a particular metric derivable from the Pythagorean Theorem. The vector perspective is particularly elegant for studying geometry in dimensions greater than 3.

Each of these perspectives was developed for high school students a generation ago, and developments at higher levels also exist, as shown in Table 1. More complete references are given at the end of the chapter.

Table 1

Perspective	High School Level	College Level	Historical Classic
Traditional	Anderson, Garon, and Gremillion, *School Mathematics Geometry*	Greenberg, *Euclidean and Non-Euclidean Geometries*	Hilbert, *Foundations of Geometry*
Transformation	Coxford and Usiskin, *Geometry—A Transformation Approach*	Martin, *Transformation Geometry: An Introduction to Symmetry*	Klein, *Elementary Mathematics from an Advanced Standpoint: Geometry*
Vector	Vaughan and Szabo, *A Vector Approach to Euclidean Geometry*	Hausner, *A Vector Space Approach to Geometry*	Choquet, *Geometry in a Modern Setting*

Any of these perspectives can be approached either synthetically or analytically. *Synthetic approaches* use numbers either sparingly or not at all in deducing properties of geometric figures. In contrast, *analytic approaches* take advantage of the properties of numbers (real or complex) to deduce geometric properties. Throughout our analysis of geometrical ideas in this book, we use (and often compare) synthetic and analytic approaches.

We begin in this chapter by briefly revisiting Euclid's development to introduce a discussion of the role of deduction and the importance of definition in geometry. Then we proceed to a transformation perspective, focusing in particular on the notion of congruence.

In Chapter 8, we explore different definitions of distance and provide a transformation perspective on similarity. Chapter 9 examines trigonometry, which is both important for its own sake and provides powerful tools to prove theorems in Euclidean geometry. In Chapter 10, we again visit similarity, examining it in relation to area and volume, and exploring these measures with respect to distance. Finally, in Chapter 11, we return to traditional geometry synthetically and analytically, reviewing many of the topics treated in preceding chapters.

Euclid did produce other scholarly works, covering a wide range of scientific knowledge and methods. Some focused on applied mathematics, including optics. Still his most famous work is the *Elements*.

In the *Elements*, Euclid compiled the fundamental results of scientific geometry up to his time. The first four books on plane geometry deal with simple constructions and proofs. Books V and VI are focused on the reformulation of the theory of proportions, taking incommensurable quantities (those whose ratio is not rational) into account. Books VII, VIII, and IX provide some fundamental notions of number theory and advanced problems in arithmetic. Book X explores the theory of incommensurables. The last three books treat 3-dimensional (solid) geometry.

The first extraordinary accomplishment in the *Elements* is the fitting of all of the geometry known at that time into a large axiomatic system. The second extraordinary aspect of the *Elements* is that it uses geometry to develop significant results in areas of mathematics that are not considered part of geometry today. For example, the basic theory of divisibility and a proof of the infinitude of primes (as in our Section 5.2.4) are found in the *Elements*. So, too, are the Law of Cosines and a geometric version of the Quadratic Formula. It is no wonder that in all of Europe for over 2000 years the *Elements* was a basic resource for anyone wishing to study mathematics or logic.

Euclid seems to have written the *Elements* as a textbook—a successor of a textbook with the same title which had been written by Theudius during the time of Aristotle. The word "Elements" is meant to convey basic principles or foundations. Although Euclid claimed no specific authorship of the results presented therein, his systematic arrangement of topics and method of logical development surely produced new discoveries for the text. However, it is very hard to establish which results were discovered by Euclid and which were the discoveries of others.

If little is known about the authorship of specific results in the *Elements*, even less is known about Euclid himself. The dates of his birth are unknown, but almost certainly he worked after Aristotle (384–322 B.C.) and he predates Archimedes (287–212 B.C.). He may have studied in Plato's school in Athens. Euclid became a teacher in Alexandria, which by this time had become the world's center for mathematical activity. It is reported that the first Ptolemy (the ruler of Egypt after the death of Alexander the Great in 323 B.C.) asked Archimedes if there were an easier way to learn geometry than by reading the *Elements* and Archimedes responded that there was no royal road to geometry. The same story is told about Alexander the Great and the mathematician Menaechmus, who responded, "O king, through the country there are roads for ordinary people and roads for royalty, but in geometry there is only one road for everybody." What precisely was this "road" that these mathematicians saw as necessary to the study of geometry? It was to read Euclid.[3]

Throughout the *Elements*, Euclid followed the same logical development of material. He proposed a statement of the proposition to be proved and gave a statement of the given data to be assumed, sometimes with a diagram. He noted how the data would be used and what additional lines or figures were required for the proof. He then constructed a proof and gave a concluding statement about what had been done. This logical structure that Euclid used in the *Elements* influenced virtually all scientific thinking from his time on.

The first book of the *Elements* opens with definitions and assumptions (common notions and postulates) that are used throughout the text. At the end of this section are listed the definitions, postulates, common notions, and propositions of Book I of the *Elements*. Every other book of the *Elements* begins with specific definitions that pertain to the subject matter to be treated therein, but no new common notions or postulates.

Looking at the definitions in the *Elements* with modern eyes, we immediately see some problems. Euclid tried to define every term, including "point" and "line".

---

[3] Heath, *Euclid's Elements*, Vol. 1, p. 2.

His definitions included ambiguous terms. For example, in Definition 1, what is meant by "part," and in Definition 2, what is meant by "breadthless length"? Today we recognize that in a mathematical system some terms must be undefined.

Looking at the propositions of Book I, you may be surprised that the first three propositions and many of the others are constructions. Constructions play an important role in the *Elements*. Notice that Postulates 1 through 3 use the language of constructions.

Euclid's diagrams and the points, lines, and figures he produced were based on straightedge-and-compass constructions, and these constructions were accepted by him as proof of the existence of these points, lines, and figures. Euclid made no provision for the existence of points, so by today's standards of logical rigor, his axioms provide an insufficient basis for the existence of the figures he constructed.

Other approaches to Euclidean geometry

Euclidean geometry is a mathematical system in which the statements that are assumed or can be deduced include all the axioms and propositions that are in Euclid's *Elements*. Today there exist many logical developments of Euclidean geometry that are somewhat different from that found in Euclid. The first to have significant use in schools was published in 1794 by Adrien-Marie Legendre, a French mathematician who had made many discoveries in number theory and analysis. Legendre was motivated by a desire to construct a postulate system that would be easier for students to learn, and wanted to clarify whether the parallel postulate of Euclid is independent of other postulates.

Legendre's approach was, in today's terms, quite similar to Euclid's own. However, to tinker at all with Euclid's approach, even two millennia after Euclid, was a novel idea. In fact, almost a century later, in *Euclid and His Modern Rivals*, Charles Dodgson (better known to us as Lewis Carroll, the author of *Alice in Wonderland*) argued that the sequence of propositions found in Euclid's *Elements* is the best possible and that no other sequence should be used. It is ironic that the rather narrow view put forth in this book is from a man who has become known for his liberal approach to the use of words (see Section 7.1.3).

As we have mentioned, many approaches to Euclidean geometry exist today. We begin a careful synthetic development in Unit 11.1, introducing a modern axiom set, and in Section 11.2.2, we point out that there is only one Euclidean geometry that emerges from these axioms. So theorems proved using synthetic or coordinate approaches, with traditional, transformation, vector, or even other approaches to Euclidean geometry, are valid in Euclidean geometry.

Geometries that are not Euclidean

Figure 1

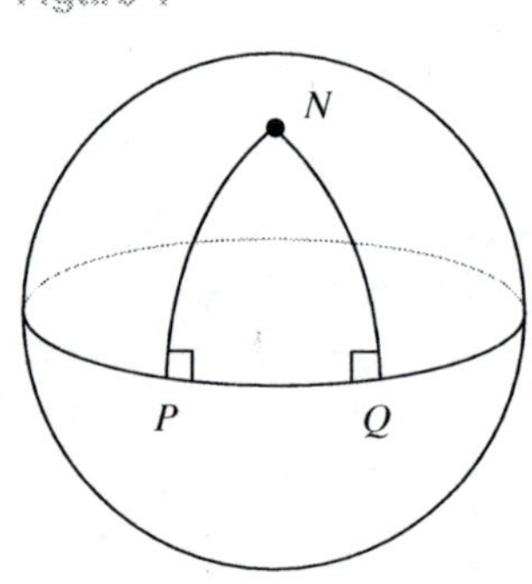

Until the 19th century, Euclidean geometry was considered to be the only true geometry. Today we recognize many geometries that are not Euclidean.

One example is the 2-dimensional geometry of the surface of Earth. In this geometry, some familiar theorems of Euclidean geometry are invalid; e.g., that the sum of the measures of the angles of a triangle is 180°. On the surface of Earth, shortest distances are along great circles, so sides of triangles must be along great circles. For instance, consider a triangle NPQ with the North Pole N as one vertex and two nonopposite points P and Q on the equator as the other two vertices, as shown in Figure 1. In this way, we can see that the geometry of the surface of Earth is not Euclidean.

Another proposition of Euclidean geometry is that lines contain infinitely many points. Consequently, a geometry with only finitely many points also cannot be Euclidean geometry. The surprise is that some of these *finite geometries* have quite a number of theorems in common with Euclidean geometry (see Section 11.1.1).

The Definitions, Postulates, Common Notions, and Propositions of Euclid's *Elements*—Book I (adapted from Heath [1956])

Definitions

1. A *point* is that which has no part.
2. A *line* is breadthless length.
3. The extremities of a line are points.
4. A *straight line* is a line which lies evenly with the points on itself.
5. A *surface* is that which has length and breadth only.
6. The extremities of a surface are lines.
7. A *plane surface* is a surface which lies evenly with the straight lines on itself.
8. A *plane angle* is the inclination to one another of two lines in a plane which meet one another and do not lie in a straight line.
9. And when the lines containing the angle are straight, the angle is called *rectilineal*.
10. When a straight line set up on a straight line makes the adjacent angles equal to one another, each of the equal angles is *right*, and the straight line standing on the other is called *perpendicular* to that on which it stands.
11. An *obtuse angle* is an angle greater than a right angle.
12. An *acute angle* is an angle less than a right angle.
13. A *boundary* is that which is an extremity of anything.
14. A *figure* is that which is contained by any boundary or boundaries.
15. A *circle* is a plane figure contained by one line such that all the straight lines falling upon it from one point among those lying within the figure are equal to one another.
16. And the point is called the *center* of the circle.
17. A *diameter* of the circle is any straight line drawn through the center and terminated in both directions by the circumference of the circle, and such a straight line also bisects the circle.
18. A *semicircle* is the figure contained by the diameter and the circumference cut off by it. And the center of the semicircle is the same as that of the circle.
19. *Rectilineal* figures are those which are contained by straight lines, *trilateral* figures being those contained by three, *quadrilateral* those contained by four, and *multilateral* those contained by more than four straight lines.
20. Of trilateral figures, an *equilateral triangle* is that which has three sides equal, an *isosceles triangle* that which has two of its sides alone equal, and a *scalene triangle* that which has its three sides unequal.
21. Further, of trilateral figures, a *right-angled triangle* is that which has a right angle, an *obtuse-angled triangle* that which has an obtuse angle, and an *acute-angled triangle* that which has its three angles acute.
22. Of quadrilateral figures, a *square* is that which is both equilateral and right-angled; an *oblong* that which is right-angled but not equilateral; a *rhombus* that which is equilateral but not right-angled; and a *rhomboid* that which has its opposite sides and angles equal to one another but is neither equilateral nor right-angled. And let quadrilaterals other than these be called *trapezia*.
23. *Parallel* straight lines are straight lines which, being in the same plane and being produced indefinitely in both directions, do not meet one another in either direction.

Postulates

1. A straight line can be drawn from any point to any point.
2. A finite straight line can be produced continuously in a straight line.

3. A circle can be described with any center and distance.
4. All right angles are equal to one another.
5. If a straight line falling on two straight lines makes the interior angles on the same sides less than two right angles, the two straight lines, if produced indefinitely, meet on that side on which are the angles less than the two right angles.

Common Notions

1. Things which are equal to the same thing are also equal to one another.
2. If equals be added to equals, the wholes are equal.
3. If equals be subtracted from equals, the remainders are equal.
4. Things which coincide with one another are equal to one another.
5. The whole is greater than the part.

Propositions of Book I

1. On a given finite straight line, an equilateral triangle can be constructed.
2. At a given extremity point (on a line) a straight line equal to a given straight line can be placed.
3. Given two unequal straight lines, from the greater a straight line equal to the lesser can be cut off.
4. If two triangles have the two sides equal to two sides respectively, and have the angles contained by the equal straight lines equal, they will also have the base equal to the base, the triangle will be equal to the triangle, and the remaining angles will be equal to the remaining angles respectively, namely, those which the equal sides subtend.
5. In isosceles triangles, the angles at the base are equal to one another, and, if the equal straight lines be produced further, the angles under the base will be equal to one another.
6. If in a triangle two angles be equal to one another, the sides which subtend the equal angles will also be equal to one another.
7. Given two straight lines constructed on a straight line (from its extremities) and meeting in a point, there cannot be constructed on the same straight line (from its extremities), and on the same side of it, two other straight lines meeting in another point and equal to the former two respectively, namely, each to that which has the same extremity with it.
8. If two triangles have the two sides equal to two sides respectively, and have also the base equal to the base, they will also have the angles equal which are contained by the equal straight lines.
9. A given rectilineal angle can be bisected.
10. A given finite straight line can be bisected.
11. A straight line can be drawn at right angles to a given straight line from a given point on it.
12. To a given infinite straight line, from a given point which is not on it, a perpendicular straight line can be drawn.
13. If a straight line set up on a straight line make angles, it will make either two right angles or angles equal to two right angles.
14. If with any straight line, and at a point on it, two straight lines not lying on the same side make the adjacent angles equal to two right angles, the two straight lines will be in a straight line with one another.
15. If two straight lines cut one another, they make the vertical angles equal to one another.
16. In any triangle if one of the sides be produced, the exterior angle is greater than either of the interior and opposite angles.

17. In any triangle two angles taken together in any manner are less than two right angles.

18. In any triangle the greater side subtends the greater angle.

19. In any triangle the greater angle is subtended by the greater side.

20. In any triangle two sides taken together in any manner are greater than the remaining one.

21. If on one of the sides of a triangle, from its extremities, there be constructed two straight lines meeting within the triangle, the straight lines so constructed will be less than the remaining two sides of the triangle, but will contain a greater angle.

22. Out of three straight lines, which are equal to three given straight lines, a triangle can be constructed if two of the straight lines taken together in any manner should be greater than the remaining one.

23. On a given straight line and at a point on it, a rectilineal angle equal to a given rectilineal angle can be constructed.

24. If two triangles have the two sides equal to two sides respectively, but have the one of the angles contained by the equal straight lines greater than the other, they will also have the base greater than the base.

25. If two triangles have the two sides equal to two sides respectively, but have the base greater than the base, they will also have the one of the angles contained by the equal straight lines greater than the other.

26. If two triangles have the two angles equal to two angles respectively, and one side equal to one side, namely, either the side adjoining the equal angles, or that subtending one of the equal angles, they will also have the remaining sides equal to the remaining sides and the remaining angle to the remaining angle.

27. If a straight line falling on two straight lines make the alternate angles equal to one another, the straight lines will be parallel to one another.

28. If a straight line falling on two straight lines make the exterior angle equal to the interior and opposite angle on the same side, or the interior angles on the same side equal to two right angles, the straight lines will be parallel to one another.

29. A straight line falling on parallel straight lines makes the alternate angles equal to one another, the exterior angle equal to the interior and opposite angle, and the interior angles on the same side equal to two right angles.

30. Straight lines parallel to the same straight line are also parallel to one another.

31. Through a given point a straight line parallel to a given straight line can be drawn.

32. In any triangle, if one of the sides be produced, the exterior angle is equal to the two interior and opposite angles, and the three interior angles of the triangle are equal to two right angles.

33. The straight lines joining equal and parallel straight lines (at the extremities which are) in the same directions (respectively) are themselves also equal and parallel.

34. In parallelogrammic areas the opposite sides and angles are equal to one another, and the diameter bisects the areas.

35. Parallelograms which are on the same base and in the same parallels are equal to one another.

36. Parallelograms which are on equal bases and in the same parallels are equal to one another.

37. Triangles which are on the same base and in the same parallels are equal to one another.

38. Triangles which are on equal bases and in the same parallels are equal to one another.

39. Equal triangles which are on the same base and on the same side are also in the same parallels.

40. Equal triangles which are on equal bases and on the same side are also in the same parallels.

41. If a parallelogram has the same base as a triangle and lies in the same parallels, the parallelogram is double the triangle.

42. In a given rectilineal angle, a parallelogram can be constructed equal to a given triangle.

43. In any parallelogram the complements of the parallelograms about the diameter are equal to one another.

44. To a given straight line, in a given rectilineal angle, a parallelogram equal to a given triangle can be applied.

45. In a given rectilineal angle, a parallelogram equal to a given rectilineal figure can be constructed.

46. On a given straight line a square can be described.

47. In right-angled triangles the square on the side subtending the right angle is equal to the squares on the sides containing the right angle.

48. If in a triangle the square on one of the sides be equal to the squares on the remaining two sides of the triangle, the angle contained by the remaining two sides of the triangle is right.

7.1.1 Problems

1. Examine the definitions of Book I of the *Elements*.

a. In today's language, how would we describe Euclid's "line", "straight line", and "finite straight line"?

b. For Euclid, does a plane figure such as a circle include the points in its interior? How can you tell?

c. What interpretation can we give to the word "extremities" in definitions 3 and 6?

d. For Euclid, is every equilateral triangle also isosceles? How do you know?

e. How does Euclid classify the various types of quadrilaterals?

2. Draw a figure for Euclid's fifth postulate and restate the postulate in modern language.

3. Give the properties of real numbers corresponding to Euclid's Common Notions 1 through 3.

4. Examine the propositions of Book I of the *Elements* listed in this section.

a. Which are constructions?

b. Which of these correspond to today's triangle congruence propositions, and to which triangle congruence proposition does each correspond?

c. Which is the Pythagorean Theorem?

d. Which conveys in modern language that the sum of the measures of the angles of a triangle is 180°?

e. In Propositions 35 through 45, what is the meaning of the word "equal"?

5. Consider the following "definitions" of *point*.

(1) an extremity that has no dimension (Posidonius)

(2) something that is indivisible (Aristotle)

(3) a monad having position (Proclus)

(4) something that has no part (Euclid)

(5) an absolutely simple space element (Friedrich Ueberweg)

(6) a sphere that does not include any other sphere (Edward Huntington)

a. If you were attempting to describe a *point* to someone, which definition would you use and why?

b. Create your own "definition" of point.

6. Consider the following "definitions" of *line*.

(1) breadthless length (Plato)

(2) a magnitude extended one way (Proclus)

(3) the path of a point when moved (Proclus)

(4) a straight line as a line which lies evenly with the points on itself (Euclid)

(5) a collection of points

a. Which definition do you prefer and why?

b. Do any of these definitions ensure that a line is not circular?

7. Consider the following "definitions" of *plane*.

(1) a surface which lies evenly with the straight lines on itself (Euclid)

(2) a surface that is stretched to the utmost (Proclus)

(3) a flat surface as that which can be laid through any three given points (Moritz Pasch)

(4) an infinite set of points equidistant from two fixed points in space (Nicholas Lobachevsky)

(5) the set of all points on all lines joining every two points of a triangle whose sides are lines—not line segments (Mario Pieri)

(6) a surface such that a straight line joining any two of its points lies entirely in the surface

(7) a collection of lines

a. Which "definition" do you feel comes closest to describing the planes you have studied in geometry?

b. Which do you feel provides the most breadth to create "nonintuitive" planes?

8. Look up the word "geometry" in a standard dictionary. Then try a mathematical dictionary and an encyclopedia. Try finding a definition of "geometry" in a mathematics textbook, and in a history of mathematics book. Collect these definitions and explain which one you prefer and why.

7.1.2 Deduction and proof

It is often said that you can be certain of the truth of a mathematical statement or the answer to a mathematics problem. There is a sense in which this is true. If you agree with the assumptions in the statement or in the problem, and if valid reasoning has been used, then you must agree with the results. This property of truth based on assumptions and valid reasoning is a result of a basic underlying aspect of mathematical thinking—the *deductive process*, or *deduction* for short. In mathematics, results that have been deduced from agreed-upon statements (axioms, definitions, or previously proved statements) using valid arguments of deduction can be thought of as *true*. We call these results **theorems**. The deductive argument itself is called a **proof**.

Before the discoveries of geometries different from Euclidean geometry, theorems proved using deduction were generally considered to be **absolute truths**, that is, true independent of the particular axioms one chose for geometry. Mathematicians occasionally formulated alternate postulate sets because they thought that some theorems might be more accessible using one logical approach than another, not because it was thought that different postulate sets led to different geometries. Now we recognize that the postulates we choose *define* a mathematical system, and that all theorems in mathematics are **relative truths**, that is, true relative to the assumptions of the system.

For example, the statement made by many a primary-school teacher to a child, "You can't take away a bigger number from a smaller one", is true in the system of natural numbers but not true in the integers. The statement "The sum of the measures of the angles of a triangle is 180°" is true in Euclidean geometry but not for triangles on the surface of Earth.

Deduction is a demanding criterion for the establishment of truth because it means we cannot assert something is true merely because we have lots of examples. For instance, Figure 2 shows an 8-by-8 grid, but with two opposite corners removed, so there are 62 squares.

Figure 2

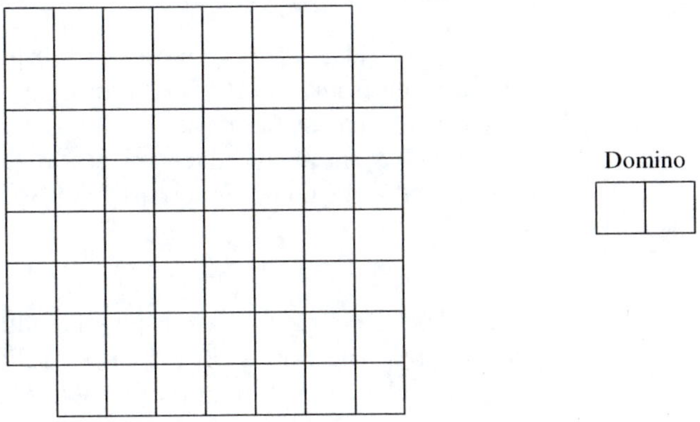

Domino

Can the grid be covered with 31 nonoverlapping dominoes? If you try to do this, you will find that no matter how you fit the dominos together they will not cover the squares. You can cover 60 of the squares, but the 2 that remain will not be next to each other. But all the unsuccessful attempts do not prove that someone could not come up with a way to cover the grid with 31 dominoes. Only deduction can do that.

Here is a proof. First, color the squares like a checkerboard (Figure 3). This does not change the problem. If dominoes can cover the original grid, they can cover this checkerboard, and vice versa.

Figure 3

Each domino always covers two adjacent squares, one black and one white. The 31 dominoes must thus cover 31 black squares and 31 white squares. But the opposite corners of an 8-by-8 checkerboard are the same color, so by removing the opposite corners, 30 remaining squares are of one color (white in Figure 3) and 32 are of the other (black in Figure 3). Hence any attempt to cover this cut-off checkerboard with 31 dominoes will be futile.

The preceding deductive argument *proves* that no matter how 30 of the 31 dominoes are placed, there will be two squares (black in Figure 3) without dominoes on them, and the remaining domino cannot cover both.

Notice that we did not show that a particular placement of dominoes would not work. Even had we shown such an example, it would not have proved the result we want. The only way to deduce the result was to show that *all* possible ways of covering with dominoes would not work.

Similar deductive arguments can be used to determine whether other checkerboards can be covered with these dominoes.

Question 1: Can a 5-by-5 grid with opposite corners removed be covered with dominoes? Can a 9-by-8 grid with opposite corners removed be covered?

These ideas apply to more than abstract problems. Any metalworker who needs to cut items from a sheet of metal, or a garment worker who needs to cut pieces of cloth from a bolt, may be faced with this kind of question—can the items be cut with no pieces left over? Money is saved if no pieces are left over. The geometry of covering the plane and its 3-dimensional counterpart, filling space, is rich in results, beauty, and application.[4]

The power of deduction

The checkerboard problem is just one illustration of the logical power of deductive reasoning. The power of deduction is illustrated whenever we deduce a property held by all members of an infinite set. Thus, when we prove in Euclidean geometry that the sum of the measures of the angles of a triangle is 180°, this proof applies to *all* triangles, not just the small ones found on a sheet of paper, not just the ones that have angles whose measures are whole numbers of degrees, and not just the ones that look close to equilateral or right. The theorem applies to Euclidean triangles formed by the earth, our moon, and the sun, to triangles in chemical bonds, to very long thin triangles, as well as to equilateral triangles. No process of observation can provide such powerful results.

---

[4] See, for example, Branko Grünbaum and G. C. Shephard, *Tilings and Patterns*, (New York: W. H. Freeman, 1987).

We know that deduction is sometimes used to verify what seems to be obvious. *The base angles of an isosceles triangle are congruent* is a theorem that we are compelled to prove even though observation makes it seem obvious. Yet what is obvious to one person may not be obvious to another. And sometimes what seems "obvious" turns out to be false. A second aspect of the power of deduction is that theorems may be proved whose truth is hard to believe. Would you believe the Pythagorean Theorem if you had never seen a proof? *Should* a person believe the Pythagorean Theorem without seeing a proof? Deduction shows that the theorem is always true, and we do not need all 370 different proofs of it found in Loomis's book *The Pythagorean Proposition*,[5] or the new proofs that were published in every volume of *The American Mathematical Monthly* from 1886 to 1899, to validate it. The power of deduction is that *one* valid proof suffices to ensure the truth of the theorem.

A third aspect of deduction is that it not only can show a statement to be true but can also indicate why it is true. For instance, the solution to the checkerboard problem shows why the smaller board cannot be covered with 2 × 1 dominoes. When we deduce the solution to an equation, we may see in our argument why no other solution could work.

A fourth aspect of the power of deduction is that it provides a universally accepted criterion for the establishment of mathematical truth. Although philosophers may debate the foundations of mathematics, and mathematicians themselves may wonder whether a proof in which computers did much of the work truly constitutes a proof, or whether a complicated proof has a gap, there is universal agreement on the principles behind deductive proof. Consequently, when a new theorem is proved, no one has to go to a laboratory to test the result. Only the argument needs to be checked. This is what makes mathematics so different from science. In science, conclusions must be confirmed by observation.

Because of its broad power, deduction is employed outside of mathematics in situations such as law and philosophy where careful argumentation is needed. It is thought by many people to be the purest form of reasoning, and it is a major reason why many people believe mathematics is important for all students to study.

The need for care

With the power of deduction comes a responsibility that comes with all forms of power—to use it carefully. In geometry in particular, because drawings often suggest theorems to be proved, care must be taken in what is assumed in and from drawings. For instance, consider the following "proof".

"Theorem" Every triangle is isosceles.

You should focus on each step, convincing yourself that the proof is legitimate from the assumptions and propositions of Euclid's *Elements*.

"Proof": Given: ΔABC with $AC \neq BC$.

1. Construct the bisector of $\angle C$. Call it ℓ. (ℓ is not perpendicular to $\overline{AB}$ because $AC \neq BC$.)
2. Construct the perpendicular bisector of segment $\overline{AB}$. Call it m and let the intersection of $\overline{AB}$ and m be D. Since ℓ and m are not both perpendicular to $\overline{AB}$, ℓ and m intersect in a point. Call it P.

[5] Elisha Loomis, *The Pythagorean Proposition*, Second Edition. Reston, VA: National Council of Teachers of Mathematics, 1972 (originally self-published in 1940).

3. Construct the perpendicular from P to $\overleftrightarrow{AC}$, intersecting it in a point E.

Figure 4

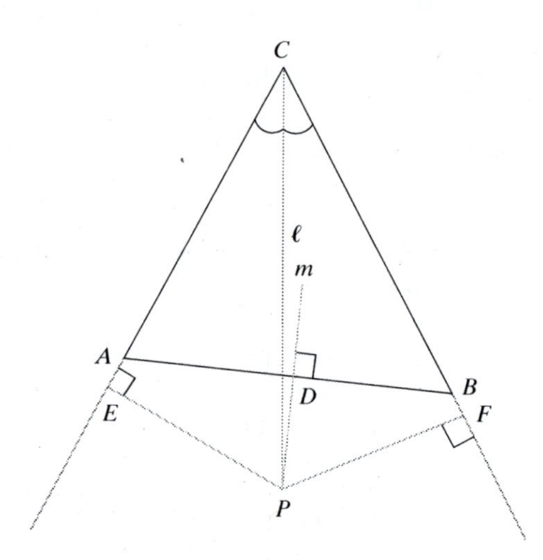

4. Construct the perpendicular from P to $\overleftrightarrow{CB}$, intersecting it in a point F (Figure 4). We now give modern reasons, but you should look at the previous section for Euclid's assumptions or propositions that justify the numbered steps.

Statement	Reason
$\angle ACP \cong \angle BCP$	$\overleftrightarrow{CP} = \ell$ is the bisector of $\angle C$.
5. $\angle CEP \cong \angle CFP$	They are both right angles since $\overline{PE}$ is perpendicular to $\overleftrightarrow{AC}$ and $\overline{PF}$ is perpendicular to $\overleftrightarrow{CB}$.
6. $\overline{CP} \cong \overline{CP}$	Reflexive property of congruence
7. $\Delta CEP \cong \Delta CFP$	AAS Congruence
8. $\overline{EP} \cong \overline{FP}$ and $\overline{CE} \cong \overline{CF}$	Corresponding parts of congruent triangles are congruent.
Now, $\overline{AP} \cong \overline{BP}$	A point on the perpendicular bisector of a line segment is equidistant from the endpoints of the line segment.
$\angle AEP \cong \angle BFP$	They are both right angles since $\overline{PE}$ is perpendicular to $\overleftrightarrow{AC}$ and $\overline{PF}$ is perpendicular to $\overleftrightarrow{BC}$.

Now we show $\overline{EA} \cong \overline{FB}$. Suppose, without loss of generality, $EA > BF$. Then there exists a point A' on $\overline{EA}$ and different from A with $\overline{EA'} \cong \overline{BF}$. Now $\angle A'EP$ is a right angle (it is identical to $\angle AEP$).

Statement	Reason
9. $\angle AA'P$ is obtuse	Exterior angle theorem
10. $\Delta FPB \cong \Delta EPA'$	SAS Congruence
11. $\overline{PA'} \cong \overline{PB}$.	Corresponding parts of congruent triangles are congruent.
12. $\overline{PA'} \cong \overline{PA}$	Transitive property of congruence

This is impossible (see step 9), so we must have $\overline{EA} \cong \overline{FB}$.

Statement	Reason
13. $\overline{AC} \cong \overline{BC}$	Since $\overline{CE} \cong \overline{CF}$ and $\overline{EA} \cup \overline{AC} = \overline{CE}$ and $\overline{FB} \cup \overline{BC} = \overline{CF}$
ΔABC is isosceles.	Definition of *isosceles*.

It appears that we have done nothing "illegal" in our "proof". Yet we know it cannot be valid. So something "illegal" must have been done. We must have used some false premise that was not given.

Question 2: Draw an accurate diagram of the construction in the preceding proof. Describe the location of points E and F relative to ΔABC.

It turns out that P must be outside the triangle *and* that one of the perpendiculars $\overline{PE}$ and $\overline{PF}$ must intersect a side of the triangle between the vertices. We can only deduce this if we have properties of "betweenness" (which Euclid did not have). Thus *one* of the points E or F is on a side of the triangle ABC and the other is outside. We falsely assumed in our proof that both E and F were outside the triangle. This destroys the "proof" that any scalene triangle is isosceles.

Using Euclid's method, but incorporating the concept of betweenness, let us examine why P must lie outside the triangle and that E or F (but not both) must lie on a side of the triangle.

Circumscribe ΔABC. (That is, draw the circle through points A, B, and C.) Then both ℓ, the bisector of $\angle C$, and m, the perpendicular bisector of $\overline{AB}$, must bisect arc $\overset{\frown}{AB}$. So point P, their intersection, is on the circle. Thus ℓ and m intersect outside ΔABC at P, as shown in Figure 5.

Figure 5

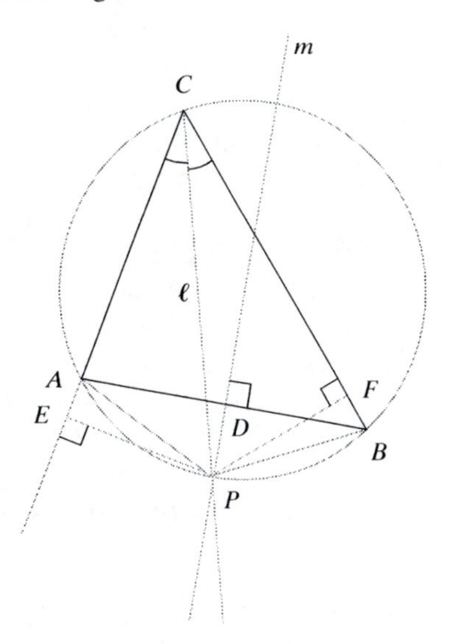

Next consider the quadrilateral $ACBP$. $\angle CAP$ is supplementary to $\angle CBP$ since they are the opposite angles of an inscribed quadrilateral. Since ΔABC is scalene, these two angles are not right angles. So one must be acute and the other obtuse. Without loss of generality, let $\angle CBP$ be acute. Then in ΔCPB the altitude on $\overline{CB}$ must be inside the triangle while in ΔCAP the altitude on $\overline{AC}$ must be outside the triangle. So point E is outside ΔABC, while point F is inside.

This example illustrates how the concept of betweenness of points can be important in a proof. In the penultimate step of our "proof", we wrongly assumed that A was between C and E, and that simultaneously B was between C and F. We made this assumption based on the diagram we drew (Figure 4), which turned out to be incorrect.

Here is another example of a difficulty caused by relying too much on a diagram. Read each step of the following "proof" carefully to convince yourself that Euclid's propositions justify each step.

"Theorem" A right angle has the same measure as an obtuse angle.

Proof: Construct a rectangle $ABCD$. Choose a point E not on the rectangle so that the segment $\overline{AD}$ is congruent to the segment $\overline{CE}$. Construct the perpendicular bisector of $\overline{AE}$. Call it ℓ. Construct the perpendicular bisector of $\overline{CD}$. Call it m.

Then ℓ and m intersect in a point. Call it P. Let M be the point of intersection of ℓ and $\overline{AE}$. Let N be the point of intersection of m and $\overline{CD}$. Draw segments $\overline{DP}, \overline{EP}, \overline{AP}$, and $\overline{CP}$.

A diagram illustrating these constructions is shown in Figure 6.

Figure 6

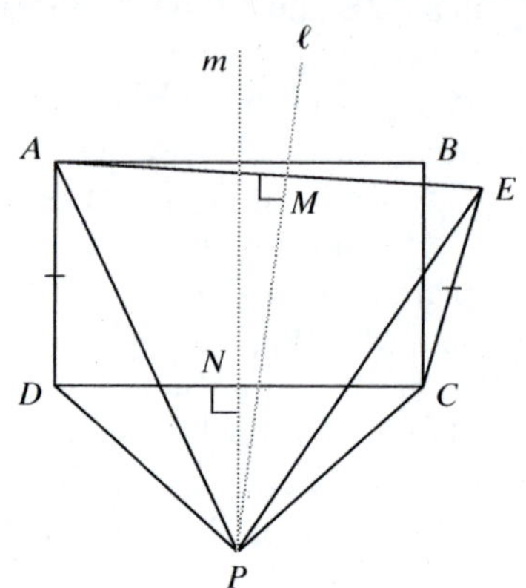

In this proof we number four missing justifications (See Problem 4).

Consider ΔECP and ΔADP. $AP = EP$ and $DP = CP$ since every point on the perpendicular bisector is equidistant from the endpoint of the line segment. Since $\overline{AP} \cong \overline{EP}$ and $\overline{DP} \cong \overline{CP}$ (because they have equal lengths), and $\overline{AD} \cong \overline{CE}$ (by construction), $\Delta ECP \cong \Delta ADP$ (1).

Therefore, $m\angle ECP = m\angle ADP$ (2).

Now we show that angles adjacent to these two angles also have the same measure. Since $\overline{DP} \cong \overline{CP}$, $\Delta DNP \cong \Delta CNP$ (3).

Thus $m\angle DCP = m\angle CDP$ (4).

But $m\angle ECP = m\angle DCP + m\angle ECD$ and $m\angle ADP = m\angle CDP + m\angle ADC$.

So $m\angle ECD = m\angle ADC$.

But $\angle ECD$ is an obtuse angle, and $\angle ADC$ is a right angle.

Thus the "theorem" is "proved". ┘

Take some time now to convince yourself that the same result is deduced if P falls on $\overline{DC}$ or inside the rectangle. What we have "proved" is obviously a fallacy. But where does the error lie?

It turns out that the diagram in Figure 6 is incorrect. $\overline{EP}$ is on the wrong side of C. To see why, notice that since P is on ℓ, $AP = EP$. And since P is on m, $BP = AP$. Thus $EP = BP$, so P must also lie on the perpendicular bisector of $\overline{BE}$, call it n. Since we constructed $\overline{BC} \cong \overline{EC}$, C must also lie on the perpendicular bisector of $\overline{BE}$, as in Figure 7.

Figure 7

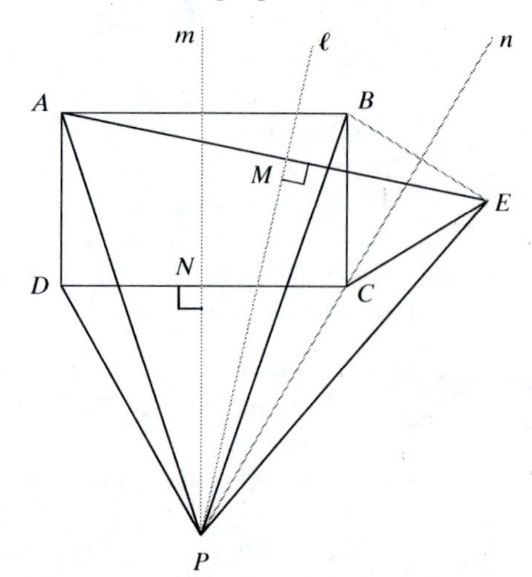

Thus ΔECP is positioned in a way that $\overline{EP}$ is on the side of C outside of the rectangle and the statement in our previous proof that $m\angle ECP = m\angle DCP + m\angle ECD$ fails.

This example illustrates how a diagram can sometimes lead us astray in a proof. Notice that the step where the error occurred was the result of drawing a conclusion from our diagram. Thus an unwarranted assumption appeared in our argument.

The moral of these examples is that, although diagrams are often useful in proofs, poor diagrams can lead to invalid assumptions and false conclusions. Examples like the two in this section convinced mathematicians that even if using our intuition and drawing diagrams based on it always yielded correct results, a proof needs to be independent of diagrams. It convinced some mathematicians to reexamine Euclid's geometry to make it independent of these drawings. It also brought attention to the need to use language carefully, both in definitions and in logical arguments. In the next section, we turn our attention to language.

7.1.2 Problems

1. Remove two opposite squares from an $n \times n$ square checkerboard. Is there any value of n for which the remaining $n^2 - 2$ squares can be covered by nonoverlapping 1×2 dominoes? Demonstrate your covering or prove that it cannot be done.

2. a. Identify a theorem you have learned for which you needed to see a proof before you would believe it.

   b. Identify a theorem you have seen proved that seemed so obvious to you that you didn't think a proof was needed.

3. Identify the assumption or proposition of Euclid that justifies the numbered steps in the "proof" that every scalene triangle is isosceles.

4. In the "proof" that an obtuse angle is equal in measure to a right angle, replace the missing justifications (1) to (4) by propositions from Euclid's *Elements*.

5. Consider the following "proof" that in Euclidean geometry there exists a triangle with two right angles.

Let two circles meet in points A and B, as in Figure 8.

Figure 8

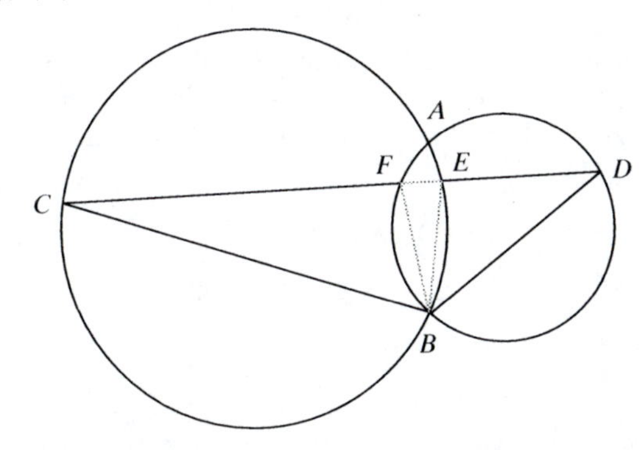

Let $\overline{BC}$ and $\overline{BD}$ be diameters of the circles. Let $\overline{CD}$ intersect the circles in F and E. Then $\angle BEC$ is a right angle and $\angle BFD$ is a right angle. So ΔBEF has two right angles. Criticize this "proof", determining where the fallacy lies.

6. Suppose "parallelogram" has just been defined as a quadrilater with two pairs of parallel sides and the following theorem is the first property of parallelograms to be deduced. There is something wrong with the proof of the following theorem. Find the error and correct the proof.

> **Theorem:** The diagonals of a parallelogram bisect each other.

Proof: Let $ABCD$ be a parallelogram with diagonals $\overline{AC}$ and $\overline{BD}$ intersecting at O. $\overline{AD}$ is parallel to $\overline{BC}$ and $\overline{AB}$ is parallel to $\overline{DC}$ since $ABCD$ is given as a parallelogram. Since alternate interior angles formed by parallel lines are congruent, $\angle CAD \cong \angle ACB$ and $\angle ADB \cong \angle DBC$. $\overline{AD} \cong \overline{BC}$, since opposite sides of a parallelogram are equal in length. Thus $\Delta AOD \cong \Delta COB$ by ASA congruence. Thus $\overline{AO} \cong \overline{OC}$ and $\overline{BO} \cong \overline{OD}$, since corresponding parts of congruent triangles are congruent.

7. Here are two valid proofs of the theorem that the base angles of an isosceles triangle are equal. Which do you like better, and why?

Proof 1: Given isosceles ΔABC with $\overline{AC} \cong \overline{BC}$. Draw the bisector of $\angle C$. Let D be the intersection of $\overline{AB}$ and this bisector. $\angle ACD \cong \angle BCD$. $\overline{CD} \cong \overline{CD}$. $\Delta ACD \cong \Delta BCD$ by SAS congruence. Thus $\angle A \cong \angle B$.

Proof 2: Given isosceles ΔABC with $\overline{AC} \cong \overline{BC}$. Then $\overline{BC} \cong \overline{AC}$ and $\angle ACB \cong \angle BCA$. Thus the parts $\overline{AC}, \overline{BC}$, and $\angle ACB$ of ΔACB are congruent, respectively, to the corresponding parts $\overline{BC}, \overline{AC}$, and $\angle BCA$ of ΔBCA. Therefore, $\Delta ACB \cong \Delta BCA$ by SAS congruence. Therefore, $\angle BAC \cong \angle ABC$.

ANSWER TO QUESTION

1. A 5-by-5 grid with opposite corners removed cannot be covered with dominoes because there are 23 squares, an odd number. But a 9-by-8 grid can be covered with its two opposite corners removed by splitting it into three easily-covered rectangles, one 7-by-8, one 1-by-8, and 1-by-6.

7.1.3 General properties of definitions

In earlier chapters in this book, you have encountered terms for which more than one definition exists. Alternate definitions are of three types. Sometimes one definition may be more convenient in one context than another, e.g., in choosing whether the natural numbers begin at 0 or at 1. Sometimes an alternate definition arises from a desire to emphasize one aspect of a concept rather than another, e.g., when a function is defined as a particular set of ordered pairs (a static, set description of a function) rather than as a mapping or correspondence (a more active, procedural notion). We then want the definitions to be equivalent. At still other times alternate definitions exist because a concept is extended beyond the point where the first definition holds (e.g., when the trigonometric functions sine and cosine are defined beyond measures of acute angles in right triangles).

Geometry is a particularly rich area for the discussion of definitions. In the next section we trace the historical development of the definition of *congruence*. Along the way, we discuss alternate definitions of *parallel lines, trapezoid*, and *isosceles trapezoid* that provide slightly different concepts. To examine why we can alter definitions and why we might wish to do this, we first discuss definitions themselves.

Purposes of definitions

Definitions serve many purposes in addition to their universal *raison d'être* of clarifying what a term or symbol means. A definition brings attention to the idea being defined. It helps distinguish that idea from other closely related ideas. And in mathematics, it serves as a statement from which properties involving the idea may be deduced, or with which something may be seen as an example of the idea.

For instance, a number is defined as *even* if and only if it equals $2k$, where k is an integer. By defining when a number is an *even number*, we bring attention to *evenness*, we help to distinguish it from an odd number, we can deduce properties of even numbers, and we can use the definition as a criterion to determine whether or not a number is even. (See Problem 4.)

Question 1: Explain why 0 is an even number.

Definitions allow more latitude than many people realize. Charles Dodgson, publishing under his pseudonym Lewis Carroll, wrote: "When I use a word—it means just what I want it to mean—neither more nor less."—Humpty Dumpty.[6] Through Humpty Dumpty, Dodgson conveys the contemporary view that definitions in mathematics are, in theory, quite arbitrary. A definition of a new term or symbol need only satisfy the following two criteria:

1. It accurately describes the idea being defined, and
2. It includes only words or symbols commonly understood, defined earlier, or purposely left undefined.

---

[6] Lewis Carroll, Through the Looking Glass, Chapter VI, in *The Annotated Alice*, with an introduction and note by Martin Gardner (New York: Bramhall House, 1960), p. 269.

A third condition is desired but not always practical or feasible:

3. It includes no more information than is necessary.

These criteria are not merely abstract conditions. Outside of mathematics, in law, in sports and games, and in many other pursuits, good definitions satisfy these same criteria, and for good reason. The definition may have serious implications, as defining when human life *begins*, or when a person is *legally dead*, or what constitutes a *hate crime*, or what is treated as *overtime* on a job.

The semantics of definitions

Every definition is, either implicitly or explicitly, an "if-and-only-if" statement, such as the following definition of *scalene triangle*:

> A triangle is *scalene* if and only if no two of its sides have the same length.[7]

This statement is of the form $p \Leftrightarrow q$ (read "p if and only if q"), where $p =$ "A triangle is scalene" and $q =$ "No two sides of a triangle have the same length." Like every other if-and-only-if statement, $p \Leftrightarrow q$ is equivalent to the logical conjunction ($p \Rightarrow q$ and $q \Rightarrow p$), read "p implies q and q implies p".

> $p \Rightarrow q$: If a triangle is scalene, then no two of its sides have the same length.

We call $p \Rightarrow q$ the **meaning** direction of a definition, for it gives a meaning for the term "scalene triangle". Contrast the meaning direction with $q \Rightarrow p$.

> $q \Rightarrow p$: If in a triangle no two sides have the same length, then the triangle is scalene.

We call $q \Rightarrow p$ the **sufficient condition** direction of a definition, for it gives a condition sufficient for a triangle to be scalene.[8]

Even when the words "if and only if" are not explicitly stated in a definition, they may be assumed. For instance, the following definition is equivalent to the preceding $p \Leftrightarrow q$ above: A *scalene triangle* is a triangle no two sides of which have the same length.

Using the word "is" in a definition (as we did in the preceding sentence) exposes an ambiguity in the English language. Consider these two statements, which have identical form.

(1) A natural number is a positive integer.

(2) A prime number is a positive integer.

Statement (1) is a suitable definition of *natural number*. In it, the word "is" is short for "is the same as". Statement (2) is not a suitable definition of prime number, because the sufficient condition direction is not true—not every positive integer is a prime number. In it, the word "is" is short for "is an example of". The ambiguity of the word "is" forces us either to call on our existing knowledge to identify which statement is a definition, or it forces us to signal that something is a definition. In this book, we attempt to use italics or boldface as a signal that we are defining a term or phrase.

---

[7] Recall from Section 2.1.1 that some people prefer to replace "if and only if" by "if" when the context makes clear that a term or phrase is being defined.

[8] A related but slightly different use of the word "sufficient" is found when $p \Rightarrow q$ and we say that p is **sufficient** for q and that q is **necessary** for p.

Alternate definitions

There exist several definitions for many of the most common ideas in mathematics because an idea can be approached in different ways or because people differ on what sufficient conditions they prefer to appear in the definition. For instance, each of the following definitions of *rectangle* can be found in books on Euclidean geometry.

a. A *rectangle* is a parallelogram with four right angles.[9]

b. A *rectangle* is a quadrilateral with four right angles.[10]

c. A *rectangle* is a parallelogram with a right angle.[11]

Definition (a) best fits the intuitive idea of rectangle but it contains more information than is necessary. Definition (b) is useful in that, with this definition, to prove a quadrilateral is a rectangle, you would not need to show that the quadrilateral is a parallelogram. Definition (c) automatically classifies the rectangle and allows us to prove that a figure is a rectangle using only one right angle provided the figure is known to be a parallelogram. It also enables us to see immediately that all properties of parallelograms apply to rectangles.

Other reasons for choosing a particular definition are its generalizability, or extendability. For instance, we might use this definition:

d. A *rectangle* is a parallelogram in which two sides are perpendicular.

Definition (d) could be extended to three dimensions by defining a *rectangular solid* to be a parallelepiped in which three faces lie in mutually perpendicular planes.

Two definitions that define exactly the same idea are called **equivalent**. Equivalence of two definitions cannot be taken for granted but must be proved by showing that the sufficient condition in each definition implies the other. Consequently, definitions can be proved equivalent only if each implies the other within a given system. For instance, with the following arguments we show that definitions (b) and (c) of a rectangle above are equivalent in Euclidean geometry. (In some non-Euclidean geometries, this theorem is not true.)

Theorem A figure is a quadrilateral with four right angles if and only if it is a parallelogram with a right angle.

Figure 9

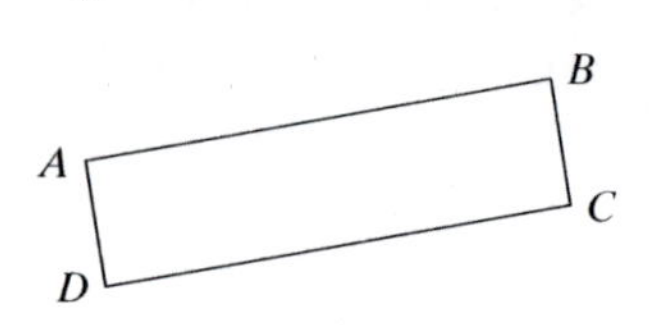

Proof: (b) can be rewritten "A quadrilateral is a rectangle if and only if it has four right angles." (c) can be rewritten "A quadrilateral is a rectangle if and only if it is a parallelogram with a right angle."

To prove (b) $\Rightarrow$ (c), we need to show that if a figure has four right angles, then it is a parallelogram (with a right angle). Suppose all the angles of $ABCD$ (Figure 9) are right angles, so its adjacent sides are perpendicular. Since two lines perpendicular to the same line are parallel, and $\overline{AB}$ and $\overline{CD}$ are both perpendicular to $\overline{BC}$, $\overline{AB} // \overline{CD}$. Since $\overline{BC}$ and $\overline{AD}$ are both perpendicular to $\overline{CD}$, $\overline{BC} // \overline{AD}$. Therefore, by the definition of parallelogram, $ABCD$ is a parallelogram.

To prove (c) $\Rightarrow$ (b), we need to show that if a figure is a parallelogram with a right angle, then it has four right angles. Let $ABCD$ be a parallelogram. Without loss of generality, assume A is the right angle. Then $\overline{AB} \perp \overline{AD}$. From the definition of parallelogram, $\overline{AB} // \overline{CD}$ and $\overline{AD} // \overline{BC}$. Since a line that is perpendicular to one of two parallel lines is perpendicular to the other, $\overline{AB} \perp \overline{BC}$ and $\overline{CD} \perp \overline{AD}$. Then, for the same reason $\overline{CD} \perp \overline{BC}$. Thus $ABCD$ has four right angles. ┘

---

[9] Laurie E. Bass et al., *Geometry* (Needham, MA: Prentice Hall, 1998), p. 90.

[10] Zalman Usiskin et al., UCSMP *Geometry* (Needham, MA: Prentice Hall, 2002), p. 316.

[11] Richard Rhoad, George Milauskas, and Robert Whipple. *Geometry for Enjoyment and Challenge*, New Edition (Evanston, IL: McDougal, Littell, 1991), p. 236.

Equivalent definitions are found throughout mathematics. For instance, in Section 5.2.2, we defined the *greatest common factor of two positive integers* a and b to be that positive integer g with the following properties:

i. g is a factor of both a and b;

ii. If d is any integer that is a factor of a and b, then d is a factor of g.

An equivalent definition is suggested by the words "greatest common factor": The *greatest common factor of two positive integers* a and b is the greatest integer that divides both a and b. In the Problems, you are asked to prove that these definitions are equivalent.

Related definitions

Some definitions serve to classify a given collection of objects into smaller collections. To this end, the following definition is useful: A **partition** of a set S is a set of disjoint subsets of S whose union is S. Then we can say, for example,

The integers are partitioned into those that are *even* and those that are *odd*.

The triangles are partitioned into *scalene triangles* and *isosceles triangles*.

The angles with measures less than 180° are partitioned into those that are *acute*, *right*, and *obtuse*.

The real numbers are partitioned into the *rationals* and the *irrationals*.

Question 2: Give a partition of all triangles based on the measures of its largest angle.

Question 3: Why can't the set of all triangles be partitioned into the set of scalene, isosceles, and equilateral triangles?

Definitions also may serve to *nest* objects. For instance, the usual definition of *rational number* is as a real number that can be written as a quotient of integers. Then, since $\frac{n}{1} = n$, every integer can be written as a rational number. In an analogous way, every even integer is an integer. These definitions create a sequence of nested sets, each of which is a subset of those that come later: the set of even integers $\subset$ the set of integers $\subset$ the set of rational numbers $\subset$ the set of real numbers.

Visually, partitions and nested sets are strikingly different, as the Venn and tree diagrams in Figure 10 show.

Figure 10

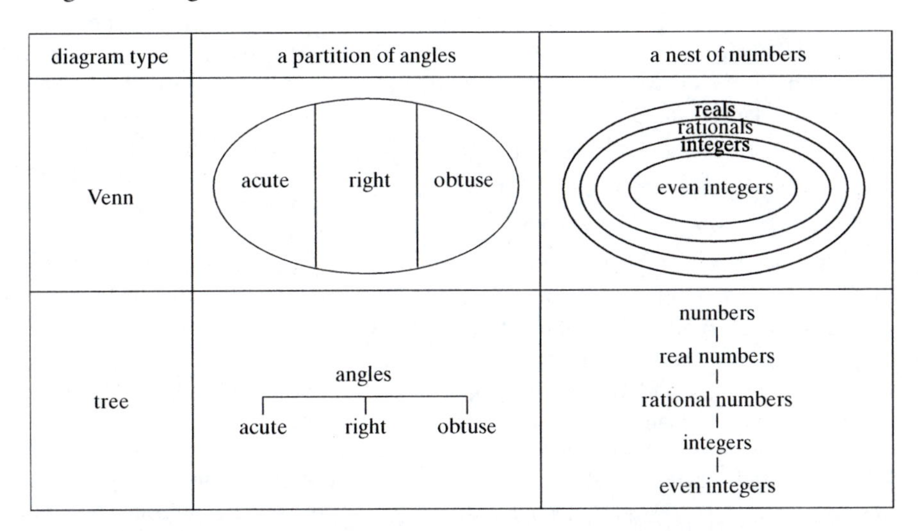

The distinction between partitions and nests can be illustrated by comparing squares and rectangles. Young children often come to believe (sometimes through instruction) that a figure with four right angles is a square if all its sides have the

same length, and is a rectangle otherwise. With this conception, rectangles and squares partition the set of figures with four right angles. On the other hand, high school and college texts invariably define *square* in such a way that every square is a rectangle. Thus squares are conceptually nested inside rectangles. Nests are particularly useful because any property of all elements of a set is a property of all elements of its subsets. Since every property of all rectangles is also a property of all squares, mathematicians prefer to consider a square as a rectangle.

7.1.3 Problems

1. a. Pick one of the terms i-iv below and discuss at least two possible definitions for it that fit the mathematical criteria for a good definition.
   b. Identify the words that are commonly understood, defined earlier (or elsewhere), or purposely left undefined.
   c. Discuss the implications of the different definitions.
      i. when human life *begins*
      ii. when a person should be considered *dead*
      iii. *hate crime*
      iv. *pornography*

2. Write the meaning and sufficient-condition parts of each definition as if-then statements.
   a. A *dodecagon* is a polygon with twelve sides.
   b. If a linear equation is of the form $Ax + By = C$, then it is said to be in *standard form*.

3. A statement is given with the word "is". Is the statement a suitable definition for the term in italics? If so, indicate the meaning direction of the definition. If not, indicate why not.
   a. A polygon with twelve sides is a *dodecagon*.
   b. A number divisible by 4 is an *even number*.
   c. A *zero of a real function f* is a number x with $f(x) = 0$.
   d. An *equilateral triangle* is an isosceles triangle.

4. Use the definition of *even number* to prove or disprove each statement for all even numbers.
   a. The sum of any three even numbers is even.
   b. The product of any even number and any integer is even.
   c. The quotient of two even numbers is even.
   d. The nth power of an even number is divisible by 2^n.

5. Can the following definition of *line* be extended to three dimensions? If so, how? A *line* is a set of points of the form $\{(x, y): ax + by = c, a, b, c \in \mathbf{R}, a^2 + b^2 \neq 0\}$.

6. Consider the following standard definition of *circle*: A *circle* is the set of points in a plane at a given distance from a given point. Extend it to three dimensions in two ways to define two different three-dimensional figures.

7. Explain why these definitions of *parallel lines* are not equivalent in Euclidean geometry.

   Definition 1: Two lines in the same plane are *parallel lines* if and only if they have no points in common.

   Definition 2: Two lines in the same plane are *parallel lines* if and only if they are both vertical or they have the same slope.

8. Consider these two definitions of *trapezoid*, both found in standard high school geometry texts.

   Definition 1: A *trapezoid* is a quadrilateral with exactly one pair of parallel sides.

   Definition 2: A *trapezoid* is a quadrilateral with at least one pair of parallel sides.

   a. Are they equivalent? Why or why not?
   b. Explain how trapezoids are related to parallelograms under each definition.

9. Consider these two definitions of *isosceles trapezoid*, both found in standard high school geometry texts.

   Definition 1: An *isosceles trapezoid* is a trapezoid with exactly one pair of parallel sides and whose nonparallel sides are congruent.

   Definition 2: An *isosceles trapezoid* is a trapezoid with one pair of parallel sides and with the other pair of sides congruent.

   a. Are they equivalent? Why or why not?
   b. Explain how isosceles trapezoids are related to rectangles under each definition.
   c. How are these definitions related to definitions 1 and 2 of *trapezoid* given in Problem 8?
   d. A quadrilateral has vertices at $(-a, 0)$, $(a, 0)$, (b, c), and (d, c). Under definition (1), what conditions must b and d satisfy if the quadrilateral is an isosceles trapezoid? Under definition 2, what conditions must b and d satisfy if the quadrilateral is an isosceles trapezoid?

10. In *Mathematique Moderne*, a text for Belgian students written by Georges Papy (1966), a *direction* is defined as a partition of the plane into lines. Then two lines are defined to be *parallel lines* if and only if they are in the same direction. Is this definition equivalent to either definition 1 or 2 of *parallel lines* given in Problem 7? Explain your answer.

11. Here are some types of natural numbers. Create a nest or a partition involving them.
    a. prime, composite, the number 1
    b. divisible by 2, divisible by 3, divisible by 6
    c. of form $2k + 1$, of form $4k + 1$, of form $4k + 3$, where k is a nonnegative integer

12. Create a nest of 5 sets whose largest set is the set of all polygons. For each set but the largest set in the nest, write a definition that precisely describes the objects in that set in terms of the objects in the next larger level.

13. In the *Merriam-Webster's Collegiate Dictionary*, tenth edition, a *rhomboid* is defined as "a parallelogram with no right angles and with adjacent sides of unequal length." Compare rhomboids with rhombi (quadrilaterals with four sides of equal length) and rectangles in terms of nesting or partitioning the set of all parallelograms.

14. Refer to the definitions of Book I of Euclid's *Elements* in Section 7.1.1.

a. Find an example of a nest and an example of a partition in those definitions.

b. Does Euclid seem to prefer nests or partitions in his related definitions?

15. Explain how congruence sets modulo n partition the integers. (Refer to Section 6.1.1.)

16. Create a Venn diagram of numbers that includes all the types in Figure 10 and also algebraic numbers, complex numbers, and numbers representable by finite decimals.

17. Prove that the two definitions of "greatest common factor" discussed on page 293 are equivalent.

ANSWERS TO QUESTIONS

1. 0 is an integer. Since $0 = 2 \cdot 0$, 0 is also twice an integer. Thus, by the definition of *even number*, 0 is an even number.

2. A triangle is an *acute triangle*, a *right triangle*, or an *obtuse triangle* based on its largest angle being acute, right, or obtuse.

3. Because equilateral triangles are also isosceles triangles, they overlap and so cannot both be in the same partition.

7.1.4 Definitions of congruence from Euclid to modern times

A good idea with which to illustrate the importance of appropriate definitions is *congruence*, which is so fundamental to Euclidean geometry. We think intuitively of congruent figures as being those that have the same size and same shape. In the language of mathematical modeling, we want congruence to be the mathematical model for the relationship between figures with the same size and shape. Examples in the real world abound: copies of the same page made by a duplicating machine; a particular part of a mass-produced product; coins minted in the same place at the same time; buttons for the same shirt; the patterns that repeat in the repeating patterns on clothes or in friezes on buildings; letters of the alphabet written in the same font and same font size; and so on.

What does the word "congruent" convey? Are congruent figures identical, but just in different positions? What is meant, then, by "identical"? Can figures of infinite extent be congruent? What, then, do we mean by "size"? Is an object congruent to its mirror image? How does that fit in with being in a different position? It is perhaps surprising that such a simple idea is not so easy to describe in precise mathematical language. Many of the ideas in Section 7.1.3 can be applied in examining the various ways in which mathematicians have dealt with the idea of congruence.

Congruence as superposition

Euclid had a general conception of congruence that implicitly assumed geometric figures could be moved without changing their size or shape. The fourth of the five common notions (axioms) in Euclid's *Elements* is "Things which coincide with one another are equal to one another." Euclid used this common notion to justify the idea that equal figures could be made to coincide. Today we use the word "congruent" in place of "equal".

This axiom was used very quickly—in the fourth theorem of the first book of *Elements*—to deduce SAS triangle congruence. The word "coincide" was interpreted to mean that two figures coincide if one could be moved onto the other with a rigid motion; that is, that figures could be moved without changing their size or shape. Since the movement was usually intended to place one figure on top of another, Euclid's method of proof is called *superposition*.

Proofs involving superposition were questioned as early as the 16th century[12] and became to be considered as invalid in the 19th century, when the postulate structure underlying geometry was called into question with the discovery of non-Euclidean geometry. In the early 20th century, they were further called into question by Einstein's theory of relativity, since under that theory figures can change size if moved.[13]

Separate definitions of congruence for each figure

Starting in the 1880s, mathematicians such as Moritz Pasch (1843–1930), Giuseppe Peano (1858–1932), and Mario Pieri (1860–1913) developed rigorous postulate systems for geometry to correct the logical defects in Euclid's work. (We discussed some of these defects briefly in Section 7.1.2. See Unit 11.1 for further discussion of the defects in Euclid's system and the modern axiomatic systems that were developed to correct these deficiencies.) The most famous of these systems was developed by David Hilbert in 1899. Hilbert treated "congruence" as an undefined term and stated postulates that dealt with congruence of segments, other postulates that dealt with congruence of angles, and still other postulates that dealt with congruence of triangles. Thus, though we use the single phrase "is congruent to" and the single symbol $\cong$, Hilbert actually had three different kinds of congruence.

Not long after Hilbert's work, Albert Einstein gave geometers a practical reason for questioning superposition, for he showed that lengths in a moving object are perceived differently by an observer standing still. Also, there could be no instantaneous movement, since objects could not move faster than the speed of light.

For a half century these developments had little effect on geometry textbooks in the United States. Until the late 1950s, Euclid's characterization of congruence as superposition appeared in almost all geometry texts and was used to prove the SAS and other triangle congruence propositions.

Like Hilbert, those who wrote the reform texts of the late 1950s chose to distinguish congruent segments, congruent angles, and congruent triangles. The geometry text of the School Mathematics Study Group (SMSG), the largest and most famous of the "new math" projects, departed from tradition by using a combination of the ideas of Hilbert and George David Birkhoff. Birkhoff's contribution was to utilize the algebraic properties of real numbers in geometry. This enabled him to define congruence of certain figures in terms of real numbers. Thus came the SMSG definitions:

> Two *segments are congruent* if and only if they have the same length.
>
> Two *angles are congruent* if and only if they have the same measure.
>
> Two *triangles are congruent* if and only if there is a correspondence between their vertices such that the three corresponding sides are congruent and the three corresponding angles are congruent.

Some books adopted the following approach to avoid congruence of segments and angles. There is only one definition of congruence in such books.

> Two *triangles are congruent* if and only if there is a correspondence between their vertices such that the three corresponding sides have the same length and the three corresponding angles have the same measure.

---

[12] See Robin Hartshorne, *Geometry: Euclid and Beyond*, p. 32.

[13] See Jim Tattersall and Shawnee McMurran, "An Interview with Dame Mary L. Cartwright, D.B.E., F.R.S.," *The College Mathematics Journal* 32(4), p. 245 (September 2001).

Since the 1960s, these two definitions of triangle congruence have been common in geometry texts in the United States.

A few books define congruence of figures other than segments, angles, and triangles. The definition for each figure is based on what seems to be the most intuitive way to think of the figure. For instance, here is a definition of *congruent circles* found in one book.

> Two *circles are congruent* if and only if their radii are congruent.[14]

The creation of separate definitions of congruence for individual figures makes it seem as if there is no general notion of congruence, for the definitions are quite different for each class of figures. As a result, in these approaches the helpful intuition gained from Euclid's superposition idea is lost. On the other hand, these individual definitions enable quick deduction of some theorems.

From correspondence to transformation

Not all geometers used Hilbert's approach. Pieri, for example, made Euclid's superposition rigorous, creating an axiom system in 1899 that allowed congruence of all figures to be described in terms of "motions" that superimpose one figure upon another. We now explore this idea of congruence.

Triangles ABC and ZWM in Figure 11 have been drawn so that the corresponding angles A and M, B and W, and C and Z have the same measure and corresponding sides $\overline{AB}$ and $\overline{MW}$, $\overline{BC}$ and $\overline{WZ}$, and $\overline{AC}$ and $\overline{MZ}$ have the same length. Thus, by the SMSG definition stated above, the triangles are congruent with the correspondence $ABC \leftrightarrow MWZ$.

Figure 11

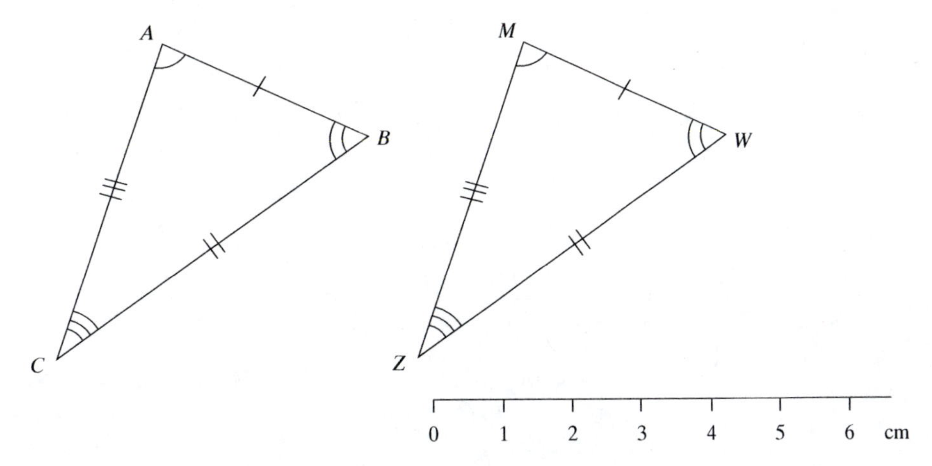

However, stating only the corresponding points disguises something quite basic about these two triangles. They are not merely congruent, but they are located relative to each other in a particular position. Every point on ΔMWZ (not just the vertices) is 5 cm to the right of a corresponding point on ΔABC. Euclid's notion of moving ΔABC to superimpose it on ΔMWZ can be characterized as a slide of 5 cm to the right. Or, if we superimpose the two triangles on a coordinate plane, we could say that for every point (x, y) on ΔABC, there is a corresponding point $(x + 5, y)$ on ΔMWZ.

We have now recast the correspondence of the three vertices as a function (or mapping) whose domain is the set of points on ΔABC and whose range is the set of points on ΔMWZ, with the rule described either in words or algebraically. Furthermore, the rule

---

[14] Alan G. Foster, Jerry J. Cummins, and Lee E. Yunker, *Merrill Geometry* (Columbus, OH: Merrill, 1984), p. 328.

for this function can be applied to all points in the plane. When this is done in such a way that the function is 1-1 and onto, we have defined a *transformation*. (Recall that a function T is 1-1 if, for all a and b in its domain, $T(a) = T(b)$ implies $a = b$.) If we call the preceding transformation T, we can write $T(\Delta ABC) = \Delta MWZ$ and we can call $T(x, y) = (x + 5, y)$ an equation for T. This equation provides an analytic description of the transformation T.

Definition A **transformation of the plane** or **plane transformation** is a 1-1 function with the plane as its domain and codomain.

When T is a transformation and $T(\alpha) = \beta$, we say the transformation **maps** α onto β. We call α the **preimage** and β the **image**. The particular transformation of the plane just used is of the type called a *slide* or a *translation*, and we call ΔMWZ the *translation image* of ΔABC. In Euclidean 2-space, a transformation of the plane is a 1-1 mapping from $\mathbf{E}^2$ onto $\mathbf{E}^2$. In the Cartesian plane, a transformation of the plane is a 1-1 mapping from $\mathbf{R}^2$ onto $\mathbf{R}^2$.

Recasting the idea of congruence in terms of transformations provides a more dynamic view of congruence that is related to superposition and can apply to all figures.

Congruence in terms of transformations

A definition of congruence should provide that a figure and its rotation image are congruent, as in Figure 12.

Figure 12

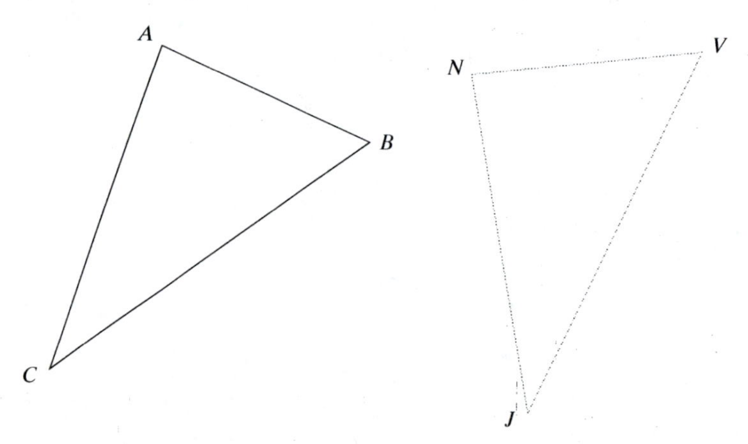

Euclid and other geometers since ancient times have also wanted triangles like ΔABC and ΔPHI in Figure 13 to be congruent. These triangles are mirror images of each other.

Figure 13

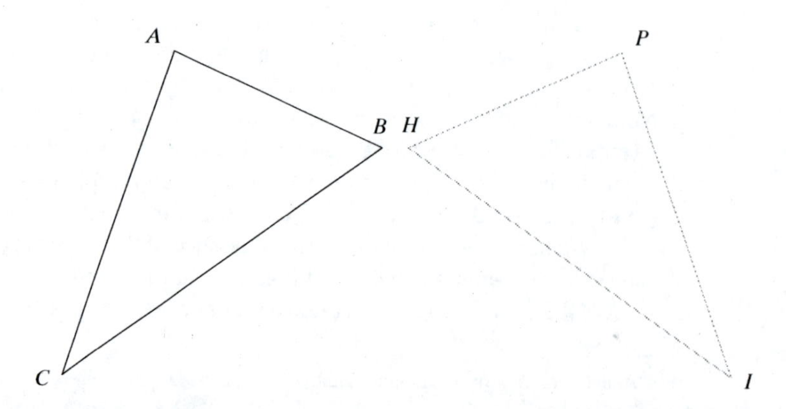

It is not as obvious why the concept of congruence should include figures that are mirror images of each other, but there are some compelling reasons. In traditional proofs that the base angles of an isosceles triangle have the same measure, and in the proofs of many other theorems, congruent triangles that are mirror images of each other are involved. A second reason is that ΔABC can be rotated in three dimensions to fall upon ΔPHI, so superposition is possible.

But the strongest reason is mathematical. Surely we would like ΔABC to be congruent to itself. Think of drawing ΔABC from Figure 13 on a storefront window. If we view the triangle from inside the store and then from outside, what we will see from one direction is ΔABC, as it looks in Figure 13. What we will see from the other direction will look more like ΔPHI.

When viewed from the same position, triangles ΔABC and ΔPHI in Figure 13 differ in *orientation*, a relation among figures that is difficult to define but easy to describe intuitively. The vertices of these triangles correspond as follows: A to P, B to H, and C to I. Imagine walking around ΔABC from A to B to C. Then the interior of the triangle will be on your right. If you imagine walking around the corresponding points P to H to I in the corresponding order, the interior of ΔPHI will be on your left. These different directions always appear when figures have different orientations.

This description of orientation is not a precise definition because it assumes we are viewing the triangle as if it is on a table below us or on a wall in front of us. However, it is possible to assign an orientation to a triangle in a plane without regard to the position of the viewer or the position of the plane in space if the plane is coordinated (see Problem 8).

In three dimensions, the right and left hands of most people can provide a good physical example of congruent figures with different orientation. Figure 14 shows two candidates for congruence with some corresponding points denoted: P corresponds to P', Q to Q', and so on.

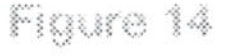
Figure 14

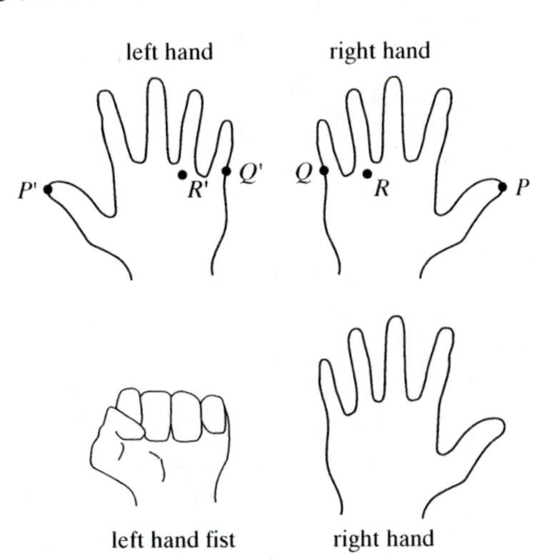

If a person has lost a finger on one hand only, then clearly the figures cannot be congruent. This indicates that the idea of a *one-to-one correspondence* could naturally be involved in a definition for congruence. Also, if one hand is in a fist while the other is open, as in the bottom two drawings in Figure 14, there is a one-to-one correspondence, yet the figures do not fit our notion of congruence. Thus the *distance* between corresponding points is critical, but the orientation of two congruent figures may be different. From these ideas comes a general definition of congruence.

Definitions A transformation T of the plane is a **congruence transformation** (or **distance-preserving transformation** or **isometry**[15]) if and only if, for any points P and Q of the plane, the distance between P and Q is equal to the distance between $T(P)$ and $T(Q)$.

Two figures α and β in the plane are **congruent**, denoted $\alpha \cong \beta$, if and only if there is a congruence transformation T such that $T(\alpha) = \beta$.

If a figure's orientation is unchanged under a congruence transformation, then the transformation is called a **direct congruence** and the figure and its image are said to be **directly congruent**. Otherwise the transformation is called an **opposite congruence** and the figures are **oppositely congruent.**

The symbols $\sim$ and $\equiv$ for similarity and congruence were first used by Leibniz in 1679. The symbol $\cong$ first appears about 100 years later.[16]

In many recent mathematics curriculum development projects in the United States and in many countries of the world, congruence is defined for students in terms of properties of transformations. Of the many reasons for using transformations, two pertain to congruence. First, transformations enable one definition to be given for congruence, a definition that applies to all figures. Second, transformations enable a myriad of applications of congruence (e.g., in frieze patterns and architectural reliefs, in mass-produced parts for machines and furniture, and to the graphs of functions and relations) that are accessible precisely because of the generality of the definition.

A plane transformation may preserve some distances in geometric figures yet not be a congruence transformation. Figure 15 pictures a *horizontal stretch* H mapping face α onto face β. Distance is not preserved because the distance between the points A and B is not equal to the distance between the corresponding points $H(A)$ and $H(B)$. So the faces are not congruent.

Figure 15

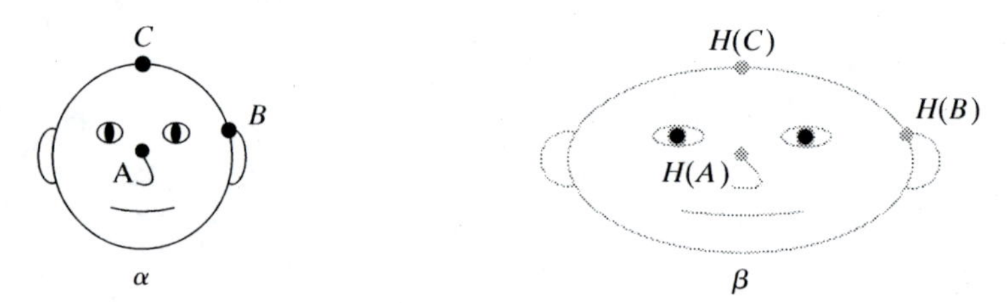

It is the case that H preserves distances between two points on the same vertical line. (Compare AC with the distance between $H(A)$ and $H(C)$.) But a transformation must keep distances equal between *all* possible pairs of preimages and images in order to be called distance-preserving. So H is not a congruence transformation.

Isometries can be characterized by the fact that they map segments onto segments of the same length. To prove this theorem, we use the definition that point C is **between** two other points A and B if and only if $AC + CB = AB$, and **line segment** $\overline{AB}$ is the set of points including A, B, and all points between A and B.

Theorem 7.1 Under an isometry, the image of a line segment is a line segment of the same length.

Proof: Let T be any isometry and $\overline{AB}$ be any segment with $T(A) = A'$ and $T(B) = B'$. Because T is an isometry, $AB = A'B'$. So we need only to show that

[15] Still another synonym is *rigid motion*.

[16] See Florian Cajori, *A History of Mathematical Notations* (LaSalle, IL: Open Court, 1928), p. 415.

$T(\overline{AB}) = \overline{A'B'}$. That is, we need to show that a point C is on $\overline{AB}$ if and only its image $T(C) = C'$ is a point on $\overline{A'B'}$. But we already know $T(A) = A'$ and $T(B) = B'$, so we need only consider points C and C' between those points.

Figure 16

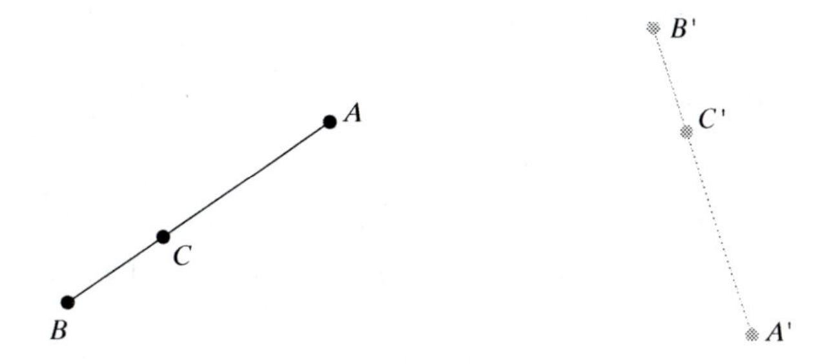

To prove $T(\overline{AB}) \subset \overline{A'B'}$, suppose C is on $\overline{AB}$ and between A and B as shown in Figure 16. Then $AC + CB = AB$. Since T preserves distance, $A'C' = AC$ and $C'B' = CB$. Consequently, $A'C' + C'B' = A'B'$, so C' is between A' and B', and so C' is on $\overline{A'B'}$.

To prove $\overline{A'B'} \subset T(\overline{AB})$, we reverse the argument. Suppose C' is on $\overline{A'B'}$ and between A' and B'. Consequently, $A'C' + C'B' = A'B'$. Since T preserves distance, it must be that its preimage C satisfies $AC + CB = AB$, and so C is between A and B. ∎

Betweenness is the property of one point being between two others. The proof of Theorem 7.1 has shown that isometries preserve betweenness.

Corollary: Isometries preserve betweenness.

Because of the corollary, under an isometry, the image of a line is a line, the image of a ray is a ray, and the image of an angle is an angle (see Problem 6).

Because a transformation is 1-1, it is always possible to determine the preimage from its image. If the correspondence is reversed, the transformation so formed is called the *inverse* of the original transformation. For example, the inverse of the horizontal stretch H maps β onto α, $H(A)$ onto A, $H(B)$ onto B, etc. We might call the inverse a horizontal squeeze! As another example, the inverse of the translation in Figure 11 is the translation in which each point on ΔMWZ is mapped onto the point on ΔABC 5 units to its left. More generally, if $\mathbf{b}$ is a given vector, then the inverse of the translation $T_{\mathbf{b}}$ is the translation $T_{-\mathbf{b}}$.

The above definition of congruence transformations generalizes directly to three dimensions. Thus, congruence describes the relationship between xerox copies of the same document, almost all of the chairs in an auditorium or classroom, and parts in industry made from the same mold, as well as the congruent figures found in all geometry courses.

Some people believe that 3-dimensional figures with different orientations should not be congruent, appealing to an intuition that says the congruent figures, in the language of Euclid, should be able to be made to coincide. Of course, no 3-dimensional figures—even those of the same orientation—can be made to coincide in the real world without destroying them. Even chairs that can be stacked cannot be made to coincide. Yet no working geometer can deal with a notion of congruence that does not allow figures of different orientations to be congruent. As shown in the next sections, both the mathematical theory and the applications of congruence would be greatly hampered by such a restriction.

7.1.4 Problems

1. Using the definition of *congruent circles* in this section and your knowledge of high school geometry, prove:

a. Congruent circles have the same area.

b. If two circles have the same area, then they are congruent.

2. Make up a definition of congruence for quadrilaterals akin to any in this section for triangles. Then, under your definition, prove or disprove each statement.

a. If two rectangles have the same length and width, then they are congruent.

b. If two trapezoids have bases of the same lengths and the same height, then they are congruent.

c. An SASAS quadrilateral theorem

3. a. Prove that $T(x, y) = (x + 5, y)$, the transformation of Figure 11, is a congruence transformation.

b. Give an equation describing T^{-1}, the inverse of T.

4. In each part, a transformation T is described by giving the coordinates of the image of each point. Prove that T is a congruence transformation or prove that it is not.

a. $T(x, y) = (x + 4, y - 10)$

b. $T(x, y) = (4x, -10y)$

c. $T(x, y) = (-y - 4, x + 10)$

5. Prove that the composite of two isometries is an isometry.

6. a. If three distinct points A, B, and C are on a line, then exactly one is between the other two. Use this statement to prove that the image of a line under an isometry is a line.

b. A *ray* $\overrightarrow{AB}$ is defined as the set consisting of $\overline{AB}$ and all points C on the line $\overleftrightarrow{AB}$ such that B is between A and C. Prove that the image of a ray under an isometry is a ray.

c. Suppose an *angle* ABC is defined as the union of the rays $\overrightarrow{BA}$ and $\overrightarrow{BC}$. Prove that the image of an angle under an isometry is an angle.

7. Let T be a transformation in Euclidean geometry defined as follows. For one given point C, $T(C) = C$, and for every other point P, $T(P)$ is the midpoint of $\overline{PC}$. Prove that this transformation is not an isometry.

8. **Defining orientation.** Suppose z_A, z_B, and z_C are three distinct complex numbers representing the vertices A, B, and C of ΔABC in the complex plane.

a. Let $b = AC$ and $c = AB$. Explain why

$$\text{Im}\left(\frac{z_C - z_A}{z_B - z_A}\right) = \pm\frac{b}{c}\sin A,$$

and the sign is positive if and only if the vertex C is to the left of the directed line from A through B.

b. Use part **a** to explain why C is to the left of the directed line from A through B if and only if

$$\text{Im}\left(\frac{z_C - z_A}{z_B - z_A}\right) > 0,$$

and use this result to define clockwise and counterclockwise orientation of ΔABC with vertices in that order.

c. Recall that $\bar{z} = x - iy$ is the complex conjugate of $z = x + iy$. Prove that

$$\text{Im}\left(\frac{z_C - z_A}{z_B - z_A}\right) = \frac{1}{c^2}\,\text{Im}((z_C - z_A)(\overline{z_B} - \overline{z_A})).$$

d. Let $z_A = x_A + iy_A$, $z_B = x_B + iy_B$, and $z_C = x_C + iy_C$. Prove that

$$\text{Im}((z_C - z_A)(\overline{z_B} - \overline{z_A})) = \text{determinant of}\begin{bmatrix} x_A & x_B & x_C \\ y_A & y_B & y_C \\ 1 & 1 & 1 \end{bmatrix}.$$

e. Explain how the determinant of part **d** tells whether a triangle in the coordinate plane is clockwise or counterclockwise oriented.

Unit 7.2 The Congruence Transformations

In this unit we examine the various types of congruence transformations both synthetically in the Euclidean plane $\mathbf{E}^2$ and analytically in the coordinate plane $\mathbf{R}^2$.

7.2.1 Translations

Terminology and notation

Informally, a translation "slides" all points in the plane through a given distance in a given direction. The direction and distance of the slide can either be specified by a vector $\mathbf{b}$ or by a directed line segment $\overrightarrow{AB}$ that represents $\mathbf{b}$, or as we shall soon see, by a complex number b. In Figure 17, β is the image of α under the translation specified by the directed line segment $\overrightarrow{AB}$. We call this translation $\mathbf{T}_{\overrightarrow{\mathbf{AB}}}$ or $\mathbf{T}_\mathbf{b}$.

Figure 17

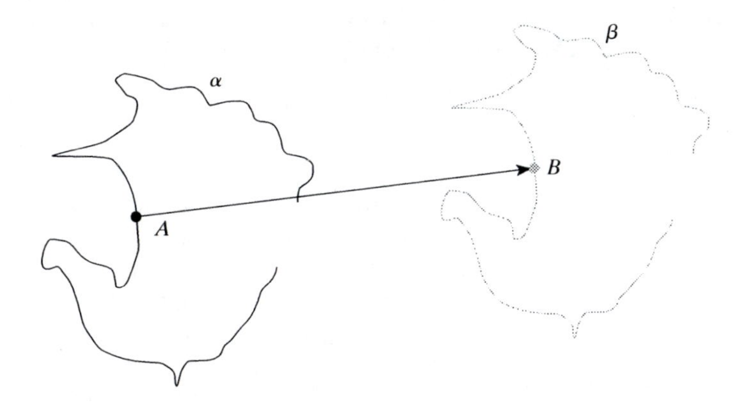

A formal definition of any particular transformation depends on identifying the image of any preimage point. Beyond this constraint, the specific definition depends on the mathematical system in which you are operating and what mathematics you have at your disposal. In this section, we provide three different definitions for translations, all deriving from the idea of translation as identified with a vector. The first definition is synthetic, using the geometry of the plane. The second is analytic, in the coordinate plane. The third is also analytic, using complex numbers.

Definition **(synthetic):** Let $T_{\overrightarrow{AB}}$ be the translation specified by the directed line segment $\overrightarrow{AB}$, and let $C' = T_{\overrightarrow{AB}}(C)$. If C is not on $\overleftrightarrow{AB}$, then C' is the point such that $BACC'$ is a parallelogram (see Figure 18). If C is on $\overleftrightarrow{AB}$, then C' is the point such that $C'C = BA$ and $C'B = CA$ (see Figure 19 and Question 2).

Figure 18

Question 1: Trace Figure 18. Describe the precise location of a point C^* other than C' with $C^*C = BA$ and $C^*B = CA$.

Question 2: Trace Figure 19, in which A, B, and C are collinear. Locate the point C' such that $C'C = BA$ and $C'B = CA$.

Figure 19

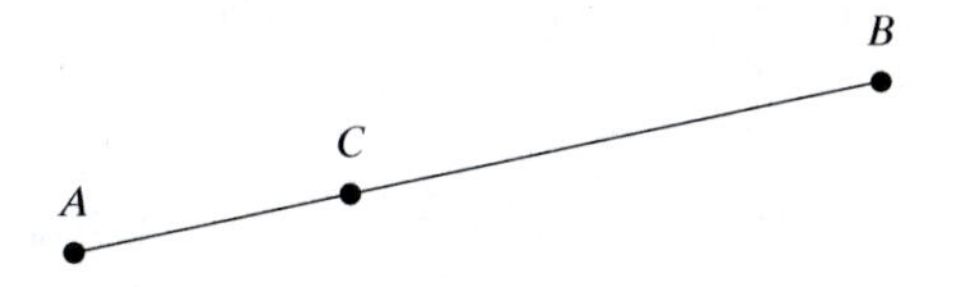

If **b** is the vector represented by the directed line segment $\overrightarrow{AB}$, then we know by its synthetic definition that the translation $T_{\mathbf{b}}$ has the property that, for any point P in the plane, $T_{\mathbf{b}}(P) = P'$ if and only if the vectors $\overrightarrow{PP'}$ and $\overrightarrow{AB}$ are parallel and equal in length. Translating this result into coordinate form, if $P = (x, y)$, $P' = (x', y')$, $A = (p, q)$, and $B = (r, s)$, then from addition of vectors, $x' - x = r - p$ and $y' - y = s - q$. From this comes an analytic definition of this translation.

Definition **(analytic):** Let $T_{\overrightarrow{AB}}$ be the translation specified by the directed line segment $\overrightarrow{AB}$, where $A = (p, q)$ and $B = (r, s)$. Then $T_{\overrightarrow{AB}}(x, y) = (x', y') = (x + h, y + k)$, where $h = r - p$ and $k = s - q$.

That is, when $\mathbf{b} = \overrightarrow{AB}$, the translation T_b is the vector form of $T_{\overrightarrow{AB}}$ with $\mathbf{b} = (h, k)$. We sometimes call this translation $T_{h,k}$; it has the effect of sliding each point of the plane h units to the right and k units up. Thus $T_{h,k}(x, y) = (x + h, y + k)$. We call h and k the **horizontal component** and **vertical component** of the translation by the vector **b**. If h is negative, then the slide is to the left. If k is negative, the slide is down. If $h = 0$, then T is a vertical translation. If $k = 0$, then T is a horizontal translation. If both $h = 0$ and $k = 0$, then T maps each point onto itself. It is the **identity transformation** or **the zero translation**.

In the complex plane, each complex number b defines a translation $T_b: \mathbf{C} \to \mathbf{C}$ by $T_b(z) = z + b$, for all $z \in \mathbf{C}$. The real and imaginary parts h and k of the complex number $b = h + ik$ play the roles that h and k played in the translation described by the vector $\mathbf{b} = (h, k)$. That is, h is the horizontal component and k the vertical component of the translation.

In Problem 5, you are asked to prove that translations are congruence transformations. As such, any figure in the plane and its image under a translation are congruent.

Composites of translations

Translations are 1-1 functions (why?) and, like all functions, they can be composed. Composition of translations is often employed to teach addition of positive and negative numbers. For example, $-5 + 13$ can be thought of as a slide of 5 units to the left composed with (followed by) a slide of 13 units to the right. The resulting composite is a slide 8 units to the right. So although translations are often first encountered after positive and negative numbers have been met, they may be more basic.

If the two-dimensional translation T_{h_1,k_1} is followed by T_{h_2,k_2}, then the composite slides h_1 and then h_2 in the horizontal direction, for a total slide of $h_1 + h_2$ in the horizontal direction. The composite slides k_1 and then k_2 in the vertical direction, for a total slide of $k_1 + k_2$ in the vertical direction. This verbal argument can be written algebraically. For any point (x, y) in the plane,

$$T_{h_2,k_2} \circ T_{h_1,k_1}(x, y) = T_{h_2,k_2}(x + h_1, y + k_1)$$ (definition of function composition and translation T_{h_1,k_1})

$$= ((x + h_1) + h_2, (y + k_1) + k_2)$$ (definition of translation T_{h_2,k_2})

$$= (x + (h_1 + h_2), y + (k_1 + k_2))$$ (associative property of addition)

$$= T_{h_1+h_2,k_1+k_2}(x, y).$$ (definition of translation T)

That is, the composite of two translations is a translation whose components are the sum of the corresponding components of the two translations.

Composition of translations is easily described with complex numbers. If we define the slides using the complex numbers b_1 and b_2, then for all complex numbers z,

$$(T_{b_2} \circ T_{b_1})(z) = T_{b_2}(T_{b_1}(z)) = T_{b_2}(z + b_1) = (z + b_1) + b_2 = T_{b_1+b_2}(z).$$

Thus, the composite of two translations determined by two complex numbers b_1 and b_2 is the translation determined by the complex number $b_1 + b_2$.

Question 3: Explain why composition of translations is commutative.

Composition of translations is also easy to describe geometrically. Let T_1 be the translation $\overrightarrow{AB}$ and T_2 be the translation $\overrightarrow{BC}$. Then $T_2 \circ T_1(A) = T_2(B) = C$. Since we know that $T_2 \circ T_1$ is a translation, and $T_2 \circ T_1$ maps A onto C, $T_2 \circ T_1$ must be the translation $\overrightarrow{AC}$. This composition is pictured in Figure 20.

Figure 20

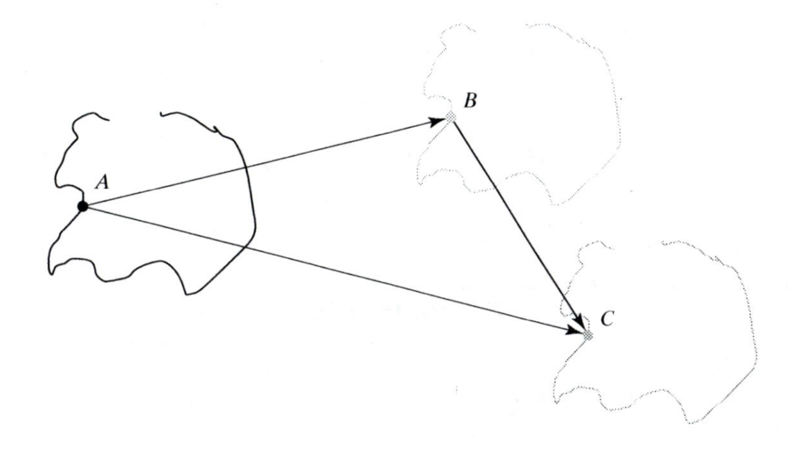

Composition of translations as vector or complex number addition

In Figure 20, the picture looks much like the addition of vectors in the plane and the addition of complex numbers in the complex plane. The reason is simple: The correspondence $T_b \leftrightarrow b$ between the system $\langle \mathbf{T}, \circ \rangle$ of all translations of the plane with the operation of composition and the system $\langle \mathbf{C}, + \rangle$ of complex numbers (or, equivalently, with the coordinate plane $\langle \mathbf{R}^2, + \rangle$), with the operation of addition is an isomorphism. That is, $T_b \leftrightarrow b$ "preserves" operations. This means that, if we are studying only translations (and not other transformations of the plane), we can think of them both as vectors in a plane and as complex numbers. The identity transformation corresponds to the zero vector or to the complex number 0, the opposite vector $-\mathbf{b}$ of a given vector $\mathbf{b}$ corresponds to the inverse of the translation $T_\mathbf{b}$, as does the opposite $-b$ of the complex number b. These isomorphisms allow us to use the rich algebraic structures of plane vectors and complex numbers to study the geometry of translations.

This discussion can be summarized in the language of structures, as follows.

Theorem 7.2 The set $\mathbf{T}$ of all translations of the plane, with composition, forms a group $\langle \mathbf{T}, \circ \rangle$ that is isomorphic to $\langle \mathbf{C}, + \rangle$, the set of complex numbers with addition, and to $\langle \mathbf{R}^2, + \rangle$, the set of all two-dimensional vectors with addition.

7.2.1 Problems

1. Let $A = (-4, 7)$, $B = (1, 0)$, and $C = (5, 6)$.

a. Find the vertices of the image of ΔABC under the translation $T_{2,-5}$.

b. To what complex number does this translation correspond?

2. Let $z = 1 + 2i$ and let $w = 3 + 5i$.

a. Find $T_z(w)$, $T_{z+\bar{z}}(w)$, and $T_{z-\bar{z}}(w)$.

b. Explain why, for any complex number z, $T_{z+\bar{z}}$ is a horizontal translation and $T_{z-\bar{z}}$ is a vertical translation.

3. If T is a translation and $T(A) = A'$, then the distance AA' is the **magnitude** of the translation.

a. What is the magnitude of the translation of Problem 1?

b. Let $\mathbf{b} = (h, k)$ be a vector. What is the magnitude of $T_\mathbf{b}$?

c. If $z = x + yi$, and x and y are real, what is the magnitude of T_z?

d. Use translations to explain why $|a| + |b| \geq |a + b|$, for any complex numbers a and b, is called the Triangle Inequality.

4. If $T_1(A) = B$, $T_2(B) = C$, and A, B, and C are collinear, then the translations T_1 and T_2 are called **parallel**.

a. Suppose $T_1(x, y) = (x + h_1, y + k_1)$ and $T_2(x, y) = (x + h_2, y + k_2)$ and T_1 and T_2 are parallel. How are h_1, h_2, k_1, and k_2 related?

b. How must the complex numbers z_1 and z_2 be related if T_{z_1} and T_{z_2} are parallel?

5. Let T be any translation.

a. Give a synthetic proof that if $T(A) = B$ and $T(C) = D$, then $AC = BD$. This shows that *translations preserve distance*, and thus that they are congruence transformations.

b. Give an analytic proof of the statement in part **a**.

c. Use the informal definition of orientation given in Section 7.1.4 to explain why translations are *direct* isometries.

ANSWERS TO QUESTIONS

1. Draw the circle with center C and radius AB. Then draw the circle with center B and radius AC. These two circles intersect at C' and C^*.

2. C' is to the right of B, as far from B as C is from A.

3. $T_{h_1,k_1} \circ T_{h_2,k_2}(x, y) = (x + h_2 + h_1, y + k_2 + k_1) = (x + h_1 + h_2, y + k_1 + k_2) = T_{h_2,k_2} \circ T_{h_1,k_1}(x, y)$ or since addition of complex numbers is commutative and $\{\mathbf{T}, \circ\}$ and $\{\mathbf{C}, +\}$ are isomorphic, composition of translations is commutative.

7.2.2 Rotations

Rotations are closely connected to turns, revolutions, and angles but are not exactly any one of them. Before deriving formulas for rotation images, we examine this related terminology.

Rotations and angles

In school geometry today, an angle is typically defined as the union of two rays with the same endpoint. No numbers are in this definition, so it is not obvious when one angle is larger than another. In the interpretation of Euclid, one angle is larger than a second angle when the opening of the first angle would contain the second. This is a *static* view of angle, quite appropriate in a synthetic development. A *dynamic* view is that one angle is larger than another when the amount of turn needed to turn one side of the angle onto the other side is greater. Teachers often employ the dynamic view when explaining angle measure. In the dynamic view, rotations and angles are inseparably connected.

Figure 21

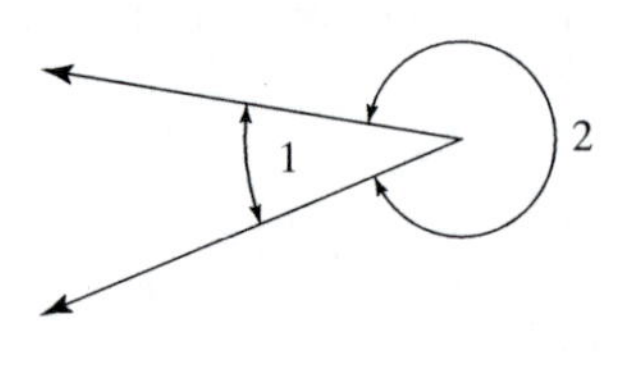

In older geometry books and in some geometry books today, two different noncollinear rays with the same endpoint are said to determine *two* angles. One of these (angle 1 in Figure 21) has measure x less than 180° and the other (angle 2 in Figure 21) has measure $(360 - x)°$. To distinguish these, the second angle is called a **reflex angle**. It is due to the desire that every union of two rays with the same endpoint have a unique measure that geometry books since the 1960s have tended to avoid mention of reflex angles.

The nonuniqueness of angle measure again appears in trigonometry. In trigonometry, an angle is typically defined as an *ordered pair* of rays with the same endpoint, with the first ray in the pair called its **initial side** and the other ray called its **terminal side**. Some call these angles **directed angles**. The trigonometric viewpoint corresponds more closely with the dynamic view of angle and, as we point out in this section and again in Section 9.1.1, is intimately connected with rotations.

Both static and dynamic views of angles are necessary to appreciate fully the concept of angle and the richness of their applications. Some applications require directed angles. You must turn the handle on a door, or spin the dial on a combination lock, or turn a car in a particular direction, not just a particular amount. But other applications do not require directed angles. The parallax of stars, the angle that creates the field of vision in a camera, and the angles measured by surveyors to create maps are undirected angles.

Turns

The word *turn* is an informal word for rotation. In driving, we speak of making a right turn or a U-turn. Like mathematical rotations, these turns change direction, but unlike mathematical rotations, these turns do not have centers. These everyday turns have something else in common with the formal mathematical definition of transformation: The focus is on the relationship between the starting and ending positions of an object and not necessarily on how the object got from start to end.

Revolution

The word *revolution* has two different meanings related to rotations. The word *revolution* is sometimes used to refer to the path our planet Earth takes about the Sun. A *revolution* is also a unit of rotation (as if things were not confused enough between the terms). One revolution $= 360° = 2\pi$ (radians).

A definition of rotation

We want a definition of rotation that has a center and allows for turns in either direction and the possibility that turns will be of more than one revolution. This is done by having a point as center and a real number that describes the amount and direction of the turn. We begin with the case where the rotation is less than a half-revolution and proceed from there.

Definition Let C be a point and ϕ be a real number such that $-\pi < \phi \leq \pi$ in radians or $-180° < \phi \leq 180°$. Then the **ϕ-rotation about C**, denoted $\mathbf{R}_{C,\phi}$, is the plane transformation that maps C onto itself, and every other point P of the plane onto the point Q such that (i) $PC = QC$ and (ii) $m\angle PCQ = |\phi|$, where if $\phi > 0$, ΔPQC is counterclockwise oriented and if $\phi < 0$, ΔPQC is clockwise oriented.

Figures 22a and 22b depict $\mathbf{R}_{C,\phi}$. We call C the **center** of the rotation.[17] If $\phi = 0$, then we define $\mathbf{R}_{C,\phi}(P) = P$ for every point of the plane. $\mathbf{R}_{C,\phi}$ is the **identity transformation** or **zero rotation** and any point can be considered its center.

Figure 22

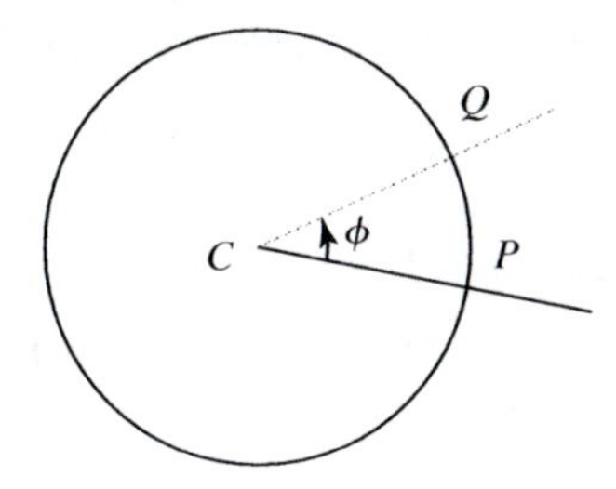

ϕ-rotation about C counterclockwise

$\phi > 0$

(a)

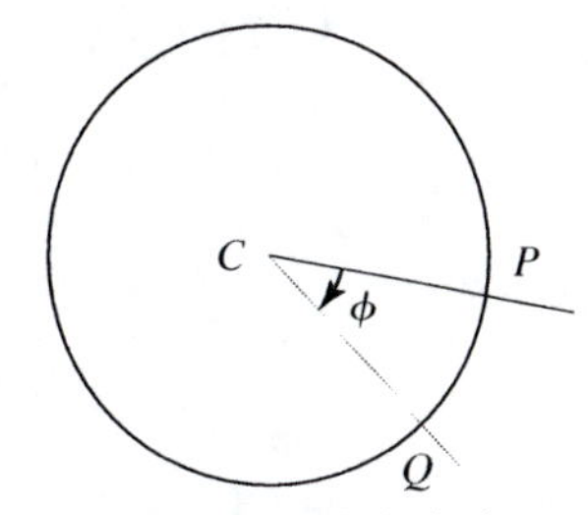

ϕ-rotation about C clockwise

$\phi < 0$

(b)

[17] ϕ is sometimes called the *angle of rotation*, sometimes the *magnitude of the rotation*.

When $|\phi| = 180°$, or $|\phi| = \pi$, $R_{C,\phi}$ is the transformation that maps C onto itself and every other point P of the plane onto the point Q such that C is the midpoint of $\overline{PQ}$. In this case, $R_{C,\phi}$ is called a **half-turn** or **point reflection** about C. Figure 23 shows a point P on a letter and its image Q on the image of that letter from the half-turn about C.

Figure 23

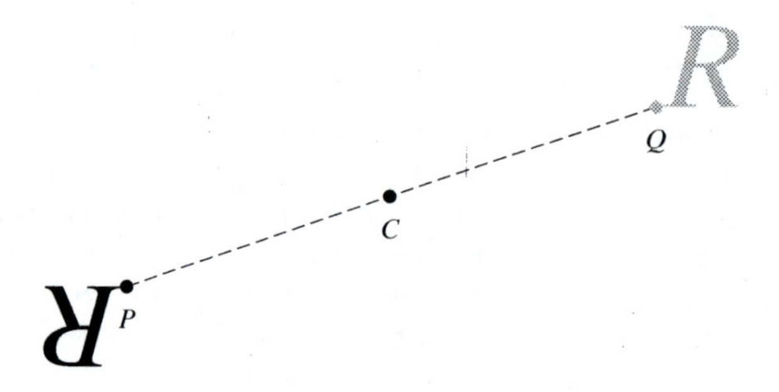

Figure 24 shows a figure I and its image II under a rotation with center C. In this figure, $m\angle ACB \approx -110°$ if we think of the direction of the rotation as clockwise. That is, just as with circular functions, we think of counterclockwise rotations as positive and clockwise rotations as negative. And just as with circular functions, two real numbers yield the same rotation if they are congruent modulo 360° or 2π. So the rotation in Figure 24 is $-110°$, $-470°$, $-830°$, etc., when considered as a clockwise rotation, or 250°, 610°, 970°, etc., when considered as a counterclockwise rotation. Calling this rotation $R_{C,x}$, then $x \equiv -110°$ Mod $(360°) \Leftrightarrow x - (-110°) = 360°n$ for some integer n; that is, if and only if there exists an integer n with $x = -110 + 360n$.

Figure 24

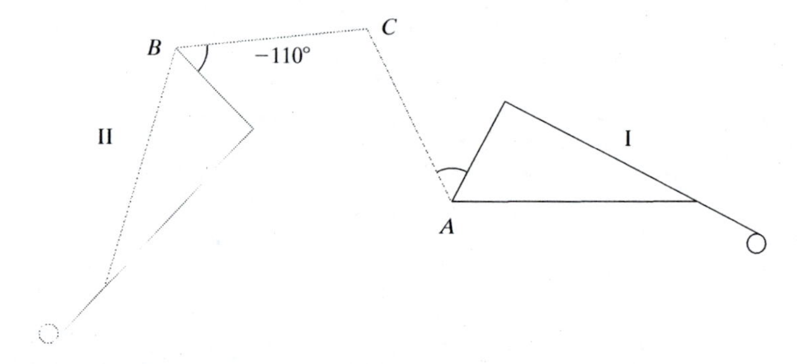

In general, when the magnitude ϕ is outside the range $(-180°, 180°]$ or $(-\pi, \pi]$ radians, then $R_{C,\phi}$ is identical to $R_{C,\phi'}$, where $\phi' \equiv \phi$ Mod(360°) and $-180° < \phi' \leq 180°$ or $\phi' \equiv \phi'$ Mod(2π) and $-\pi < \phi \leq \pi$ radians. That is, adding integer multiples of 360° or 2π radians to ϕ does not change the transformation. For example, if $\phi = -\frac{11\pi}{4}$ radians $= -495°$, then $\phi' = -\frac{3\pi}{4} = -135°$, and $R_{C,-495°}$ and $R_{C,-135°}$ are the same transformation of the plane.

Question: Describe algebraically all possible k if $R_{C,k} = R_{C,\pi/6}$.

In Problem 10, you are asked to prove that rotations are congruence transformations. Hence any geometric figure in the plane and its image under a rotation are congruent.

Composites of rotations

A basic property of angles, often taken as a postulate, is Angle Addition, more accurately "angle measure" addition: If ray $\overrightarrow{CD}$ is between rays $\overrightarrow{CA}$ and $\overrightarrow{CB}$, then

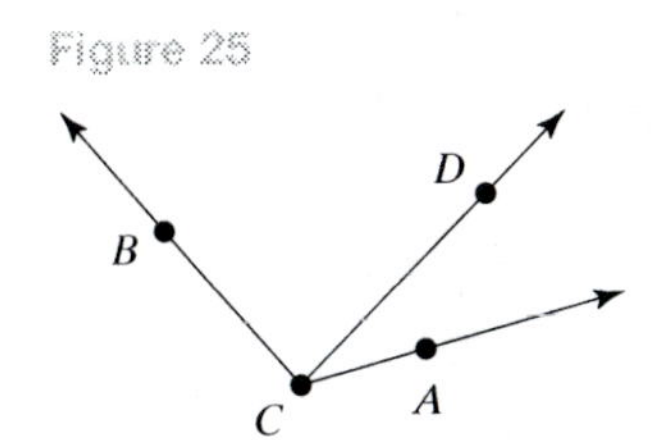

Figure 25

$m\angle ACD + m\angle DCB = m\angle ACB$ (Figure 25). When teachers explain the Angle Addition Property to children, arcs are made and turns are mentioned. Thus, the intuition of combining turns is basic to the idea of angle addition.

If two rotations have the same center, then their composite is easy to describe. If $R_{C,\theta}$ is followed by $R_{C,\phi}$, the composite is $R_{C,\theta+\phi}$, as in Figure 26.

Figure 26

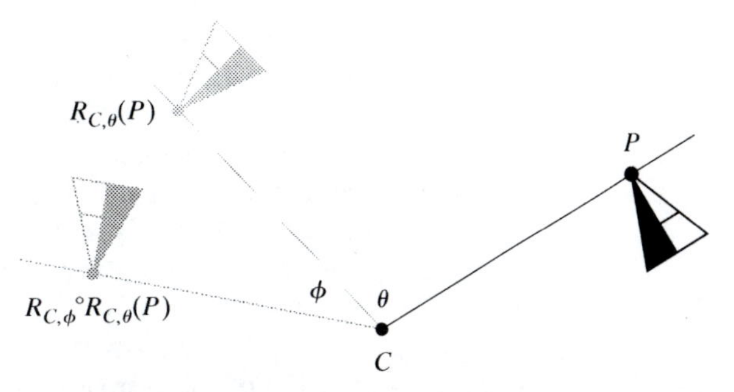

The addition property of rotations is true whether the rotations are clockwise or counterclockwise. For instance, the composite of a 70° rotation followed by a −153° rotation with the same center is the −83° rotation with that center. Of course, if $\theta = 0$, then $R_{C,\theta}$ maps each point onto itself. So composing $R_{C,0}$ with $R_{C,\phi}$ leaves $R_{C,\phi}$ unchanged. That is, $R_{C,0} \circ R_{C,\phi} = R_{C,\phi} \circ R_{C,0} = R_{C,\phi}$. So $R_{C,0}$ is an identity transformation under composition. Also, since $R_{C,-\phi} \circ R_{C,\phi} = R_{C,0}$, the rotations $R_{C,-\phi}$ and $R_{C,\phi}$ are inverse transformations. This is a mathematical way of stating that turning counterclockwise undoes turning clockwise by the same amount, and vice versa.

These observations are often described in the language of abstract algebra by the statement of Theorem 7.3.

Theorem 7.3 With composition, the set of all rotations with the same center is a group.

In the language of abstract algebra, the connection between rotations and the real numbers modulo 360° or 2π is as follows: The group of all plane rotations with center C under composition is isomorphic to the group of real numbers under addition modulo 360° or modulo 2π.

If two rotations have different centers, then in most cases their composite is a rotation, but in some cases the composite is a translation. This result is left as Problem 5 in Section 7.2.5.

Rotations on the coordinate plane

When C is the origin, we use the symbol $\mathbf{R}_\phi$ for $R_{C,\phi}$. In polar coordinates, when the pole is the center of rotation, $R_\phi[r, \theta]$, the image of $[r, \theta]$ under a rotation with magnitude ϕ is $[r, \theta + \phi]$, as shown in Figure 27.

Figure 27

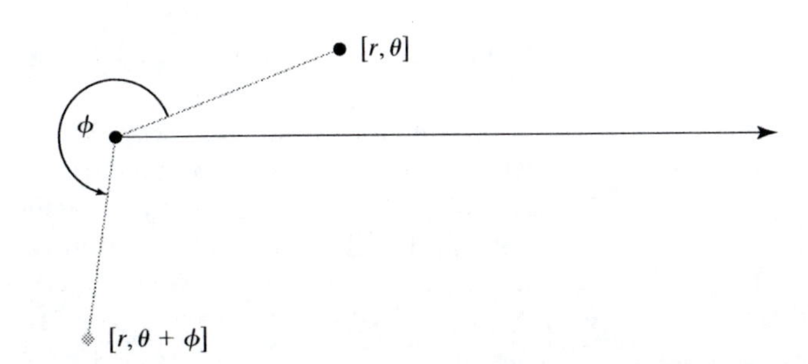

Formulas for images of points under R_ϕ in Cartesian coordinates usually require trigonometry. However, for a few values of ϕ, formulas can be obtained without trigonometry.

Theorem 7.4 For all points (x, y) in the coordinate plane, $R_{90^\circ}(x, y) = (-y, x)$.

Proof: We need to show that the image of $P = (x, y)$ under R_{90° is $(-y, x)$ regardless of the values of x and y. Clearly the formula for the image holds if both $x = 0$ and $y = 0$.

Now suppose $y = 0$ but $x \neq 0$. Then $P = (x, y)$ is on the x-axis at a distance $|x|$ from the center $(0, 0)$. If $x > 0$, its image $Q = R_{90^\circ}(x, y)$ is on the positive ray of the y-axis at a distance $|x|$ from the center and must be $(0, x)$. If $x < 0$, then Q is on the negative ray of the y-axis and must again be $(0, x)$. Thus in these positions, $R_{90^\circ}(x, y) = (-y, x)$.

For all other points P, $y \neq 0$. Let $Q = (r, s)$. Q is not on the y-axis, so $r \neq 0$. From the definition of rotation, $QO = PO$. Thus

$$r^2 + s^2 = x^2 + y^2. \tag{1}$$

Since the magnitude of the rotation is 90°, ΔPQO is a right triangle, so $QO^2 + PO^2 = PQ^2$. In terms of coordinates

$$(x - r)^2 + (y - s)^2 = x^2 + y^2 + r^2 + s^2. \tag{2}$$

From equation (2), $2rx + 2sy = 0$, so an equivalent equation is

$$rx + sy = 0.$$

Consequently,
$$\frac{s}{r} = \frac{-x}{y}.$$

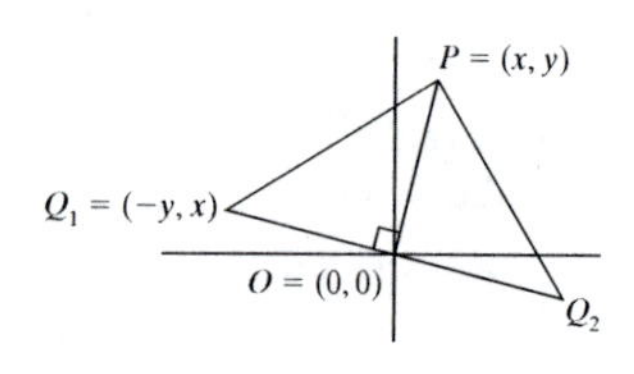

Figure 28

Thus $(r, s) = (ky, -kx)$ for some $k \neq 0$. Then, substituting for r and s in equation (1),

$$k^2y^2 + k^2x^2 = x^2 + y^2.$$

So $k = \pm 1$, and so $(r, s) = (-y, x)$ or $(r, s) = (y, -x)$. These coordinates determine the two points Q_1 and Q_2 in Figure 28.

That is, we have obtained two possible images of (x, y) under rotations of 90°. Q_1 is the image under the counterclockwise rotation because the signs of the coordinates agree with where the rotation image should be. Specifically, when x and y are both positive, (x, y) is in the first quadrant and the image $(-y, x)$ is in the second quadrant. When x is negative and y is positive (second quadrant), then $-y$ is negative and so $(-y, x)$ is in the third quadrant. You should try the other two possibilities, and also try the cases where (x, y) is on a coordinate axis. You should also check that Q_2. is the image of (x, y) under R_{-90°. ┘

From Theorem 7.4, formulas for the image of (x, y) under R_{180° and R_{270° are easily established (see Problem 5).

A formula for the image of any point under R_ϕ comes directly from a definition of the sine and cosine functions in terms of the unit circle. The point $(\cos\phi, \sin\phi)$ is the image of $(1, 0)$ under the rotation R_ϕ (Figure 29). The point $(-\sin\phi, \cos\phi)$ is the image of $(0, 1)$ under the same rotation because it is 90° further around the circle from $(\cos\phi, \sin\phi)$. From these two images, we can prove the following theorem. A proof is given in Section 9.3.1.

Theorem 7.5 For all points (x, y) in the coordinate plane,

$$R_\phi(x, y) = (x \cos\phi - y \sin\phi, x \sin\phi + y \cos\phi).$$

Figure 29

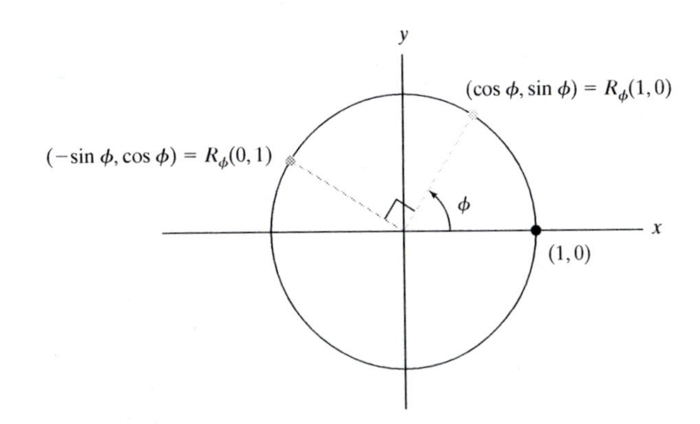

In matrix form, if (x', y') is the image of (x, y), then under R_ϕ, Theorem 7.5 becomes

$$\begin{bmatrix} x' \\ y' \end{bmatrix} = \begin{bmatrix} \cos\phi & -\sin\phi \\ \sin\phi & \cos\phi \end{bmatrix} \begin{bmatrix} x \\ y \end{bmatrix}.$$

We say that *the matrix* $\begin{bmatrix} \cos\phi & -\sin\phi \\ \sin\phi & \cos\phi \end{bmatrix}$ *represents the transformation* R_ϕ. Most people seem to find the matrix form easier to remember than the form of Theorem 7.5.

A formula for rotations with centers at points other than (0, 0)

The formula of Theorem 7.5 can be extended to centers other than the origin by using the following strategy. To obtain $R_{C,\phi}(P)$, employ the following three transformations, in this order.

1. Translate P by the slide T_{-C} to a point P'.
2. Apply the rotation R_ϕ to P' to obtain a point Q'.
3. Translate the point Q' by the slide T_C to obtain Q.

This procedure for implementing the rotation $R_{C,\phi}$ is described by the following composite:

$$R_{C,\phi} = T_C \circ R_\phi \circ T_{-C}$$

This is a **basic transformation strategy**, used not only in mathematics but in many endeavors. We first perform a transformation to bring the object to be transformed to a position where we can deal with it. In the new position, we deal with the object. Then we transform back. In mathematical modeling, the first transformation is a translation from the real world into mathematics. Then the mathematics is applied and finally there is a translation back from the mathematical solution to the real-world solution. In appliance repair, we first bring our appliance to a repair shop. The appliance is repaired there. Then we bring the appliance back. The classic transformation strategy requires that the first transformation not disrupt the object being transformed. That is, if the act of bringing the appliance to a repair shop would change the appliance, then the strategy would not work. This is why a guitar player would generally not bring a guitar to a shop merely to be tuned; transporting the guitar causes a change in tune.

To return to the specific problem of rotating a point P a magnitude ϕ about a center C, let P be represented by the complex number $p = (x, y) = x + iy$. Let c be the complex number corresponding to the center C of the rotation. Recall from

Section 2.2.2 that $z_\phi = \cos\phi + i\sin\phi$ is the complex number corresponding to the rotation R_ϕ. Thus the complex number q representing the image $Q = R_{C,\phi}(P)$ can be computed from the complex number p representing P by the formula

$$q = z_\phi(p - c) + c.$$

Theorem 7.6 In the complex plane, the image of p under the ϕ-rotation with center, $R_{C,\phi}(p)$, is given by

$$R_{C,\phi}(p) = z_\phi(p - c) + c,$$

where $z_\phi = \cos\phi + i\sin\phi$.

Theorem 7.6 provides a straightforward procedure for finding any rotation image of any point. For example, if the point $P = (-1, 2)$ is to be rotated $\frac{\pi}{4}$ about the point $C = (3, 5)$, then $c = 3 + 5i$, $\phi = \frac{\pi}{4}$, and $p = -1 + 2i$. Letting q be the complex number corresponding to the image Q,

$$\begin{aligned} q &= \left(\cos\frac{\pi}{4} + i\sin\frac{\pi}{4}\right)(-1 + 2i - (3 + 5i)) + 3 + 5i \\ &= \left(\frac{\sqrt{2}}{2} + i\frac{\sqrt{2}}{2}\right)(-4 - 3i) + 3 + 5i \\ &= \left(3 - \frac{\sqrt{2}}{2}\right) + \left(5 - \frac{7\sqrt{2}}{2}\right)i. \end{aligned}$$

This shows that the image Q is the point with coordinates $\left(3 - \frac{\sqrt{2}}{2}, 5 - \frac{7\sqrt{2}}{2}\right)$.

We can employ complex numbers in this way because, for a given point C, the correspondence

$$R_{C,\phi} \longleftrightarrow z_\phi$$

is an isomorphism between the set of all rotations around C in the plane with the operation of composition and the set of complex numbers on the unit circle with the operation of multiplication.

7.2.2 Problems

1. Draw a triangle ABC and select a fourth point O in the exterior of ΔABC.

a. Draw the image of ΔABC under the rotation of magnitude $-42°$ with center O.

b. Draw the image of ΔABC under the rotation of $-42°$ with center A.

c. Prove that the two images in parts **a** and **b** are congruent.

d. Generalize the result of part **c**.

2. What rotations are identical to $R_{C,k}$, if k is measured in

a. degrees?

b. radians?

3. In the group of all rotations with the same center C, give the magnitude of each transformation.

a. the identity

b. the inverse of $R_{C,\pi/4}$

4. Use the formula for $R_{90°}$ to explain why perpendicular lines that intersect at the origin have slopes whose product is -1.

5. a. By composing the transformation $R_{90°}$ with itself, derive formulas for $R_{180°}$ and $R_{270°}$.

b. Use your answers to part **a** to show analytically that $R_{180°}$ is its own inverse, and that $R_{90°}$ and $R_{270°}$ are inverses.

6. a. Solve the system $x + y = 0$; $x^2 + y^2 = 1$.

b. The answer to part **a** gives the cosines and sines of what numbers?

7. Give a formula for $R_{5\pi/6}(x, y)$ and check your formula in some way.

8. Check the result of the example following Theorem 7.6 in two ways.
a. with a graph and decimal approximations to the coordinates of the image
b. by showing that $PC = QC$ exactly
c. Is either check a sure check?

9. Consider the triangle PQR, where $P = (2, 5), Q = (-1, 4)$, and $R = (6, 0)$.
a. Rotate this triangle 40° about the point $(0, 1)$. Give both exact values and approximations for the image.
b. By calculation using the exact values, verify that the image is congruent to the original.

10. Prove that every rotation is an isometry:
a. using congruent triangles
b. using Theorem 7.5

11. Let $\bar{z}$ be the complex conjugate of z. Prove: For all ϕ, $\overline{z_\phi} = z_{-\phi}$.

12. Find a formula for $R_{C,\phi}(x, y)$ when $C = (a, b)$.

13. Let $w = \cos\frac{2\pi}{3} + i\sin\frac{2\pi}{3}$.
a. Explain why w is a cube root of 1.
b. Find the other nonreal cube root of 1.
*c. Prove: If a, b, and c are complex numbers representing the vertices of an equilateral triangle ABC in clockwise order, then $aw^2 + bw + c = 0$. (*Hint*: First prove the result for the special case where triangle ABC is inscribed in the unit circle.)

ANSWER TO QUESTION

1. $\frac{\pi}{6} + 2\pi n$ for some integer n or $m \equiv \frac{\pi}{6}\,\mathrm{Mod}(2\pi)$

7.2.3 Reflections

Reflections may seem a little more complicated than either translations or rotations, because they switch orientation. But reflections are easier to define synthetically than either translations or rotations. And in this section we show that there is a sense in which reflections are the basic building blocks of the congruence transformations.

Definitions and Terminology

Definition Let m be a line in a plane. For each point P in that plane the **reflection image of P over line** m, denoted by $\mathbf{r_m(P)}$, is the point Q such that (i) $Q = P$ if P is on m, and (ii) m is the perpendicular bisector of $\overline{PQ}$ if P is not on m.

Figure 30

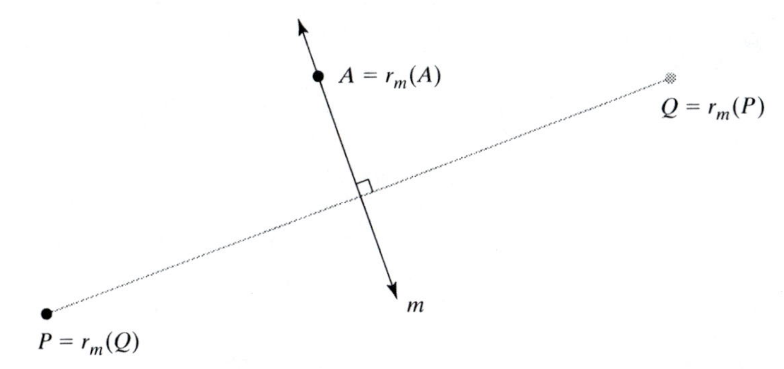

From the definition of reflection, since $\overline{PQ}$ and $\overline{QP}$ are the same segment, if $r_m(P) = Q$, then $r_m(Q) = P$ (see Figure 30). That means that, considered as a function, the reflection r_m is its own inverse.

In the coordinate plane, reflection images over some lines are very easy to find. Figure 31 displays a point (a, b) and its reflection images over the x-axis, the y-axis, and the line $y = x$. In the drawing, a is negative and b is positive. That the coordinates of these images are correct follows from the geometry theorem that if two points P and Q are equidistant from the endpoints of a segment $\overline{AB}$, then $\overleftrightarrow{PQ}$ is the perpendicular bisector of $\overline{AB}$.

Figure 31

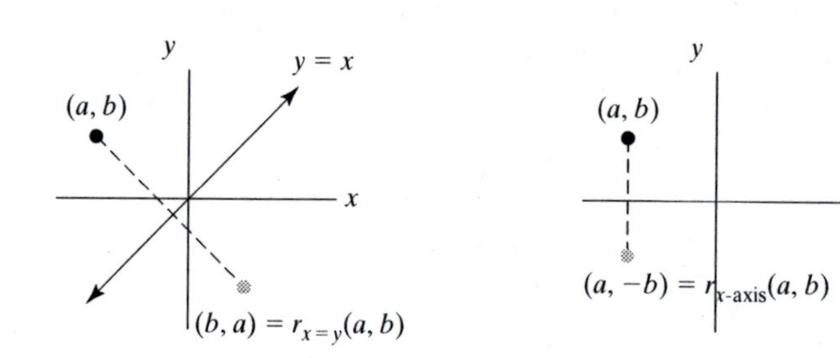

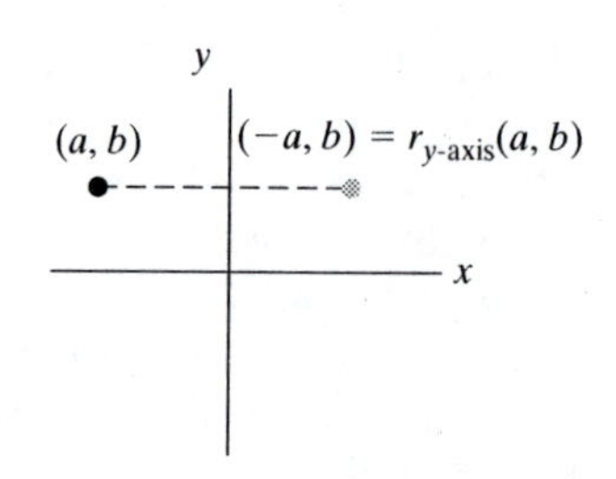

For instance, every point on the line $y = x$ has coordinates (a, a). The distance from (a, a) to (x, y) is $\sqrt{(a - x)^2 + (a - y)^2}$, which is also the distance from (a, a) to (y, x). These arguments prove the following theorem.

Theorem 7.7 For all points (x, y),

a. $r_{x\text{-axis}}(x, y) = (x, -y)$ **b.** $r_{y\text{-axis}}(x, y) = (-x, y)$ **c.** $r_{x=y}(x, y) = (y, x)$.

Each part of Theorem 7.7 can be written in matrix form. Call the image (x', y'). Then, for part (a),

$$\begin{bmatrix} x' \\ y' \end{bmatrix} = \begin{bmatrix} x \\ -y \end{bmatrix} = \begin{bmatrix} 1 & 0 \\ 0 & -1 \end{bmatrix} \begin{bmatrix} x \\ y \end{bmatrix}.$$

This tells us that $\begin{bmatrix} 1 & 0 \\ 0 & -1 \end{bmatrix}$ is a matrix representing $r_{x\text{-axis}}$.

Properties of Reflections

In some treatments of geometry, some properties of reflections are assumed and from them the properties of congruence of triangles are deduced. In the following discussion, we use familiar properties of congruent triangles to derive properties of reflections. In later sections, we will reverse the process and use properties of reflections to deduce properties of congruence.

Theorem 7.8 Every reflection is a congruence transformation. (Reflections preserve distance.)

Proof: Let r_m be a reflection with $r_m(A) = A'$ and $r_m(B) = B'$. To prove that r_m is a distance-preserving transformation, we must show $AB = A'B'$. There are four cases: (1) A and B are both on m; (2) A is on m and B is not; (3) A and B are on the same side of m; (4) A and B are on opposite sides of m. Figure 32 shows the cases and what can be deduced immediately from the definition of reflection. We leave it to you as Problem 2 to finish this proof.

Figure 32

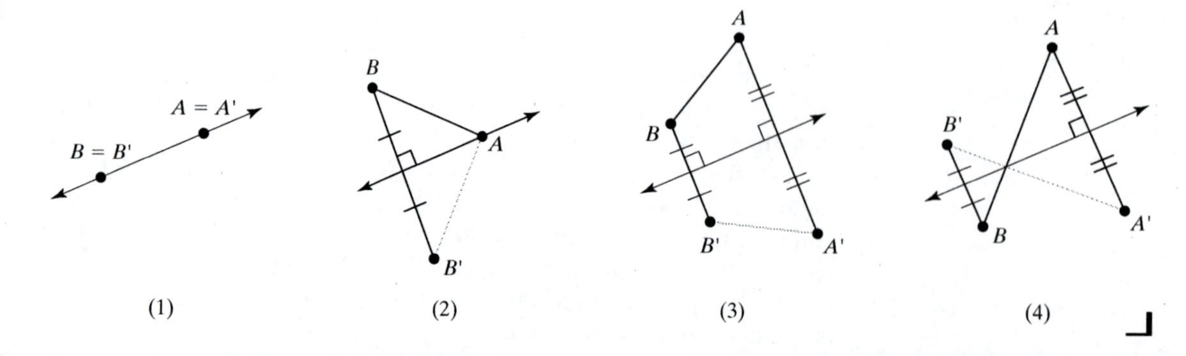

Applying Theorem 7.1, the following corollary is immediate.

Corollary 1: Under a reflection, the image of a line segment is a line segment of the same length.

Now we can apply Problem 6 of Section 7.1.4 to conclude that, under a reflection, the image of a line is a line, the image of a ray is a ray, the image of an angle is an angle, and the image of a polygon is a polygon with corresponding sides of the same length as the preimage.

What about the measures of the angles? Using triangle congruence theorems, the next corollary follows. We ask you to supply a proof, using Figure 33.

Corollary 2: Under a reflection, the image of an angle is an angle with the same measure.

Figure 33

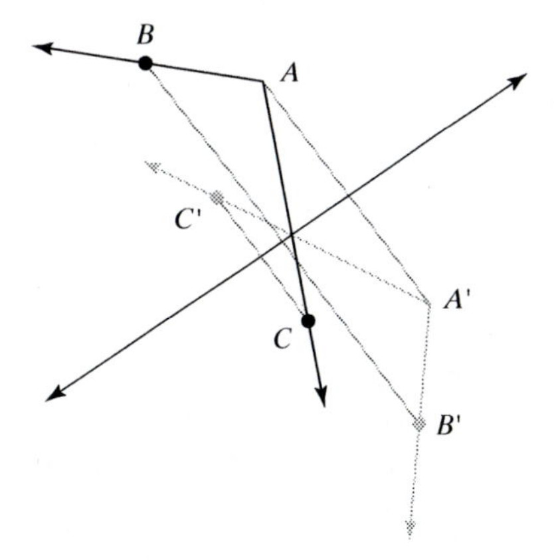

Because of these corollaries, we say that *reflections preserve betweenness, collinearity, and angle measure.*

Composites of reflections

The preservation properties of reflections are important because of the surprising relationship among composites of reflections, rotations, and translations. Whatever is preserved by the transformations in a set is also preserved by a composite of those transformations.

In a clothes store, you have probably noticed mirrors connected to one another. These mirrors are like intersecting planes, and they are the 3-dimensional counterpart to intersecting reflecting lines. If you look at the image in one mirror of your image in the other mirror, you will see a *rotation* image of yourself. You can verify this by raising a hand; you will see what looks like the opposite hand raised in the image but the same hand raised in the image of the image. This phenomenon is an instance of the following theorem.

Theorem 7.9(a) **(Two-Reflection Theorem for Rotations):** If m intersects n, then $r_n \circ r_m$ is the rotation with center the intersection of m and n, and twice the measure of the directed angle from m to n.

Before the proof, let us understand the theorem. Figure 34 shows a figure F, its reflection image F^* over line m (the first reflecting line), and the image F' of F^* over line n (the second reflecting line). Now block out F^* from the picture and consider only the relationship between F and F'. If you need to, turn the page keeping the point O fixed. You should see that F can be rotated into F'. To prove that this is actually the case, we need to show that the angle formed by any point on F, O, and its image on F' has a constant measure, and that any point and its image are the same distance from O.

Figure 34

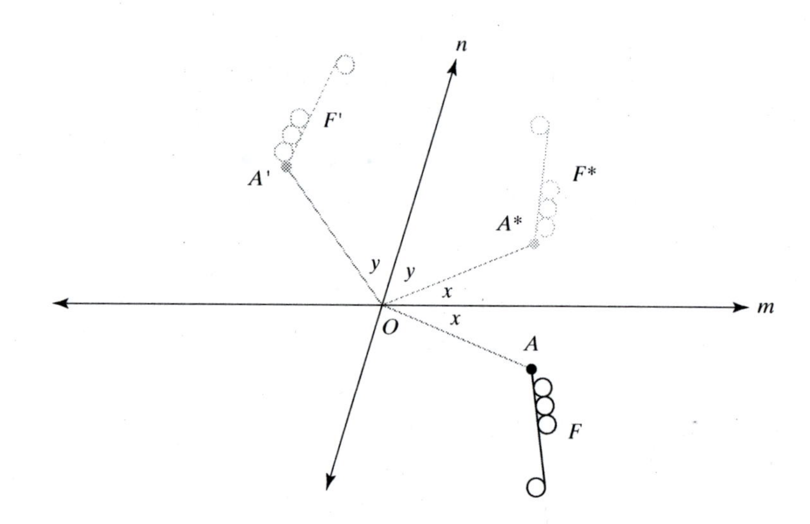

Proof: Suppose m and n are lines intersecting at O. Let $A^* = r_m(A)$ and let $A' = r_n(A^*)$. Then $A' = r_n \circ r_m(A)$. Since O is on both lines, O coincides with its image under both reflections. That is, $O = r_m(O) = r_n \circ r_m(O)$. Now, because reflections preserve distance, $OA = OA^* = OA'$. This means that A and A' are on the same circle with center O, which implies that A' is the image of A under some rotation with center O.

To find the measure of $\angle A'OA$, notice that m bisects $\angle AOA^*$ and n bisects $\angle A^*OA'$. Consequently, if the measure of the angle from line m to line n is $x + y$, as shown in Figure 34, then $m\angle A'OA = 2x + 2y$. This proves the theorem for the case in which the point A is located as we have in Figure 34. You should prove the theorem for other locations. ┘

As surprising as the relationship between reflections and rotations might be, there is an analogous relationship between reflections and translations. You have encountered this relationship if you have been in a room with parallel mirrored walls. In such a room, if you look into one of the mirrors, you can see image after image of the room, one behind the other, seemingly forever.

Theorem 7.9(b) **(Two-Reflection Theorem for Translations):** If $m // n$, then $r_n \circ r_m$ is the translation with direction perpendicular to the lines m and n, and with magnitude twice the distance between m and n, in the direction from m to n.

Not only are the Two-Reflection Theorems for Translations and Rotations analogues of each other, so are their proofs. Figure 35 and the following proof have been written to emphasize the analogy.

Figure 35

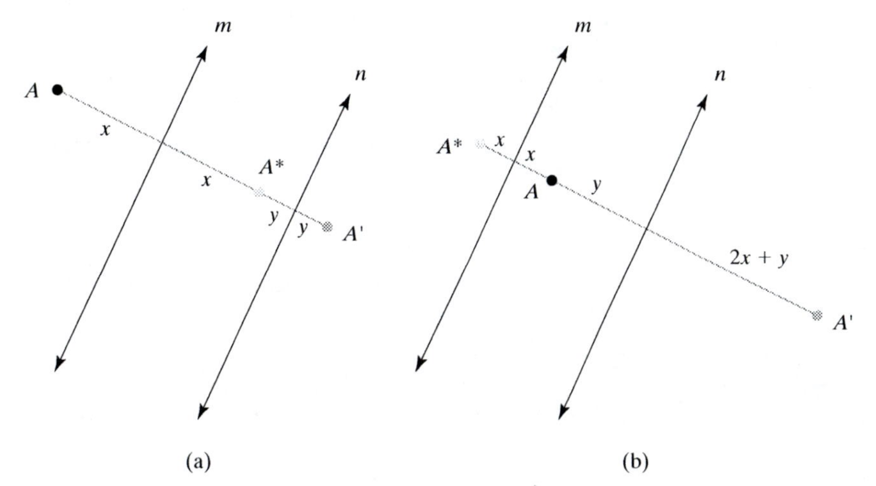

(a) (b)

Proof: Suppose m and n are parallel lines. Let $A^* = r_m(A)$ and let $A' = r_n(A^*) = r_n \circ r_m(A)$. Since $\overline{AA^*}$ and $\overline{A^*A'}$ are perpendicular to parallel lines, the three points A, A^*, and A' are collinear. So $\overleftrightarrow{AA^*} \perp m$.

To find the length AA', notice that m and n are the perpendicular bisectors of $\overline{AA^*}$ and $\overline{A^*A'}$. Consequently, if the distance between lines m and n is $x + y$, as shown in Figure 35, then $AA' = 2x + 2y$. This proves the theorem for the cases in which the point A is located as we have in Figure 35. You should prove the theorem for other locations. ┘

Consequences of the Two-Reflection Theorems

Theorems 7.9(a) and 7.9(b) enable any translation or rotation to be "decomposed" into a composite of two reflections. Therefore, we could have started this unit by deducing properties of reflections and defining translations and rotations synthetically. Then, because of the Two-Reflection Theorems, all properties preserved by reflections would also be preserved by rotations and translations. So, from the properties of reflections, we can quickly deduce that rotations and translations preserve distance, segments, rays, angles, and angle measure. In this sense, reflections are more basic than translations or rotations.

We mentioned in Section 7.1.4 that mirror images switch orientation. From this property, applying one reflection after another will switch orientation and switch it back, and so rotations and translations preserve orientation.

Another consequence of the Two-Reflection Theorems is that the composite of many pairs of reflecting lines can yield the same rotation or translation. If you wish to rotate a figure 70° about a point M, for example, you can pick any line through M as the first reflecting line. Then the second line is determined as the line that contains M and makes a 35° angle (counterclockwise) with the first line. We call these "Flexibility Theorems" because they enable a rotation or translation to be written in particular different ways. These ways are applied in Section 7.2.5.

Theorem 7.10(a) **(Rotation Flexibility Theorem):** Suppose lines m' and n' are images of intersecting lines m and n under a rotation whose center is the intersection of m and n. Then $r_{n'} \circ r_{m'} = r_n \circ r_m$.

Theorem 7.10(b) **(Translation Flexibility Theorem):** Suppose lines m' and n' are images of parallel lines m and n under a translation. Then $r_{n'} \circ r_{m'} = r_n \circ r_m$.

Here is another interpretation of Theorem 7.10: (a) $R_{C,\phi} = r_n \circ r_m$, where m and n are any two lines that intersect at C such that the directed angle from m to n is congruent to $\frac{\phi}{2}$ Mod(360°). (b) $T_{\overrightarrow{AB}} = r_n \circ r_m$, where m and n are any two parallel lines perpendicular to $\overleftrightarrow{AB}$ and such that the directed line segment from a point on m to a point on n and perpendicular to n is $\frac{1}{2}\overrightarrow{AB}$.

Reflections on the coordinate plane

The Two-Reflection Theorem for Rotations enables a formula to be found for any reflection over a line through the origin in the coordinate plane. Although the formula is messy, its derivation is quite elegant.

Theorem 7.11 Let L be the line containing (0, 0) and forming an angle ϕ with the positive ray of the x-axis. Then $r_L(x, y) = (x\cos 2\phi + y\sin 2\phi, x\sin 2\phi - y\cos 2\phi)$.

Figure 36

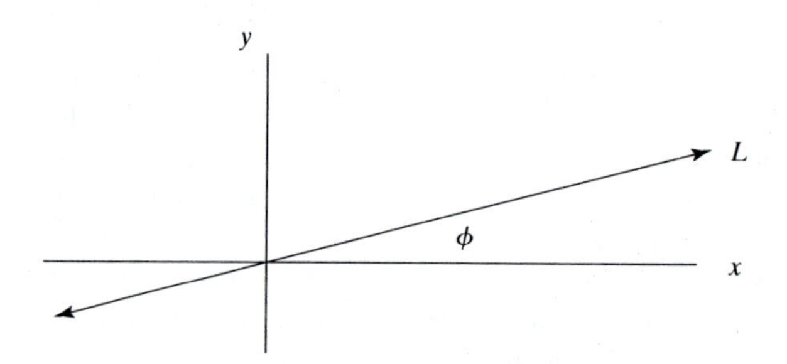

Proof: Refer to Figure 36. Identify the reflection over the x-axis as r_x. By the Two-Reflection Theorem for Rotations, since the measure of the angle formed by L and x is ϕ,

$$r_L \circ r_x = R_{2\phi}.$$

In matrix form, this equation becomes

$$\text{Matrix for } r_L \cdot \begin{bmatrix} 1 & 0 \\ 0 & -1 \end{bmatrix} = \begin{bmatrix} \cos 2\phi & -\sin 2\phi \\ \sin 2\phi & \cos 2\phi \end{bmatrix}.$$

To solve for the matrix for r_L, we multiply both sides of this equation on the right by the inverse of $\begin{bmatrix} 1 & 0 \\ 0 & -1 \end{bmatrix}$. But, just like the transformation r_x, that matrix is its own inverse. Consequently,

$$\text{Matrix for } r_L = \begin{bmatrix} \cos 2\phi & -\sin 2\phi \\ \sin 2\phi & \cos 2\phi \end{bmatrix} \cdot \begin{bmatrix} 1 & 0 \\ 0 & -1 \end{bmatrix}$$
$$= \begin{bmatrix} \cos 2\phi & \sin 2\phi \\ \sin 2\phi & -\cos 2\phi \end{bmatrix}.$$

The matrix representation means that if (x, y) is any point in the coordinate plane, its image (x', y') under r_L satisfies

$$\begin{bmatrix} x' \\ y' \end{bmatrix} = \begin{bmatrix} \cos 2\phi & \sin 2\phi \\ \sin 2\phi & -\cos 2\phi \end{bmatrix} \begin{bmatrix} x \\ y \end{bmatrix}$$
$$= \begin{bmatrix} x\cos 2\phi + y\sin 2\phi \\ x\sin 2\phi - y\cos 2\phi \end{bmatrix}.$$

From this, the theorem follows. ⌟

Reflections in the complex plane

In the complex plane, there is a very important reflection. It is defined by

$$r_R(z) = \bar{z},$$

where $\bar{z}$ is the complex conjugate $a - bi$ of the complex number $a + bi$. Because the point representing $\bar{z}$ in the complex plane is the reflection image over the real axis of the point representing z, r_R is *reflection over the real axis* in the complex plane.

One way of obtaining analytic descriptions of other reflections is to transform them to reflection in the real axis by the same sort of fundamental transformation strategy used with rotations in Section 7.2.2. Suppose we want to reflect over a line ℓ containing the origin. Let ϕ be the measure of the angle from the positive real axis to ℓ. (The slope of line ℓ is $\tan \phi$.) Then, for any point P of the plane, $Q = r_\ell(P)$ can be found by the following procedure, pictured in Figure 37:

1. Apply the rotation $R_{-\phi}$ centered at the origin to P to obtain a point P'. (Note that the image of ℓ under $R_{-\phi}$ is the real axis.)
2. Apply r_R to P' to obtain a point Q'.
3. Apply the rotation R_ϕ centered at the origin to Q' to obtain Q.

Figure 37

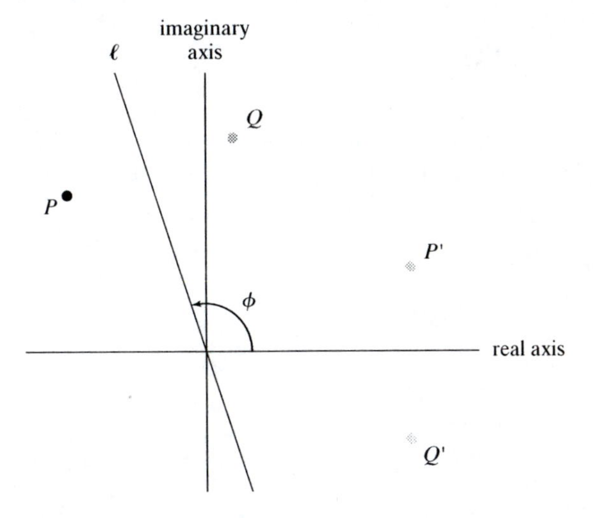

This procedure is summarized by the following composite of transformations:

$$r_\ell = R_\phi \circ r_R \circ R_{-\phi}.$$

The reason that this transformation strategy works so well is that each of the three transformations on the right side of the equation can be accomplished using a unary operation with a complex number! In particular, R_ϕ can be performed by multiplying by $z_\phi = \cos \phi + i \sin \phi$, and r_R is done by taking the conjugate. Consequently, for any complex number z,

$$\begin{aligned} r_\ell(z) &= R_\phi \circ r_R \circ R_{-\phi}(z) \\ &= R_\phi \circ r_R(z_{-\phi} z) \\ &= R_\phi(\overline{z_{-\phi} z}) \\ &= z_\phi(\overline{z_{-\phi} z}). \end{aligned}$$

For all complex numbers, the conjugate of a product equals the product of conjugates: $\overline{z_1}\,\overline{z_2} = \overline{z_1 z_2}$. So

$$r_\ell(z) = z_\phi \overline{z_{-\phi}}\, \bar{z}.$$

Problem 11 of Section 7.2.2 shows that the complex numbers corresponding to the rotations of ϕ and $-\phi$ about the origin are conjugates of one another. Hence $\overline{z_{-\phi}} = z_\phi$, and so

$$\begin{aligned} r_\ell(z) &= z_\phi z_\phi \bar{z} \\ &= z_{2\phi}\bar{z}. \end{aligned}$$

Consequently, the image of z under a reflection over the line making an angle ϕ with the positive ray of the real axis is the product of the conjugate of z and the number $z_{2\phi} = \cos 2\phi + i \sin 2\phi$. This yields the result of Theorem 7.10 in the language of complex numbers.

Theorem 7.11 **(complex number form):** In the complex plane, the image of z under the reflection r_ϕ over the line through the origin and making an angle ϕ with the positive ray of the real axis is given by

$$r_\phi(z) = z_{2\phi}\bar{z}.$$

If the line ℓ does not contain the origin, then you can employ the classic transformation strategy again. First translate ℓ so that its image does contain the origin. Then reflect over the image. Then translate back. The result is the following generalization of Theorem 7.11.

Theorem 7.12 In the complex plane, the image of z under the reflection $r_{C,\phi}$ over the line containing point C and making an angle ϕ with the positive ray of the real axis, is given by

$$r_{C,\phi}(z) = z_{2\phi}\overline{(z - c)} + c.$$

7.2.3 Problems

1. Give the matrix for each transformation. (Use the results of Theorem 7.7.)

a. $r_{y\text{-axis}}$

b. $r_{x=y}$

c. $r_{y=-x}$

2. Finish the proof of Theorem 7.8.

3. Prove Corollary 2 to Theorem 7.8.

4. Use the Two-Reflection Theorem for Rotations and formulas for the image of (x, y) under the transformations $r_{x\text{-axis}}$ and $r_{y=x}$ to obtain the formula for the image of (x, y) under a rotation of 90° about the origin.

5. Which of the Two-Reflection Theorems for Rotations and Translations could apply when the reflecting lines m and n are the same line? Describe the composite transformation.

6. Find equations for a pair of lines m and n so that $r_n \circ r_m$ is the given transformation.

a. the translation that maps the origin onto (5, 0)

b. the rotation of 180° about a point (x_0, y_0)

c. the rotation of 40° about the origin

d. the translation T with $T(x, y) = (x + 3, y + 4)$

7. If you go to a hair salon or barber shop to have someone cut your hair, there is often a mirror on the wall in back of you. To see the back of your head, a hand mirror may be placed in front of you. Explain what transformation maps the back of your head onto its image in the hand mirror.

8. Figure 38 is a picture of an image in a kaleidoscope. The preimage is what lies in the sector containing point P between the mirrors a and b, which are radii of the circle. All the

images result from reflections or composites of reflections over those mirrors. The point P and its five images Q, R, S, T, and U are identified. Indicate the composite of reflections that maps P onto each of these points.

Figure 38

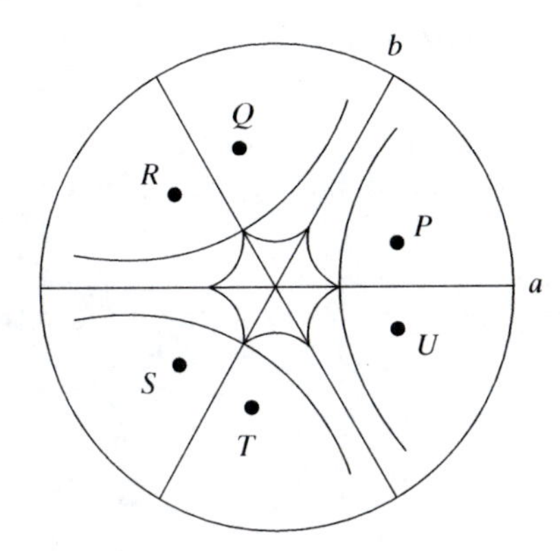

9. Let k and ℓ be the perpendicular bisectors of $\overline{AB}$ and $\overline{BC}$, and let k and ℓ intersect at O.

a. Explain why O is the center of a circle containing A, B, and C.

b. Explain why $r_\ell \circ r_k$ is a rotation with center O, and why the measure of arc $\widehat{AC}$ is the magnitude of this rotation.

c. Use part **b** and the Two-Reflection Theorem for Rotations to explain why the measure of the inscribed angle ABC is half of its intercepted arc $\widehat{AC}$. (This is a theorem in Euclid's *Elements* that we discuss in Section 8.2.4.)

10. Let m be the line with equation $y = 3x + 5$ and n be the line with equation $y = 3x + 9$.

a. Describe, as specifically as possible the transformation $r_n \circ r_m$.

b. Give equations for two other lines u and v such that $r_v \circ r_u$ is the same transformation.

11. Find a formula for (x', y'), the image of (x, y) under the reflection over the line $y = mx$.

12. Find a formula for the reflection image z' of the complex number z over the line $y = kx$.

*13. A rectangular mirror is placed on a wall so that its top is 72″ off the floor and its bottom is 30″ off the floor. A person has a height of h inches, eyes e inches off the ground, and stands d inches from the mirror.

a. Give particular values of h, e, and d for which the person can see the image of his shoes in the mirror.

b. Give particular values of h, e, and d for which the person can see the image of his entire front in the mirror.

c. Does standing closer to the mirror increase or decrease the ability to see your feet in a mirror?

7.2.4 Glide reflections

In this section, we consider a congruence transformation that is different from reflections, rotations, and translations. It is the transformation that maps one footstep onto the next (Figure 39). For this reason, we give it the informal name *walk*, but its formal name is *glide reflection*.

Figure 39

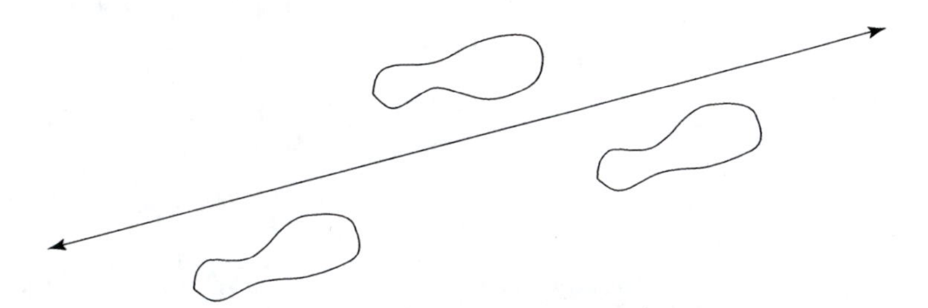

A glide reflection has a reflecting line and a directed magnitude. The reflecting line centers the walk and changes the feet, while the magnitude gives the length and direction of the stride in the walk. Thus a glide reflection shares some of the characteristics of reflections and translations. This is evident in its definition.

Definition Suppose that m is a line in the plane and that $\mathbf{b}$ is a vector parallel to m. Then a **glide reflection** is the composite $r_m \circ T_\mathbf{b}$ or $T_\mathbf{b} \circ r_m$ of the reflection r_m and the translation by $\mathbf{b}$.

Because the translation $T_\mathbf{b}$ in a glide reflection is itself the composite of two reflections, every glide reflection is the composite of three reflections. Thus the existence of glide reflections answers (in the affirmative) the question whether new transformations can be created as a result of three reflections.

Question 1: Why is a glide reflection an isometry?

Question 2: Can a glide reflection have fixed points?

The study of glide reflections is made simpler by the fact that in a given glide reflection, either the reflection or the translation can be performed first.

Theorem 7.13 The glide reflection $r_m \circ T_{\mathbf{b}}$ is the same transformation as the glide reflection $T_{\mathbf{b}} \circ r_m$.

We call either $r_m \circ T_{\mathbf{b}}$ or $T_{\mathbf{b}} \circ r_m$ the glide reflection $\mathbf{g}_{m,\mathbf{b}}$.

Figure 40

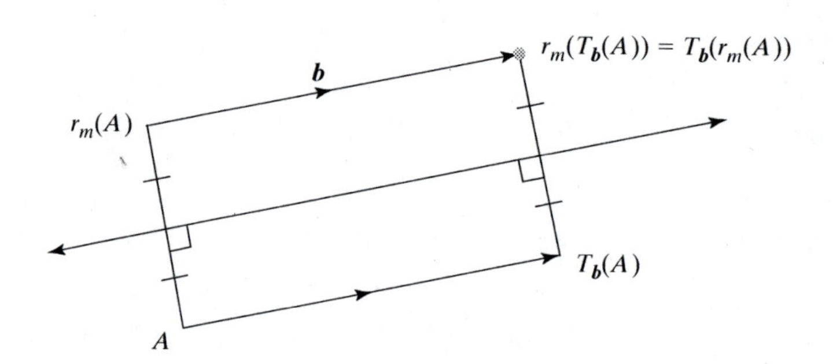

Proof: Figure 40 shows all the given information that can be deduced from the definition of G and the definitions of translation and reflection. For any point A not on the glide reflection line m,

$$r_m \circ T_{\mathbf{b}}(A) = T_{\mathbf{b}} \circ r_m(A).$$

If A is on m, then

$$r_m(A) = A,$$

so

$$r_m \circ T_{\mathbf{b}}(A) = T_{\mathbf{b}}(A) = T_{\mathbf{b}}(r_m(A)) = T_{\mathbf{b}} \circ r_m(A).$$

Consequently, $r_m \circ T_{\mathbf{b}}(A) = T_{\mathbf{b}} \circ r_m(A)$ for all points A. So $r_m \circ T_{\mathbf{b}} = T_{\mathbf{b}} \circ r_m$. ∎

Keep in mind that, although the definition of *glide reflection* is as a composite of a translation and a reflection, a glide reflection is a single transformation. When you walk, you do not think of separating the switching of your feet from your stride forward; you do both at the same time. Similarly, in a glide reflection, you should think of the reflection and translation as being done together.

Analytic descriptions of glide reflections

Throughout this book you have seen the advantages of multiple descriptions of mathematical objects. A property that is difficult to prove with one description may be easier to prove with another. A problem difficult to solve with one description may be easier to solve with another. We have provided a synthetic definition of glide reflections. Now we seek an analytic description.

Every glide reflection is the composite of a reflection and a translation. Therefore, because an analytic description for any reflection can be found, it can be combined with the analytic description of a translation to obtain a description for a glide reflection. As with reflections, a very nice description of any glide reflection can be found using complex numbers. In fact, all that is required is to affix a translation to the formula of Theorem 7.12.

Theorem 7.14 Let m be a line in the complex plane that contains a point c and that makes an angle ϕ with the positive ray of the real axis, and let b be the complex number representing a vector parallel to m. Then the image of z under the glide reflection $g_{m,\mathbf{b}} = T_{\mathbf{b}} \circ r_m$ is given by

$$g_{m,b}(z) = z_{2\phi}(\overline{z - c}) + c + b,$$

where $z_{2\phi} = \cos(2\phi) + i\sin(2\phi)$.

7.2.4 Problems

1. a. Find the image of (x, y) under the glide reflection $r_{y\text{-axis}} \circ T_{0,-3}$.

   b. Draw a scalene triangle and its image under the transformation of part **a**.

2. a. Explain why every reflection can also be thought of as a glide reflection.

   b. Explain why no translation can be thought of as a glide reflection.

3. a. Find a formula for the image z' of the complex number z under the glide reflection $r_m \circ T_{2,6}$, where m is the line with equation $y = 3x$.

   b. Find a formula for the image (x', y') of the point (x, y) under the glide reflection of part **a**.

   c. Repeat part **a** if the line has equation $y = 3x - 5$.

   d. Check your answer to part **c** by a process different from that used to find the image.

4. Find a formula for the image of (x, y) under the glide reflection $g_{m,\mathbf{b}} = T_{\mathbf{b}} \circ r_m$, where m contains the point (x_0, y_0) and $\mathbf{b} = (h, k)$ is a vector parallel to m.

5. Give a geometrical description of the transformation G of the complex plane defined by $G(z) = i\bar{z} + (1 + i)$.

6. A point F is a **fixed point** of a plane transformation T if $T(F) = F$.

   a. Find all fixed points, if any, of the following plane transformations:

      i. the rotation $R_{C,\phi}$

      ii. the translation $T_{\mathbf{b}}$ by the vector **b**

      iii. the reflection r_m

   b. For given complex numbers a and b with $|a| = 1$, define the plane transformation $G(a, b)$ of the complex plane by $G(a, b)(z) = a\bar{z} + b$. Prove that $G(a, b)$ has fixed points if and only if $a\bar{b} + b = 0$. (*Hint*: If $f = a\bar{f} + b$, then $\bar{f} = \bar{a}f + \bar{b}$ by properties of complex conjugation.)

   c. If $a\bar{b} + b = 0$, show that $a\bar{z} + b = a\overline{\left(z - \frac{b}{2}\right)} + \frac{b}{2}$.

   d. Use parts **b** and **c** to conclude that if $G(a, b)$ has fixed points, then $G(a, b)$ is a reflection.

   e. Prove that if $G(a, b)$ does not have fixed points, then it is a reflection followed by a translation.

ANSWERS TO QUESTIONS

1. Because it is the composite of two isometries.
2. Only if the translation is the identity transformation. Then points on the glide reflecting line are mapped onto themselves. The glide reflection is then a reflection.

7.2.5 Are there other congruence transformations?

Let us return to the intuitive idea behind congruence. Two figures are congruent if we can reflect or rotate or translate one onto the other. In Section 7.2.3, we showed that every rotation and every translation is a composite of two reflections. Consequently, we can formally restate the intuitive notion as follows: Two figures are congruent if and only if we can get from one to the other by a reflection or composite of reflections.

But in Section 7.2.4, we saw that some composites of reflections give rise to a new transformation, the glide reflection. Are there others? The answer is no. In this section, we show that every congruence transformation is either a translation, a rotation, a reflection, or a glide reflection. For this reason, reflections can be regarded as the basic building blocks of congruence transformations in the sense that any congruence transformation can be expressed as the composite of one, two, or three reflections.

Composites of three reflections

Because two lines are either intersecting or parallel, we have already considered all composites of two reflections. So now we consider all composites of three reflections $r_c \circ r_b \circ r_a$ by examining all possible configurations of three lines. There are three cases to consider. The lines may have no points of intersection (Case 1, Figure 41), 1 point of intersection (Case 2, Figure 42), or 2 or more points of intersection (Case 3, Figure 43). In all cases, a', b', and c' are images of a, b, and c, respectively. In Case 3, b'' is the image of b'.

Figure 41

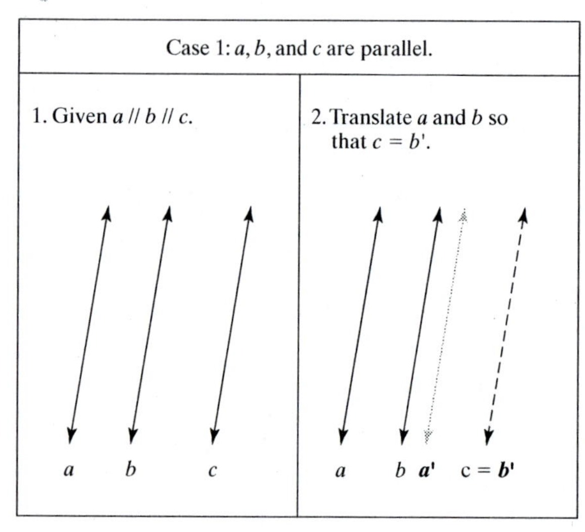

Figure 42

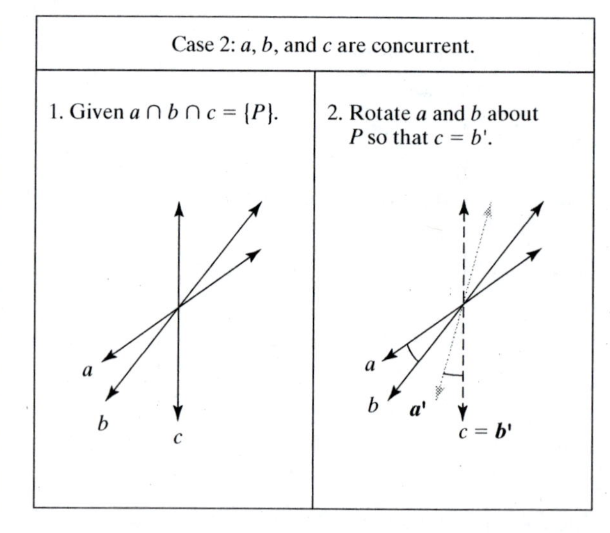

In the first two cases, the algebra is the same. Because of the Rotation and Translation Flexibility Theorems 7.10(a) and 7.10(b),

$$\begin{aligned} r_c \circ r_b \circ r_a &= r_c \circ (r_{b'} \circ r_{a'}) & \\ &= r_c \circ (r_c \circ r_{a'}) & \text{(since } c = b'\text{)} \\ &= (r_c \circ r_c) \circ r_{a'} & \text{(since composition is associative)} \\ &= r_{a'}. & \text{(since } r_c \circ r_c = I\text{)} \end{aligned}$$

So the composite of three reflections over parallel or concurrent lines is a single reflection.

In Case 3, two lines a and b intersect at a point P and the third line c does not contain P nor is it parallel to a or b, so there is a second point of intersection (but we do not need to use the second point).

Figure 43

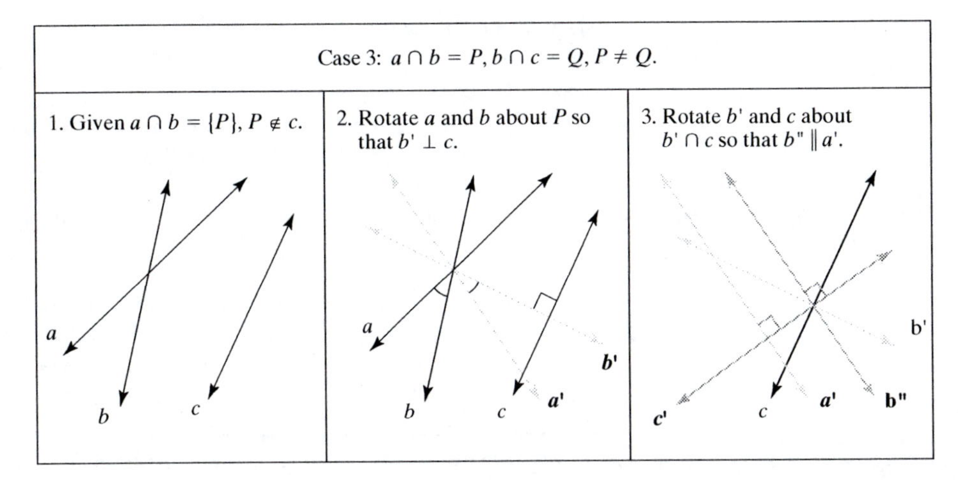

In Case 3, because of the Rotation Flexibility Theorem,

$$
\begin{aligned}
r_c \circ r_b \circ r_a &= r_c \circ (r_{b'} \circ r_{a'}) \\
&= (r_c \circ r_{b'}) \circ r_{a'} && (b' \perp c) \\
&= (r_{c'} \circ r_{b''}) \circ r_{a'} && (b'' \perp c' \text{ and } b'' /\!/ a) \\
&= r_{c'} \circ (r_{b''} \circ r_{a'}) \\
&= r_{c'} \circ T, \text{ where } T \text{ is a translation (since } b'' /\!/ a').
\end{aligned}
$$

Since $b'' \perp c'$, the direction of T is parallel to c'. Consequently, $r_{c'} \circ T$ is a glide reflection. These three cases exhaust the possibilities, proving the following theorem.

Theorem 7.15 The composite of three reflections is either a reflection or a glide reflection.

Composites of four reflections

What about the composite of four reflections $r_d \circ r_c \circ r_b \circ r_a$? Any such composite can be considered as the composite of r_d following the reflection or glide reflection $r_c \circ r_b \circ r_a$. If $r_c \circ r_b \circ r_a$ is a single reflection, then $r_d \circ (r_c \circ r_b \circ r_a)$ is the composite of two reflections, and so it must be a rotation or translation. If $r_c \circ r_b \circ r_a$ is a glide reflection, say $r_{c'} \circ T$, then $r_d \circ (r_c \circ r_b \circ r_a) = r_d \circ (r_{c'} \circ T) = r_d \circ r_{c'} \circ T$. Now only two cases remain.

If $r_d \circ r_{c'}$ is a translation, then $r_d \circ r_{c'} \circ T$ is the composite of two translations. In Section 7.2.1, we showed that the composite of two translations is a translation.

The other case is that $r_d \circ r_{c'}$ is a rotation following the translation $T = r_{b'} \circ r_{a'}$, as shown in Figure 44.

$$
\begin{aligned}
(r_d \circ r_{c'}) \circ (r_{b'} \circ r_{a'}) &= (r_d \circ r_{c'}) \circ (r_{b''} \circ r_{a''}) && \text{(Rotation Flexibility Theorem)} \\
&= (r_d \circ r_{c'} \circ r_{b''}) \circ r_{a''} && (\circ \text{ is associative})
\end{aligned}
$$

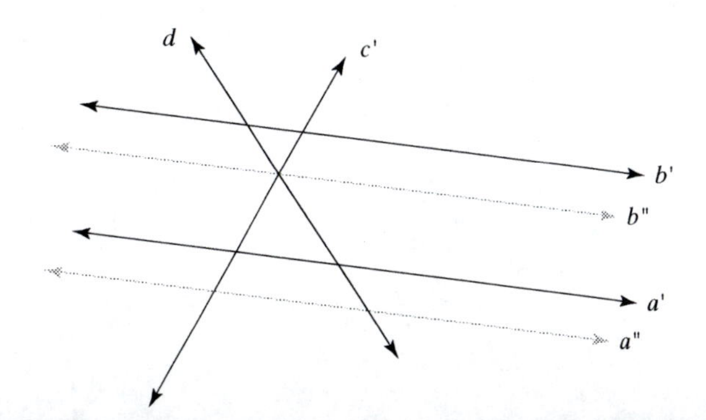

Translate a' and b' so that b'' contains the intersection of c' and d. Since d, c', and b'' are concurrent, the composite $r_d \circ r_{c'} \circ r_{b''}$ is a single reflection. Thus the original composite of four reflections can be reduced to a composite of two reflections, which means the original composite is either a rotation or translation.

The result of all this is the following theorem.

Theorem 7.16 The composite of four reflections is either a translation or a rotation.

Since any composite of four reflections can be reduced to a composite of two reflections, any composite of five can be reduced to three, any composite of six can be reduced to four, and so on. This shows that the four types of transformations we have discussed exhaust all the possibilities. There are no other composites of reflections.

Theorem 7.17 Every composite of reflections is either a reflection, a rotation, a translation, or a glide reflection.

Distance-preserving transformations

Remember that we have defined two figures to be *congruent* if and only if they are related by a distance-preserving transformation. Although we know that reflections, rotations, translations, and glide reflections preserve distance, and these are the only composites of reflections, we have not yet shown that there are no other distance-preserving transformations. Now, we prove that there are no congruence transformations other than the ones we have seen. Notice in these proofs that the images of only three noncollinear points are enough to determine all possible distance-preserving transformations. This is why triangle congruence is enough to determine the congruence of any figures in a plane.

Recall that a point P is a fixed point under a transformation T if and only if $T(P) = P$, that is, T maps P onto itself.

Theorem 7.18 A distance-preserving transformation with a fixed point is either a reflection or a rotation.

Proof: Suppose T is distance-preserving, A is a fixed point under T, and T is not the identity. Now consider a triangle ABC, and let $B' = T(B)$ and $C' = T(C)$. $AB = AB'$ and $AC = AC'$, because T preserves distance. This means that B and B' are on the same circle with center A. Once the location of B' is known, because $BC = B'C'$, C' must be on a circle with center B' and radius BC. Also, since $AC = AC'$, C and C' are on the same circle with center A. This means that there are only two possible locations for C', because C' is at the intersection of two known circles with different centers. These two possibilities are shown in Figure 45.

Figure 45

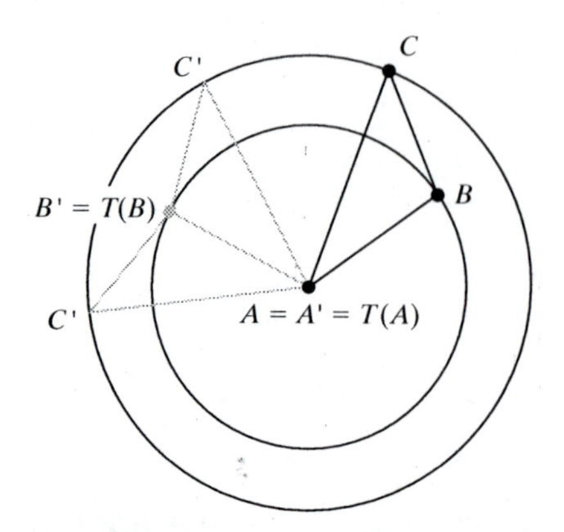

If ΔABC and $\Delta AB'C'$ have the same orientation, then C' is the image of C under the rotation with center A and magnitude $m\angle BAB'$. Notice that the center and magnitude of this rotation are independent of C. On the other hand, if ΔABC and $\Delta AB'C'$ have opposite orientation, then C' is the image of C under the reflection over the perpendicular bisector of $\overline{BB'}$. This reflecting line contains A because $\Delta ABB'$ is isosceles with base $\overline{BB'}$, and its location is seen to be independent of C. Thus, other than the identity, there are only two possible distance-preserving transformations with a fixed point A a rotation with center at A or a reflection about a line through A. ┘

Because a distance-preserving transformation with a fixed point is either a reflection or a rotation, it can be viewed as the composite of at most two reflections. Now we are very close to the conclusion that we desire.

Theorem 7.19 Every distance-preserving transformation T is a composite of reflections.

Proof: If T has a fixed point, then Theorem 7.18 shows that T is a composite of reflections. If T has no fixed point, then suppose $T(A) = B$, with $B \neq A$. Now let m be the perpendicular bisector of $\overline{AB}$. We create a transformation with a fixed point by composing T and r_m. By definition of reflection,

$$r_m(B) = A.$$
$$r_m \circ T(A) = r_m(B) = A.$$

So $r_m \circ T$ has the fixed point A. So $r_m \circ T$ equals either a rotation R or a reflection r. Now we do a cute thing. We compose r_m with $(r_m \circ T)$. The result is T. That is,

$$T = r_m \circ (r_m \circ T) = r_m \circ R$$

or

$$T = r_m \circ (r_m \circ T) = r_m \circ r.$$

In either case T is a composite of reflections. ┘

Corollary 1: Every distance-perserving transformation is either a reflection, a translation, a rotation, or a glide reflection.

Thus the following sets of transformations in a plane are identical:

the set of all composites of reflections
the set of all composites of three or fewer reflections
the set of all composites of reflections, rotations, and translations,

and they are identical to the following sets, which are equal by their definitions

the set of all distance-preserving transformations
the set of all congruence transformations
the set of all isometries.

Corollary 2: If two figures are directly congruent, then either one is the image of the other under a rotation or translation. If two figures are oppositely congruent, then either one is the image of the other under a reflection or glide reflection.

Composition is always associative. The sets named above are closed under composition because the composite of a composite of reflections is also a composite of reflections. Their identity is in each case the identity transformation. The inverse of any distance-preserving transformation is certainly distance-preserving. Thus each of these sets, with $\circ$, form the same group. It is called the **congruence group**.

Theorem 7.20 With composition, the set of all congruence transformations is a group.

7.2.5 Problems

1. Which of the four types of composites of reflections have the indicated property?
   a. preserve orientation
   b. switch orientation
   c. can be the composites of 2001 reflections

2. For what values of n does a composite of n reflections preserve orientation?

3. Trace Figures 46 and 47. Find a line z so that $r_z \circ r_y = r_x \circ r_w$.

a.

Figure 46

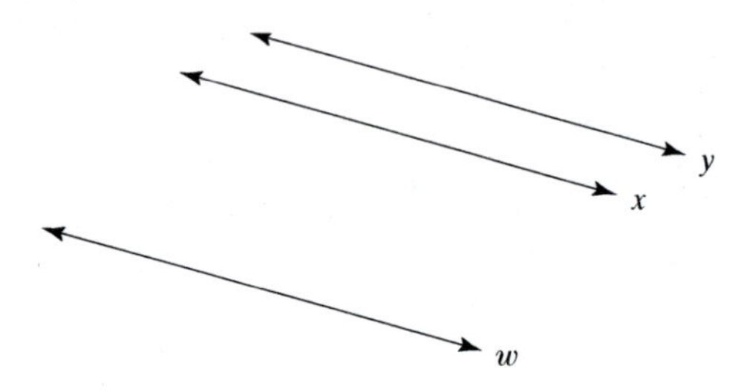

b.

Figure 47

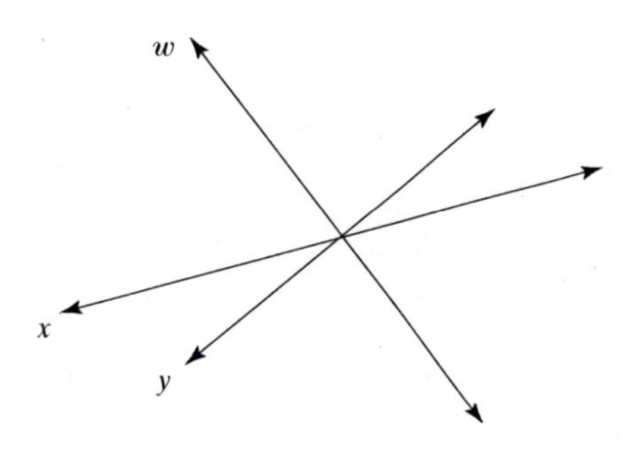

4. Prove: The composite of a translation following a rotation is a rotation.

*5. Prove: The composite of two rotations with magnitudes ϕ and θ and different centers is a translation if $\phi + \theta \equiv 0 \text{ Mod } (2\pi)$, and a rotation otherwise.

6. Classify the composite as one of the four types of transformations we have discussed.
   a. reflection over the x-axis followed by rotation of 80° with center $(5, 0)$
   b. reflection over the line $y = x$ followed by the translation $T_{7,1}$
   c. rotation of 75° with center $(0, 0)$ followed by a rotation of 75° with center $(10, 0)$
   d. rotation of 90° with center $(0, 0)$ followed by the translation $T_{\pi,0}$

7. Recall that a rotation of 180° is also called a *half-turn*. Give a full description of the composite of the two half-turns with centers A and B. (*Hint*: Do drawings.)

8. Examine a chapter on triangle congruence proof in a traditional school geometry textbook. A number of pairs of congruent triangles will be pictured in the text and problems. How many of these pairs are related by reflections? How many by rotations? How many by translations? How many by glide reflections?

9. In Corollary 1 to Theorem 7.19, the assertion is made that six sets of transformations are identical. Explain why this assertion is true. (You may use any theorems from this chapter.)

7.2.6 Congruent graphs

In this section, we prove four powerful theorems about graphs. Although most of the graphs we consider are of relations in two variables, the ideas can be applied in contexts involving one variable as well as contexts involving three or more variables.

Question: The graph of $39x - 31y = 700$ is a line. Two lattice points on this graph are $(41, 29)$ and $(10, -10)$. With this information, you can quite easily find two points on the graph of $39(x + 8) - 31(y - 5) = 700$. Find these points before reading on.

The question is easy to answer if you see the calculations that need to be made in the first equation to get the left side to equal 700. Then, for each point on the line $39x - 31y = 700$, you must *subtract* 8 from the x-coordinate and *add* 5 to the y-coordinate to get a point on the graph of the second line in order to compensate for the addition and subtraction in the equation. Corresponding to the point $(41, 29)$ on the first line is the point $(33, 34)$ on the second; corresponding to $(10, -10)$ is the point $(2, -5)$.

In general, for each point (a, b) on the graph of the first line, there is a point $(a - 8, b + 5)$ on the graph of the second line. This indicates that the second line is the image of the first under the translation T with $T(x, y) = (x - 8, y + 5)$. This is shown in Figure 48.

Figure 48

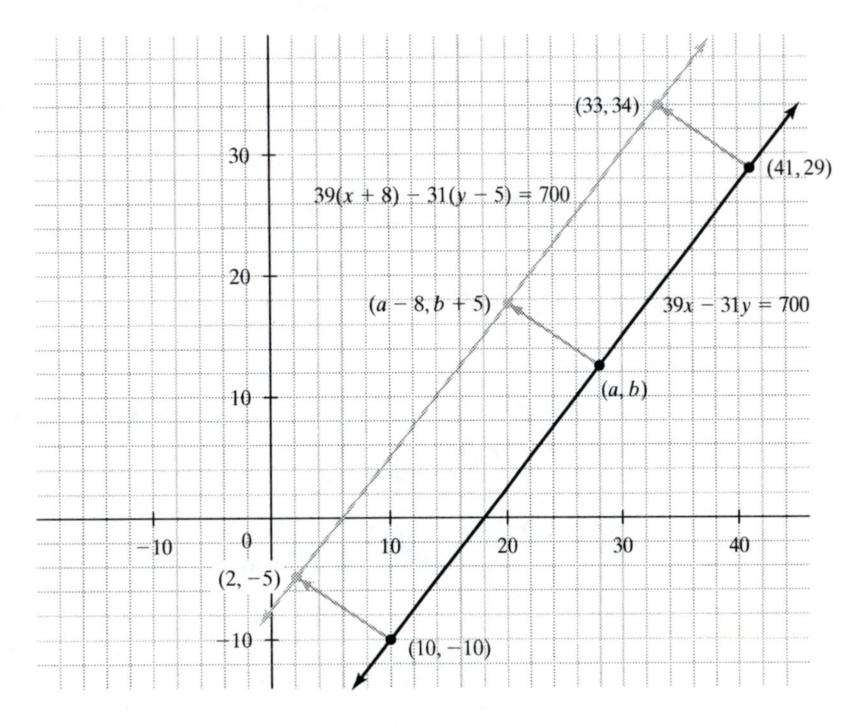

The argument above illustrates a powerful theorem about graphing.

Theorem 7.21 **(Graph Translation Theorem):** In an equation or inequality in x and y with graph G, the following two processes yield the same Cartesian coordinate graph congruent to G.

1. applying the translation $T_{h,k}(x, y) = (x + h, y + k)$ to G;
2. replacing x by $x - h$, and y by $y - k$.

Proof: Let (x', y') be a point on the image graph G' from the translation in (1). Then $x' = x + h$ and $y' = y + k$. Solving these equations for x and y, $x = x' - h$ and $y = y' - k$. Thus, by substitution, we obtain an equation or inequality in $x' - h$ and $y' - k$. Because we customarily drop the "primes" in x' and y' in equations or inequalities, (2) follows. G' and G are congruent because translations are congruence transformations. ┘

The Graph Translation Theorem has many familiar applications in school mathematics, some of which are given here as corollaries. In the corollaries, we use the letters most commonly found in textbooks. Because different letters are used for the variables and because the equations or inequalities for the graphs may be in slightly modified form, you may not have realized that these are all special cases of one theorem.

Corollary 1: The graph of $y = a(x - h)^2 + k$ is a parabola congruent to the graph of $y = ax^2$, with vertex (h, k), line of symmetry $x = h$, and opening upward.

Proof: The graph of $y = ax^2$ is a parabola with vertex $(0, 0)$, line of symmetry $x = 0$ (the y-axis), and opening upward. Translating that graph using $T_{h,k}(x, y) = (x + h, y + k)$, the image will be congruent with everything translated h units to the right and k units up. Thus the image is a parabola with vertex (h, k), line of symmetry $x = h$, and opening upward. From the Graph Translation Theorem, its equation is $y - k = a(x - h)^2$, which becomes the given equation by adding k to both sides. ⌟

Corollary 2 (Point-Slope Formula for a Line): An equation for the line through (x_0, y_0) with slope m is $y - y_0 = m(x - x_0)$.

Proof: The line $y = mx$ contains $(0, 0)$ and has slope m. Translating this line using T_{x_0, y_0}, we replace y by $y - y_0$ and $x - x_0$. ⌟

Corollary 3 (Slope-Intercept Formula for a Line): An equation for the line with slope m and y-intercept b is $y = mx + b$.

Corollary 4 (Phase Shift): The graph of $y = A\sin(x - B) + C$ is the sine wave $y = A\sin x$ shifted B units to the right and C units up.

Proof: $y = A\sin(x - B) + C \Leftrightarrow y - C = A\sin(x - B)$. By the Graph Translation Theorem, the graph of $y - C = A\sin(x - B)$ is the image of the graph of $y = A\sin x$ under $T_{B,C}$. ⌟

We have focused on equations for relations. In function language, if $y = f(x)$, replacing x by $x - h$ and y by $y - k$ means considering a new image function f' with equation $y = f(x - h) + k$. The function $f'\colon x \to f(x - h) + k$ can be viewed as the composite $T_{0,k} \circ f \circ T_{h,0}$ of three functions. That is, to achieve f' we do a horizontal translation, then apply f, then apply the vertical translation.

There are theorems analogous to the Graph Translation Theorem for other transformations. Reflections are somewhat special since every reflection is its own inverse, and you may have encountered them in your earlier study.

Theorem 7.22 **(Graph Reflection Theorem):** In an equation or inequality in x and y with graph G, the following processes yield the same Cartesian coordinate graph congruent to G:

a. **1.** reflecting G over the y-axis; and
2. replacing x by $-x$.

b. **1.** reflecting G over the x-axis; and
2. replacing y by $-y$.

c. **1.** reflecting G over the line $x = y$; and
2. switching x and y.

Proof (a): Let (x', y') be a point on the image graph. For reflection over the y-axis, $x' = -x$ and $y' = y$. Solving these equations for x and y, $x = -x'$ and $y = y'$. Thus, by substitution, the original equation or inequality is replaced by the same one with $-x'$ in place of x and y' in place of y. Because we customarily drop the "primes" in x' and y' when we give equations for relations, the theorem follows.

Proofs (b) and (c): These are left to you. ┘

The analytic description of the general rotation image is more complicated than that of the reflections mentioned in Theorem 7.22 or any translation. Consequently, to rotate a graph requires a more complicated substitution.

Theorem 7.23 **(Graph Rotation Theorem):** In an equation or inequality in x and y with graph G, the following processes yield the same Cartesian coordinate graph congruent to G:

1. applying the rotation R_ϕ to G;
2. replacing x by $x \cos \phi + y \sin \phi$, and y by $-x \sin \phi + y \cos \phi$.

Proof: Let (x', y') be a point on the image graph G'. Then $(x', y') = R_\phi(x, y) = (x \cos \phi - y \sin \phi, x \sin \phi + y \cos \phi)$ by Theorem 7.5. This yields the system

$$x' = x \cos \phi - y \sin \phi$$
$$y' = x \sin \phi + y \cos \phi.$$

Solving this system for x and y will tell us what substitution should be made in the original equation or inequality. These details and the rest of the proof are left for you. (*Hint*: Follow the ideas in the proofs of Theorems 7.21 and 7.22.) ┘

EXAMPLE Find an equation for the image of the graph of $y = x^2$ under $R_{30°}$ and describe the image.

Solution To find an equation for the rotation image of this graph under a 30° rotation, we make the substitution $x \cos 30° + y \sin 30°$ for x and $-x \sin 30° + y \cos 30°$ for y. Since $\cos 30° = \frac{\sqrt{3}}{2}$ and $\sin 30° = \frac{1}{2}$, an equation for the image is

$$-\frac{x}{2} + \frac{y\sqrt{3}}{2} = \left(\frac{x\sqrt{3}}{2} + \frac{y}{2}\right)^2.$$

Multiplying both sides by 4 and then expanding the right side,

$$-2x + 2\sqrt{3}y = 3x^2 + 2\sqrt{3}xy + y^2.$$

In standard form, this equation is $3x^2 + 2\sqrt{3}xy + y^2 + 2x - 2\sqrt{3}y = 0$. The graph of $y = x^2$ is a parabola with vertex at $(0, 0)$, opening up, and with the y-axis as an axis of symmetry (Figure 49). The image graph will be a parabola with the same vertex. Its axis of symmetry is the image of the y-axis under a 30° rotation. This is the line that contains $(\cos 120°, \sin 120°)$, or $\left(-\frac{1}{2}, \frac{\sqrt{3}}{2}\right)$, so an equation for it is $y = -\sqrt{3}x$.

Figure 49

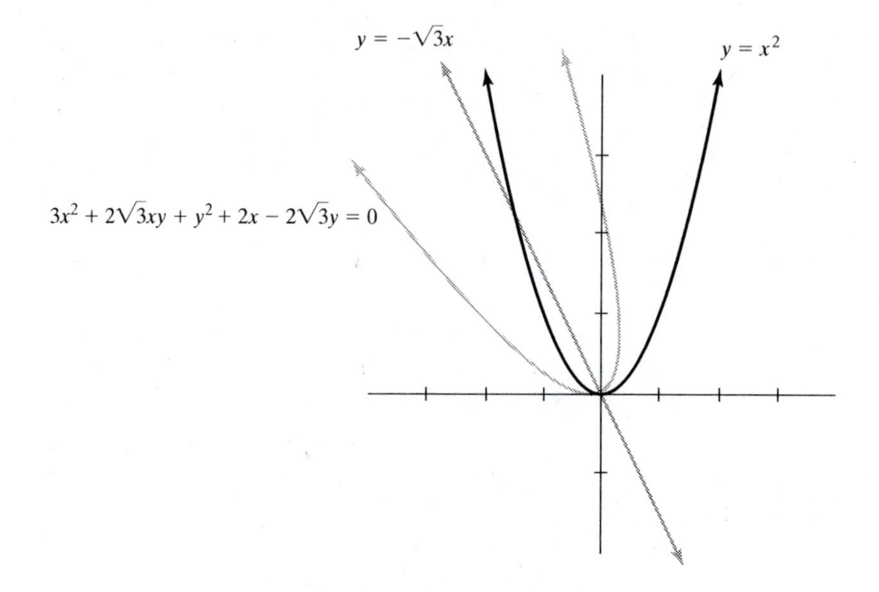

There is an evident pattern to Theorems 7.21 through 7.23 and their proofs. The substitutions for x and y in order to determine the equation for the image of a graph under a transformation T are what x' and y' respectively equal in the formula for T^{-1}.

Theorem 7.24 **(General Graph Transformation Theorem):** In an equation or inequality in x and y with graph G, the following processes yield the same Cartesian coordinate graph:

1. applying a transformation T to G;
2. replacing (x, y) by $T^{-1}(x, y)$.

7.2.6 Problems

1. a. Find two lattice points on $61x + 39y = 2000$.

b. Find two lattice points on $61(x + 14) + 39(y - 167) = 2000$.

2. Explain how the graph of $y = mx + b$ is related to the graph of $y = mx$, and thus prove Corollary 3 to the Graph Translation Theorem.

3. Describe the graphs of the two relations in x and y.

a. $x^2 + y^2 = r^2$, $(x - h)^2 + (y - k)^2 = r^2$

b. $x^2 + y^2 < r^2$, $(x - h)^2 + (y - k)^2 < r^2$

c. $y = a|x|$; $y = a|x - h| + k$

d. $\frac{x^2}{a^2} + \frac{y^2}{b^2} = 1$; $\frac{(x-h)^2}{a^2} + \frac{(y-k)^2}{b^2} = 1$

e. $\frac{x^2}{a^2} - \frac{y^2}{b^2} = 1$; $\frac{(x-h)^2}{a^2} - \frac{(y-k)^2}{b^2} = 1$

4. Prove that the graphs of $y = \sin x$ and $y = \cos x$ are congruent.

5. Use the Graph Translation Theorem and the fact that $ax^2 = k$ has the solutions $\pm\sqrt{\frac{k}{a}}$ to find all solutions to $a(x - h)^2 = k$. (This shows why completing the square is a very effective procedure for solving quadratic equations.)

6. Prove that the graphs of $y = \log kx$ are congruent for all positive values of k.

7. a. There are 222 units of a quantity, and it is being reduced by 0.4 units each year. How many units will there be n years from now?

b. The world record in the mile has been changing by about 4 seconds a decade ever since 1920. The world record in 1999 was 3 minutes, 42 seconds. From this information, find an equation approximately relating the year Y and the record time t for the last four-fifths of the 20th century.

8. In October 1999, the world population passed 6 billion and was growing at a rate of 1.3% a year. If this growth rate were

to continue, then n years after 1999, the world population P (in billions) would be given by $P = 6(1.013)^n$. Modify this formula so that it gives the population P in the year Y.

9. In Section 3.1.2, the equation $t = \dfrac{3}{\frac{w}{50} - 1}$ is graphed. Show that this graph is a translation image of one branch of the graph of an equation $wt = c$, where c is a constant, and find c.

10. In a 3-dimensional Cartesian coordinate system, the sphere with center $(0, 0, 0)$ and radius r has equation $x^2 + y^2 + z^2 = r^2$. What is an equation for the sphere with center (a, b, c) and radius r?

11. In the complex plane, $\{z: |z| = k, k \in \mathbf{R}, k > 0\}$ is the circle with center 0 and radius k. Give an equation for the circle with center z_0 and radius k.

12. Suppose a relation is described in polar coordinates $[r, \theta]$ and $\theta - \phi$ is substituted for θ. What happens to the graph?

13. a. Prove part (b) of the Graph Reflection Theorem.

b. Prove part (c) of the Graph Reflection Theorem.

14. Complete the proof of the Graph Rotation Theorem.

15. Find an equation for the image of the graph of the given relation under the indicated rotation.

a. $x^2 - y^2 = 1$, $R_{45°}$

b. $x^2 - y^2 = 1$, $R_{60°}$

c. $y = x^2$, $R_{90°}$

d. $y = x^2$, $R_{45°}$

e. $y = x^2$, $R_{-90°}$

f. $\frac{x^2}{4} + \frac{y^2}{9} = 1$, $R_{90°}$

16. Prove that the graphs of $y = \tan x$ and $y = \cot x$ are congruent.

17. Prove that the graphs of $f(x) = e^x$ and $g(x) = 1 - e^{-x}$ are congruent.

18. a. What substitution should be made for (x, y) in a relation in order to reflect the graph of the relation over the line $x = 3$?

b. Verify your answer to part **a** with a relation of your own choosing.

Unit 7.3 Symmetry

Symmetry is a rich topic of study with important applications to biology, to art and architecture, to chemistry, and to physics ranging from the everyday to the esoteric. The mathematical study of symmetry also is broad, ranging from tasks of paper-folding to the study of group theory.

Without any study of mathematics, people can recognize by folding when a geometric figure possesses reflection symmetry with respect to a vertical or horizontal line. Consequently, some schoolbooks treat symmetry as somewhat of a trivial phenomenon, meant for the slower student as a recreation. Or symmetry is treated as an algebraic phenomenon. In the study of functions and relations in high school, students may be asked to determine algebraically whether a graph is symmetric to the x-axis, y-axis, or line $x = y$.

Curiously, these most common instances of symmetry in the study of mathematics usually do not include the fundamental relationship between symmetry and congruence. A gap is consequently created between the geometric intuition by which we recognize congruent parts within a symmetric figure and the mathematical theory that we employ to deduce that those parts are congruent. In this unit, we discuss the elementary mathematics of symmetry and show how it can be used to deduce results about geometric objects.

7.3.1 Reflection symmetry

It is not uncommon for students of mathematics to learn that certain figures possess symmetry, particularly reflection symmetry. They learn that an isosceles triangle has a line of symmetry and that a square has four lines of symmetry. But these results are all from *sight* or have been found informally by drawing and folding. In this section, we *deduce* the reflection-symmetry properties of many of the common figures of elementary geometry. The statements of many of the theorems in this section are likely to be known to you, but you may never have seen proofs of them. Possibly you have never even thought of how they could be proved.

The defining condition for a figure to possess symmetry is exceedingly simple. Reflection symmetry and all the other symmetries are only special cases of the following definition.

Definition A figure F is **symmetric** with respect to a transformation T if and only if $T(F) = F$.

Usually T is a congruence transformation. If T is the reflection r_m, F is **reflection-symmetric** or **symmetric with respect to m**, or **symmetric to m**, and m is called a **symmetry line** or **axis of symmetry** for F. Humans and most mammals are roughly *bilaterally symmetric*, the biologic term for reflection-symmetric. Perhaps this is the reason that it is so easy for us to recognize reflection symmetry with respect to a vertical line.

Many of the common figures studied in geometry possess reflection symmetry. In order to prove that a figure is reflection-symmetric, the following theorem is very handy. It enables us to establish reflection symmetry simply by showing that the reflection image of a figure is a subset of the original figure.

Theorem 7.25 Let F be any figure and let r_m be the reflection over line m. If $r_m(F) \subset F$, then F is reflection-symmetric with respect to m.

Proof:

$r_m(F) \subset F$	Given
$r_m(r_m(F)) \subset r_m(F)$	For any sets, $A \subset B \Rightarrow f(A) \subset f(B)$.
$F \subset r_m(F)$	$r_m \circ r_m$ is the identity transformation.
$r_m(F) = F$	For any sets, $A \subset B$ and $B \subset A \Rightarrow A = B$.
F is symmetric to m.	Definition of reflection symmetry

We begin our study of the reflection symmetry of plane figures with the circle. It is surprisingly easy to show that a circle has infinitely many lines of symmetry.

Theorem 7.26 Every circle is reflection-symmetric with respect to any line through its center.

Figure 50

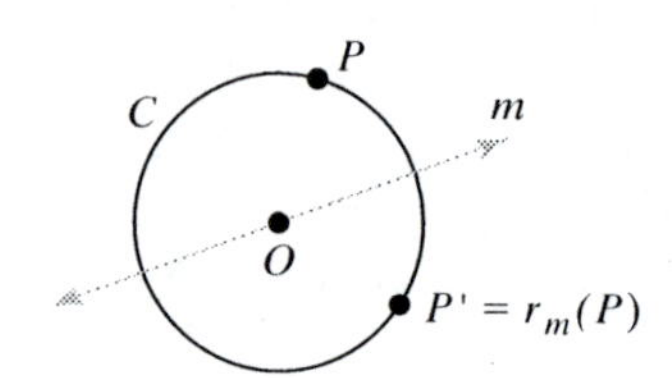

Proof: Let C be a circle with center O. Because of Theorem 7.25, all we need to show is that for any point P on C, $r_m(P)$ is on C (see Figure 50).

Let $P' = r_m(P)$. Since m contains O, $r_m(O) = O$. Because reflections preserve distance,

$$OP' = OP.$$

Hence P' is on circle C. (Now we are essentially done.) Since for all P, $r_m(P)$ is on C,

$$r_m(C) \subset C.$$

Thus, from Theorem 7.25, $r_m(C) = C$ and so C is symmetric to m.

Notice that, other than the definition of symmetry (which obviously must be involved to prove a figure symmetric), the proof of Theorem 7.25 involves only general properties of sets and the fact that every reflection is its own inverse. The proof of Theorem 7.26 uses only distance, which is the defining characteristic of a circle. These proofs are brief and elegant because the characterization of symmetry using transformations gets to the essence of symmetry, and no extraneous ideas are needed.

To display more of this elegance, we consider the symmetry of the ellipse. Often in schoolbooks, this symmetry is introduced only when coordinates are present and there exists an equation for the ellipse. When this is done, a proof is analytic and the ellipse has to be located in a convenient position. With properties of reflections in hand, a synthetic proof is possible regardless of the location of the ellipse.

Recall standard definitions of the ellipse and the hyperbola as a locus of points: Given two points F_1 and F_2 and a real number $k > 0$, the **ellipse with foci F_1 and F_2 and focal constant k** is the set of points P such that $PF_1 + PF_2 = k$. The **hyperbola with foci F_1 and F_2 and focal constant k** is the set of points P such that $|PF_1 - PF_2| = k$.

Theorem 7.27 Every ellipse and every hyperbola has two symmetry lines: the line through its foci and the perpendicular bisector of the segment connecting its foci.

Proof: We prove the theorem for ellipses.

Figure 51

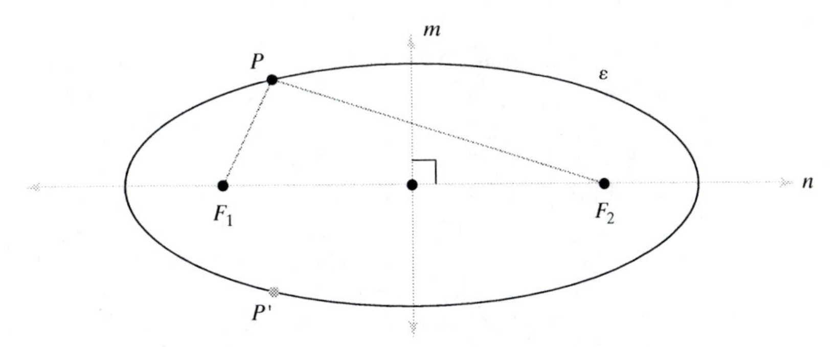

Call the ellipse ε, and let its foci be F_1 and F_2 and its focal constant be k as in the definition of ellipse. So for any point P on ε, $PF_1 + PF_2 = k$.

First consider the line n through F_1 and F_2 (see Figure 51). Proceed as in the proof of circle symmetry, letting P be any point on ε, with $P' = r_n(P)$. We wish to show that P' is on ε. Since n contains F_1 and F_2, $r_n(F_1) = F_1$ and $r_n(F_2) = F_2$. Since reflections preserve distance, $P'F_1 = PF_1$ and $P'F_2 = PF_2$. Adding, $P'F_1 + P'F_2 = PF_1 + PF_2 = k$. Consequently, P' is on ε. Thus $r_n(\varepsilon) \subset \varepsilon$, and ε is symmetric with respect to n.

The proof that the perpendicular bisector of $\overline{F_1F_2}$ is a symmetry line proceeds similarly. Let m be this perpendicular bisector, let Q be any point on ε, and let $Q' = r_m(Q)$. By the definition of reflection, $r_m(F_1) = F_2$ and $r_m(F_2) = F_1$. Since reflections preserve distance, $Q'F_1 = QF_2$ (notice the switch) and $Q'F_2 = QF_1$. Adding, $Q'F_1 + Q'F_2 = QF_2 + QF_1 = k$. So Q' is on ε, and using the same argument as before, from this ε is symmetric with respect to m.

We leave the proof for the hyperbola to you. Figure 52 pictures the situation.

Figure 52

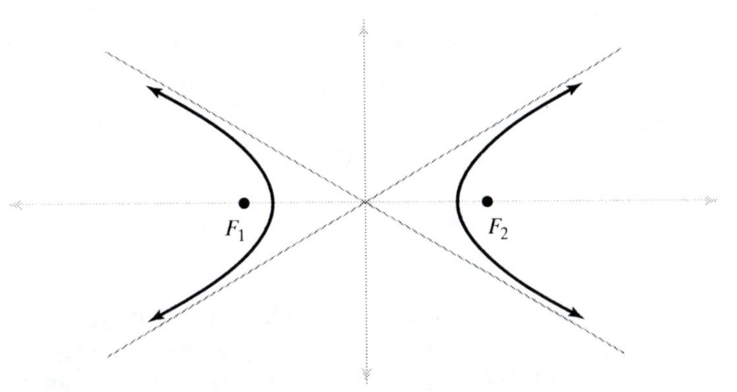

∎

The only other conic section, the *parabola*, has one line of symmetry. You are asked to show this in Problem 3.

Now let us turn to the figures usually studied in plane geometry. One of the most basic, the line segment, has the same symmetry as the ellipse and the hyperbola.

Theorem 7.28 Every line segment has two symmetry lines: itself and its perpendicular bisector.

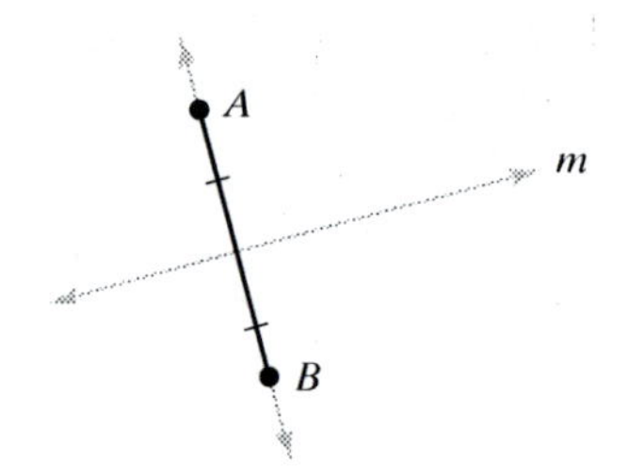

Figure 53

Proof: First consider the segment $\overline{AB}$ itself as a reflecting line (see Figure 53). Since each point of the segment is on the reflecting line, $r_{\overleftrightarrow{AB}}(\overline{AB}) = \overline{AB}$. Consequently, $\overleftrightarrow{AB}$ is a line of symmetry for $\overline{AB}$.

Now let m be the perpendicular bisector of $\overleftrightarrow{AB}$. By the definition of reflection, $r_m(A) = B$ and $r_m(B) = A$. Since reflections preserve betweenness, the image of any point between A and B will be between B and A. Consequently, $r_m(\overline{AB}) \subset \overline{AB}$. Thus $\overline{AB}$ is symmetric to m. ┘

Notice that the proof of Theorem 7.28 is quite similar to the proof of Theorem 7.27. The major difference is the use of distance in Theorem 7.27 compared to the use of betweenness in Theorem 7.28. This difference results from the fact that *line segment* is defined in terms of betweenness while *ellipse* is defined in terms of distance. In addition to the idea that reflections preserve betweenness, the other ideas used in Theorem 7.28 are the definitions of symmetry, reflection, and line segment. The lack of need for additional concepts again demonstrates the basic connection between symmetry and transformations.

The preservation of angle measure by reflections can be used to deduce that an angle has exactly one symmetry line.

Theorem 7.29 Every acute, right, or obtuse angle has exactly one symmetry line.

Figure 54

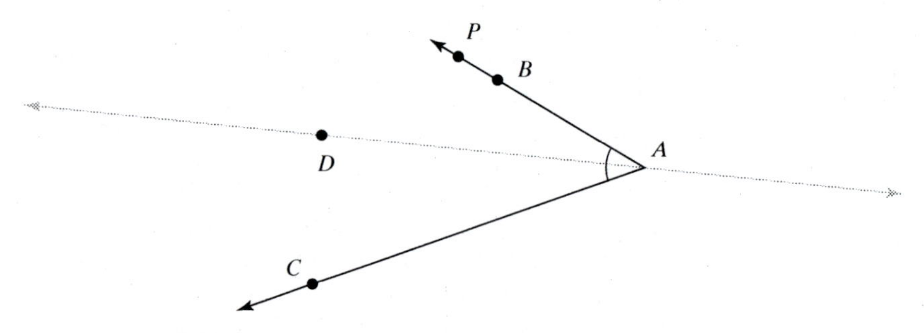

Proof: Let $m = \overrightarrow{AD}$ be the bisector of angle BAC and let P be a point on side $\overrightarrow{AB}$ of the angle as in Figure 54. $r_m(\angle PAD)$ is an angle of the same measure as $\angle PAD$, $r_m(\overrightarrow{AD}) = \overrightarrow{AD}$, and $r_m(P)$ is on the other side of $\overleftrightarrow{AD}$ from P, so $r_m(\angle PAD) = r_m(\angle PAD) = \angle DAC$. This implies $r_m(\overrightarrow{AB}) = \overrightarrow{AC}$. Consequently, $r_m(\overrightarrow{AC}) = \overrightarrow{AB}$, and so (taking the reflection image of the union of its sides), $r_m(\angle BAC) = \angle CAB$, which means $\angle BAC$ is symmetric to m. No other symmetry line exists, because the line of symmetry must bisect the angle, and an angle has exactly one bisector. ┘

We are now ready to consider reflection symmetry in polygons, because a line of symmetry for a polygon can only be either a perpendicular bisector of a side or a bisector of an angle. The simplest polygon with symmetry is the isosceles triangle.

Theorem 7.30 Every isosceles triangle is symmetric to the bisector of its vertex angle.

Figure 55

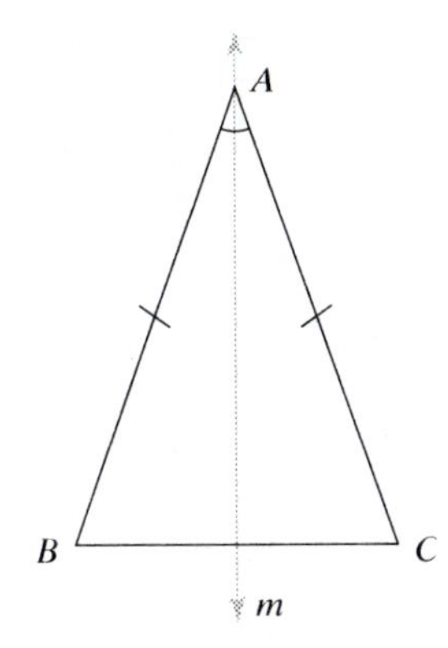

Proof: Let ΔABC be isosceles with $AB = AC$, and let m be the bisector of its vertex $\angle BAC$. $r_m(A) = A$ because A is on m. Now we need to show that $r_m(B) = C$. Because of the symmetry of $\angle BAC$ to m (Theorem 7.29), $B' = r_m(B)$ is on $\overrightarrow{AC}$. Since $AB = AC$ (the triangle is isosceles) and $AB = AB'$ (reflections preserve distance), we have $AB' = AC$. Thus $r_m(B)$ is on $\overrightarrow{AC}$ and the same distance from A as C is, so we must have $r_m(B) = C$. From this, $r_m(C) = B$, and so $r_m(\Delta ABC) = \Delta ACB$. Consequently, ΔABC is symmetric to m. ┘

Corollary: The bisector of the vertex angle of an isosceles triangle is also an altitude, perpendicular bisector, and median of the triangle.

Why prove these theorems about symmetry? One reason is that the symmetry of a figure leads rather automatically to many of its properties. For instance, in Figure 55, since $r_m(A) = A$, $r_m(B) = C$, and $r_m(C) = B$, we have $r_m(\angle ABC) = \angle ACB$. That is, *from the symmetry of the isosceles triangle*, we can deduce that the base angles have the same measure. This symmetry argument is closer to our intuition about isosceles triangles, in which we see the congruence of base angles as a property of the entire figure and not as a result of congruent triangles within the figure.

In the proof of the symmetry of the angle and of the isosceles triangle, we used the fact that the reflection image of the union of two sets equals the union of the reflection images of those sets. This is a general property of one-to-one functions, true also of intersections of sets. That is, if f is any 1-1 function and A and B are subsets of its domain,

$$f(A \cup B) = f(A) \cup f(B)$$
$$f(A \cap B) = f(A) \cap f(B).$$

Figure 56

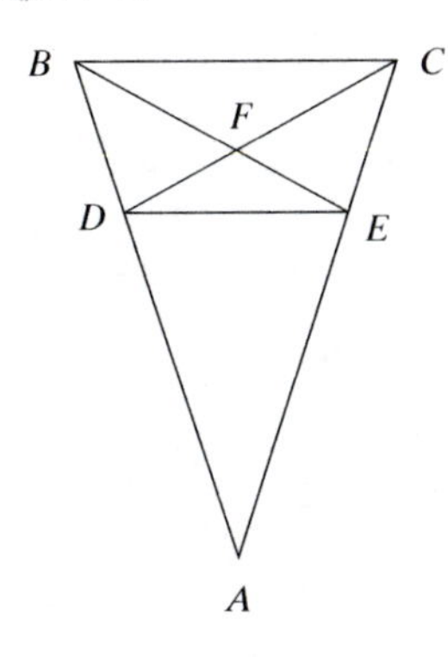

These properties are useful because many geometric figures are combinations of unions or intersections of others.

For instance, consider Figure 56, where $AB = AC$ and $AD = AE$. Since these two isosceles triangles share a common vertex angle, they have a common line of symmetry, the bisector of $\angle A$. Since $\overline{BE} \cap \overline{CD} = \{F\}$ and $r(F) = r(\overline{BE} \cap \overline{CD}) = r(\overline{BE}) \cap r(\overline{CD}) = \overline{CD} \cap \overline{BE} = \{F\}$, F must be on this bisector. Thus the entire figure is symmetric to that line, and as a result any angle and its reflection image over that line have equal measure. Also, any segment has the same length as its reflection image over that line. This provides a symmetry argument proving, for instance, $m\angle CDA = m\angle BEA$ and $DC = BE$.

Put two isosceles triangles together with the same base and remove their intersection, and a *kite* appears (Figure 57). A quadrilateral is a **kite** if and only if it has

Figure 57

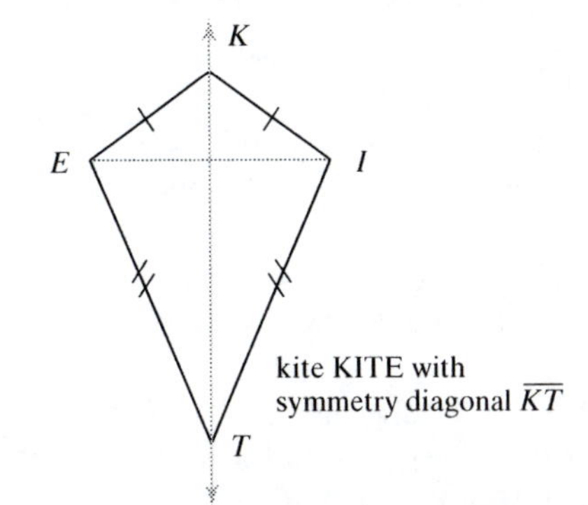

two mutually exclusive pairs of congruent adjacent sides. The perpendicular bisector of the removed base must bisect the vertex angles of the original kite (Corollary to Theorem 7.30), so it contains a diagonal of the kite and is a line of symmetry for the kite. It is called the **symmetry diagonal** of the kite.

Theorem 7.31 Every kite has a symmetry line, the line containing the common vertices of its pairs of congruent sides.

Recall the definition of *rhombus*: A **rhombus** is a quadrilateral with four congruent sides. Consequently every rhombus is a kite.

Corollary: Every rhombus has two symmetry lines that are perpendicular to each other.

The kite's symmetry guarantees that a pair of opposite angles have the same measure. (This could also be deduced by splitting the kite back into its original isosceles triangles and adding up the measures of their base angles.) Because a rhombus is a kite in two different ways, both pairs of opposite angles are congruent.

Caution: The diagrams in Figures 55 through 57 have vertical symmetry lines, but the theorems apply for the figures regardless of their tilt.

7.3.1 Problems

1. A figure is the union of two circles. What are the possible numbers of symmetry lines for the figure?

2. Prove Theorem 7.27 for hyperbolas.

3. Given a point F not on a line m, the **parabola** with focus F and directrix m is the set of points P such that the distance from P to F equals the distance from P to m. From this definition, prove that a parabola has a line of symmetry.

4. Explain why an equilateral triangle has three symmetry lines.

5. Explain how the Corollary to Theorem 7.30 follows from Theorem 7.30.

6. Prove: If F is any figure and m any line, $F \cap m = r_m(F) \cap m$. That is, prove that a figure and its reflection image intersect the reflecting line at the same points.

7. Use the result of Problem 6 and the theorems of this section to prove that the medians of an isosceles triangle are concurrent.

8. In the drawing of Figure 56, let $F = \overline{BE} \cap \overline{CD}$. Use a symmetry argument (not congruent triangles) to explain why $DF = EF$.

9. A figure F is **convex** if and only if for all points $A \in F$ and $B \in F$, $\overline{AB} \subset F$. Some kites are nonconvex. Does Theorem 7.31 hold for nonconvex kites? If so, why? If not, why not?

10. A set of points (x, y) described by an equation or inequality is reflection-symmetric with respect to the line $y = x$ if replacing x by y yields the same equation or inequality. Give similar conditions for a set of points (x, y) to be reflection-symmetric with respect to the given line.

   a. x-axis

   b. y-axis

   c. $y = -x$

   d. $x = 10$

   *e. $x + 2y = 5$

   *f. $y = 4x - 3$

11. Prove that the graph of $y = 3x^2 + 6x - 10$ has the line of symmetry $x = -1$.

12. Let F be a figure. Prove: With composition, the set of congruence transformations that map F onto itself is a group. This is the **symmetry group** of F.

13. Let f be a 1-1 function and A and B be subsets of its domain.

   a. Show that $f(A \cup B) \subset f(A) \cup f(B)$ and $f(A) \cup f(B) \subset f(A \cup B)$, and thus prove that $f(A \cup B) = f(A) \cup f(B)$.

   b. Prove that $f(A \cap B) = f(A) \cap f(B)$ and give an example of such a function f and sets A and B.

7.3.2 Other congruence transformation symmetries

Land animals are likely to possess external reflection symmetry, perhaps because balance is so important in walking and running. Flowers and some sea animals possess rotation symmetry because they need to be able to attract pollinators or see predators coming from any direction. These symmetries different from reflection symmetry are also possessed by common geometric figures.

Rotation symmetry

Pinwheels and circular saw blades are two of the many figures that possess rotation symmetry without possessing reflection symmetry (Figure 58). On the other hand, regular polygons and some gears possess both kinds of symmetry (Figure 59).

Figure 58

Figure 59

Definitions A figure F is **rotation-symmetric** if and only if there exists a rotation R that is not the identity such that $R(F) = F$. The center of the rotation is the **center of symmetry**, or just the **center** of F.

All figures that have a point called the *center*—circles, ellipses, hyperbolas, and regular polygons—have rotation symmetry.[18] This shows how intuitive the idea of rotation symmetry is, for the word *center* arose in these figures independently of the study of symmetry.

The existence of rotation symmetry in some figures with reflection symmetry is due to the following theorem.

Theorem 7.32 If a figure is reflection symmetric with respect to two intersecting lines m and n, then it has rotation symmetry.

Proof: Let F be a figure with intersecting symmetry lines m and n. Thus $r_m(F) = F$ and $r_n(F) = F$. From this, $r_n \circ r_m(F) = r_n(r_m(F)) = r_n(F) = F$. But, by the Two-Reflection Theorem, $r_n \circ r_m$ is a rotation. Thus F is the image of itself under a rotation, and so F has rotation symmetry. ┘

Corollary 1: Every circle possesses rotation symmetry.

If the rotation $R_{C,\phi}$ maps a figure F onto itself, then $R_{C,\phi} \circ R_{C,\phi}$, which is $R_{C,2\phi}$ also maps F onto itself. Continuing this idea, the composite of any number k of rotations $R_{C,\phi}$ maps the figure onto itself. If ϕ is the smallest positive number for which

[18] Figures may have a center of gravity without having a center.

$R_{C,\phi}(F) = F$, and n is the smallest value of $k > 1$ for which $R_{C,k\phi}$ maps F onto itself and also maps each point of F onto itself, then F is said to have **n-fold symmetry**.

Corollary 2: Every ellipse, hyperbola, segment, rhombus, and rectangle has 2-fold symmetry (Figure 60).

Figure 60

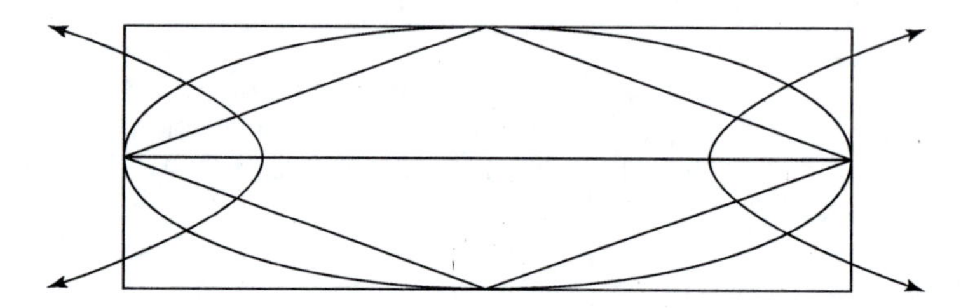

Proof: All these figures have perpendicular symmetry lines. So a rotation of 180° maps each figure onto itself. Composing that rotation with itself yields the rotation of 360°, which is the identity transformation, and that transformation maps each point onto itself. ┘

Recall that a rotation of 180°, under which all of the figures mentioned in Corollary 2 can be mapped onto themselves, is called a *half-turn*. When a figure possesses the 2-fold symmetry given to it by a half-turn, it is said to be **symmetric to** (or **about**) **a point**, or **point symmetric** with respect to the point that is its center.

Half-turns have some special properties. The following are quite useful.

Theorem 7.33 In every half-turn,

a. The center is the midpoint of the segment joining any point other than the center to its image.

b. Any line is parallel to its image.

Figure 61

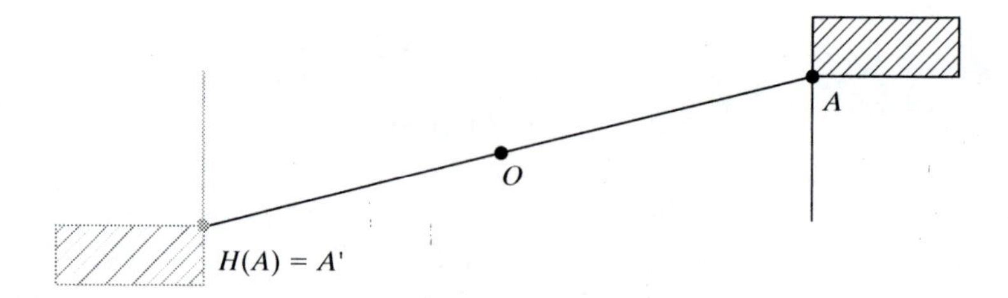

Proof:

a. Let H be a half-turn with center O and let $H(A) = A'$ (Figure 61). Then, by the definition of half-turn, $m\angle AOA' = 180°$, so A, O, and A' are collinear. But also by the definition of rotation, $AO = A'O$. So O is the midpoint of $\overline{AA'}$.

b. Let H be a half-turn with center O and let ℓ be any line. H is the composite of any two reflections over perpendicular lines intersecting at O, so choose one of those lines to be the perpendicular to ℓ through O, and call it m. Then the second line, call it n, is parallel to ℓ. Because $\ell \perp m$, $r_m(\ell) = \ell$. And because $n // \ell$, $H(\ell) = r_n(r_m(\ell)) = r_n(\ell) // \ell$ (see Figure 62).

Figure 62

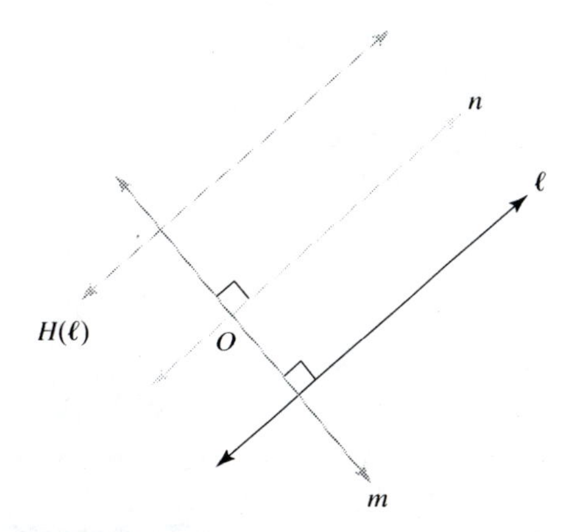

With Theorem 7.33, we can deduce the rotation symmetry of the parallelogram, the figure with rotation symmetry but not reflection symmetry that is most studied in standard treatments of Euclidean geometry. *If* we know that the diagonals of a parallelogram have a common midpoint, then the 2-fold symmetry of the parallelogram follows immediately from the definition of a half-turn. But in the proof given below we assume no such knowledge. All we assume known about the parallelogram is from its definition: Its opposite sides are parallel.

Theorem 7.34 Every parallelogram possesses 2-fold symmetry about the point that is the common midpoint of its diagonals.

Figure 63

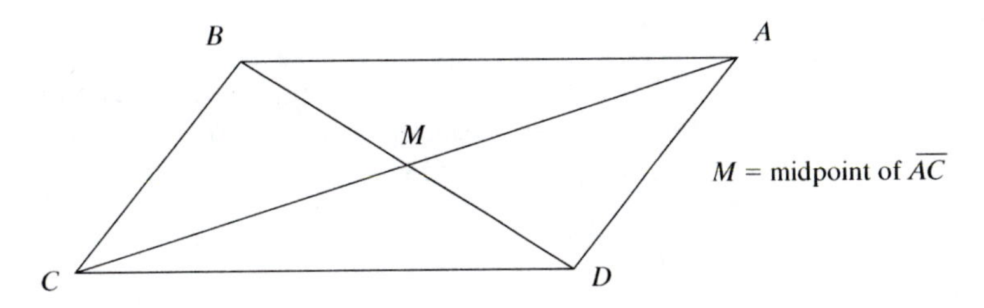

Proof: Let $ABCD$ be a parallelogram with M the midpoint of $\overline{AC}$, as in Figure 63. Consider the half-turn H with center M. Then, by the definition of half-turn, $H(A) = C$ and $H(C) = A$. Now we want to show that $H(B) = D$. Since B is on $\overleftrightarrow{BC}$, $H(B)$ is on $H(\overleftrightarrow{BC})$, which is the line parallel to $\overleftrightarrow{BC}$ through A. So $H(B)$ is on $\overleftrightarrow{AD}$. Likewise, since B is on $\overleftrightarrow{BA}$, $H(B)$ is on $H(\overleftrightarrow{BA})$, which is the line parallel to $\overleftrightarrow{CD}$ through C. So $H(B)$ is on $\overleftrightarrow{CD}$. Since $H(B)$ is on both $\overleftrightarrow{AD}$ and $\overleftrightarrow{CD}$, it must be that $H(B) = D$. Consequently, M is the midpoint of $\overline{BD}$ and $H(D) = B$. So $H(ABCD) = CDAB$ and the parallelogram possesses 2-fold symmetry. ∎

The converse of Theorem 7.34 is also true; every quadrilateral with rotation symmetry is a parallelogram (see Problem 5).

Corollary: Opposite sides of a parallelogram are congruent.

Proof: Opposite sides are the images of each other under the distance-preserving transformation H. ∎

Translation symmetry

Repeating designs, such as those found in wallpaper (Figure 64), in tartans or other textile patterns, or on friezes, possess *translation symmetry*. Translation symmetry is associated with *periodicity*.

Figure 64

Definition A figure F is **translation-symmetric** if and only if there is a translation T not equal to the identity and with $T(F) = F$.

No polygons or conic sections possess translation symmetry. But graphs of periodic functions and some other functions do.

Symmetry of sine waves

Every sine wave possesses not only reflection, rotation, and translation symmetry, but also *glide-reflection symmetry*. All these symmetries are related to algebraic properties of the sine function (see Figure 65).

Figure 65

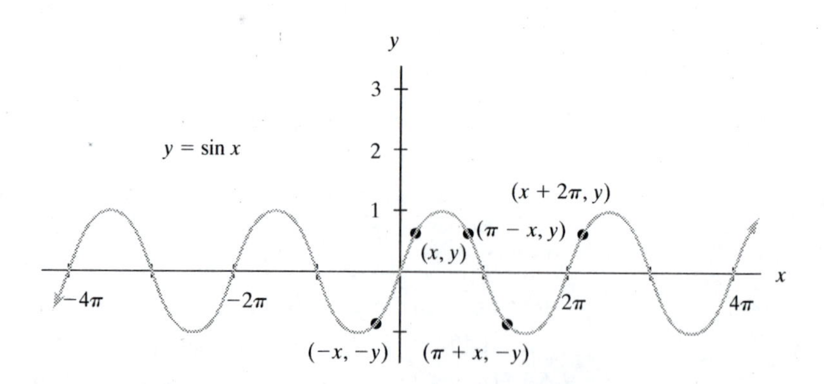

Translation symmetry: A sine wave repeats. The sine function, in particular, is periodic with period 2π. That means that when (x, y) is on the graph, so is $(x + 2\pi, y)$. This is because of the identity, for all x,

$$\sin(x + 2\pi) = \sin x.$$

Rotation symmetry: A sine wave has many centers of symmetry, every point that is halfway between consecutive maximum and minimum points. The sine function has a center of symmetry at $(0, 0)$, meaning that when (x, y) is on its graph, so is $(-x, -y)$. So when $y = \sin x$, then $-y = \sin(-x)$, from which we have the identity

$$\sin(-x) = -\sin x.$$

Reflection symmetry: A sine wave is reflection-symmetric to any line perpendicular to the direction of translation symmetry that contains a maximum or minimum point. For the graph of $y = \sin x$, one of these symmetry lines is $x = \frac{\pi}{2}$. Now the reflection image of (x, y) over this line is $(\pi - x, y)$. So if $y = \sin x$, then also $y = \sin(\pi - x)$. The identity is

$$\sin(\pi - x) = \sin x.$$

Glide-reflection symmetry: Every sine wave has glide-reflection symmetry. For the graph of the sine function, the glide-reflection line is the x-axis. One glide reflection maps the maximum point $\left(\frac{\pi}{2}, 1\right)$ onto the next minimum point $\left(\frac{3\pi}{2}, -1\right)$, and in general maps (x, y) onto $(x + \pi, -y)$. The corresponding identity is

$$\sin(x + \pi) = -\sin x.$$

7.3.2 Problems

1. Draw a figure with 5-fold rotation symmetry and no reflection symmetry.

2. The uncompleted (and unnumbered) black and white square grids for daily crossword puzzles found in most newspapers usually contain a particular type of symmetry. Examine a few such puzzles to find out what this is. (This symmetry is one of the key aids in solving diagramless puzzles.)

3. Prove that every equilateral triangle possesses 3-fold rotation symmetry.

4. Prove that every square possesses 4-fold rotation symmetry.

5. Prove: If a quadrilateral has rotation symmetry, then it is a parallelogram.

6. Many household items possess symmetry, sometimes for decorative purposes, but often for flexibility or to match the symmetry of a natural object. For each item, describe its typical symmetry, give a reason for the symmetry, and if possible name another item with the same symmetry for the same reason.

a. light bulb

b. bathtub

c. doorknob

d. 4-sided card table

7. The graph of $y = \lfloor x \rfloor$ has translation symmetry.

a. Describe the translation with smallest magnitude that maps this graph onto itself.

b. What is this magnitude?

8. Give an identity for each kind of symmetry of the sine function that is different from the one mentioned in this section.

9. All translation symmetries of the graph of the sine function can be described in the identity $\sin x = \sin(x + 2n\pi)$, for all integers n. In a similar manner, describe for the sine function all the symmetries of the given type.

a. rotation symmetries

b. reflection symmetries

c. glide-reflection symmetries

10. Give an example of each of the four kinds of symmetry for the graph of the cosine function, and indicate the corresponding identity.

11. Repeat Problem 10 for the graph of the tangent function.

12. Repeat Problem 10 for the graph of the cosecant function.

13. Find all the symmetries of the graph of each function.

a. $f(x) = ax^3$

b. $g(x) = a + \frac{b}{x-c}$

c. $h(x) = \sin x + \tan x$

d. $j(x) = ax^2 + bx + c$

*e. $k(x) = ax^3 + bx^2 + cx + d, a \neq 0$

Unit 7.4 Traditional Congruence Revisited

In this unit, we return to some familiar theorems about congruence, but along the way develop some theorems that may not be so familiar.

7.4.1 Sufficient conditions for congruence

The purpose of this section is to use the definition of congruence via transformations to deduce the defining properties of congruent segments and angles, and also to deduce all the triangle congruence propositions. We assume basic properties of lengths of segments and measures of angles that do not require congruence. We utilize the definitions of transformations given in Sections 7.2.1 to 7.2.3 and Theorem 7.8 and its corollaries of Section 7.2.3 as if they are postulates. You may wish to review those sections at this time.

We began Section 7.1.4 by contrasting two types of definitions of congruence. One type, popularized by the axiomatic treatment of geometry constructed by David Hilbert a century ago, involves separate characterizations of congruence for segments, angles, and triangles, and still other definitions for other kinds of figures as they might need to be introduced. (See Unit 11.1 for an in-depth look at Hilbert's idea.) In this approach, the SAS congruence proposition (or some equivalent proposition) needs to be assumed, and from it the other congruence theorems can be deduced.

The second type involves a single definition of congruence for all figures. This is accomplished by using transformations: Two figures α and β are congruent if and only if there is a distance-preserving transformation mapping α onto β. In this approach, which can be traced back to Euclid (though Euclid's approach had gaps), the SAS and all other congruence propositions are theorems because they can be deduced from properties of transformations.

Congruence of segments and angles

We offer two proofs of the basic theorem about segment congruence. These proofs exhibit the flexibility that is available in future proofs of this type.

Theorem 7.35 If two segments have the same length, then they are congruent.

Figure 66

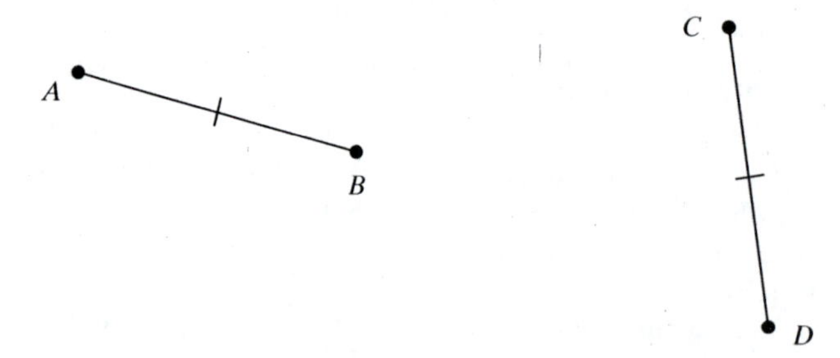

Proof: Suppose $\overline{AB}$ and $\overline{CD}$ have the same length, as in Figure 66. Given the general notion of congruence that we have been working with, to deduce congruence, we need to show that there is a distance-preserving transformation that maps $\overline{AB}$ onto $\overline{CD}$.

Proof 1 (using a translation and a rotation): Let T be the translation associated with the vector $\overrightarrow{AC}$. This translation maps $\overline{AB}$ onto a segment $\overline{A'B'} = \overline{CB'}$ of the same length. Then let R be the rotation with center C and magnitude $m\angle B'CD$ in the appropriate direction. Because $CD = AB = A'B'$, $R(\overline{A'B'}) = \overline{CD}$. Thus $R \circ T(\overline{AB}) = \overline{CD}$, and so $\overline{AB} \cong \overline{CD}$ (Figure 67).

Figure 67

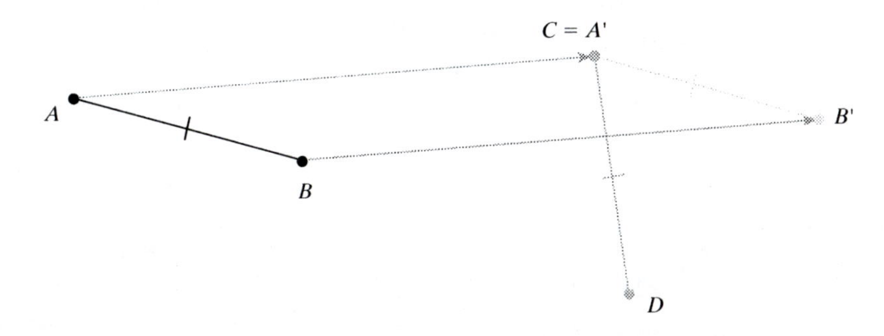

Proof 2 (using only reflections): Let m be the perpendicular bisector of $\overline{AC}$. Then r_m maps $\overline{AB}$ onto a segment $\overline{A'B'} = \overline{CB'}$ of the same length. Then let n be the bisector of $\angle B'CD$. Because $CD = AB = A'B' = CB'$, $\Delta B'CD$ is isosceles, and so $r_n(\overline{CB'}) = \overline{CD}$. Thus $r_n \circ r_m(\overline{AB}) = \overline{CD}$, and so $\overline{AB} \cong \overline{CD}$ (Figure 68).

Figure 68

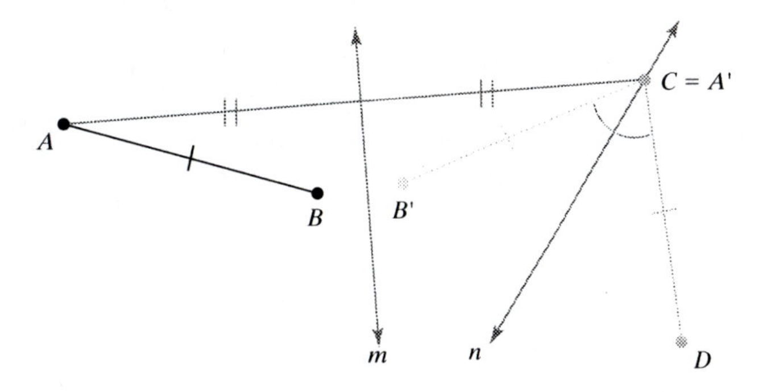

Proof 2 shows that any segment can be mapped to any other of the same length with at most two reflections. If m intersects n, as in the situation shown in Figure 68, then the composite is a rotation. If $m // n$, the composite is a translation.

The converse of Theorem 7.35 is true because if figures are congruent, then they are related by a congruence transformation, and every congruence transformation preserves distance. Combining the theorem and its converse yields a proposition that is often taken as a definition of congruence of segments.

Corollary: Two segments have the same length if and only if they are congruent.

The treatment of angles is quite similar to that for segments.

Theorem 7.36 If two angles have the same measure, then they are congruent.

Proof: This is left to you. (See Problems 1 and 2.)

Corollary: Two angles have the same measure if and only if they are congruent.

The two corollaries of this section enable one to substitute "equal length" or "equal measure" for "congruent".

Congruence of triangles

The three traditional triangle congruence propositions—SAS, SSS, and ASA—can all be deduced from the general definition of congruence and the preservation properties of the congruence transformations in much the same way that the theorems for congruence of segments and angles are deduced previously.

One substantive feature of these proofs distinguishes them from the proofs involving segment or angle congruence. Congruent triangles may have different orientation.

Theorem 7.37 **(SAS Congruence):** If two sides and the included angle of one triangle are congruent to two sides and the included angle of a second triangle, then the triangles are congruent.

Proof: Let ΔABC and ΔXYZ be two triangles with $\overline{AB} \cong \overline{XY}, \overline{AC} \cong \overline{XZ}$ and $\angle BAC \cong \angle YXZ$. This means, of course, that the measures of these segments and angles are equal. We need to show that there is a distance-preserving transformation that maps ΔABC onto ΔXYZ. We offer three proofs, one that is intuitive, and two that might be considered elegant.

Proof 1 (using a translation and a rotation): If ΔABC and ΔXYZ have different orientation, apply a reflection (any reflection!) to ΔABC, yielding an image $\Delta A^*B^*C^*$ that has the same orientation as ΔXYZ. Then proceed as follows. Suppose ΔABC (or $\Delta A^*B^*C^*$) and ΔXYZ have the same orientation.

Figure 69

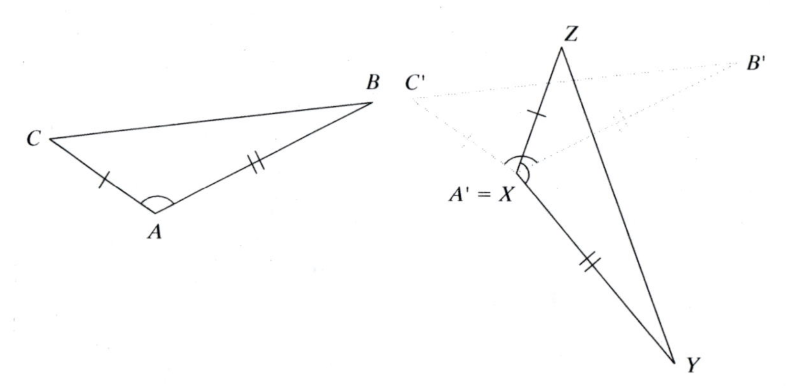

Apply the translation T that maps A onto X, so that $T(\Delta ABC) = \Delta XB'C'$, as shown in Figure 69. From the given information, because of transitivity, $XB' = XY$, $XC' = XZ$, and $m\angle B'XC' = m\angle YXZ$.

Now apply the rotation R with center X and magnitude $m\angle B'XY$ to $\Delta XB'C'$. Because $\angle B'XC' \cong \angle YXZ$, this rotation maps $\angle B'XC$ onto $\angle YXZ$. But it does more than that. Because $XB' = XY$, $R(B') = Y$, and because $XC' = XZ$, $R(C') = Z$. Consequently, $R \circ T(\Delta ABC) = R(\Delta XB'C') = \Delta XYZ$. (Notice that all three parts of the *SAS* condition come into play at the same time.) Thus, by the definition of congruence, $\Delta ABC \cong \Delta XYZ$.

Proof 2 (using only reflections): We outline this proof. Begin as in Proof 1, applying a reflection to ΔABC if its orientation is different from that of ΔXYZ. Now consider the reflection over m, the perpendicular bisector of $\overline{AX}$. This reflection maps A onto X, so that $r_m(\Delta ABC) = \Delta XB'C'$. (See Figure 70.) Now there is a reflecting line n with $r_n(\Delta XB'C') = \Delta XYZ$. We leave it to you to describe the location of this line. Then, because $r_n(r_m(\Delta ABC)) = \Delta XYZ$, the triangles are congruent.

Figure 70

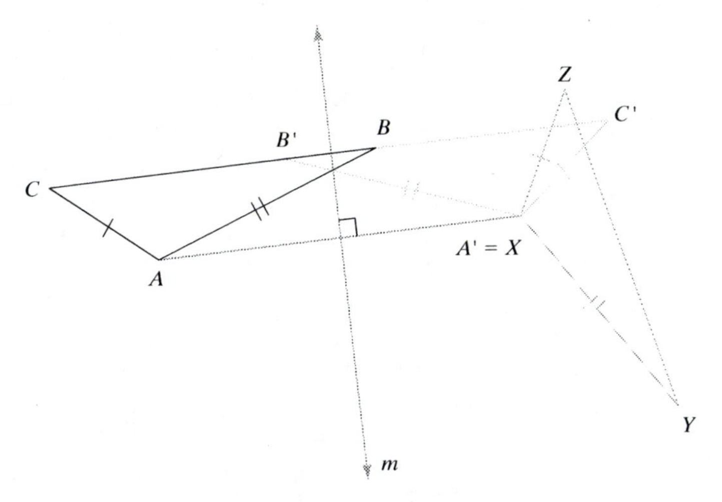

Proof 3 (using angle congruence): Since $\angle BAC \cong \angle YXZ$, there is a congruence transformation that maps *angle BAC* onto *angle YXZ*. Call this transformation T and apply it to ΔBAC. $T(\overrightarrow{AB})$ is either $\overrightarrow{XY}$ or $\overrightarrow{XZ}$.

If $T(\overrightarrow{AB}) = \overrightarrow{XY}$, then also $T(\overrightarrow{AC}) = \overrightarrow{XZ}$. Because $AB = XY$, $T(B) = Y$, and because $AC = XZ$, $T(C) = Z$. Thus $T(\Delta ABC) = \Delta XYZ$.

If $T(\overrightarrow{AB}) = \overrightarrow{XZ}$, then reflect $T(\Delta ABC)$ over the bisector of $\angle XYZ$. Then the images of B and C under $r \circ T$ are Y and Z, respectively, so $r \circ T(ABC) = \Delta XYZ$.

In either case, ΔXYZ is the image of ΔABC under a congruence transformation, and so they are congruent. ┘

Any of the proofs of the SAS Congruence Theorem can be adapted to prove the ASA congruence proposition. For all the ASA proofs, the goal is to map the included side of one triangle onto the included congruent side of the other, whereas in the proofs of SAS the goal was to map the included angle of one onto the included congruent angle of the other. We leave it to you to supply the details.

Theorem 7.38 **(ASA Congruence):** If two angles and the included side of one triangle are congruent respectively to two angles and the included side of another triangle, then the triangles are congruent.

The *SSS* congruence proposition poses a slightly more difficult problem because there is no included side or angle on which to pivot. The proof we supply makes use of properties of kites.

Theorem 7.39 **(SSS Congruence):** If the three sides of one triangle are congruent to the three sides of another triangle, then the triangles are congruent.

Figure 71

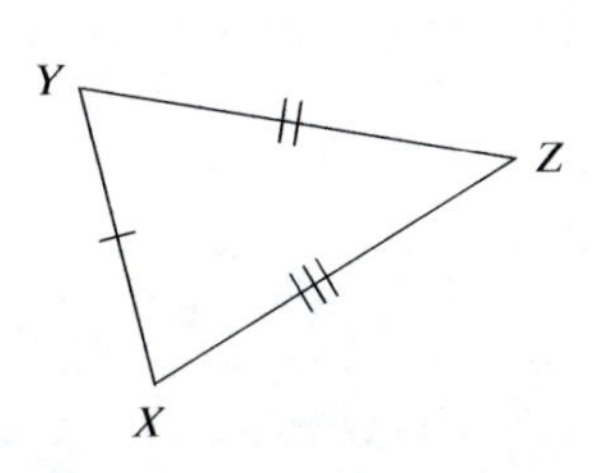

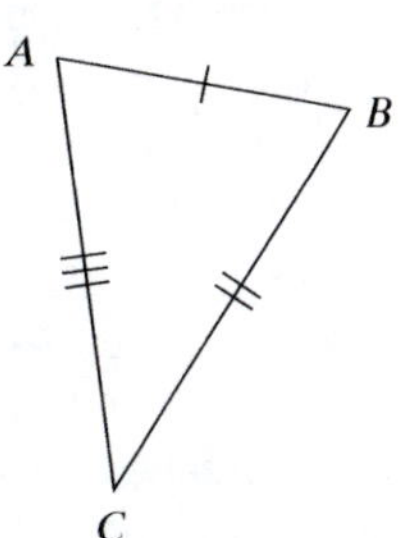

Proof: Again suppose the triangles are ΔABC and ΔXYZ, but this time with $\overline{AB} \cong \overline{XY}, \overline{AC} \cong \overline{XZ}$, and $\overline{BC} \cong \overline{YZ}$ as in Figure 71. Because $\overline{AB} \cong \overline{XY}$, there is a congruence transformation T with $T(\overline{AB}) = \overline{XY}$. This transformation T can be chosen so that $T(A) = X$ and $T(B) = Y$. (If the first transformation T does not have that property, then reflect $T(\Delta ABC)$ over the perpendicular bisector of $\overline{XY}$, and consider a new T including that reflection.) Furthermore, this transformation T can be chosen so that $T(C)$ is on the other side of $\overleftrightarrow{XY}$ from Z. We call this point C'. (If $T(C)$ is on the same side of $\overleftrightarrow{XY}$ as Z, then reflect $T(\Delta ABC)$ over $\overleftrightarrow{XY}$ and consider a still newer T including that reflection.) We are left with the situation pictured in Figure 72.

Figure 72

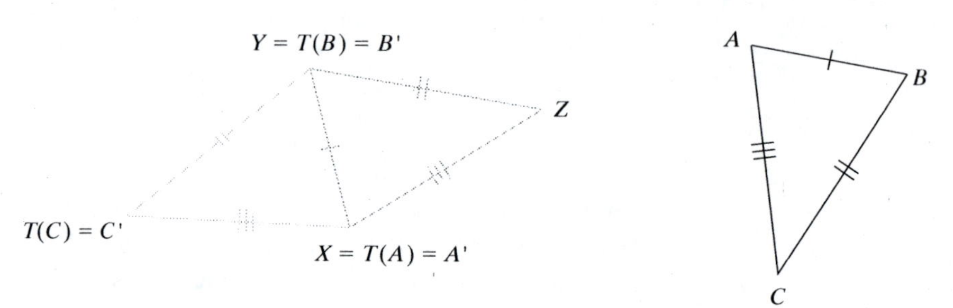

Now we use the other given congruent sides. Because $\overline{AC} \cong \overline{XZ}$ and $\overline{BC} \cong \overline{YZ}$, quadrilateral $XC'YZ$ is a kite, and $\overleftrightarrow{XY}$ is a symmetry line for the kite. Let r be the reflection over $\overleftrightarrow{XY}$. Then $r(\Delta XYC') = \Delta XYZ$, and so $r \circ T(\Delta ABC) = \Delta XYZ$. Thus $\Delta ABC \cong \Delta XYZ$. ┘

The SsA Congruence Theorem

The condition in which two triangles have two sides of one congruent to two sides of the other, and a *nonincluded* angle of one congruent to a corresponding angle of the other is called the SSA condition. It is well-known that two triangles can satisfy the SSA condition and yet not be congruent. Such a situation is pictured in Figure 73. This situation is typically encountered first by students in the study of the Law of Sines. In trigonometry, it is often called the *ambiguous case*, because there are two values for the third side and for each other angle of the triangle. This name is unfortunate, for the situation is no more ambiguous than having two solutions to a quadratic or absolute-value equation in algebra.

Figure 73

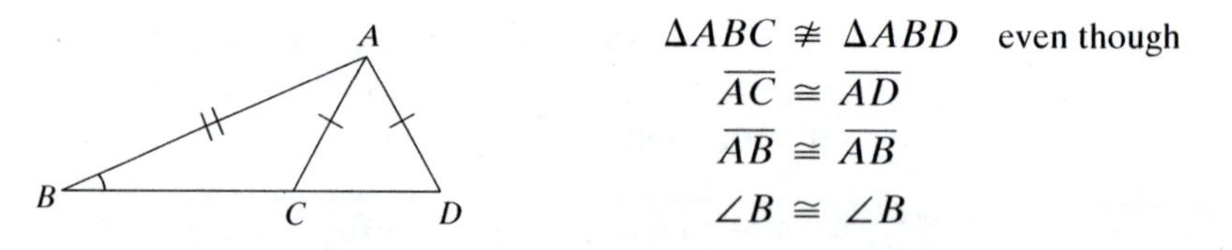

There are situations in which the SSA condition is not ambiguous. A common one is the HL Congruence Theorem: If the hypotenuse and leg of one right triangle are congruent to the hypotenuse and leg of a second right triangle, then the triangles are congruent. The HL Congruence Theorem is a special case of the following theorem, which deserves to be more widely known.

Theorem 7.40 **(SsA Congruence):** If two sides and the angle opposite the longer of the two sides in one triangle are congruent, respectively, to two sides and the corresponding angle in another triangle, then the triangles are congruent.

We write this as the SsA condition (rather than SSA) to emphasize that the larger sides cannot be adjacent to the congruent angles. HL Congruence is a special case because the hypotenuse is opposite the right angle and always larger than a leg.

Figure 74

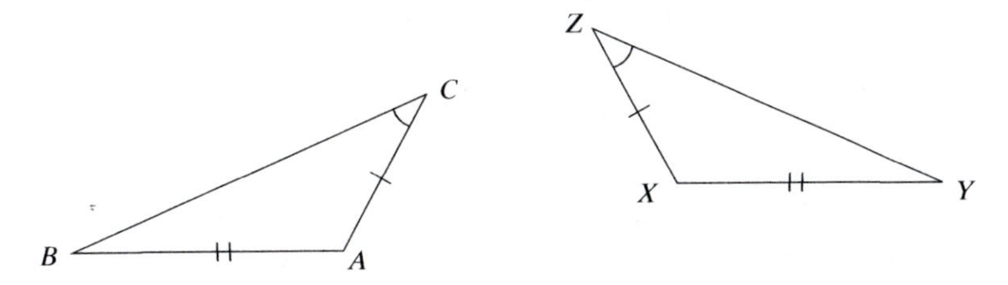

Proof: Suppose the given triangles are ΔABC and ΔXYZ, as in Figure 74, with $\overline{AB} \cong \overline{XY}, \overline{AC} \cong \overline{XZ}$, $XY > XZ$, and $\angle ACB \cong \angle XZY$. This is a long proof, so we number some steps. We begin as in the proof of SSS Congruence, by mapping ΔABC onto a conveniently located congruent image $\Delta A'B'C'$.

Figure 75

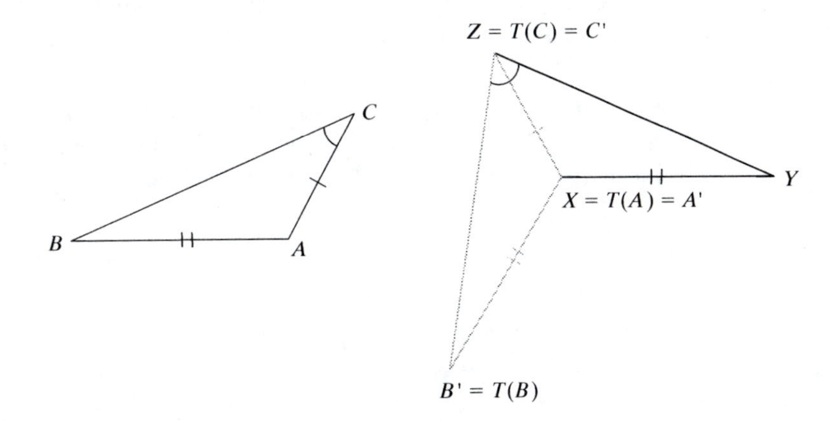

1. Since $\overline{AC} \cong \overline{XZ}$, there is a congruence transformation T with $T(\overline{AC}) = \overline{XZ}$. T can be chosen so that $T(A) = A' = X$ and $T(C) = C' = Z$. Furthermore, T can be chosen so that $T(B) = B'$ is on the other side of $\overline{XZ}$ from Y (as shown in Figure 75). Then $\Delta A'B'C'$ is the image of ΔABC and the two triangles are congruent.
2. We need to show that $\Delta A'B'C' \cong \Delta XYZ$. We will do this by showing that ΔXYZ is the reflection image of $\Delta A'B'C'$ over $\overleftrightarrow{ZX}$. Call this reflection r. Since $C' = Z$ and $A' = X$ and they are on the reflecting line, $r(C') = Z$ and $r(A') = X$. All we need to do now is to show that $r(B') = Y$.
3. $\angle ACB \cong \angle XZY$ is given and $\angle ACB \cong \angle A'C'B'$ from the congruence transformation T, so $\angle XZY \cong \angle A'C'B'$. This makes $\overleftrightarrow{ZX}$ the bisector of $\angle B'ZY$. Due to the symmetry of the angle to its bisector, $r(B')$ is on ray $\overrightarrow{ZY}$.
4. Now consider the circle with center X and radius XY (Figure 76). Z is in the interior of this circle since $XZ < XY$. Thus the ray $\overrightarrow{ZY}$ intersects the circle in exactly one point.
5. B' is on this circle because $XB' = XY$. $r(B')$is on this circle because $\overleftrightarrow{ZX}$ is a symmetry line for the circle. Consequently, $r(B') = Y$, the only point of intersection of the circle and $\overrightarrow{ZY}$. So $r(\Delta A'B'C') = \Delta XYZ$. Consequently, $r \circ T(\Delta ABC) = \Delta XYZ$, from which $\Delta ABC \cong \Delta XYZ$.

Figure 76

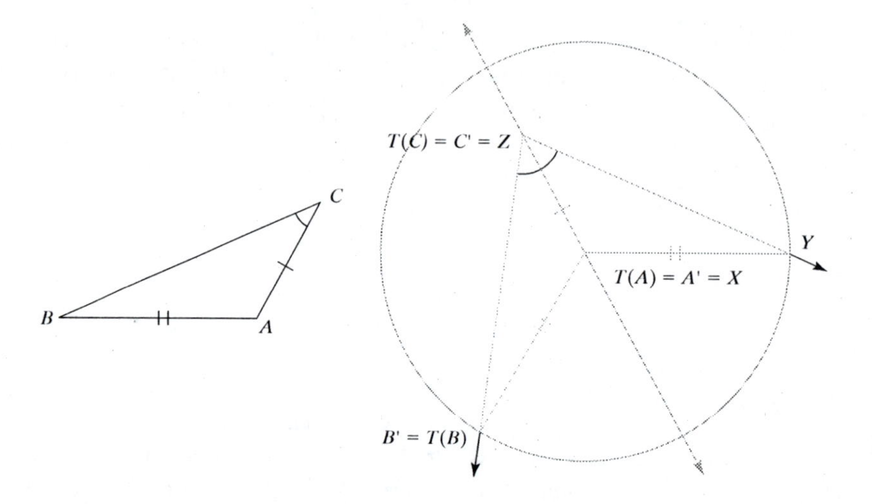

Corollary (HL Congruence): If the hypotenuse and leg of one right triangle are congruent to the hypotenuse and leg of another right triangle, then the triangles are congruent.

The SsA Theorem is not found in Euclid's *Elements* (SAS, SSS, and ASA are), nor is it found in many geometry textbooks. But it and its HL Corollary explain very nicely the various cases of the Law of Sines (see Section 9.1.2).

7.4.1 Problems

1. Figure 77 shows two angles of the same measure. Prove that they are congruent using the idea in Proof 1 of Theorem 7.35.

Figure 77

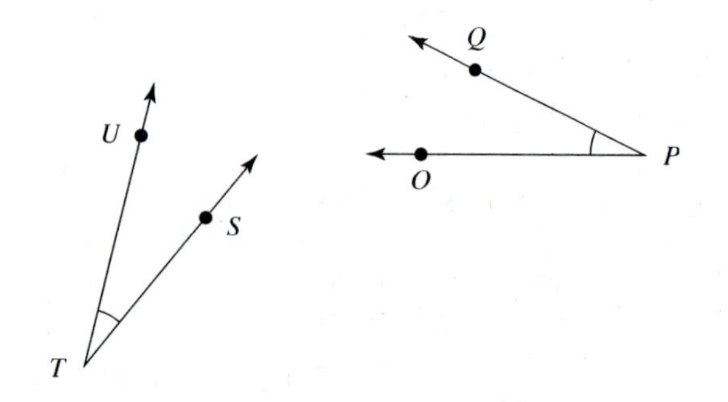

2. Prove that the two angles in Figure 77 are congruent using the idea in Proof 2 of Theorem 7.35.

3. Draw a picture of Proof 2 of the SAS Congruence Theorem and identify the line n.

4. Prove the ASA Congruence Theorem by using one of the methods of the proof of the SAS Congruence Theorem.

5. The triangles in Figure 78 are congruent. Trace the figure and find reflecting lines ℓ, m, and n such that the transformation $r_n \circ r_m \circ r_\ell$ maps one triangle onto the other.

Figure 78

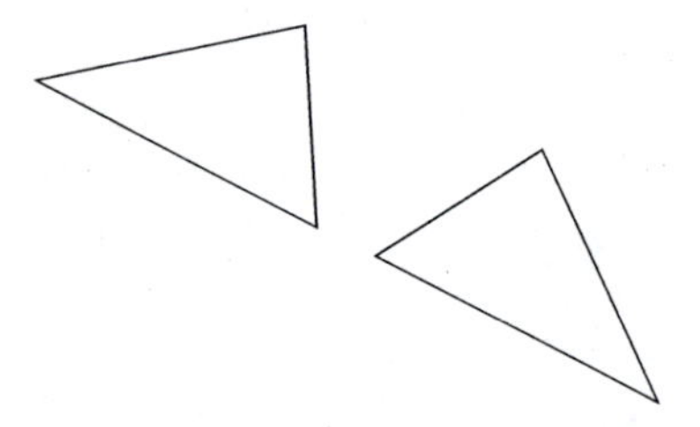

6. If $AB = 10$, $AC = 6$, and $m\angle B = 30°$, in ΔABC, use trigonometry to find all possible values of BC, $m\angle BCA$, and $m\angle BAC$.

7. In Step 4 of the proof of the SsA Theorem, the given $XZ < XY$ is used.

a. Why does the argument not work when $XZ > XY$?

b. Does the argument work if $XZ = XY$?

8. In ΔPQR, $PR = 10$, $m\angle P = 40°$, and $QR = x$.

a. For what values of x is no triangle possible?

b. For what values of x are all possible triangles congruent?

c. For what values of x do there exist noncongruent triangles?

d. In part **c**, how many different noncongruent triangles exist?

9. Prove or disprove: If in quadrilaterals $ABCD$ and $EFGH$, angles A, C, E, and G are right angles, $AB = EF$, and $BC = FG$, then the quadrilaterals are congruent.

7.4.2 Concept analysis: analyzing a geometric figure

You are familiar with various kinds of triangles: isosceles, right, acute, obtuse, equilateral. In this book we have mentioned seven kinds of quadrilaterals: isosceles trapezoids, kites, parallelograms, rectangles, rhombi, squares, trapezoids. But these are not the only kinds of triangles or quadrilaterals, and polygons with more than 4 sides also contain some special cases. The study of Euclidean geometry is to a great extent a study of the special properties of a type of figure.

What kinds of properties are there?

For the moment let us restrict ourselves to a polygon. You can begin by examining the main aspects of the polygon.

Consider its *angles*. Are any congruent? supplementary? complementary? What is the sum of their measures?

Consider its *sides*. Are any congruent? parallel? perpendicular? Are there relationships among the lengths of sides (such as the Pythagorean Theorem)? Are there relationships connecting lengths of sides and measures of angles (such as the Law of Sines or Law of Cosines)?

Consider its *diagonals*. How many are there? Are any congruent? What kinds of figures do the diagonals determine? What are their lengths?

Does it possess any *symmetry*? If so, what are its symmetry lines or what is its center of symmetry? Does it *tesselate* (that is, can congruent copies of it and the region it encloses cover the plane with no overlap)?

How does it relate to other known types of polygons? Is it a special case? A more general figure?

Consider its *area*. Is there a nice formula?

In what manufactured flat objects does its shape occur? In what manufactured solid objects is its shape one face?

Here are some other kinds of questions that mathematicians have explored about polygons.

Are there properties of the *midpoints* of its sides? Of the *perpendicular bisectors* of its sides? Of the *bisectors* of its angles?

Are key points determined by the polygon collinear or concyclic? Are key lines concurrent?

If equilateral triangles or other regular polygons are constructed on its sides, do their centers have any properties?

The list can be endless, limited only by a person's creativity, and you can easily modify this list to apply to curves and 3-dimensional geometric figures.

A different type of property is one that determines the figure. What *sufficient conditions* other than the defining conditions determine this polygon? A good place to start is with any of the properties. For instance, here are some of the properties of all parallelograms.

Both pairs of opposite sides parallel (usual defining property)
Both pairs of opposite sides congruent
Both pairs of opposite angles congruent
Midpoints of diagonals coincide

Each one of these properties is also a sufficient condition for a quadrilateral to be a parallelogram. But here is another property of all parallelograms.

Diagonal splits it into two congruent triangles

This is not a sufficient condition for a quadrilateral to be a parallelogram.

An example: cyclic quadrilaterals

A **cyclic quadrilateral** is a quadrilateral that can be inscribed in a circle. That is, if $ABCD$ is a cyclic quadrilateral, then A, B, C, and D are points in order on the same circle, as shown in Figure 79.

Figure 79

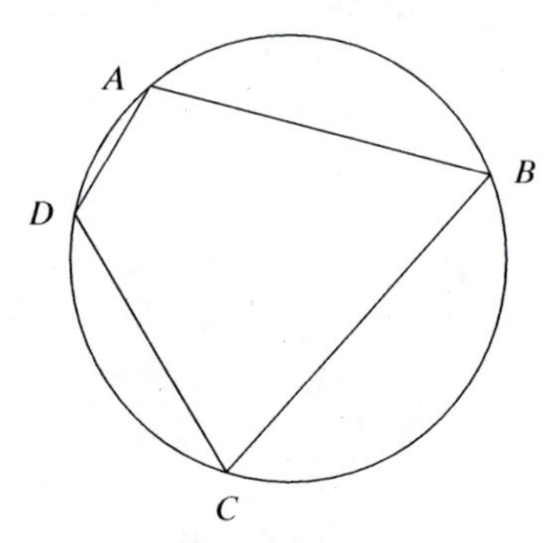

Cyclic quadrilaterals were not studied by Euclid. So they offer a nice example of an interesting type of figure that has been studied by geometers over the centuries since Euclid but that is not typically studied in school geometry courses. We present some of the properties of cyclic quadrilaterals here without proof. You are asked for some proofs in Problem 3.

Angles: A basic property of cyclic quadrilaterals is that both pairs of opposite angles are supplementary. This is also a sufficient condition for a quadrilateral to be cyclic.

Sides and diagonals: The Greek mathematician Ptolemy proved the following relationship among the sides and diagonals of a cyclic quadrilateral $ABCD$: $AC \cdot BD = AD \cdot BC + AB \cdot CD$. That is, the sum of the products of the lengths of the opposite sides of a cyclic quadrilateral equals the product of the lengths of its diagonals.

Symmetry: The four vertices of a cyclic quadrilateral can be any points on a circle, so there need be no symmetry.

Relationship to other polygons: Every square is a cyclic quadrilateral because its center is equidistant from its four vertices. It may not be as obvious that every isosceles trapezoid (see Problem 3) is a cyclic quadrilateral. Cyclic quadrilaterals share a property with all triangles and all regular polygons: they can be inscribed in a circle. This last property is the origin of their name.

Area: There exist several formulas for the area of a cyclic quadrilateral. One was discovered by Brahmagupta in the 7th century. Let the sides of a cyclic quadrilateral be a, b, c, and d, and let s be half its perimeter. Then its area is $\sqrt{(s-a)(s-b)(s-c)(s-d)}$. (See Problem 6 in Section 10.1.2.) Another formula is shown in Problem 5 of this section.

Thousands of theorems about polygons have been deduced over the years. But no figures have been studied more than the triangle and circle. If you are interested in learning about them, you might consult the books by Johnson and Altshiller-Court listed in the bibliography for this chapter. Each book contains hundreds and hundreds of theorems, many of them quite astounding. If you are interested in learning about the properties of curves, then the books of Lockwood and Yates are very fine resources.

7.4.2 Problems

1. Define a **trapezoid** to be a quadrilateral with at least one pair of parallel sides. Define an **isosceles trapezoid** to be a trapezoid with congruent base angles.

a. The network in Figure 80 represents a hierarchy of quadrilaterals. The top node represents quadrilaterials. Node x is connected to node y below it if and only if y is a special type of x, with one stipulation: When y is a special type of x, and z is a special type of y, we do not connect x to z. Put the correct name of figures in the nodes A–H: cyclic quadrilaterals, isosceles trapezoids, kites, parallelograms, rectangles, rhombi, squares, trapezoids.

Figure 80

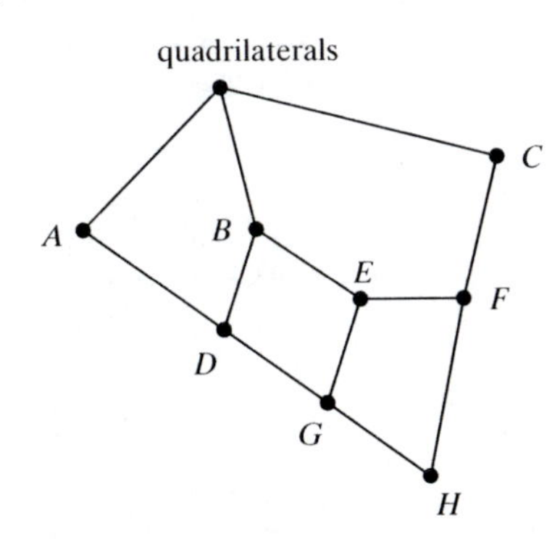

b. Prove that every isosceles trapezoid has a symmetry line.

c. Describe the implications of part **b** for rectangles and squares.

d. How are parts **a, b,** and **c** affected if *trapezoid* is defined to be a quadrilateral with exactly one pair of parallel sides?

2. The diagonals of a trapezoid partition its interior into 4 triangular regions. Is it possible for two of these to be congruent without the trapezoid being a parallelogram or isosceles trapezoid?

3. Deduce the following properties of cyclic quadrilaterals.

a. The perpendicular bisectors of the four sides are concurrent.

b. Both pairs of opposite angles are supplementary.

c. Every isosceles trapezoid is a cyclic quadrilateral.

d. If two opposite angles of a quadrilateral are supplementary, then the quadrilateral is cyclic.

4. By considering a rectangle as a cyclic quadrilateral, show that the Pythagorean Theorem can be viewed as a special case of Ptolemy's Theorem mentioned in this section.

5. a. Use a dynamic geometry program to confirm the following theorem. If a cyclic quadrilateral with consecutive sides a, b, c, and d is inscribed in a circle of radius R, then its area is $\frac{\sqrt{(ab+cd)(ac+bd)(ad+bc)}}{4R}$.

b. Prove that the statement of part **a** is true for a rectangle.

6. Define a *parallexagon* to be a hexagon whose three pairs of opposite sides are parallel.

a. Deduce at least one property of this hexagon different from its defining property.

b. How are parallexagons related to equiangular, equilateral, and regular hexagons?

7. A *60°-triangle* is a triangle with at least one 60° angle.

a. Suppose x, y, and z are sides of a 60° triangle with z opposite the 60° angle. How are x, y, and z related?

b. Deduce some other property of this type of triangle.

8. Repeat Problem 7 for a *120°-triangle.*

9. The sides of a parallelogram are the sides of four squares drawn outside the parallelogram. Give some properties of the quadrilateral whose vertices are the centers of these four squares.

10. An object or theorem is identified. Use books or the Internet to find out what this object or theorem is. Also find out why it is so-named, and draw an accurate picture.

a. Napoleon's Theorem

b. the *Euler line*

c. the *Nagel point*

d. the *Nine-Point Circle Theorem*

11. Examine either the book by Johnson, the book by Altshiller-Court, or some other book with many theorems on Euclidean geometry. Pick out a theorem that is particularly interesting to you. Draw a picture and give its proof.

12. Examine either the book by Lockwood, the book by Yates, or some other book with many properties of curves. Pick out a theorem that is particularly interesting to you. Draw a picture and give its proof.

13. Invent your own type of figure and give some of its properties.

7.4.3 General theorems about congruence

Congruence is so associated with triangles in the minds of many people who have studied geometry that if you asked them about congruent *figures* they might not understand that you are talking about *any* figures, including triangles and other polygons, but also including circles, curves, drawings and any other objects that might be considered as sets of points in Euclidean geometry.

In addition to the definition of congruence via transformations, you saw in Corollary 2 to Theorem 7.19 two theorems that apply to congruent figures of all kinds in a plane.

If two figures are directly congruent, then either one is the image of the other under a rotation or translation.

If two figures are oppositely congruent, then either one is the image of the other under a reflection or glide reflection.

The purpose of this section is to prove two general theorems that follow from these two. We believe these theorems are quite surprising and demonstrate the power of proof both to demonstrate *that* a statement is true and also to demonstrate *why* a statement is true.

We need two theorems before the ones that are the goal of this section. The first is found in many schoolbooks. We provide two distinctly different proofs: one employs triangle congruence, the other symmetry.

Theorem 7.41 The perpendicular bisector of one leg of a right triangle contains the midpoint of the hypotenuse of that triangle.

Proof: Let ΔABC be a triangle with a right angle at B, let M be the midpoint of $\overline{AB}$, and let the perpendicular to $\overline{AB}$ at M intersect $\overline{AC}$ at N (Figure 81). We wish to prove $AN = NC$.

Figure 81

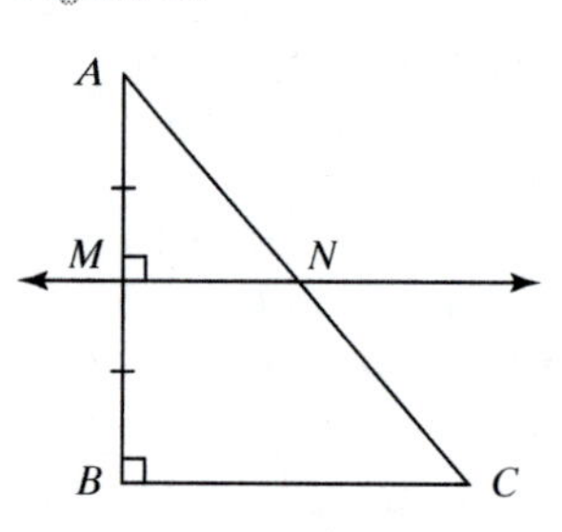

Figure 82

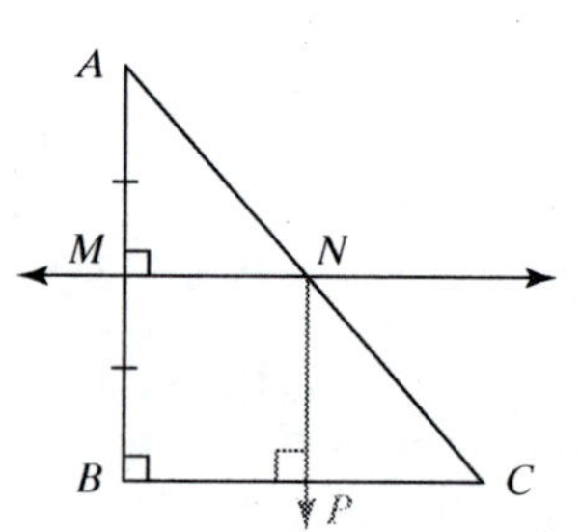

Figure 83

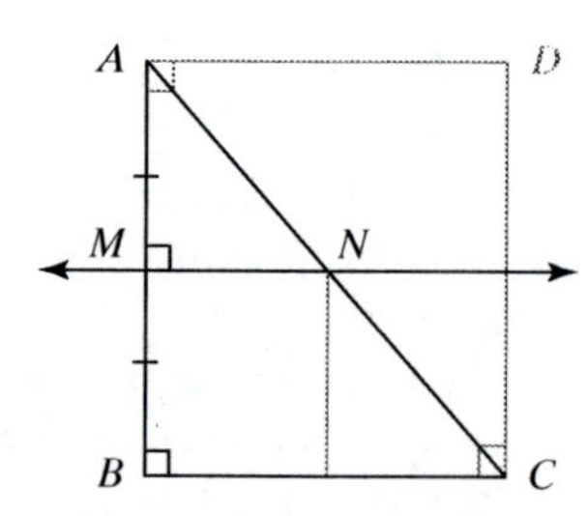

Proof 1 (triangle congruence): Let the perpendicular from N to $\overline{BC}$ intersect $\overline{BC}$ at P (Figure 82). Since quadrilateral $BMNP$ has four right angles, it is a rectangle. So its opposite sides $NP = MB$. Thus $NP = AM$. From the parallel lines, $m\angle MNA = m\angle BCN$ and the right angles AMN and NPC are congruent. So $\Delta AMN \cong \Delta NPC$, from which $AN = NC$.

Proof 2 (symmetry): Let the perpendicular to $\overline{AB}$ at A and the perpendicular to $\overline{BC}$ at C intersect at D, forming rectangle $ABCD$. $\overleftrightarrow{MN}$, being the perpendicular bisector of $\overline{AB}$, is a symmetry line for the rectangle (Figure 83). It intersects the second symmetry line for $ABCD$ at the center of rotation for $ABCD$. This rotation is a half-turn mapping A onto C, so its center is the midpoint of $\overline{AC}$. Thus N is the midpoint of $\overline{AC}$. ∎

Our reason for discussing Theorem 7.41 here is to deduce the following general theorem about glide reflections.

Theorem 7.42 In a glide reflection G, where $G = r_m \circ T$, the midpoint of the segment connecting any point and its image lies on m.

Figure 84

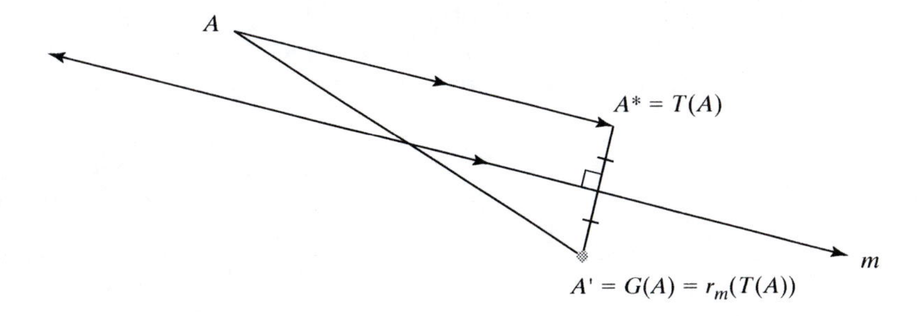

Proof: If A is a point on m, then $T(A)$ is on m because T is parallel to m. Then $G(A) = r_m \circ T(A) = T(A)$. Thus $G(A)$ is on m.

If A is not on m, then the triangle with vertices A, $T(A)$, and $G(A)$ is a right triangle. This is the situation pictured in Figure 84, with $A^* = T(A)$ and $A' = G(A)$. m is the perpendicular bisector of $\overline{A^*A'}$, and so from Theorem 7.39, it contains the midpoint of $\overline{AA'}$, which was to be shown. ■

Theorems 7.41 and 7.42 supply the machinery to deduce the two theorems which are the reason for this section. Theorems 7.43 and 7.44 are extraordinary because (1) they apply to *all* plane figures, (2) they are very simple to state, (3) they are easy to prove, and (4) they would seem to be something that would be obvious to anyone who has studied congruent figures, and yet they are not obvious.

Theorem 7.43 If α and β are *any figures in the plane* and α is directly congruent to β, then the perpendicular bisectors of all segments connecting corresponding points on α and β are either parallel or concurrent.

Proof: Because α is directly congruent to β, then β is the image of α under either a single translation or a single rotation.

If β is the image of α under a translation, then all segments connecting corresponding points on α and β are parallel, so their perpendicular bisectors are parallel.

Figure 85

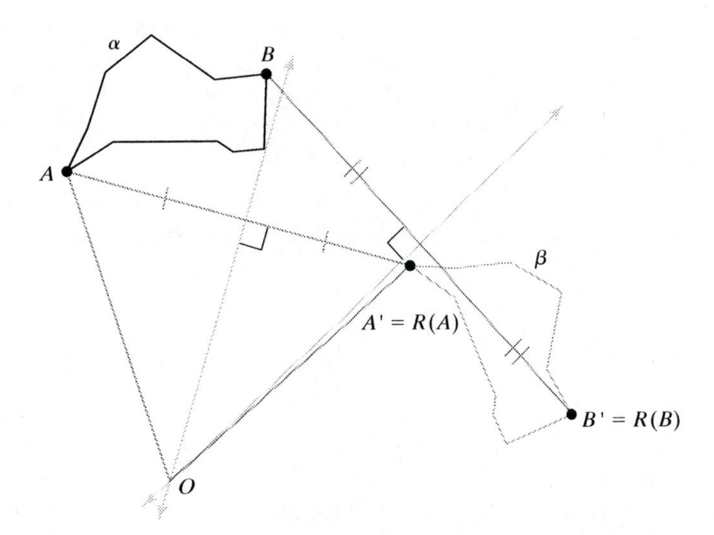

If β is the image of α under a rotation (Figure 85), call the rotation R and let O be its center. Any point A and its image $A' = R(A)$ are equidistant from O, because R is a rotation. So $\Delta OAA'$ is an isosceles triangle with vertex O. Thus the perpendicular bisector of $\overline{AA'}$ contains O. The perpendicular bisector of the segment $\overline{BB'}$ connecting any other point and its image will contain O because the rotation has only one center. ┘

Question: Figure 85 pictures the case in Theorem 7.43 where β is a rotation image of α. Picture the situation of Theorem 7.43 when β is a translation image of α.

When two congruent figures are oppositely congruent, the relationship is even simpler to state.

Theorem 7.44 If α and β are *any figures in the plane* and α is oppositely congruent to β, then the midpoints of all segments connecting corresponding points on α and β are collinear.

Figure 86

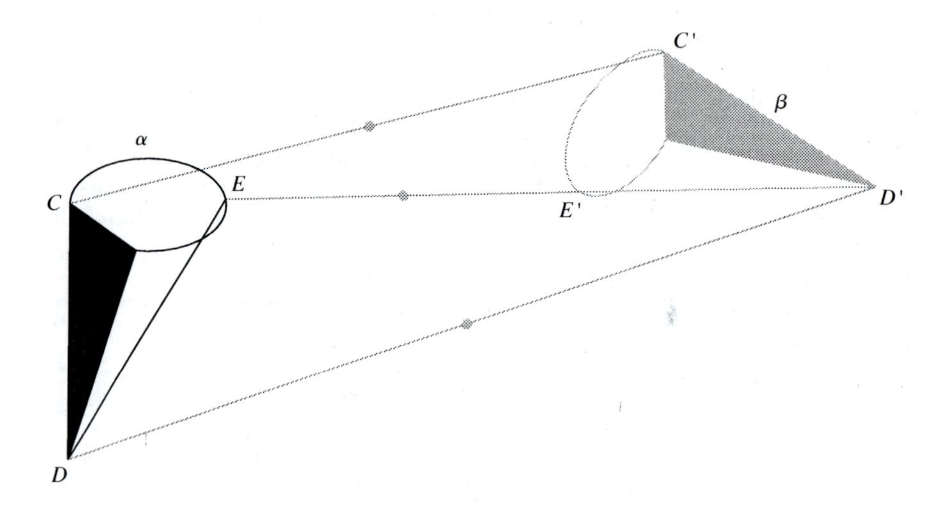

Proof: When α and β are oppositely congruent, β is the image of α under either a reflection or a glide reflection.

If β is the image of α under a reflection, then the midpoints of all segments connecting corresponding points on α and β lie on the reflecting line.

If β is the image of α under a glide reflection, then from Theorem 7.42 the midpoints of all segments connecting corresponding points on α and β lie on the glide reflecting line (Figure 86). ┘

Theorems 7.43 and 7.44 are not in Euclid's *Elements*. They seem relatively recent, but we do not know who first discovered them.

7.4.3 Problems

1. Give an analytic proof for Theorem 7.41.
2. Generalize Theorem 7.41 and prove your generalization.
3. The graphs of the relations $y = x^2$ and $x = (y - 3)^2$ can be viewed as either directly congruent or oppositely congruent due to the symmetry of either graph. Viewing them as directly congruent, what is the center of the rotation that maps one onto the other?
4. Take two congruent copies of a small picture. Place the photocopies on a page so that one is not a translation image of the other. By identifying points on the pictures and drawing appropriate lines, show how to obtain the center of the rotation mapping one onto the other.
5. If $A = (0, 0)$, $B = (5, 0)$, $C = (5, 12)$, $D = (6, 0)$, $E = (18, 0)$, and $F = (6, -5)$, then ΔABC and ΔDEF are oppositely congruent. Determine the correspondence and find an equation for the glide reflecting line.

ANSWER TO QUESTION

See Figure 87.

Figure 87

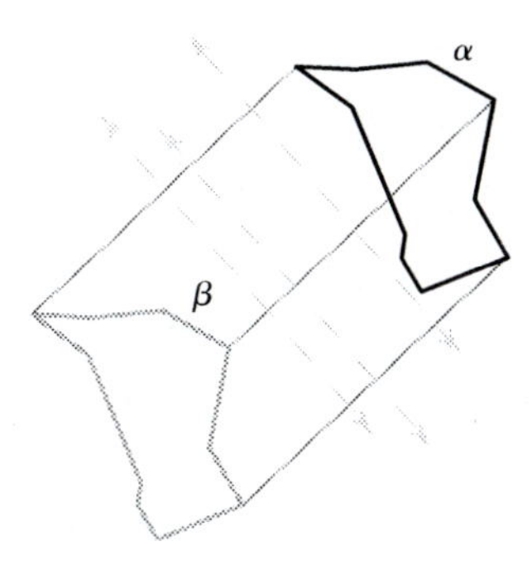

Chapter Projects

1. Euclidean congruences in the complex plane. Suppose that T is a transformation of the complex plane.

a. Explain why T is a congruence transformation if and only if $|T(z) - T(w)| = |z - w|$ for all complex numbers z and w. Use properties of complex numbers to prove that T transforms lines into lines.

b. Suppose that a and b are complex numbers with $|a| = 1$. Prove that the transformations h and k of the complex plane defined by $h(z) = az + b$, for all complex numbers z, and $k(z) = a\bar{z} + b$, for all complex numbers z, are congruence transformations.

2. Direct and opposite congruences. Suppose that T is a congruence transformation of the complex plane. Show that there are complex numbers a and b with $|a| = 1$ such that either (1) $T(z) = az + b$, for all complex numbers z, or (2) $T(z) = a\bar{z} + b$, for all complex numbers z, by completing the following parts:

a. If $T(0) = b$ and $T(1) = c$, then the transformations g and h of the complex plane defined by $g(z) = (c - b)z + b$, for all complex numbers z, and $h(z) = (c - b)\bar{z} + b$, for all complex numbers z, are congruence transformations such that $g(0) = h(0) = T(0)$ and $g(1) = h(1) = T(1)$.

b. Show that for each complex number z, the point $T(z)$ must be on the circle C_0 centered at $b = T(0)$ of radius $|z|$, and also on the circle C_1 centered at $c = T(1)$ of radius $|z - 1|$. Conclude that $T(z)$, $g(z)$, and $h(z)$ must be points of intersection of C_0 and C_1.

c. Show that if z is a real number, then the circles C_0 and C_1 in part **b** have only one point of intersection and that point is on the line determined by b and c. Conclude that $T(z) = g(z) = h(z)$.

d. Show that if z is not a real number, then the circles C_0 and C_1 in part **b** have two points of intersection, and that one of them is $h(z)$ while the other is $g(z)$.

e. Suppose that T is not equal to g or h. Show that it follows that there are complex numbers u and v such that $T(u) = g(u) \neq h(u)$; $T(v) = h(v) \neq g(v)$. Use this to prove that $|u - v| = |u - \bar{v}| = |\bar{u} - v|$. Conclude that u and v are real numbers and show that this would contradict the result of part **c**. The congruences of the complex plane of type (1) are direct congruences, and the congruences of type (2) are opposite congruences.

3. The four kinds of isometries. Let a and b be complex numbers with $|a| = 1$. Prove the following.

a. The direct congruence of the complex plane $T(z) = az + b$ is a translation if $a = 1$, and a rotation centered at $\frac{b}{1-a}$ of magnitude $\text{Ang}(a)$ if $a \neq 1$.

b. The opposite congruence of the complex plane $T(z) = a\bar{z} + b$ is a reflection in the line through $\frac{b}{2}$ that makes an angle $\frac{\text{Ang}(a)}{2}$ with the positive real axis if $a\bar{b} + b = 0$, and a glide reflection in the line through $\frac{b}{2}$ that makes an angle $\frac{\text{Ang}(a)}{2}$ with the positive real axis with a glide of $\frac{a\bar{b}+b}{2}$ if $a\bar{b} + b \neq 0$. (See Problem 6 in Section 7.2.4.)

4. Isometries in space. Every isometry in E^3 is of one (or more) of these types: rotation about an axis, translation, reflection over a plane, glide reflection, screw displacement, or rotatory reflection.

a. Find a source that describes these transformations synthetically, and describe each.

b. Give an expression for the image of (x, y, z) under an example of each type of transformation.

c. Which of these are direct isometries, which opposite?

d. Explain why every isometry in E^3 is the composite of at most four reflections over planes.

5. Frieze patterns. A **frieze pattern** is a plane figure that has translation symmetry in one direction and is bounded in the perpendicular direction. Frieze patterns may be categorized by their symmetries other than the translation symmetry. There are only seven "essentially different" frieze patterns. Find a source that exhibits these patterns and give a detailed explanation of why there are only seven "essentially different" ones.

Bibliography

Unit 7.1 References

Anderson, Richard D., Jack W. Garon, and Joseph G. Gremillion. *School Mathematics Geometry*. Boston: Houghton Mifflin, 1966.

Carroll, Lewis. *Euclid and His Modern Rivals*, New York: Dover, 1973.
This edition is the unabridged republication of the second edition of the work of Charles Dodgson originally published in 1885. Through a drama akin to a Platonic dialogue, Dodgson argues that all modern approaches to Euclidean geometry are faulty compared to Euclid's own approach.

Choquet, Gustave. *Geometry in a Modern Setting*. New York: Houghton Mifflin, 1969. (Translated from the original *L'enseignement de la géométrie*, Paris: Hermann, 1964.)

Coxford, Arthur F., and Zalman Usiskin. *Geometry - A Transformation Approach*. River Forest, IL: Laidlaw, 1971.

Greenberg, Marvin J. *Euclidean and non-Euclidean Geometries*. 3rd edition. New York: W. H. Freeman, 1993.

Hausner, Melvin. *A Vector Space Approach to Geometry*. Englewood Cliffs, NJ: Prentice Hall, 1965.

Heath, Thomas. The Thirteen Books of *Euclid's Elements*. Volumes I–III. New York: Dover, 1956.
This definitive English-language reference for Euclid's Elements should be in every high school professional's library.

Hilbert, David. *The Foundations of Geometry*. Translated from the German by E. J. Townsend. LaSalle, IL: Open Court, 1902.
A rigorous synthetic traditional approach to Euclidean geometry by one of the greatest mathematicians of his time.

Klein, Felix. *Elementary Mathematics from an Advanced Standpoint: Geometry*. New York: Dover, 1939. (Translated from the 1925 3rd German edition by E. R. Hedrick and C. A. Noble.)

Kramer, Edna E. *The Nature and Growth of Modern Mathematics*. Princeton, NJ: Princeton University Press, 1981.
Chapter 17, "The Unification of Geometry", details the geometric work of Felix Klein and his association with Sophus Lie.

Martin, George E. *Transformation Geometry: An Introduction to Symmetry*. New York: Springer-Verlag, 1982.
A mathematical development of transformations and symmetry. A resource for also for Unit 7.3.

Vaughan, Herbert E., and Steven Szabo. *A Vector Approach to Euclidean Geometry*. Volumes 1 and 2. New York: Macmillan, 1971.

Unit 7.2 References

Chrestenson, H. E. *Mappings of the Plane with Applications to Trigonometry and the Complex Numbers*. San Francisco: W. H. Freeman, 1966.
This book presents a detailed and readable account of isometries and similarities of the plane and their coordinate representations.

Silvester, J. R. *Geometry: Ancient and Modern*. New York: Oxford University Press, 2001.
This book, a resource also for Units 7.3, 7.4, 8.2, 9.1, and Chapter 11, has many excellent problems.

Thomas, David. *Active Geometry*. Pacific Grove, CA: Brooks/Cole, 1998.
A lab manual of inquiry-based, technology-rich activities into the study of geometry.

Yaglom, I. M. *Geometric Transformations*. Translated from the Russian by Allen Shields. New Mathematical Library, Volume 8. Washington, DC: Mathematical Association of America, 1962.
This first of three Yaglom volumes on transformations in the MAA series is devoted to congruence. It contains numerous theorems provable using properties of transformations.

Unit 7.3 References

Costa, Antonio, and Bernardo Gomez. *Arabesques and Geometry* (video). New York: Springer Videomath, 1999.

Crowe, David. Symmetry, Rigid Motions and Patterns. *UMAP Module 4*. Cambridge, MA: Consortium for Mathematics and Its Applications (COMAP), 1986.
This is a unit written for college students but accessible to high school students.

Kinsey, L. C., and T. E. Moore. *Symmetry, Shape and Space: Introduction to Geometry*. New York: Springer-Verlag, 2001.

Rosen, Joe. *Symmetry Discovered*. New York: Cambridge University Press, 1975.
A very nice book with applications of symmetry in many areas.

Shubnikov, A., and V. Koptsik. *Symmetry in Science and Art*. New York: Plenum Press, 1974.

Weyl, Hermann. *Symmetry*. Princeton, NJ: Princeton University Press, 1952.
A long essay written by a top-flight mathematician.

Unit 7.4 References

Altshiller-Court, Nathan. *College Geometry: An Introduction to the Modern Geometry of the Triangle and the Circle*. Second edition. New York: Barnes & Noble, 1952.
A source for hundreds of theorems of Euclidean geometry not typically found in high school texts.

Coxeter, H. S. M., and Samuel Greitzer. *Geometry Revisited*. Washington, DC: Mathematical Association of America, 1967.

A wonderful book with many beautiful theorems proved in a variety of ways.

Johnson, Roger A. *Advanced Euclidean Geometry: An Elementary Treatise on the Geometry of the Triangle and the Circle*. New York: Dover, 1960.
Similar to the Altshiller-Court reference above, but with even more theorems.

Lockwood, E. H. *A Book of Curves*. Cambridge, England: Cambridge University Press, 1961.
Concept analysis of a large number of types of curves, including the conics, cycloid family, various spirals, etc.

Posamentier, Alfred, and Charles T. Salkind. *Challenging Problems in Geometry*. New York: Dover, 1996.
Excursions in advanced Euclidean geometry.

Hahn, L. S. *Complex Numbers and Geometry*. Washington, DC: Mathematical Association of America, 1994.

Usiskin, Zalman, Arthur F. Coxford Jr., and Daniel Hirschhorn. *UCSMP Geometry*. Needham, MA: Prentice Hall, 2002.
A high school textbook that employs many of the mathematical ideas in this chapter.

Yates, Robert C. *Curves and Their Properties*. Classics in Mathematics Education, No. 4. Reston, VA: National Council of Teachers of Mathematics, 1974.
Similar to the Lockwood reference, and similarly organized by curve.

Chapter

8

DISTANCE AND SIMILARITY

The fundamental figures of geometry—points, lines, planes, circles, spheres, polyhedra—are abstract models of natural and human-made physical objects or their components. To apply the concepts of geometry to the objects of our physical world, we measure quantities such as the distance between points, and the area and volume of a great variety of geometric figures. By integrating such measurements with the theorems of geometry, notably those on congruence and similarity, we enrich our understanding of these abstract concepts, providing practical applications of the geometrical results.

Most physical objects are too large, too small, too unwieldy, or otherwise inappropriate for direct study. We study them through photographs or other images; with telescopes, microscopes, or other instruments; using maps or other drawings; or with models or other physical likenesses. In all cases, we are able to learn much about the original object through the careful study of an object similar to it.

Unit 8.1 discusses and explores the idea of distance in different environments and examines what assumptions we make in calculating distances. We then move on to similarity. Although these may seem like strange partners, distance is basic to similarity. Figures are similar if and only if the ratio of distances between any corresponding pairs of points on the figures is constant.

When we assert that a length or distance is x, we may omit any mention of units, but with real lengths or distances there must be a unit. We are then again faced with ratios, for if a distance or length is measured in two different units (e.g., miles and kilometers),

$$AB = x \text{ unit1} = y \text{ unit2},$$

then all other distances and lengths in those units, say

$$CD = a \text{ unit1} = b \text{ unit2},$$

are such that $\frac{y}{x} = \frac{b}{a}$. Thus when we study similar figures we are also implicitly studying what happens when we change units of length in a single figure.

Unit 8.2 consists of an exploration of similarity in a variety of contexts. After describing the general idea, we examine similar graphs, similar polygons, similar arcs, and similarity transformations. In these applications, we think usually of the similarity of two different figures. But in Unit 8.3, we examine the presence of similarity in single figures. This does not end our look at distance and similarity. The trigonometry of Chapter 9 and the theorems of area and volume in Chapter 10 build on the ideas of this chapter.

Unit 8.1 Distance

8.1.1 What is distance?

Both in mathematical discourse and in everyday parlance, the words *distance* and *length* are sometimes but not always identical. For instance, the distance between two points in Euclidean geometry is the same as the length of the segment connecting those points. However, we also apply the idea of length more generally when we have a route in mind, a set of points along which the length is being measured. We speak of the length of a segment, or the length of a path, or the length of an arc. In contrast, when we speak mathematically of distance, we tend to mean *shortest distance.* Thus we may speak of the "length of a curve from A to B", and this may be different from the "distance from A to B". We muddle the distinction between the words by speaking of the "distance along the curve".

In everyday life, the word *distance* can take on different meanings depending upon the context in which it is used. For example, we might say that "the town of Centerville is 21 miles from the town of Oak Grove along the Interstate" but then we might add "but Centerville is 15 miles from Oak Grove as the crow flies". We also often describe distance between towns by the amount of time it will take to get from one town to another, rather than by the number of miles that separate them. For example, the distance from Centerville to Oak Grove might be described as "20 minutes by car". Thus, although we think of distance as a quantity that measures *how far apart* two things are from one another, we recognize that there are often different measures of "apartness" that make sense and that are useful.

Question 1: For each of the three measures (along a road, as the crow flies, time) of the distance from Centerville to Oak Grove, describe a situation in which it might be most appropriate.

Distance is also fundamentally important in mathematical and scientific contexts, where it ranges from measures as small as those found in atoms to distances as large as those between galaxies. Mathematically, we think of a particular measure of distance as a function that maps an ordered pair of points P and Q into the set of real numbers.

In different contexts, there are different formulas for distance. In this section, we examine Euclidean distance and taxicab distance in the coordinate plane and Hamming distance in data sets to gain an intuitive understanding of the characteristics that a mathematical definition of distance should satisfy.

Euclidean distance

In this chapter, we are concerned primarily with *Euclidean distance functions* in the Cartesian plane $\mathbf{R}^2$ and Cartesian space $\mathbf{R}^3$. We represent these functions as d_E. Formulas for Euclidean distance can be deduced from the Pythagorean Theorem. Sometimes they are simply taken as given. If $P = (x_1, y_1)$ and $Q = (x_2, y_2)$ are points in the Cartesian plane $\mathbf{R}^2$, then $d_E(P, Q)$, the **(2-dimensional) Euclidean distance** between them, is given by

$$d_E(P, Q) = \sqrt{(x_1 - x_2)^2 + (y_1 - y_2)^2}.$$

If $P = (x_1, y_1, z_1)$ and $Q = (x_2, y_2, z_2)$ are points in Cartesian space $\mathbf{R}^3$, then $d_E(P, Q)$, the **(3-dimensional) Euclidean distance** between them, is given by

$$d_E(P, Q) = \sqrt{(x_1 - x_2)^2 + (y_1 - y_2)^2 + (z_1 - z_2)^2}.$$

From these distance formulas, we can derive equations for figures in $\mathbf{R}^2$ and $\mathbf{R}^3$.

Question 2:

a. Find an equation for the sphere S in Cartesian space $\mathbf{R}^3$ with center $Q = (1, -2, 3)$ and radius 5.

b. Find an equation for the intersection of the sphere S in part **a** with the xy-plane and describe that intersection.

We are also sometimes concerned with distance on a sphere. The **spherical distance** between two points P and Q on a sphere S, $d_S(P, Q)$, is defined as the length of the shortest arc of a great circle on S passing through P and Q.

Question 3: In Cartesian space $\mathbf{R}^3$, explain why the points $P = (1, 0, 0)$ and $Q = (0, 1, 0)$ are on the sphere S of radius 1 centered at the origin, and compare the Euclidean and spherical distance between P and Q.

In your previous work in mathematics, you have used the Euclidean distance d_E extensively in working with geometric objects in the plane. We now examine a different way of measuring distance in the Cartesian plane $\mathbf{R}^2$ that is based on an everyday experience (if you live in a city) and that produces some very interesting results in comparison with Euclidean distance.

Taxicab distance

Assume you live in a place that is flat and in which all streets are north-south or east-west. Suppose you want to measure distance in terms of taxi fares—stipulating that the distance between two points is the lowest taxi fare that will get you from one point to another. If the taxi fare is \$1 per mile, how can the "taxicab distance" be calculated?

To answer this question, envision the Cartesian plane as a grid of streets along every horizontal and vertical line. Suppose you want to go from a point $P = (x_1, y_1)$ and a point $Q = (x_2, y_2)$. Then you use the Euclidean distance formula to calculate the taxi fare only if the route you take is entirely along a horizontal line (east-west), or entirely along a vertical line (north-south), because traveling diagonally cannot be done. Let $\mathbf{d_T(A, B)}$ be the **taxicab distance** between A and B. Then $d_T(P, Q) = d_T(P, L) + d_T(L, Q)$, where $L = (x_2, y_1)$.

Figure 1

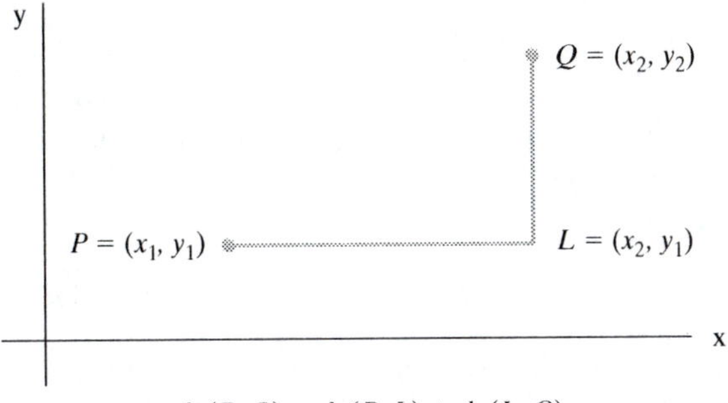

$d_T(P, Q) = d_T(P, L) + d_T(L, Q)$

So the **taxicab distance formula** is $d_T(P, Q) = |x_1 - x_2| + |y_1 - y_2|$.

Comparing Euclidean distance with taxicab distance

Even though distance using d_T is different from d_E, the Euclidean distance, the context in which we are working is the same as when we use d_E; namely, the Cartesian plane or $\mathbf{R}^2$.

Question 4: Given any two points P, Q of the Cartesian plane, what is the relationship between $d_E(P, Q)$ and $d_T(P, Q)$? When are they equal? When are they unequal?

To answer these questions, we appeal to the triangle inequality, namely, that the length of any side of a triangle is less than the sum of the lengths of the other two sides. First notice that on any line that is either horizontal or vertical, $d_E = d_T$. In other words, segments on horizontal or vertical lines have the same length whether we use d_E or d_T.

EXAMPLE Show that the Euclidean distance is always less than or equal to the taxicab distance.

Solution Use Figure 1. Since the length of the hypotenuse of a triangle cannot exceed the sum of the lengths of its legs,

$$d_E(P, Q) \leq d_E(P, L) + d_E(L, Q) = d_T(P, L) + d_T(L, Q) = d_T(P, Q).$$

So we always have

$$d_E(P, Q) \leq d_T(P, Q).$$

The example confirms our intuition that the path of a taxicab can never be shorter than the distance "as the crow flies".

Now that we know how taxicab distance compares with Euclidean distance, what other comparisons can we make? Would plane geometry be the same if we used the taxicab distance instead of the Euclidean distance? The answer is no. All we need to do is produce one counterexample to a known theorem of Euclidean geometry. We choose the triangle inequality.

Figure 2

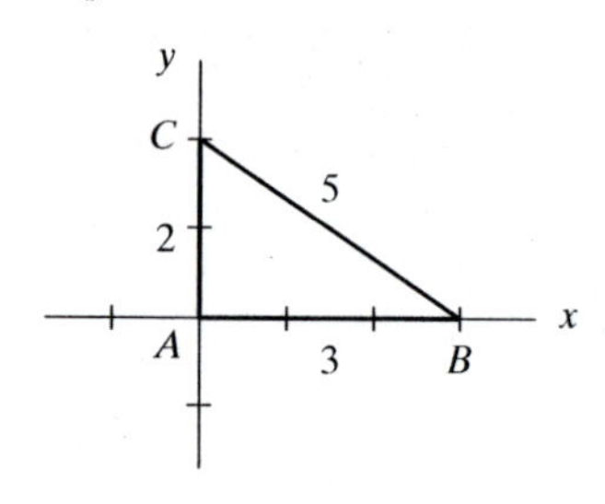

Consider triangle ABC, where $A = (0, 0)$, $B = (3, 0)$, and $C = (0, 2)$, as shown in Figure 2. Notice that $d_T(A, B) = |3 - 0| + |0 - 0| = 3$, and $d_T(A, C) = |0 - 0| + |0 - 2| = 2$. Yet $d_T(B, C) = |3 - 0| + |0 - 2| = 5$. So with taxicab distance, the sum of the lengths of two sides of a triangle is not necessarily greater than the length of the third side. This means that we cannot use taxicab distance to measure the length of segments in Euclidean geometry. It also suggests that geometric figures defined in terms of distance might have different shapes if taxicab distance were used.

Circles from Euclidean distance and taxicab distance

Circles are usually defined in terms of distance. We use the standard definition: A **circle** is the set of points Q that are at a fixed distance (its radius) from a fixed point O (its center). With d_E, the circle with center $(0, 0)$ and radius 1 is the set of points (x, y) satisfying

$$1 = d_E((x, y), (0, 0)) = \sqrt{(x - 0)^2 + (y - 0)^2} = \sqrt{x^2 + y^2}.$$

This gives rise to the familiar equation for the unit circle (Figure 3): $x^2 + y^2 = 1$.

In contrast, with taxicab distance,

$$1 = d_T((x, y), (0, 0)) = |x - 0| + |y - 0| = |x| + |y|.$$

What does this taxicab unit circle look like when we use d_T instead of d_E? In Quadrant 1, since $x > 0$, and $y > 0$, the set of points on the circle is given by $\{(x, y)\colon x + y = 1\}$; in Quadrant 2 by $\{(x, y)\colon -x + y = 1\}$, in Quadrant 3 by $\{(x, y)\colon -x - y = 1\}$ and in Quadrant 4 by $\{(x, y)\colon x - y = 1\}$. Putting these four parts together, as in Figure 4, it turns out that the unit "circle" defined by taxicab distance is a square!

Figure 3

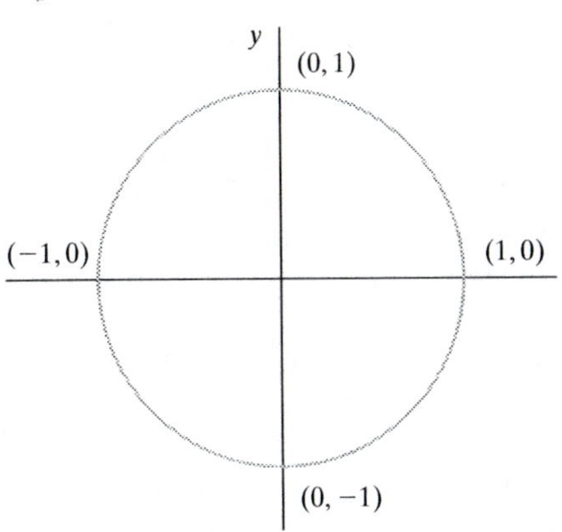

Figure 4

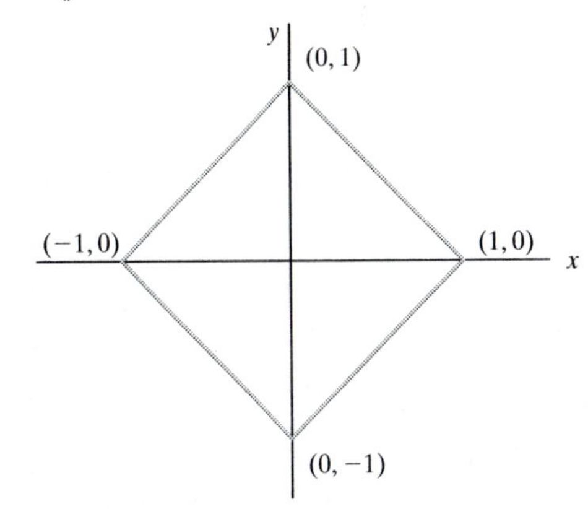

Notice that the unit circles in Figures 3 and Figure 4 both have equations of the form $|x^p| + |y^p| = 1$, where p is a positive real number. If $p = 1$, we get the unit square, or, as we have seen, the "taxicab circle". If $p = 2$, we get the traditional Euclidean unit circle. We ask you in Problem 8 to explore on a computer what happens as p becomes larger.

Hamming distance for data transmission

Aside from the Euclidean plane and space, there are many other mathematical contexts in which distance is meaningful and important. Here is an example from outside geometry.

In computer science, information is often stored and transmitted electronically as *binary strings* of some fixed length; that is, as finite sequences $s = (s_1, s_2, s_3, \ldots, s_n)$ with n terms of 0's and 1's. We write these sequences without commas but separate them with parentheses. For example, the set S_3 of all strings of length 3 is

$$S_3 = \{(000), (100), (010), (001), (110), (011), (101), (111)\}.$$

A piece of data or information is encoded as a series of strings of a given length and then stored or transferred electronically to another location where it is stored or decoded. If mistakes occur in transferring the data, then the piece of data that arrives at the destination will be a different series of binary strings than those describing the original message. Of course, if the transferred piece of data is "close" to the original, it may be possible to recognize and even correct the transferred piece of data. But what does "close" mean in this context?

In 1950, Richard W. Hamming introduced the following distance on the set S_n of binary strings of length n. For two binary strings $s = (s_1 s_2 s_3 \ldots s_n)$ and $t = (t_1 t_2 t_3 \ldots t_n)$,

$$d_H(s, t) = \text{the number of indices } k \text{ such that } s_k \neq t_k.$$

That is, $d_H(s, t) =$ the number of places in which the strings s and t differ. This is called the **Hamming distance** between the two strings. For example, consider two strings s and t of length 8, where $s = (01011110)$ and $t = (01110110)$. These strings are at a Hamming distance of 2 from each other because s and t differ in the third and fifth places only. Here $d_H(s, t) = 2$.

Error detection and correction capabilities can be added to the process of transmitting information with binary strings of length n by using a subset C_k of S_n such that the Hamming distance between any two strings in C_k is at least k, where $k > 1$. For example, for binary strings of length 3, the subset

$$C_2 = \{(000), (011), (101), (110)\}$$

of S_3 consists of strings separated by a Hamming distance of 2, while the strings in the subset

$$C_3 = \{(001), (110)\}$$

are separated by a Hamming distance of 3. Notice that each time we increase the Hamming distance between strings by 1, we cut the number of strings by one-half.

Suppose that information is encoded using only the binary strings of length n that are in a subset C_k of S_n for some $k > 1$. If there are transmission errors, then when the information is transmitted, some of the strings that arrive at the end of the transmission channel are different from those in the encoded message. These transmission errors can be detected if the received information includes strings from S_n that are not in the subset C_k of S_n that was used to encode the message.

Under the reasonable assumption that the probability of making a transmission error of a given binary digit in a binary string is small and is equally likely at any place in the string, then, of the possible errors that might occur in transmitting a given string, the most likely error is that one binary digit is incorrect, the next most likely error is that two binary digits were transmitted incorrectly, and so on. Thus, if $k \geq 2$, the most likely transmission errors can be detected by the arrival of strings that are not in the encoding subset C_k of S_n. If $k \geq 3$, these most likely errors can be corrected by replacing these detected strings by the string in C_k of S_n that is closest to the transmitted string in Hamming distance. Also, if $k \geq 3$, then it will be possible to detect the next most likely type of incorrectly transmitted string, one with two incorrect binary digits. Thus, by choosing k large, we can encode information in such a way that the most likely transmission errors can be detected and even corrected. Of course, by choosing k large, we make C_k a relatively small subset of S_n. So we must make the string length n large enough to provide enough strings in C_k to encode the necessary information.

The general idea of distance

We have now considered several different examples of distance functions in different contexts. In each of these examples, the distance function provides a reasonable measure of the amount of separation or apartness of two objects. But what properties should we require of a distance function in any given context?

First we want to have shorter and longer distances. So the set of distances must be ordered, whether we are considering the distance between two points as the length of a segment, time elapsed in traversing, or cost of making the trip. We can therefore specify that the distance between an ordered pair of objects P and Q be a real number.

All of the examples we have considered assign *undirected* distances to pairs of objects; the distance from P to Q is the same as the distance from Q to P. Moreover, the number assigned as the distance between P and Q is a nonnegative real number that is equal to 0 if and only if $P = Q$. Also, all of the examples of distances that we have considered have the following *triangle inequality property*: For any three objects P, Q, and R, the distance from P to R is less than or equal to the sum of the distance from P to Q and the distance from Q to R.

For Euclidean distance in a plane or in space, the triangle property follows from the fact that the sum of the lengths of two sides of a triangle always is greater than the length of the third. This feature also holds for the sides of any spherical triangle, so the spherical distance function has the triangle inequality property. You are asked to verify the triangle inequality property for the taxicab and Hamming distances in the Problems.

These considerations lead us to the following formal definition of a *distance function*.

Definition

A **distance function** or **metric** on a set S is a function $d: S \times S \to R$, such that for all points P, Q, and R in S:

D1: $d(P, Q) \geq 0$ [nonnegativity property]

D2: $d(P, Q) = 0$ if and only if $P = Q$ [nondegeneracy property]

D3: $d(P, Q) = d(Q, P)$ [symmetric property]

D4: $d(P, Q) + d(Q, R) \geq d(P, R)$ [triangle inequality]

In some fields of mathematics, distance functions are only required to satisfy D1, D2, and D3. The definition of a metric stated here was first given in 1906 by the French mathematician Maurice Fréchet.

8.1.1 Problems

1. The authors claim that standard fares for a given airline often do not satisfy the criteria for a distance function. More precisely, if S is the set of all cities served by a certain airline at a certain time and if $d(P, Q)$ = standard coach airline fare between cities P and Q for that airline at that time, then d may have properties D1, D2, and D3 of a distance function, but property D4 generally fails for at least one choice of three cities P, Q, and R. Probe the Internet to find an example to verify this claim.

2. In this section, formulas for Euclidean distance are given for $\mathbf{R}^2$ and $\mathbf{R}^3$. Give a formula for a distance function d_E in $\mathbf{R}^1$.

3. Prove that taxicab distance has all of the properties required of a distance function.

4. In Euclidean geometry the set of all points equidistant from two given points P and Q is the perpendicular bisector of the segment $\overline{PQ}$. In the geometry of the taxicab metric, this is rarely the case. Construct two examples to show that the set of all points equidistant from two given points P and Q with the taxicab metric can have more than one shape.

5. Different distance functions produce different sets of allowable isometries (distance-preserving transformations) for the plane. For $\{\mathbf{R}^2, d_E\}$, all of the familiar translations, rotations, reflections, and glide reflections of Euclidean geometry are allowable under d_E, as you would expect. Show that for $\{\mathbf{R}^2, d_T\}$ rotations of 0, 90, 180 degrees (or any multiple of these) are allowable under d_T, but all other rotations are not. *Hint*: Show a rotation that maps a vertical line to a line that is not parallel to either coordinate axis and indicate that it will increase the taxicab distance rather than preserve it.

6. a. Construct a taxicab circle with center $P = (0, 1)$ and radius $r = 1.8$.

   b. Generalize part **a**.

7. Identify the centers and radii of two different taxicab circles that have more than two points in common.

8. A distance formula that was put forth by Hermann Minkowski (1864–1909) is

$$d_p(P, Q) = \sqrt[p]{|x_1 - x_2|^p + |y_1 - y_2|^p}.$$

It is called the **Minkowski p-distance** from the point (x_1, y_1) to the point (x_2, y_2) on $\mathbf{R}^2$.

   a. For what values of p does $d_p(P, Q)$ correspond to $d_E(P, Q)$ and $d_T(P, Q)$?

   b. Consider the equation $|x|^p + |y|^p = 1$, which defines a "unit circle" under Minkowski distance. Graph the Minkowski circle for $p = 1$.

   c. Try increasing values of p and describe what happens.

   d. Prove that $\lim_{p\to\infty} \{(x, y): |x|^p + |y|^p = 1\}$ is the square with horizontal and vertical sides defined by the four lines $x = 1$, $x = -1$, $y = 1$, and $y = -1$.

9. Let distance be defined for the Cartesian plane using the following function d_M: Given points P, Q, with coordinates (x_1, y_1) and (x_2, y_2), respectively, define

$$d_M(P, Q) = \max(|x_1 - x_2|, |y_1 - y_2|).$$

   a. Interpret $d_M(P, Q)$ geometrically.

b. Show that d_M is a metric.

c. Show that the Minkowski distance d_p converges to d_M as p goes to infinity.

10. Explain why the Hamming distance has all of the properties required of a distance function.

11. Display the binary strings of length 3 as vertices of a cube in such a way that the Hamming distance between two binary strings of length 3 is equal to the smallest number of cube edges needed to join the vertices representing the two strings.

12. In Section 3.1.3, we exhibited a geometric model of the real number system R on a semicircle of radius 1 centered at the point $C = (0, 1)$ in the Cartesian plane. In this model (see Figure 5), each real number r corresponds to the point r' on this semicircle that is the point of intersection with this semicircle of the line segment joining C to the point $(r, 0)$ on the x-axis.

This model of the real number system is sometimes called the **finite semicircle model of R**. Given two real numbers r and s, define the **chordal distance $\mathbf{d_c(r, s)}$** to be the length of the chord on the semicircle that joins the points r' and s' representing r and s in the finite semicircle model of **R**.

a. Explain why d_c is a distance function on the set **R** of all real numbers.

b. How are the values of the chordal distance function related to the usual distance function $d(r, s) = |r - s|$?

c. Explain why $d_c(r, s) < 2$ for all real numbers r and s.

Figure 5

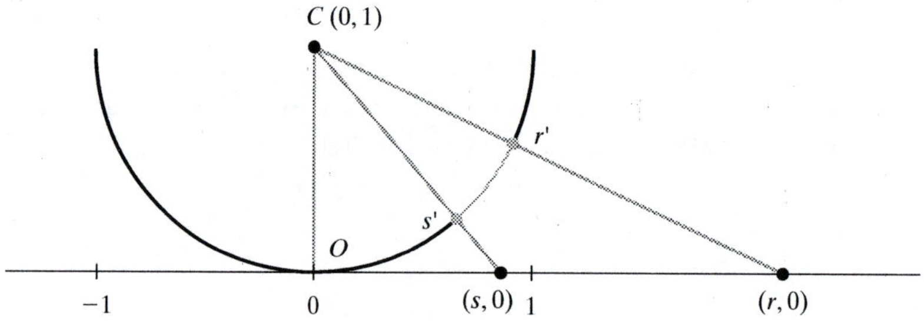

$C\,(0, 1)$ r' s' O -1 0 $(s, 0)$ 1 $(r, 0)$

ANSWERS TO QUESTIONS

1. Sample answers: If low on gas and driving from Centerville to Oak Grove, we might want to know the distance along the highway. But if taking a helicopter, we would prefer the distance as the crow flies. And if we were in a hurry, we would want to know how long it would take.

2. a. $(x - 1)^2 + (y + 2)^2 + (z - 3)^2 = 25$;

b. $(x - 1)^2 + (y + 2)^2 = 16$, a circle.

3. The distance from P to $(0, 0, 0)$ is 1 as is the distance from Q to $(0, 0, 0)$. $d_E(P, Q) = \sqrt{2}$ and $d_S(P, Q) = \frac{\pi}{2}$.

8.1.2 Minimum distance problems

The geometric distance functions of Section 8.1.1 all share the property that the distance between two points is the length of the shortest path between those points in the given system. For instance, the taxicab distance is the minimum distance along horizontal and vertical lines. The Euclidean distance between P and Q in both two and three dimensions is no longer than the length of any path between P and Q.

We now generalize this idea to consider the distance between a point and a set. From here on we are in the Euclidean plane E^2 or Euclidean space E^3.

The distance between a point and a set

Suppose that d is a metric defined for all pairs of points in E^2 (E^3) and that S is a set in E^2 (or E^3). We define the **distance d(P, S) between a point P outside S and the set S** as the greatest lower bound (glb) of the distances between P and any point Q in S. Symbolically,

$$d(\text{point } P, \text{set } S) = \text{glb}\{d(P, Q) : Q \in S\}.$$

Sometimes there is a point Q in S such that the greatest lower bound is the distance between P and Q. This is the situation in Question 1. In Question 2, there is no such point Q.

Question 1: Suppose that P is the point $(2, 2)$ and that S is the unit circle together with its interior; that is, $S = \{(x, y): x^2 + y^2 \leq 1\}$. Find the Euclidean and taxicab distances from P to S.

Question 2: If S is the interior of the unit circle in the Cartesian plane; that is, if $S = \{(x, y): x^2 + y^2 < 1\}$. Find the Euclidean and taxicab distances from P to S.

The distance from a point to a line

Depending on the shape of the set S, the distance between a point and S may be easy or quite difficult to determine. The following result shows that the distance from a point P to a line ℓ is easy to determine for the Euclidean distance function.

Theorem 8.1 For the Euclidean distance function, the distance $d(P, \ell)$ between a point P and a line ℓ not containing P is the length of the line segment $\overline{PQ}$ perpendicular to ℓ at Q.

Proof: The theorem is proved if we can show that the hypotenuse of a right triangle is longer than either leg. Let P be a point not on line ℓ. Let Q be the foot of the perpendicular from P to ℓ. Let R be any other point on ℓ as shown in Figure 6. We need to prove that $PR > PQ$.

Figure 6

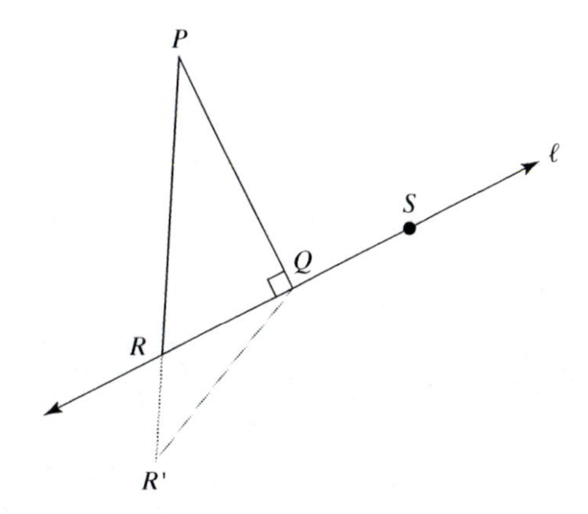

The proof is indirect: Suppose that $PR < PQ$. Then extend $\overline{PR}$ to R' so that $PR + RR' = PR' = PQ$. Then $\Delta PR'Q$ is isosceles, so $m\angle PR'Q = m\angle PQR'$. But

$$m\angle PQR' = m\angle PQR + m\angle RQR' > m\angle PQR = 90°.$$

So $\Delta PR'Q$ has two obtuse angles, which is impossible, since the sum of the measures of the angles of a triangle is 180°. Consequently, we cannot have $PR < PQ$. Also, PR cannot equal PQ. (Why?) So $PR > PQ$.

∎

This proof of Theorem 8.1 relies on the property that the sum of the measures of the angles of a triangle is 180°. It is also the case that Theorem 8.1 follows from the Pythagorean Theorem. In the problems, you are asked to provide a proof using the Pythagorean Theorem.

The distance from a point to a plane

A 3-dimensional analogue of the shortest distance from a point to a line in E^2 is the shortest distance between a point and a plane in E^3. It seems that the shortest distance should be the length of the perpendicular from the point to the plane. Since we have not defined what a perpendicular from a point to a plane means, and you may not have encountered this in your other courses, we pause to clarify terms.

Put a pencil at an oblique angle to the floor. Notice that even when the pencil is at such an angle, you can draw a line on the floor that is perpendicular to the pencil. For this reason, a line is not perpendicular to a plane merely because it is perpendicular to one line in the plane through the intersection of the line and plane. The requirement is stiffer. The pencil must be perpendicular to *every* line in the floor that contains the point of intersection of the pencil and floor.

Definition **A line ℓ is perpendicular to a plane at a point P** if and only if ℓ and the plane both contain P, and ℓ is perpendicular to *every* line in the plane through P.

In many mathematical and physical contexts, a line perpendicular to a plane, or to a curve, is called a **normal** to the plane or curve.

It would be difficult to prove that a line is perpendicular to a plane were it not for the following theorem, which reduces the number of lines required for perpendicularity from an infinity to two.

Theorem 8.2 A line is perpendicular to a plane if it is perpendicular to two lines in the plane through the point of intersection.

Figure 7

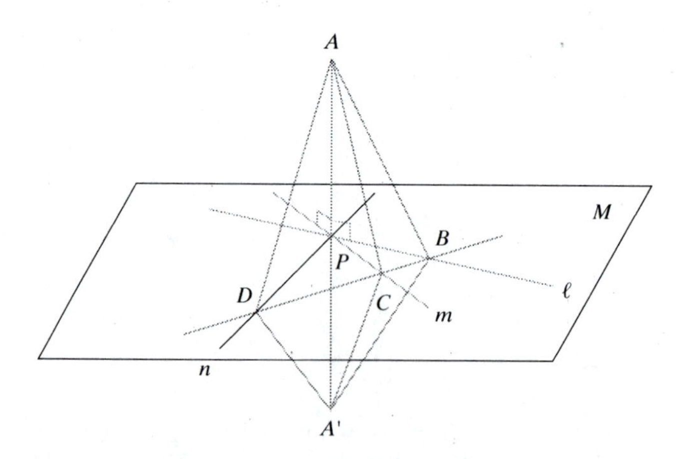

Proof: Let $\overleftrightarrow{AP}$ intersect plane M at P. Let lines ℓ and m containing P be in plane M with $\overleftrightarrow{AP} \perp \ell$ and $\overleftrightarrow{AP} \perp m$. We wish to show that for any other line n in plane M through P, $\overleftrightarrow{AP} \perp n$. Refer to Figure 7. Extend $\overline{AP}$ to point A' so that $A'P = AP$. Now consider a line in M that intersects ℓ at B, m at C, and n at D. $\Delta PAC \cong \Delta PA'C$ and $\Delta PAB \cong \Delta PA'B$ by SAS, so $CA = CA'$ and $BA = BA'$. Thus $\Delta BAC \cong \Delta BA'C$ by SSS. Consequently, $m\angle DBA = m\angle DBA'$. This, with what we know, forces $\Delta BAD \cong \Delta BA'D$ by SAS. So $DA = DA'$. This implies $\Delta DAP \cong \Delta DA'P$ by SSS, from which $m\angle DPA = m\angle DPA'$. Since these angles are also supplementary, they must be right angles. So $\overline{DP} \perp \overline{PA}$, which was to be shown. ∎

Theorem 8.2 explains why the intersection of two walls of a room, being perpendicular to the intersections of those walls with the floor of the room, is also perpendicular to every other line in the floor through the point of intersection. It also shows why, in a 3-dimensional coordinate system, because the z-axis is perpendicular to the x-axis and y-axis, the z-axis is perpendicular to every line in the xy-plane through the point of origin.

Theorem 8.3 The shortest distance between a point P and a plane M is the length of the perpendicular from P to M.

Shortest paths

One natural path connecting two points is the single line segment connecting them. However, as we saw in the discussion of the taxicab metric, there are more complicated kinds of paths between two points. The remainder of this section is devoted to problems related to the lengths of certain paths between points.

Definition If $P_1, P_2, \ldots, P_n (n \geq 2)$ are distinct points, the union of segments $\overline{P_1P_2} \cup \overline{P_2P_3} \cup \ldots \cup \overline{P_{n-1}P_n}$ is called a **path** and is said to connect P_1 and P_n.

The paths we explore are often called **piecewise linear paths**. The **length** of a path is the sum of the lengths of its segments. Piecewise linear paths are often used to approximate curves and their lengths.

Hero's problem

Hero (also called Heron) the Elder of Alexandria (c. 100 B.C.) was a practical surveyor and prolific writer on mathematics and mechanics who is celebrated for his invention of the steam-powered engine, among other things, and is known for his extensive commentary on Euclid's *Elements*. Hero made the following discovery based on the fact that a light ray from a point A meeting a plane mirror in a point C is reflected in the direction of a point B in such a way that $\overline{AC}$ and $\overline{BC}$ form equal angles with the mirror. He found that the actual piecewise linear path of light ACB is the shortest possible path from A to B by way of the mirror. In other words, for any other point C' of the mirror, $AC' + C'B > AC + CB$.

Theorem 8.4 **(Hero's theorem):** Let ℓ be a line and A and B be two points on the same side of ℓ. Then the shortest path from A to ℓ to B is $\overline{AC} \cup \overline{BC}$, where C is the point of intersection of ℓ and the segment $\overline{AB'}$, with B' being the reflection image of B over ℓ.

Figure 8

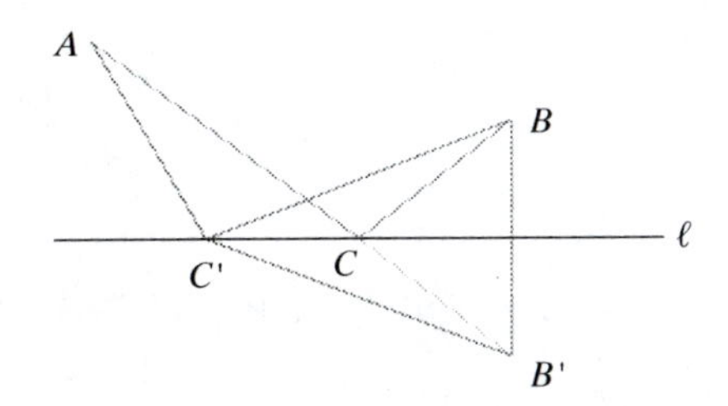

Proof: Refer to Figure 8. We must prove $AC + CB$ is less than $AC' + C'B$, where C' is any other point on line ℓ. Since B' is the reflection image of B over ℓ, $CB = CB'$ and $C'B = C'B'$. Then

$$\begin{aligned} AC' + C'B &= AC' + C'B' && \text{(Substitution)} \\ &> AB' && \text{(Triangle Inequality)} \\ &= AC + CB' && \text{(Betweenness Property)} \\ &= AC + CB. && \text{(Substitution)} \end{aligned}$$

Versions of Hero's problem have ℓ being a river and A and B be starting and ending points on a journey by horse. If the horse needs to drink from the river, then the shortest path is to ride towards C, drink from the river there, then ride directly to the endpoint. Figure 8 also displays the path of a billiard ball shot from A to point C off the side ℓ of the billiard table, assuming no spin.

The Fermat point

Sometimes shortest distance problems involve more than two points. One such problem was posed by Pierre Fermat (1601–1665) and first solved by a student of Galileo, Evangelista Torricelli (1608–1647). The proof we outline below was given by J. E. Hofmann in 1929 and makes use of the fact that rotations preserve distance.

> **Fermat's problem:** In a given acute-angled triangle ABC, locate a point P whose Euclidean distances from A, B, and C have the smallest possible sum.

The solution to Fermat's problem is called the **Fermat point** of ΔABC. This problem has some practical significance. If A, B, and C are locations and roads or pipelines need to be built from an unknown central point P to A, B, and C, then locating P where the sum $PA + PB + PC$ is smallest may be the least expensive site.

Let P be an arbitrary point inside the triangle. We show that if P minimizes $AP + BP + CP$, then $m\angle BPC = 120°$. A similar argument will then show that $m\angle CPA = m\angle APB = 120°$.

Solution Draw $\overline{AP}$, $\overline{BP}$, and $\overline{CP}$. Rotate ΔAPB 60° about B to obtain $\Delta C'P'B$ (Figure 9). Because of the magnitude of the rotation, $\Delta ABC'$ and $\Delta PBP'$ are equilateral.

Figure 9

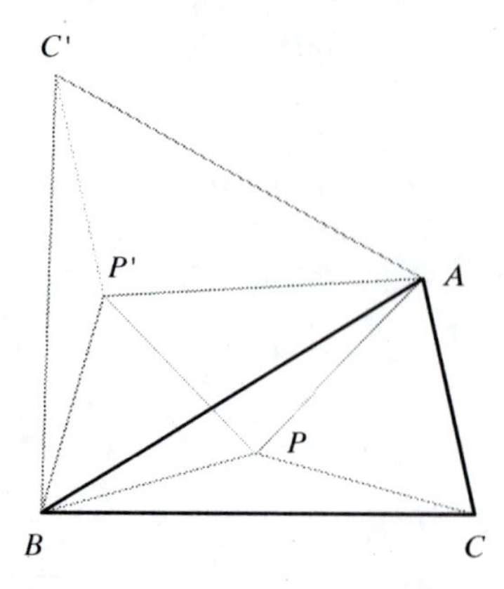

$$\begin{aligned} AP + BP + CP &= C'P' + BP' + CP, \text{ since rotations preserve distance.} \\ &= C'P' + P'P + PC, \text{ since } \Delta PBP' \text{ is equilateral.} \end{aligned}$$

$\overline{C'P'} \cup \overline{P'P} \cup \overline{PC}$ is a path of three segments connecting C' to C. By the Triangle Inequality, we know this path has the shortest length when it is straight, in which case $\angle BPC$ and $\angle BPP'$, as well as $\angle C'P'B$ and $\angle PP'B$, form linear pairs, so they are supplementary. Then $m\angle BPC = 180° - m\angle BPP' = 120°$. Similarly, $m\angle BP'C = m\angle BPA = 120°$. Thus, the desired point P, for which $AP + BP + CP$ is the smallest, is the point from which each of the sides $\overline{BC}$, $\overline{CA}$, and $\overline{AB}$ of ΔABC subtends an angle of 120 degrees.

Notice that the point C' in the solution to Fermat's problem is independent of the location of point P, for it is the third vertex of an equilateral triangle constructed on $\overline{AB}$. This enables an easy construction with ruler and compass of the Fermat point for any given triangle. *Construct an equilateral triangle on each side of the given triangle. Join the new vertex of each of these equilateral triangles to the opposite vertex of the given triangles. These segments are concurrent at the Fermat point.*

We began our solutions with an acute-angled triangle. You are asked in the Problems to explore why this was the case, and whether Fermat's solution applies in other cases.

Fagnano's problem

The following problem was first proposed and solved, using calculus, in 1775, by J. F. Toschi di Fagnano (1715–1797). It speaks of one triangle inscribed in another. This occurs when the vertices of the first triangle are on the sides of the second triangle.

> **Fagnano's problem:** In a given acute-angled triangle ABC, inscribe a triangle UVW whose perimeter is as small as possible.

The solution we give was discovered by Leopold Fejér (1880–1959) and bears a remarkable resemblance to our solution to Fermat's problem.

Solution Select arbitrary points U, V, and W such that U is on $\overline{BC}$, V is on $\overline{CA}$, and W is on $\overline{AB}$, as in Figure 10. Let U' and U'' be the reflection images of U over $\overleftrightarrow{CA}$ and $\overleftrightarrow{AB}$, respectively. Then

$$UV + VW + WU = U'V + VW + WU'',$$

since reflections preserve distance.

Figure 10

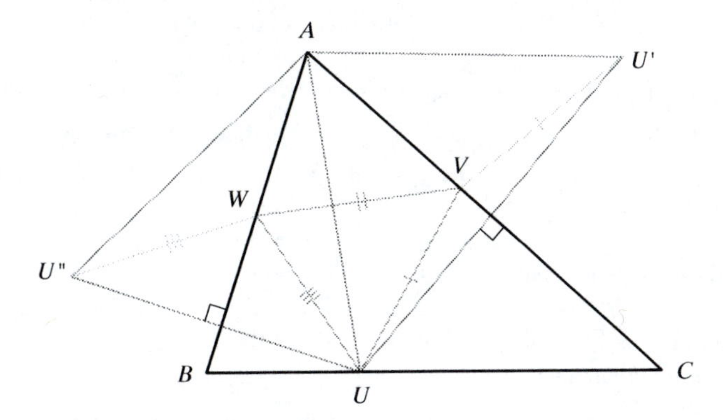

$\overline{U'V} \cup \overline{VW} \cup \overline{WU''}$ is a path from U' to U'', which, by the Triangle Inequality, is minimal in length when the path is a straight line segment. Therefore, among all inscribed triangles with a given vertex U on $\overline{BC}$, the one with the smallest perimeter occurs when V and W lie on the line $\overleftrightarrow{U'U''}$. In this way we obtain a definite triangle UVW for each choice of U on $\overline{BC}$. The problem will be solved when we have chosen U so as to minimize $U'U''$, which is equal to the perimeter of triangle UVW.

Since $\overline{AU'}$ and $\overline{AU''}$ are reflection images of $\overline{AU}$ over $\overleftrightarrow{CA}$ and $\overleftrightarrow{AB}$, they are congruent segments and $m\angle U'AU'' = 2m\angle BAC$. Thus $\Delta AU'U''$ is an isosceles triangle whose angle at A has measure independent of the choice of U. The base $\overline{U'U''}$ of $\Delta AU'U''$ has the smallest length when the equal sides of $\Delta AU'U''$ are smallest, i.e., when $\overline{AU}$ is minimal, in other words, when AU is the shortest distance from the given point A to the given line $\overleftrightarrow{BC}$. Thus, *the desired location of U is such that $\overline{AU}$ is perpendicular to $\overleftrightarrow{BC}$*. Thus $\overline{AU}$ is the altitude from A. This determines the desired triangle. A similar argument shows that *V must be the foot of the altitude from B to $\overline{AC}$, and W must be the foot of the altitude from C to $\overline{AB}$. Then ΔUVW is the unique triangle whose perimeter is smaller than that of any other inscribed triangle.*

The triangle UVW that solves Fagnano's problem is called the **pedal triangle**, or **orthic triangle** of ΔABC.

8.1.2 Problems

1. What is the shortest distance between a point and a circle? Consider the three possibilities and prove your result in each case.
   a. the point is in the interior of the circle
   b. the point is in the exterior of the circle
   c. the point is on the circle

2. a. What is the shortest distance between a point and a sphere?
   b. What is the shortest distance between a sphere and a plane that does not intersect it?
   c. What is the shortest distance between two circles? Consider all possibilities and discuss the generalization to spheres.

3. Prove Theorem 8.1, making use of the Pythagorean Theorem.

4. Prove that the distance from the point (x_0, y_0) to the line $Ax + By + C = 0$ is $\frac{|Ax_0+By_0+C|}{\sqrt{A^2+B^2}}$.

5. Using Euclidean distance, find the length of the shortest path from $(-1, 7)$ to $(8, 5)$ that intersects the x-axis.

6. In Figure 8, let $\overrightarrow{CD}$ be the ray perpendicular to line ℓ on the same side of ℓ as points A and B. When $\overline{AC}$ and $\overline{CB}$ are viewed as parts of the path of a light ray or sound wave reflected off of ℓ, $\angle ACD$ is called the **angle of incidence** and $\angle DCB$ is the **angle of reflection**.
   a. From the geometry of the situation, deduce the law of physics that states that these two angles have equal measure.
   b. Suppose that $A = (0, 4)$, $B = (6, 2)$, and ℓ is the x-axis. Locate point C.

7. Use a dynamic geometry program to construct Figure 9. Print out the figure for several locations of P, thus demonstrating that the location of C' is independent of P.

8. Draw an acute triangle and construct the Fermat point for that triangle.

9. a. Show that the solution to Fermat's problem given in this section breaks down if $m\angle ABC > 120°$.
   b. Show that the solution to Fermat's problem is valid in triangles for which there is no angle greater than 120 degrees.
   c. Solve Fermat's problem for a triangle ABC where $m\angle ABC > 120°$.

10. Let $C = (0, 0)$, $A = (3, 0)$, and $B = (0, 4)$. Locate the Fermat point of this triangle and determine the sum of its distances from the three vertices. Pick some other point on or inside the triangle and show that the sum of its distances from the three vertices is greater.

11. Suppose that equilateral triangles are drawn on the three sides of an arbitrary ΔABC external to the triangle. Let the triangles so formed be ABC', BCA', and CAB'. Prove that $\overline{AA'}$, $\overline{BB'}$, and $\overline{CC'}$ are concurrent.

12. Use an atlas to help obtain answers to these questions.
   a. The Tri-Cities Regional Airport serves Kingsport, Johnson City, and Bristol, Tennessee. How well located is this airport with respect to the Fermat point of the triangle determined by the centers of these cities?
   b. Repeat part **a** for the Tri-City Airport serving Midland, Bay City, and Saginaw, Michigan.

13. Consider the problem of finding the shortest total length of a set of segments connecting the vertices of a polygon.
   a. Use Figure 9. Let P be the Fermat point of ΔABC. Prove that $PA + PB + PC < AB + AC$. (This inequality shows that the three segments connecting the Fermat point to the vertices of the triangle solve this problem for a triangle.)

b. Let $ABCD$ be a square with diagonals intersecting at E. Show that $AE + BE + CE + DE < AB + BC + CD$.

c. Let $ABCD$ be a square as shown in Figure 11a, where the angles at E and F all measure 120°. Show that $AE + ED + EF + FC + FB$ is shorter than the lengths of the diagonals of $ABCD$.

d. State the theorem suggested by Figure 11b and calculate the shortest total length.

16. Prove that the sum of the lengths of the three altitudes of a triangle is less than the perimeter of the triangle.

17. Prove that the point in a quadrilateral for which the sum of its Euclidean distances to the vertices is the smallest is the point of intersection of the diagonals of the quadrilateral. (*Hint*: Choose any other interior point of the quadrilateral and compare its sum of distances to the vertices to that of the intersection point of the diagonals.)

Figure 11

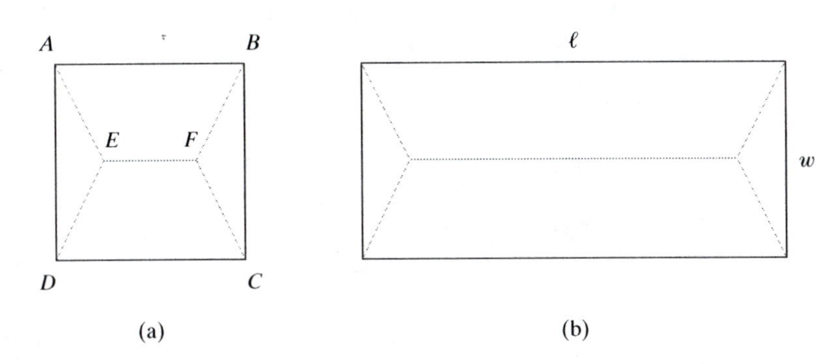

14. Construct Figure 10 using a dynamic geometry program. Have the program calculate the perimeter of ΔUVW and the length of the path $U'V + VW + WU''$ for various positions of U. Show that the length of the path is minimized when U is the foot of the perpendicular from A to $\overline{BC}$.

15. How does Fagnano's problem collapse when we try to apply it to a triangle ABC in which angle A is obtuse?

18. Consider the triangle with vertices $A = (-2, 0)$, $B = (2, 0)$, and $C = (1, 3)$.

a. Determine the vertices of the inscribed triangle whose perimeter is smallest.

b. What is that perimeter?

19. What is the smallest perimeter possible for a triangle that is inscribed in an equilateral triangle of side 5?

ANSWERS TO QUESTIONS

1. Euclidean distance $2\sqrt{2} - 1$; taxicab distance $4 - \sqrt{2}$.

2. Answers are the same as for Question 1.

8.1.3 Extended analysis: locus problems

An important set of problems in mathematics deals with points that satisfy given conditions regarding their position in a plane or in space. Historically, these are called **locus problems**, from the Latin word "locus" (plural, "loci") meaning "place". Here are answers to some common locus problems in early plane geometry. In recent years, the word "locus" is often replaced by "set", so we provide both wordings.

1. The locus (set) of points in a plane equidistant from two given points in the plane is a line, the perpendicular bisector of the segment connecting them.
2. The locus (set) of points in a plane equidistant from two given intersecting lines in the plane is the union of two lines, the bisectors of the four angles determined by the given lines.
3. The locus (set) of points in a plane at a fixed distance from a given point in the plane is a circle with center at the point and radius the fixed distance.

The third of these loci is so familiar to us that it would seem that no interesting problems could ensue from it. On the contrary, one classic problem is particularly rich.

The dog-on-leash problem

The following problem appears in many places in many forms, almost always with an animal of some type on a leash tied to the side of a rectangular-shaped object with given dimensions.

> **Problem 1:** A dog on a leash is tied to a rectangular-shaped barn 20 meters long and 10 meters wide. The leash is 8 meters long and fastened 6 meters from the end of the longer side of the barn (Figure 12). In what region can the dog roam, and what is the area of that region?

Figure 12

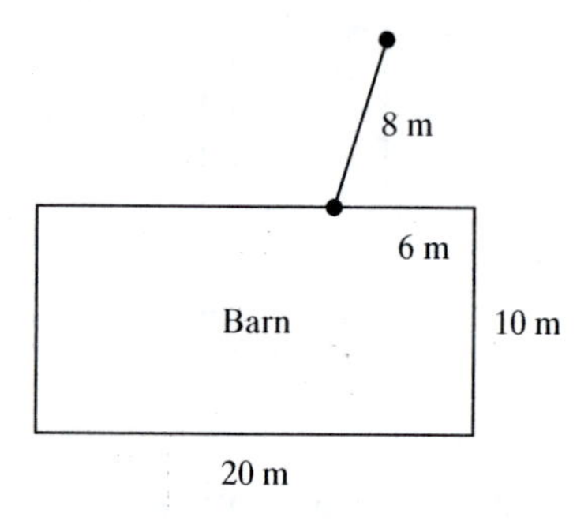

Question 1: Before reading on, take time to solve Problem 1.

As you solve this problem, you might first wish that the leash had been shorter than the distance to the end of the barn. Then the region is semicircular, as in Figure 13a below. But because the leash is more than 6 m long, the region is the union of the interiors of a semicircle and a quarter circle.

There are some similarities between this problem and the problem of maximizing the volume of a box discussed in Section 3.2.3 and the catch-up problem of Section 3.1.2. We may begin with rectangles of any shape. We may ask what happens as the length of the leash becomes longer and longer, or shorter and shorter. Four specific lengths are given in the problem, and we may wonder if there is a way to generalize the problem. We explore one direction here. In the Problems, you are asked to explore other possibilities.

> **Problem 2:** Without changing the values of any of the other parameters in Problem 1, where should the leash be tethered to maximize the area for the dog?

Figure 13

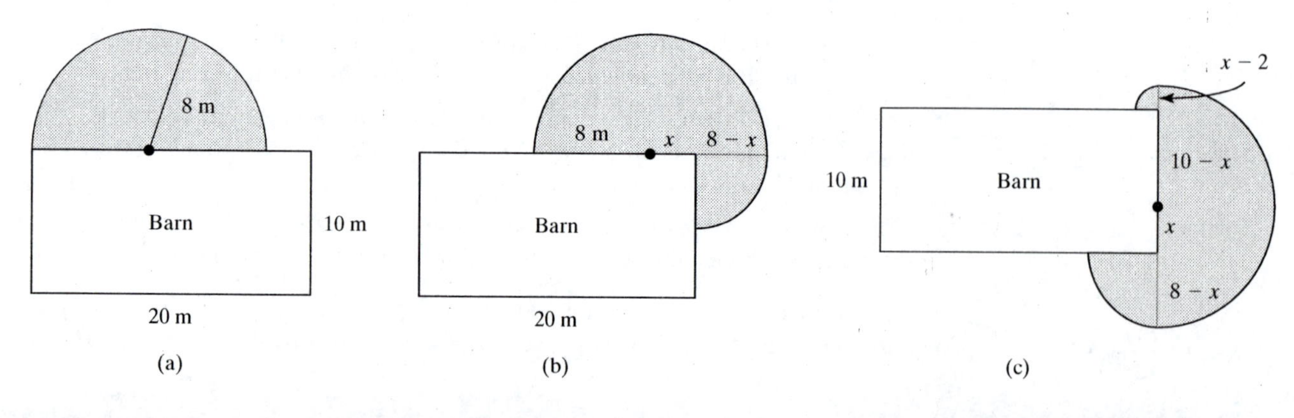

Question 2: Without reading on, and without doing any calculations or drawing an additional picture, where does your intuition lead you to think the dog should be tethered?

There are four possibilities to consider:

1. Tether on the long side in a place where there is no overlap. Then the dog can roam in a semicircle of radius 8 meters, for an area of 32π square meters (Figure 13a).
2. Tether on the long side close enough to the corner so that the dog can turn the corner. (See Figure 13b.) If the leash is attached x meters from the corner, then the dog still can roam in a semicircle of radius 8 meters, but now the dog can roam in an additional quarter circle of radius $(8 - x)$ meters, for a total area of $\left(32\pi + \frac{\pi}{4}(8 - x)^2\right)$ square meters. This expression is clearly maximized when $x = 0$, that is, when the dog is at the corner.
3. Tethering at the corner allows the dog to roam in a 270° sector of a circle with radius 8, for an area of 48π square meters.
4. Tethering on a shorter side of the rectangle is the only other possibility. Certain positions allow the dog the possibility of looking over two edges. If the dog is x m from the nearest corner, where $2 < x \le 5$, then the region for roaming includes the 8-m semicircle and two additional quarter circles, as shown in Figure 13c. This region can be shown to have less area at all times than tethering at a corner. So tethering at a corner is the best strategy. (Did your intuition guide you in the right direction?)

The locus of points at a fixed distance from a line

Now we turn to a different locus problem. What is the locus of points at a fixed distance from a fixed line? The answer is simply stated and just as easily proved.

Theorem 8.5 In a plane, the set of points at a fixed distance from a given line ℓ is the union of two lines, one on each side of ℓ parallel to ℓ and at the fixed distance from ℓ.

Figure 14

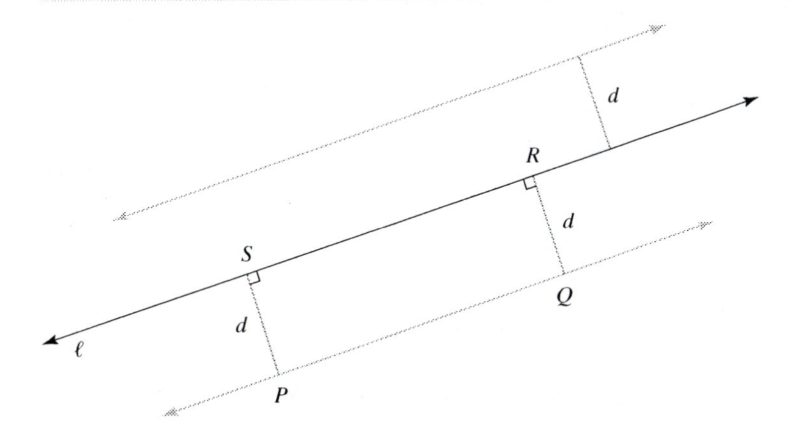

Proof: Let d be the fixed distance. Let two points P and Q be on the same side of ℓ at a distance d from it (Figure 14). Then drop the perpendiculars from P and Q to ℓ, intersecting ℓ at S and R, respectively. By Theorem 8.1, $PS = QR = d$. Since $\overline{PS} // \overline{QR}$ (both being perpendicular to ℓ), $PQRS$ is a parallelogram. Thus $\overline{PQ} // \overline{RS}$. Since there is only one parallel to ℓ through P, all points on that side of ℓ at a distance d from it are on that parallel. Similarly, there is a line on the other side of ℓ at a distance d from ℓ. ┘

The locus of points at a fixed distance from a set of points or region

Disputes over boundaries of territory have led to wars between nations and bickering between neighbors over fence lines. International conventions and zoning ordinances, treaties, titles, and plats are all written attempts to clarify boundaries. The boundaries themselves are almost always written in a technical language that is quite mathematical, sometimes identifying points and indicating that the boundary connects them, or identifying a natural barrier (such as a river) and indicating that the boundary is the middle (determined in some way), or identifying a region and indicating that the boundary is a locus of points at a given distance from that region. It is the last of these that we discuss here.

In the United States, a home, apartment building, or other structure is typically built on a parcel of land owned by an individual, organization, or the government. The boundary of the parcel is typically specified by latitude and longitude, or distances from fixed natural objects. In towns and cities, where land parcels are small, people are not allowed to build to the boundaries of the lot. An ordinance might say that no building is to be built within 3 feet of the boundary. There may also be ordinances forbidding certain identified activities within fixed distances of others, such as no selling of alcohol within 250 feet of a school.

In international law, the baseline for the boundary of a country on an ocean is the low-water line along the coast, except for ports, harbors, and bays that are internal waters of the country. The ocean within 3 miles of the baseline is within a country's jurisdiction unless another country is closer, and a country can claim land up to a distance of 12 miles from the baseline. Any land over 12 miles from the baseline is in the high seas, that is, in international waters and under international law. These distances can have personal significance: For instance, if a couple wishes to get married on a ship by the ship's captain, they must be in international waters.

All of these situations involve the locus of points at a certain distance from a given set. To study the nature of this problem, we again begin with a specific condition on a rectangle.

Problem 3: What is the locus of points 1 unit from an 8×13 rectangle?

Solution The answer combines two fundamental locus propositions, involving distance from a point and from a line. These are pictured in Figure 15.

Figure 15

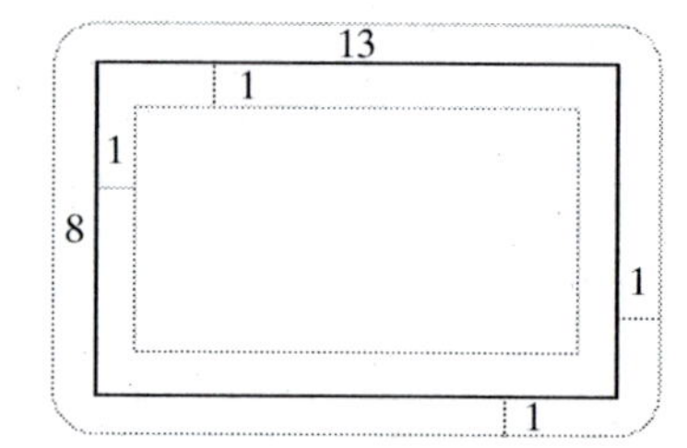

In the interior of the rectangle, the locus is a rectangle with dimensions 6×11. This might lead us to believe that in the exterior of the rectangle, the locus is also a rectangle, one with dimensions 10×15. This is the case except at the corners, where points are closer to the corners than to the sides, so quarter-circles result.

You are asked to extend and generalize Problem 3 in the problems that follow.

8.1.3 Problems

1. At the beginning of this section, answers to three plane geometry locus problems are given. Give the answers if these problems are considered in 3 dimensions rather than 2.

2. In Problem 1 in this section, let the length of the leash be x meters and let all other given dimensions be the same. Assume that the barn is in the middle of a flat field of infinite

extent in all directions. Let $A(x)$ = the area (in square meters) of the region in which the dog can roam.

a. Because of the nature of the problem, there is no one simple algebraic expression for $A(x)$. For what values of $x \leq 30$ does the expression change?

b. Give expressions for $A(x)$ for all values of $x \leq 30$.

c. On the interval $0 \leq x \leq 30$, is A continuous?

d. On the interval $0 \leq x \leq 30$, is A differentiable?

e. As $x \to \infty$, what function does A approach?

3. a. How do the answers to Problem 2 immediately above change if the leash is tethered at the upper right vertex of the rectangle?

b. Do the answers to part **a** change if the leash is tethered at one of the other vertices of the rectangle?

4. In Problem 1 in this section, let the dog be tethered on the long side of the rectangle d meters from the upper-right vertex (generalizing from the stated value of 6 meters), where $0 \leq d \leq 20$. Now let $B(d)$ = the area (in square meters) of the region in which the dog can roam.

a. Give a formula for $B(d)$ in terms of d.

b. Describe the graph of B.

5. In Problems 1 and 2 of this section, for maximum roaming area, is it always best to tether at a vertex of the rectangle, regardless of the dimensions of the rectangle or length of the leash? Explain your answer.

*6. Assumed in the analyses of Problems 1 and 2 in this section is that the leash is tied to the barn at the same height as it is fastened to the dog's collar. Suppose in Problem 1 that the leash is tied 1 meter higher than the dog's collar. How does that affect the solution?

7. Give equations for the two lines at a distance 3 units from the line with equation $x + 2y = 5$.

8. What is the locus of points 1 unit from a $\frac{1}{2} \times \frac{3}{4}$ rectangle?

9. Generalize Problem 3 of this section in some way and answer the question of your generalization.

ANSWER TO QUESTION

The region is the union of a semicircle with radius 8 and a quarter-circle to the right of the rectangle with center at the top right vertex and radius 2. The total area is 33π m^2.

8.1.4 Distance on the surface of a sphere

The Euclidean distance formula $\mathbf{R}^3$ is not often usable on the surface of Earth because the locations of points on Earth are not usually described by their relationship to the center of Earth. Instead, the most common way of locating points uses latitude and longitude. Our desire is to be able to determine the distance between two locations on the surface of Earth given their latitude and longitude. For this, we assume Earth is a sphere whose center is the point O.

Latitude and longitude

Both latitude and longitude are measures of arcs, and to determine these we review some facts about planes and spheres. If a plane contains a point in the interior of a sphere, then its intersection with the sphere is a circle. If the circle's center is the center of the sphere, then it is a **great circle**. If not, it is a **small circle**. (See Problem 1.)

Figure 16

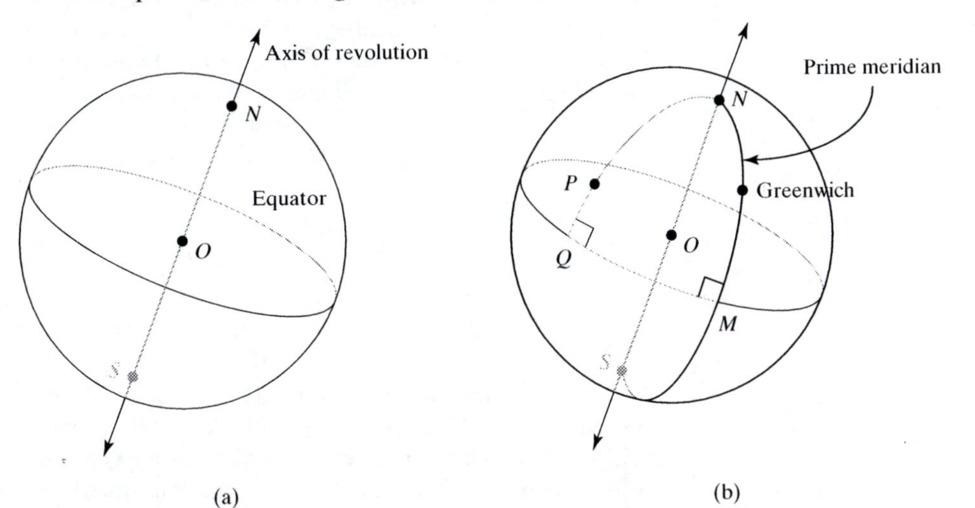

Earth rotates about a line called the **axis of revolution**. This line contains the center O of Earth and intersects the surface in two points, the **north pole** and **south pole**, which we call N and S. (In Figures 16a and 16b, S is on the back side of Earth from the viewer.) The set of points on the surface of Earth equidistant from N and S is the intersection of Earth's surface with a plane that contains the center O of Earth, so it is a great circle. This great circle is the **equator**. The equator splits Earth and its surface into northern and southern hemispheres.

Since N, O, and S are collinear, there are infinitely many great circles that contain N and S. Each semicircle with endpoints N and S is called a **meridian**. Exactly one of these infinitely many meridians passes through each point on Earth other than the poles. The meridian that contains the observatory in Greenwich, England is called the **prime meridian**.

Let P be a point on the surface of Earth other than N or S. Let the meridian containing P intersect the equator in Q. (See Figure 16b.) The **latitude** of P is the degree measure of the arc $\overset{\frown}{PQ}$ and is identified as north or south depending on which hemisphere P lies. Consequently, latitudes range from 0 to 90° in each direction. Notice that all points with the same latitude lie on the same *small* circle. The north and south poles are said to have latitude 90°N and 90°S, respectively, for they are 90° from the equator regardless of the meridian used to measure the arc.

Let M be the intersection of the prime meridian with the equator. The **longitude** of P is the degree measure of the arc $\overset{\frown}{MQ}$ and is identified as east or west depending on the direction in which it is measured from M. Thus longitudes range from 180°W to 180°E. Notice that all points on the same meridian have the same longitude. The meridian with longitude 180°W or 180°E is the other half of the circle containing the prime meridian. The north and south poles do not have a unique longitude. You can think of them either as having no longitude or as having every possible longitude.

The dimensions of Earth

Earth is not a sphere, even ignoring its mountains and valleys. The equator has a diameter of about 7,928 miles (12,759 km), while the meridians have diameters of about 7,901 miles (12,715 km). The fraction $\frac{7928-7901}{7928}$ is about $\frac{1}{294}$ (and is closer to $\frac{1}{297}$ if the diameters are measured more accurately) and is used as a measure of Earth's *oblateness*, the extent to which it differs from a sphere. This difference is so small, and the mountains and valleys of Earth are so small compared to its diameter (Mt. Everest's height of 29,028 feet (8,848 m) is less than .001 of Earth's diameter), that if an accurate bowling-ball-sized model of Earth was crafted, you would have a difficult time discerning that it is not a sphere and you would likely not feel any of the peaks and valleys. For this reason, it is quite reasonable to treat Earth as a sphere, particularly for such things as latitude and longitude which do not pertain to physical properties such as gravity. For our purposes, we take the diameter of this sphere to be 7920 miles, or 12,740 km.

From the value for the diameter, you can calculate the length of a degree of arc of a great circle to be about 69.1 miles or 111.0 km.

The spherical law of cosines

A problem of both practical and theoretical importance is to calculate the shortest distance between two points P and Q along the Earth's surface given their latitudes and longitudes. If P and Q are endpoints of a diameter, then we know the distance PQ from the above discussion. If they are not endpoints of a diameter, then there is exactly one plane that contains them and the center O of the earth. The shortest distance between P and Q can be proved to be the length of the arc of the great circle that is the intersection of Earth and this plane. To find this distance, we apply plane geometry and plane trigonometry to arrive at a fundamental theorem about spherical triangles.

Definition Given three points A, B, and C on a sphere but not all on the same great circle, the **spherical triangle ABC** is the union of the minor arcs AB, BC, AC of the three great circles that contain them.

Figure 17

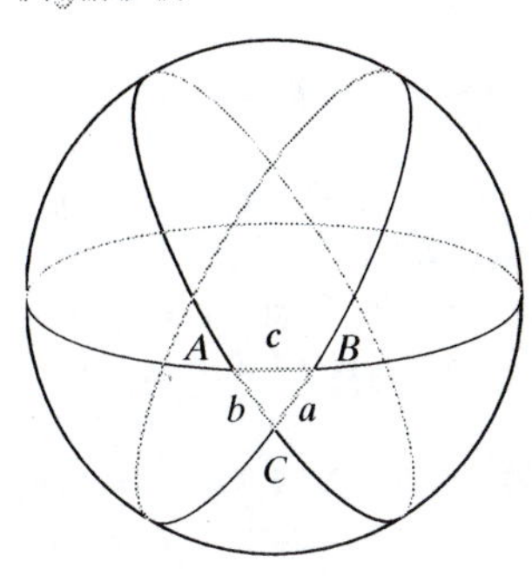

We will find a formula for the length of side c in spherical ΔABC in terms of the lengths of sides a and b and the measure of angle C. For triangles in the plane, this goal is achieved in the Law of Cosines. Consequently, the corresponding formula for spherical triangles is called the *Spherical Law of Cosines.*

Figure 18 shows spherical ΔABC in a sphere with center O. In the plane of arc AC, the tangent to arc AC at C meets $\overrightarrow{OA}$ at E. In the plane of arc BC, the tangent to arc BC at C meets $\overrightarrow{OB}$ at D. The **measure of the spherical angle ACB** is defined as the measure of the plane angle ECD between the tangents to the arcs CA and CB at C. (See Figure 18.) Notice that the three sides of the spherical triangle are arcs with measures equal to those of central angles: $a = m\angle BOC$, $b = m\angle AOC$, and $c = m\angle AOB$.

Figure 18

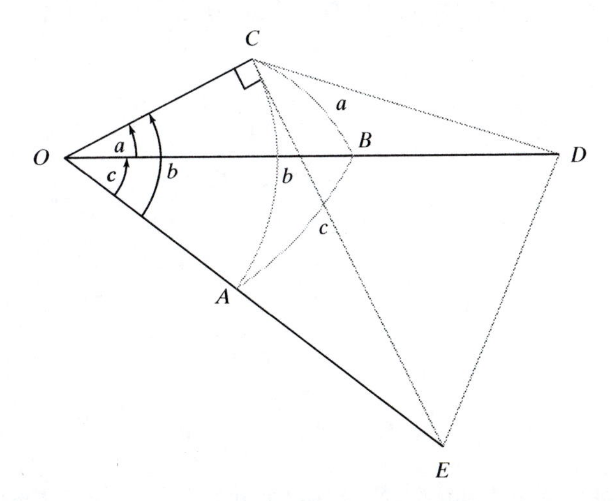

Theorem 8.6 **(Spherical Law of Cosines):** In any spherical ΔABC,

$$\cos c = \cos a \cos b + \sin a \sin b \cos C.$$

Proof: This proof has three major steps. First we find two expressions for DE^2. Then we set them equal to each other and solve for $\cos c$, which is in one of the expressions. The third step is to use right triangle trigonometry to put the answer in the form stated in the theorem.

Refer to Figure 18. By the plane Law of Cosines applied in ΔODE and ΔCDE,

$$DE^2 = OE^2 + OD^2 - 2OD \cdot OE \cos c$$

and

$$DE^2 = CE^2 + CD^2 - 2CD \cdot CE \cos C.$$

Subtract the second equation from the first.

$$0 = (OE^2 - CE^2) + (OD^2 - CD^2) - 2OD \cdot OE \cos c + 2CD \cdot CE \cos C$$

ΔOCE and ΔOCD are right triangles due to $\overline{CE}$ and $\overline{CD}$ being tangents to the great circles. So, by the Pythagorean Theorem, each expression in parentheses equals OC^2. Make this substitution and divide both sides by 2.

$$0 = OC^2 - OD \cdot OE \cos c + CD \cdot CE \cos C$$

Solve for $\cos c$ and use the trigonometry of right triangles.

$$\cos c = \frac{OC}{OD} \cdot \frac{OC}{OE} + \frac{CD}{OD} \cdot \frac{CE}{OE} \cos C$$
$$= \cos a \cos b + \sin a \sin b \cos C$$

An example

Let P and Q be two places on Earth not on the equator and consider spherical ΔNPQ, where N is the north pole. Let n, p, and q be the sides of this triangle opposite the vertices N, P, and Q. From the latitudes of P and Q, we can determine the degree measures of arcs NP and NQ. The difference in their longitudes determines the degree measure of spherical angle PNQ. With these two sides and the included angle, the Spherical Law of Cosines can be applied to find the degree measure of side n. The distance is then $\frac{n}{360}(7920\pi)$ miles, or $\frac{n}{360}(12{,}740\pi)$ km.

For instance, to find the distance between New York City, and New Delhi, India, let P be New York, latitude 40°45′N, longitude 74°0′W, and let Q be New Delhi, with latitude 28°38′N, longitude 77°12′E. Form spherical ΔNPQ as shown in Figure 19.

Figure 19

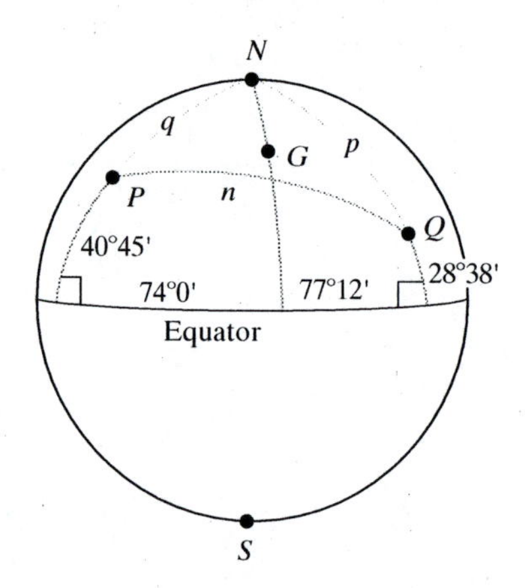

From the Spherical Law of Cosines,

$$\cos n = \cos \widehat{PN} \cdot \cos \widehat{QN} + \sin \widehat{PN} \cdot \sin \widehat{QN} \cdot \cos \angle PNQ$$
$$\cos n = \cos 49°15' \cdot \cos 61°22' + \sin 49°15' \cdot \sin 61°22' \cdot \cos 151°12'$$
$$\approx -.2699$$

From this,

$$n \approx \cos^{-1}(-.2699) \approx 105.66°.$$

This tells us that the arc of the great circle from New York to New Delhi is approximately $\frac{105.66}{360}$ of the circle. To find the distance between these cities, we multiply the circumference of Earth by this number. The desired distance

$$d \approx \frac{105.66}{360} \cdot 7920\pi\text{miles} \approx 7302 \text{ miles.}$$

8.1.4 Problems

1. In this section it is asserted that the intersection of a sphere with a plane that contains an interior point of the sphere is a circle. Prove this theorem synthetically.

2. a. In $\mathbf{R}^3$, what is an equation for the sphere with center (h, k, j) and radius r?

   b. Every plane in $\mathbf{R}^3$ has an equation of the form $ax + by + cz = d$. Use this fact to provide an analytic proof that if a plane and the sphere with center $(0, 0, 0)$ have at least two points in common, then their intersection is a circle.

3. Assuming that the intersection of two nonparallel distinct planes is a line, prove that two great circles of a sphere intersect in two points which are endpoints of a diameter.

4. Explain why, in a great circle of Earth, an arc with measure $x°$ is about $22\pi x$ miles long.

5. a. Explain why the proof of the Spherical Law of Cosines given in this section fails if one or both of the known sides of the spherical triangle has measure 90°.

b. Does the Spherical Law of Cosines hold for a triangle with exactly one known side of measure 90°?

c. Does the Spherical Law of Cosines hold for a triangle with both known sides having measure 90°?

6. We took our information for the latitude and longitude of New York and New Delhi from *The World Almanac and Book of Facts 2000*. In that same source, the air distance between the cities of New York and New Delhi is given as 7318 miles. Give some possible reasons for the disparity between this figure and the one calculated in the lesson.

7. In flying a nonstop great circle route from Chicago (41°52N, 87°38′W) to Tokyo, Japan (35°45′N, 139°45′E), demonstrate that you will fly quite near Anchorage, Alaska (61°10′N, 149°59′W). About what percent of the flight will be finished when you are near Anchorage?

8. Use latitude and longitude to determine the distance between Paris, France (48°50′N, 2°20′E) and Rio de Janiero, Brazil (22°54′S, 43°13′W).

9. Beijing, China and Philadelphia, Pennsylvania are both at approximately 40°N latitude. Beijing's longitude is 116°28E, while Philadelphia's longitude is 75°9′W.

a. What is the great circle distance between these cities?

b. If a person were to fly due east from Beijing to Philadelphia, what distance would be traveled?

c. If a person were to fly due west from Beijing to Philadelphia, what distance would be traveled?

10. a. Assuming that Earth is a sphere of radius 3960 miles, find a formula for the circumference of the circle containing all points at a latitude $L°$.

b. Use part **a** to find the distance *along the circle of latitude* between places on the same latitude whose longitudes differ by $d°$.

c. Find a formula for the *great circle distance* between two points on the same latitude whose longitudes differ by $d°$.

*d. Use parts **b** and **c** to prove that the distance between two points along the circle of latitude is never less than the great circle distance between those points. When is it equal?

11. Let ABC be a spherical triangle with right angle at C and with no leg of measure π (radians). Prove:

a. $\cos c = \cos a \cos b$

b. $\cos A = \frac{\tan b}{\tan c}$

c. $\sin A = \frac{\sin a}{\sin c}$

d. $\tan A = \frac{\tan a}{\sin b}$

e. $\cos c = \cot A \cot B$

f. $\cos A = \cos a \sin B$

12. Let $d_S(P, Q)$ be the length of the smaller arc of the great circle passing through two points P and Q on the surface of a sphere with radius r. Show that d_S satisfies the triangle inequality with respect to any third point R on the sphere.

13. Let S be a surface in three-dimensional Euclidean space. Define $d_m(P, Q)$ as the minimum length of the paths passing along the surface S and joining the points P and Q. You may assume that for each pair of points on S there exists such a path. Show that d_m is a distance function.

Unit 8.2 Similar Figures

In the development of a negative of a picture taken with a camera, there is almost always a choice of size of the resulting photo. There are even more choices when portrait photos are taken by students as class pictures or for graduation. We usually think of the results as being the *same picture* even though their sizes are different.

Mathematics tends to take the same view of its objects. Types of figures are not distinguished by their size. A large figure does not have a different name than a small one of the same shape. For instance, the name *ellipse* is given both to Earth's orbit around the sun and to a picture of the orbit drawn on a sheet of paper, even though the former is over a trillion times as large. The name "similar" that we use for figures with the same shape has the same linguistic root as the word "same".

Because the study of Euclidean geometry normally places congruence before similarity, there is a natural tendency to think of similarity as a less important extension of congruence. Perhaps the first to recognize that this was not the case was the German mathematician Felix Klein, who in 1872 defined Euclidean geometry as "the science that studies those properties of geometric figures that are not changed by similarity transformations".

In an essay entitled "What Is Geometry?", the 20th-century Russian mathematician Isaac Moisevitch Yaglom commented on Klein's view.

"Klein's definition says that, in a definite sense, not only does geometry not distinguish between congruent figures, but it does not even distinguish between similar figures; indeed, in order to assert that two triangles are congruent, and not merely similar, we must have fixed a definite unit of measurement, once and for all, with which to measure the sides of both triangles. It is this very "indistinguishability" of similar figures that enables us to represent figures of large dimensions in a picture; the teacher uses this principle when he tells the students to reproduce 'accurately' in their notebooks the figure he is drawing on the blackboard and which, of course, could not possibly fit into their notebooks without being reduced in size."[1]

Indeed, if students in a class are each asked to draw an accurate picture of a 3-4-5 right triangle, and if the drawings are then compared, the drawings may not be congruent but they will be similar. All the theorems of Euclidean geometry that deal with length are theorems that relate ratios of lengths, not simply lengths. For instance, we cannot prove that the diagonal of a square has a particular length, but we can prove that the ratio of the length of a diagonal of a square to the side of that square is $\sqrt{2}$. When we deduce that the opposite sides of a parallelogram are congruent, we have proved that the ratio of their lengths is 1; we have not found a precise length for the sides.

8.2.1 When are two figures similar?

In 1938, George David Birkhoff and Ralph Beatley, both professors at Harvard University, developed a high school geometry course that takes advantage of the power of similarity in Euclidean geometry. Working from a set of postulates for geometry that Birkhoff had developed some years earlier and that assumed all properties of real numbers, their approach utilizes only five geometric postulates. There is no postulate of congruence, but they assume *SAS Similarity*: If in two triangles, two sides of one are proportional to two sides of the other and the included angles have the same measure, then the triangles are similar. From this and the other four postulates, they are able to develop all the theorems of Euclidean geometry. Such is the power of similarity.

Informally, we think of two figures as similar when they have the same shape. As with congruence, the definition of similar figures found in many books is restrictive, applying only to convex polygons. Here is a standard definition: Two convex polygons are *similar* if and only if there is a correspondence between their vertices such that corresponding angles are congruent and corresponding sides are proportional. (*Proportional* means that the ratios of lengths of corresponding sides are equal.)

Question 1: Two similar polygons are shown in Figure 20. What is the correspondence between their vertices? By measuring, determine the constant ratio of lengths of corresponding sides of the larger to the smaller.

---

[1] I. M. Yaglom, *Geometric Transformations II*, p. 6.

Figure 20

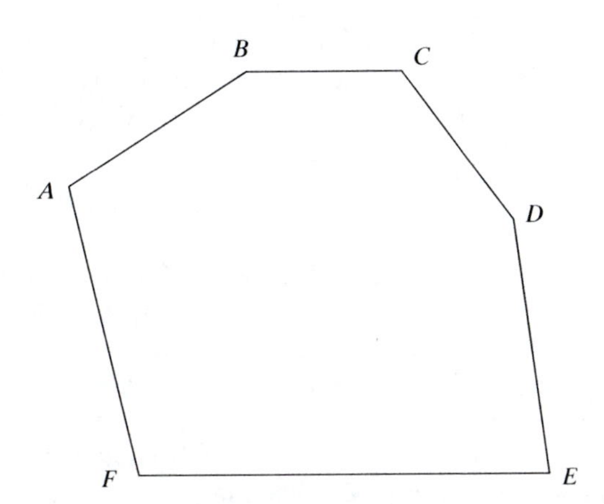

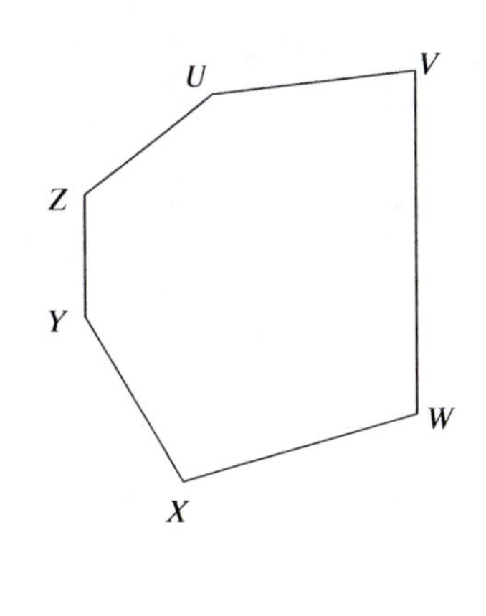

The interiors of all triangles are *convex*. That is, all points between any two points A and B in the interior also lie in the interior. But polygons with 4 or more sides can be *nonconvex* (Figure 21). In nonconvex figures, how angles are measured affects whether the standard definition of similarity applies.

Figure 21

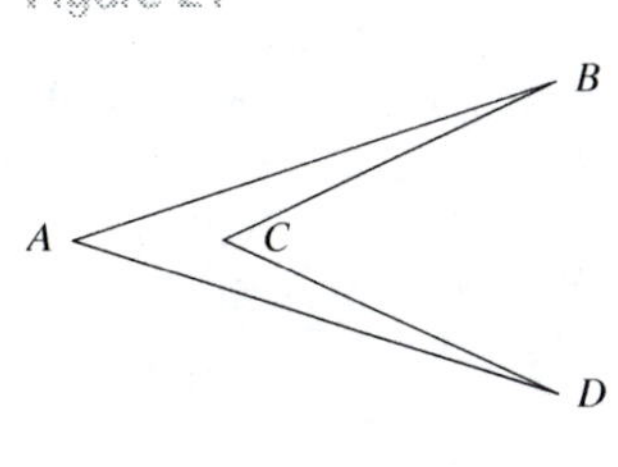

Consider the quadrilateral in Figure 21. We would like to think of it as having four angles: $\angle DAB$, $\angle ABC$, $\angle BCD$, and $\angle CDA$. Of these, $\angle BCD$ presents a difficulty. If angles are restricted to having measures from 0° to 180°, then the interior of $\angle BCD$ is in the exterior of the polygon. But then the quadrilateral $ABCD$ has four acute angles and the sum of its angle measures is quite small, far from the sum of 360° for all convex quadrilaterals.

We could consider $\angle BCD$ as a reflex angle. If we allow reflex angles, $\angle BCD$ (and every other angle that is not a straight angle) has two measures. This presents difficulties in discussions of angles, as we pointed out in Section 7.2.2. With reflex angles or with directed angles, the sum of the measures of the interior angles of ABCD can be seen to be 360°. But reflex angles and directed angles are needed for so little of geometry that most geometry schoolbooks today avoid them.

Figure 22

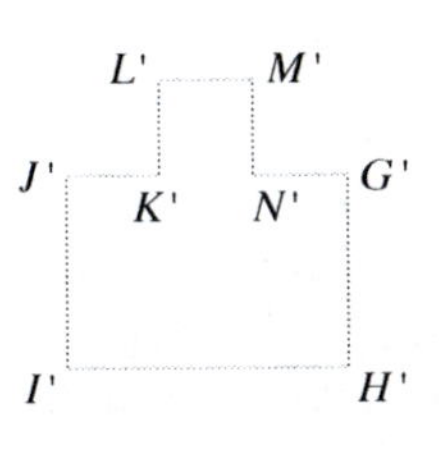

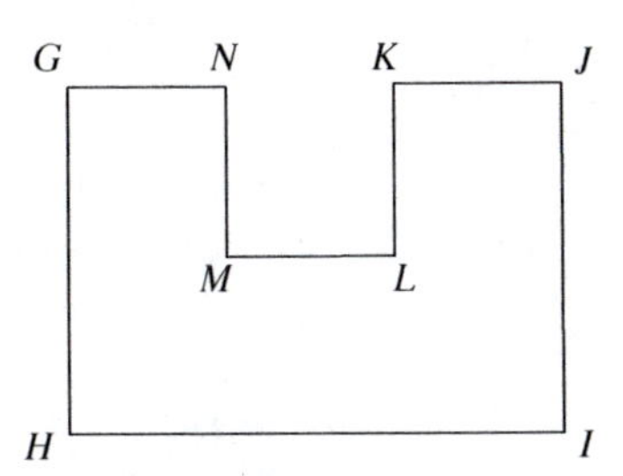

Figure 22 shows another difficulty if measures of angles are unique and restricted to being from 0° to 180°. Here the correspondence between vertices of the octagons is $GHIJKLMN \leftrightarrow G'H'I'J'K'L'M'N'$. Corresponding sides are proportional since the sides of $G'H'I'J'K'L'M'N'$ are half the length of the corresponding sides of $GHIJKLMN$. If angles are always measured from 0° to 180°, all angles of both octagons are right angles, so corresponding angles have equal measures. Thus, under this restriction of angle measure and the standard definition of similar polygons, these polygons are similar. Because of this difficulty, you will often see the definition of similar polygons restricted to convex polygons.

To avoid the restriction of the definition of similar figures to convex polygons, or even polygons more generally, we employ transformations. Refer back to Section 7.1.4 to compare these definitions with the corresponding definitions for congruence.

Definitions A transformation T of the plane is a **similarity transformation** if and only if there is a positive real number k (the **ratio of similitude**) such that for any points P and Q of the plane, if $T(P) = P'$ and $T(Q) = Q'$, then $P'Q' = k \cdot PQ$.
Two figures α and β in the plane are **similar**, written $\alpha \sim \beta$, if and only if there is a similarity transformation T such that $T(\alpha) = \beta$.

Some books call k the **ratio of similarity**. We call k a ratio because $P'Q' = k \cdot PQ$ implies $\frac{P'Q'}{PQ} = k$. If $k > 1$, then β is larger than α, and T is called an **expansion**. If $k < 1$, then β is smaller than α, and T is a **contraction.** If $k = 1$, then T preserves distance, and so T is an isometry and $\alpha \cong \beta$. That is, every isometry is a similarity transformation of magnitude 1.

Two similar figures α and β are **directly similar** if they have the same orientation; otherwise, they are **oppositely similar.**

Recall that the definition of congruent figures in terms of transformations required only one criterion—the preservation of distance. This contrasts with the two criteria—equal side lengths and equal angle measures—in the customary definition of congruent triangles. Similarly, the definition of similar figures in terms of transformations requires only one criterion—the multiplication of distance—for figures to be similar. This single criterion contrasts with the two criteria—proportional side lengths and equal angle measures—in the customary definition of similar polygons. This criterion is the fundamental property of similarity because from it all other properties of similarity can be developed. Notice its generality: It deals with all corresponding distances, not just corresponding side lengths. We restate this defining criterion in terms of ratios, as follows.

Fundamental Property of Similarity: If two figures a and b are similar, then all ratios of corresponding distances on a and b are equal.

Depending on the order in which the ratios are taken, the equal ratios will either equal the ratio of similitude or its reciprocal.

Before we discuss any specific similarity transformation, we can deduce some properties that apply to all similarity transformations.

Suppose B is between A and C. Then $AB + BC = AC$. Multiplying both sides by $k \neq 0$, we have $k \cdot AB + k \cdot BC = k \cdot AC$. If T is a similarity transformation with this magnitude k, the images B', A', and C' of B, A, and C will be such that $A'B' = k \cdot AB$, $B'C' = k \cdot BC$, and $A'C' = k \cdot AC$ (Figure 23). Consequently, $A'B' + B'C' = A'C'$, so B' is between A' and C'. This argument shows that similarity transformations preserve betweenness.

Figure 23

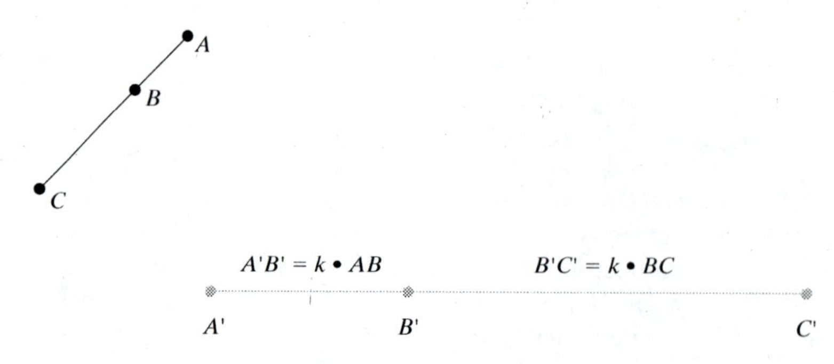

Recall from Section 7.1.4 that the preservation of betweenness by isometries leads immediately to the preservation of collinearity and the fact that the image of a line is a line, the image of a line segment is a line segment, and so on for the basic figures. The same arguments lead to the following theorem.

Theorem 8.7

a. Every similarity transformation preserves betweenness.

b. Every similarity transformation preserves collinearity.

c. Under a similarity transformation, the image of a line is a line, the image of a line segment is a line segment, the image of a ray is a ray, the image of an angle is an angle, and the image of a polygon is a polygon with the same number of sides.

Transformations more general than similarities possess all the properties identified in Theorem 8.7. In particular, *affine transformations*—defined in $\mathbf{R}^2$ as those transformations that map a point (x, y) onto a point $(ax + by + e, cx + dy + f)$—have all these properties. Affine transformations include *horizontal stretches* (those that map (x, y) onto (ax, y)) and *vertical stretches* (those that map (x, y) onto (x, by)), and all affine transformations are composites of these stretches and similarity transformations.

Question 2: In Figure 22, think of $G'H'I'J'K'L'M'N'$ as the image of $GHIJKLMN$ under a transformation. Find two pairs of corresponding vertices such that the distance between the vertices in the two figures does not equal the ratio of lengths of corresponding sides. (This question shows that a transformation mapping $GHIJKLMN$ onto $G'H'I'J'K'L'M'N'$ cannot be a similarity transformation.)

Unlike affine transformations, which may distort a figure by stretching it in one direction and not others, similarity transformations expand or contract figures uniformly in all directions. This can be seen from the following theorem.

Theorem 8.8 Under any similarity transformation, the image of any circle is a circle.

Figure 24

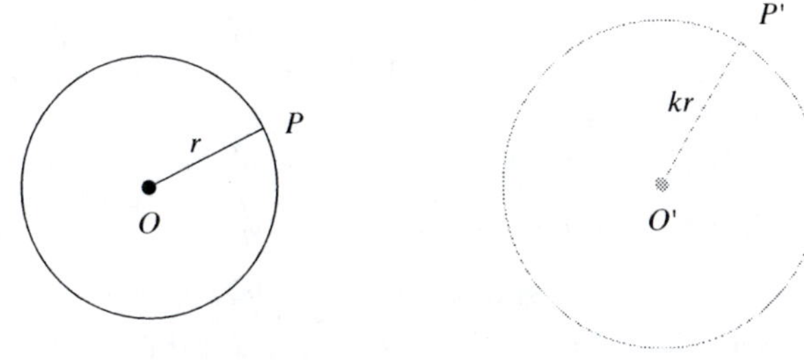

Proof: Let O be the center of a circle C with radius r, as shown in Figure 24. Then, from the definition of a circle, $C = \{P: OP = r\}$. (Read this as: C is the set of points P whose distance from O equals r.) Call the similarity transformation S, let its magnitude be k, and use primes to denote images. Since $O'P' = k \cdot OP = kr$, every image point P' of a point P on C is at a distance kr from O'. So every point on the image C' is on the circle with center O' and radius kr. And if a point P' is on the circle with center O' and radius kr, then its preimage must have been at a distance r from O. So C' is a circle. ∎

Now we turn to descriptions of specific similarity transformations. The most basic type of similarity transformation fixes a point O and expands or contracts the preimage by corresponding its points with images that are k times as far from O.

Definition Given a point O in a plane and a real number $k > 0$, the transformation S that maps each point P of the plane onto the point $S(P) = P'$ on $\overrightarrow{OP}$ such that $OP' = k \cdot OP$ is the **size change** (or **size transformation**)[2] with **center** O and **magnitude** k.

Figure 25 shows a preimage nonconvex pentagon $ABCDE$ and its images under three size changes with center O, with magnitudes $\frac{2}{3}$, $\frac{4}{3}$, and 2.

Figure 25

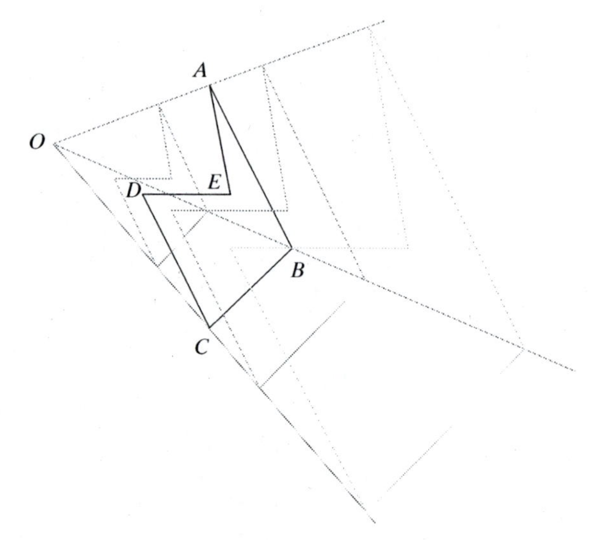

To deduce properties of size changes, it is often easier to use coordinates than to deal with the above synthetic definition. When the center of the size transformation is the origin, the image of any point is remarkably easy to find. We call this transformation S_k.

Theorem 8.9 Under the size change S_k with center $(0, 0)$ and magnitude $k > 0$, $S_k(a, b) = (ka, kb)$.

Figure 26

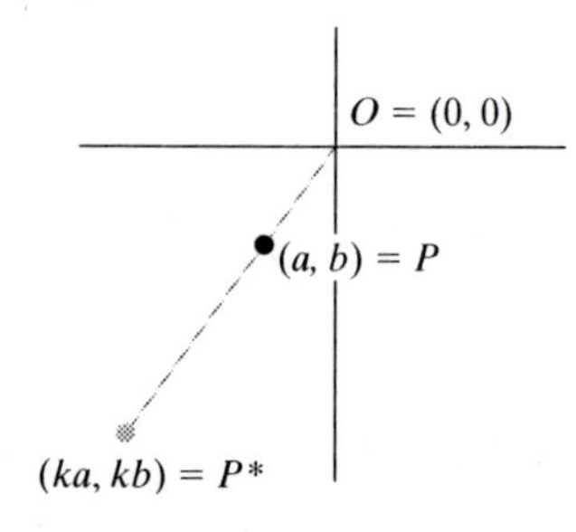

Proof: Let $O = (0, 0)$, $P = (a, b)$, and $P^* = (ka, kb)$, as shown in Figure 26. We apply the definition of size change to this situation to show that P^* is on $\overrightarrow{OP}$ and that $OP^* = k \cdot OP$.

By the Euclidean distance formula,

$$OP^* = \sqrt{(ka - 0)^2 + (kb - 0)^2} = \sqrt{k^2a^2 + k^2b^2} = |k|\sqrt{a^2 + b^2}.$$

Since $k > 0$, this last expression equals $k\sqrt{a^2 + b^2}$, which is $k \cdot OP$.

To show that P^* is on $\overrightarrow{OP}$, consider the three distances OP, PP^*, and OP^*.

$$PP^* = \sqrt{(ka - a)^2 + (kb - b)^2} = |k - 1|\sqrt{a^2 + b^2}.$$

If $k - 1 > 0$, then $PP^* = (k - 1)\sqrt{a^2 + b^2}$, so $OP^* = OP + PP^*$. This implies that P is between O and P^*, so P^* is on $\overrightarrow{OP}$. In this case the size change is an expansion. The situation with $k - 1 < 0$ is left to you. ┘

Because any point in the plane can be the origin for a coordinate system, any synthetic properties of S_k apply to all size transformations. Theorem 8.10 includes the most important of these properties.

---

[2]No single term is consistently used to identify this transformation. Various authors call this transformation a *homothety*, *homothetic transformation*, *dilation*, or *dilatation*. Other authors use those terms to refer to a more inclusive set of transformations. We use the term *homothetic* on page 393–4.

Theorem 8.10 Let S_k be the size change with magnitude $k > 0$ and center $(0, 0)$.

a. S_k is a similarity transformation.
b. Under S_k, the image of any line is a line parallel to it.
c. Under S_k, an angle and its image have the same measure.
d. The inverse of S_k is $S_{\frac{1}{k}}$.

Proof:

a. All that needs to be shown is that the distance between two images is k times the distance between their preimages. This is left to you.

b. Let $(x', y') = S(x, y)$. Then $(x', y') = (kx, ky)$, which implies that $x = \frac{x'}{k}$ and $y = \frac{y'}{k}$.

Either the preimage line has equation $y = mx + b$ or the preimage is the vertical line $x = a$. If the preimage line is $x = a$, then by substitution, the image has equation $\frac{x'}{k} = a$, from which $x' = ka$, another vertical line. So the preimage and image are parallel. If the preimage line has equation $y = mx + b$, then the image has equation $\frac{y'}{k} = m\left(\frac{x'}{k}\right) + b$, from which $y' = mx' + kb$, a line with the same slope. So the preimage and image are parallel.

c. Because of (a) and Theorem 8.7, the image of an angle is an angle. Because of (b), the sides of an angle and the sides of its image are parallel. Consequently, the angle and its image have the same measure.

d. Let $P = (x, y)$ and $Q = (kx, ky)$. Then for all points P and Q, $S_k(P) = Q$ and $S_{\frac{1}{k}}(Q) = P$. Consequently $S_{\frac{1}{k}}$ is the inverse of S_k. ┘

Figure 25 exhibits Theorem 8.10(b), which helps greatly in drawing images of figures. These properties of S_k are important because size change transformations are building blocks for similarity transformations.

Theorem 8.11 Every similarity transformation is the composite of a single size change and an isometry.

Proof: Let T be the given similarity transformation. Suppose the ratio of similitude of T is r. Then T multiplies distances by r. Now let

$$D = S_{\frac{1}{r}} \circ T.$$

Because $S_{\frac{1}{r}}$ multiplies distances by $\frac{1}{r}$, D multiplies distances by 1, so D is an isometry. Now solve the transformation equation for T by composing each side with S_r. Thus

$$S_r \circ D = S_r \circ S_{\frac{1}{r}} \circ T.$$

So

$$S_r \circ D = T.$$ ┘

Actually, the proof of Theorem 8.11 establishes an even stronger result than stated in the theorem.

Corollary 1: Every similarity transformation is the composite of a size change centered at $(0, 0)$ and an isometry.

Because of Theorem 8.11, many of the properties of similarity transformations can be found by examining the properties of S_k. In particular, if S_k has a property that is also a property of all isometries, then this will be a property of all similarity transformations. For this reason, under any similarity transformation, the images of segments, rays, and lines are segments, rays, and lines; the images of angles are angles with the same measure; and the images of polygons are polygons. This is why similar figures look so much like their preimages.

Corollary 2: If two figures are similar, corresponding angles have the same measure.

Not all properties of S_k extend to all similarity transformations. For instance, under some isometries, lines are not parallel to their images. So there are similarity transformations in which lines are not parallel to their images. In other words, in similar figures corresponding lines or segments or rays do not have to be parallel. Figure 20 illustrates this idea.

8.2.1 Problems

1. Deduce SAS triangle congruence from SAS Similarity.

2. Suppose, in the definition of *similar* given in this section, the condition $P'Q' = k \cdot PQ$ was replaced by $\frac{P'Q'}{PQ} = k$. Does this make any difference?

3. Explain why, under the definition of *similar* given in this section, the polygons of Figure 22 are not similar.

4. Prove that if two figures α and β are congruent, then they are similar.

5. Suppose you have a 3″ by 5″ photo that you wish to enlarge as much as possible yet have it fit on an 8.5″ by 11″ piece of paper.
   a. What is the magnitude of the similarity transformation you should use and what will be the dimensions of the enlargement?
   b. Generalize part **a** to apply to any dimensions of a rectangular photo and a rectangular piece of paper.
   c. Suppose you have a 6″ by 9″ photo and you wish to contract it as little as possible yet have it fit on a 3″ by 5″ card. What is the magnitude of the similarity transformation you should use and what will be the dimensions of the smaller image?

6. If T_1 and T_2 are similarity transformations with magnitudes k_1 and k_2, prove that $T_2 \circ T_1$ is a similarity transformation and determine its ratio of similitude.

7. Finish the proof of Theorem 8.9 for the case where $k - 1 < 0$.

8. What is the inverse of the size change with center $(5, -7)$ and magnitude 5?

9. Find an equation for the image of the line $3x + 2y = 4$ under S_{100}.

10. Prove Theorem 8.10(a).

11. Prove: All circles are similar.

12. Use Theorem 8.9 to extend the definition of S_k to allow negative values of k.
    a. Using this extended definition, draw the image of quadrilateral $ABCD$ under $S_{-1.5}$ when $A = (4, 6)$, $B = (4, -2)$, $C = (-3, -8)$, and $D = (-6, 0)$.
    b. Let O be the **centroid** (the intersection of the medians) of ΔABC and let L, M, and N be the midpoints of $\overline{BC}$, $\overline{AC}$, and $\overline{AB}$, respectively. Then ΔLMN is the image of ΔABC under a size change. What are the center and magnitude of the size change?

13. Suppose that S is a size change with center A, $S(B) = C$, and $B \neq C$.
    a. Prove: If $S(D) = E$, then $\frac{AB}{BC} = \frac{AD}{DE}$.
    b. Prove: If $\frac{AB}{BC} = \frac{AD}{DE}$, A, D, and E are collinear, and E is on $\overleftrightarrow{AD}$, then $\overleftrightarrow{BD} \parallel \overleftrightarrow{CE}$.

14. a. Describe the size change S_k, with $k > 0$, as a function in the complex plane **C**.
    b. ℓ is a line in the complex plane **C** if and only if there exist complex numbers c and z_0 with $\ell = \{z\colon z = c + mz_0, m \in \mathbf{R}\}$. Use this fact and part **a** to provide an alternate proof of Theorem 8.10b.

ANSWERS TO QUESTIONS

1. $ABCDEF \leftrightarrow XYZUVW$; $\frac{3}{2}$ or 1.5

2. Possible answer: IM and $I'M'$.

8.2.2 Similarity of graphs

To emphasize the generality of the definition of *similar* given in Section 8.2.1, we consider here the similarity of graphs on the coordinate plane.

An example

Under the size change transformation S_2, every point (a, b) in $\mathbf{R}^2$ is mapped to the point $(2a, 2b)$. This transformation expands any geometric figure to a similar figure twice its size.

$$S_2(a, b) = (2a, 2b) \tag{1}$$

S_2 also expands the graph of any function to a larger graph that is similar to the original graph. This larger graph is the graph of a different function. How are the formulas for these two graphs related? Answering this question will give us a way of looking at different functions and saying when these functions have graphs that are similar.

As an example, consider the cubic function f defined by

$$f(x) = x(1 - x^2) \qquad 0 \le x \le 1. \tag{2}$$

Figure 27

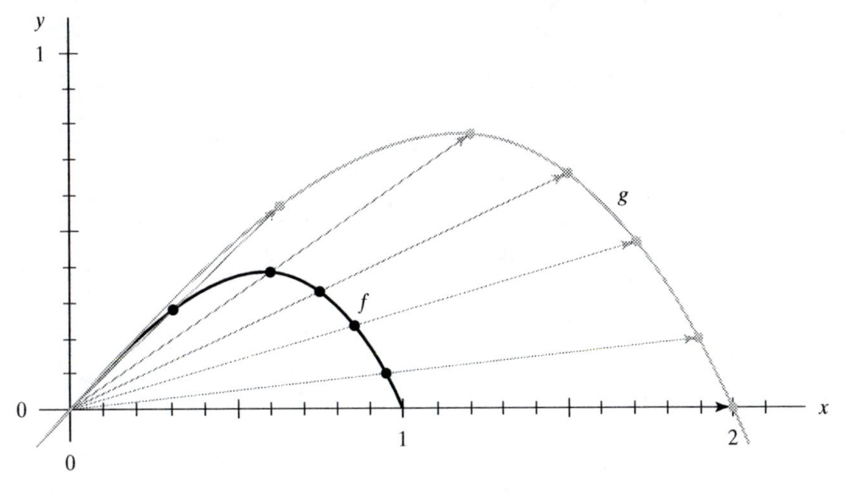

Part of the graph of (2) is the smaller curve shown in Figure 27. The image of this part of the graph is the larger curve in Figure 27. The larger curve is similar to the original. This larger curve is clearly the graph of some function g. There is a simple procedure for obtaining a formula for the function g from the formula for f. We state the procedure first, and return to its justification later.

Let (x', y') be the image of (x, y) under the transformation S_2 in (1). Since $(x', y') = (2x, 2y)$,

$$x = \frac{x'}{2} \quad \text{and} \quad y = \frac{y'}{2}. \tag{3}$$

Now substitute for x and y in the equation (2) for f and in the inequality defining the domain of f.

$$\frac{y'}{2} = \frac{x'}{2}\left(1 - \left(\frac{x'}{2}\right)^2\right) \qquad 0 \le \frac{x'}{2} \le 1$$

From this,

$$y' = x'\left(1 - \frac{x'^2}{4}\right) \qquad 0 \le x' \le 2 \tag{4}$$

Equation (4) is a formula for the function g. Now we write this formula in terms of x, not x', but it still expresses the same relationship defined in (4).

$$(5) \qquad g(x) = x\left(1 - \frac{x^2}{4}\right) \qquad 0 \le x \le 2$$

You can check that the graph of g seems to be twice the size of the graph of f by graphing both functions.

Generalizing the result

A justification of the procedure we used above results in a theorem for size changes that corresponds to Theorems 7.21 to Theorem 7.24 for isometries.

As when working with those theorems, you should be careful not to fall into a common trap. Under the size change transformation S_k, $S_k(x, y) = (kx, ky)$. So you might take a literal way of applying this definition and think that the image of $y = f(x)$ has equation $ky = f(kx)$, or, equivalently, $y = \frac{1}{k}f(kx)$. You would be wrong. As with those earlier theorems, the appropriate substitution is the inverse.

Theorem 8.12 **(Graph Size Change Theorem):** Under the size change S_k, the image of the graph of a function f with formula $y = f(x)$ is the graph of the function g with formula $g(x) = kf\left(\frac{x}{k}\right)$.

Proof: Consider a point $P = (a, b)$ on the graph of f (as in Figure 28), and its image $S_k(P) = (ka, kb)$. We know that $g(ka) = kb$. Expressing b in terms of f, $f(a) = b$, so that $g(ka) = kf(a)$. To get a general formula for $g(x)$ we set $x = ka$ and express a in terms of x as $a = \frac{x}{k}$. Hence $g(x) = kf\left(\frac{x}{k}\right)$.

Figure 28

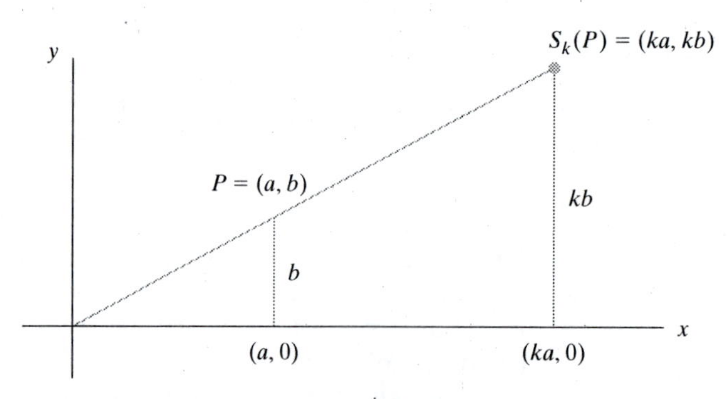

Since a size change transformation S_k maps a figure onto one similar to it, Theorem 8.12 implies the following about graphs of functions.

Corollary 1: For all functions f, the graph of $y = f(x)$ is similar to the graph of $y = kf\left(\frac{x}{k}\right)$.

Applying S_k to the parabola that is the graph of the function with equation $y = x^2$ results in an image parabola that is the graph of the function with equation $y = k\left(\frac{x}{k}\right)^2$, or, more simply, $y = \frac{x^2}{k}$. Thus every parabola with an equation of the form $y = ax^2$, $a > 0$, is a size change image of every other parabola with an equation of this form. Any other parabola in the plane is congruent to one of these parabolas. The result is a theorem that surprises many seeing it for the first time.

Corollary 2: All parabolas are similar.

Question 1:

a. What is the value of k for which S_k maps the graph of $y = x^2$ onto the graph of $y = 5x^2$?

b. What is the value of k for which S_k maps the graph of $y = 6x^2$ onto the graph of $y = 17x^2$?

Similarity of graphs and units on the axes

The ideas about similarity of graphs that we have been exploring have direct application to situations involving functions where we need to change units.

Suppose that wrapping a cube of side x yards in a certain way requires a length y yards of string, where

(6) $$y = 6x - 2. \quad (x \text{ and } y \text{ in yards})$$

What is a formula for the length y *in feet* of the string required to wrap a cube of side x *in feet*?

The relationship between yards (x or y) and feet (x' or y') units is that x yards $= 3x$ feet. So, if x yards $= x'$ feet, then y yards $= y'$ feet, and we obtain the conversions

(7) $$x = \frac{x'}{3} \quad \text{and} \quad y = \frac{y'}{3}.$$

Notice how this argument parallels the above argument with the graphs of functions. Substituting these conversions in (6), we get first $\frac{y'}{3} = 6\frac{x'}{3} - 2$, and then, after simplifying, we obtain $y' = 6x' - 6$. This means that the relationship (6) expressed in feet is given by

(8) $$y = 6x - 6 \quad (x \text{ and } y \text{ in feet})$$

In this way, we can apply the transformation S_k to change formulas for functions when the units on both axes are the same. Specifically, when $S_k(x, y) = (x', y') = (kx, ky)$, S_k changes formulas given in (x, y) units to formulas with measurements in (x', y') units, where the (x', y') units are $\frac{1}{k}$ times the size of the (x, y) units. In the example just given, the (x, y) units are yards and the (x', y') units are feet, and $k = 3$.

Corollary 3 (Unit conversion in formulas): The unit change transformation S_k given by $S_k(x, y) = (x', y') = (kx, ky)$ takes any formula $y = f(x)$ in (x, y) units to a formula $y' = kf\left(\frac{x'}{k}\right)$ in (x', y') units.

Thus the special type of similarity relationship of graphs under the transformation S_k occurs when we change units on the x- and y-axes. We give a special name to S_k for this application to unit conversion. The transformation S_k is called a *k-scaling*. Specifically, given a function f defined by a formula $y = f(x)$, we say that the function g defined by $g(x) = kf\left(\frac{x}{k}\right)$ is a **k-scaling** of f. If g is a k-scaling of f, then, by Corollary 3, if f expresses a relationship involving certain units, then g expresses the same relationship in units $\frac{1}{k}$ as large. Moreover, by Theorem 8.12, the graph of g is related to the graph of f by a size change of magnitude k. So the graphs of f and g are similar. This kind of similarity has a special name, corresponding to k-scaling. Two graphs are **homothetic** if and only if they are related by a size change transformation.

Clearly, graphs that are homothetic are also similar. On the other hand, two graphs may be similar without being homothetic. For example, if one graph is a translation or a rotation or a reflection image of another graph, they are similar; in fact, they are congruent. Still, they are not homothetic.

Another example with linear functions

As geometric figures, any line segment is similar to any other line segment. However, graphs of two linear functions are homothetic *only if their slopes are the same.* This follows since, under the size change $x' = kx$, $y' = ky$, the formula $f(x) = mx + b$ becomes $g(x') = mx' + bk$, a function with the same slope. Yet, even when their slopes are the same, segments might not be homothetic.

Consider the following linear functions f and g.

$$f(x) = 10 - x, \quad \text{when} \quad -10 \leq x \leq 10$$
$$g(x) = 120 - x, \quad \text{when} \quad -120 \leq x \leq 120$$

When 12-scaled, f becomes g. By defining f on a finite interval, we see that when the expansion transformation operates, it expands the interval of definition as well. The arrows from the graph of f to the graph of g (Figure 29) illustrate the expansion of the graph.

Figure 29

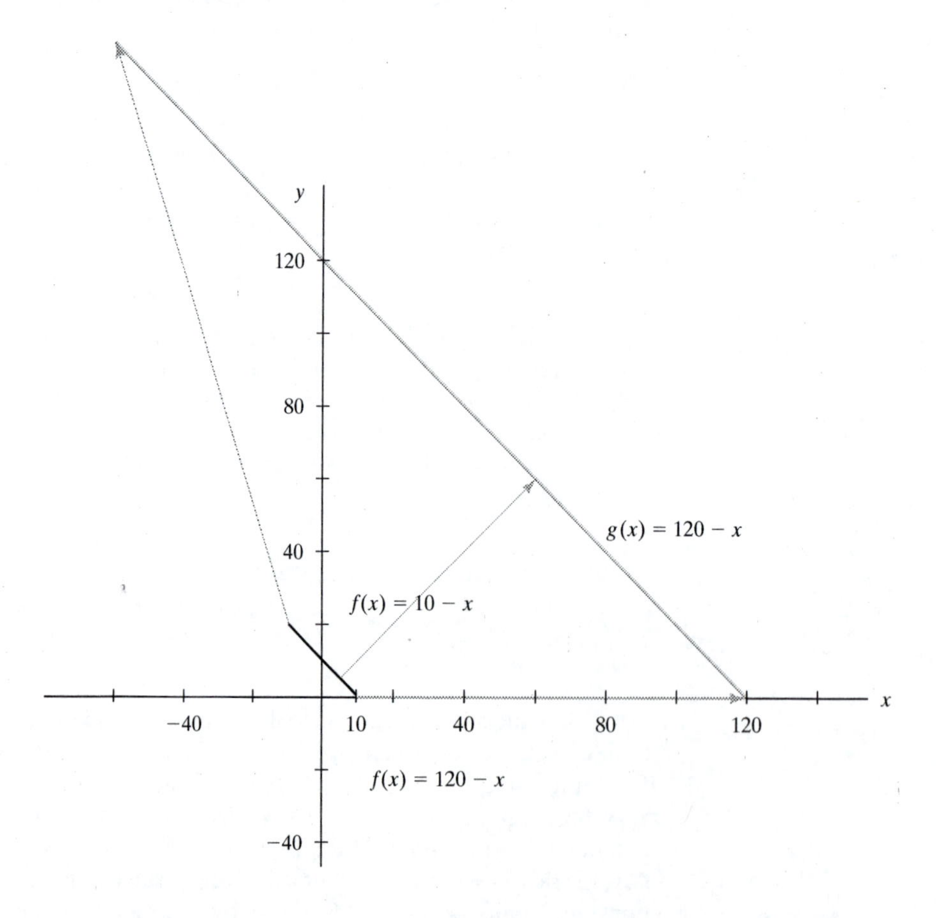

In applications, this sort of relationship of graphs is entirely natural. For example, if we use x feet of a string of length 10 feet, the number of feet left is given by $f(x) = 10 - x$, a function meaningful on $0 \leq x \leq 10$. Using the conversion

$x' = 12x$, which converts x in feet to x' in inches, we obtain the function $g(x) = 120 - x$. This function gives the number of inches left and is meaningful on $0 \le x \le 120$. The slope of each function is -1. For f this slope has the meaning 1 foot less remaining for every 1 foot used. For g this slope has the meaning 1 inch less remaining for every 1 inch used.

Similarity of graphs of exponential functions

It may seem that exponential functions with different bases could not have similar graphs. However, that is not the case. For example, the transformation $x' = 3x$, $y' = 3y$ carries the function f below onto the function g. These functions are graphed on their respective intervals in Figure 30.

$$f(x) = 5 \cdot 2^{0.1x} \qquad 0 \le x \le 20$$

$$g(x) = 15 \cdot (\sqrt[3]{2})^{0.1x} \qquad 0 \le x \le 60$$

Figure 30

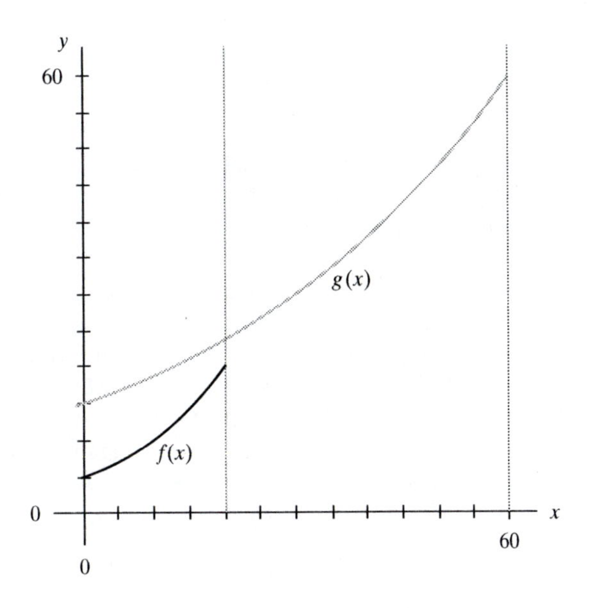

Pairs of similar graphs might arise if the function f gave a value in yards in terms of x given in yards. The function g would describe the same situation, where g would give a value in feet in terms of x given in feet. The general statement for exponential functions provides another corollary to the theorem of this section.

> **Corollary 4:** The graph of the exponential function $f: x \rightarrow b^x$ is similar to the graph of the function $g: x \rightarrow kb^{\frac{x}{k}}$.

8.2.2 Problems

1. a. For what value of k does S_k map the graph of $y = ax^2$ onto the graph of $y = bx^2$, where $a > 0$ and $b > 0$?
   b. Why does the graph $y = \frac{1}{10}x^2$ look so much wider than the graph of $y = x^2$ if the two graphs are similar?

2. The total length y (in mm) and tail length x (in mm) of the snake species *lampropeltis polyzona* are approximately related by $y = 7.44x + 8$, for $200 < x < 1400$ mm. Convert this into a formula where x and y are measured in inches.

3. Show that the graph of any quadratic function of the form $f(x) = ax^2 + bx$ is homothetic to the graph of $g(x) = x^2 + bx$, but that the graph of $f(x) = ax^2 + bx$ is not homothetic to the graph of $h(x) = x^2 + cx$ if $c \neq b$.
4. The formula $y = x^3$ defines a cubic. Find formulas for other cubics whose graphs are similar to the graph of $y = x^3$.
5. Show that the graphs of functions f with formulas of the form $f(x) = \frac{c}{x}$ are all similar. Does this imply that all hyperbolas are similar?

6. Is the graph of $f(x) = \sin x$ similar to the graph of every function g with a formula of the form $g(x) = \cos(bx + c)$?

7. In all of the examples of S_k in this section, k is positive. Which results might not apply if k were negative?

8. Discuss the similarity of graphs of logarithm functions of different bases.

9. The graph of $f(x) = 3x^2$ is a horizontal translation image of the graph of $g(x) = 3(x - 2)^2$, which means that the two graphs are congruent. However, they are not homothetic. Is this a difficulty with the definition of *homothetic*? That is, do you think the definition of homothetic should be extended to say that two graphs are homothetic if they are related by the composite of a size change transformation and a translation? (Some authors do define *homothetic* in this broader way.)

10. In what way is the theorem of this section related to zooming in on graphs, as can be done with a calculator or computer graphing utility?

ANSWER TO QUESTION

1. a. $\frac{1}{5}$ b. $\frac{6}{17}$

8.2.3 Similar polygons

When two polygons are known to be similar, because one is the image of the other under a similarity transformation, any pair of corresponding angles is congruent and all corresponding lengths and distances are proportional. So, for instance, if two triangles are similar, then corresponding medians are proportional to corresponding sides, as are corresponding altitudes and corresponding angle bisectors. Without transformations, results like these need to be deduced individually. With transformations, they all come at once.

Because of the power of similarity, it is useful to know when two polygons are similar. Figure 31 shows two triangles. If they were faces or other familiar objects, we could probably tell whether or not they were similar by sight, by examining features of one that are not in the other. But triangles are very difficult figures to analyze by sight. So we are forced to use the given information, the lengths of their sides. ΔIMS has sides 4, 5, and $\sqrt{17}$; ΔJNT sides 12, 15, and $\sqrt{153}$.

Figure 31

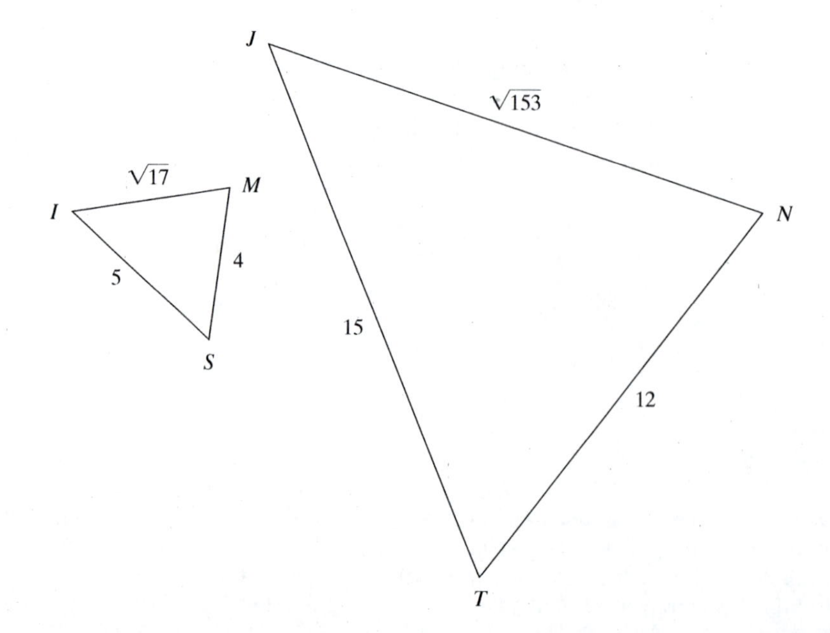

Examination of the side lengths shows that the sides of ΔJNT are 3 times as long as the sides of ΔIMS.

$$\frac{15}{5} = \frac{12}{4} = \frac{\sqrt{153}}{\sqrt{17}} = 3$$

Consequently, the three sides of ΔJNT are proportional to the three sides of ΔIMS.

In Section 8.2.1, we proved that every similarity transformation is the composite of a single size change and a congruence transformation. Now we apply this idea to show that the triangles *IMS* and *JNT* are similar. Apply a size change of magnitude 3 to ΔIMS. (The center of this size change can be at any point.) The result is an image triangle $\Delta I'M'S'$ (not drawn) with sides of length 12, 15, and $3\sqrt{17}$. These side lengths imply that $\Delta I'M'S'$ and ΔJNT are congruent by SSS Congruence. Since $\Delta I'M'S' \cong \Delta JNT$, there is an isometry mapping $\Delta I'M'S'$ onto ΔJNT. So there is a similarity transformation mapping ΔIMS onto ΔJNT, and the two triangles are similar.

The SSS similarity condition is one of the three conditions sufficient for triangle similarity presented in Euclid's *Elements*. These are still studied in most school geometry courses.

Theorem 8.13 **(Sufficient Conditions for Triangle Similarity):** Two triangles are similar if

a. (SSS similarity condition) the lengths of three sides of one are proportional to the lengths of three sides of the other;

b. (SAS similarity condition) the lengths of two sides are proportional to the lengths of two sides of the other and the included angle of one is congruent to the included angle of the other; or

c. (AA similarity condition) two angles of one are congruent to two angles of the other.

Proof: The proof of each part follows the specific idea mentioned above. Namely, find an appropriate magnitude for a single size change to be applied to one triangle so that the image is congruent to the second. Then apply an appropriate triangle congruence theorem.

a. Suppose that the lengths of the sides of ΔABC are a, b, and c. Since the sides are proportional, the lengths of the sides of the other triangle, call it ΔDEF, can be considered to be ka, kb, and kc, as in Figure 32. Now apply the size change S_k to ΔABC.

Figure 32

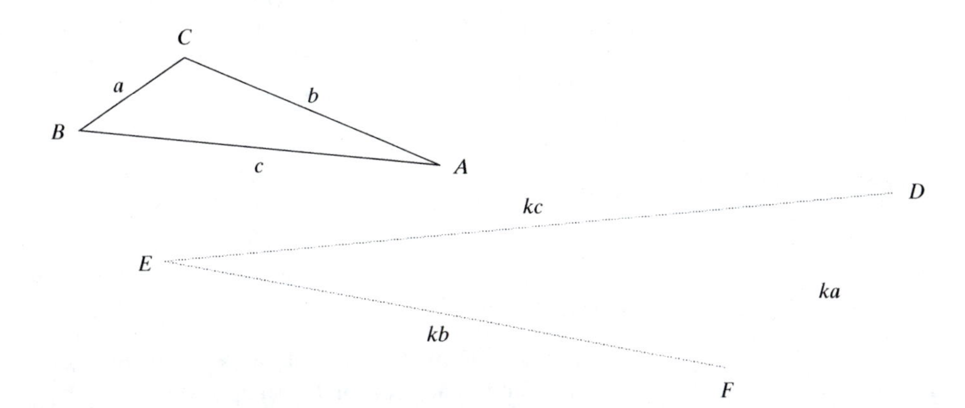

Since this size change multiplies lengths by k, the lengths of the sides of the image $\Delta A'B'C'$ (not drawn) are ka, kb, and kc. So $\Delta A'B'C' \cong \Delta DEF$ by SSS Congruence. This means that there is an isometry T mapping $\Delta A'B'C'$ onto ΔDEF. Now we have $T \circ S_k(\Delta ABC) = T(\Delta A'B'C') = \Delta DEF$. The transformation $T \circ S_k$, being the composite of a size change and an isometry, is a similarity transformation. So $\Delta ABC \sim \Delta DEF$.

b. In triangles ABC and DEF, suppose that $m\angle C = m\angle F$ and that $\frac{FD}{CA} = \frac{FE}{CB}$. Then if $CA = b$ and $CB = a$, there is a positive number k with $FD = kb$ and $FE = ka$. Consequently, we apply a size change of magnitude k to ΔABC. The image $\Delta A'B'C'$ is congruent to ΔDEF. (Which congruence theorem justifies this?) So, as in part (a), ΔABC can be mapped onto ΔDEF by the composite of a size change and an isometry. So the triangles are similar.

c. Let the triangles be ABC and DEF with $m\angle A = m\angle D$ and $m\angle B = m\angle E$. Since the sum of measures of the angles of the triangles is 180°, $m\angle C = m\angle F$. Pick any pair of corresponding sides, say $\overline{AB}$ and $\overline{DE}$. Then apply a size change of magnitude $\frac{DE}{AB}$ to ΔABC. The rest of the proof is quite a bit like the proofs in parts (a) and (b) and is left to you to finish. ⌟

Similar polygons with more than 3 sides

Theorem 8.13 shows that either having all sides proportional (SSS similarity condition) or all angles congruent (AA similarity condition) is sufficient for triangles to be similar. This is not true for quadrilaterals, even familiar ones. Rectangles have all their angles congruent but may not have proportional sides (Figures 33a and 33b). Rhombuses have all their sides congruent, so any two rhombuses have proportional sides, but their angles may be different (Figures 33b and 33c).

Figure 33

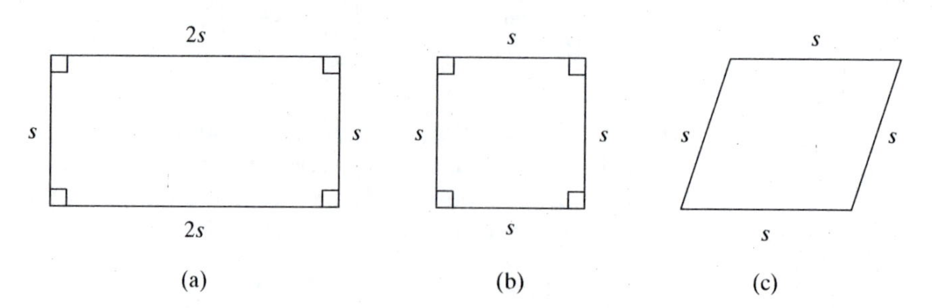

These examples show that if we multiply all the side lengths in a polygon by the same factor, we do not necessarily get a similar polygon. We need to keep the angle measures the same. But if we only keep angle measures the same, we do not necessarily get a similar polygon either. The lengths need to be multiplied by the same factor. In almost all cases, with polygons having more than three sides, both conditions (proportional lengths and equal angle measures) must be checked to ensure similarity.

One type of polygon, however, needs not be checked. The proof of this theorem is left to you.

Theorem 8.14 All regular polygons with the same number of sides are similar to each other.

In Section 7.1.2, you saw that an argument should be independent of the drawing that pictures it.

Question 1: Draw a picture of the following theorem and use your picture to determine the missing reasons in the proof.

Theorem 8.15 Suppose A, B, C, and D are any four points on a circle such that $\overleftrightarrow{AB}$ and $\overleftrightarrow{CD}$ intersect at E. Then $AE \cdot BE = CE \cdot DE$.

Proof: $\angle CAE \cong \angle BDE$. (Why?) Also, $\angle CEA \cong \angle BED$. (Why?) Consequently, $\Delta ACE \sim \Delta DBE$ by AA Similarity. So

$$\frac{AE}{DE} = \frac{CE}{BE} = \frac{AC}{DB}.$$

The theorem follows from the first equality. ┘

Question 2: In your drawing from Question 1, was point E located inside the circle? If so, modify your drawing so that point E is located outside the circle. If it was located outside the circle, modify your drawing so that the point E is located inside. Modify your answer to Question 1 based on the new figure.

The product $AE \cdot BE$ in Theorem 8.15 is called the **power of point E with respect to the circle.**[3] The theorem shows that the power of a point with respect to a given circle is independent of the choice of lines through that point. It also shows that two theorems separated in many geometry texts are, in a very strong way, two cases of the same theorem. We return to this theorem in Section 8.2.5.

8.2.3 Problems

1. Could there be an SSA condition sufficient for triangle similarity? If so, what would the condition be? If not, why not?

2. Prove or disprove.

a. All equilateral triangles are similar.

b. All isosceles triangles are similar.

c. All right triangles are similar.

d. All isosceles right triangles are similar.

e. If one triangle has sides of length 2 and 3, and a second has sides of length 20 and 30, and each triangle has a 120° angle, then the triangles are similar.

3. Prove or disprove: If corresponding bases of two isosceles trapezoids are congruent, then the isosceles trapezoids are similar.

4. Explain why the word "if" in Theorem 8.13 can be replaced by "if and only if", and the resulting statement is true.

5. By Theorem 8.13, all triangles with sides of length $3k$, $4k$, and $5k$ are similar. Prove or disprove: All quadrilaterals with consecutive sides of length $3k$, $4k$, $5k$, and $6k$ are similar.

6. Prove Theorem 8.14.

7. State, in words, the two theorems that Theorem 8.15 represents.

8. Make a drawing for Theorem 8.15 under each circumstance given here. Is the theorem still true? Why or why not?

a. $E = B$

b. $A = B$ and $C \neq D$

c. $A = B$ and $C = D$

9. Give the power of each point with respect to the circle with center $(0, 0)$ and radius r.

a. $(2r, 2r)$

b. $(0, 0)$

c. $(0, -r)$

*d. (a, b)

10. What is the power of a vertex of an equilateral triangle with side s with respect to the inscribed circle inscribed in that triangle?

11. Line m_1 intersects circle C_1 at points P_1 and Q_1 and circle C_2 at points R_1 and S_1. Line m_2 intersects circle C_1 at points P_2 and Q_2 and circle C_2 at points R_2 and S_2. Lines m_1 and m_2 intersect at E.

a. Prove: $EP_1 \cdot EQ_1 \cdot ER_1 \cdot ES_1 = EP_2 \cdot EQ_2 \cdot ER_2 \cdot ES_2$.

b. If $Q_1 = R_1$ and $Q_2 = R_2$, does the result in part **a** still hold? Why or why not?

[3] As directed distances from point E, when E is inside the circle, AE and BE have opposite signs, so their product is negative. When E is outside the circle, AE and BE have the same sign, so their product is positive. In some work, this distinction is important. But in school geometry, where lengths are rarely allowed to be negative, the distinction is not made.

12. A point P is in the exterior of a given circle C with center O.

a. Use straightedge and compass or a geometric construction program to construct the tangent lines to C that pass through P. (*Hint*: Why are the points of tangency on C the points of intersection of C with the circle C' whose diameter is the segment $\overline{OP}$?)

b. Prove that your construction in part **a** works.

13. Suppose that C_1 and C_2 are circles with different radii such that neither circle is contained within the other.

a. Use a geometric construction program to construct the common external tangent lines to C_1 and C_2. (*Hint*: For a given point Q_1 on C_1, construct the point Q_2 on C_2 such that the radial segments $\overline{O_1Q_1}$ and $\overline{O_2Q_2}$ are parallel, as shown in Figure 34. Then the point P of intersection of the lines $\overleftrightarrow{O_1O_2}$ and $\overleftrightarrow{Q_1Q_2}$ is independent of the choice of Q_1. [Why?] This point is called the **external center of similitude** of circles C_1 and C_2. Then apply the construction in Problem 12.)

b. Prove that your construction in part **a** works and that point P is the external center of similitude for the two circles.

c. Modify the construction in part **a** to produce the common *internal* tangents of circles C_1 and C_2 whose interiors do not intersect. These tangents intersect at the **internal center of similitude** of circles C_1 and C_2.

14. Suppose that C_1, C_2, and C_3 are circles with different radii such that each lies exterior to the other two.

a. Use a geometric construction program to verify **Monge's Theorem**: The external centers of similitude P_1, P_2, and P_3 of the three possible pairs of the circles C_1, C_2, and C_3 are collinear.

b. Show by example that the corresponding statement to Monge's Theorem for internal centers of similitude is not true.

c. Modify Monge's Theorem to apply to combinations of internal and external centers of similitude.

Figure 34

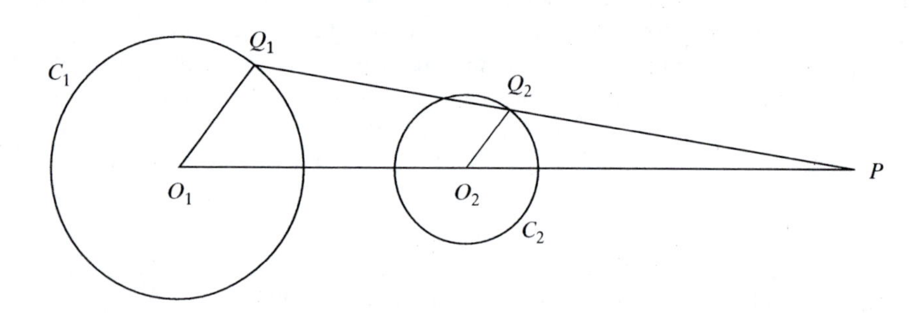

ANSWERS TO QUESTIONS

1 and 2. The two circles in Figure 35 exhibit the two cases.

Figure 35

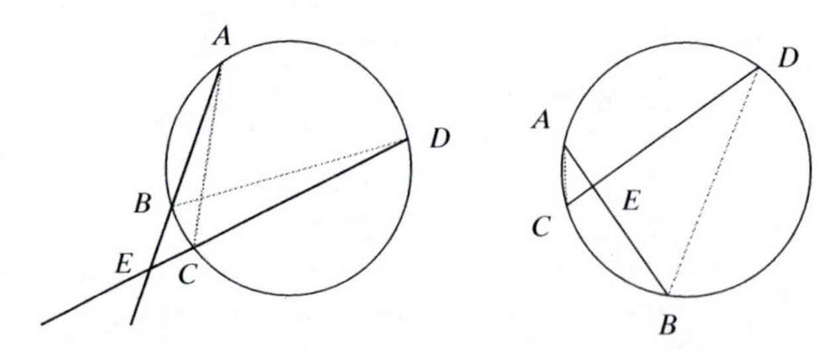

$m\angle CAE = m\angle BDE$ because each equals $\frac{1}{2}m\,\widehat{BC}$. $m\angle CEA = m\angle BED$ because they are the same angle (in the configuration at the left) or vertical angles (at the right).

8.2.4 Similar arcs

The definition of similarity is broad enough to apply to all figures. That is, any figure can be a preimage under a similarity transformation. In the previous three sections, preimages have been polygons, graphs, or subsets of graphs of functions. In this brief section, we deal with preimages that are arcs of circles. Our purpose is to demonstrate the similarity of all circular arcs with the same degree (or radian) measure. This demonstration also serves to point out the fundamental difference between arc length and arc measure, two ideas often confused by young students.

Arc measure

We begin with some definitions of familiar terms. A **central angle** of a circle O is an angle whose vertex is O. Suppose that a central angle intersects the circle at points A and B. Points A and B and the intersection of circle O with the interior of $\angle AOB$ constitute the **minor arc $\overset{\frown}{AB}$**. The **measure of the minor arc** $\overset{\frown}{AB}$ is defined as $m\angle AOB$. From the definition, the measure of a minor arc is between 0° and 180°, or between 0 and π radians.

Figure 36

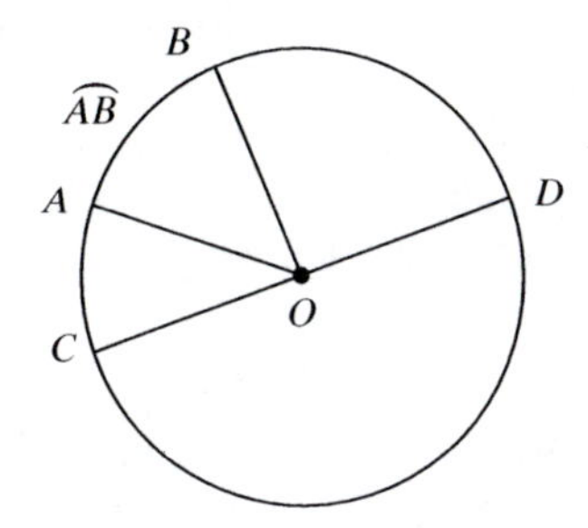

Points A and B and the rest of the circle constitute the **major arc $\overset{\frown}{AB}$**. The **measure of a major arc** is defined as 360° or 2π radians less the measure of the corresponding minor arc $\overset{\frown}{AB}$. The measure of a major arc is between 180° and 360°, or between π and 2π radians. In writing, we may distinguish major arcs from minor arcs by using three letters to represent the major arc. In Figure 36, $\overset{\frown}{ACB}$ is a major arc.

In between minor arcs and major arcs are *semicircles*. A **semicircle** of a circle O consists of the endpoints of a diameter $\overline{CD}$ of circle O and the points on circle O on one side of the line $\overleftrightarrow{CD}$. The measure of a semicircle is 180° or π radians. Every minor arc is contained in many semicircles, and every major arc contains many semicircles.

In the preceding three paragraphs, we have used a static definition of *arc* and *arc measure*. But an arc can also be thought of as a dynamic figure. Think of an arc $\overset{\frown}{AB}$ as the part of a circle swept out by a point as it moves around the circle from A to B. This dynamic conception of arc can be formalized using a parametric analytic description of a circle. The circle with center (h, k) and radius r can be described by the parametric equations

$$\begin{cases} x = h + r\cos t \\ y = k + r\sin t. \end{cases}$$

In this context, an arc is defined as the set of points on the circle determined by all values of t in a given interval, and the arc measure of the arc is the length of that interval Mod 360° or Mod 2π. Regardless of its radius, an entire circle has arc measure 360° or 2π.

Arc length

When you walk around a circular fountain or other circular structure, you can identify how far you have traveled in two ways. If you say you have walked 90° around the circle, or, equivalently, $\frac{1}{4}$ of the way around, you are using *arc measure*. If you say you have walked 50 feet or 25 steps, you are measuring *arc length*. Arc measure tells how much you have turned; arc length tells how far you have traveled.

In Section 10.3.3, we point out a fundamental mathematical difference between arc length and arc measure. Arc measure (like angle measure) is dimensionless, or 0-dimensional; arc length (like segment length) is 1-dimensional.

When we calculated the distance between two points on the surface of Earth in Section 8.1.4, we were calculating arc length. We did this by multiplying the circumference of the circle (an arc length) by the portion of the circle occupied by the arc. That proportion is found by dividing the degree measure of the arc by 360°, or by dividing the radian measure by 2π.

That process illustrates how arc length and arc measure are related. In a circle of radius r units, an arc of measure $x°$ has a length of $\frac{x}{360} \cdot 2\pi r$ units. In the same circle, an arc of measure x radians has a length of $\frac{x}{2\pi} \cdot 2\pi r$, or xr units. The simplicity of calculating arc length from arc measure in radians provides one reason why radian measure is simpler in many applications. (Other reasons are given in Section 9.1.1.)

Congruent arcs

Two arcs are congruent if and only if one can be mapped onto the other by an isometry, by the definition of congruence. Consequently, two arcs with the same measure may not be congruent. The arcs must also be subsets of circles with the same radius, for only such circles are images of each other under isometries.

The endpoints of congruent arcs determine congruent chords (Figure 37a), but the converse is not necessarily true (Figure 37b). Congruent chords determine congruent minor arcs only in congruent circles.

Figure 37

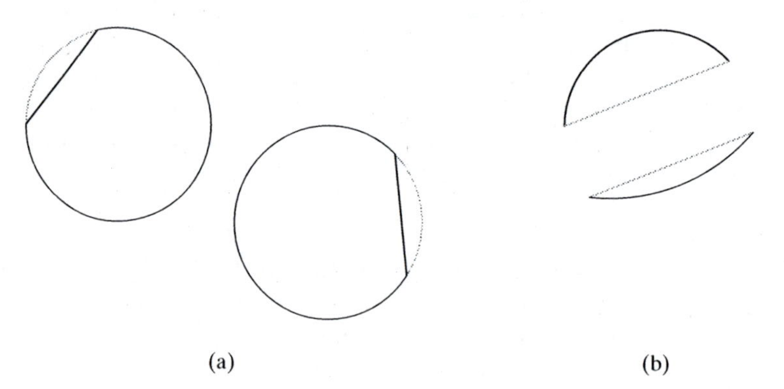

(a) (b)

Similar arcs

With the equivalence of congruent arcs and congruent chords only in congruent circles, it may seem that there is little that can be said about similar arcs. Theorem 8.19 shows that the opposite is the case: Similar arcs may be found in any circles.

Theorem 8.16 Two arcs are similar if and only if they have the same arc measure.

Proof:

($\Rightarrow$) Suppose $\widehat{AB}$ and $\widehat{CD}$ are similar arcs in circles O and P, respectively (Figure 38a). Then there is a similarity transformation T such that $T(\widehat{AB}) = \widehat{CD}$. T must map circle O onto circle P, so $T(O) = P$. Consequently, since T preserves angles, $T(\angle AOB) = \angle CPD$. Since T preserves angle measure, $m\angle AOB = m\angle CPD$.

($\Leftarrow$) Suppose $\widehat{AB}$ and $\widehat{CD}$ are arcs in circles O and P, respectively, and that they have the same arc measure. Then, by the definition of arc measure, $m\angle AOB = m\angle CPD$.

Figure 38

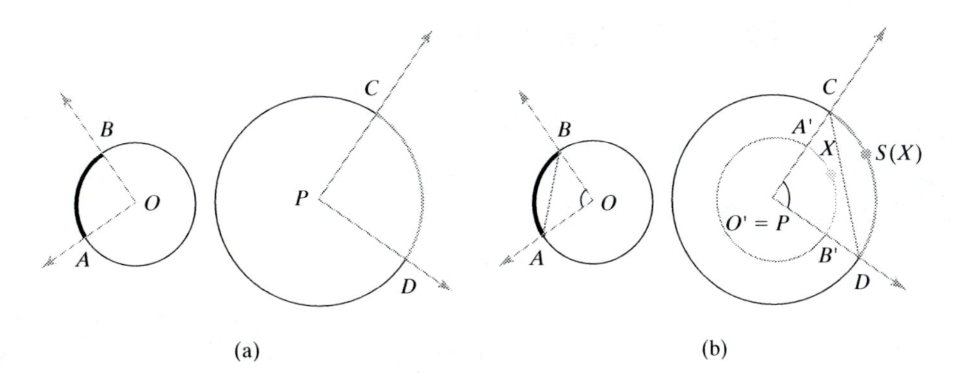

(a) (b)

Consequently, there is an isometry T that maps $\angle AOB$ onto $\angle CPD$. Let T(circle O) = circle O', $T(A) = A'$ and $T(B) = B'$. (Circle O' has center P, as shown in Figure 38b.) Now $AO = BO$ and $CP = DP$ since they are radii from the same circle, and $A'P = B'P = AO = BO$ since isometries preserve distance.

The size change S with center P and radius $\frac{PC}{PA'}$ maps A' onto C and B' onto D. If a point X is on arc $A'B'$, then $XP = A'P$, and so $S(X)$ will be at a distance PC from P. So $S(X)$ is on arc $\overset{\frown}{CD}$. Reversing the argument, every point on the arc $\overset{\frown}{CD}$ has a preimage from the arc $\overset{\frown}{A'B'}$. Consequently, $S(\overset{\frown}{A'B'}) = \overset{\frown}{CD}$.

Thus $S \circ T(\overset{\frown}{AB}) = \overset{\frown}{CD}$, implying that $\overset{\frown}{CD}$ is the image of $\overset{\frown}{AB}$ under a similarity transformation. ■

The transformation $S \circ T$ in the second part of the proof of Theorem 8.16 maps the chord $\overline{AB}$ onto the chord $\overline{CD}$. Since similarity transformations multiply distances, and lengths are defined in terms of distances, $\frac{\text{length of } \overset{\frown}{AB}}{AB} = \frac{\text{length of } \overset{\frown}{CD}}{CD}$. This proves a relationship between chord length and arc length that was used by the mathematicians who first studied trigonometry (see Section 9.1.2).

Theorem 8.17 If two arcs are similar, then the ratios of their lengths to the lengths of their chords are equal.

8.2.4 Problems

1. Which idea, arc length or arc measure, is applied in the following situations?

a. measuring your waist

b. measuring how far you can turn your neck

c. creating a circle graph

d. calculating the difference in longitude for two cities at the same latitude

2. What is the radius of a circle for which the radian measure of an arc equals its length?

3. Consider the parametric equations for a circle given in this section.

a. As t increases from 0 to 1 radian, what part of the circle will be graphed?

b. As t increases from 0 to 1 radian, what is the length of the arc that is graphed?

4. What is the length of a 212° arc in a circle of radius 14?

5. In a circle of radius 2, suppose a chord $\overline{AB}$ has length 1.

a. What are the degree measures of the two arcs $\overset{\frown}{AB}$ in that circle?

b. Generalize part **a**.

6. Show by example that a 45° arc can have the same length as a 90° arc.

7. In a circle with radius r, what is the length of a chord of an arc with measure s?

8.2.5 When many theorems become one

Theorem 8.15, proved in Section 8.2.3, is usually stated as two theorems, which we now identify as Theorems 8.15(a) and 8.15(b) and depict in Figures 39a and 39b.

Theorem 8.15 **(restatement):**

a. If two chords intersect in a circle, the product of the segments of one chord equals the product of the segments of the other.

b. If two secants are drawn to a circle from a point outside it, the product of the first secant and its external segment equals the product of the second secant and its external segment.

Figure 39

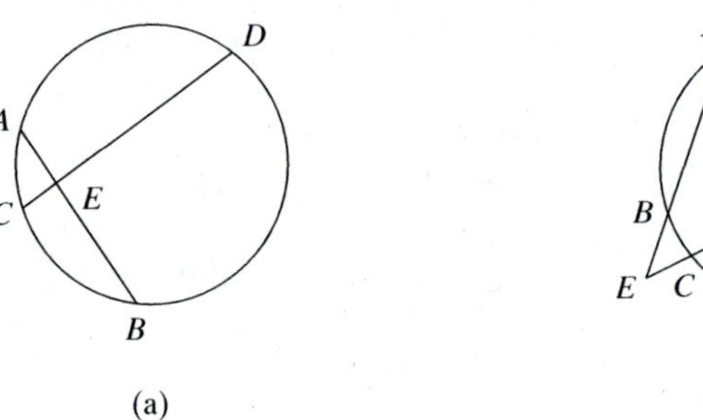

(a) (b)

The pictures and the wordings of Theorems 8.15(a) and 8.15(b) look different enough that most students do not realize that the theorems have much in common. But the statement of Theorem 8.15 applies to both: If A, B, C, and D are any four points on a circle, and if $\overleftrightarrow{AB} \cap \overleftrightarrow{CD} = \{E\}$, then $AE \cdot BE = CE \cdot DE$. In this way, two theorems become one.

A third traditional theorem can also be subsumed in our original statement of Theorem 8.15. It is closely related to Theorem 8.15(b):

Theorem 8.15

c. If a secant and a tangent are drawn to a circle from a point outside it, the square of the length of the tangent equals the product of the secant and its external segment.

This theorem, pictured in Figure 39c, can be viewed as the situation of Theorem 8.15 when E is external to the circle and points A and B coincide.

Figure 39

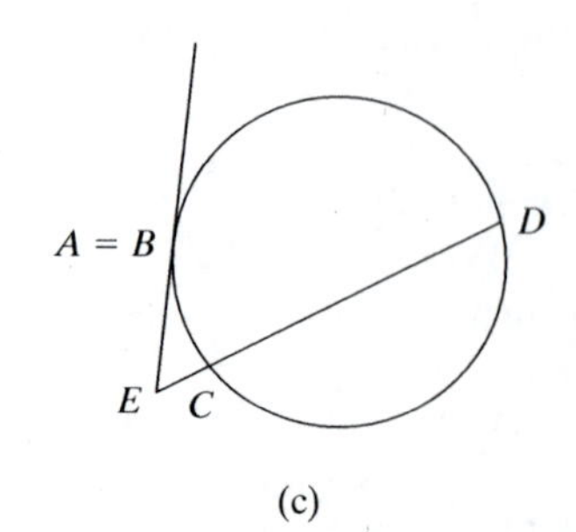

(c)

Thus our original statement of Theorem 8.15 was broad enough to cover three theorems usually stated separately.

Six theorems and a definition

Theorem 8.15 applies to segment lengths in a circle. Can theorems about measures of angles in circles likewise be combined into one? For this, we recall that if a circle intersects one side of an angle in point A and the other side in point B, and not both A and B are the vertex of the angle, then the angle **intercepts** arc AB. It is possible for an angle to intercept two arcs of a circle, as in Figures 40d, 40e, and 40f.

The **measure of a central angle** of a circle is defined to be the measure of its intercepted arc. From this definition, it is customary to deduce six theorems. We do not show their proofs since they are not difficult and can be found in many geometry textbooks.

Theorem 8.18 In a circle:

a. The measure of an inscribed angle equals half the measure of its intercepted arc.

b. The measure of the angle formed by a chord and a tangent equals half the measure of its intercepted arc.

c. The measure of an angle formed by two chords equals half the sum of the measures of the arcs intercepted by the angle and its vertical pair.

Figure 40

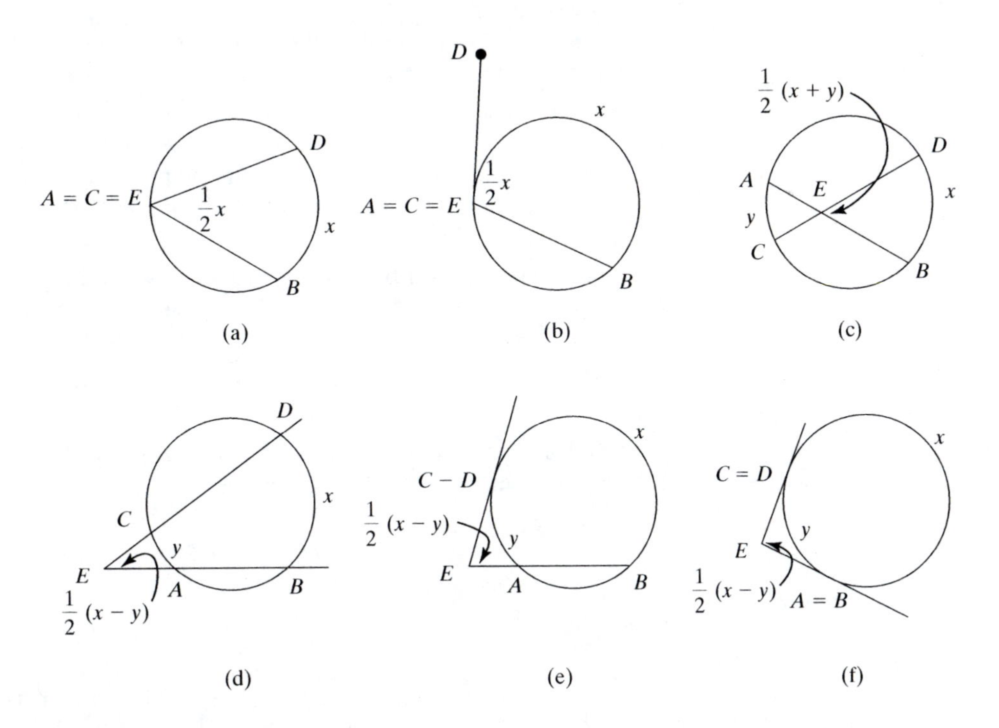

Theorem 8.18 In a circle:

d. The measure of an angle formed by two secants equals half the difference of the measures of the arcs intercepted by the angle.

e. The measure of an angle formed by a secant and a tangent equals half the difference of the measures of the arcs intercepted by the angle.

f. The measure of an angle formed by two tangents equals half the difference of the measures of the arcs intercepted by the angle.

Connecting the six parts to each other

We have lettered the six diagrams 40a–40f so that in each case two lines $\overleftrightarrow{AB}$ and $\overleftrightarrow{CD}$ intersect at E. In some cases, $A = B$ or $C = D$ so that E is needed in order to determine the line. The angle whose measure is being found is $\angle BED$. And, in each case $x = m(\overset{\frown}{BD})$ and $y = m(\overset{\frown}{AC})$.

Begin by examining Figure 40c, for it is the paradigm for all six situations. Notice that if E is the center of the circle, then $\angle BED$ is a central angle, so $m\angle BED = x$ and $m\angle CEA = y$. Because these angles are vertical angles, $x = y$ and so $m\angle BED = \frac{1}{2}(x + y) = \frac{1}{2}(x + x) = x$. Thus the definition of the measure of a central angle can be viewed as a special case of Theorem 8.18(c).

Now move C in Figure 40c along the circle so that it coincides with A. The result is that $y = 0$ and we have Figure 40a. Rotate ray $\overrightarrow{AD}$ in Figure 40a about A until it is tangent to the circle and we achieve Figure 40b. In this way, we see that Figures 40a and 40b are limiting cases of Figure 40c, and so Theorems 8.18(a) and 8.18(b) are limiting cases of Theorem 8.18(c). In these limiting cases, $y = 0$, and so $\frac{1}{2}(x + y)$ has become $\frac{1}{2}x$.

By beginning with Figure 40d and moving D along the circle until it coincides with C, we obtain Figure 40e. Similarly, moving B in Figure 40e along the circle until it coincides with A produces Figure 40f. This shows that Theorems 8.18(e) and 8.18(f) are limiting cases of Theorem 8.18(d). For this reason, the formulas are the same in all three situations.

Now we wish to connect the first three figures with the last three. We do this by comparing Figures 40c and 40d. Move A in Figure 40c along the circle so that it is between C and B. Now E, the intersection of $\overleftrightarrow{AB}$ and $\overleftrightarrow{CD}$, is outside the circle. Furthermore, the order of points around the circle has been changed. In Figures 40a–40c, the points are in the order *B-D-A-C* with the possibility that neighboring points might coincide. In Figures 40d–40f, the points are in the order *B-D-C-A* with the possibility that neighboring points might coincide. This change in order implies that the *directed arc* $\overset{\frown}{AC}$ has been replaced by its opposite arc $\overset{\frown}{CA}$, and it is the reason that $\frac{1}{2}y$ is subtracted from $\frac{1}{2}x$ rather than added to $\frac{1}{2}x$ to obtain $m\angle BED$. That is, in all six cases,

$$m\angle BED = \frac{1}{2}(m(\overset{\frown}{BD}) + m(\overset{\frown}{AC})),$$

with $\overset{\frown}{BD}$ and $\overset{\frown}{AC}$ considered as directed arcs. In Theorems 8.18(d), 8.18(e), and 8.18(f), we have

$$m\angle BED = \frac{1}{2}(m(\overset{\frown}{BD}) + m(\overset{\frown}{CA})),$$

from which

$$m\angle BED = \frac{1}{2}(m(\overset{\frown}{BD}) - m(\overset{\frown}{AC})).$$

A second way of connecting the six parts

Another way of interpreting the six parts as one is to define a **complete angle** as the union of an angle with its vertical pair. The measure of a complete angle is the sum of the measures of the two angles, which is twice the measure of either part. Then consider the measure of any intercepted arc as positive if the arc is convex as viewed from the vertex of the angle, and negative if the arc is concave as viewed from the vertex of the angle. For instance, in Figure 40c, arcs $\overset{\frown}{BD}$ and $\overset{\frown}{AC}$ are both convex when viewed from point E. Consequently, $m\angle BED = \frac{1}{2}(m(\widehat{BD}) + m(\widehat{AC}))$, but in Figures 40d, 40e, and 40f, $\widehat{BD}$ with measure x is positive but $\widehat{AC}$ with measure y is negative, so $m\angle BED = \frac{1}{2}(m(\widehat{BD}) - m(\widehat{AC}))$. Lastly, in Figures 40a and 40b, the complete angle BED intercepts only the one arc with measure x, so the angle's measure has to be $\frac{1}{2}x$.

With these conventions, Theorem 8.18 can be restated in a single sentence.

Theorem 8.18 **(restatement):** If a circle intersects both sides of a complete angle, then the measure of the complete angle equals the sum of the measures of its intercepted convex and concave arcs.

The combining of theorems

Combining many theorems into one general theorem is not limited to geometry. For instance, Theorem 4.9, *$f(x) = g(x)$ is equivalent to $h(f(x)) = h(g(x))$ if and only if h is 1-1*, combines several propositions students typically learn as different properties. Theorem 4.9 and Theorem 8.18 share the characteristic that a new idea, or a modification of an old idea, is needed in order to obtain the generalization. With Theorem 4.9, the new idea is the rewriting of an equation into function language. With Theorem 8.18, the new ideas are the complete angle and the introduction of the criterion of convexity.

In fact, combining many results into one is a fundamental characteristic of mathematics, for virtually every theorem is a generalization. Theorem 4.9 and the theorems of this section may be distinguished from many other generalizations by the fact that the similarities are not obvious, and generalizations rather than specifics are being combined in one statement. Is this similar to combining the statements

1. Multiplication of natural numbers is commutative.
2. Multiplication of fractions is commutative.
3. Multiplication of decimals is commutative.
4. Multiplication of real numbers is commutative.

into the single statement

Multiplication of real numbers is commutative?

It would be similar if the learner sees statements (1) to (4) as separate and unrelated ideas and does not realize they are aspects of a single generalization.

8.2.5 Problems

1. In Figure 39b, suppose that AB is 10 units longer than EB and that CD is 12 units longer than EC. Find possible lengths for these segments.
2. In Figure 39c, suppose that CD is 15 units longer than EC and that EB, EC, and ED are all of integer length. Find at least two possible triples (EB, EC, ED).
3. Prove that an angle that intercepts a semicircle is a right angle.
4. Use Figure 40f.
   a. Find $m\angle E$ in terms of x.
   b. What is the range of possible values of x?
5. a. Assume that when a regular polygon is inscribed in a circle, its vertices are equally spaced on the circle. Use Theorem 8.18(a) to deduce a formula for the measure of each interior angle of a regular n-sided polygon.
   b. A polygon is **circumscribed about a circle** when each of its sides is tangent to the circle. Assume that when a regular polygon is circumscribed about a circle, its vertices are equally spaced on the circle. Use Theorem 8.18(f) to deduce a formula for the measure of each interior angle of a regular n-sided polygon.

8.2.6 Types of similarity transformations

In Section 7.2.5 we proved that if two figures are congruent, regardless of their complexity, one figure is the image of the other under a rotation, translation, reflection, or glide reflection. If the congruence transformation preserves orientation, then it is either a rotation or a translation. If the congruence transformation switches orientation, then it is either a reflection or glide reflection.

In this section, we examine the transformations that map figures onto similar figures. These transformations are the ratio-preserving transformations or, equivalently, the distance-multiplying transformations. In Section 8.2.1, we saw that any similarity transformation can be thought of as the composite of a congruence transformation and the transformation S_k. But now we wish to think of these similarity transformations not as composites but as single entities. Obviously, the similarity transformations include the size changes and all congruence transformations. But what else do they include?

The result may surprise you. The similarity transformations, like their congruence counterparts, can be classified into four types! Furthermore, analogous to the situation with congruence transformations; two types preserve orientation and two types reverse it. If the similarity transformation preserves orientation, then it is either a spiral similarity or a translation. If the similarity transformation switches orientation, then it is either a reflective similarity or a glide reflection. The rest of this section is devoted to clarifying what these results mean and proving them.

First, let us clarify how we determine whether similar figures have the same or opposite orientation. Suppose α and β are similar, with ratio of similitude k, so that lengths on β are k times corresponding lengths on α. Then apply a size change of magnitude k to α, calling the image α'. Now α' and β are congruent. If α' and β have the same orientation, then we say α and β have the same orientation and are **directly similar**. If α' and β have opposite orientation, then so do α and β, which are then **oppositely similar**. These definitions imply that size changes preserve orientation.

Next, recall from Chapter 2 that if z_1 and z_2 are complex numbers, and, in polar form, $z_1 = [r_1, \theta_1]$ and $z_2 = [r_2, \theta_2]$, then $z_1 z_2 = [r_1 r_2, \theta_1 + \theta_2]$. When we compare the product $z_1 z_2$ to z_2, we see that multiplication by z_1 has given us a product whose graph is r_1 times as far from the origin and, at the same time, has its argument increased by θ_1. That is, multiplication by z_1 yields the composite of a size change with center $(0, 0)$ and magnitude r_1 and a rotation with center $(0, 0)$ and magnitude θ_1. This transformation is a *spiral similarity*. In general, a **spiral similarity** is the composite of a rotation and a size change with the same center. It may be helpful to think of a spiral similarity as emulating the process of the development of the spiral of a ram's horn or of the shell of a chambered nautilus; simultaneously as the transformation rotates a figure the transformation also enlarges it. Figure 41 pictures the spiral similarity $R_{120°} \circ S_{\frac{2}{3}}$ with center O.

Figure 41

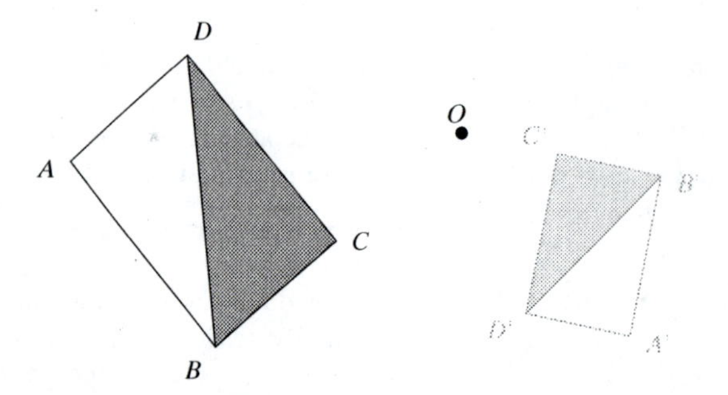

$R_{120°} \circ S_{\frac{2}{3}}(ABCD) = A'B'C'D'$

Finally, we need some way of indicating when two similarity transformations are the same without having to determine the images of all points. Fortunately, the situation with similarity is like that with congruence. If we know the images of all three vertices of a triangle, then we can locate the image of every point in the plane. But if we know whether the transformation preserves or switches orientation, then we need only the image of two of the vertices. This is true of congruence transformations (though we did not mention it earlier) and now we show its truth for similarity transformations.

Theorem 8.19 A similarity transformation is uniquely determined by the images of any two of its points and whether or not it preserves or switches orientation.

Proof: Suppose that S is a similarity transformation, and A and B are different points with $A' = S(A)$ and $B' = S(B)$. S is determined by the images of each of its points, so let us consider the image of a third point C. If C is on $\overleftrightarrow{AB}$, then $C' = S(C)$ is on $\overleftrightarrow{A'B'}$, since S preserves collinearity. Furthermore, S will preserve the betweenness relationship of the three points. Now, because S is a similarity transformation, $\frac{A'C'}{AC} = \frac{A'B'}{AB}$, so $A'C' = A'B' \cdot \frac{AC}{AB}$, which uniquely locates C' on the line (Figure 42a). So the images of A and B uniquely determine the images of any other point on $\overleftrightarrow{AB}$ whether S preserves orientation or not.

Figure 42

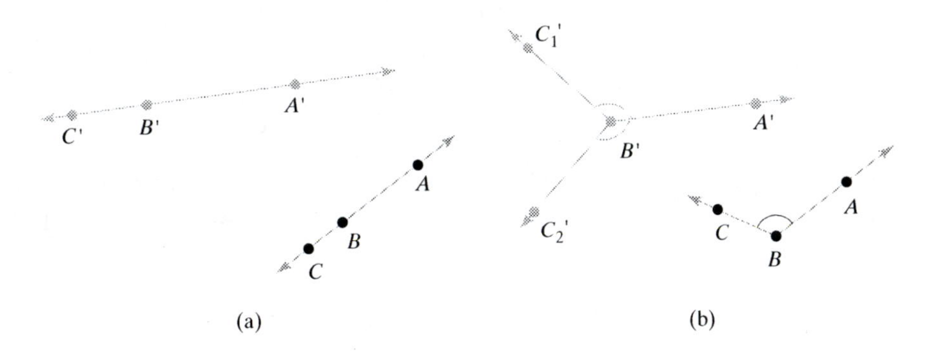

If C is not on $\overleftrightarrow{AB}$, then since $m\angle A'B'C' = m\angle ABC$, C' must be on a ray with end point A' that makes an angle of $m\angle ABC$ with $\overrightarrow{B'A'}$. There are only two such rays, and they are reflection images of each other over $\overleftrightarrow{A'B'}$. Since $\frac{A'C'}{AC} = \frac{A'B'}{AB}$, the distance from A' to C' is also determined by the given information. So C' has only two possible locations (called C_1' and C_2' in Figure 42b), and they are reflection images of each other over $\overleftrightarrow{A'B'}$. Thus if we know whether S preserves or switches orientation, we can determine which is the location of C'. ┘

With this language and machinery, we are able to deal with the classification of similarity transformations.

Theorem 8.20 If α and β are directly similar, then they are related by a spiral similarity or a translation.

Proof: Let P and Q be points on α, and let $\overline{P'Q'}$ be the image of $\overline{PQ}$ under the similarity transformation. Clearly P' and Q' are on β. Either $\overline{P'Q'} \parallel \overline{PQ}$ or not and either $\overline{P'Q'} \cong \overline{PQ}$ or not. So there are four possibilities.

1. $\overline{P'Q'} \cong \overline{PQ}$ and $\overline{P'Q'} // \overline{PQ}$. Then quadrilateral $PP'Q'Q$ is a parallelogram (Figure 43a). So the translation in the direction $\overrightarrow{PP'}$ and with magnitude PP' maps α onto β. This fits the desired conclusion of the theorem.

Figure 43

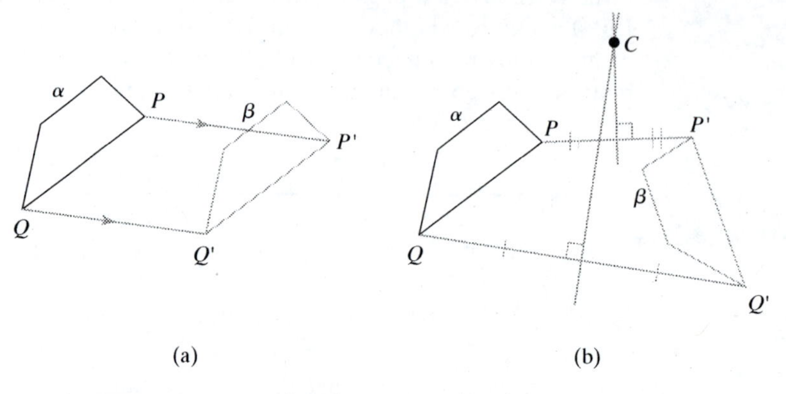

2. $\overline{P'Q'} \cong \overline{PQ}$ and $\overleftrightarrow{P'Q}$ intersects $\overleftrightarrow{PQ'}$ at a single point (Figure 43b). Then a rotation with center at the intersection C of the perpendicular bisectors of $\overline{PP'}$ and $\overline{QQ'}$ maps α onto β (Theorem 7.41). This is a spiral similarity because it equals the composite of a rotation and a size change (with magnitude 1) each with center C.
3. $\overline{P'Q'} // \overline{PQ}$ and $\overline{P'Q'}$ has a length unequal to PQ. Then let O be the point of intersection of $\overline{PP'}$ and $\overline{QQ'}$. Then the size change with center O and magnitude $\frac{OP'}{OP}$ maps P onto P' and maps Q onto Q'.
4. $\overline{P'Q'}$ is not congruent to $\overline{PQ}$ and $\overline{P'Q'}$ is not parallel to $\overline{PQ}$. Then let C be the intersection of $\overleftrightarrow{P'Q'}$ and $\overleftrightarrow{PQ}$ (Figure 43c). The circles circumscribed about $\Delta PP'C$ and $\Delta QQ'C$ contain C and intersect at a second point O. In fact, this point O is the center of the spiral similarity, because we can prove that $m\angle POP' = m\angle QOQ'$ and $\frac{OP'}{OP} = \frac{OQ'}{OQ}$. A proof is left to you.

Figure 43

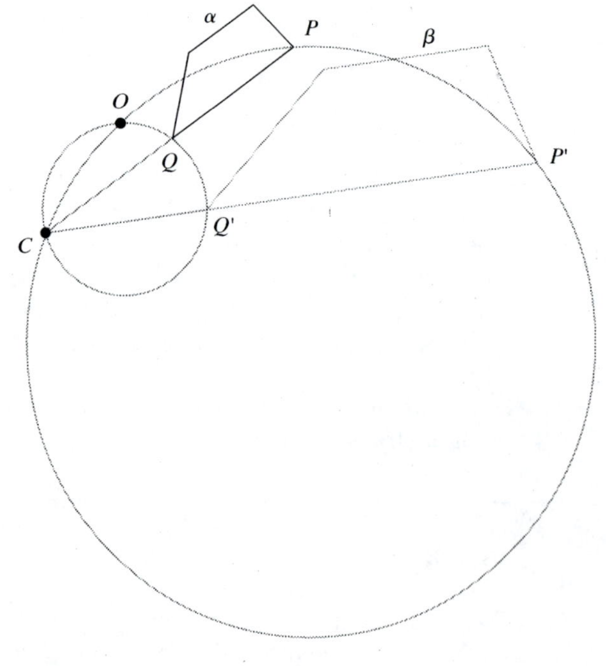

(c)

In order to classify the similarity transformations relating oppositely similar figures, we need to define the fourth type of similarity transformation, the *dilative reflection*. A **dilative reflection** is the composite of a size change with center C and a reflection over a line m that contains C. The line m and the center C are the **reflecting line and center of the dilative reflection**. Successive images under iterations of the same dilative reflection are like a walk in which each footprint is a fixed percent larger or smaller than the previous and the distances between the footprints become larger or smaller in proportion to the sizes of the footprints. Figure 44 shows a figure α and two consecutive images under a dilative reflection with magnitude 2.

Figure 44

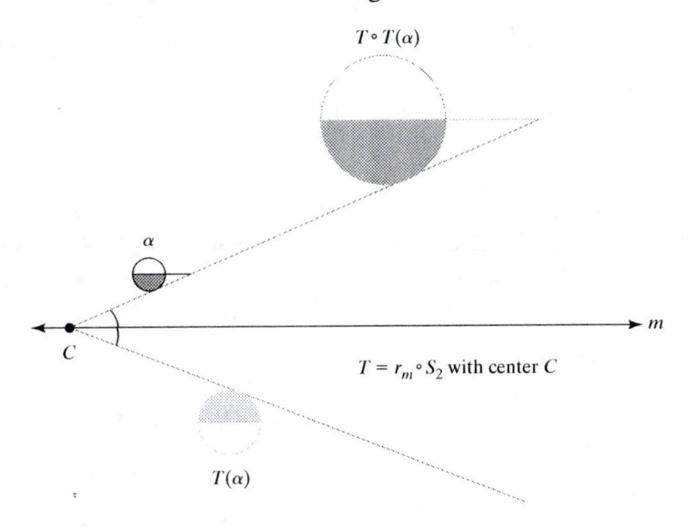

Theorem 8.21 If α and β are oppositely similar, then they are related by a dilative reflection or a glide reflection.

Proof: Let P' and Q' be the image of P and Q under the similarity transformation that maps α onto β. Now let Q^* be the translation image of Q' under the translation that maps P' onto P (Figure 45). Then m, the line of the dilative reflection, is the unique line that is parallel to the bisector of $\angle Q^*PQ$ and intersects $\overline{PP'}$ at a point C that is $\frac{QQ'}{PP'}$ of the way from P to P'. The center of the size change is C. A full proof can be found in Yaglom (1968). (See the References for this chapter.)

Figure 45

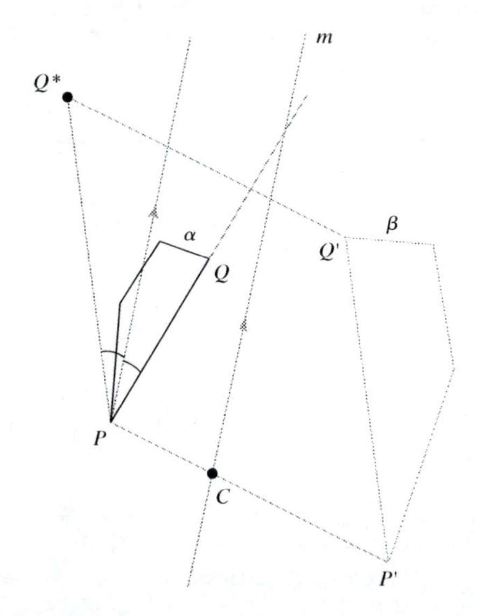

Notice that each of the four types of similarity transformations is either already a congruence transformation or is a generalization of one. This should be expected, since similarity generalizes congruence.

Table 1

Orientation	Congruence Transformation	Similarity Transformation
Preserved	*Rotation* Translation	*Spiral similarity* Translation
Switched	*Reflection* Glide reflection	*Dilative reflection* Glide reflection

Those transformations that are in italics in Table 1 have at least one fixed point.

The congruence group from an analytic viewpoint

In Chapter 7, we found that each type of congruence transformation had a simple analytic description in the complex plane. Let $\bar{z}$ be the conjugate of z. Also, let $z_\phi = \cos\phi + i\sin\phi$, the complex number that graphs as the point $[1, \phi]$ on the unit circle. Then the isometries can be described as follows.

Translations: If b is the complex number representing a vector in the plane, then

$$T_b(z) = z + b, \qquad \text{for all complex numbers } z.$$

Rotations: If ϕ is an angle and C is a point in the plane represented by the complex number c, then

$$R_{C,\phi}(z) = z_\phi(z - c) + c, \qquad \text{for all complex numbers } z.$$

Reflections: If m is the line in the plane inclined at an angle ϕ to the horizontal axis and passing through the point C, and if c is the complex number representing C, then

$$r_m(z) = z_{2\phi}\overline{(z - c)} + c, \qquad \text{for all complex numbers } z.$$

Glide reflections: If $G_{m,b} = r_m \circ T_b = T_b \circ r_m$, where m and b are described as above, then

$$G_{m,b}(z) = z_{2\phi}\overline{(z - c)} + c + b, \qquad \text{for all complex numbers } z.$$

Since every isometry is one of these transformations, and since $|z_\phi| = 1$ for all ϕ, these analytic descriptions fall into one of two forms, mentioned earlier in Projects 1, 2, and 3 for Chapter 7.

Theorem 8.22 Suppose that $T: C \to C$ is a transformation of the complex plane. Then

a. T is a direct congruence if and only if there are complex numbers a and b with $|a| = 1$ such that

$$T(z) = az + b, \qquad \text{for all complex numbers } z.$$

b. T is an opposite congruence if and only if there are complex numbers a and b with $|a| = 1$ such that

$$T(z) = a\bar{z} + b, \qquad \text{for all complex numbers } z.$$

We also noted in Chapter 7 that the set of all congruence transformations of the plane is a group under composition. This means that the set contains composites of all its elements, the identity transformation, and the inverse of any of its elements.

The similarity group from an analytic viewpoint

Every congruence transformation is a similarity transformation. The most basic similarity transformation that is not a congruence transformation is the size change $S_{C,k}$ of magnitude $k > 0$ with $k \neq 1$, centered at the point C.

Theorem 8.23 If c is the complex number representing the center C of the size change $S_{C,k}$ of magnitude $k > 0$ with $k \neq 1$, then

$$S_{C,k}(z) = k(z - c) + c, \qquad \text{for any complex number } z.$$

Proof: The proof uses the transformation strategy introduced in Section 7.2.2 and is left to you. ∎

When $c = 0$, the formula of Theorem 8.23 yields the formula for the size change S_k centered at the origin: $S_k(z) = kz$, for any complex number z. This agrees with the earlier discussion in Section 8.2.1.

Corollary 1 to Theorem 8.11 tells us that any similitary transformation T of the plane is the composite $T = S_k \circ S$, where S is a congruence transformation. From this, a complex-number description of all similarity transformations follows quite quickly.

Theorem 8.24 Suppose that $T: C \to C$ is a transformation of the complex plane. Then

a. T is a direct similarity (translation or spiral similarity) if and only if there are complex numbers a and b with $a \neq 0$ such that

$$T(z) = az + b, \qquad \text{for all complex numbers } z.$$

b. T is an opposite similarity (glide reflection or dilative reflection) if and only if there are complex numbers a and b with $a \neq 0$ such that

$$T(z) = a\bar{z} + b, \qquad \text{for all complex numbers } z.$$

The set of all similarity transformations of the plane is a group under composition that contains the group of congruence transformations as a subgroup. The identity transformation I is described analytically by

$$I(z) = z, \qquad \text{for all complex numbers } z.$$

To compute the inverse of a transformation T, solve the formula $w = T(z)$ for z in terms of w and interchange the roles of z and w to obtain a formula for $w = T^{-1}(z)$. For example, the inverse of the direct similarity given by

$$T(z) = 2iz + 4 + i, \qquad \text{for all complex numbers } z$$

is obtained by letting $w = 2iz + 4 + i$. Solving for z in terms of w, $z = \frac{w-4-i}{2i} = -\frac{i}{2}w + \frac{-1+4i}{2}$. Now interchange the roles of z and w to obtain $w = -\frac{i}{2}z + \frac{-1+4i}{2}$ as a rule for the transformation T^{-1}. That is, $T^{-1}(z) = -\frac{i}{2}z + \frac{-1+4i}{2}$.

To determine synthetic descriptions for T and T^{-1}, you can take any triangle in the plane (with its vertices represented by complex numbers) and find its image. The relationship between the preimage and image determines the transformation, just as we saw that three noncollinear points and their images determine a congruence transformation.

The ease with which all similarity transformations can be described by complex numbers makes these analytic descriptions a powerful tool in deducing properties of similarity transformations. You are asked to explore some of these properties in the Problems.

8.2.6 Problems

1. Determine whether the figures are directly or oppositely similar.

a. the hexagons of Figure 20 in Section 8.2.1

b. the triangles of Figure 28 in Section 8.2.2

c. the triangles of Figure 32 in Section 8.2.3

2. Find the image of the triangle with vertices $(0, 0)$, $(1, 0)$, and $(0, 2)$ under each similarity transformation. Draw both preimage and image.

a. the spiral similarity $S_5 \circ R_{90°}$

b. the dilative reflection $S_3 \circ r_{y=x}$

3. A spiral of isosceles right triangles is created in the following way. Begin with isosceles right triangle ABC with right angle at C. Find the image of ΔABC under the spiral similarity with center A that maps leg $\overline{CA}$ onto hypotenuse $\overline{AB}$. Use A as the center of a second spiral similarity that maps a leg of the image onto the hypotenuse of the image. Continue this process until seven images have been drawn.

a. Sketch the original triangle and the seven images.

b. If $AB = x$, what is the length of the hypotenuse of the last image?

4. Finish part (4) of the proof of Theorem 8.20.

5. Trace the drawing of hexagons in Figure 20 of Section 8.2.1. Locate the center of the spiral similarity that maps one hexagon onto the other.

6. A quadrilateral has vertices $A = (1, 1)$, $B = (0, 2)$, $C = (1, 3)$, and $D = (4, 2)$. $ABCD$ is rotated 90° about $(0, -3)$ and then S_2 is applied to the image $A^*B^*C^*D^*$ to obtain a final image $A'B'C'D'$. Find the center and magnitude of the spiral similarity that maps $ABCD$ onto $A'B'C'D'$.

7. Repeat Problem 6 when the rotation has magnitude 180°.

8. Let $\overline{CD}$ be the altitude to the hypotenuse of right triangle ABC. Describe the similarity transformation that maps ΔADC onto ΔACB.

9. The hexagon with vertices $H = (4, 5)$, $E = (0, 5)$, $X = (-2, 2)$, $A = (-2, -2)$, $G = (4, -2)$, and $N = (6, 0)$ is a preimage under the transformation $T_2 \circ T_1$, where T_1 is the reflection over the line $x = 6$ and T_2 is the size change with magnitude $\frac{1}{2}$ and center $(0, 0)$. Find the center and reflecting line of the dilative reflection mapping $HEXAGN$ onto $T_2 \circ T_1(HEXAGN)$.

10. In a dilative reflection, show that the order of the reflection and size change make no difference.

11. Prove Theorem 8.23.

12. Classify the transformation T as one of the four types of similarity transformations from its formula in the complex plane, and find the ratio of similitude.

a. $T(z) = \left(-\frac{1}{2} + \frac{i\sqrt{3}}{2}\right)\bar{z} + [1 + i]$

b. $T(z) = \left(\frac{1}{\sqrt{2}} - \frac{i}{\sqrt{2}}\right)z + i$

c. $T(z) = (2 + 2i)z + (1 - i)$

d. $T(z) = -2i\bar{z} + i$

13. a. Describe, in specific geometric terms, the transformation $T(z) = 2iz + 4 + i$ mentioned in this section.

b. Describe T^{-1} in specific geometric terms. Verify your description by using the formula for T^{-1} in this section.

14. a. If $z_1 = 1 + i$, $z_2 = 2 - i$, $w_1 = -1 + 2i$, $w_2 = 3 - i$, find a direct similarity D and an opposite similarity O such that both D and O map z_1 onto w_1, and z_2 onto w_2.

b. Explain why D and O exist for any two pairs of distinct points in the complex plane.

15. Use the complex number description of size changes to prove: The composite of two size changes $S_{C,k} \circ S_{D,m}$ is the size change with magnitude km centered at the point E that divides the line segment $\overline{CD}$ in the ratio $\frac{1-m}{m(1-k)}$. (E is negative when it does not lie on $\overline{CD}$.)

Unit 8.3 Distances within Figures

From the definition of similar figures, lengths on one of two similar figures determine lengths on the other if a ratio of similitude is known. This is why we can use the scale of a map in an atlas to find the actual distance between two places identified on the map. From the power of a microscope, we can determine the length of a small object. These applications are ubiquitous and accessible to anyone who can solve proportions, and they exhibit the power of similarity to determine *distances in one figure from distances in another figure.*

It is not as obvious that the theorems of similarity allow us to deduce relationships among *distances within a single figure.* For instance, the Pythagorean Theorem describes a relationship among the sides of a right triangle. In this unit, we discuss other relationships in a single figure that can be determined through similarity. Like the Pythagorean Theorem, these relationships are interesting and important.

8.3.1 Geometric means

Figure 46a illustrates a well-known geometric version of the distributive property $a(c + d) = ac + ad$. Figure 46b shows $(a + b)(c + d) = ac + bc + ad + bd$. These figures do more than merely picture the algebraic properties. They provide ways of understanding the properties that are different from logical proofs but are just as compelling.

Figure 46

c d
a ac ad

(a)

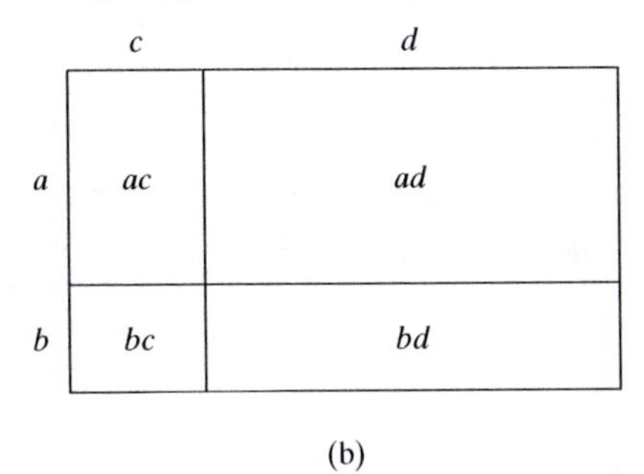

(b)

In this section, we utilize some geometry theorems to substantiate algebraic relationships that are themselves theorems. The relationships involve the various kinds of means that have already been discussed in this book: geometric, arithmetic, and harmonic.

Recall that the **geometric mean** (or **mean proportional**) of two positive numbers x and y is $\sqrt{xy}$. In the problems of Section 4.3.4 are some algebraic properties of the geometric mean. But, as its name suggests, this number received its name from its geometric properties. In particular, there are a number of situations in geometry where three segments have the lengths x, y, and $\sqrt{xy}$.

Geometric means in circles

The following theorem follows from Theorem 8.15c of Section 8.2.5.

Theorem 8.25 Let $\overline{PT}$ be tangent to circle O at T, and let a line through P intersect the circle at points A and B. Then PT is the geometric mean of PA and PB.

Figure 47

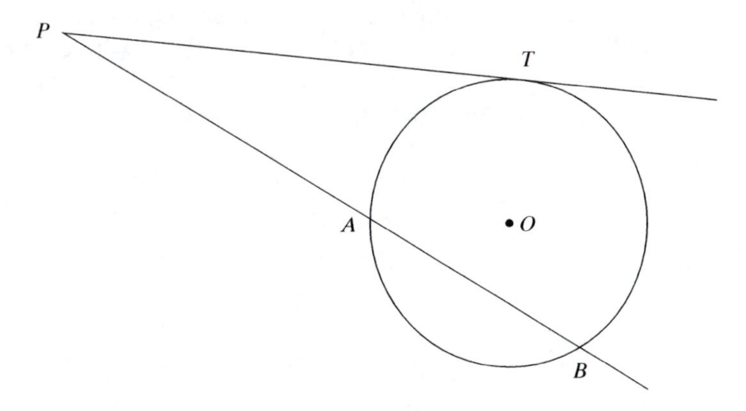

Refer to Figure 47. If $\overline{PT}$ remains fixed while $\overline{PA}$ moves across the circle from T to the other point of tangency, then we see that $PA \leq PT \leq PB$ at all times, with equality only if A and B are the same point. This justifies calling PT a "mean", for it is always somewhere between PA and PB.

Figure 48

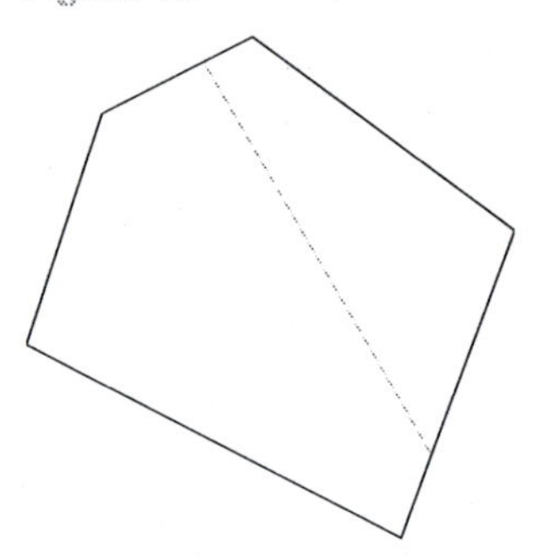

Geometric means in right triangles

When a segment is drawn from a point on one side of a convex polygon to a point on another side (as shown in Figure 48), two new convex polygons are formed. Each new polygon contains the segment as one side and part of the original polygon as its other sides. The two convex polygonal regions that are formed in this way do not constitute a precise partition of the original polygonal region, since they now share a segment. So we use informal terminology; we say that the segment "splits" the polygon into two parts. In this section, we discuss the triangles formed by splitting an original triangle into two or more parts.

Perhaps the most "meaningful" situation in geometry is a result of splitting a right triangle by the altitude to its hypotenuse (Figure 49).

Theorem 8.26 The altitude to the hypotenuse of a right triangle splits the triangle into two triangles, similar to each other and to the original right triangle.

Figure 49

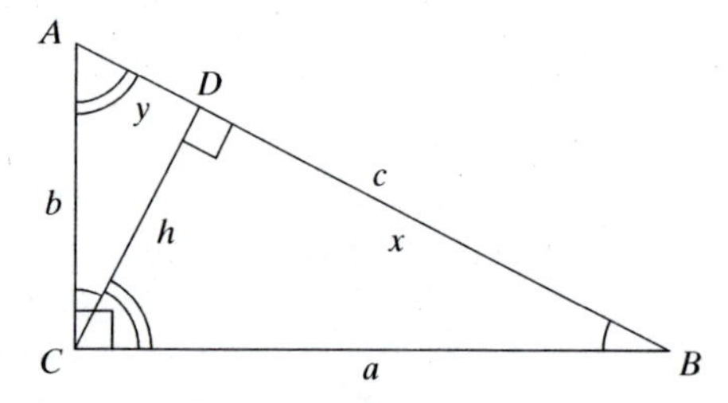

Each pair of similar triangles leads to a situation with a geometric mean. For instance, $\Delta ABC \sim \Delta ACD$ implies $\frac{AB}{AC} = \frac{BC}{CD} = \frac{AC}{AD}$ from which Theorem 8.27(a) follows. Proofs of the other parts are left to you.

Theorem 8.27 Let D be the foot of the altitude to the hypotenuse $\overline{AB}$ of right triangle ABC. Then

a. AC is the geometric mean of AB and AD.
b. BC is the geometric mean of BA and BD.
c. DC is the geometric mean of DA and DB.

We have stated the theorem using a particular triangle to show a mnemonic for remembering it. Each segment emanating from C is the geometric mean of the two other segments emanating from its other endpoint.

Place two segments of lengths x and y next to each other and draw a semicircle using the union of the segments as a diameter. Next draw the perpendicular from the common endpoint of the segments to the circle, as in Figure 50a.

Figure 50

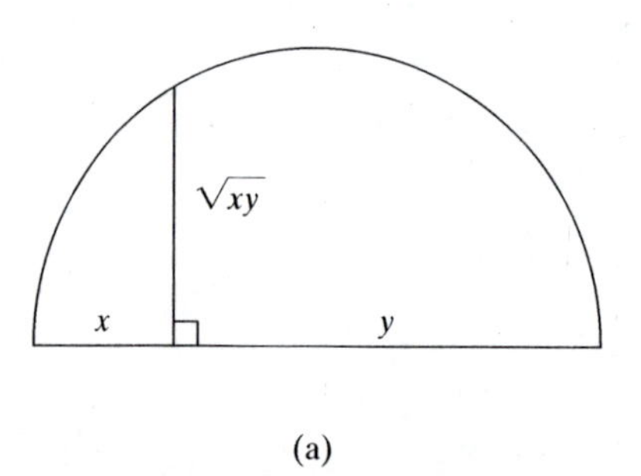

(a)

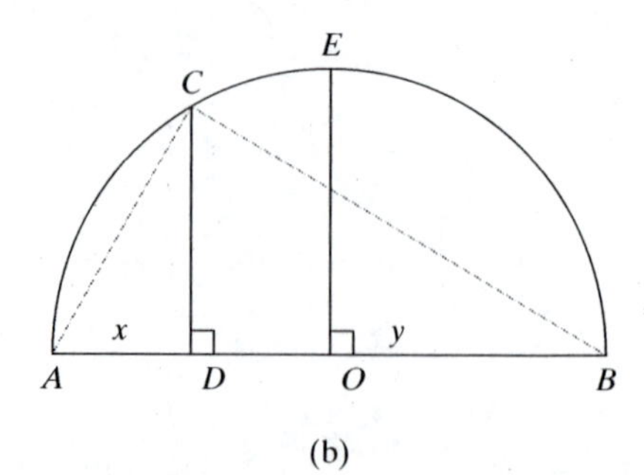

(b)

The length of this perpendicular is the geometric mean of x and y, and Figure 50b suggests the proof. $\angle ACB$, being an inscribed angle with a semicircle (the half of the circle not drawn) as its intercepted arc, is a right angle. So ΔACB is a right triangle and

$\overline{CD}$ is the altitude to its hypotenuse. By Theorem 8.27(c), CD is the geometric mean of AD and BD.

In Figure 50b, notice that $OE = \frac{x+y}{2}$ because it is the radius of circle O. Now think of $\overline{CD}$ as moving along the diameter from A to B, perpendicular to $\overline{AB}$. As it moves, CD is never larger than OE, and it equals OE only when C coincides with E, that is, when $x = y$. This is an elegant geometric proof that the geometric mean of x and y is less than or equal to the arithmetic mean of x and y.

The situation in Theorem 8.26 allows a simple proof of the Pythagorean Theorem using similarity: $\frac{x}{a} = \frac{a}{c}$ implies $x = \frac{a^2}{c}$. In the same way, $y = \frac{b^2}{c}$. Thus $c = x + y = \frac{a^2+b^2}{c}$, from which $c^2 = a^2 + b^2$.

We are accustomed to using the Pythagorean Theorem to find any one of a, b, or c given the other two. It turns out that any two of a, b, c, h, x, and y are sufficient to determine all the others.

Corollary: In the right triangle with legs a and b and hypotenuse c, with the altitude to the hypotenuse h splitting the hypotenuse into segments of lengths x and y, two of these six lengths determine the other four uniquely.

Proof: There are 15 ways of choosing a pair of lengths from six, so at first glance it might seem that there are 15 separate analyses to be done. But some of these choices are essentially identical. For instance, suppose a and c are given. Then b can be found by the Pythagorean Theorem. Since $\Delta ABC \sim \Delta ACD$, $\frac{h}{b} = \frac{a}{c}$. This obtains h. From b and h, the Pythagorean Theorem can be used to obtain y, and in the same way a and h determine x. We leave it to you to demonstrate the other parts of the theorem. (See Problem 5b.) ┘

Question: Which other pairs yield essentially the same analysis as the one in the preceding proof?

Harmonic means

In Section 4.3.4, you were introduced to the *harmonic mean* and several of its applications. Recall that the harmonic mean of two numbers a and b is the reciprocal of the arithmetic mean of the reciprocals of a and b. Algebraically, let **H(a, b)** be the harmonic mean of a and b, and **A(a, b)** their arithmetic mean. Then

$$H(a,b) = \frac{1}{A\left(\frac{1}{a},\frac{1}{b}\right)} = \frac{1}{\frac{\left(\frac{1}{a}\right)+\left(\frac{1}{b}\right)}{2}} = \frac{2}{\frac{1}{a}+\frac{1}{b}} = \frac{2ab}{a+b}.$$

We now introduce the notation **G(a, b)** for the geometric mean of a and b. Figure 50b pictures the relationship between $G(a,b)$ and $A(a,b)$. It is natural to ask whether there is a geometric picture of the relationship among all three means, but not natural to expect that there is one!

A picture does exist. Start with a point P outside a circle so that a and b are the shortest and longest distances from P to some point on the circle. This is shown in Figure 51. $\overleftrightarrow{PB}$ is the secant that contains the center of circle O and intersects the circle at points A and B. So $a = PA$ and $b = PB$. $\overline{PT}$ is tangent to the circle at T, and H is the foot of the perpendicular from T to $\overleftrightarrow{PB}$.

Figure 51

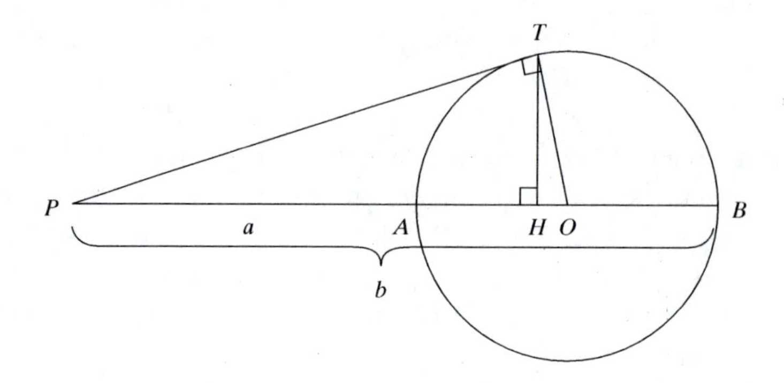

The arithmetic and geometric means of a and b are rather easily seen in Figure 51.

$$A(a, b) = PO$$
$$G(a, b) = PT$$

We now show that PH is the harmonic mean of a and b. It is not as hard as it may seem at first. By AA Similarity, $\Delta PHT \sim \Delta PTO$, so $\frac{PH}{PT} = \frac{PT}{PO}$. Thus $PH = \frac{PT^2}{PO} = \frac{ab}{\frac{a+b}{2}} = \frac{2ab}{a+b} = H(a, b)$. In this way, Figure 51 elegantly shows a relationship between these three means, when $a < b$.

Theorem 8.28 If $0 < a < b$, then $H(a, b) < G(a, b) < A(a, b)$. That is, for any two positive unequal numbers, their geometric mean is greater than their harmonic mean and less than their arithmetic mean.

Figure 54 for Problem 13 provides another elegant demonstration of Theorem 8.28.

8.3.1 Problems

1. a. The length of a tangent to circle O from a point P outside it is 15. If the radius of the circle is 10, find OP.
   b. Generalize part **a** in some way.
2. Prove Theorem 8.26, indicating which sides of the three triangles correspond.
3. In Figure 49, suppose $a = 8$ and $h = 3$. Find $b, c, x,$ and y.
4. a. Suppose $b = 75$ and $x = 80$ in Figure 49. Find a, c, h, and y.
   b. Prove or disprove: If b and x are rational, so are $a, c, h,$ and y.
5. a. Finish the proof of Theorem 8.27.
   b. Finish the proof of the corollary to Theorem 8.27.
6. Trigonometry enables us to examine the result of Theorem 8.26 in a different manner. From AAS Congruence, a right triangle is determined by its hypotenuse and one of its acute angles. Suppose in Figure 49 that $c = 1$ and angle B is known. Prove all of the following: $a = \cos B$, $b = \sin B$, $h = \cos B \sin B$, $x = \cos^2 B$, and $y = \sin^2 B$.
7. In right triangle ABC shown in Figure 52, D is the foot of the perpendicular from C to $\overline{AB}$, and M is the midpoint of $\overline{AB}$. Explain why this provides another elegant demonstration that the geometric mean of x and y is never greater than the arithmetic mean of x and y, and the two means are equal only when $x = y$.

Figure 52

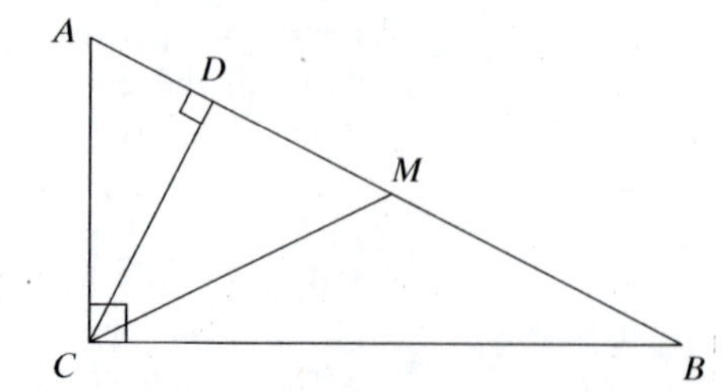

8. Refer to Figure 51. Explain why PO is the arithmetic mean of PA and PB.

9. Suppose $a = b$ in Figure 51. How does this affect PO, PT, and PH and what implication does this have for Theorem 8.28?

10. In Figure 53, $\overline{AB}$ and $\overline{CD}$ are chords of the circle and P is the midpoint of $\overline{CD}$. Two lengths and their geometric mean are in the figure. Identify them.

Figure 53

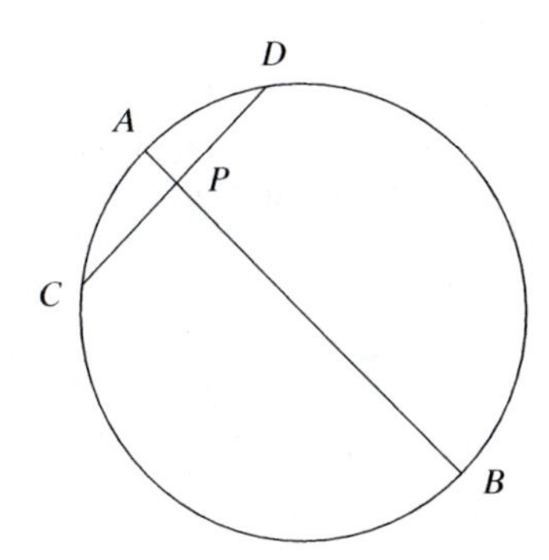

11. Let $A(x, y)$ be the arithmetic mean of x and y. Suppose that $\overrightarrow{BD}$ is the bisector of $\angle EBC$, and that P is a point on the same side of $\overleftrightarrow{BE}$ as C and the opposite side of $\overrightarrow{BC}$ as E. Prove or disprove that $m\angle PBD = A(m\angle PBE, m\angle PBC)$.

12. Let $A(x, y)$ be the arithmetic mean of x and y. Suppose that M is the midpoint of $\overline{PQ}$ and that $\angle PQR$ is a right angle. Prove or disprove that $RM = A(RP, RQ)$.

13. Figure 54, which pictures the three means of this section, is due to Pappus. O is the center of a circle with diameter $\overline{AB}$. P is on the circle, $\overline{PQ} \perp \overline{AB}$, and $\overline{QR} \perp \overline{OP}$. Find three lengths on the figure that are (a) the arithmetic mean, (b) the geometric mean, and (c) the harmonic mean of the lengths AQ and BQ.

Figure 54

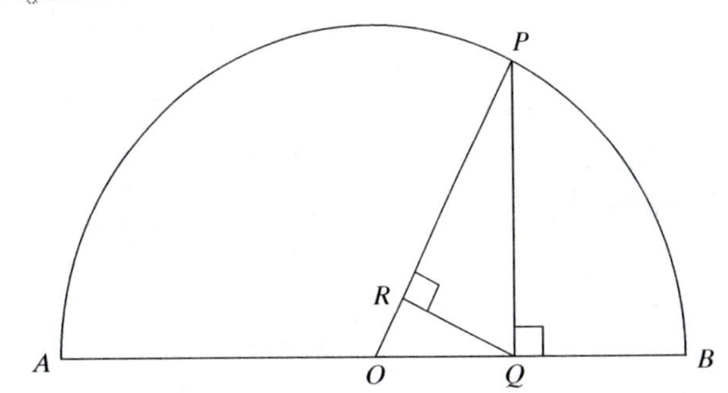

8.3.2 Similarity and parallel lines

One of the basic theorems in the study of similarity deals with the lengths of segments formed when a line intersects two sides of a triangle and is parallel to the third side, as in Figure 55.

Theorem 8.29 Suppose line ℓ is parallel to $\overline{AB}$ and intersects $\overline{OA}$ and $\overline{OB}$ in C and D, respectively. Then $\frac{OC}{OA} = \frac{OD}{OB} = \frac{DC}{AB}$.

Figure 55

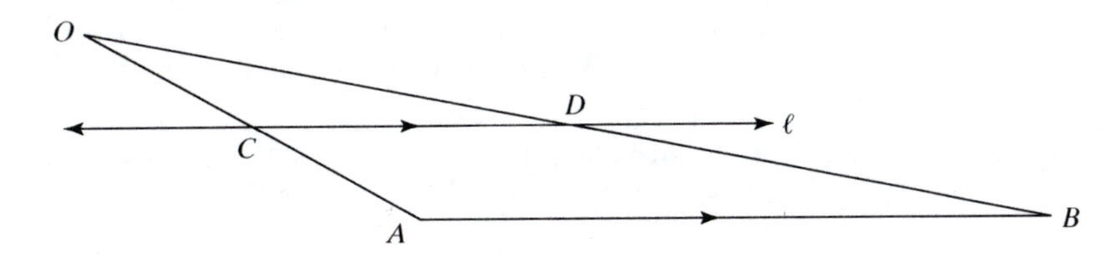

Proof: One proof involves similar triangles. We offer a different proof, using size changes. The size change with center O and magnitude $\frac{OC}{OA}$, by its definition, maps O onto O, A onto C, and B onto a point on ray $\overrightarrow{OB}$. The image of $\overleftrightarrow{AB}$ contains C and is parallel to $\overline{AB}$ (Theorem 8.10), so it must be $\overleftrightarrow{CD}$. Consequently, the image of B is on $\overleftrightarrow{CD}$ and on $\overrightarrow{OB}$, so the image of B is D. Thus the magnitude of the size change is $\frac{OD}{OB}$, and since the size change has only one magnitude, $\frac{OC}{OA} = \frac{OD}{OB} = \frac{DC}{AB}$. ∎

From the single proportion

$$\frac{OC}{OA} = \frac{OD}{OB} \quad (1)$$

comes many others. Switch the means of the proportion (why can this be done?).

$$\frac{OC}{OD} = \frac{OA}{OB} \tag{2}$$

Apply the reciprocal function to both sides of (1).

$$\frac{OA}{OC} = \frac{OB}{OD} \tag{3}$$

Subtract 1 from each side (in the forms of $\frac{OC}{OC}$ and $\frac{OD}{OD}$).

$$\frac{CA}{OC} = \frac{DB}{OD} \tag{4}$$

Equation (4) shows that the four parts of the sides $\overline{OA}$ and $\overline{OB}$ are proportional. We might say that (1) and (3) involve part-whole ratios on each segment, (2) involves part-part and whole-whole ratios of different segments, and (4) involves part-part ratios on each segment.

The converse of Theorem 8.29 is also true.

Theorem 8.30 If a line intersects $\overline{OA}$ and $\overline{OB}$ of a triangle OAB in points C and D such that any of the four proportions (1) to (4) holds, then $\overline{CD} \parallel \overline{AB}$.

Proof: There is only one line through C parallel to $\overline{AB}$. This line must intersect $\overline{OB}$ at some point (due to the postulate of Pasch—see Theorem 11.2). The point of intersection makes all of the proportions (1) to (4) true. Since there is only one point that splits the segment in a particular ratio, that point must be D. ∎

Theorem 8.30 is basic because it has many variants and many applications, some of which we now explore.

Representing proportional relationships graphically

$$d = 90t \tag{5}$$

Suppose a vehicle takes t hours to travel d kilometers at $90\frac{\text{km}}{\text{h}}$. Then $d = 90t$. The variables d and t are proportional; multiplying t by any amount multiplies d by the same amount. Any simple proportional relationship such as that between distance d and time t in (5) can be represented in a diagram as in Figure 56 by choosing length units for t and d. We can imagine this diagram as varying. As t grows, d grows in proportion, as indicated by the arrows, and we have a dynamic representation of the proportional relationship (5). This representation is equivalent to the typical graph of $d = 90t$, where t is the variable on the horizontal axis and d is the variable on the vertical

Figure 56

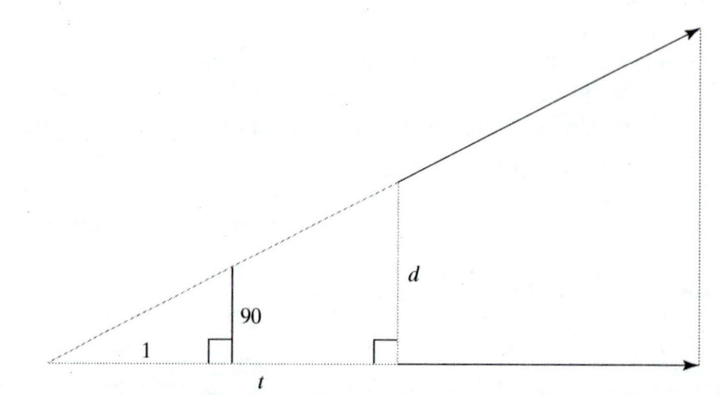

axis. Clearly, the scales are different in the two directions, since "1" in the horizontal direction is longer than "90" in the vertical direction.

In Figure 56 the distance d and time t are represented in perpendicular directions. It is also possible to represent them in parallel directions, as in Figure 57, another representation of (5). In this figure we need two different scales, a distance scale associated with the small triangle, and a time scale associated with the large triangle.

Figure 57

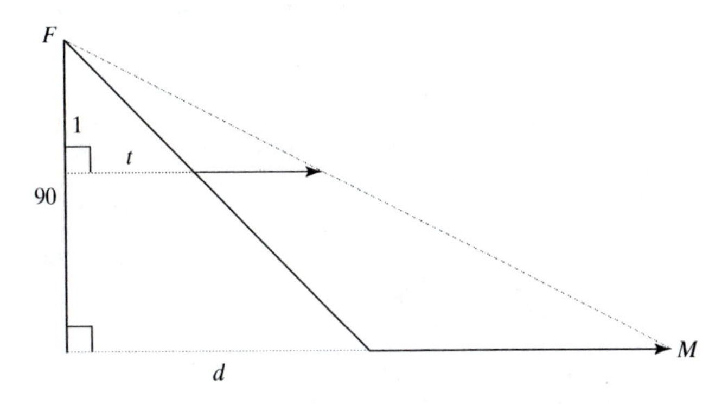

As t increases (indicated by the arrows), point M moves to the right. This causes the line segment $\overline{FM}$ to pivot at F, thus increasing d proportionally.

The situation in Figures 56 and 57 is very simple, but we can use the same ideas to look at more complex cases. For example, suppose a car has been driven a total distance D for a time T at a constant speed of $90\,\frac{km}{hr}$. If the trip time was split between two drivers as $T = t_1 + t_2$, then the distance was split, too. We represent this split as $D = d_1 + d_2$ and show it in Figure 58.

Figure 58

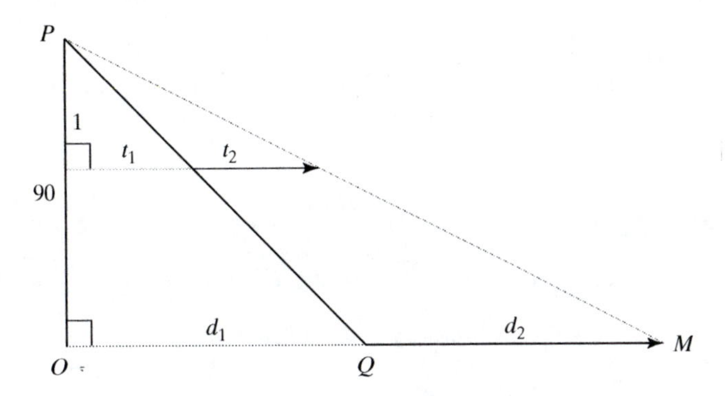

Intuitively it seems that these distance pieces d_1 and d_2 must be proportional to the time pieces t_1 and t_2. This is in fact true: Since the speed is constant, $90 = \frac{d_1}{t_1} = \frac{d_2}{t_2}$, which is equivalent to

$$\frac{d_1}{d_2} = \frac{t_1}{t_2}. \tag{6}$$

Figure 58 can be interpreted dynamically in a different way. Triangle OPM is fixed, but point Q can move from O to M, indicating different partitions of T into t_1 and t_2. At the same time, line segment $\overline{QP}$ pivots at P and creates the appropriate corresponding partition of D into d_1 and d_2. The proportional relationship (6) shown in Figure 58 can be proved directly using Theorem 8.29, but we prove a more general theorem that illustrates an interesting and useful fact about proportional relationships. Refer to Figure 59.

Theorem 8.31 In $\Delta PTT'$, let D and D' be points on $\overrightarrow{PT}$ and $\overrightarrow{PT'}$ so that $\overline{DD'} \parallel \overline{TT'}$. Let rays from the point P partition $\overline{DD'}$ and $\overline{TT'}$ into n parts with lengths d_i and t_i, respectively, so that $DD' = \sum_{i=1}^{n} d_i$ and $TT' = \sum_{i=1}^{n} t_i$. Then the elements of the two partitions formed in this way are proportional. That is,

a. $\dfrac{d_i}{t_i} = \dfrac{d_j}{t_j} = \dfrac{DD'}{TT'}$ for all i and j where $1 \le i, j \le n$

b. $\dfrac{d_i}{DD'} = \dfrac{t_i}{TT'}$ for all i where $1 \le i \le n$

c. $\dfrac{d_i}{d_j} = \dfrac{t_i}{t_j}$ for all i and j where $1 \le i, j \le n$

Figure 59

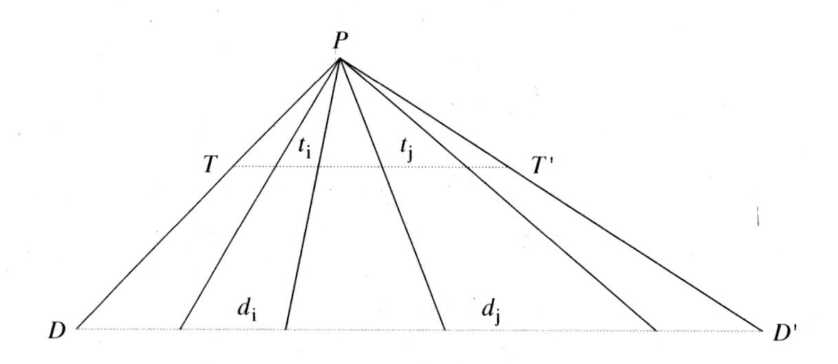

Proof: A proof follows from Theorem 8.29 and is left to you. ∎

Three or more parallels

Another generalization of Theorem 8.29, shown in Figure 60, involves more than two parallel lines.

Theorem 8.32 Let parallel lines partition $\overline{OA}$ and $\overline{OB}$ into n parts with lengths a_i and b_i, respectively, so that $OA = \sum_{i=1}^{n} a_i$ and $OB = \sum_{i=1}^{n} b_i$. Then the elements of the two partitions formed in this way are proportional.

Figure 60

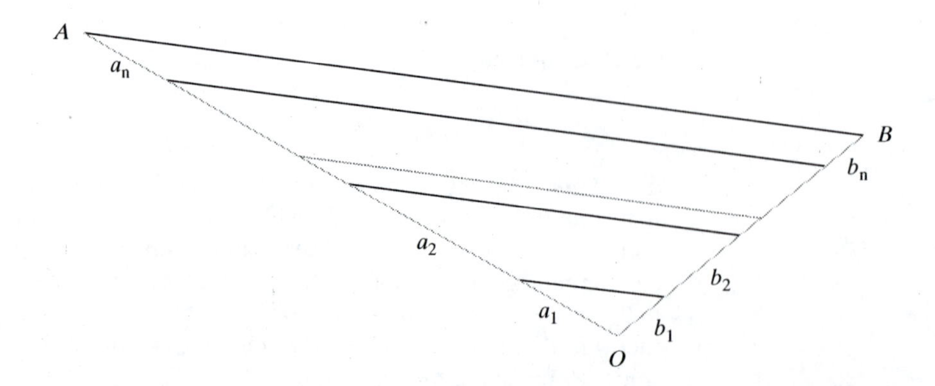

Proof: In the Problems, you are asked to describe what it means for the elements of the partitions to be proportional by following the pattern in Theorem 8.31, and then to prove the theorem. ┘

Receding telephone poles

We conclude this section with a nontrivial application of Theorems 8.30 and 8.31 to the two-dimensional representation of three-dimensional scenes in perspective drawings. Figure 61 shows a picture of a row of telephone poles that line the side of a highway as a photograph taken with a simple camera might show. Only 7 poles are shown, but you should imagine that there are many more receding off into the distance. We assume that the poles along the road are of the same height and the same distance apart and that the road is straight and level.

Figure 61

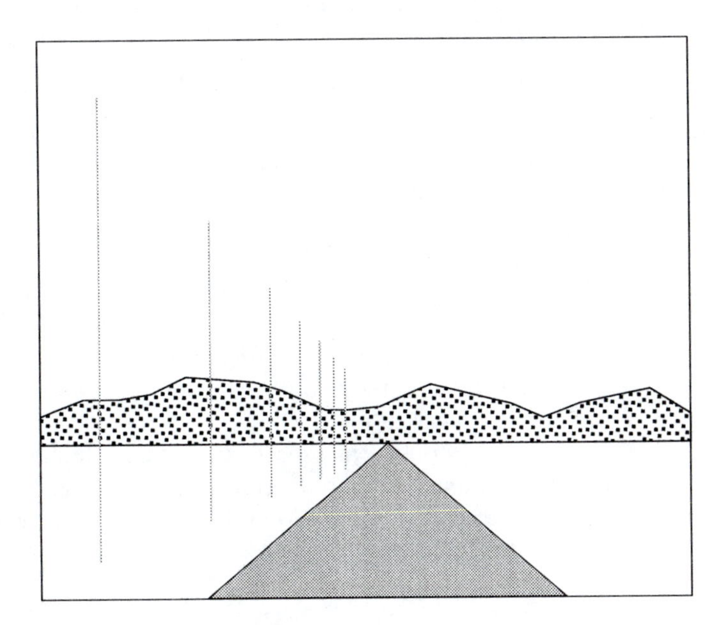

Suppose you were an artist who wished to depict these poles but you had no photograph to assist you. How could you determine in a perspective drawing the appropriate distances between the poles and the appropriate height for each? We begin by describing an approach to the first question.

Figure 62 gives a view from above of a row of poles $P_1, P_2, P_3, \ldots, P_n$ along the ray $\overrightarrow{PP_1}$. The poles are a distance s apart and the line of poles is at a distance w from the center of the road. Imagine you are viewing this scene through a vertical transparent screen SMS' placed perpendicular to the line of poles. (This is called a "picture plane".) Your eye is at a fixed point O behind that screen (at eye level and at the middle of the road). The screen intersects the line of poles at P such that P is at a distance q from P, while your eye is at a distance c behind the screen. Segment $\overline{OM}$ is above and parallel to the middle of the road.

Imagine that the images of the poles as they appear to you are "etched" onto the screen. To identify the locations of these images, let y_n be the distance from the middle M of the road on the screen to the point P_n' where your line of sight $\overrightarrow{OP_n}$ crosses the screen. That is, $y_n = MP_n'$. Thus, pole P_1 is etched at point P_1', and $y_1 = MP_1'$. The point M represents the place on the screen where the line of poles appears to converge. This is the "vanishing point" of the drawing.

Figure 62

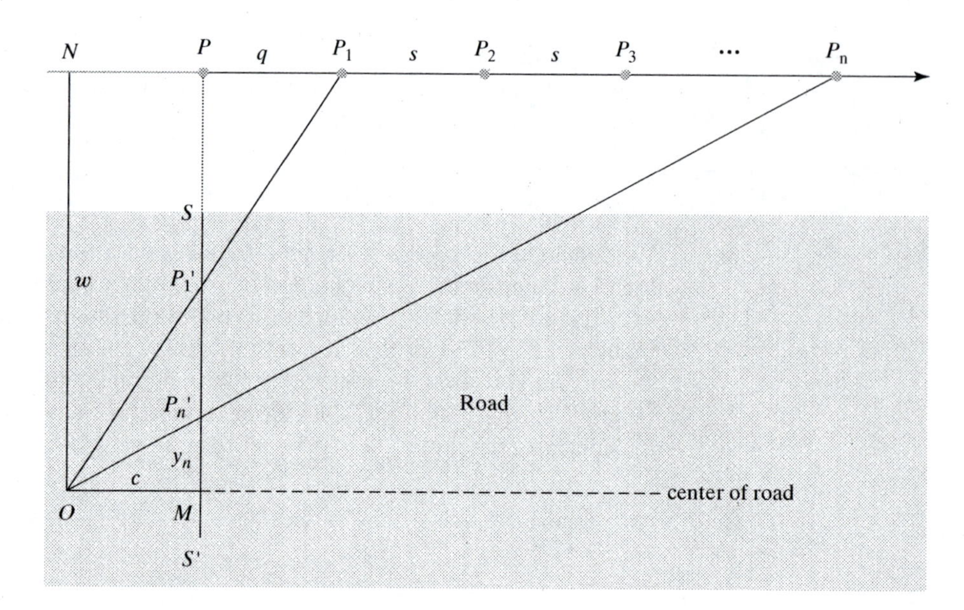

From the similar triangles $OMP_n{}'$ and ONP_n in Figure 62, $\frac{MP_n{}'}{OM} = \frac{ON}{NP_n}$, so $\frac{y_n}{c} = \frac{w}{c+q+(n-1)s}$, from which the distance y_n of the nth pole from the center of the picture is given by

$$(7) \qquad y_n = \frac{wc}{c + q + (n - 1)s}.$$

This gives us a precise way of constructing the spacing of poles in such a perspective drawing in terms of the parameters c, q, s, and w that represent the location of the poles in relation to your eye.

Looking at a special case of this situation can give a better overall picture of what is going on in a perspective drawing. Specifically, assume that the setback q of the transparent screen from the first pole and the setback c of your eye from the screen are both equal to s, the separation of the poles. Then $c = q = s$. With these assumptions, (7) becomes

$$(8) \qquad y_n = w\left(\frac{1}{n + 1}\right).$$

That is, the distances $y_1, y_2, \ldots, y_{n-1}$ of the pole images from the vanishing point follow the harmonic sequence $\left\{\frac{1}{n}\right\}$, scaled by the factor w. Actually, the more general sequence (7) is itself always quite close to being part of a harmonic sequence. This follows once we notice that there are constants r and k such that (7) can be written as

$$(9) \qquad y_n = r \cdot \frac{1}{k + n}.$$

Thus, the locations y_n of perspective images of receding equally spaced objects from the vanishing point are versions of the harmonic sequence that are scaled (by a constant r) and displaced (by a constant k).

What about the heights of the poles? Figure 63 gives a view from the side of the row of poles, numbered $1, 2, 3, \ldots, n, \ldots$ and the screen SMS'. Your eye is at O, and ray $\overrightarrow{OO'}$ is horizontal, above the middle of the road. The poles have a height $h + b$, of which h is above eye level and b below. Your line of sight to the top of pole n hits the screen at a distance h_n above the center M of the line $\overleftrightarrow{OO'}$, while your line of sight to the bottom of pole n hits the screen at a distance b_n below the center M. The apparent heights of the poles can be read off this diagram.

Figure 63

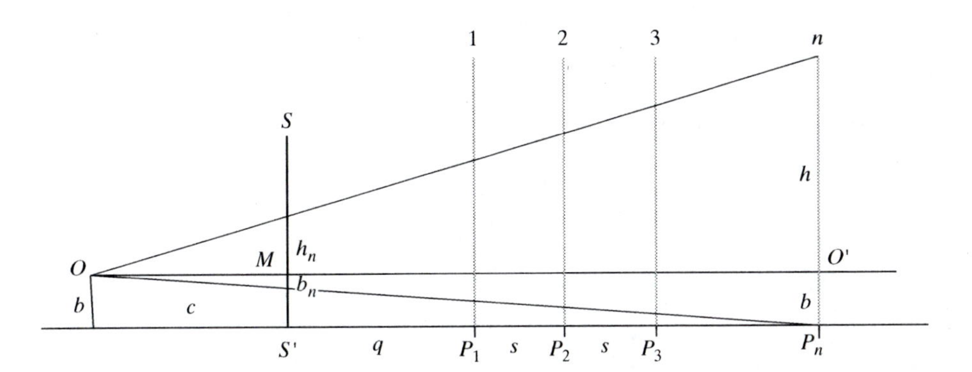

Question 1: By analyzing this diagram, find the distances h_n and b_n from point M on the screen SS' representing the images of the top and bottom of the nth pole.

Question 2: Find the sum of the distances h_n and b_n in the special case $c = q = s$.

In this example we have produced a perspective drawing by recording a distant scene on a nearby transparent screen. The general relationship between a length of a distant object parallel to such a screen and its image on the screen is represented diagrammatically in Figure 64.

Figure 64

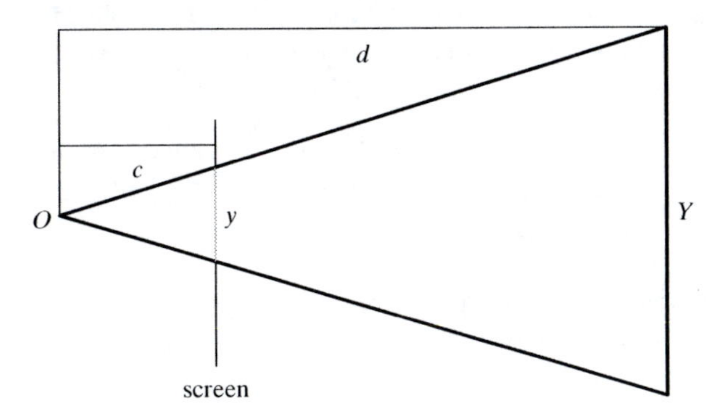

From this diagram we have the basic relationship that underlies all perspective drawings.

$$y = c \cdot \frac{Y}{d} \tag{10}$$

The relationship (10) says that the length y of an object recorded on the screen is directly proportional to its actual length Y and inversely proportional to its distance d from the observation point O. Thus, the same object twice as far away appears half as tall, which is what we would expect. The length y is also directly proportional to the distance c of the screen from the observation point. This is also natural. We can make the size y of the image as small as we like by making c small. On the other hand, we can also make y as large as we want by making c large. If c equals d, the image has become the same size as the object.

We use the basic relationship (10) in Problems 7 to 10 to further explore perspective situations. Specifically, Problem 7 asks you to analyze a perspective drawing of a simple scene using the ideas of the section. Problem 8 asks you to establish a more general method of analyzing perspective drawings by constructing a 2-dimensional coordinate system for the drawing that is related in a systematic way to a representation of the actual scene in 3-dimensional coordinates. Problem 9 is an application of this coordinate method to the receding poles situation of the text. Finally, Problem 10 is an application of the coordinate method to a new problem.

8.3.2 Problems

1. Prove Theorem 8.29 using similar triangles.

2. Let ΔABC have points B' and C' on sides $\overline{AB}$ and $\overline{AC}$ so that $\overline{B'C'} // \overline{BC}$, as shown in Figure 65. Let M be the midpoint of $\overline{AC}$ and P be the image of C' under the half-turn with center M. Prove that $\frac{BB'}{BA} = \frac{AP}{AC}$ (See Project 5 of this chapter for an application of this result.)

Figure 65

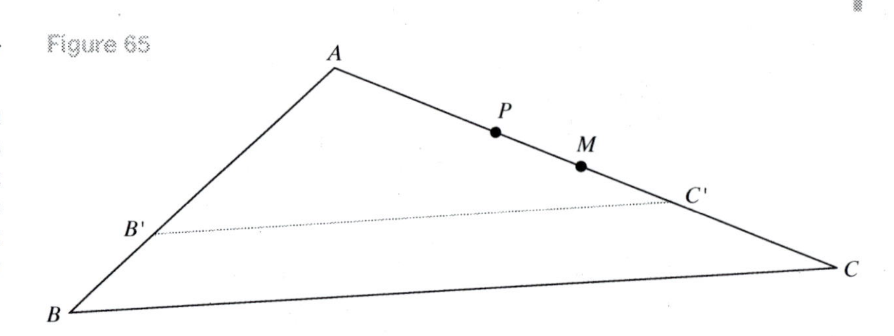

3. What algebraic property or properties enable switching the means of a proportion, as was done in going from equation (1) to equation (2) in this section?

4. Which part of Theorem 8.32 involves
   a. part-part ratios on each segment?
   b. part-whole ratios on each segment?
   c. part-part and whole-whole ratios of different segments?

5. Interpret Theorem 8.31(a) as a statement about a constant-speed relationship between distance and time. What is the value of the constant speed?

6. A large tree is between a sidewalk and the curb, 5 feet from the curb and 5 feet from the sidewalk. You are driving in a car parallel to the curb and 20 feet from it. When a person is walking on the sidewalk in the direction opposite your driving direction, it is often the case that if the person is almost completely blocked by the tree when you first look, the person continues to be blocked by the tree as you drive along. Explain why.

7. Figure 66 represents a simple photograph taken looking down a hallway. The camera is in the middle of the hallway, halfway between floor and ceiling. The width of the hallway is 18 feet, the height is 12 feet, and the distance from the camera to the near end of the hallway is 15 feet. How long is the section of hallway shown in the photograph? Draw a diagram to illustrate your answer. (*Hint*: Measuring on the diagram shows that objects at the far end of the hallway appear one-third the size of congruent objects at the near end of the hallway.)

Figure 66

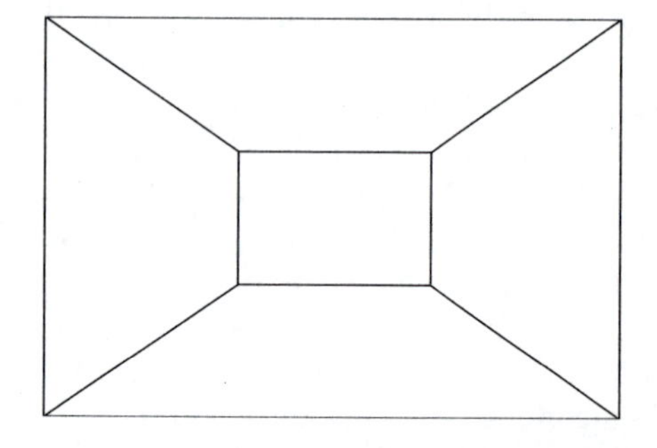

8. This problem generalizes Problem 7 by establishing a relationship between coordinates in three dimensions of a point in a scene to coordinates in two dimensions of the corresponding point in a perspective drawing of the scene. Figure 67 imposes a coordinate plane on a perspective drawing of a hallway. The origin is at the vanishing point. Figure 68 shows a horizontal section through this hallway. The width of the hallway is w. The height (not shown in this diagram) is h. We establish a three-dimensional x-y-z coordinate system for the hallway scene so that the observation point O is the origin $(0, 0, 0)$. Let d be the distance from O to the plane of P, Q, and R. Then point P has coordinates $\left(\frac{w}{2}, \frac{h}{2}, d\right)$. Let d' be the distance from O to the plane of P', Q', and R'. Then point R' has coordinates $\left(-\frac{w}{2}, 0, d'\right)$.

Figure 67

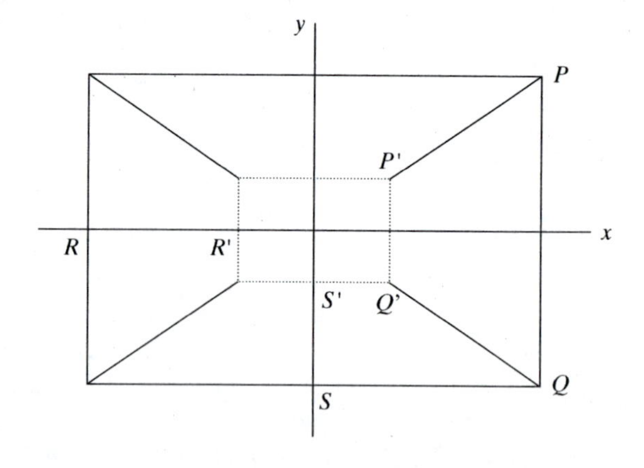

(Note that the numbers used as coordinates are all positive. This means that the z-direction in this system has an orientation opposite to the usual Cartesian x-y-z system.) Suppose the point (X, Y, Z) in the real scene corresponds to the point (x, y) in perspective coordinates. (Uppercase letters X, Y, and Z represent distances in the real scene, while lowercase letters x and y represent coordinates on the diagram.) Let c be the distance from O to the screen.

   a. Express x and y in terms of X, Y, Z, and c. Illustrate with a diagram.

Figure 68

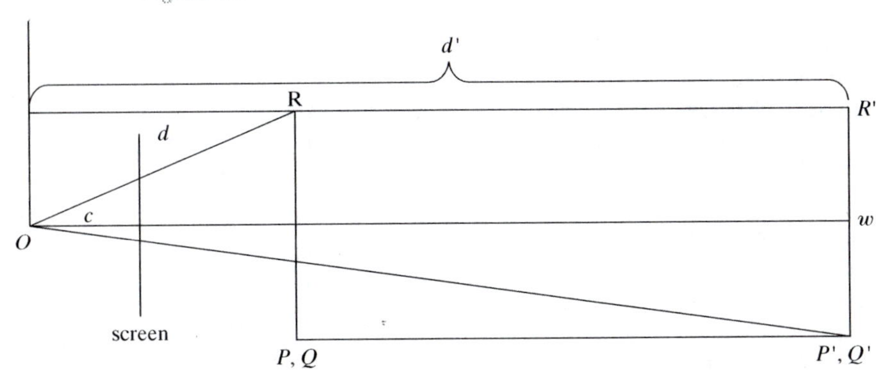

b. Show that X, Y, and Z cannot be expressed in terms of x, y, and c alone.

c. Express X and Y in terms of x, y, Z, and c

9. Analyze the receding poles situation in terms of the coordinate approach of Problem 8. Specifically, find the coordinates in a perspective drawing of the top and the bottom of the nth pole. Identify the parameters you are using.

10. Figure 69 shows a simple photograph of railroad tracks receding into the distance. The tracks appear to converge, forming an angle in the photograph. Suppose the measure of this angle is α.

Figure 69

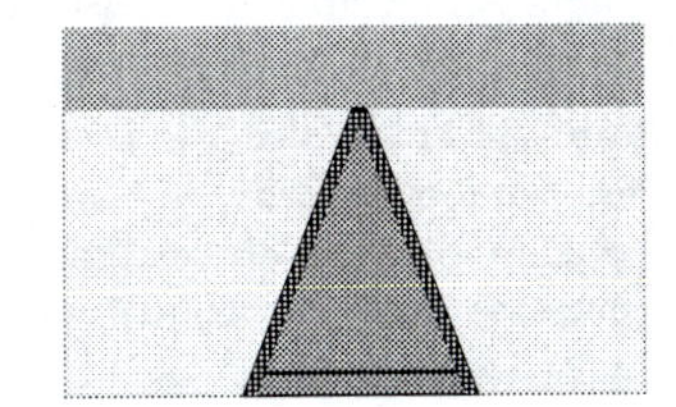

a. Show how the angle α is determined by the two relevant parameters of the situation, the distance w between the tracks, and the height h of the camera above the center of the tracks. (*Note*: A direct approach to this problem using the methods of the text section will lead to a result, though the complexity of the analysis is a little greater than the example in the text. An approach using the coordinate method of Problem 8 will yield a result much more readily. Finally, it is possible to write down the answer directly "by inspection" if you look at the situation in a certain, simple way.)

b. Suppose the distance between the tracks is 4 feet 8.5 inches (the standard U.S. gauge). What is the height h for this photograph?

c. Suppose the distance between the tracks is 4 feet 8.5 inches and that the railroad ties are 1 foot apart. If the first tie visible in the photo is at the bottom edge of the photo, and if the camera is above a point 15 feet back from this tie, show how to locate accurately the next two ties, and add them to a copy of the photo.

ANSWERS TO QUESTIONS

1. $h_n = h \cdot \frac{c}{c+q+(n-1)s}$ and $b_n = b \cdot \frac{c}{c+q+(n-1)s}$

2. $h_n + b_n = (h + b)\left(\frac{1}{n+1}\right)$

Chapter Projects

1. Global positioning. There are reasons for analyzing grids of city streets using taxicab distance. Analysis can help city planners decide where to locate parks, phone booths, traffic lights, bus stops, etc. Government space agencies and commercial airlines also benefit from use of it for the Global Positioning System developed by the United States Department of Defense. This system uses satellites orbiting the earth in a gridlike pattern in six different planes oriented at 55° to the equator. Four satellites on each plane send radio signals to earth. The taxicab distance formula is useful for evaluating distances and positions of the satellites. Prepare a presentation describing some of the uses of the taxicab metric.

2. The Absolute. In 1859 the great English mathematician Arthur Cayley derived a distance function related to a fixed conic in the plane, called the Absolute. When the Absolute consists of the pair of circular points at infinity, the Cayley distance reduces to the ordinary Euclidean distance function. Read about Cayley's theory of distance and write an essay on it.

3. Voronoi diagrams. Given $n \geq 2$ points in the plane, a Voronoi diagram is a partitioning of the plane into n regions each of whose interiors contain those points closer to one of the given points than any other. Read about Voronoi diagrams and describe them when the given points are vertices of common triangles, quadrilaterals, other polygons, and points at random.

4. Splitting polygons into similar polygons. The altitude to the hypotenuse of a right triangle splits that triangle into two triangles similar to it.

   a. Prove that no triangle other than a right triangle can be split into two triangles similar to it.

   b. Prove that every triangle can be split into four triangles similar to it.

   c. Use part **b** to explain why every triangle can be split into 7, 10, 13, ..., $3k + 1$, ... triangles similar to it, when k is an integer ≥ 2.

   d. Prove that every triangle can be split into 9, 12, ... $3k$, ... triangles similar to it, when k is an integer ≥ 3. Use the splitting into 9 similar triangles to show that every triangle can be split into 6 similar triangles.

   e. Which parallelograms can be split into similar parallelograms? Can any kites that are not rhombi be split into similar kites? What about trapezoids that are not parallelograms? Work on one of these questions or a related problem.

5. Suppose that P_0, P_1, and P_2 are noncollinear points in the plane, and let $\mathbf{p}_0$, $\mathbf{p}_1$, and $\mathbf{p}_2$ be the corresponding vectors as shown in Figure 70. For any real number t such that $0 \leq t \leq 1$, define the vectors

$$\mathbf{p}_{01}(t) = (1 - t)\mathbf{p}_0 + t\mathbf{p}_1; \quad \mathbf{p}_{12}(t) = (1 - t)\mathbf{p}_1 + t\mathbf{p}_2;$$
$$\mathbf{b}_{02}(t) = (1 - t)\mathbf{p}_{01} + t\mathbf{p}_{12}$$

Figure 70

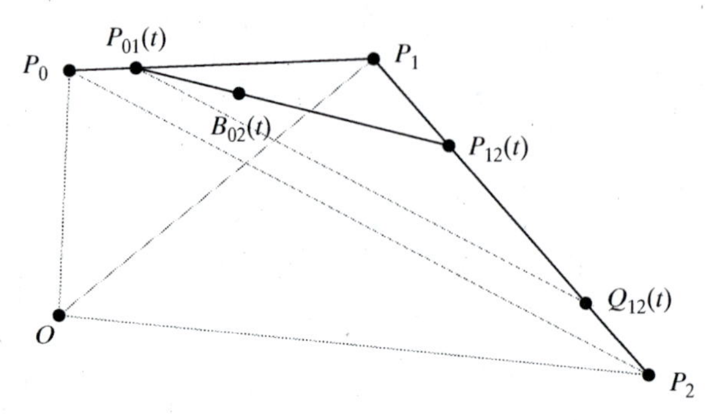

   a. Use the result of Problem 2 of Section 8.3.2 to prove that

$$\frac{P_0P_{01}}{P_0P_{01}} = \frac{P_1P_{12}}{P_1P_2} = \frac{P_{01}B_{02}}{P_{01}P_{12}} = 1 - t.$$

   b. Prove that as t increases from 0 to 1, the point $B_{02}(t)$ varies along a parabolic path from P_0 to P_2.

   c. Use a dynamic geometry program to trace the locus of points $B_{02}(t)$ for $0 \leq t \leq 1$ for three given noncollinear points P_0, P_1, and P_2. This locus is a parabolic curve called the **second-degree Bezier curve** for the points P_0, P_1, and P_2, and the points P_0, P_1, and P_2 themselves are called **control points** for the curve.

   d. Bezier curves are used extensively in computer graphics and in computer-aided design programs to generate smooth, aesthetically pleasing curves that can be varied smoothly by simply moving the control points. Find an example of the use of a Bezier curve.

6. The traditional centers of a triangle. Each of the first three parts of this project constructs and explores one of the centers of a triangle that are traditional to Euclidean geometry. You will need a dynamic geometry construction program. For each part, begin with the vertices of a ΔABC as free points. Display your results by printing out the diagrams for two quite different triangles.

   a. Display the midpoints A', B', and C' of the sides opposite the vertices A, B, and C, and the medians $\overline{AA'}$, $\overline{BB'}$, and $\overline{CC'}$. Verify these results dynamically and with a proof.

      i. The medians are concurrent at a point (the **centroid** of ΔABC).

      ii. Calling the centroid G, $\frac{AG}{GA'} = \frac{BG}{GB'} = \frac{CG}{GC'} = 2$.

   b. Display the lines containing the perpendicular bisectors of the three sides of the triangle. Verify these results dynamically and with a proof.

      i. The lines are concurrent at a point (the **circumcenter** of ΔABC).

      ii. The circumcenter O is the center of a circle (the **circumcircle**) that contains the three vertices of ΔABC.

c. Display the lines containing the three altitudes of the triangle. Verify dynamically that these lines are concurrent at a point (the **orthocenter**) of ΔABC. Call this point H.

d. Draw the triangle $A'B'C'$ (suitably named) connecting the midpoints of the sides of ΔABC and display the lines of parts **b** and **c**. Notice that $\Delta ABC \sim \Delta A'B'C'$. Prove that G is the centroid of $\Delta A'B'C'$ and that G, H, and O are collinear with $2GH = GO$. (The line containing G, H, and O is the **Euler line** of ΔABC.)

7. Excenters of a triangle. Every triangle has six exterior angles. Create a dynamic geometry drawing that begins with the vertices of a ΔABC as free points and constructs the bisectors of the three interior angles and the six exterior angles. Verify these results dynamically that and with a proof.

a. The interval angle bisectors are concurrent at a point (the **incenter** of ΔABC).

b. The incenter I is the center of a circle (the **incircle**) that is tangent to the three sides of ΔABC.

c. Each internal bisector is concurrent with two external angle bisectors. These points of concurrency X, Y, and Z are the **excenters** of ΔABC.

d. Each excenter is the center of a circle that is tangent to a side of the triangle and extensions of the other two sides. These are the **excircles** of the triangle.

e. Find a numerical relationship among the reciprocals of the radii of the three excircles and the reciprocal of the radius of the incircle.

8. Napoleon triangles and concurrencies. Create a dynamic geometry drawing that begins with the vertices of a ΔABC as free points.

a. Construct equilateral triangles LBC, MAC, and NAB such that L, M, and N are exterior to ΔABC. Prove that $\overline{AL}$, $\overline{BM}$, and $\overline{CN}$ are concurrent at the Fermat point of ΔABC. (*Hint*: Use properties of the Fermat point described in Section 8.1.2.)

b. Let P, Q, and R be the centers of the equilateral triangles LBC, MAC, and NAB. Verify dynamically that $\overline{AP}$, $\overline{BQ}$, and $\overline{CR}$ are concurrent at a point (the **Napoleon point** or **Torricelli point** of ΔABC).

c. Construct equilateral triangles $L'BC$, $M'AC$, and $N'AB$ that overlap ΔABC. Then construct segments corresponding to those constructed in parts **a** and **b**. Do there exist corresponding points of concurrency?

9. The Erdös-Mordell Theorem. This theorem was conjectured by Paul Erdös in 1935 and proved by Louis Mordell in the same year. If, from a point P inside a given triangle ABC, perpendiculars $\overline{PD}$, $\overline{PE}$, and $\overline{PF}$ are drawn to its sides, then

$$PA + PB + PC \geq 2(PD + PE + PF).$$

Equality holds if and only if ΔABC is equilateral.

Study the proof of this theorem that is given by Leon Bankoff in the *American Mathematical Monthly* 65 (1958), page 521 ("An Elementary Proof of the Erdös-Mordell Theorem"). Then check out the proof by Mordell and Barrow, which uses trigonometry, in the *American Mathematical Monthly* 44 (1937), pages 252–254 (Problem 3740). Compare these two proofs and describe which, in your estimation, is preferable, and why.

Bibliography

Unit 8.1 References

Curtis, H. J. "A Note on the Taxicab Geometry." *American Mathematical Monthly* 60 (1953), 416–417.
This article shows that using the taxicab distance on a Cartesian grid with integer points of intersection, a "circle" centered at one of these points is a square. It furthermore shows that a "circle" not centered at one of these intersection points is a nonconvex figure.

Krause, Eugene. *Taxicab Geometry*. Reading, MA: Addison-Wesley, 1975. Reprint New York: Dover, 1986.
This entire short paperback is devoted to a treatment of taxicab geometry designed for high school students.

North, J. D. *The Measure of the Universe*. Oxford: Clarendon Press, 1965.
Primarily a book on twentieth-century cosmology, in Chapter 15 it examines the fundamental features of distance as they relate to the development of cosmological models.

Pattern, P. "Finding the Shortest Distance on the Earth's Surface from Here to Timbuktu." UMAP Module 562, Cambridge, MA: COMAP, 1986.

Rademacher, Hans, and Otto Toeplitz. *The Enjoyment of Mathematics*. Princeton, NJ: Princeton University Press, 1957.
Chapter 3 considers maximum area problems for polygons inscribed in circles, and Chapters 5 and 6 develop two proofs of Fagnano's Problem in Section 8.1.2.

Rubinstein, M. *Tools for Thinking and Problem Solving*. New Jersey: Prentice Hall, 1986.
Sections 3.14 and 3.15 provide a detailed discussion of Hamming distance and code.

Schattschneider, Doris. "The Taxicab Group." *American Mathematical Monthly* 91 (1984), 423–425.
This article examines the Euclidean isometries that preserve taxicab distance and uses them to characterize the group of isometries of the plane.

Unit 8.2 References

Birkhoff, George David, and Ralph Beatley. *Basic Geometry.* New York: Scott Foresman, 1940.
The authors use SAS similarity as one of five basic postulates in a development of Euclidean geometry intended for high school students.

Usiskin, Zalman, Arthur Coxford, and Daniel Hirschhorn. *Geometry*. Needham, MA: Prentice Hall, 2002.
Chapters 12 and 13 of this high school text treat similarity through a transformation perspective.

Klein, Felix. *Elementary Mathematics from an Advanced Standpoint: Geometry*. Translated from the German by E. R. Hedrick and C. A. Noble. New York: Dover, 1939.
The text includes a treatment of rotations and expansions using quaternion multiplication.

Yaglom, I. M. *Geometric Transformations II.* Translated from the Russian by Allen Shields. New Mathematical Library No. 21, 1968. New York: Random House. Now published by the Mathematical Association of America.
This second volume in a series of three classifies similarity transformations and offers a large collection of related solved problems. Theorem 8.21 is proved on pp. 54–55.

Unit 8.3 References

Cusmariu, Adolf. "A Proof of the Arithmetic Mean—Geometric Mean Inequality." *American Mathematical Monthly* 88 (1981), 192–194.
A proof of the inequality between these means for n numbers.

Nelson, Roger B. *Proofs Without Words: Exercises in Visual Thinking*, Washington DC: Mathematical Association of America, 1993.

Nelson, Roger B. *Proofs Without Words II: More Exercises in Visual Thinking*, Washington, DC: Mathematical Association of America, 2000.
These volumes illustrate many proofs in geometry, algebra, and trigonometry by showing a picture or a diagram that helps the observer to see *why* a particular statement is true, and also to see *how* one might begin to construct a more formal proof. Pages 49–57 of the first volume illustrate proofs about inequalities involving means.

Chapter

9

TRIGONOMETRY

The word "trigonometry" literally means "the measure of triangles". However, from its inception what we today call "trigonometry" has been utilized to deal with a wide variety of problems. From its roots in Greece around 300 B.C. until the 17th century, developments in trigonometry were driven by problems in astronomy involving the locations and relative motions of the sun, moon, planets, and stars, as well as by earth-bound problems of surveying, navigation, map making, and construction. These applications required the measurement of angles, the lengths of line segments and circular arcs, and areas of geometric figures in the plane and on the surface of a sphere. As a result, plane and spherical trigonometry developed simultaneously with many fruitful connections between them, but with spherical trigonometry preeminent.

In the 17th century, mathematicians and scientists began to realize that the algebraic, graphic, and analytic properties of the trigonometric functions could be exploited to solve problems in new areas of pure and applied mathematics. For example, trigonometry played a basic role in the development of coordinate geometry and calculus. These developments were stimulated by such problems as modeling periodic behavior, representing and approximating numbers and functions, analyzing wave phenomena and mechanical vibrations and motion problems, and representing and analyzing curves and surfaces. In this setting, the connections between plane trigonometry and the analysis of functions predominated. Though spherical trigonometry continued to play an important supporting role in certain applications, plane trigonometry's role became larger.

The primary emphasis of trigonometry, as it is currently taught in high school and college, is on plane trigonometry, and that is the focus of our discussion in this chapter. Because trigonometric content is typically presented in several different courses in high school and college, we provide here a more global perspective of the basic principles and concepts of plane trigonometry as well as a sample of the diversity of its applications. We also discuss in more detail the fascinating historical and conceptual evolution of trigonometry. This historical and evolutionary perspective is essential to understanding and teaching trigonometry within the broader context of mathematics and science.

Unit 9.1 Angle Measure and the Trigonometric Ratios

9.1.1 Angle measure and arc length

In this section, we discuss the common units for measuring angles and arcs. You may wish to refer to the discussions of angles and rotations in Section 7.2.2 and of arcs in Section 8.2.4.

Units of angle and arc measure

Degree measure, introduced by the Babylonians between the years 2000 and 1600 B.C., is not only the oldest surviving system for measuring angles, but also the most familiar. A **degree** can be defined as the measure of an angle $\angle ACB$ that corresponds to a rotation of the initial side $\overrightarrow{CA}$ to the terminal side $\overrightarrow{CB}$ of magnitude $\frac{1}{360}$ of a revolution around C. More briefly,

$$1 \text{ degree} = \frac{1}{360} \text{ of a revolution.}$$

The other common measure is the **radian**. This word was introduced around 1870 by Thomas Muir and James Thompson, Sr., to stand for "radial angle". Given an angle $\angle ACB$, and a circle of radius r centered at the vertex C of the angle, **the radian measure of $\angle$ACB** is $\frac{s_r}{r}$, where s_r is the length of the arc with central angle $\angle ACB$ on the given circle.

Because the circumference of a circle of radius r is $2\pi r$, this definition implies that

$$2\pi \text{ radians} = 1 \text{ revolution.}$$

This definition raises a mathematical question, "Does the radian measure of an angle depend on the radius r of the given circle in that definition?" The following result shows that it does not.

Theorem 9.1 **(Arc-to-Radius Similarity Principle):** Suppose two arcs have the same measure on circles with radii r_1 and r_2 and a common center C. If the arcs have lengths L_1 and L_2, then $\frac{L_1}{r_1} = \frac{L_2}{r_2}$.

Figure 1

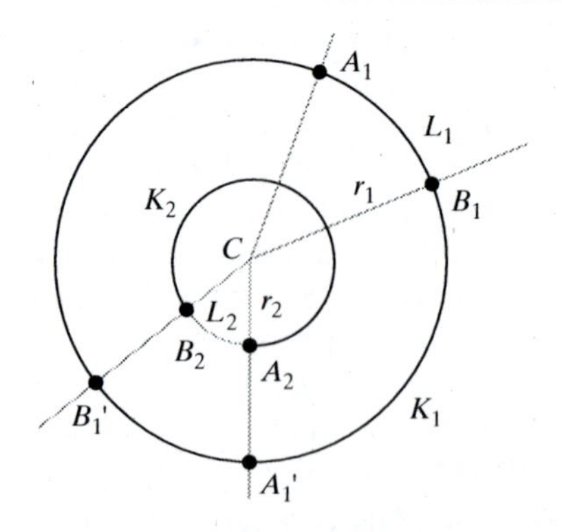

Proof: Let the arcs $\overset{\frown}{A_1B_1}$ and $\overset{\frown}{A_2B_2}$ have the same measure in the circles K_1 and K_2, both centered at C, as shown in Figure 1. Then a suitable rotation around C maps the central angle $\angle A_1CB_1$ onto the congruent central angle $\angle A_2CB_2$ and the arc $\overset{\frown}{A_1B_1}$ onto an arc $\overset{\frown}{A_1'B_1'}$ on K_1. The length of $\overset{\frown}{A_1'B_1'}$ is equal to L_1, the length of $\overset{\frown}{A_1B_1}$ because a rotation is a congruence. Then a size change transformation centered at C of magnitude $\frac{r_2}{r_1}$ maps the circle K_1 onto the circle K_2 and image arc $\overset{\frown}{A_1'B_1'}$ on K_1 onto the arc $\overset{\frown}{A_2'B_2'}$ with length L_2. This size change is a similarity transformation, so $L_2 = \frac{r_2}{r_1}L_1$. Consequently, $\frac{L_1}{r_1} = \frac{L_2}{r_2}$. ∎

Degree measure is directly related to radian measure by the radian-degree conversion formula

$$\text{measure of angle in radians} = \tfrac{\pi}{180} \cdot (\text{measure of angle in degrees}),$$

because 2π radians $= 360°$.

The formula for the length s of an arc of radius r with a central angle of measure θ in radians is especially simple and elegant. The fraction $\frac{\theta}{2\pi}$ represents how much of the circle's circumference has been traversed. Multiply that by the circumference $2\pi r$ and we arrive at the following theorem.

Theorem 9.2 Let s be the length of an arc of a central angle with measure θ radians in a circle with radius r. Then $s = r\theta$.

If the central angle θ is measured in degrees rather than radians, we can still compute the length s of the arc but the formula is not as elegant. Using the radian-degree conversion formula, we obtain this corollary.

Corollary: Let s be the length of an arc of a central angle with measure θ degrees in a circle of radius r. Then $s = \frac{r\pi\theta}{180}$.

In many applications, either Theorem 9.2 or its corollary can be used, as we choose. However, the simplicity of the formula in radians leads to a corresponding simplicity in analytic formulas for the trigonometric functions. For example, the following familiar formulas from calculus are valid only if θ is measured in radians.

$$\frac{d}{d\theta}(\sin\theta) = \cos\theta \qquad \frac{d}{d\theta}(\cos\theta) = -\sin\theta$$

$$\sin\theta = \theta - \frac{\theta^3}{3!} + \frac{\theta^5}{5!} - \cdots + \frac{(-1)^k\theta^{2k+1}}{(2k+1)!} + \cdots$$

$$\cos\theta = 1 - \frac{\theta^2}{2!} + \frac{\theta^4}{4!} - \cdots + \frac{(-1)^k\theta^{2k}}{(2k)!} + \cdots$$

These formulas are more complicated if θ is measured in degrees (see Problem 8).

Other angle measure systems

Although radian and degree measure are the most commonly used systems for angle measure, other units are useful in certain contexts. The following are illustrations of such systems.

In **grad measure**, which is used in some engineering contexts,

$$1 \text{ grad} = \frac{1}{400} \text{ of a revolution.}$$

Grad measure is better suited to the decimal system than degree measure because a right angle has a measure of 100 grads, a straight angle is 200 grads and a full revolution is 400 grads. The formulas

$$\text{measure of angle in radians} = \frac{\pi}{200} \cdot (\text{measure of angle in grads})$$

$$\text{measure of angle in degrees} = \frac{9}{10} \cdot (\text{measure of angle in grads})$$

relate grad measure to degree and radian measure.

In **mil measure**, which was developed for use in some military contexts,

$$1 \text{ mil} = \frac{1}{6400} \text{ of a revolution.}$$

The apparently odd choice of the fraction $\frac{1}{6400}$ is explained by the calculation

$$\frac{2\pi}{6400} = .00098175 \approx \frac{1}{1000}.$$

Thus, an angle of 1 mil subtends an arc of length very nearly 1 yard at a distance of 1000 yards. This approximation makes it possible to use quick mental arithmetic to calculate the distance to objects whose size is known.

9.1.1 Problems

1. What is the radius of a circle for which the radian measure of an arc equals its length?

2. What is the length of a $212°$ arc in a circle of radius 14?

3. In a circle of radius 2, suppose a chord $\overline{AB}$ has length 1.

a. What are the degree measures of the major and minor arcs $\overset{\frown}{AB}$ in that circle?

b. Generalize part **a**.

4. Unique among the measure systems mentioned in this section, the degree contains subunits of minutes and seconds that are often used in contexts such as latitude and longitude and surveying. Specifically, $1° = 60$ minutes, denoted $60'$, and 1 minute $= 60$ seconds, denoted $60''$.

a. Convert $37°6'52''$ to a decimal number of degrees.

b. Convert $37°6'52''$ to radians.

c. Find $\sin^{-1} 0.4$ to the nearest second.

5. Consider a watch with an hour hand, a minute hand, and a second hand. In a time interval of x seconds, what is the measure of the arc traversed by each of these hands? Give an answer in degrees and an answer in radians.

6. Find the grad and mil measures of the angles with degree measure $30°$, $45°$, and $60°$.

7. An observer sees a tank in the distance and measures the angle of elevation from ground level to the top of the tank to be 2 mils. If he knows that this type of tank is 9 ft tall, approximately how far away is the tank?

8. The following three calculus formulas require the measure of the angle θ to be in radians.

i. $\frac{d}{d\theta}(\sin\theta) = \cos\theta$

ii. $\lim_{\theta\to 0} \frac{\sin\theta}{\theta} = 1$

iii. $\sin\theta = \theta - \frac{\theta^3}{3!} + \frac{\theta^5}{5!} - \cdots + \frac{(-1)^k\theta^{2k+1}}{(2k+1)!} + \cdots$

a. Find the corresponding formulas for degree measure.

b. Find the corresponding formulas for grad measure.

9. The radian measure of an angle θ is defined in a calculus book[1] as follows: Given an angle $\angle ACB$, and a circle of radius r centered at the vertex C of the angle, the radian measure of $\angle ACB$ is $\frac{2S_r}{r^2}$, where S_r is the area of the sector with central angle $\angle ACB$ on the given circle. Explain why this definition is equivalent to the definition given in this section.

10. Prove the converse to Theorem 9.1.

9.1.2 The trigonometric ratios

The trigonometric ratios for the sine, cosine, and tangent are so important that they are introduced in textbooks at grade levels from before high school to college. Before the appearance of hand-held calculators, tables of their values from $0°$ to $90°$ appeared as appendices in the backs of these textbooks.

Trigonometric ratios of acute angles

Recall that these ratios are defined for acute angles as follows: Let ΔABC be a triangle with right angle at C, as shown in Figure 2.

[1] Apostol, Tom. *Calculus*, Volume 1 (New York, NY: Blaisdell, 1964).

Figure 2

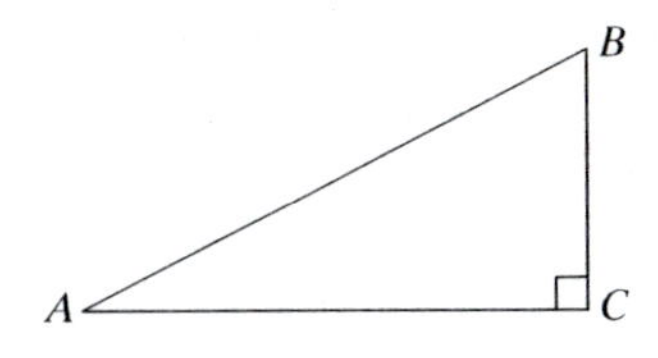

The **sine of the acute angle A**, abbreviated **sin A**, is defined to be

$$\sin A = \frac{BC}{AB} = \frac{\text{length of opposite side}}{\text{length of hypotenuse}}.$$

Similarly, the **cosine of the acute angle A**, abbreviated **cos A**, is defined as

$$\cos A = \frac{AC}{AB} = \frac{\text{length of adjacent side}}{\text{length of hypotenuse}}.$$

The remaining four trigonometric ratios of different sides in a right triangle, the **tangent**, the **cotangent**, the **secant**, and the **cosecant** of angle A are abbreviated and defined for the ΔABC as

$$\tan A = \frac{BC}{AC}, \qquad \cot A = \frac{AC}{BC}, \qquad \sec A = \frac{AB}{AC}, \qquad \csc A = \frac{AB}{BC}.$$

It is straightforward to verify that the value of any one of these trigonometric ratios for a given ΔABC determines the values of the remaining five trigonometric ratios.

Question: If angle A is acute and $\sin A = k$, give the values of the other five trigonometric ratios in terms of k.

From the triangle congruence theorems applied to right triangles, all side lengths and all angle measures for a given right triangle ΔABC are determined if the given information about ΔABC includes either the lengths of two sides (by SAS or SsA, or by the Pythagorean Theorem), or the length of one side and the measure of one of the acute angles (by ASA or AAS). Determining these lengths and angle measures is called **solving the triangle**. The following example is a very simple illustration of this point. More substantial applications of this procedure are discussed in Section 9.1.3.

EXAMPLE 1 Suppose that ΔABC is a right triangle with $\angle C$ as the right angle, and suppose that $AC = 10$ ft and $BC = 13$ ft. Solve the triangle. That is, find the measures of the other sides and angles of the triangle.

Solution By the Pythagorean Theorem,

$$AB = \sqrt{AC^2 + BC^2} = \sqrt{269}\text{ ft} \approx 16.4\text{ ft}.$$

Now any of the trigonometric ratios can be used to determine an angle. We choose to use the tangent.

$$\tan \angle A = \frac{BC}{AC} = 1.3, \quad \text{so} \quad m\angle A \approx \tan^{-1} 1.3 \approx 52.4°.$$

From this, $m\angle B \approx 90° - 52.4° = 37.6°$.

Similarity and the trigonometric ratios

The answers to the question above demonstrate that the value of any one of the six trigonometric ratios determines the values of the other five for a given right triangle. However, to apply these ratios to solve *any* right triangle, we need the following theorem. It lets us use the *measure of the angle* as opposed to the angle itself as the argument for the sine and other trigonometric ratios. Although it is a consequence of the Fundamental Property of Similarity mentioned in Section 8.2.1, it can also be proved without appealing to that result.

Theorem 9.3 **(Right Triangle Similarity Principle):** Let ΔABC be a right triangle with its right angle at C. Then any ratio of two side lengths of ΔABC determines (1) all other ratios of two corresponding side lengths and (2) both of the corresponding acute angles for any triangle $\Delta A'B'C'$ that is similar to ΔABC.

Proof:

(1) We noted above that the value of any one of the six trigonometric ratios for any given triangle determines the values of the other five for that triangle.

(2) Suppose that $\Delta ABC \sim \Delta A'B'C'$. Also suppose that the ratio $\frac{BC}{AC}$ is known. Because these triangles are similar, there is a constant k such that

$$A'C' = k \cdot AC \qquad A'B' = k \cdot AB \qquad B'C' = k \cdot BC.$$

$$\text{Consequently, } \tan A' = \frac{B'C'}{A'C'} = \frac{k \cdot BC}{k \cdot AC} = \frac{BC}{AC} = \tan A.$$

This determines $m\angle A$ and $m\angle A'$, and $\angle B$ and $\angle B'$ are their complements. (If another ratio of sides were given, we would have used a different ratio in place of the tangent ratio.) ∎

Theorem 9.3 enables us to define the sine of the *measure* of the acute angle A to equal the ratio defined above for any angle with that measure.

Now we connect the measures of arcs to the measures of angles. Recall the usual procedure for measuring an angle with a protractor. We place the protractor so that its center point is at the vertex C of the angle $\angle ACB$ and so that the ray $\overrightarrow{CA}$ is along the straightedge of the protractor, as shown in Figure 3.

Figure 3

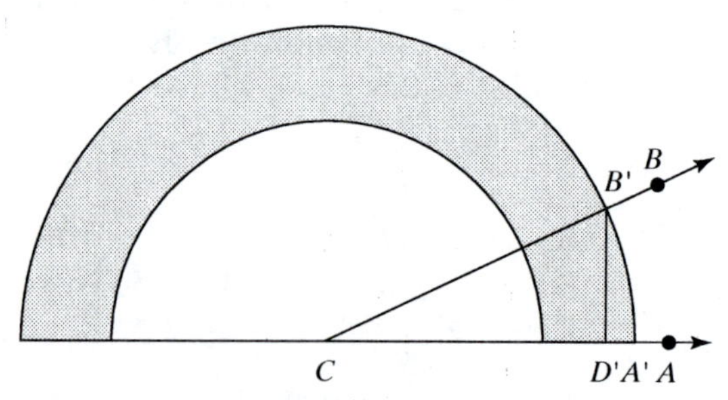

We then read the measure of the angle in degrees from the marks on the arc $A'B'$ on the protractor. Thus, we assign the degree measure to $\angle ACB$ not by a measurement at C, but by a measurement of the length of an arc with center C at some distance from C. Also, note from Figure 3 that the trigonometric ratios for the angle $\angle ACB$ can be computed for the right triangle $\Delta D'CB'$. What we need to show is that the ratio of the length of the leg $\overline{B'D'}$ to the length of the arc $\overset{\frown}{A'B'}$ is independent of the size of our protractor. The following result, which is a consequence of Theorems 8.16 and 8.17, provides one way to obtain the numerical correspondence between acute angles and their measures. It is an analogue to Theorem 9.1.

Theorem 9.4 **(Arc-to-Chord Similarity Principle):** Suppose two arcs have the same measure, and chords with lengths c_1 and c_2. If the arcs have lengths L_1 and L_2, then $\frac{L_1}{c_1} = \frac{L_2}{c_2}$.

Proof: By Theorem 8.16, two arcs have the same measure if and only if they are similar. Suppose arcs $\overset{\frown}{A_1B_1}$ and $\overset{\frown}{A_2B_2}$ are similar in circles of radii r_1 and r_2, respectively, as shown in Figure 4. Then $\frac{r_2}{r_1}$ is the magnitude of the similarity transformation mapping circle O_1 onto circle O_2. So the ratio of the lengths of the chords, $\frac{A_2B_2}{A_1B_1} = \frac{r_2}{r_1}$.

Figure 4

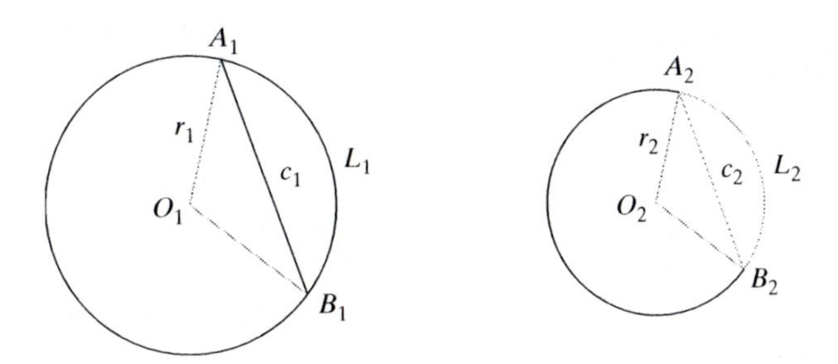

From the definition of arc measure, the central angles O_1 and O_2 of arcs $\overset{\frown}{A_1B_1}$ and $\overset{\frown}{A_2B_2}$ have the same measure. Call that measure θ. By Theorem 9.2, the length of $\overset{\frown}{A_1B_1} = r_1\theta$ and the length of $\overset{\frown}{A_2B_2} = r_2\theta$. Consequently, $\theta = \frac{\text{length of } \overset{\frown}{A_1B_1}}{r_1} = \frac{\text{length of } \overset{\frown}{A_2B_2}}{r_2}$. Now, by SAS Similarity, $\Delta A_1O_1B_1 \sim \Delta A_2O_2B_2$. Thus $\frac{r_1}{r_2} = \frac{c_1}{c_2}$, and the theorem follows. ❑

In the historical evolution of trigonometry, in order to obtain measures of angles and arcs, and lengths of sides of figures, Hipparchus and Ptolemy computed tables of the *chord lengths* for a circle of fixed radius for central angles having measures between 0° and 180°. Hipparchus computed these values for multiples of 7.5°. We do not know the radius of the circle he used. By the time of Ptolemy, the Greeks were using a sexagesimal (based on 60) numeration system, so Ptolemy used a circle with radius 60. He was able to calculate these measures for multiples of 0.5°.[2]

Figure 5

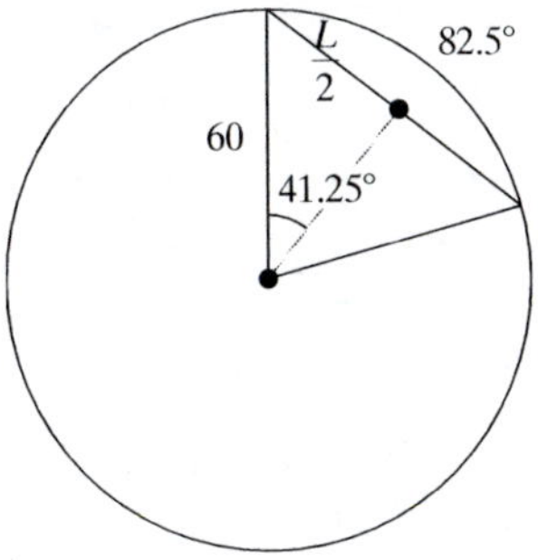

Here is an example of how Ptolemy's table of chords relates to today's values of sines. Suppose an arc has measure 82.5° in a circle of radius 60 and we want to know the length L of its chord. Then the triangle formed by a radius of the circle to an endpoint of the arc and to the midpoint of the chord of that arc (Figure 5) has a central angle with measure 41.25°. So $\frac{\frac{L}{2}}{60} = \sin 41.25°$. From this we see that $L = 120 \sin 41.25°$. In general, the chord lengths in these ancient tables for a given arc measure are 120 times the values of sines of half that arc measure.

Trigonometric ratios of obtuse angles

So far, we have defined the trigonometric ratios for the acute angles of a right triangle. To solve all triangles, trigonometric ratios of obtuse angles are also needed.

The usual method for making the transition for trigonometric ratios from acute to obtuse angles is to begin by placing the acute angle A on rectangular coordinate axes, as shown in Figure 6a. In this position, we see that $\cos A = \frac{x}{r}$ and $\sin A = \frac{y}{r}$, where x, y, and r are the lengths of sides of the right triangle. Also, most significantly, x and y are the coordinates of the point B.

Figure 6

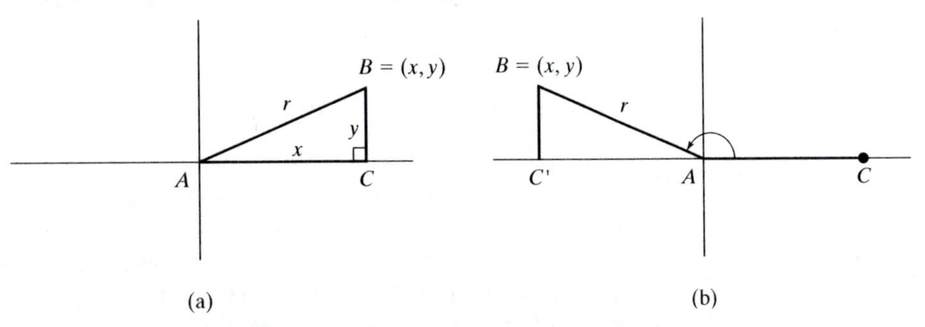

[2] See Asger Aaboe, *Episodes from the Early History of Mathematics*, p. 112ff.

Next we consider an obtuse angle with one side in common with the acute angle (Figure 6b) and the other side and point B now in the second quadrant. Here again $\angle A = \angle CAB$. Now we *define* $\sin A = \frac{y}{r}$ and $\cos A = \frac{x}{r}$. This definition is clearly consistent with the definition of sine and cosine for acute angles. Notice, however, that the leg $\overline{AC'}$ of the right triangle $\Delta ABC'$ in Figure 6b has length $-x$ (since x is negative). Sometimes this length is thought of as a directed length so that it can be negative. Consequently, for obtuse angles, the cosine ratio is negative while the sine ratio is positive.

The remaining four trigonometric ratios are now defined for both acute and obtuse angles by

$$\tan A = \frac{y}{x}, \quad \cot A = \frac{x}{y}, \quad \sec A = \frac{r}{x}, \quad \text{and} \quad \csc A = \frac{r}{y}.$$

Because x is negative while y and r are positive for an obtuse angle A, the ratios $\tan A$, $\cot A$, and $\sec A$ are negative while $\csc A$ is positive.

To obtain $\sin A$ and $\cos A$ when A is a right angle, we can adapt either Figure 6a or Figure 6b to the situation where $x = 0$ and $y > 0$. Then B is on the positive ray of the y-axis, and $\angle CAB$ is a right angle. Since $y = r$, $\sin A = 1$, $\cos A = 0$, $\tan A$ is undefined, and so on.

The solution of oblique triangles

In Section 9.1.3, we discuss several examples in which we solve triangles that are not right triangles. The important tools in these solutions are the trigonometric ratios and two theorems, the *Law of Cosines* and the *Law of Sines*.

Theorem 9.5 **(Law of Cosines):** For any ΔABC,

$$c^2 = a^2 + b^2 - 2ab\cos C.$$

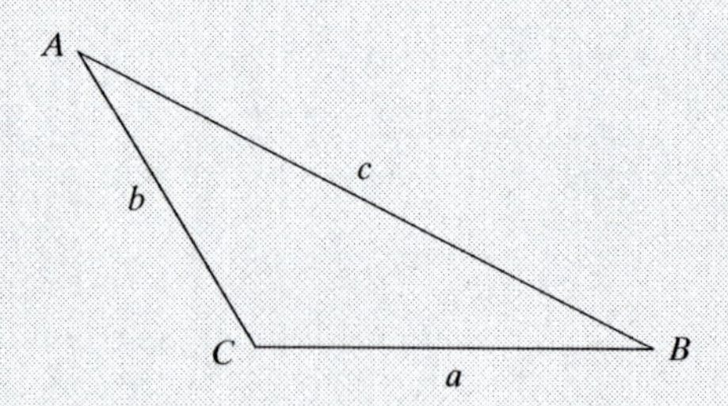

Proof: See Problem 6. ∎

Theorem 9.6 **(Law of Sines):** For any ΔABC,

$$\frac{\sin A}{a} = \frac{\sin B}{b} = \frac{\sin C}{c}.$$

Proof: See Problem 7. ∎

The Law of Cosines enables the third side of a triangle to be determined given two sides and an included angle. In this regard, it generalizes the Pythagorean Theorem to all triangles. It also makes it possible to determine any angle of a triangle given the three sides. Thus it enables a triangle to be solved given SAS or SSS. By applying the Law of Sines a second side of a triangle can be found given two angles and a side, and also a second angle can be found given two sides and an angle. So it enables a triangle to be solved given AAS, ASA, or SSA.

Care must be exercised in the last case, since triangles are not necessarily congruent with SSA. For example, in Figure 7, if the angle A and the sides c and a are given and if a is less than c, then the angle C as well as the side length $b = AC$ are not uniquely determined. This is usually called the *ambiguous case for SSA*.

Figure 7

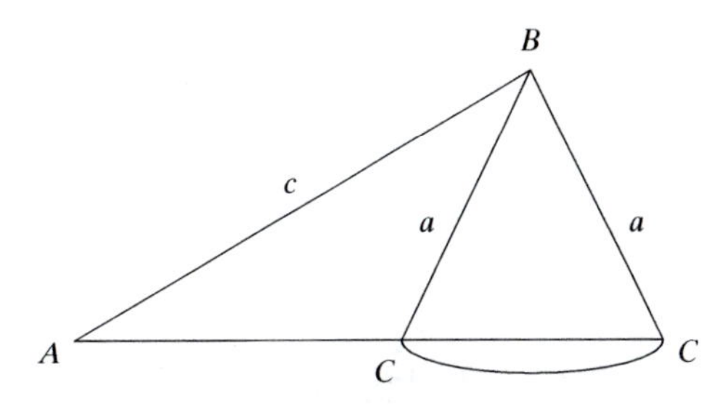

However, if $a \geq c$, then angles C and B are uniquely determined. This is the case of SsA congruence discussed in Section 7.4.1.

9.1.2 Problems

1. Show that the Pythagorean Theorem is a special case of the Law of Cosines.

2. Prove: In ΔABC, angle C is obtuse if and only if $a^2 + b^2 < c^2$.

3. a. Explain why the constant K in the Law of Sines

$$K = \frac{a}{\sin A} = \frac{b}{\sin B} = \frac{c}{\sin C}.$$

is the diameter of the circumcircle of ΔABC. (*Hint:* If S is the circumscribed circle of the triangle ΔABC, construct a second triangle $\Delta A'B'C'$ in which $A = A'$, $\overline{AB'}$ is a diameter of S, and $\angle B'AC' \cong \angle BAC$.)

b. Use the result of part **a** to conclude that if ΔABC is inscribed in a circle of diameter 1, then $a = \sin A$, $b = \sin B$, and $c = \sin C$.

4. Explain why the following is an equivalent definition of the sine of an angle: The sine of an angle between 0° and 180° is equal to the length of the chord of the arc subtended by inscribing the angle ($\angle ACB$ in Figure 8) in a circle of diameter 1. (This definition is very closely related to the definition Hipparchus and Ptolemy used to construct the first sine tables.)

Figure 8

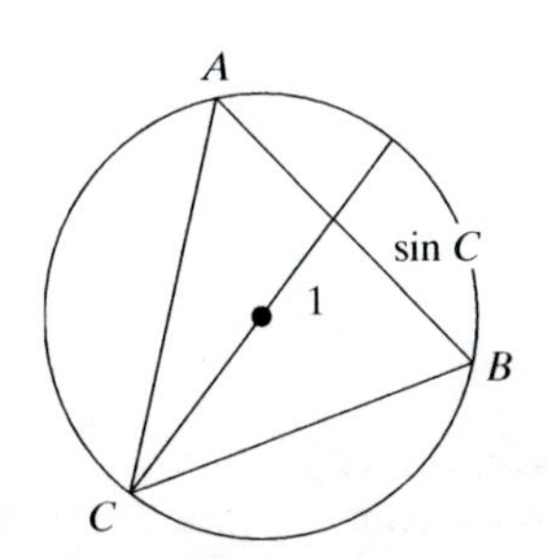

5. Show that Ptolemy's procedure gives tables like sine tables. Let ΔABC be a right triangle inscribed in a circle with diameter $AB = 120$. Prove: If L_{2A} is the length of the chord (from his table of chords) corresponding to the central angle with measure $2A$, then $\sin A = \frac{L_{2A}}{120}$.

6. Prove the Law of Cosines using the following procedure. Place the triangle ΔABC from Figure 9 so that C is at $(0, 0)$ and side $\overline{CB}$ lies along the positive x-axis. Identify the coordinates of the vertices A and B in terms of the side lengths a, b and the sine and cosine of C. Then apply the distance formula to compute the square c^2 of the length of the side $\overline{AB}$ and simplify the result.

Figure 9

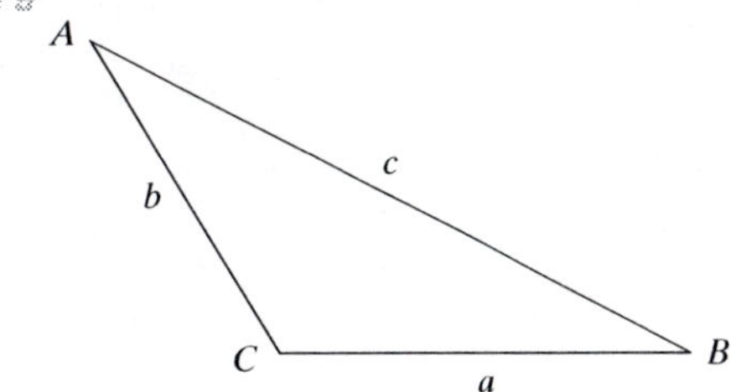

7. a. Use Figure 10. Prove: $\frac{\sin A}{a} = \frac{\sin B}{b}$.

Figure 10

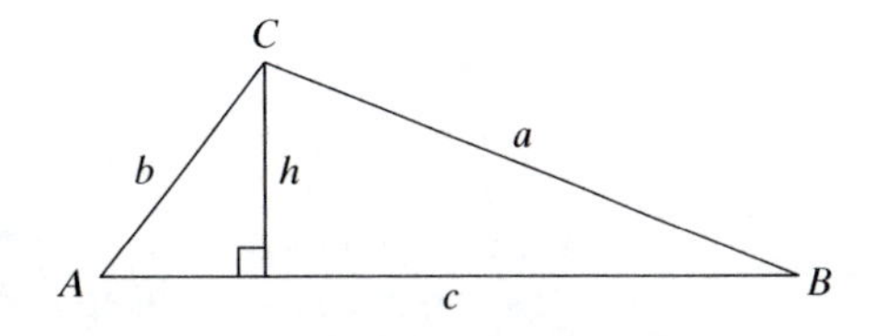

b. Explain how one can go from the result of part **a** to the full statement of the Law of Sines.

8. Suppose sides a and b and $m\angle A$ of ΔABC are given.

a. Prove that there is exactly one triangle satisfying these conditions if $a \geq b$.

b. For what values of $a < b$ is there exactly one triangle satisfying these conditions?

c. When is there no triangle satisfying these conditions?

ANSWER TO QUESTION:

$\csc A = \frac{1}{k}$; $\cos A = \sqrt{1 - k^2}$; $\sec A = \frac{1}{\sqrt{1-k^2}}$; $\tan A = \frac{k}{\sqrt{1-k^2}}$; $\cot A = \frac{\sqrt{1-k^2}}{k}$

9.1.3 Extended analysis: indirect measurement problems

The need to determine distances, angles, and arcs that could not be measured directly led to the study of trigonometry in ancient Egypt and Babylonia. The Rhind Papyrus, a scroll of mathematical problems written in Egypt about four thousand years ago, includes several problems (Problems 56–60) that deal with measurements of angles and distances related to the pyramids. Babylonian astronomers recorded the motions of the planets on the celestial sphere, an imaginary sphere centered at Earth whose surface they believed contained all the stars and planets.

Although these ancient measurements were recorded with meticulous care and detail, they were observational rather than analytical. The development of tables by Hipparchus and Ptolemy and the development of the basic theorems of plane and spherical trigonometry took place between the second century B.C. and the second century A.D. This made it possible to use trigonometry to calculate indirect measurements.

Triangulation

Ferdinand Magellan's circumnavigation of the earth (1519–1522) provided the first direct proof that the earth is roughly spherical. It prompted questions such as whether the earth was actually an oblate sphere flattened at the poles, and stimulated interest in the development of accurate maps of continents and oceans on the earth's surface. As a result, a number of geodetic surveys were conducted based on the *method of triangulation*. This method begins with a baseline segment of known length joining two points A and B on the earth's surface that are a known distance apart. For a given visible distant landmark, A' in Figure 11, $\angle A'AB$ and $\angle A'BA$ are measured, and the Law of Sines is used to compute the distances AA' and BA'. This process can be repeated for another visible landmark B' to compute the distances AB' and BB' as well as the length $A'B'$ of a new baseline $\overline{A'B'}$, and so on.

Figure 11

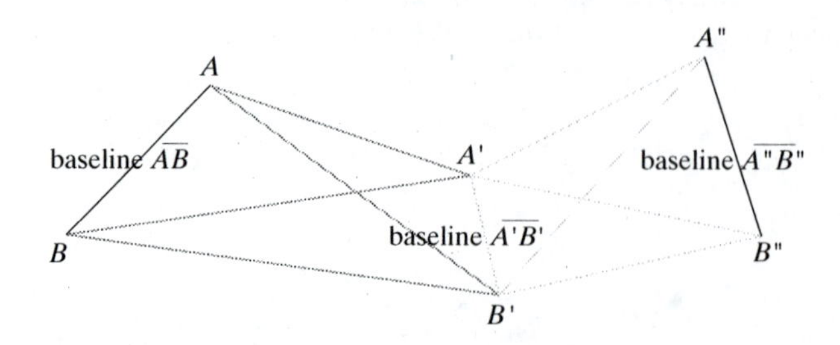

For relatively small regions of the earth's surface, the points in this triangulation grid are assumed to lie in a plane, but for larger regions the tools of spherical trgonometry are needed. Using the triangulation method, Abbé Jean Picard (1620–1682) started with a baseline consisting of a 7-mile stretch of road from Paris to Fontainebleau, and completed a detailed survey of a region extending to a section of the French coastline. This survey was completed for all of France by Giovanni Cassini (1625–1712), an Italian astronomer and cartographer, his son Jacques (1677–1756), his grandson Cesar Francois (1714–1784), and his great grandson Jean Dominique (1748–1848). In the course of these successes in cartography, the Cassinis persuaded King Louis XV to authorize two expeditions to Peru and Lapland that used the triangulation method to determine that the earth is indeed an oblate sphere as Newton had suspected.

Problems in indirect measurement are usually the first types of applications of trigonometry encountered in school. The Pythagorean Theorem, the trigonometric ratios for acute and obtuse angles, and the Law of Sines and Law of Cosines are all that is needed. Consider the following sequence of four related and progressively more complicated indirect measurement problems.

Problem 1: We want to determine the height h of a certain flagpole $\overline{QP}$ (see Figure 12) and we cannot measure its height by climbing it. Instead, we measure $\angle PAQ$, the angle of elevation to the top of the flagpole from a point A on level ground 350 feet from the base of the pole, and find that it is 20°. Find the height h of the flagpole.

Figure 12

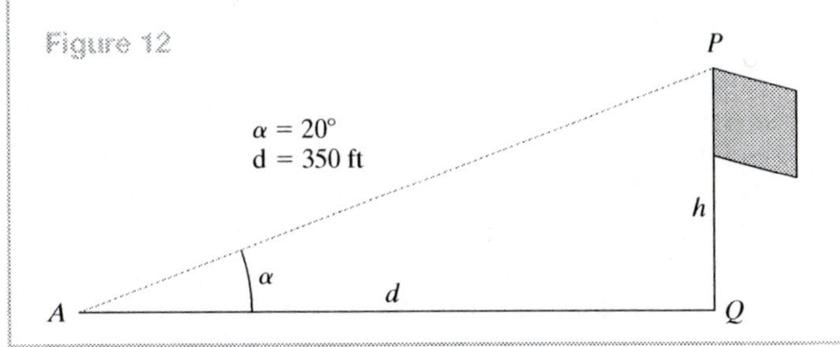

Question 1: Solve Problem 1. Then generalize your solution by solving the problem when the length from the base of the flagpole is d and the angle of elevation is α.

Problem 2: Suppose that the flagpole is located on the other side of a stream so that it is not possible to measure directly the distance from the observation point A to the base Q of the flagpole. Assume that the ground around point A is level and at the same level as the base Q of the flagpole. Move back from A along $\overrightarrow{QA}$ to point B and measure the angle of elevation at B, and the length m of the baseline $\overline{AB}$ (see Figure 13). Find the height h of the flagpole in terms of the distance m, the angle of elevation α at A, and the angle of elevation β at B.

Figure 13

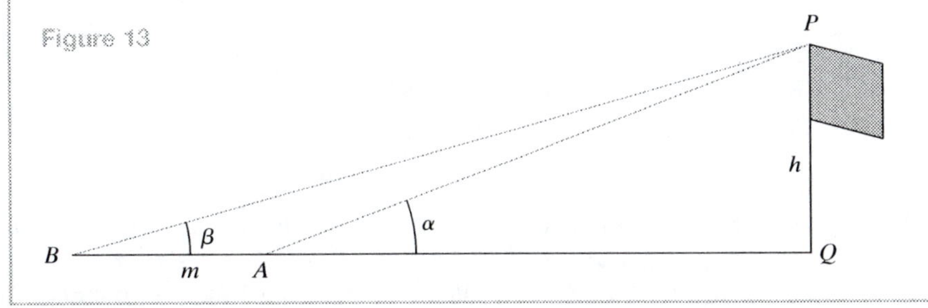

Question 2: Solve Problem 2.

Problem 2 assumes that the measured baseline $\overline{AB}$ is collinear with a line to the base of the flagpole. But in some situations the easiest baseline to measure may be *perpendicular* to a line to the base of the flagpole. Problem 3 addresses this case.

Problem 3: Suppose that the flagpole is located on the other side of a stream. You measure the angle α of elevation of P, the top of the flagpole from a point A. Then you walk along the stream bank to a second point B at a distance b from A along a line perpendicular to the line joining A to the flagpole base Q, and measure the angle $\beta = m\angle ABQ$, as shown in Figure 14. Assume that the ground on your side of the stream is level and at the same level as the base Q of the flagpole. Find the height h of the flagpole.

Figure 14

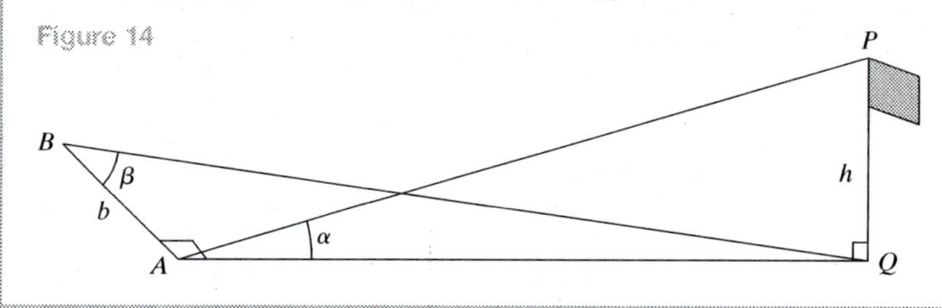

Question 3: Solve Problem 3.

Each of the Problems 1–3 involves finding the height of a flagpole by knowing the length of a baseline and the angle of inclination of the top of the flagpole from one end of that baseline. The following problem generalizes these separate cases. It asks for the height of a flagpole from the length of *any* baseline and the angles to the top and bottom of the flagpole from each end of that baseline. This problem is more challenging than the right triangle Problems 1–3. Its solution requires a multistep analysis of more than one oblique triangle, and consequently involves the Law of Sines and the Law of Cosines.

Problem 4: Suppose that there is a baseline $\overline{AB}$ of known length b, but that the baseline is not necessarily in the plane perpendicular to the flagpole at its base. The points P, Q, A, and B are all visible from one another, but do not lie in the same plane (Figure 15). Find a formula for the height h of the flagpole in terms of angles measured at the observation points A and B and the length b of the baseline.

Figure 15

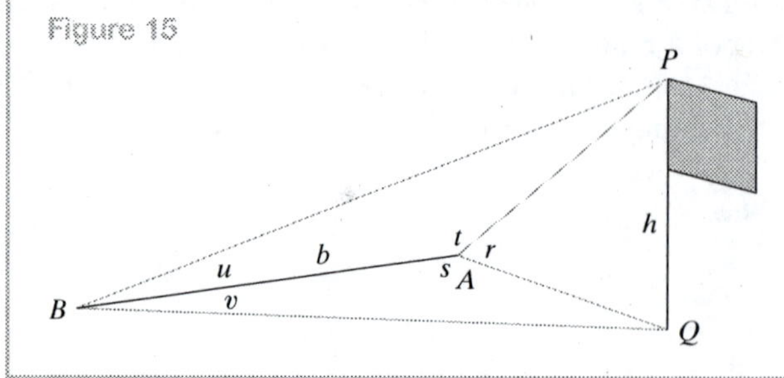

Solution For simplicity, we identify the various angles that can be measured at the observation points A and B. Let $u = m\angle ABP$, $t = m\angle PAB$, $v = m\angle ABQ$, $s = m\angle QAB$, and $r = m\angle PAQ$. Using the Law of Sines on ΔAPB and ΔAQB allows us to find AP and AQ. Then using the Law of Cosines on ΔPAQ allows us to find PQ. The details are left to you as Problem 4a.

9.1.3 Problems

1. a. Generalize Problem 1 in this section to a situation where a (vertical) flagpole is on a hill: Find the height h in terms of the distance d to the base of the pole and the angles α and γ (Figure 16), where γ is the angle between $\overrightarrow{AP}$ and the vertical.

Figure 16

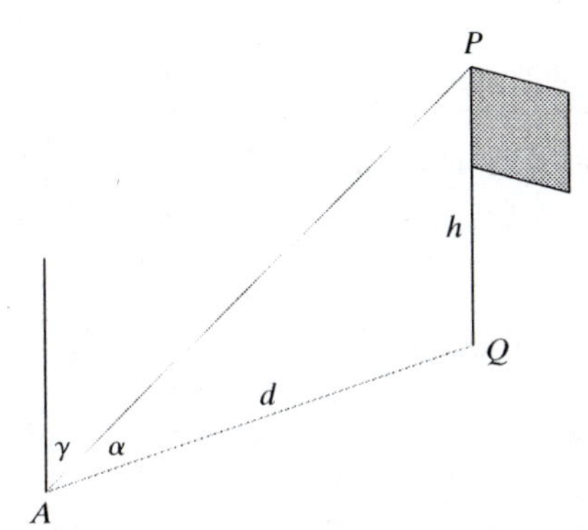

b. Show that the formula you found for h in part **a** holds also for the case of a flagpole down in a valley $\left(\gamma > \frac{\pi}{2}\right)$, and in fact for any case where $\alpha + \gamma < \pi$.

c. Derive the result of Question 1 as a special case of this formula.

2. Generalize Problem 2 in this section to a situation where the base of the flagpole is on a hill above the level with the observation point A, and the line segment BAQ makes an angle ε with the horizontal.

3. Generalize Problem 3 in this section to a situation where $\angle BAQ$ and $\angle AQP$ are not necessarily right angles. Specifically, assume that $\beta = m\angle ABQ$, $\gamma = m\angle BAQ$, and $\varepsilon = m\angle AQP$ are known, as well as the length $b = AB$.

4. a. Complete the derivation of the formula for the height of the flagpole in Problem 4 in this section.

b. The measure $w = m\angle PBQ$ is not needed to solve Problem 4, even though this angle could be measured from point B. (Note that this angle is *not* in general the sum of angles $u = m\angle ABP$ and $v = m\angle ABQ$, since A might not lie in the plane of triangle PBQ.) Show that w is in fact determined from the values r, s, t, u, v, and b.

5. Suppose we want to know the height h of a flagpole on a hill that we can see from a building. We measure the angle of elevation to the bottom from a low floor and an angle of depression to the top from a higher floor of the building, and measure the baseline distance b between these two floors, as shown in Figure 17. Is this information sufficient to determine the height of the flagpole or how far away it is? If it is, find these. If not, why not?

6. Consider the triangle with sides 13, 14, and 15.

a. Find the measures of the three angles in the triangle using the Law of Cosines.

Figure 17

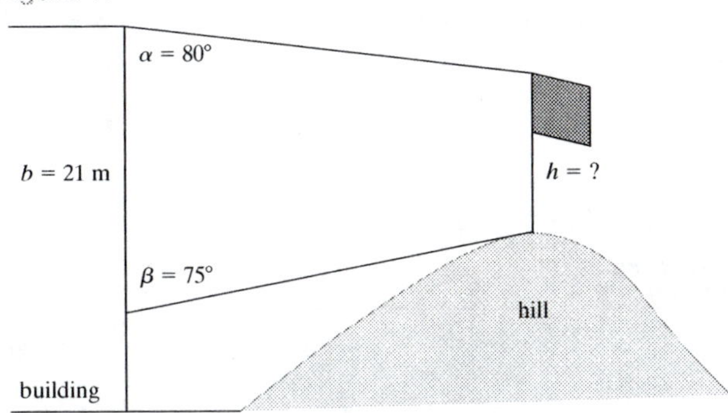

b. Find the sine of the largest angle in this triangle and use the Law of Sines to find the measures of the other angles.

7. **Belt problems.** Suppose that a belt is stretched tightly over two pulleys P_1 and P_2 of radii r_1 and r_2 and whose centers are d units apart, as in Figure 18.

Figure 18

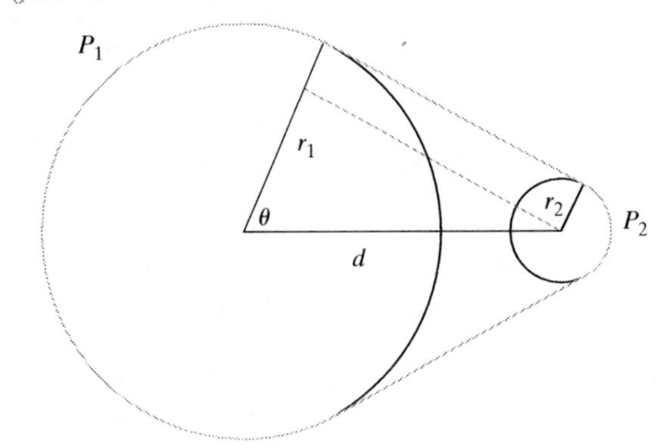

Find a formula for the total length L of the belt in terms of d, r_1, and r_2 in each case.

a. $d = 6$, and $r_1 = r_2 = 3$

b. $r_1 = r_2$ and $d = r_1 + r_2$

c. $r_1 = 12$, $r_2 = 3$, and $d = 18$

d. $r_1 > r_2$, and $d > r_1 + r_2$

8. **A Law of Cosines for Quadrilaterals:** If we know two sides and the included angle of a triangle, then SAS triangle congruence tells us that the triangle is fully determined and the Law of Cosines tells us how to find the third side.

a. Prove that if we know *three* sides and the *two* included angles of a plane quadrilateral, then the quadrilateral is fully determined, and hence that there is a SASAS quadrilateral congruence theorem.

b. Find a formula for the fourth side of a quadrilateral in terms of the other sides a, b, and c and the included angles A and B.

ANSWERS TO QUESTIONS:

1. For the particular case given, $\tan(20°) = \frac{h}{350\text{ ft}} \Rightarrow h = 350\tan(20°) \approx 127.40$ ft. For the general case, the height h is a function of d and α given by $h(d, \alpha) = d\tan(\alpha)$, where $0 < \alpha < \frac{\pi}{2}$ and $d > 0$.

2. Let $d = AQ$. Then $\frac{h}{d} = \tan\alpha$ and $\frac{h}{d+m} = \tan\beta$, from which $h = \frac{m\tan\alpha\tan\beta}{\tan\alpha - \tan\beta}$.

3. Let $d = AQ$. Then, from the diagram, $d = b\tan\beta$ and $h = d\tan\alpha$, so $h = b\tan\alpha\tan\beta$.

Unit 9.2 The Trigonometric Functions and Their Connections

In Unit 9.1, we defined the six trigonometric ratios, sine, cosine, tangent, cosecant, secant, and cotangent for all angles with measures between 0 and π radians (or 0° and 180°). As the problems of Section 9.1.3 demonstrate, these ratios together with the Laws of Sines and Cosines are powerful tools for solving a variety of indirect measurement problems in plane trigonometry. Another class of applications of plane trigonometry involves the analysis and modeling of periodic phenomena. For this class of applications, the trigonometric ratios are useful but not sufficient. These problems require that we extend the six trigonometric ratios to define functions for all real arguments. These real functions and their complex counterparts have a variety of remarkable algebraic, geometric, and analytic properties that are powerful tools for representing and analyzing periodic phenomena. We define these functions in Section 9.2.1 and discuss some of their applications in Section 9.2.2 and Unit 9.3.

9.2.1 The trigonometric functions

Some sources distinguish between the *trigonometric functions* and the *circular functions*, reserving the former term for functions of angles (or of their measures) with the domains we discussed in Unit 9.1. The **circular functions** refer to the sine, cosine, and other derived functions defined over all the real numbers for which they have meaning. However, in advanced mathematics, these (circular) functions and their extensions that include complex number domains are called the **trigonometric functions** and that is the wording we use here.

Angles, directed angles, and their measures

Recall that an *angle* in a plane is defined as the union of two rays, its *sides*, with a common initial point, its *vertex*.

In contrast, a **directed angle** is an *ordered pair* of rays with the same endpoint (its **vertex**). One side of the directed angle is prescribed as the **initial side**, the other side as the **terminal side**. The angle notation $\angle ACB$ is also used for directed angles, sometimes with the understanding that ray $\overrightarrow{CA}$ is the initial side and $\overrightarrow{CB}$ is the terminal side. Some authors use a special symbol (e.g., ∡) to distinguish angles from directed angles, but we do not.

Angles and directed angles differ significantly in the measures allowed for them. We have discussed angle measure in some detail in Section 8.2.5, so we turn now to directed angles. Whereas an (undirected) angle is usually considered to have a unique measure (and has two measures only when reflex angles are being

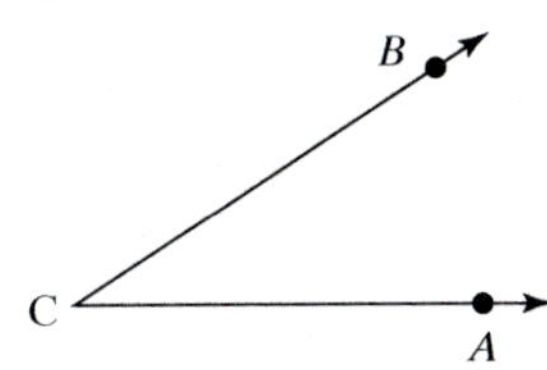

Figure 19

considered), a directed angle has infinitely many measures. These measures are the possible magnitudes of the rotations mapping the initial side of the angle onto its terminal side. If the rotation is counterclockwise, the directed angle is **positive**. If the rotation is clockwise, the directed angle is **negative**, and if the initial and terminal sides coincide, then the directed angle is a **zero angle**. Thus if an angle formed by two rays has undirected measure m, then a directed angle formed by the same rays may have any of the measures $(m \pm 360n)^\circ$ or $(-m \pm 360n)^\circ$ for any integer n, depending on which side of the directed angle is chosen to be the initial side. For example, if $m\angle ACB = 30^\circ$ in Figure 19, then the directed $\angle ACB$ has possible magnitude of $(30 \pm 360)^\circ$, while the directed angle $\angle BCA$ has possible magnitudes of $(-30 \pm 360)^\circ$.

The motivation behind the multiple measures for a directed angle is to provide a means of indicating turns that are more than one revolution, and to indicate the direction of such turns. Another way to describe the magnitude and direction of such turns is to consider *directed arcs*.

Recall that a *central angle* of a circle K is an angle whose vertex is at the center of that circle. If $\angle ACB$ is a central angle of circle K, and A and B are points on the circle, then A, B, and the points on arc $\overset{\frown}{AB}$ in the interior of the angle constitute the *arc subtended by* $\angle ACB$. Similarly, a **directed central angle** of a circle K is a directed $\angle ACB$ whose vertex is at the center of that circle. If A and B are on the circle, we think of the circle not only as containing the path of a point from A to B as the initial side of $\angle ACB$ is rotated to its terminal side, but we allow the path to wind around the circle more than once. This path on K traced by the point A as it is rotated to the point B by the rotation corresponding to directed $\angle ACB$ is called the **directed arc** $\overset{\frown}{\mathbf{AB}}$ **with central angle** $\angle \mathbf{ACB}$. For example, in the coordinate plane $\mathbf{R}^2$, if $C = (0,0)$, $A = (2,0)$, and $B = (0,2)$, and if the magnitude of the directed angle $\angle ACB$ is 450°, then the directed arc $\overset{\frown}{AB}$ is the path that starts at $(2,0)$ and proceeds around the circle of radius 2 centered at the origin for 1.25 revolutions counterclockwise and ends at the point $(0,2)$. The length of the directed arc $\overset{\frown}{AB}$ is 1.25 times the circumference of the circle, so it is $1.25 \cdot 2\pi \cdot 2 = 5\pi$.

For a given initial side of a directed central angle in a given circle, radian measure sets up a one-to-one correspondence between the set of directed angles and the set $\mathbf{R}$ of real numbers. Degree measure for directed angles is directly related to radian measure by the radian-degree conversion formula

$$\text{measure of directed angle in radians} = \tfrac{\pi}{180} \cdot (\text{measure of directed angle in degrees}),$$
$$\text{because } 2\pi \text{ radians} = 360^\circ.$$

The unit circle and wrapping function definitions of the sine and cosine functions

We begin our discussion of the trigonometric functions by recalling two ways in which they are often defined—with the unit circle and the wrapping function.

Let t be a real number, and let $\angle AOP$ be the directed angle of radian measure t with its vertex at the origin in the xy-plane, its initial side on the positive x-axis, and its terminal side intersecting the unit circle $x^2 + y^2 = 1$ at the point $P = (x, y)$. Then we *define* $\cos t = x$ and $\sin t = y$, as shown in Figure 20. Another way of conceptualizing this definition is to realize that the directed angle is itself measured by the magnitude of a rotation mapping the initial side onto its terminal side. So an equivalent definition is to define $\cos t$ and $\sin t$ to be the first and second coordinates of the image of $(1, 0)$ under a rotation of t radians.

Figure 20

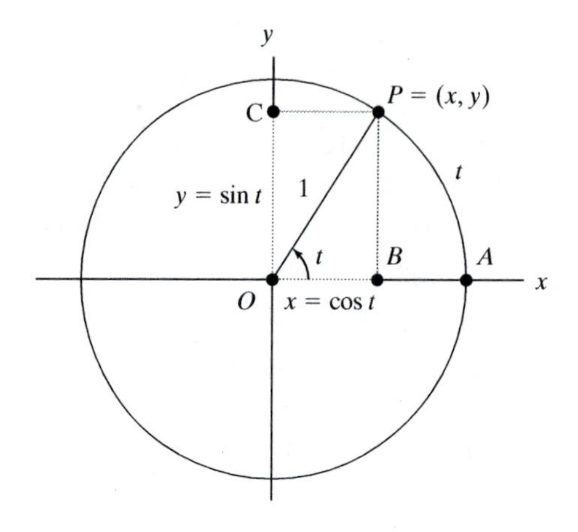

A major advantage of using the unit circle is that, when $-\pi < m < AOP \leq \pi$, due to Theorem 9.2, $m\angle AOP$ in radians is numerically equal to the length of the arc of the circle subtended by $\angle AOP$. This way of defining the trigonometric functions is usually referred to as the **unit circle definition.**

Figure 21

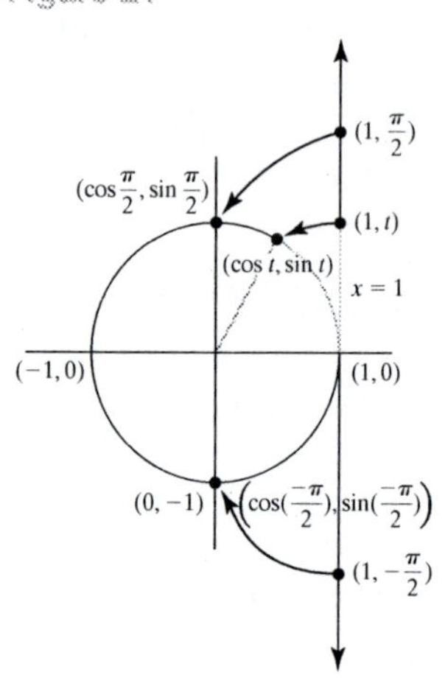

In the late 1940s, a **wrapping function definition** of the trigonometric functions, was proposed. It is also based on the unit circle but in an interestingly different way.

The wrapping function definition can be described dynamically as follows: Begin with a unit circle and the line $x = 1$, as shown in Figure 21. Imagine wrapping this line around the circle in two directions. The ray above the x-axis is wrapped counterclockwise around the circle. Many points on the line map onto the same point of the circle. To find the image of a point on the line, we match the distance of this point above the x-axis with the length of an arc of the unit circle as measured from counterclockwise $(1, 0)$. For instance, the point $\left(1, \frac{\pi}{2}\right)$ on the line is $\frac{\pi}{2}$ above the x-axis. Since the circumference of the unit circle is 2π, $\frac{\pi}{2}$ is a quarter of the circumference. So its image will be a quarter-way around the circle, at the point $(0, 1)$. In analogous fashion, the image of $\left(1, \frac{\pi}{4}\right)$ on the line is the point $\left(\frac{\sqrt{2}}{2}, \frac{\sqrt{2}}{2}\right)$. The ray below the x-axis is mapped clockwise around the circle. So the image of $\left(1, -\frac{\pi}{2}\right)$ on the line is the point $(0, -1)$ on the circle. In general, the images of all points of the form $(1, k + 2n\pi)$, where n is an integer, map onto the same point of the circle.

With this conception, $\cos x$ and $\sin x$ are defined as the first and second coordinates of the image of $(1, x)$. This definition is far removed from triangles, but it has some advantages. It gets us straight to thinking in terms of radians rather than degrees. The periodicity of the sine and cosine functions is an immediate consequence (it also is an immediate consequence of the definition in terms of arcs and rotations). And, by identifying the real number x with the length x, the trigonometric functions are seen as functions of real numbers, enabling them to be treated as real functions.

There are, however, a number of pedagogical disadvantages of the wrapping function approach. The lack of connection with triangle trigonometry leads some students to think these are different sines and cosines than those they have seen before. The lack of experience with arc length makes it difficult for some students to identify values of the functions. And the lack of experience with the idea of wrapping makes the definition difficult to apply.

Both the unit circle and wrapping function definitions clearly establish the sine and cosine functions as real functions that are periodic with a period of 2π. By way of contrast, the sine and cosine trigonometric ratios are defined only for acute and obtuse angles.

The unit circle and wrapping function definitions of the sine and cosine for $0 < t < \frac{\pi}{2}$ and $\frac{\pi}{2} < t < \pi$ can be shown to be consistent with the ratio definitions of these functions, as follows: Refer to Figure 20. There $t = m\angle POA$. If $0 < t < \frac{\pi}{2}$, then from the ratio definitions $\sin t = \frac{PB}{OP} = PB = y$ and $\cos t = \frac{OB}{OP} = OB = x$. For obtuse angles, $\sin t$ and $\cos t$ are defined as the ratios $\frac{y}{r}$ and $\frac{x}{r}$, respectively (see Section 9.1.1). Here $r = 1$, so again we have $\sin t = y$ and $\cos t = x$ as in both the unit circle and wrapping function definitions.

The familiar graphs (Figure 22) of the sine and cosine functions—*sine waves* of infinite extent—are only possible when the domain of these functions is the set **R**.

Figure 22

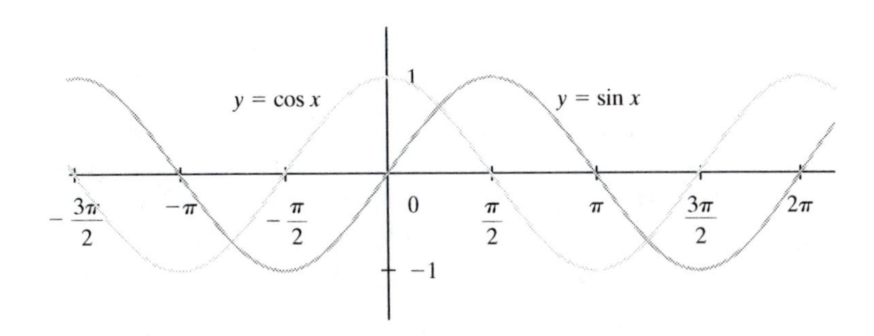

Once $\cos t$ and $\sin t$ have been defined for all real numbers t, the other four trigonometric functions are defined either as their reciprocals or their quotients.

$$\tan t = \frac{\sin t}{\cos t}, \quad \cot t = \frac{\cos t}{\sin t}, \quad \sec t = \frac{1}{\cos t}, \quad \text{and} \quad \csc t = \frac{1}{\sin t},$$

for all real numbers t for which the denominators are not zero. As an immediate consequence of being the reciprocals of the sine and cosine functions, the secant and cosecant functions are periodic of period 2π. The tangent and cotangent functions are also periodic, but their period is π (see Problem 1).

Obtaining values of the trigonometric functions

We could obtain the values of trigonometric functions from the values of the trigonometric ratios in the interval $\left[0, \frac{\pi}{2}\right]$. But in practice, this is difficult to do. We would have to obtain certain values of the functions $\left(\text{those for } \frac{\pi}{2}, \frac{\pi}{3}, \frac{\pi}{4}, \text{ and } \frac{\pi}{6}\right)$ from theorems of geometry, employ a variety of trigonometric identities to obtain values for the sums, differences, and multiples of these numbers, and interpolate between values to determine other values to the desired accuracy. This is how Ptolemy constructed his tables of lengths of chords, which preceded later tables of sines. Problem 6 in Section 9.3.1 illustrates how trigonometric identities can be used to obtain exact values of $\sin \frac{\pi}{2^k}$ and $\cos \frac{\pi}{2^k}$ for any integer $k \geq 2$, values which are critical in this process.

An easier way to obtain all values to any desired accuracy is by the representation of these functions as power series by using the Taylor series expansion. Recall from calculus that, for a given real function f with derivatives of all orders on its domain, the Taylor series centered at point a in the domain of f is

$$\begin{aligned} f(x) &\approx \sum_{n=0}^{\infty} \frac{f^{(n)}(a)}{n!}(x-a)^n \\ &= f(a) + f^{(1)}(a)(x-a) + \frac{f^{(2)}(a)}{2!}(x-a)^2 + \cdots + \frac{f^{(n)}(a)}{n!}(x-a)^n + \cdots. \end{aligned}$$

When $a = 0$, the Taylor series is called the **Maclaurin series for f**.

From calculus, the Maclaurin series expansion of many functions can be computed rather easily. Principal among these are the Maclaurin series for the cosine and sine functions and for the exponential function defined by $\exp(x) = e^x$.

$$\cos(x) = 1 - \frac{x^2}{2!} + \cdots + (-1)^n \frac{x^{2n}}{(2n)!} + \cdots = \sum_{n=0}^{\infty} (-1)^n \frac{x^{2n}}{(2n)!}$$

$$\sin(x) = x - \frac{x^3}{3!} + \cdots + (-1)^n \frac{x^{2n+1}}{(2n+1)!} + \cdots = \sum_{n=0}^{\infty} (-1)^n \frac{x^{2n+1}}{(2n+1)!}$$

$$\exp(x) = e^x = 1 + x + \frac{x^2}{2!} + \frac{x^3}{3!} + \cdots + \frac{x^n}{n!} + \cdots = \sum_{n=0}^{\infty} \frac{x^n}{n!}$$

By using a convergence test such as the Ratio Test or some other convergence test, it can be shown that all three of these series converge for all real x. In fact, these series converge to their corresponding function values for all real x.

In fact, these series *converge absolutely* for all real x. Recall that a power series $\sum_{n=0}^{\infty} a_n x^n$ converges absolutely if the corresponding series $\sum_{n=0}^{\infty} |a_n||x|^n$ converges. Also recall that if a power series is absolutely convergent, it is necessarily convergent, but that the converse is not true in general. For example, the power series $\sum_{n=1}^{\infty} \frac{(-1)^n}{n} x^n$ converges when $x = 1$ (it becomes an alternating series with terms decreasing to 0) but it does not converge absolutely because the series $\sum_{n=1}^{\infty} \frac{1}{n} = 1 + \frac{1}{2} + \frac{1}{3} + \cdots + \frac{1}{n} + \cdots$, the harmonic series, diverges).

Extending the domain of the trigonometric functions to include complex numbers

It is unfortunate that nearly all modern calculus texts fail to include any discussion of complex numbers. There are many fruitful connections between real and complex functions that are relatively easy to develop. Here is one important example. Suppose that we replace the real variable x in the power series representations of $\sin x$, $\cos x$ and e^x by a complex variable z. Then, because $|z|$ is a real number for any complex number z, and because each of these series converges absolutely for all real numbers, the power series for $\sin z$, $\cos z$ and e^z converge absolutely for all complex z. Just as for real series, absolute convergence implies convergence, so the corresponding complex power series formulas can be regarded as *definitions* of $\sin z$, $\cos z$ and e^z for all complex numbers z,

$$\cos(z) = 1 - \frac{z^2}{2!} + \cdots + (-1)^n \frac{z^{2n}}{(2n)!} + \cdots = \sum_{n=0}^{\infty} (-1)^n \frac{z^{2n}}{(2n)!},$$

$$\sin(z) = z - \frac{z^3}{3!} + \cdots + (-1)^n \frac{z^{2n+1}}{(2n+1)!} + \cdots = \sum_{n=0}^{\infty} (-1)^n \frac{z^{2n+1}}{(2n+1)!},$$

$$\exp(z) = e^z = 1 + z + \frac{z^2}{2!} + \frac{z^3}{3!} + \cdots + \frac{z^n}{n!} + \cdots = \sum_{n=0}^{\infty} \frac{z^n}{n!}.$$

These are complex extensions of the real cosine, sine, and exponential functions. That is, their values agree with the values of these real functions when z is a real number. As complex functions, the sine and cosine functions are still periodic with period 2π, and interestingly enough, the exponential function is periodic with period $2\pi i$ (see Problem 6).

Euler's Formula

One of the most remarkable consequences of the extension of the sine, cosine, and exponential functions to the complex numbers is the following result, which relates trigonometric and exponential functions.

Theorem 9.7 **(Euler's Formula):** For any real number θ,

$$e^{i\theta} = \cos\theta + i\sin\theta.$$

Proof: For any real number θ and any positive integer k,

$$(i\theta)^{4k} = \theta^{4k}, (i\theta)^{4k+1} = i\theta^{4k+1}, (i\theta)^{4k+2} = -\theta^{4k+2}, (i\theta)^{4k+3} = -i\theta^{4k+3}.$$

Therefore,

$$\begin{aligned} e^{i\theta} &= 1 + i\theta + \frac{(i\theta)^2}{2!} + \frac{(i\theta)^3}{3!} + \frac{(i\theta)^4}{4!} + \frac{(i\theta)^5}{5!} + \cdots \\ &= \left(1 - \frac{\theta^2}{2!} + \frac{\theta^4}{4!} - \cdots + \frac{(-1)^k\theta^{2k}}{(2k)!} + \cdots\right) + i\left(\theta - \frac{\theta^3}{3!} + \frac{\theta^5}{5!} - \cdots + \frac{(-1)^k\theta^{2k+1}}{(2k+1)!} + \cdots\right) \\ &= \cos\theta + i\sin\theta \end{aligned}$$

The most famous consequence of Euler's Formula relates the important mathematical numbers e, π, i, 1, and 0.

Corollary: $e^{i\pi} = -1$, or, equivalently, $e^{i\pi} + 1 = 0$.

9.2.1 Problems

1. Explain why the tangent and cotangent functions have period π while the sine and cosine functions have period 2π.

2. The identities in this problem are known as the Pythagorean identities.

a. Explain why $\sin^2 A + \cos^2 A = 1$ for any acute or obtuse angle A using the trigonometric ratio definition of sine and cosine.

b. Explain why $\sin^2 x + \cos^2 x = 1$ for any x using the unit circle definitions of the sine and cosine functions.

c. Divide both sides of the identity in part **b** by appropriate expressions to obtain two more identities similar to that one.

3. The identities $\sin(-x) = -\sin x$ and $\cos(-x) = \cos x$ are not meaningful for the trigonometric ratio definition of sine and cosine. Derive these identities from one of the other definitions of sine and cosine.

4. a. In the wrapping function, name two points on the line $x = 1$ whose image is $\left(-\frac{1}{2}, \frac{\sqrt{3}}{2}\right)$.

b. What specific values of sines and cosines are obtained from the information in part **a**?

5. Figure 23 suggests how four of the trigonometric functions got their names. O is a circle of radius 1 and t is the measure of the acute angle POQ. Each of the six trigonometric functions of t has length equal to the length of a segment on the figure. Find the segment for each function.

Figure 23

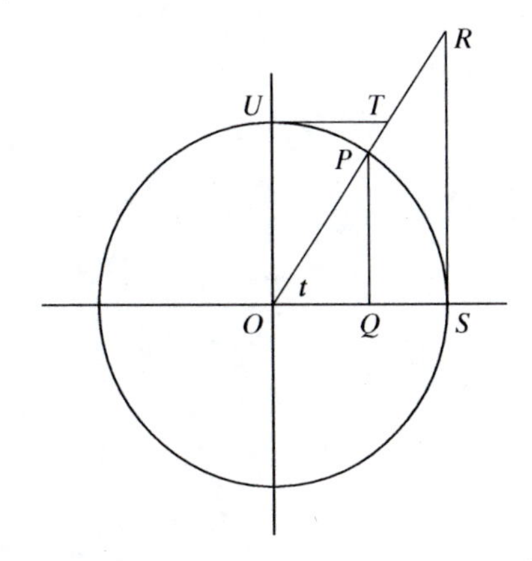

6. Prove that the exponential function defined on C, $\exp(z)$, is periodic with period $2\pi i$.

7. Prove the corollary to Theorem 9.7.

8. By applying Theorem 9.7 to the exponential $e^{i(\theta+\phi)}$, derive the familiar formulas for $\sin(\theta + \phi)$ and $\cos(\theta + \phi)$.

9. Use Theorem 9.7 to deduce formulas for $\sin x$ and $\cos x$ in terms of e and x. (*Hint:* Replace θ by $-\theta$ to deduce a corollary to the theorem.)

9.2.2 Modeling with trigonometric functions

Many phenomena, including sound, electromagnetic and water waves, mechanical vibrations, seasonal, biological and economic cycles, tides, and the motion of celestial bodies and orbiting spacecraft, are driven by motions that are periodic or approximately periodic. Manufactured objects such as electric generators and motors, internal combustion engines, and axles of cars and trucks all rotate while in use, so are described by periodic functions. The rotation of an alternating current generator produces a current with voltage, or AC. It is the reason why the complex exponential function (Problem 6 of Section 9.2.1) is so important in electrical engineering. The trigonometric functions are perhaps the most familiar examples of periodic functions.

Modeling periodic phenomena

Recall that a function f is **periodic** if and only if there exists a positive number p such that $f(x + p) = f(x)$ for all x. The smallest such number p (if there is a smallest number) is the **period** of the function.

From the definitions of $\cos x$ and $\sin x$ in terms of rotations, $\cos(x + 2\pi) = \cos x$ for all x. Furthermore, the cosine function is decreasing on the interval $[0, \pi]$ and increasing on the interval $[\pi, 2\pi]$, so there is no number p less than 2π for which $\cos(x + p) = \cos x$ for all x. This implies that 2π is the period of the cosine function.

Question: Show that if $c > 0$, the function f defined by $f(x) = \sin(cx)$ for all real numbers x is periodic with period $\frac{2\pi}{c}$.

The length of daylight

The following example illustrates how trigonometric functions can model periodic phenomena. The earth tilts on its axis at a fairly constant angle of $\gamma \approx 23.5°$ to the perpendicular to the plane of its orbit around the sun. As a result of this tilt, the length of the daily period of daylight in the northern hemisphere varies throughout the year from the longest period of daylight near the summer solstice (≈June 21) to the shortest period of daylight near the winter solstice (≈December 21) (see Figure 24).

Figure 24

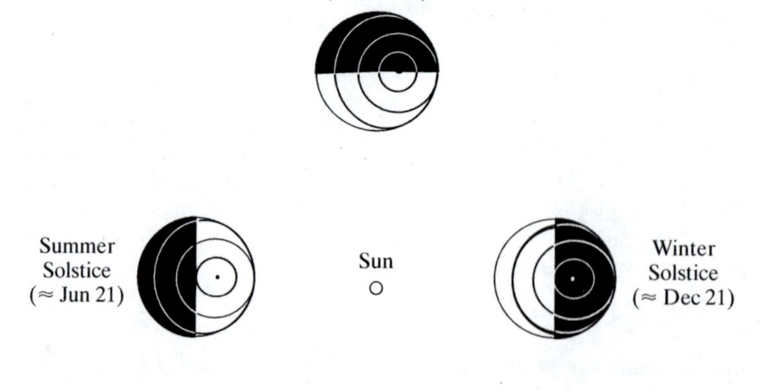

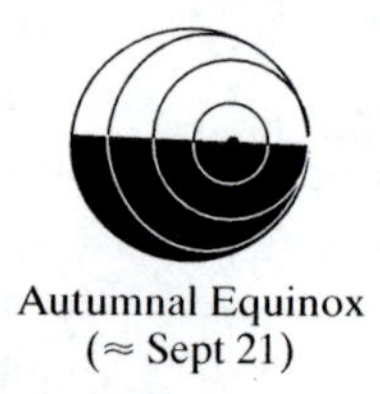

If the north pole N is the upper end point of the earth's axis, then Figure 25a describes the position of Earth at the summer solstice. Imagine the Sun at the right (over 11,600 Earth diameters away!), so the shaded part is night. The equator and the circle for latitude α are shown. From this diagram, we see that the period of daylight on the longest day of the year varies with the latitude. For instance, at the equator half the circle of latitude is in daylight, so half the day is in daylight. At the latitude α drawn, about $\frac{2}{3}$ of the day would be in daylight. Figure 25b shows a cross-section viewed from the plane of the Equator; the Sun is then 23.5° above the horizontal.

Figure 25

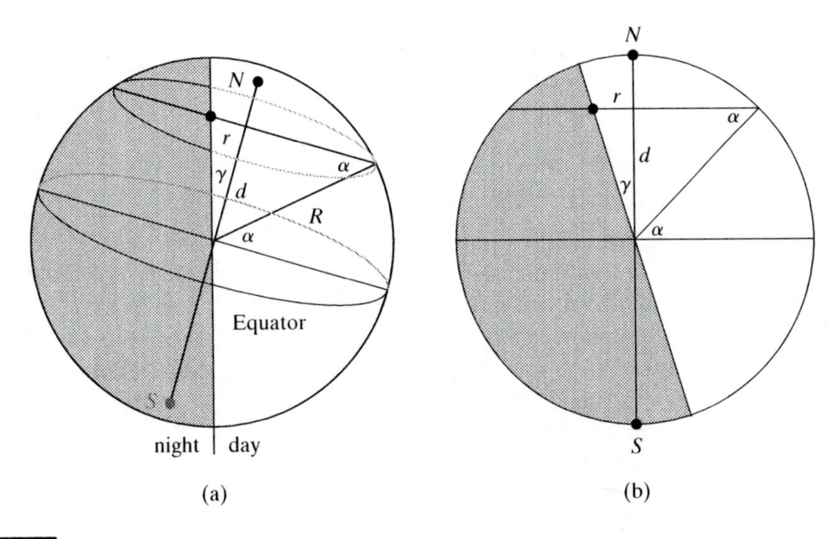

EXAMPLE 1 Find the length of the longest day as a function of the latitude α in degrees.

Solution We need only right-triangle trigonometry. Refer to Figure 25. Let d be the distance from the plane of the circle of latitude α to the center of Earth. Then $\sin\alpha = \frac{d}{R}$, and also $\tan\gamma = \frac{r}{d}$. Solving each equation for d and equating the solutions, we find that the distance r and Earth's radius R are related by

$$r = R\sin\alpha\tan\gamma.$$

If we slice through the earth at latitude α and show a view of the resulting circle from Figure 25, we obtain Figure 26. The radius of that circle is $R\cos\alpha$.

Figure 26

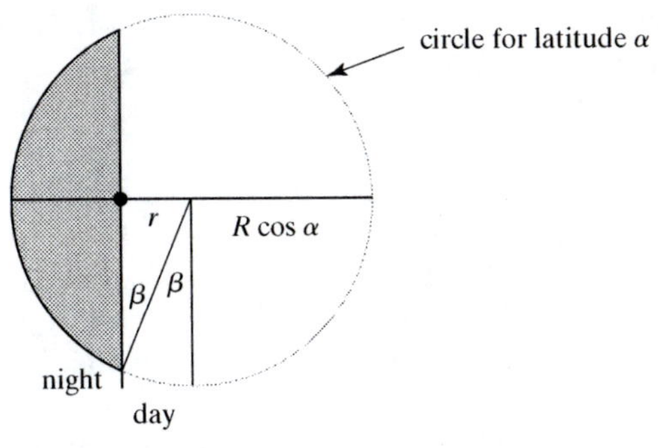

In Figure 26, $\sin\beta = \frac{r}{R\cos\alpha} = \tan\alpha\tan\gamma$, from which $\beta = \sin^{-1}(\tan\alpha\tan\gamma)$. (Notice that R, the earth's radius, cancels out.)

The portion of the day that is daylight at this latitude equals the portion of the arc of the circle that is blue. From Figure 26, that portion is $\frac{2\beta+180°}{360°}$. Since there are 24 hours in a day, the length D of the longest day in hours is given by $D = 24\frac{2\beta + 180°}{360°}$.

Since $\gamma \approx 23.5°$, $\tan\gamma \approx \tan 23.5° \approx 0.435$, and $\beta \approx \sin^{-1}(0.435\tan\alpha)$.

Evaluating D at various latitudes α, we have the following:

Latitude°	Hours in Longest Day
0°	12 h = 12.00 h
10°	12 h 35 min ≈ 12.58 h
20°	13 h 12 min = 13.20 h
30°	13 h 56 min ≈ 13.93 h
40°	14 h 51 min ≈ 14.85 h
50°	16 h 10 min ≈ 16.17 h
60°	18 h 31 min ≈ 18.52 h
66.5°	24 h

Do these values of D obtained theoretically agree with actual longest days at the various latitudes? Table 1 lists data drawn from the *World Almanac and Book of Facts 2000* for 40° North latitude on the 21st day of each month of the year 2000. We have purposely picked the 21st days of months because they provide (within a minute or so) the shortest and longest days of the year and the solstices.

Table 1 40° North Latitude Actual Data

Month	Jan.	Feb.	Mar.	April	May	June
Day of Yr	21	52	81	112	142	173
Sunrise	7:17	6:46	6:02	5:13	4:40	4:31
Sunset	17:05	17:42	18:13	18:35	19:14	19:32
Day length (hours)	9.83	10.93	12.18	13.53	14.57	15.02
Month	**July**	**Aug.**	**Sept.**	**Oct.**	**Nov.**	**Dec.**
Day of Yr	203	234	265	295	326	356
Sunrise	4:49	5:17	5:47	6:17	6:52	7:18
Sunset	19:24	18:48	17:58	17:12	16:39	1638
Day length (hours)	14.58	13.52	12.18	10.92	9.78	9.33

At a given latitude α, the length of the daily period of daylight varies from its maximum at the summer solstice to its minimum at the winter solstice and the period of this variation is roughly 365 days (actually, 365.2425 days—see Section 6.1.2). Our calculations produce a length of 14 hr 51 min for the longest day, but the actual longest day is 15 hr 1 min. The difference of 10 minutes is due mainly to the width of the Sun, which is about a half degree, or 1/720 of a circle. This is verified by the fact that on the equinoxes, the length of the day is not exactly 12 hours (the theoretical value), but about 11 minutes longer. If we judge the length of day from the time the center of the Sun rises to the time the center sets, then our calculations are well within the accuracy of the value of γ used for the tilt of Earth. The oblateness of Earth and refraction of the atmosphere at the horizon also affect the length of the day.

The data in the table are periodic. If you consult an almanac for the year in which you read this page, you will find values for the days of the year that are very close to the ones given. Furthermore, graphing the values suggests that a sine wave will fit the data reasonably well.

Sine waves are images of the graph of $y = \sin x$ under stretches and translations. By the Graph Translation Theorem (Theorem 7.21), a translation image of the wave that is rotation symmetric about (x_0, y_0) rather than $(0, 0)$ has equation

$y - y_0 = \sin(x - x_0)$. From the Question of this section, if we wish the period of the sine wave to be p rather than 2π, and still maintain that symmetry, we may write

$$y - y_0 = \sin\left(\frac{x - x_0}{\frac{p}{2\pi}}\right).$$

Finally, if we wish the amplitude of the sine wave to be A rather than 1, we can use the equation

$$\frac{y - y_0}{A} = \sin\left(\frac{x - x_0}{\frac{p}{2\pi}}\right).$$

EXAMPLE 2 Let $L(d)$ be the length of the day d of the year 2000 at 40°N latitude. Using the data in Table 1, find an equation for a sine wave that approximates the graph of the function L.

Solution The length of day has symmetry about the equinoxes, so we pick March 21, the 81st day of the year, as the center of rotation symmetry. Then $x_0 = 81$. On this day, $L(d) = 12.18$, so $y_0 = 12.18$. The amplitude A is half of the difference of the day length (in hours) at the summer and winter solstices. Thus

$$A = \frac{1}{2}(15.02 - 9.33) = 2.85.$$

The year 2000 was a leap year, so $p = 366$ days. Therefore,

$$\frac{L(d) - 12.18}{2.85} = \sin\left(\frac{d - 81}{\frac{366}{2\pi}}\right).$$

Consequently, $L(d) = 12.18 + 2.85\sin\left[\frac{2\pi}{366}(d - 81)\right]$.

This model produces day length data for the 21st of each month in the year 2000 (Table 2) that are close to but not identical to the data in Table 1.

Table 2 40° North Latitude Model Values

Month	Jan.	Feb.	Mar.	April	May	June
Day of Yr	21	52	81	112	142	173
Day length (hours)	9.83	10.93	12.18	13.53	14.57	15.02
L(d) (hours)	9.75	10.82	12.18	13.62	14.63	15.02
Month	**July**	**Aug.**	**Sept.**	**Oct.**	**Nov.**	**Dec.**
Day of Yr	203	234	265	295	326	356
Day length (hours)	14.58	13.52	12.18	10.92	9.78	9.33
L(d) (hours)	14.63	13.58	12.13	10.74	9.70	9.34

Describing plane curves with angles as parameters

Many important plane curves can be described most conveniently and completely by parametric equations in which the parameter is a suitably selected directed angle. In such cases, the tools of trigonometry often play an important role in the development of the description.

Theorem 9.8(a) The circle of radius r centered at the point (h, k) has parametric equations

$$x = h + r\cos\theta, \qquad y = k + r\sin\theta,$$

where the parameter θ is the directed angle with vertex at (h, k) with initial side parallel to the positive x-axis and terminal side joining (h, k) to the point (x, y). As θ increases from 0 to 2π, the corresponding point (x, y) traces out the circle with the rectangular equation $(x - h)^2 + (y - k)^2 = r^2$ once in the counterclockwise direction starting from the point $(h + r, k)$.

Proof: In Problem 5, you are asked to show that any point satisfying the parametric equations is on the circle with the indicated rectangular equation, and vice versa.

By stretching the circle of Theorem 9.8(a), we obtain a more general theorem involving ellipses.

Theorem 9.8(b) The ellipse centered at (h, k) with semimajor axis of length a parallel to the x-axis, and semiminor axis of length b parallel to the y-axis, has parametric equations

$$x = h + a\cos\theta, \qquad y = k + b\sin\theta,$$

with θ defined as in Theorem 9.8a. As θ increases from 0 to 2π, (x, y) traces out the ellipse with the rectangular equation $\frac{(x-h)^2}{a^2} + \frac{(y-k)^2}{b^2} = 1$ once in the counterclockwise direction starting from the point $(h + a, k)$.

Proof: A proof is again left to you (see Problem 6).

The next problem is concerned with finding parametric equations involving an angular parameter for a famous plane curve called a *cycloid*. The solution of this problem makes significant use of concepts and methods from trigonometry.

EXAMPLE 3 A wheel of radius r rolls along a straight level track. The point P on the rim of the wheel that is in contact with the track when the wheel began to roll describes a plane curve C as the wheel rolls. If t is the central angle in radians through which the wheel has rolled to place the wheel in its current position $P(x, y)$, find equations for x and y in terms of the parameter t.

Figure 27

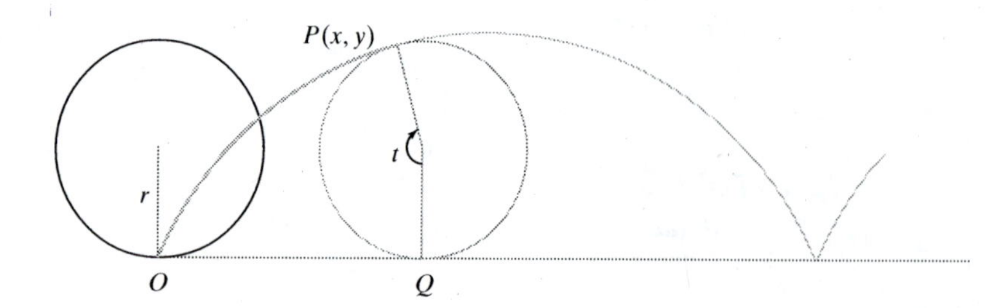

Solution: In Figure 27, the length of $\overline{OQ}$ is equal to the length of the circular arc $\widehat{QP}$ because both represent the distance that the point P has traveled as the central angle t increased from 0 to t. Therefore, because the radius of the wheel is r, the x-coordinate of Q is tr if t is measured in radians. Therefore, the x-coordinate of the point P is $tr - r\sin t$ and the y-coordinate of the point P is $r - r\cos t$. (Note that $\cos t < 0$ for the position of P in Figure 27.) Thus, the cycloid is described by the parametric equations

$$\begin{cases} x = r(t - \sin t) \\ y = r(1 - \cos t) \end{cases}.$$

We could cite many other types of problems for which plane trigonometry is useful. The samples that we have selected in this section and in Section 9.1.3 give some indication of the diversity of these problems as well as the tools and concepts of trigonometry that are used for their solution. In the applications that we have considered, there are other trigonometric concepts and tools that have not been used such as the addition formulas, the formulas for rotation of axes, polar coordinates, and the trigonometric form of complex numbers. Their widespread use attests to the fundamental importance of trigonometry.

9.2.2 Problems

1. Give the amplitude and period of the function f defined by $f(x) = 3\sin(4x) + 5$.
2. Give a center of rotation symmetry for the function g with $g(x) = A\sin(B(x + C)) + D$.
3. Can a sine wave always be described by an equation involving the cosine as the only trigonometric function? Why or why not?
4. Using information from Example 1, find an equation for $L(d)$, the length of day d of the year 2000 on Earth at latitude 30°S.
5. Prove Theorem 9.8a.
6. Prove Theorem 9.8b.
7. a. Explain why $\begin{cases} x = a\sec\phi + h \\ y = b\tan\phi + k \end{cases}$ are parametric equations for the hyperbola with the rectangular equation $\frac{(x-h)^2}{a^2} - \frac{(y-k)^2}{b^2} = 1$.

   b. Let $a = b = 1$ and $h = k = 0$. Graph this hyperbola using a function grapher. Explain why the order of tracing out the hyperbola is as it is.
8. Many periodic functions are not trigonometric functions. For example, the **square wave function** s is defined by

$$s(x) = \begin{cases} 1 \text{ if } & 2k\pi < x \le (2k+1)\pi \\ -1 \text{ if } & (2k+1)\pi < x \le 2(k+1)\pi \end{cases} \text{ for any integer } k.$$

   a. Sketch the graph of the square wave function on the interval $[-3\pi, 3\pi]$. Then use a graphing calculator or computer to plot the same function on the same interval. Explain any differences that you see between the resulting graph and the graph that you sketched.

   b. Use a graphing calculator or computer to graph the function

$$p(x) = \frac{4}{\pi}\left[\sin(x) + \frac{1}{3}\sin(3x) + \frac{1}{5}\sin(5x) + \frac{1}{7}\sin(7x)\right]$$

   together with the square wave function on the interval $[-3\pi, 3\pi]$. Discuss the closeness of the two graphs.
9. Suppose that a circle C_a of radius a is fixed with center at the origin O and that a circle C_b inside of C_a with radius $b < a$ and tangent to the circle C_a at the point $A = (a, 0)$ begins to roll in a counterclockwise direction around the inside of C_a (Figure 28a). Then the point $P = (x, y)$ on the circle that was initially at A describes a curve inside of the circle C_a called a **hypocycloid** (Figure 28b). We wish to find an equation for the hypocycloid in terms of t, where $t = m\angle AOC$ (in radians).

Figure 28

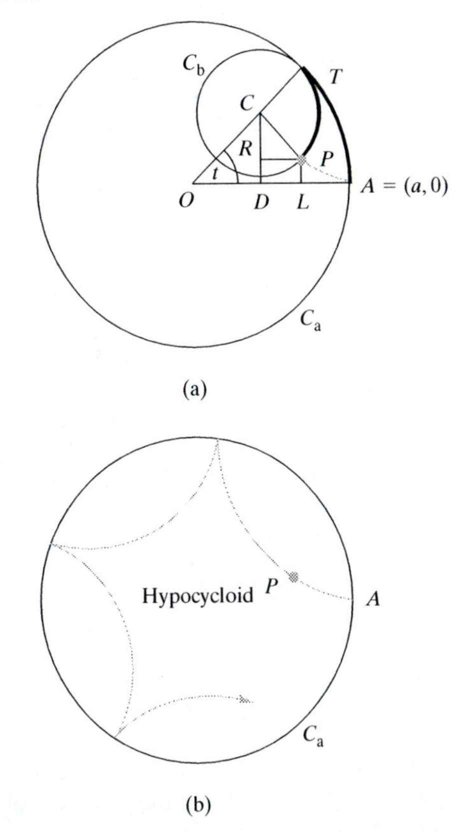

Let C be the (moving) center of circle C_b, let T be the point of tangency of the two circles, and let $s = m\angle PCT$.

a. Explain why $s = \frac{at}{b}$ and $s - t = \frac{a-b}{b}t$.

b. Show that parametric equations for the hypocycloid are given by

$$\begin{cases} x = (a-b)\cos(t) + b\cos\left(\dfrac{a-b}{b}t\right) \\ y = (a-b)\sin(t) + b\sin\left(\dfrac{a-b}{b}t\right) \end{cases}.$$

c. If $b = \frac{a}{4}$, show that these parametric equations for the hypocycloid simplify to

$$\begin{cases} x = a\cos^3(t) \\ y = a\sin^3(t) \end{cases}$$

and that a rectangular equation for this hypocycloid is

$$x^{2/3} + y^{2/3} = a^{2/3}.$$

(*Hint*: Use identities for $\sin(3t)$ in terms of $\sin t$ and $\cos(3t)$ in terms of $\cos t$.)

10. If the circle C_b in Problem 9 rolls along the *outside* of the circle C_a, a point P on the circumference of C_b describes a curve in the plane called an **epicycloid**. Assuming that the point $P = (x, y)$ is initially at the point $(a, 0)$ and that C_b rolls counterclockwise around C_a, show that

$$\begin{cases} x = (a+b)\cos(t) - b\cos\left(\dfrac{a+b}{b}t\right) \\ y = (a+b)\sin(t) - b\sin\left(\dfrac{a+b}{b}t\right) \end{cases},$$

where $t = m\angle AOC$ (in radians), as in Problem 9.

11. Use integration to prove that the area under one arch of the cycloid

$$x = r(t - \sin t) \qquad y = r(1 - \cos t)$$

is $3\pi r^2$ and that the length of this arch is $8r$.

ANSWERS TO QUESTION:

1. For all $x \in \mathbf{R}$, $f(x) = \sin(cx) = \sin(cx + 2\pi) = \sin(c(x + \frac{2\pi}{c}))$, and $\frac{2\pi}{c}$ is the smallest number p for which $\sin(cx) = \sin(c(x + p))$, so $\frac{2\pi}{c}$ is the period of f.

9.2.3 The historical and conceptual evolution of trigonometry

Babylonian and Greek contributions

The Babylonians were first to introduce degree measure and a spherical coordinate system (c. 2000 to 1600 B.C.). The rationale for the division of a circle into 360 parts (degrees) is not known for certain. It is often theorized to be 360 because 360 is close to the number of days of the year (and so one degree is close to the amount a star moves in the sky each day) and to the fact that 360 has so many factors. The choice of 360 may be related to the fact that the Babylonians used a sexagesimal (base 60) number system and a circle is divided neatly into 6 parts by the vertices of an inscribed regular hexagon, with the resulting chord lengths equal to the radius of the circle. It also may be that the base system was related to the choice of the degree.

The Greeks adopted and refined both degree measure and the work with spherical coordinates. Hipparchus of Nicaea (c. 180–125 B.C.) prepared the first tables of lengths of chords subtending given circular arcs, and used these tables to calculate, among other things, longitudes and latitudes on the celestial sphere. Greek trigonometry reached a high point with Menelaus (98 A.D.). In his seminal work *Sphaerica*, Menelaus introduced the concept of a spherical triangle and proved theorems about spherical triangles analogous to those Euclid had for plane triangles.

Claudius Ptolemy (c. 100–178 A.D) wrote a comprehensive and definitive account of Greek astronomy based on the earlier work of Hipparchus and Menelaus. The first volume of this 13-volume treatise was largely devoted to an exposition of the mathematics essential to his discussion of astronomy, primarily content from what is now called plane and spherical trigonometry. Ptolemy's books had a profound influence on subsequent developments in astronomy throughout the world until the sixteenth century. Islamic writers referred to these books as the *Almagest* (the greatest), a name that has now been generally adopted for Ptolemy's treatise. The tables of Hipparchus and Ptolemy (discussed in Section 9.1.2) were precursors of later tables of sines. Until the advent of calculators in the 1970s, tables of sines and tangents were essential for anyone who wished to apply trigonometry.

Hindu and Arab contributions

In the ninth and tenth centuries, Hindu and Arab mathematicians further refined the trigonometric concepts and methods developed in Ptolemy's *Almagest*. In particular, they introduced the modern sine concept from the Greek work on chords and arcs. The historical development is quite similar to the way that Hindu-Arabic decimal notation became the worldwide standard. Indian mathematicians defined the cosine, tangent, and cotangent (of course, not using those names). The Muslim mathematician and astronomer Mohammed ibn Jabir ibn Sinan Abu Abdullah al-Battani (850–929) developed the spherical Law of Cosines to calculate the measure of the arc in space between a planet and the Sun. It was given in his astronomical treatise whose title translates as *On the science and number of stars and their motions*. He computed tables of sines, tangents, and cotangents for angles from 0° to 90°. The Persian Abu'l-Wafa (940–998) first discovered the spherical Law of Sines, and Al Buruni (973–1048) later established the corresponding result for plane triangles. This order of discovery may seem curious to us now, but it should be remembered that a major impetus for the study of trigonometry from ancient times through the sixteenth century came from astronomy, and for astronomy, spherical trigonometry was more critical than plane trigonometry.

Almost none of these achievements of the Indian and Islamic mathematicians were known in the medieval West. Europe was struggling through the Dark Ages, and it wasn't until the 12th century that Latin translations were made of ancient mathematical treatises. The Hindu mathematician Aryabhata first used the half-chord, which he called "jyardha", later shortened to "jya". The Arabs translated this word as *jiba*, which meant "nothing" in Arabic, and which is written (as all Arabic is written) without vowels as *jb*. Later readers thought that *jb* stood for the word *jaib* (also written as *jb*), which means "inlet or bay" but also can mean "bosom". (With a little imagination, you can picture a half-chord and half-arc as outlining a mother's arm as she is cradling a child.) Robert of Chester around 1140 made a Latin translation of a treatise by al-Khwarizmi and translated the word *jaib* into the Latin word *sinus* (which means "inlet, bay, and bosom"). Some books say that *jaib* (or *jayb*) had the meaning of "chord of an arc" but that the Europeans thought it meant "fold of a garment", for which the Latin was again "sinus". Some sources believe that Plato of Tivoli first introduced the word "sinus", around 1116–1136; others credit Gherardo of Cremona, around 1150.

European contributions

Although Fibonacci, in his *Practica Geometriae* of 1220, had initiated the use of plane trigonometry, until 1450 the focus in trigonometry was on spherical trigonometry. In the late fifteenth century, because of the need for accurate navigation and surveys of land, plane trigonometry became very important.

Trigonometry was studied by most mathematicians of the Renaissance. By the sixteenth century it began to be treated as a "subject" in the literature, and acquired the status of a branch of mathematics. Georg Peurbach (1423–1461) corrected the Latin translations of Ptolemy's *Almagest* and produced more accurate trigonometric tables. The first systematic treatment of trigonometry as a branch of geometry was given by Regiomontanus (1436–76) in his book *De Triangulis Omnimodus*, in which he gave an axiomatic development of both plane and spherical trigonometry within Euclid's framework. Regiomontanus's book continues the work of Peurbach. Others in the fifteenth and sixteenth centuries continued to work on more accurate trigonometric tables, including George Joachim Rheticus (1514–1613), Nicolaus Copernicus (1473–1543), and Bartholomaus Pitiscus (1561–1613). Rheticus had published chapters of Copernicus's famous *De revolutionibus orbium Coelestium*. These dealt with planar and spherical trigonometry and served as a compendium of the trigonometry pertinent to the astronomy at that time.

The word "trigonometry" was introduced by Pitiscus in 1595 in his book *Trigonometria*. The word "goniometry" was introduced to refer to that part of trigonometry dealing only with angles. Pitiscus also introduced the formulas for the sine and cosine of the sum and difference of two angles, which up to then had been calculated on the basis of tables. But although mathematicians had the formulas, they didn't have the language of algebra to work with them.

François Viète (1540–1603) brought the algebra of trigonometry into being with several treatises he wrote in the late 16th century. He also extended the tables of Rheticus and added to the trigonometric identities that had been established by Ptolemy. In 1615 his versions of formulas for $\sin(nA)$ as functions of $\sin A$ and $\cos A$ were published posthumously.

As functions became central to algebra, many new trigonometric formulas were developed. Important relationships were established in the seventeenth and eighteenth centuries between the trigonometric functions and powers and multiples of angles, series, continued fractions, polar coordinates, and other ideas. The names associated with these developments are quite familiar. Isaac Newton (1642–1727) used polar coordinates in connections with tangents, curvature, and rectification of curves. He derived a formula for the radius of curvature using polar coordinates and trigonometric ratios, and he connected the trigonometric functions to infinite series.

Leonhard Euler (1707–1783) wrote a textbook, *Introductio in analysin infinitorum*, in 1748, which has been compared to the *Elements*, doing for analysis what Euclid did for geometry. In this textbook, Euler popularized the definition of the trigonometric functions as ratios and derived their series expansions using binomial series and a limiting argument. We have noted earlier that Euler produced the formula $e^{ix} = \cos x + i \sin x$, unifying the power function with the trigonometric functions and the imaginary number i. Euler is responsible for expanding much of the theory of trigonometry in the later part of the 18th century. He analyzed all the trigonometric functions, systematically listed all the usual formulas of goniometry, emphasized the periodicity of the trigonometric functions, etc. He connected trigonometry with coordinate geometry and calculus. He expanded the distance function in mathematical astronomy as a series involving trigonometric ratios. He looked back at the spherical trigonometry of Menelaus and showed how the various theorems could be derived algebraically one from the other.

In 1798–1799, Sylvestre-François Lacroix (1765–1843) published an influential textbook on trigonometry called *Traite elementaire de trigonometrie rectiligne et spherique et application de l'algebre a la geometrie*. This textbook went through many editions and was translated into many languages. Trigonometry books in the United States in the early twentieth century were similar to this one. For most of the twentieth century, however, plane and spherical trigonometry were separated and only plane trigonometry was taught in schools.

9.2.3 Problems

1. a. If A and B are points on a circle O of radius 1, prove that $\frac{1}{2}AB = \sin\frac{1}{2}(\angle AOB)$. (This equation shows how a table of chords in a circle can determine a table of sines.)

   b. Does the formula of part **a** work if $\angle AOB$ is measured by its major rather than its minor arc? Why or why not?

*2. Ptolemy was the first to solve the following problem. Given two stars in the sky whose longitude and latitude on the celestial sphere are known, find the angle between them. Ptolemy sliced the celestial sphere by a plane through the two stars and the center of the earth, obtaining a great circle. If C and D represent the two stars, and the diameter of the circle is $\overline{AB}$, so that A, B, C, and D are in order on the circle (see Figure 29), then Ptolemy was able to prove, using similar triangles, that $AB \cdot CD + AD \cdot BC = BD \cdot AC$. Derive his theorem. (*Hint:* Choose point E on $\overline{AC}$ so that $\angle ABE \cong \angle EBC$. Then find two pairs of similar triangles.)

Figure 29

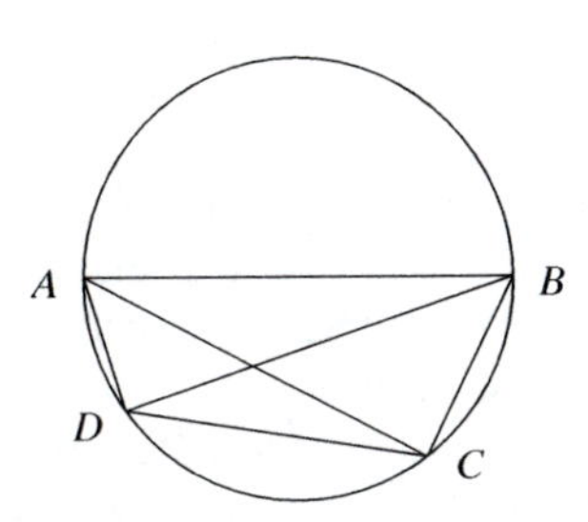

3. **Earth–Moon distance.** Aristarchus of Samos (c. 310–230 B.C.) used trigonometry to estimate the distance from Earth to the Sun. His procedure was to measure the angle $\angle MES$ between the Moon and the Sun at the moment when the moon was a half moon. (To exhibit the geometric setting, Figure 30 is not drawn to scale.) At that moment, $\angle SME$ was a right angle. It was difficult for Hipparchus (or for anyone else for that matter!) to determine when the Moon is exactly half full from observation on Earth. He estimated the angle $\angle MES$ to be about 87°. Because Hipparchus did not know the distance from Earth to the Moon, he expressed his estimate of the Earth–Sun distance as a multiple of the Earth–Moon distance.

Figure 30

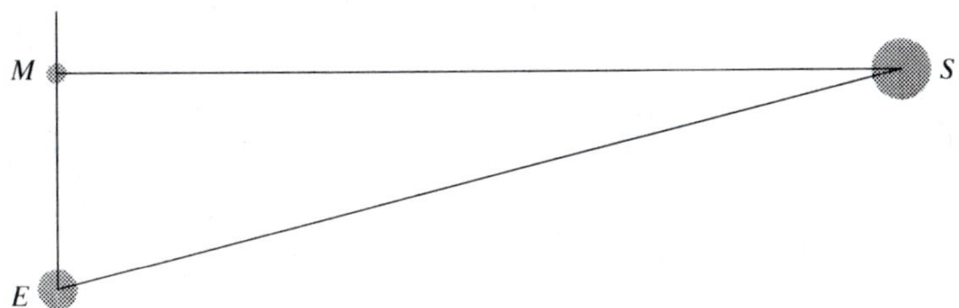

a. Based on the estimate of 87° that Hipparchus made for $\angle MES$ and an Earth–Moon distance of 238,800 miles, find the corresponding estimate of the distance from Earth to the Sun in miles.

b. The actual average distance from Earth to the Sun is about 93 million miles. What is the corresponding degree measure of the angle $\angle MES$?

c. Explain why small errors in the measurement of $\angle MES$ result in relatively large errors in measurement of the Earth–Sun distance.

4. **Circumference of Earth.** Eratosthenes (275–194 B.C.) of Cyrene (a city in what is now Libya) was generally regarded as second only to Archimedes among mathematicians of his era. Among his many accomplishments, he was the first to measure the size of Earth on the basis of Earth observations. His procedure can be described as follows. At noon of the summer solstice (June 21), it was known that the Sun's rays struck the bottom of a deep well in the Egyptian town of Syene (now Aswan), due south of Alexandria. At the same time, Eratosthenes observed that the Sun made an angle of 7.2° to the vertical in Alexandria. Taking Earth to be a sphere, Eratosthenes concluded that the circumference of Earth was 50 times the distance between Alexandria and Syene.

a. Explain this conclusion.

b. Given that the surface distance from Alexandria to Syene is about 500 miles, what is the corresponding circumference and radius of Earth?

c. Suppose Eratosthenes was off by .2° in his observations. What then would be possibilities for his estimate for Earth's circumference?

Unit 9.3 Properties of the Sine and Cosine Functions

The trigonometric functions are among the most interesting functions in all of mathematics. An analysis of them using the ideas of Section 3.2.1 reveals a host of special properties. In this unit we separate those special properties into three types: algebraic, by which we mean the trigonometric identities that relate the values of these functions; geometric (or graphical), including examination of their graphs and their relationships to physical phenomena; and analytical, those properties that relate these functions and their derivatives and the consequent applications.

9.3.1 Algebraic properties of the trigonometric functions

Most courses in trigonometry place considerable emphasis on deriving and manipulating trigonometric identities. There are the *defining identities* that relate the basic values of the six trigonometric functions. For all values of t for which the denominators do not equal 0,

$$\tan t = \frac{\sin t}{\cos t}, \quad \cot t = \frac{\cos t}{\sin t}, \quad \sec t = \frac{1}{\cos t}, \quad \text{and} \quad \csc t = \frac{1}{\sin t}.$$

You have also seen the *Pythagorean identities*. For all t,

$$\sin^2 t + \cos^2 t = 1, \quad 1 + \tan^2 t = \sec^2 t, \quad \text{and} \quad 1 + \cot^2 t = \csc^2 t.$$

In Section 9.2.2, we discussed several examples in which the trigonometric functions are used to model periodic phenomena and to describe and analyze motion problems and geometric curves. Other applications of the trigonometric functions to the analysis of mechanical vibrations are discussed in Section 9.3.2. Trigonometric identities are at the heart of all such applications. The trigonometric identities govern and direct the use of the trigonometric functions in the same way that algebraic properties such as the commutative and distributive properties govern and direct the use of numbers.

The possible trigonometric identities are too numerous to list. In this section we content ourselves with a derivation of the most basic of the identities. Then in the Problems we ask you to use these to derive some of the other important identities.

A formula for rotation images

The definition of $\cos t$ and $\sin t$ as the x-coordinate and the y-coordinate of the image of $(1, 0)$ under a rotation centered at the origin of magnitude t makes it natural that trigonometric functions would be involved in formulas for rotation images in $\mathbf{R}^2$. Here is a proof of a theorem found in Section 7.2.2.

Theorem 9.9 **(Rotation Image Formula):** Let R_ϕ be the rotation with center $(0, 0)$ and magnitude ϕ. Then

$$R_\phi(x, y) = (x \cos \phi - y \sin \phi, x \sin \phi + y \cos \phi).$$

Proof: Because $R_\phi(1, 0) = (\cos \phi, \sin \phi)$, it follows that

$$R_\phi(x, 0) = (x \cos \phi, x \sin \phi), \text{ for all real numbers } x.$$

Also, $R_\phi(0, y) = (-y \sin \phi, y \cos \phi)$, for all real numbers y (see Problem 1). The vector $\mathbf{c} = (x, y)$ is the sum of the vectors $\mathbf{a} = (x, 0)$, and $\mathbf{b} = (0, y)$, and the points $(0, 0)$, $(x, 0)$, (x, y), and $(0, y)$ are successive vertices of a rectangle (see Figure 31). When the three vectors **a, b,** and **c** are rotated, the image of **c** is the sum of the images of **a** and **b**, since the image of a rectangle under a rotation is a congruent rectangle.

From the formulas, above,

$$R_\phi(\mathbf{a}) = R_\phi(x, 0) = (x \cos \phi, x \sin \phi)$$

and

$$R_\phi(\mathbf{b}) = R_\phi(0, y) = (-y \sin \phi, y \cos \phi).$$

Figure 31

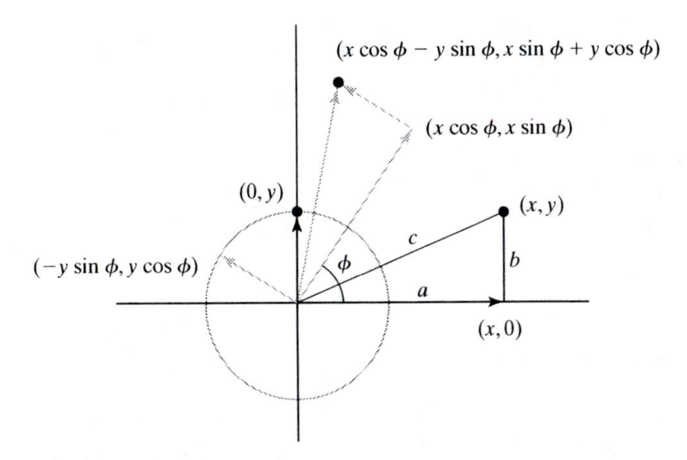

So

$$R_\phi(\mathbf{c}) = R_\phi(x, y) = (x\cos\phi - y\sin\phi, x\sin\phi + y\cos\phi),$$

which completes the proof.

The product of the complex numbers $x + iy$ and $\cos\phi + i\sin\phi$ is $(x\cos\phi - y\sin\phi) + i(x\sin\phi + y\cos\phi)$. This once again shows that we can think of multiplying $x + iy$ by the complex number $\cos\phi + i\sin\phi$ as equivalent to applying R_ϕ to the point (x, y).

Formulas for $\cos(x \pm y)$ and $\sin(x \pm y)$

In Theorem 9.9, we used ϕ to identify the argument of the trigonometric functions because x and y are so identified with coordinates of points in $\mathbf{R}^2$. Now we use the Rotation Image Formula to derive identities involving the sum and difference of arguments. The letters ϕ and θ are sometimes used for these arguments, but more often these formulas are remembered using the letters x and y, as we show here. The symbol $\mp$, when used in conjunction with $\pm$, means that there are two sentences being written as one, with the top signs ($-$ and $+$) being used for one identity, and the bottom signs ($+$ and $-$) being used for the other.

Theorem 9.10 **(Sum and Difference Formulas):** For all real x and y, for which the expressions are defined

a. $\sin(x \pm y) = \sin x\cos y \pm \cos x\sin y$

b. $\cos(x \pm y) = \cos x\cos y \mp \sin x\sin y$

c. $\tan(x \pm y) = \dfrac{\tan x \pm \tan y}{1 \mp \tan x\tan y}$

Proof:

a. and **b.** (simultaneously!): Think of $(\cos(x + y), \sin(x + y))$ as the image of $(1, 0)$ under R_{x+y}. Then separate out the two rotations and apply the Rotation Image Theorem.

$$\begin{aligned}(\cos(x + y), \sin(x + y)) &= R_{x+y}(1, 0) && \text{(def. of cos and sin)}\\ &= R_x \circ R_y(1, 0) && \text{(angle addition)}\\ &= R_x(\cos y, \sin y)\\ &= (\cos x\cos y - \sin x\sin y, \sin x\cos y + \cos x\sin y)\end{aligned}$$

Equating the components of the first and last ordered pairs provides formulas for both $\cos(x + y)$ and $\sin(x + y)$. The corresponding formulas for $\cos(x - y)$

and $\sin(x-y)$ follow from those for $\cos(x+y)$ and $\sin(x+y)$ because for all x, $\cos(-x) = \cos x$ and $\sin(-x) = -\sin x$.

c. $\tan(x \pm y) = \frac{\sin(x\pm y)}{\cos(x\pm y)}$. Now use the formulas found in parts (a) and (b) and divide each by $\cos x \cos y$. See Problems 9a and 12.

The proof of parts **a.** and **b.** shows how intimately rotations, sines, and cosines are related. Four identities are deduced in a few lines. There is even a more elegant form of the addition half of these identities. Since $R_{x+y} = R_x \circ R_y$,

$$\begin{bmatrix} \cos(x+y) & -\sin(x+y) \\ \sin(x+y) & \cos(x+y) \end{bmatrix} = \begin{bmatrix} \cos x & -\sin x \\ \sin x & \cos x \end{bmatrix} \begin{bmatrix} \cos y & -\sin y \\ \sin y & \cos y \end{bmatrix}$$

$$= \begin{bmatrix} \cos x \cos y - \sin x \sin y & -\sin x \cos y + \cos x \sin y \\ \sin x \cos y - \cos x \sin y & \cos x \cos y - \sin x \sin y \end{bmatrix}.$$

The sum and difference formulas are significant because from them so many other formulas can be derived. By letting x = y, the *double-angle formulas* for $\cos 2x$ and $\sin 2x$ follow. From the double-angle formulas, formulas for $\cos 3x$, $\sin 3x$, $\cos 4x$, $\sin 4x$, etc., can be derived. Also, from the double-angle formulas, formulas for $\cos\left(\frac{x}{2}\right)$ and $\sin\left(\frac{x}{2}\right)$ can be deduced. By adding or subtracting the formulas for $\cos(x+y)$ and $\cos(x-y)$ and other pairs of sums and/or differences, the *product-to-sum identities* can be obtained. Before logarithms were discovered, the product-to-sum identities were used by mathematicians to perform difficult multiplications. These ideas are explored in the Problems.

9.3.1 Problems

1. Explain why, for all real numbers y, the image of $(0, y)$ under a rotation of magnitude ϕ around the origin is $R_\phi(0, y) = (-y \sin\phi, y\cos\phi)$.

2. Use the Pythagorean identities and the Sum Formulas to derive the following **double-angle formulas**.
   a. $\cos(2x) = \cos^2 x - \sin^2 x$
   b. $\cos(2x) = 1 - 2\sin^2 x$
   c. $\cos(2x) = 2\cos^2 x - 1$
   d. $\sin(2x) = 2\sin x \cos x$

3. Use the identities in Problem 2 to derive the following **half-angle formulas**.
   a. Prove: For all x, $\left|\cos\left(\frac{x}{2}\right)\right| = \sqrt{\frac{1+\cos x}{2}}$.
   b. Find and prove a formula for $\left|\sin\left(\frac{x}{2}\right)\right|$ in terms of $\cos x$.
   c. Your formulas and proofs for parts **a** and **b** should be similar. Explain why there is no similar formula and proof for a formula for $\left|\sin\left(\frac{x}{2}\right)\right|$ in terms of $\sin x$.

4. Use the Sum Formulas to find **multiple-angle formulas**.
   a. Find a formula for $\cos(3x)$ in terms of $\cos x$ and $\sin x$, and verify your formula with a specific value of x.
   b. Find a formula for $\cos(3x)$ in terms of $\cos x$.
   c. Find a formula for $\sin(3x)$ in terms of $\sin x$ and $\cos x$, and verify your formula with a specific value of x.
   d. Find a formula for $\cos(4x)$ in terms of $\cos x$ and $\sin x$.
   e. Find a formula for $\cos(4x)$ in terms of $\cos x$.
   f. Find a formula for $\sin(4x)$ in terms of $\sin x$ and $\cos x$.
   g. Prove that $\cos(nx)$ can be expressed as a polynomial of degree n in $\cos x$.

5. a. Derive an expression for $\frac{\sin 6x}{\sin 2x}$ in terms of $\cos x$.
   b. Generalize part **a** in some way.

6. Use the fact that $\sin\left(\frac{\pi}{4}\right) = \frac{\sqrt{2}}{2} = \cos\left(\frac{\pi}{4}\right)$ and identities in Problem 3 to determine the following sines and cosines.
   a. $\sin\left(\frac{\pi}{8}\right) = \frac{\sqrt{2-\sqrt{2}}}{2}$; $\cos\left(\frac{\pi}{8}\right) = \frac{\sqrt{2+\sqrt{2}}}{2}$
   b. $\sin\left(\frac{\pi}{16}\right) = \frac{\sqrt{2-\sqrt{2+\sqrt{2}}}}{2}$; $\cos\left(\frac{\pi}{16}\right) = \frac{\sqrt{2+\sqrt{2+\sqrt{2}}}}{2}$
   c. $\sin\left(\frac{\pi}{32}\right) = \frac{\sqrt{2-\sqrt{2+\sqrt{2+\sqrt{2}}}}}{2}$; $\cos\left(\frac{\pi}{32}\right) = \frac{\sqrt{2+\sqrt{2+\sqrt{2+\sqrt{2}}}}}{2}$
   d. Do you think that the pattern established in parts **a**–**c** persists? Why or why not?

7. In 1593, Vieté proved that

$$\frac{2}{\pi} = \sqrt{\frac{1}{2}} \cdot \sqrt{\frac{1}{2} + \frac{1}{2}\sqrt{\frac{1}{2}}} \cdot \sqrt{\frac{1}{2} + \frac{1}{2}\sqrt{\frac{1}{2} + \frac{1}{2}\sqrt{\frac{1}{2}}}} \cdot \ldots.$$

He found this formula by computing areas of regular polygons with 4, 8, 16, ..., 2^n sides inscribed in a circle of radius 1. This problem asks you to carry out the steps of this derivation.
   a. Find the area of a regular polygon of 2^2 sides inscribed in a circle of radius 1.
   b. Find the area of a regular polygon of 2^3 sides inscribed in a circle of radius 1, and show that this area can be written in the form $\frac{2}{\sqrt{\frac{1}{2}}}$.

c. Show that as the number of sides is doubled from n to $2n$, an area equal to $n\left(\sin\frac{\pi}{n}\right)\left(1 - \cos\frac{\pi}{n}\right)$ is added.

d. Find the area of a regular polygons of 2^4 sides inscribed in a circle of radius 1, and show that this area can be written in the form $\dfrac{2}{\sqrt{\frac{1}{2}}\cdot\sqrt{\frac{1}{2}+\frac{1}{2}\sqrt{\frac{1}{2}}}}$.

e. Based on a pattern in parts **a–d**, make a conjecture for the area of a regular polygon of $n = 2^k$ sides inscribed in a circle of radius 1.

f. Show how your conjecture leads to Vieté's formula.

8. Use the Sum and Difference Formulas to find the **product-to-sum formulas**. Before the invention of logarithms, these formulas were used to perform complicated multiplications.

a. Prove that $\cos x \cos y = \frac{1}{2}(\cos(x + y) + \cos(x - y))$.

b. Assume you have a table of cosines (use a calculator for this) that enables you to find inverse cosines as well. Multiply 95632 by 61807 in the following way. Determine $x = \cos^{-1}(.95632)$ and $y = \cos^{-1}(.61807)$ with your "table". Find $\cos(x + y)$ and find $\cos(x - y)$, again using your table. Divide by 2 and put the decimal point in the proper place.

c. Prove that $\sin x \sin y = \frac{1}{2}(\cos(x - y) - \cos(x + y))$ and perform the same multiplication as in part **b** using that identity.

9. a. Prove that $\tan(x \pm y) = \frac{\tan x \pm \tan y}{1 \mp \tan x \tan y}$.

b. By dividing the formula for $\sin(2x)$ by the formula for $\cos(2x)$ in Problem 2, prove that $\tan(2x) = \frac{2\tan x}{1 - \tan^2 x}$.

c. Deduce a formula for $\tan(3x)$ in terms of $\tan x$.

d. Deduce a formula for $\tan\left(\frac{x}{2}\right)$ in terms of $\cos x$.

e. Deduce a formula for $\tan\left(\frac{x}{2}\right)$ in terms of $\sin x$ and $\cos x$ that contains no radicals.

10. a. Deduce a formula for $\cot(2x)$ in terms of $\cot x$.

b. Deduce a formula for $\cot\left(\frac{x}{2}\right)$ in terms of $\sin x$ and $\cos x$ that contains no radicals.

11. It can be proved that if x is a nonzero rational number, then $\cos x$ is irrational.[3] Use this result in the following problems.

a. Prove that π is irrational.

b. Prove that $\sin x$, $\tan x$, $\cot x$, $\sec x$, and $\csc x$ are irrational whenever x is a nonzero rational number for which the function is defined. (*Hint*: Use formulas for $\cos 2x$.)

c. Prove that nonzero values of $\sin^{-1} x$, $\cos^{-1} x$, and $\tan^{-1} x$ are irrational whenever x is a rational number in their domains.

12. Use a formula for $\tan(x - y)$ (see Problem 9**a**) to deduce a formula for the tangent of the acute angle formed by two nonvertical and nonperpendicular lines with slopes m_1 and m_2.

9.3.2 Geometric properties of the sine and cosine functions

We know that, by definition, the sine and cosine functions are *periodic with period* 2π. That is, if $p = 2\pi$, then

$$\sin(x + p) = \sin(x) \quad \text{and} \quad \cos(x + p) = \cos(x), \qquad \text{for all real numbers } x,$$

and no smaller positive value of p has this property. More generally, for each positive number c, the functions

$$x \to \sin(cx) \quad \text{and} \quad x \to \cos(cx)$$

have period $p = \frac{2\pi}{c}$. In some applications, the period p is called the **wave length**. We also say that these functions have **frequency** $f = \frac{c}{2\pi}$. Thus, for example, the real function $x \to \sin(\pi x)$ has period 2 and frequency $\frac{1}{2}$ (see Figure 32), while $x \to \cos(2x)$ has period π and frequency $\frac{1}{\pi}$ (see Figure 33).

Figure 32

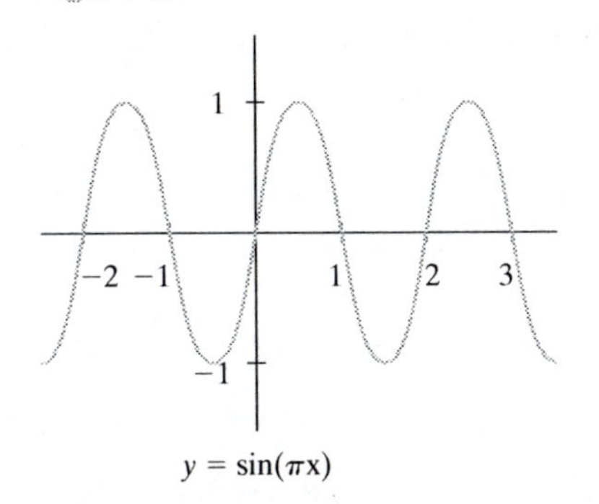

$y = \sin(\pi x)$

Figure 33

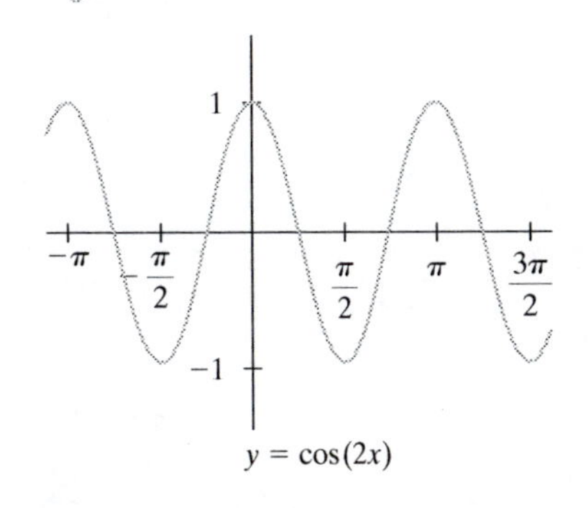

$y = \cos(2x)$

[3]For a proof, see Ivan Niven, *Irrational Number*, Carus Monograph No. 11 (Washington, DC: Mathematical Association of America and John Wiley, 1956), 17.

Sum of multiples of a sine and a cosine function with the same period

Combinations of sine and cosine functions often have interesting but predictable graphical features, as the examples in this section illustrate. It is best if you have a graphing utility as you read this section, so that you can construct the graphs shown here while you read.

EXAMPLE 1

a. Graph the function f defined by

$$f(x) = -3\cos(2x) + 4\sin(2x)$$

on the window $-2\pi < x < 2\pi$, $-6 < y < 6$.

b. Analyze why the graph looks the way that it does.

Solution

a. Your graph should look like the graph of a positive multiple of the sine or cosine function that has been shifted along the x-axis (see Figure 34). The values of the function appear to range from -5 to 5 and the period should appear to be about π, which is the same as the period of the given sine and cosine terms.

Figure 34

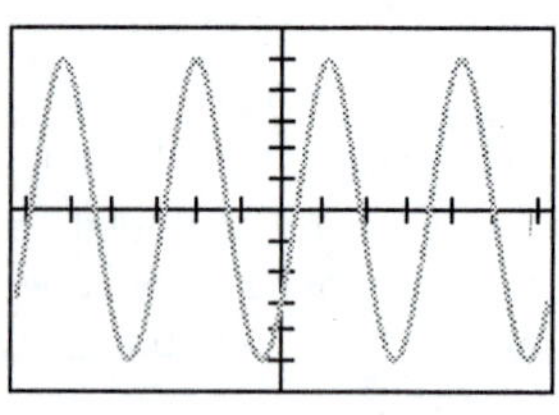

$-2\pi < x < 2\pi$, x-scale $= 1$
$-6 < y < 6$, y-scale $= 1$

b. It seems that $f(x)$ is expressible in the form $f(x) = A\cos(c[x - B])$ or $f(x) = A\sin(c[x + B])$ for appropriate constants A and B. Either of these forms of $f(x)$ would explain the apparent nature of the graph of $f(x)$ shown in Figure 34. To obtain the $A\cos(c[x - B])$ form, we multiply and divide the given expression for $f(x)$ by $5 = \sqrt{(-3)^2 + 4^2}$ to obtain

$$f(x) = -3\cos(2x) + 4\sin(2x) = 5\left(\frac{-3}{5}\cos(2x) + \frac{4}{5}\sin(2x)\right).$$

We have chosen the multiplier so that the coefficients, $-\frac{3}{5}$ and $\frac{4}{5}$ are coordinates of a point $\left(-\frac{3}{5}, \frac{4}{5}\right)$ on the unit circle. Consequently, there is an angle θ such that

$$\cos\theta = -\frac{3}{5} \quad \text{and} \quad \sin\theta = \frac{4}{5}.$$

(This angle has measure approximately 2.214 radians or 126.9°.) Therefore, we can express $f(x)$ in the form

$$\begin{aligned} f(x) &= 5(\cos\theta\cos(2x) + \sin\theta\sin(2x)) \\ &= 5\cos(2x - \theta) \\ &= 5\cos\left[2\left(x - \frac{\theta}{2}\right)\right]. \end{aligned}$$

This form of $f(x)$ shows that the graph of $f(x)$ is a cosine curve with an amplitude of 5, period π, that is shifted to the right by $\frac{\theta}{2}$.

Example 1 can be generalized into a theorem, whose proof we leave to you.

Theorem 9.11 Every function f of the form

$$f(x) = a\cos(cx) + b\sin(cx),$$

where a, b, and c are real numbers, can also be expressed in the form

$$f(x) = A\cos[c(x - B)],$$

where $A = \sqrt{a^2 + b^2}$, $\cos(Bc) = \frac{a}{A}$, and $\sin(Bc) = \frac{b}{A}$.

Either of the forms $f(x) = A\cos[c(x - B)]$ or $f(x) = A\sin[c(x + B)]$ is called a **phase-amplitude form** of $f(x)$ because the constant A represents the *amplitude* and B the horizontal shift or *phase shift* from the standard position of the corresponding sine or cosine function. As you will see later, the phase-amplitude form of a function is quite useful in analyzing vibrational phenomena.

Sums of multiples of two sine and/or cosine functions with the same amplitude and different periods

When multiples of sine and cosine functions have different periods, adding them may result in a function whose graph is not a pure sine wave. Still, a simple equation for the sum can be obtained and its graph analyzed.

EXAMPLE 2

a. Graph the function g defined by $g(x) = 2\cos(8x) - 2\cos(10x)$ on the window

$$-\pi < x < \pi, -6 < y < 6.$$

b. Analyze why the graph looks the way that it does.

Solution

Figure 35

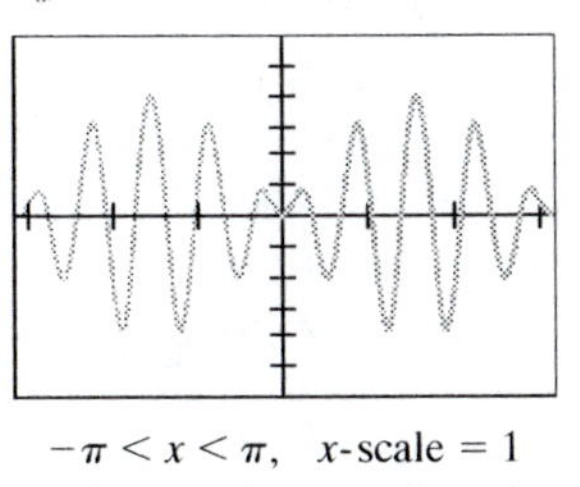

$-\pi < x < \pi$, x-scale = 1
$-6 < y < 6$, y-scale = 1

a. Your graph should look like that of a sine or cosine function with a small period that has been "pinched" along its length near $x = -\pi$, $x = 0$ and $x = \pi$ (see Figure 35). The graph appears to be symmetric to the y-axis.

If you graph this function over a larger interval, for x such as $-4\pi < x < 4\pi$, you may find that the graph breaks up badly due to the pixel limitations of your calculator screen, but you should still see that the pinching effect persists in intervals of apparent length 2π.

b. Note that the period of the first term is $\frac{2\pi}{8} = \frac{\pi}{4}$ while that of the second term is $\frac{2\pi}{10} = \frac{\pi}{5}$. Also note that

$$\begin{aligned} g(x) &= 2\cos(8x) - 2\cos(10x) \\ &= 2\cos(9x - x) - 2\cos(9x + x). \end{aligned}$$

Figure 36

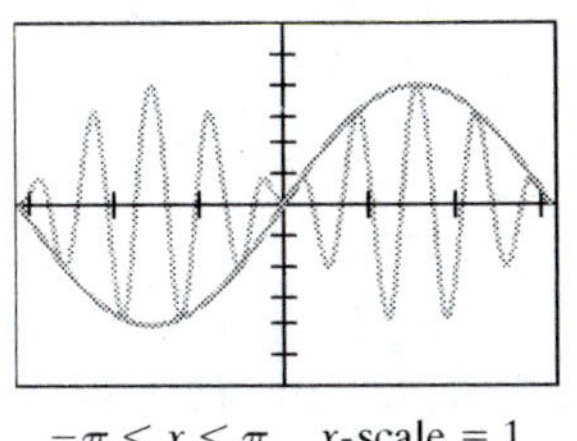

$-\pi < x < \pi$, x-scale = 1
$-6 < y < 6$, y-scale = 1

This is in the form of the identity of Problem 8**c** in Section 9.3.1. From that identity,

$$g(x) = 4\sin(9x)\sin x.$$

This expression for $g(x)$ reveals why its graph has its unusual appearance. The $4\sin(9x)$ term is periodic with period $\frac{2\pi}{9}$ and amplitude 4. It is multiplied by the factor $\sin x$, which has period 2π, and has values ranging from -1 to 1. Thus, the graph of the $h(x) = \sin x$ factor provides an outline for the graph of g. This is seen in Figure 36.

The most critical features of the function g in Example 2 are that the coefficients of the two terms are equal and the frequencies of the two terms are close in size. Thus their difference is much smaller than their sum. The fact that the two functions being added are cosine functions is not critical. (See Problem 7 following this section.)

An application to the analysis of mechanical vibrations.

The analysis of mechanical vibrations is important to a diversity of real-world phenomena, from situations as large as the design of machinery and earthquake-resistant structures to the analysis of atomic and molecular vibrations, among others. As with many applications of mathematics, there is a simple physical model that is used as a basis for mathematical models of the application. For example, in probabilistic or

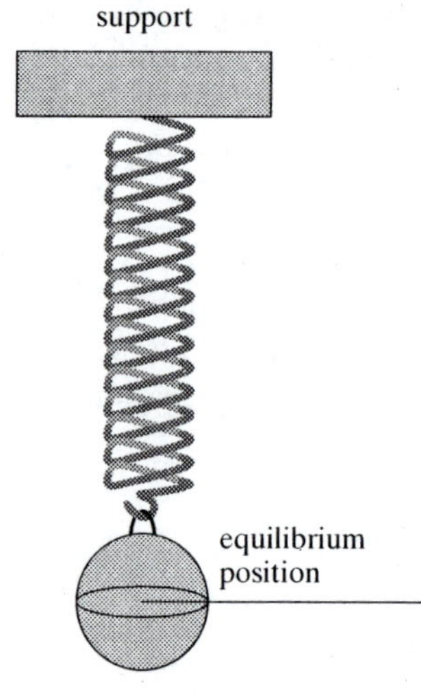

Figure 37

statistical applications, the physical model of drawing cards from a deck or balls from an urn is a simple but useful model. For problems in mechanical vibrations, the *spring-mass* system pictured in Figure 37 is useful.

When the mass at the base of the spring (shown in Figure 37 as a sphere) is disturbed in the vertical direction at a certain time $t = 0$, by positioning it above or below the equilibrium position and/or by imparting an initial velocity upward or downward, the mass will oscillate in the vertical direction for $t > 0$. If $x(t)$ is the directed distance of the center of the mass from its equilibrium position at time $t \geq 0$, then $x(t)$ depends not only on the initial position $x_0 = x(0)$ and the initial velocity $v_0 = x'(0)$ given to the mass but also on the mass m and the physical characteristics of the spring.

Experience suggests that the spring oscillates and, in the idealized absence of friction, would oscillate forever. But it is a wonderful surprise that the height of the mass is given by a sine or cosine function. Here is why.

Within their elastic limits, stretching and compressing physical springs typically satisfy Hooke's Law: *The magnitude $|F|$ of the force F that is required to stretch or compress a spring a distance D is proportional to D; that is, $|F| = kD$ for some positive constant k.* The constant k is called the **spring constant** for the spring. Springs with large spring constants are stiff; those with small spring constants stretch and compress more easily.

Suppose that a spring-mass system consists of a mass m suspended from a spring with spring constant k. If the internal friction of the spring and other forces acting on the spring are assumed to be negligible, then as a consequence of **Newton's Second Law of Motion** $ma = F$; that is,

$$(\text{mass})(\text{acceleration}) = \text{force}.$$

Because acceleration is the second derivative of position, and force is proportional to the second derivative, the function $x(t)$ describing the position of the mass at time t must satisfy the differential equation

$$m\frac{d^2x(t)}{dt^2} = -kx(t) \qquad \text{for all } t \geq 0 \tag{1}$$

and the initial conditions:

$$x(0) = x_0 \qquad \text{(initial position)} \tag{2}$$

$$\frac{dx(0)}{dt} = v_0 \qquad \text{(initial velocity)}. \tag{3}$$

Equations (1), (2), and (3) are sometimes called the **equations of motion** for the spring-mass system.

EXAMPLE 3 Suppose that suspending a 6-lb weight from a certain spring stretches it 6 in. At time $t = 0$, the weight is moved 4 in. above its equilibrium position and given an initial velocity of 2 ft per second upward. Find the equations of motion of this spring-mass system.

Solution In the ft-lb-sec system, mass is measured in slugs, and mass m in slugs of an object is related to its weight w in pounds by $w = 32m$. Also, because the spring is stretched 6 in. $= \frac{1}{2}$ ft, the spring constant k is given by $6\text{ lb} = k \cdot \frac{1}{2}$ ft or $k = 12$ lb per ft. Therefore, the directed distance $x(t)$ of the weight above the equilibrium position satisfies the equation $\frac{6}{32}\frac{d^2x(t)}{dt^2} = -12x(t)$, which simplifies to

$$\frac{d^2x(t)}{dt^2} = -64x(t) \qquad \text{for all } t \geq 0. \tag{1'}$$

By adding the information about the initial position and velocity, we obtain (2′) and (3) of the equations of motion.

(2′) $$x(0) = \frac{1}{3}\text{ ft} \qquad \text{(initial position)}$$

(3′) $$\frac{dx(0)}{dt} = 2\text{ ft per sec} \qquad \text{(initial velocity)}$$

It is possible to find the motion function $x(t)$ from the equations of motion of the spring-mass system in Example 3. This requires, quite obviously, use of the formulas for differentiation of the sine and cosine functions derived in calculus. Recall that if $f(t) = \sin t$, then $f'(t) = \frac{df(t)}{dt} = \cos t$, and if $f(t) = \cos t, f'(t) = \frac{df(t)}{dt} = -\sin t$. From this, we can check that some specific functions satisfy the equations of motion.

Question 1: Verify that each of the functions

$$x_1(t) = \cos\left[\sqrt{\frac{k}{m}}t\right] \quad \text{and} \quad x_2(t) = \sin\left[\sqrt{\frac{k}{m}}t\right]$$

satisfy equation (1) of the equations of motion. More generally, show that if a and b are any real numbers, the function

$$x_3(t) = a\cos\left(\sqrt{\frac{k}{m}}t\right) + b\sin\left(\sqrt{\frac{k}{m}}t\right) \qquad \text{for all } t \geq 0$$

satisfies equation (1).

Question 2: Show that if $a = x_0$ and $b = v_0\sqrt{\frac{m}{k}}$ in the function x_3 of Question 1, which satisfies equation (1) of the equations of motion, the resulting function also satisfies equations (2) and (3).

EXAMPLE 4 Consider the spring-mass system described in Example 3.

a. Use the results of Question 2 to find the motion function $x(t)$ from the equations of motion of the system.

b. Write $x(t)$ in phase-amplitude form and use it to describe the graphical characteristics of the motion function $x(t)$ of the system.

c. Use a graphing utility to confirm the conclusions from part (b).

Solution

a. $x(t) = \cos x_0\left(\sqrt{\frac{m}{k}}t\right) + v_0\sqrt{\frac{m}{k}}\sin\left(\sqrt{\frac{m}{k}}t\right)$. Here $k = -12\frac{lb}{ft}$ and $m = \frac{6}{32}$ slugs. Thus

$$x(t) = \frac{1}{3}\cos(8t) + \frac{1}{4}\sin(8t).$$

b. To write $x(t)$ in the phase-amplitude form $A\cos[8(t - B)]$, note that

$$A = \sqrt{\left(\frac{1}{3}\right)^2 + \left(\frac{1}{4}\right)^2} = \frac{5}{12}; \quad \cos(8B) = \frac{\frac{1}{3}}{\frac{5}{12}} = \frac{4}{5}; \quad \sin(8B) = \frac{\frac{1}{4}}{\frac{5}{12}} = \frac{3}{5}$$

and so $8B \approx .6435$ radians or 36.8°. Therefore, $B \approx .08$ radians or 4.61°. Therefore, a phase-amplitude form of $x(t)$ is

$$x(t) \approx \frac{5}{12}\cos[8(t - .08)],$$

Figure 38

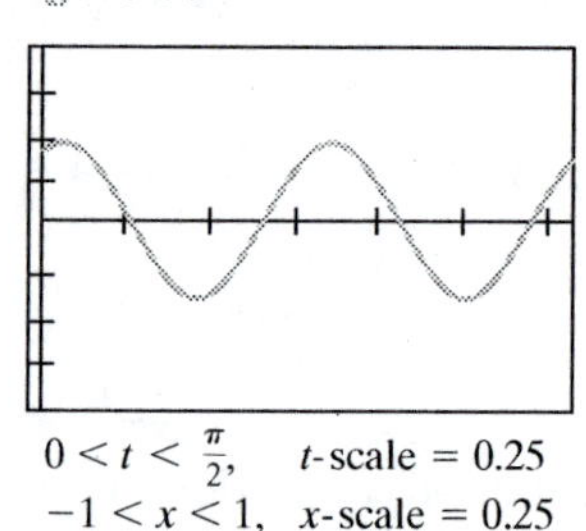

$0 < t < \frac{\pi}{2}$, t-scale = 0.25
$-1 < x < 1$, x-scale = 0.25

which shows that the amplitude of the oscillation of the mass is $\frac{5}{12}$ ft or 5 in. from the equilibrium position.

c. In the window $0 < t < \frac{\pi}{2}, -1 < x < 1$, the graph of $x(t)$ looks like the graph in Figure 38. This graph shows that the mass, which is initially 4 in. above its equilibrium, continues upward for another inch because the initial velocity is upward, before beginning its downward movement toward its equilibrium. The frequency of the motion is $\frac{8}{2\pi}$ seconds or 1.27 seconds, so the motion pictured in Figure 38 takes place in a little more than 2.5 seconds.

In the preceding discussion, it was assumed that no outside forces and no damping forces were acting on the spring-mass system. If an outside force is acting on the system, then the resulting motion would obviously be affected. For example, suppose that an outside force is acting on the spring-mass system in Example 1 so that it is itself vibrating according to the function

$$f(t) = \frac{1}{4}\sin(2\pi t).$$

(This might be accomplished by moving the support up and down with an amplitude of $\frac{1}{4}$ ft starting at $t = 0$.) Then it can be shown that the motion function $x(t)$ of the mass would be a function of the form

$$x(t) = A\cos[8(t - B)] + C\sin[2\pi t],$$

where A, B, and C are constants determined by the initial conditions, the mass m and the spring constant k. This is the sort of function that we considered in Examples 3 and 4. Consequently, a motion of the sort displayed in Example 3 may result. In the analysis of mechanical vibrations, such motions are called **beats**. Most of us have experienced beats in sound vibrations. For example, the sound that we hear from an electric motor running at a constant speed might have a periodically varying amplitude because the support on which the motor is mounted is vibrating at frequency close to the frequency of the sound produced by the motor itself.

9.3.2 Problems

1. Show that the function f with $f(t) = -3\cos(2t) + 4\sin(2t)$ in Example 1 can also be expressed as $f(t) = A\sin[c(t + B)]$.

2. Prove Theorem 9.11.

3. Prove the analogue to Theorem 9.11 suggested by Problem 1 of this set.

4. Graph the function g with $g(x) = \sin x + \cos x$, and explain why the graph has the shape it has.

5. Graph $h(x) = 7\cos x - 24\sin x$ and explain why the graph has the period and amplitude it has.

6. a. Graph $f(x) = \sin x - \cos x$ and explain why its graph looks the way it does.

   b. Repeat part **a** for the function $g(x) = (\sin x - \cos x)e^{-x}$.

7. Suppose that c_1 and c_2 are real numbers such that $c_1 > c_2 > 0$ such that $c_1 + c_2$ is much larger than $c_1 - c_2$. Prove: If a is any nonzero real number and if $f(x)$ has any of the following forms,

$$f(x) = \begin{cases} a[\cos(c_1x) \pm \cos(c_2x)] \\ a[\sin(c_1x) \pm \cos(c_2x)] \\ a[\cos(c_1x) \pm \sin(c_2x)] \\ a[\sin(c_1x) \pm \sin(c_2x)] \end{cases}$$

then f has a graph somewhat like that of the function in Example 2 in this section.

ANSWERS TO QUESTIONS

1. If $x(t) = \cos\left(\sqrt{\frac{k}{m}}t\right)$, then $x'(t) = -\sqrt{\frac{k}{m}}\sin\left(\sqrt{\frac{k}{m}}t\right)$, and $x''(t) = -\frac{k}{m}\cos\left(\sqrt{\frac{k}{m}}t\right)$, shows that equation (1) holds. If $x(t) = \sin\left(\sqrt{\frac{k}{m}}t\right)$, then $x'(t) = \sqrt{\frac{k}{m}}\cos\left(\sqrt{\frac{k}{m}}t\right)$, and $x''(t) = -\frac{k}{m}\sin\left(\sqrt{\frac{k}{m}}t\right)$, so equation (1) again follows. If $x(t) = a\cos\left(\sqrt{\frac{k}{m}}t\right) + b\sin\left(\sqrt{\frac{k}{m}}t\right)$, then $x'(t) = -a\sqrt{\frac{k}{m}}\sin\left(\sqrt{\frac{k}{m}}t\right) + b\sqrt{\frac{k}{m}}\cos\left(\sqrt{\frac{k}{m}}t\right)$, so $x''(t) = -a\frac{k}{m}\cos\left(\sqrt{\frac{k}{m}}t\right) - b\frac{k}{m}\sin\left(\sqrt{\frac{k}{m}}t\right)$, and again equation (1) holds.

2. $x_3(t) = x_0\cos\left(\sqrt{\frac{k}{m}}t\right) + v_0\sqrt{\frac{m}{k}}\sin\left(\sqrt{\frac{k}{m}}t\right)$, so $x_3(0) = x_0\cos\left(\sqrt{\frac{k}{m}}\cdot 0\right) + v_0\sqrt{\frac{m}{k}}\sin\left(\sqrt{\frac{k}{m}}\cdot 0\right) = x_0\cdot 1 + 0 = x_0$, so equation (2) is satisfied. $\frac{dx(0)}{dt} = \frac{dx_3(0)}{dt} = -x_0\sqrt{\frac{k}{m}}\sin\left(\sqrt{\frac{k}{m}}\cdot 0\right) + v_0\sqrt{\frac{m}{k}}\cdot\sqrt{\frac{k}{m}}\cos\left(\sqrt{\frac{k}{m}}\cdot 0\right) = v_0$. So equation (3) is satisfied.

9.3.3 Analytical properties of the sine and cosine functions

The formulas that you learned in calculus for the derivatives of the sine and cosine functions, for x in radians,

$$\frac{d}{dx}(\sin x) = \cos x \quad \text{and} \quad \frac{d}{dx}(\cos x) = -\sin x,$$

and the corresponding second derivative formulas

$$\frac{d^2}{dx^2}(\sin x) = -\sin x \quad \text{and} \quad \frac{d^2}{dx^2}(\cos x) = -\cos x,$$

are remarkable in their simplicity and yet powerful in a variety of applications in pure and applied mathematics.

In calculus, these formulas are often derived analytically by applying the limit definition of the derivative, the addition formulas for the sine and cosine, and the special limit:

$$\lim_{\theta\to 0}\frac{\sin\theta}{\theta} = 1 \qquad \theta \text{ in radians} \qquad \text{(see Problem 3)}.$$

Although these derivations are straightforward, they give little insight into the nature of these functions as tools for modeling periodic behavior.

We now show that analytic properties of the sine and cosine functions embodied in the differentiation formulas can be understood dynamically by using the sine and cosine functions as models of periodic behavior. Specifically, we will explain these formulas in terms of a mathematical model for *uniform circular motion in a plane.*

Uniform circular motion of an object in a plane

A uniform circular motion model is appropriate for physical situations such as (a) the motion of a communications satellite in a circular orbit around earth, or (b) motion of an object that you are spinning rapidly in a circle by holding one end of a string with the other end tied to the object. A precise description of uniform circular motion of an object requires statements about its path, its velocity, and its speed.

(P) Path of the object: The object travels in a path T that is a directed circle (i.e., a circle with a specified direction) with a radius r and with a center at a point O in that plane.

(V) Velocity and speed of the object: The speed s of the object along the path is constant, and the velocity vector of the object at any point P of the path is tangent to the path T at P and points in the direction of the path.

We need to say more than this to describe the situations (a) and (b) mentioned above. For instance, if the string in situation (b) would break, the object would fly off in a direction tangent to the circle at the point P where the object is located when the string breaks. Your hand must exert (centripetal) force to keep the object moving along the circular path T. This force F is exerted on the object in the direction from the object to your hand and follows Newton's law $F = ma$. With this example in mind, we can add the following statement to the description of uniform motion.

(A) Acceleration of the object: The object is accelerated toward the center of the circular path T and the magnitude of the acceleration is constant. (The centripetal force on the object accounts for the acceleration.)

We assume that all forces other than the centripetal force are negligible relative to the centripetal force.

Mathematical models of uniform circular motion

Now we create a mathematical model for uniform circular motion. First, we need a convenient location and description for the directed path T. Since T is to travel along a circle of radius R, suppose that we locate that circle in the xy-plane with its center at the origin. Then the path T is located on the circle

$$x^2 + y^2 = R^2.$$

However, this equation is not a suitable description of the path. Motion means that the position P of the object depends on time. The equation $x^2 + y^2 = R^2$ does not allow us to determine the position of the object at a given time.

What we need is a description of the path in which the time t is a parameter and that gives the location of the point P along the circular path at any time t. This can be done with an equation for the position vector $\mathbf{r(t)}$ joining O to P at time t,

$$\mathbf{r(t)} = g(t)\mathbf{i} + h(t)\mathbf{j},$$

where $\mathbf{i} = (1, 0)$ and $\mathbf{j} = (0, 1)$ are the standard unit vectors, and $g(t)$ and $h(t)$ are real functions defined for all real numbers t. Any such path has a *natural direction*—the direction of increasing t values—and a *natural initial point*—the point that corresponds to $t = 0$.

We now identify conditions on the functions $g(t)$ and $h(t)$ that correspond to the path, velocity, and acceleration requirements (P), (V), and (A) above for uniform circular motion on a circle of radius R centered at the origin O. Recall that velocity is defined as the derivative of position with respect to time, and the derivative of velocity with respect to time is acceleration. Thus, since the position vector $\mathbf{r}$ joining O to P (Figure 39) is given by the vector function

$$\mathbf{r(t)} = g(t)\mathbf{i} + h(t)\mathbf{j},$$

Figure 39

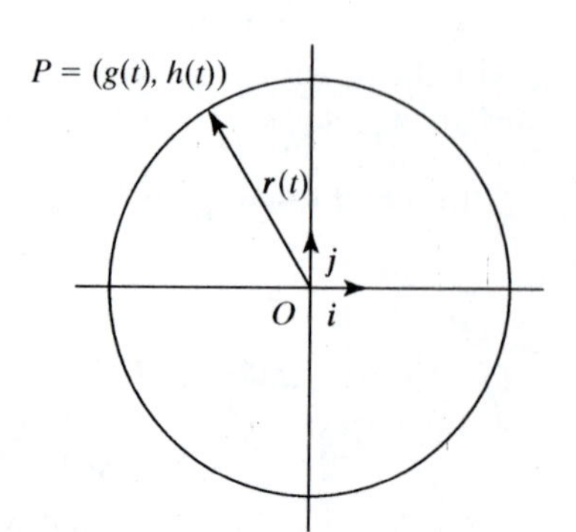

the velocity vector **v** at P is

$$\mathbf{v(t)} = g'(t)\mathbf{i} + h'(t)\mathbf{j}.$$

The speed $s(t)$ at time t is the magnitude of the velocity vector. That is,

$$s(t) = |\mathbf{v(t)}| = \sqrt{g'(t)^2 + h'(t)^2}.$$

The acceleration vector **a** is given by

$$\mathbf{a(t)} = g''(t)\mathbf{i} + h''(t)\mathbf{j}.$$

To assure that this path is on the circle of radius R centered at the origin as required in condition (P), we need to require that $g(t)$ and $h(t)$ satisfy the two conditions shown here as (P*).

(P*) **i.** $g(t)^2 + h(t)^2 = R^2$ for all t, and

ii. for all (x, y) with $x^2 + y^2 = R^2$, there is a t such that $x = g(t)$ and $y = h(t)$.

According to the velocity condition (V), the speed of the object must be constant for uniform circular motion. Consequently, $g(t)$ and $h(t)$ must also satisfy the velocity condition (V*).

(V*) The functions $g(t)$ and $h(t)$ are differentiable and there is a constant K such that $g'(t)^2 + h'(t)^2 = K^2$ for all t.

Condition (V) for uniform circular motion also requires that the velocity vector **v(t)** at any point P along the circular path T must be perpendicular to the position vector **v(t)** at P. (The tangent to a circle is perpendicular to the radius at its endpoint on the circle.) In Problem 4, you are asked to verify that $\mathbf{r(t)} = g(t)\mathbf{i} + h(t)\mathbf{j}$ and $\mathbf{v(t)} = g'(t)\mathbf{i} + h'(t)\mathbf{j}$ are perpendicular vectors if and only if the following condition is satisfied,

$$g(t)g'(t) + h(t)h'(t) = 0 \text{ for all } t;$$

that is, if their dot product is zero. However, this condition is automatically satisfied for differentiable functions $g(t)$ and $h(t)$ that satisfy condition (P) because

$$g(t)^2 + h(t)^2 = R^2 \text{ for all } t \Rightarrow 0 = \frac{d}{dt}(R^2) = 2g(t)g'(t) + 2h(t)h'(t) \text{ for all } t.$$

Consequently, conditions (P*) and (V*) together assure that the requirements (P) and (V) for uniform circular motion on a circle of radius R are satisfied.

Finally, we consider condition (A) on the motion. This requires that the acceleration of the object is directed toward the center of the circular path and that it has a constant magnitude. Because the acceleration is directed toward the center of the circle, its direction is opposite to that of the position vector $\mathbf{r(t)} = g(t)\mathbf{i} + h(t)\mathbf{j}$. Because its magnitude is constant, it is simply a negative multiple of the position vector. Consequently, requirement (A) for uniform circular motion is met if the functions $g(t)$ and $h(t)$ satisfy the following condition.

(A*) There is a positive constant M such that $\mathbf{a(t)} = -M\mathbf{r(t)}$; that is,

$$g''(t) = -Mg(t) \quad \text{and} \quad h''(t) = -Mh(t) \text{ for all } t.$$

In summary, if we can find a pair of twice differentiable functions $g(t)$ and $h(t)$ that satisfy conditions (P*), (V*), and (A*), then the vector function

$$\mathbf{r(t)} = g(t)\mathbf{i} + h(t)\mathbf{j}$$

provides a mathematical model for uniform circular motion.

Notice that the requirements

$$g(t)^2 + h(t)^2 = R^2 \quad \text{and} \quad g'(t)^2 + h'(t)^2 = K^2 \qquad \text{for all } t$$

in (P*) and (V*) bear a strong resemblance to the Pythagorean identity $\cos^2 x + \sin^2 x = 1$, and that the requirement

$$g''(t) = -Mg(t) \quad \text{and} \quad h''(t) = -Mh(t) \qquad \text{for all } t$$

is reminiscent of the derivative formulas

$$\frac{d^2}{dx^2}(\sin x) = -\sin x \quad \frac{d^2}{dx^2}(\cos x) = -\cos x \qquad x \text{ in radians.}$$

These similarities suggest that vector function

$$\mathbf{r(t)} = \mathrm{R}(\cos t)\mathbf{i} + \mathrm{R}(\sin t)\mathbf{j}$$

might satisfy all the conditions (P*), (V*), and (A*). This is the case.

Theorem 9.12(a) The vector function

$$\mathbf{r(t)} = R\cos(t)\mathbf{i} + R\sin(t)\mathbf{j}$$

describes uniform circular motion for an object P on a circle of radius R centered at the origin O with a constant speed R and a constant magnitude of acceleration R.

Proof: Condition (P*) can be verified for this position function as follows:

i. $\sqrt{(R\cos t)^2 + (R\sin t)^2} = \sqrt{R^2(\cos^2 t + \sin^2 t)} = R,$

ii. If (x, y) satisfies $x^2 + y^2 = R^2$, then $\left(\frac{x}{R}, \frac{y}{R}\right)$ is a point on the unit circle and so $\frac{x}{R} = \cos t, \frac{y}{R} = \sin t$ for some t with $0 \leq t < 2\pi$. Since

$$\frac{d}{dx}(\sin x) = \cos x \quad \text{and} \quad \frac{d}{dx}(\cos x) = -\sin x,$$

and

$$\frac{d^2}{dx^2}(\sin x) = -\sin x \quad \text{and} \quad \frac{d^2}{dx^2}(\cos x) = -\cos x,$$

conditions (V*) and (A*) are also satisfied by

$$\mathbf{r(t)} = R(\cos t)\mathbf{i} + R(\sin t)\mathbf{j},$$

with the constant K equal to R and the constant M equal to 1. Thus, this position function is a model for uniform circular motion on a circle of radius R provided that the speed of the object is also R and provided that the acceleration vector and the position vector have equal magnitudes. ┘

A good mathematical model of uniform circular motion on a circle of radius R should not require the speed of the object to be numerically equal to the radius of the path. That might be the case for a particular motion, but we need a model with more flexibility—one that allows us to specify the speed of the object as well as the radius of its circular path. That can be accomplished by introducing a parameter k into our model, as in the following generalization of the preceding theorem and proof. The proof is left to you as Problem 1.

Theorem 9.12(b) Suppose that k is a nonzero real number. Then the vector function

$$\mathbf{r_k(t)} = R\cos(kt)\mathbf{i} + R\sin(kt)\mathbf{j}$$

describes uniform circular motion for an object P on a circle of radius R centered at the origin O with a constant speed kR and a constant magnitude of acceleration k^2R.

Applying the models for uniform circular motion

We now return to one of the problems that motivated this discussion.

EXAMPLE 1 Determine the speed that must be attained by a launch vehicle to place a communications satellite in a circular orbit above the equator at a distance of 5000 miles above the center of Earth.

Solution Assume that Earth is a sphere of radius 3960 miles and that the satellite's orbit is near enough to the earth so that the acceleration due to Earth's gravity is essentially the same as that on Earth's surface, namely,

$$g = 32.2\frac{\text{ft}}{\text{sec}^2} \approx 79000\frac{\text{mi}}{\text{hr}^2}.$$

Also assume that other forces acting on the satellite, such as atmospheric drag and the gravitational attraction of the moon and other celestial bodies, are negligible compared to the gravitational attraction of the earth. (These assumptions limit the radius R of the orbit to the range from roughly 4100 to about 20,000 miles.)

By Theorem 9.12(b), we can model the uniform circular motion of the satellite so that its position at time t is given by

$$\mathbf{r(t)} = 5000\cos(kt)\mathbf{i} + 5000\sin(kt)\mathbf{j}.$$

The velocity $\mathbf{v(t)}$ of the satellite at time t is the derivative of $r(t)$.

$$\mathbf{v(t)} = -5000k\sin(kt)\mathbf{i} + 5000k\cos(kt)\mathbf{j}$$

Its speed $s(t)$ at time t is the magnitude of the velocity vector, so $s(t) = 5000k$, a constant. We call the constant speed s. Its acceleration at time t, $a(t)$ is the derivative of velocity.

$$\mathbf{a(t)} = -5000k^2\cos(kt)\mathbf{i} - 5000k^2\sin(kt)\mathbf{j}$$

The acceleration has magnitude $|\mathbf{a(t)}| = 5000k^2 = \dfrac{s^2}{5000}$.

Since the satellite's acceleration must equal the acceleration due to gravity g,

$$\frac{s^2}{5000} \approx 79000\frac{\text{mi}}{\text{hr}^2},$$

from which $s \approx \sqrt{5000 \cdot 79000} \approx 19{,}900\frac{\text{mi}}{\text{hr}}$, and $k \approx 3.98\frac{1}{\text{hr}}$.

The period p of the orbit is the period of the position vector, so $p = \frac{2\pi}{k} \approx$ 1.58 hours.

Are there other mathematical models for uniform circular motion on a circle of radius R that differ significantly from the model given in Theorem 9.12(b)? The answer is no. Under reasonable differentiability restrictions on the pair of functions g and h, the converse of Theorem 9.12(a) can be proved. That is, if g and h satisfy the three conditions

$$g(t)^2 + h(t)^2 = 1, \quad \frac{d}{dt}g(t) = -h(t), \quad \text{and} \quad \frac{d}{dt}h(t) = g(t),$$

then $g(t) = \cos t$ and $h(t) = \sin t$. Thus, not only are the sine and cosine functions useful for modeling uniform circular motion, but the mathematical model of uniform circular motion also explains why the sine and cosine functions have the first and second derivatives that they do.

9.3.3 Problems

1. Prove Theorem 9.12(b).

2. Explain why the model in Theorem 9.12(b) has the following property: Once the radius R and the speed s of the motion are specified, the constant k and the magnitude of the acceleration are determined.

3. a. Use a geometric argument based on Figure 40 to verify the limit formula

$$\lim_{\theta\to 0}\frac{\sin\theta}{\theta} = 1, \text{ where } \theta \text{ is in radians.}$$

b. Apply the limit definition of the derivative

$$\frac{df(x)}{dx} = \lim_{h\to 0}\frac{f(x+h)-f(x)}{h}$$

to derive the formula $\frac{d}{dx}(\sin x) = \cos x$, where x is in radians.

4. Verify that $\mathbf{r(t)} = g(t)\mathbf{i} + h(t)\mathbf{j}$ and $\mathbf{v(t)} = g'(t)\mathbf{i} + h'(t)\mathbf{j}$ are orthogonal vectors if and only if the following condition is satisfied:

$$g(t)g'(t) + h(t)h'(t) = 0 \qquad \text{for all } t.$$

5. Compute the orbital insertion speed and altitude to establish a stationary orbit above the equator for a communications satellite.

Figure 40

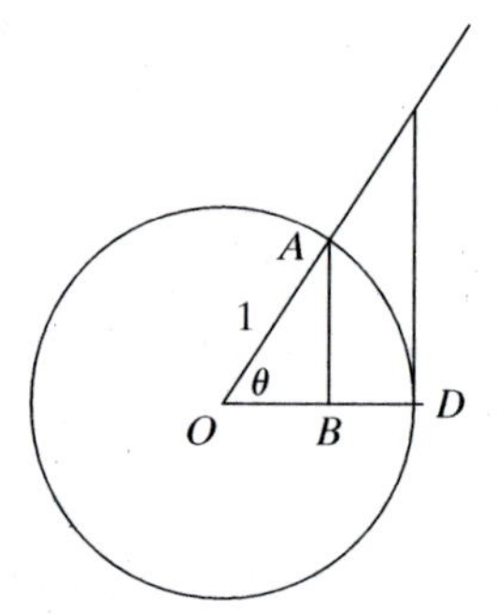

Chapter Projects

1. Trigonometry and musical instruments. Read "The Mathematics of Musical Instruments", by Rachel W. Hall and Kresimir Josic, *The American Mathematical Monthly* 108: 347–357 (April 2001). Write a summary of this article that could be used for presentation to a class.

2. Regiomontanus's problem. Imagine you are looking at a vertical sign that is above your eye level, as shown in Figure 41.

Assume that the vertical dimension ℓ of the sign and the height h of its bottom above eye level are known. Let r be the distance of your eye from the projection of the sign on the ground. We want to know the best place from which to read the sign. For best viewing of the sign we would want to make the measure of the angle $\theta_2 - \theta_1$ as large as possible. The sign then

Figure 41

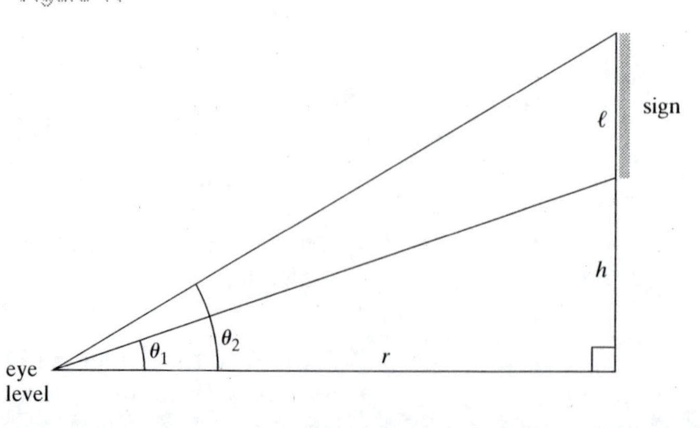

occupies the greatest portion of your field of vision. Consequently, this problem can be formulated as follows: Given fixed values of ℓ and h, find the value of r that maximizes $\theta_2 - \theta_1$.

a. Show that the maximum viewing angle occurs when the distance $r = \sqrt{h(h + \ell)}$.

b. Interpret the result in part **a** by showing that the maximum viewing angle occurs when a line of sight from the eye elevated at a 45° angle points at the "middle" of the sign, where here the "middle" is the geometric mean of the height of the bottom of the sign and the height of the top of the sign above eye level, as shown in Figure 42.

Figure 42

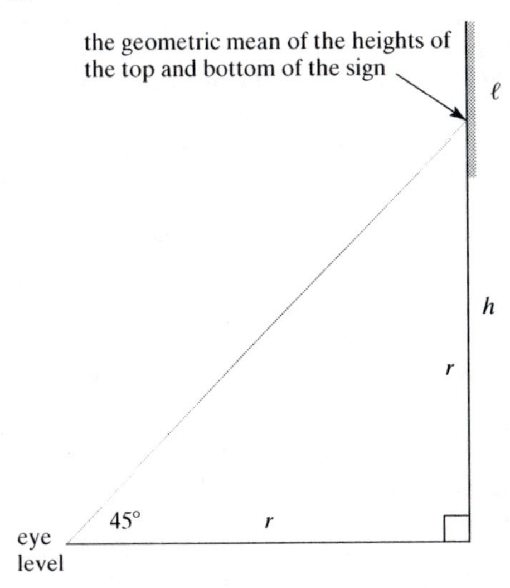

c. Interpret the result in another way by showing that the maximum viewing angle occurs when the viewing point P is the point of tangency with eye level of a circle in a plane perpendicular to the ground and to the sign, tangent to the ground, and which contains the top and bottom of the sign (see Figure 43).

Figure 43

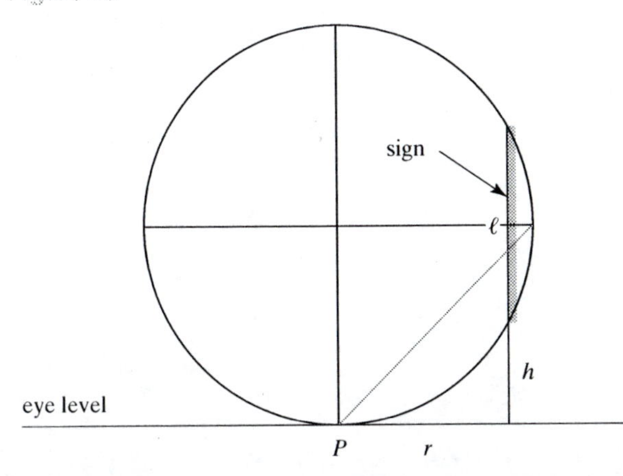

3. Constructing a table of sines and cosines. Assume you have a calculator that has no keys for the trignometric functions but can calculate sums, products, differences, quotients, and square roots. Construct an approximate table of sines and cosines for angle measures from 0° to 90° in increments of 1° with the following steps.

a. From the known values of these functions for 30° and 45°, using the identities in Section 9.3.1, obtain exact values for $\sin 15n°$ and $\cos 15n°$ for $n = 1, 4,$ and 5. Then record decimal approximations to these values.

b. Prove that $\sin 18° = \frac{\sqrt{5}-1}{4}$.

c. Use the values from parts **a** and **b** and the identities in Section 9.3.1 and its problems to obtain decimal approximations to $\sin 3n°$ and $\cos 3n°$ for $1 \le n \le 29$.

d. Interpolate from the values you found in part **c** to obtain decimal approximations for all the other integer degree measures from 0° to 90°.

e. Compare the values you obtained in part **d** with the values given by a calculator. Where you are farthest off, use the half-angle formulas and more interpolation to obtain better estimates.

4. Remarkable equalities. Values of the trigonometric functions are related to each other in many and wondrous ways. Here are eight identities and four relationships among specific values. Deduce as many as you can.

a. For any x, $\sin x + \sin(2x) + \sin(3x) + \cdots + \sin(nx) = \dfrac{\sin\frac{1}{2}(n+1)x \cdot \sin\frac{1}{2}(nx)}{\sin\frac{1}{2}x}$.

b. For any x, $\cos x + \cos(2x) + \cos(3x) + \cdots + \cos(nx) = \dfrac{\cos\frac{1}{2}(n+1)x \cdot \sin\frac{1}{2}(nx)}{\sin\frac{1}{2}x}$.

c. For any x, $\sin x + \sin(3x) + \sin(5x) + \cdots + \sin((2n-1)x) = \frac{\sin^2(nx)}{\sin x}$.

d. For any x, $\cos x + \cos(3x) + \cos(5x) + \cdots + \cos((2n-1)x) = \frac{\sin(2nx)}{2\sin x}$.

In identities **e–h**, A, B, and C are angles in any triangle ABC.

e. $\sin A + \sin B + \sin C = 4\cos\frac{1}{2}A \cdot \cos\frac{1}{2}B \cdot \cos\frac{1}{2}C$

f. $\cos A + \cos B + \cos C = 1 + 4\sin\frac{1}{2}A \cdot \sin\frac{1}{2}B \cdot \sin\frac{1}{2}C$

g. $\tan A + \tan B + \tan C = \tan A \cdot \tan B \cdot \tan C$

h. $\cot\frac{1}{2}A + \cot\frac{1}{2}B + \cot\frac{1}{2}C = \cot\frac{1}{2}A \cdot \cot\frac{1}{2}B \cdot \cot\frac{1}{2}C$

i. $\sin 20° \cdot \sin 40° \cdot \sin 60° \cdot \sin 80° = \frac{3}{16}$

j. $\cos 20° \cdot \cos 40° \cdot \cos 60° \cdot \cos 80° = \frac{1}{16}$

k. $\sin 6° \cdot \sin 42° \cdot \sin 66° \cdot \sin 78° = \frac{1}{16}$

l. $\sin 12° \cdot \sin 24° \cdot \sin 48° \cdot \sin 96° = \frac{1}{16}$

5. Modeling with trigonometric functions. In Section 9.2.2, the length of a day in the year 2000 at 40°N latitude over a year is approximated by a trigonometric function. Find data for your location for at least one day in each month of the current year for each of these three times: sunrise, sunset, and the length of a day. (*Note*: For sunrise and sunset, you may need to interpolate from values given in an almanac to take into account the longitude of your location.) Model each of these times with a trigonometric function and compare the values given by your model with the actual data for your location.

Bibliography

Unit 9.1 References

Aaboe, A. *Episodes from the Early History of Mathematics*. New Mathematical Library Volume 13. Washington, DC: Mathematical Association of America, 1964.
This volume includes a detailed exposition on how Ptolemy constructed a trigonometric table in his *Almagest*.

Berggren, John Lennart. *Episodes in the Mathematics of Medieval Islam*. New York: Springer-Verlag, 1986.
Chapter 5 discusses tables of chords and the sine, introduction of the six trigonometric functions, proofs of the addition formulas, among other things. Chapter 6 discusses spherical trigonometry.

Gelfand, I. M., and Mark Saul. *Trigonometry*. Boston: Birkhäuser, 2001.
This short text treats in a simple and elegant fashion all of the trigonometric results usually seen in high school, and many others as well, often with interesting historical information added. Calculus is not used, but the discussion is much broader and fuller than is typical in most precalculus texts.

Unit 9.2 References

Dunham, William. *Euler: The Master of Us All*. Washington, DC: Mathematical Association of America, 1999.
Euler's work with complex numbers and trigonometry is found in Chapter 5.

Katz, Victor. *A History of Mathematics*. New York: HarperCollins, 1993.
Attention to the history of trigonometry, including non-European work, is found in Chapter 4 and Sections 6.6, 7.4, and 10.3.

Kennedy, Edward S. "The History of Trigonometry—An Overview." In *Historical Topics for the Mathematics Classroom* (Washington, DC: National Council of Teachers of Mathematics, 1969), pp. 333–359. Reprinted in *Studies in the Islamic Exact Sciences*. Beirut: American University of Beirut, 1983, pp. 3–29.
A readable history from ancient times to the 18th century, followed by several short histories of individual trigonometric ideas.

Maor, Eli. *Trigonometric Delights*. Princeton, NJ: Princeton University Press, 1998.
This very readable and engaging book discusses the historical evolution of trigonometry.

Unit 9.3 References

Hobson, Ernest William. *A Treatise on Plane Trigonometry*. New York: Cambridge University Press, 1921.
This classic treatise on trigonometry written by an outstanding mathematician of that time demonstrates that the subject has a richness and depth that is not at all evident in most modern textbooks.

Chapter

10

AREA AND VOLUME

In two-dimensional space, area measures the space covered or enclosed by a figure. The calculation of area of land is one of the oldest problems that we know of in mathematics, originating in part from problems of fair apportionment of land inherited by children from their parents.

The problem of calculating such areas immediately becomes difficult. Boundaries of land are often not straight; they may include parts of rivers or edges of foothills or swamps. The land itself may include small bodies of water. The land may be hilly. One method of finding the area of irregular pieces of land is not very far from the approach taken in calculus. If the land is not too large, it is photographed from above, thus making it flat. A grid of congruent squares with known dimensions is placed over the photo (Figure 1a). The number of squares entirely inside the boundary plus half the number crossed by the boundary provides an estimate for the area. If a more precise estimate is desired, then smaller squares are used (Figure 1b).

Figure 1

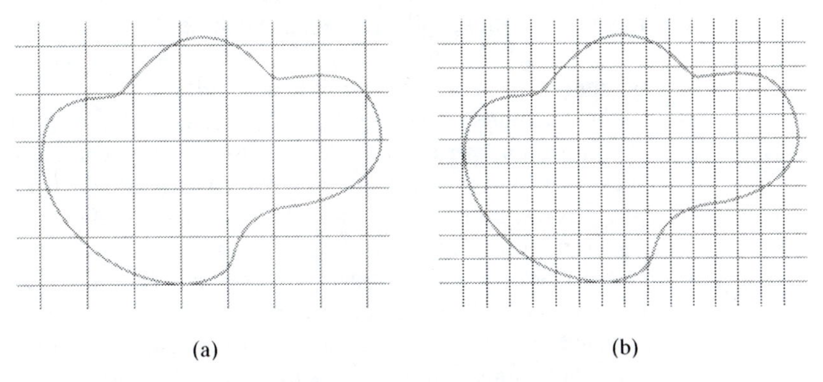

This approach is quite similar to the approach we take in Unit 10.1 to develop and discuss the area of plane figures. It applies also to the calculation of the surface area of common three-dimensional figures. That is, we can trace everything ultimately back to the areas of squares because, as we will show, we can rearrange the parts of any polygonal region to form a square.

In Unit 10.2 we discuss volume. The volume of a 3-dimensional figure is a measure of the space occupied or enclosed by it. The approach that is used for area would suggest that the volume of a polyhedron can be determined by rearranging its parts to form a cube. It turns out that, although we can and do measure volume in cubic units, we cannot reduce all three-dimensional polyhedra to cubes the way that we can reduce all two-dimensional polygons to squares. Nonetheless, we can still obtain formulas for the volume of the three-dimensional figures commonly studied in school.

Thus, despite important similarities, area and volume possess significant differences that make them interesting to study and compare.

Unit 10.1 Area

Many students leave their study of mathematics associating area with formulas (rather than size) and thinking that they need to have the formula for the area of a figure before they can obtain its area. In this unit, we show many ways of arriving at the area of a figure, and many ways in which area formulas are related. We begin at the logical beginning, with a definition of area.

In comparison with distance as linear measure, area can be conceived of as surface measure—one that gives the plane or surface "content" of a figure. And, like distance, the calculation of area is subject to the context in which it is applied. In geometry we calculate area using formulas based on the sides and angles of figures. In calculus, we define area as the limit of a sum and calculate it as a definite integral. Where did these methods come from? What assumptions do we make when we apply them? These are some of the questions we address in this unit.

10.1.1 What is area?

Suppose we want to calculate the area of a plane region F and we do not know any formulas for calculating the area of F. How could we proceed? One possible method is to compare F with a region whose area we do know. If we can superimpose that region on F, we can be confident both have the same area.

Another method we could use is to choose a region and designate it as "the unit of area". Then we calculate the area of F by decomposing it into nonoverlapping units that completely cover the region. This decomposition method presupposes that we can decompose F into a finite number of nonoverlapping units. We then add the areas of these units to obtain the area of F. This is the approach we follow.

Area is a property of 2-dimensional figures. The basic regions of elementary geometry for which we calculate area fall into two types: those bounded by line segments and those bounded by curves. Of course, we can combine these regions to obtain regions with both characteristics. Our treatment of area will extend to all these figures, but we begin with the first type, that is, polygonal regions. Because polygonal regions can be decomposed into triangular regions, our definition of an area function needs to cover only triangular regions at first.

Specifically, a **triangular region** is the set of all points interior to or on the sides of a triangle. The area of a polygon is then found by adding up the areas of its constituent triangular regions. Before we do this, however, we need to choose a unit of area. We choose the square as the unit of area for several reasons. It has congruent edges, so its area calculation is the same no matter which edge you choose, unlike the triangle whose area calculation depends on what side is its base. Also, it turns out (as we show in Section 10.1.5) that any union of triangular regions can be decomposed and rearranged into a single square. And because we customarily measure in *square units*, we can transfer the theory based on the square as a unit of area directly to practical problems.

Definition Let F be the union of a finite collection of triangular regions in E^2. An **area function** α is a function that assigns to each such F a positive real number $\alpha(F)$ such that:

1. If $F_1 \cong F_2$, then
$$\alpha(F_1) = \alpha(F_2). \qquad \text{(Congruence property)}$$
2. If the triangular regions making up F_1 and F_2 have no interior points in common, then
$$\alpha(F_1 \cup F_2) = \alpha(F_1) + \alpha(F_2). \qquad \text{(Additive property)}$$
3. If F is a square region with side x, then
$$\alpha(F) = x^2. \qquad \text{(Area of square)}$$

Some books replace (3) by the area $\alpha(F) = ab$ of a rectangular region with adjacent sides a and b. We could also replace (3) by the area $\alpha(F) = \frac{1}{2}x^2$ of an isosceles right triangle with leg x (i.e., half a square), which puts the entire definition in terms of triangular regions. We begin with the area of a square because area is typically measured in square units and because we can derive these other formulas quite easily from it, as you will see.

Part (2) of this definition rests on a tacit assumption: that if a region is split into triangular regions in two different ways, then both the areas calculated are the same. That assumption relies on a further assumption, that the pieces of a triangular region can be rearranged into a square, and if there is more than one rearrangement into squares, then all such squares have the same size.[1] You might think of this in the following way: Suppose you calculate the area of a triangle using the familiar formula $A = \frac{1}{2}bh$, that is, area = half the product of a length of a side and the altitude to that side. Will you get the same area if you pick another side of the same triangle? The answer is yes, but this is not something that a complete mathematical theory could take for granted. In Section 10.1.2, we return to this question.

The theory of area technically requires that we always refer to the areas of regions because area is defined in terms of unions of sets. But in practice, we speak of the area of a square or triangle, not the area of a square region or triangular region. There is nothing mathematically wrong with this. We are merely referring to a region by the boundary of that region. Moreover, it is natural to speak of the area of a polygon or other simple closed curve when we think of how much space it encloses, and to speak of the area of the corresponding plane region when we think of how much space the region covers.

From squares to polygons and beyond

We have mentioned that some books begin by assuming that the area of a rectangle is length times width. They then proceed in the following sequence:

area of rectangle ⇒ area of right triangle ⇒ area of any triangle ⇒ area of trapezoid.

In contrast, we begin with the more basic area formula for a square and go as far as area calculations using calculus. Along the way we stop to gaze at the landscape and deduce some broadly applicable theorems not found in all high school books. Our development is described schematically in Table 1.

[1] For a detailed treatment, see *Geometry: A Metric Approach with Models*, by Richard S. Millman and George D. Parker (New York: Springer-Verlag, 1981).

Table 1

Area of square (def. of area function)	⇒	Area of rectangle (Theorem 10.1) ⇓	⇒	Riemann sums (Section 10.1.4)		
		Area of right triangle (Corollary) ⇓	⇒	SASΔ formula (Theorem 10.8)	⇒	ASAΔ formula (Theorem 10.9)
Area of quadrilateral with ⊥ diagonals (Theorem 10.4)	⇐	Area of triangle (Theorem 10.3) ⇓	⇒	Area of circumscribed polygon (Theorem 10.6) ⇓	⇒	Hero's formula (Theorem 10.7)
		Area of trapezoid (Theorem 10.5) ⇓		Area of circle (Theorem 10.10) ⇓		
		Trapezoidal rule		Area of ellipse		

From squares to rectangles

The proof of Theorem 10.1 shows how the formula for the area of a rectangle follows from the definition of an area function. The proof is short and involves only elementary algebra, yet it uses all three parts of the definition.

Theorem 10.1 The area of a rectangle is the product of the lengths of two adjacent sides.

Proof: Without loss of generality, let $ABCD$ be the rectangle and let $AB = b$ and $BC = h$. Extend $\overline{BA}$ by a line segment of length h to the point E, and extend $\overline{BC}$ by a segment of length b to the point F. $\overline{BE}$ and $\overline{BF}$ have length $(h + b)$ and are the sides of a square $EBFG$, as shown in Figure 2.

Figure 2

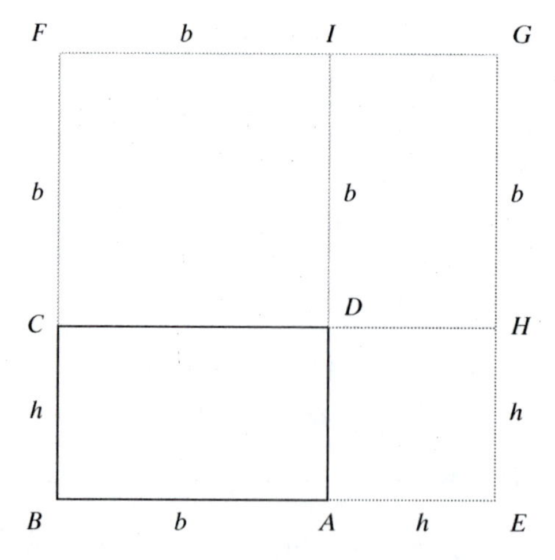

Rectangles $ABCD$ and $DHGI$ are congruent. By the congruence property (1) of the definition of area function, they have the same area. By property (3) of the definition of area function, $\alpha(EBFG) = (h + b)^2$, $\alpha(EADH) = h^2$, and $\alpha(\text{DCFI}) = b^2$. ┘

Question: Complete the proof, indicating where the additive property (2) of the area function is used.

In the statement of Theorem 10.1, the word "side" refers to a segment. But often we use that word to refer to the segment's length. For instance, we speak of a rectangle with sides a and b. The dual use of a word to refer either to a segment or its length is found throughout geometry, for example, with "radius", "diameter", "leg", "hypotenuse", "diagonal", etc.

Since a rectangle with dimensions a and b can be formed from two congruent right triangles with legs a and b, a formula for the area of a right triangle follows immediately from Theorem 10.1.

Corollary: The area of a right triangle is $\frac{1}{2}$ the product of its legs.

Other measures of two-dimensional regions

Figure 3

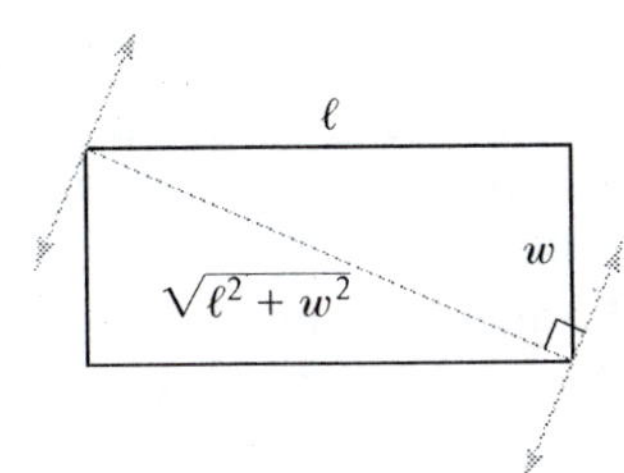

Area is not the only measure of two-dimensional regions. For example, the **width** of a two-dimensional region in a particular direction is the length of the longest segment joining two points of the region parallel to that direction. For instance, since the longest segments joining two points on a rectangle are its diagonals, the longest width of a rectangle with dimensions ℓ and w is $\sqrt{\ell^2 + w^2}$, in the direction of either diagonal (Figure 3). A circle with radius r has constant width $2r$ and is one of many figures with constant width. Three such figures are displayed in Figures 4a–c.

Figure 4a

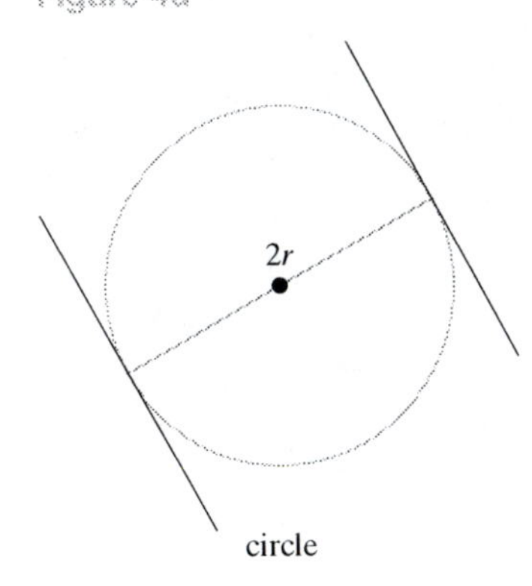

Figure 4b

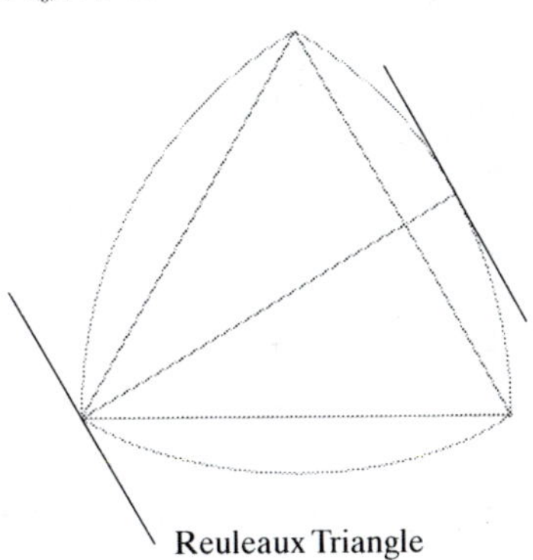

Figure 4c

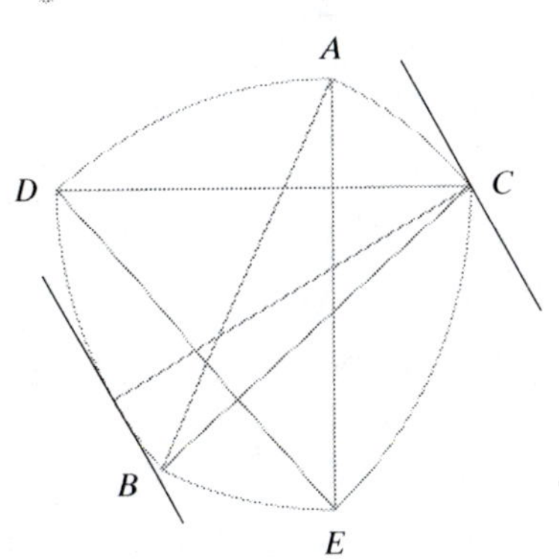

And of course there is *perimeter*. The **perimeter** $p(R)$ of any polygonal region R is the sum of the lengths of its sides. This defines a function p that students often confuse with area. One reason for the confusion is that the units of the function are ignored in the mathematical definition, so that the ranges of the area and perimeter functions are identical. However, the units are never the same: The unit of perimeter is a 1-dimensional unit segment and the unit of area is the 2-dimensional unit square. A second reason for the confusion is that the numbers used for dimensions of rectangles and triangles in schoolbooks are often small integers, and for these integers the numerical values of area and perimeter are of the same general size. A third reason for the confusion may be that these topics are taught without reference to physical examples. With examples, the difference is significant: A lake's area indicates how much room there is for fishing, while its perimeter gives the amount of shoreline; a room's area tells how much flooring is needed, while its perimeter gives the amount of baseboard; and so on.

The following theorem establishes that area does not determine perimeter even in simple figures. The figure of Theorem 10.1 suggests a geometric proof for Theorem 10.2 that we encourage you to produce. Here we provide an elegant algebraic proof. The key to the proof of Theorem 10.2 is to let x be the amount by which a side differs from $\frac{1}{4}$ the perimeter p. When the rectangle is a square, $\frac{1}{4}p$ is the length of the each side.

Theorem 10.2 Of all rectangles with a given perimeter, the square has the greatest area.

Proof: Let the rectangle have perimeter p, length ℓ, width w, and area A. By the definition of perimeter, $p = 2\ell + 2w$, so $\frac{p}{2} = \ell + w$. We know then that ℓ is some number between 0 and $\frac{p}{2}$. So we can let $\ell = \frac{p}{4} + x$. Then $w = \frac{p}{4} - x$. So $A = \ell w = \left(\frac{p}{4} + x\right)\left(\frac{p}{4} - x\right) = \frac{p^2}{16} - x^2$. Since x^2 is always nonnegative, the area A is maximized when $x = 0$: that is, when $\ell = w$ and the rectangle is a square. ┘

Area formulas quickly deduced

From the area formula for a right triangle (Corollary to Theorem 10.1), we can deduce the familiar area formula for any triangle.

Theorem 10.3 The area of a triangle is $\frac{1}{2}$ the product of a side and the altitude to that side.

Proof: Let the triangle be ABC, let $b = AC$, and let $h =$ the length of the altitude $\overline{BD}$ from B to $\overline{AC}$. We wish to show that $\alpha(\Delta ABC) = \frac{1}{2}hb$. There are three cases.

Case 1: If $D = A$ or $D = C$ (Figure 5a), then ΔABC is a right triangle and its area is $\frac{1}{2}hb$ from the Corollary to Theorem 10.1.

Figure 5a

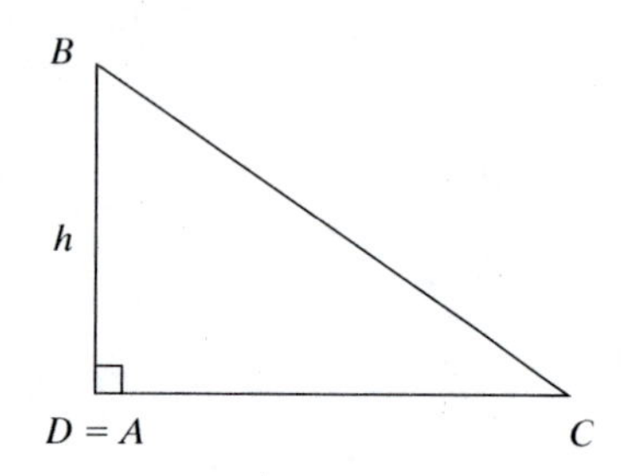

Figure 5b

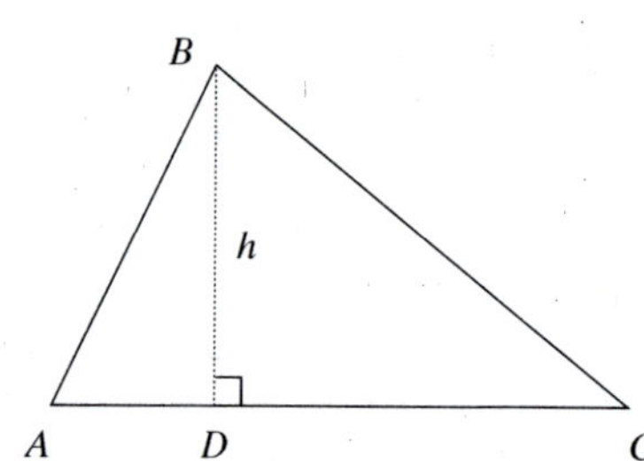

Figure 5c

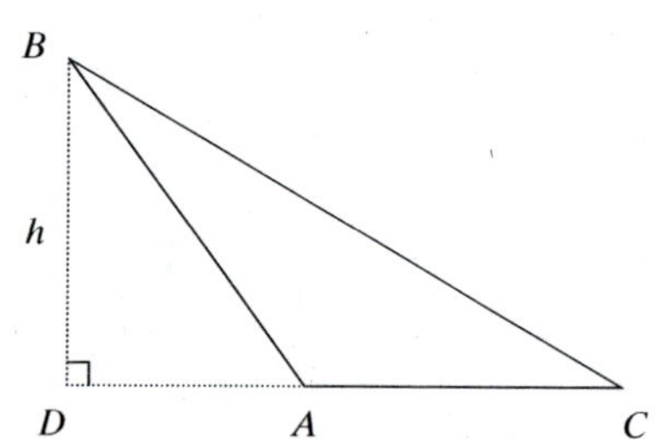

Case 2: If D lies on $\overline{AC}$ (Figure 5b), then, using the Corollary to Theorem 10.1.

$$\alpha(\Delta ABC) = \frac{1}{2}h \cdot AD + \frac{1}{2}h \cdot DC \qquad \text{(from the additive property of the area function)}$$

$$= \frac{1}{2}h(AD + DC) = \frac{1}{2}hb.$$

Case 3: If D lies outside of $\overline{AC}$, say on $\overrightarrow{CA}$ as shown (Figure 5c), then, again from the additive property of the area function,

$$\alpha(\Delta ABC) + \frac{1}{2}h \cdot AD = \frac{1}{2}h \cdot DC.$$

Thus

$$\alpha(\Delta ABC) = \frac{1}{2}h \cdot DC - \frac{1}{2}h \cdot AD$$

$$= \frac{1}{2}h(DC - AD)$$

$$= \frac{1}{2}hb.$$

If D lies outside of $\overline{AC}$ on $\overrightarrow{AC}$, the proof is similar to that in (3). ┘

From either the formula for the area of a rectangle or for the area of a triangle, a formula for the area of any quadrilateral with perpendicular diagonals can be obtained. This formula is useful because it applies to all kites and thus to all rhombi and squares.

Theorem 10.4 If a quadrilateral has perpendicular diagonals of lengths d_1 and d_2, then its area is $\frac{1}{2}d_1 d_2$.

Proof: Let $ABCD$ be a quadrilateral with perpendicular diagonals $\overline{AC}$ and $\overline{BD}$ intersecting at E, as shown in Figure 6a. From the additive property of the area function, and from Theorem 10.3,

$$\alpha(ABCD) = \alpha(\Delta ADB) + \alpha(\Delta CDB) = \frac{1}{2}\cdot AE\cdot BD + \frac{1}{2}\cdot EC\cdot BD = \frac{1}{2}\cdot AC\cdot BD.$$

Figure 6

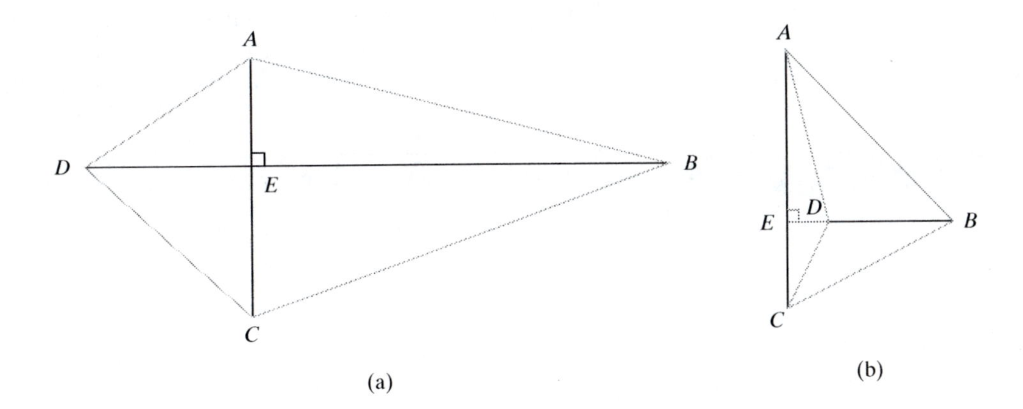

If the diagonals do not intersect, then the quadrilateral is not convex. Extend the diagonals as needed to again intersect at E (Figure 6b).

Now

$$\alpha(\Delta ABC) = \alpha(ABCD) + \alpha(\Delta ADC).$$

So

$$\frac{1}{2}\cdot AC\cdot BE = \alpha(ABCD) + \frac{1}{2}\cdot AC\cdot DE$$

$$\frac{1}{2}\cdot AC\cdot BE - \frac{1}{2}\cdot AC\cdot DE = \alpha(ABCD)$$

$$\frac{1}{2}\cdot AC\cdot BD = \alpha(ABCD).$$

From the formula for the area of a triangle, we can obtain a formula for the area of any trapezoid. By the shortness of its proof, it could be considered a corollary to Theorem 10.3. Its importance leads us to call it a theorem.

Theorem 10.5 The area of a trapezoid with bases b_1 and b_2 and height h is $\frac{1}{2}h(b_1 + b_2)$.

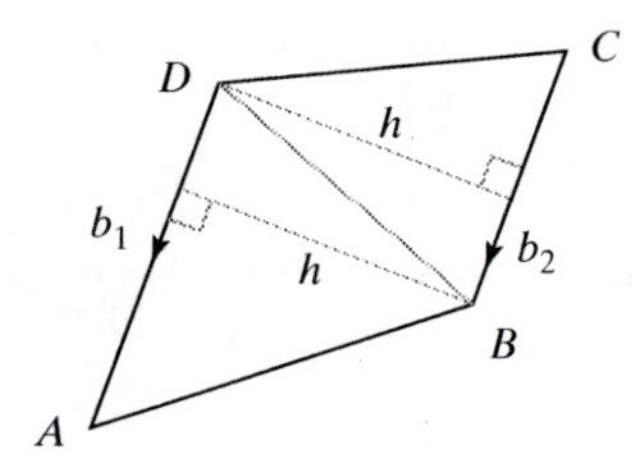

Figure 7

Proof: Refer to Figure 7. Suppose $\overline{AD}$ and $\overline{BC}$ are bases of trapezoid $ABCD$, h is the distance between $\overline{AD}$ and $\overline{BC}$, $b_1 = AD$, and $b_2 = BC$. Then

$$\begin{aligned} \alpha(ABCD) &= \alpha(\Delta ABD) + \alpha(\Delta DBC) && \text{by property (2) of the area function} \\ &= \frac{1}{2}b_1h + \frac{1}{2}b_2h \\ &= \frac{1}{2}h(b_1 + b_2). \end{aligned}$$

The trapezoid area formula is significant because it applies to all parallelograms, rhombi, rectangles, and squares, and in calculus is used to develop the Trapezoidal Rule that approximates areas under curves. (In calculus, the bases are typically vertical, so some students at first do not recognize that the figures are trapezoids.)

In Section 8.3.2, in two ways we extended a theorem about a line intersecting a side of a triangle. One way was by increasing the number of parallel lines; the second was by increasing the number of rays from one vertex. Theorem 10.5 and the next theorem roughly arise applying the same types of generalization to the area formula $A = \frac{1}{2}hb$ for a triangle. The proof of Theorem 10.5 uses triangles with the same height between parallel lines. The proof of Theorem 10.6 uses triangles with the same height emanating from the same vertex. Recall that a polygon is circumscribed about a circle, or, equivalently, a circle is inscribed in a polygon, if and only if each of the polygon's sides is tangent to the circle.

Theorem 10.6 The area of a polygon with perimeter p circumscribed about a circle with radius r is $\frac{1}{2}rp$.

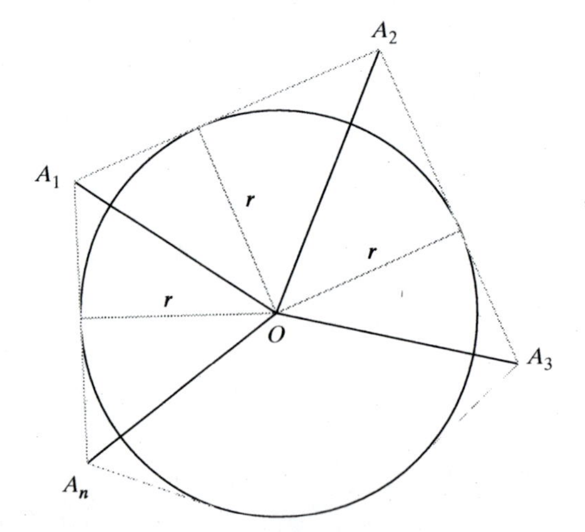

Figure 8

Proof: The proof is straightforward. Let $A_1A_2A_3 \ldots A_n$ be a polygon circumscribed about circle O with radius r (see Figure 8). The polygonal region $A_1A_2A_3 \ldots A_n$ is the union of the triangular regions $OA_1A_2, OA_2A_3, \ldots, OA_nA_1$. Each triangular region has height r because a radius of a circle is perpendicular to any tangent to the circle at the point of tangency. So

$$\begin{aligned} \alpha(A_1A_2A_3 \ldots A_n) &= \alpha(OA_1A_2) + \alpha(OA_2A_3) + \cdots + \alpha(OA_nA_1) \\ &= \frac{1}{2}r(A_1A_2) + \frac{1}{2}r(A_2A_3) + \cdots + \frac{1}{2}r(A_nA_1) \\ &= \frac{1}{2}r(A_1A_2 + A_2A_3 + \cdots + A_nA_1) \\ &= \frac{1}{2}rp. \end{aligned}$$

Theorem 10.6 applies to all regular polygons, and also to any triangle, since a circle can be inscribed in any of these figures. The application to triangles, discussed in the next section, is particularly productive.

10.1.1 Problems

1. a. Estimate the area of the region in Figure 1a given that each square is 1 unit on a side.

b. Estimate the area of the region in Figure 1b given that each square is $\frac{1}{2}$ unit on a side.

2. In the proof of Theorem 10.1, if the original rectangle $ABCD$ has dimensions 7″ and 10″, what are the dimensions of the two squares $HEAD$ and $FIDC$? Explain how the area of the rectangle can be derived from the areas of these two squares.

3. Figure 9 suggests a proof of the triangle area formula $A = \frac{1}{2}bh$ from the area formula for a rectangle for the case where the angles including the base of the triangle are both acute.

Figure 9

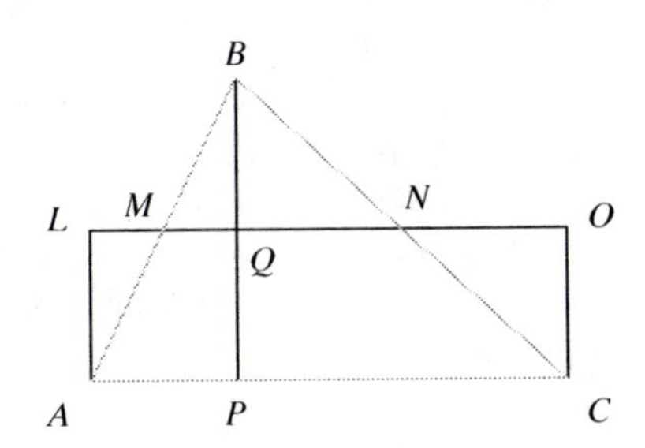

a. Provide the proof.

b. Draw the corresponding figure for the case in which a base angle is obtuse and prove the theorem for this case.

4. A rectangle has dimensions x and y, with $x \geq y$.

a. What is its minimal width?

b. What is its maximal width?

c. If this rectangle is the frame of a door, what is the radius of the largest circular table top that can be passed through the door? (Ignore the thickness of the table top.)

5. Let ℓ and m be parallel lines containing the vertices A and C of the rectangle $ABCD$ and not containing any other points of the rectangle. Prove that the distance between ℓ and m is greatest when either line is perpendicular to $\overline{AC}$.

6. Let ΔABC be equilateral. Consider the region bounded by the circular arc $\widehat{BC}$ with center A, the circular arc $\widehat{AC}$ with center B, and the circular arc $\widehat{AB}$ with center C. Prove that this region has constant width. (The union of the three arcs $\widehat{BC}$, $\widehat{AC}$, and $\widehat{AB}$ is known as a **Reuleaux triangle** and is shown in Figure 4b of this section.)

7. Provide a proof for the situation of Figure 6a of Theorem 10.4 that avoids Theorem 10.3 by separating the quadrilateral into four right triangles.

8. Let $ABCD$ be a trapezoid with $\overline{AB} \parallel \overline{CD}$. The segment joining the midpoints of the other sides $\overline{AD}$ and $\overline{BC}$ is a **median** of the trapezoid $ABCD$. Prove that the area of a trapezoid is the product of its height and the length of the median perpendicular to that height.

9. What is the area of an isosceles trapezoid with sides of length 20, 25, 25, and 28?

10. An isosceles trapezoid has bases of length 9 and 15 and it can be circumscribed about a circle. Can its area be determined? If so, find the area. If not, tell why the area cannot be determined.

11. Let $ABCD$ be an isosceles trapezoid with bases $\overline{AB}$ and $\overline{CD}$ and $AB < CD$ (Figure 10). Let E be the intersection of $\overline{AC}$ and $\overline{BD}$.

Figure 10

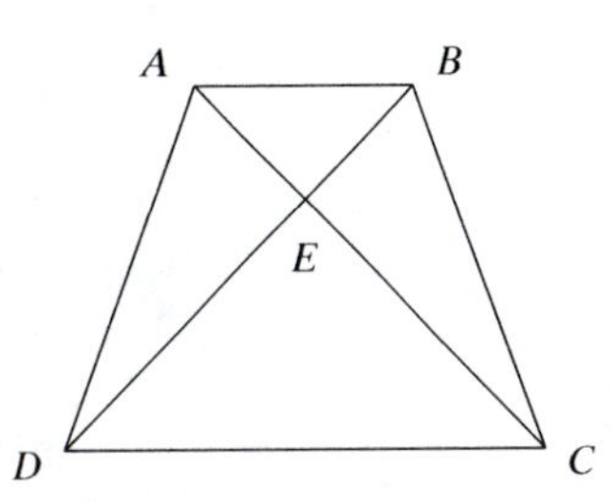

a. Prove: $\alpha(\Delta BEC) = \alpha(\Delta AED)$.

b. Prove: $\alpha(\Delta ABE) < \alpha(\Delta EDC)$.

12. A rhombus has diagonals of lengths a and b. Show three different ways of finding its area.

13. Let $ABCD$ be a parallelogram and let E be a point on $\overline{AC}$. Let the parallel to $\overline{AD}$ through E intersect $\overline{AB}$ at H and $\overline{CD}$ at I. Let the parallel to $\overline{AB}$ through E intersect $\overline{AD}$ at F and $\overline{BC}$ at G. Prove that parallelograms $HEGB$ and $FDIE$ have equal area. (This is a theorem that Euclid exploited in his development of area.)

14. Let E, F, G, and H be the midpoints of the sides $\overline{AB}, \overline{BC}, \overline{CD}$, and $\overline{DA}$ of convex quadrilateral $ABCD$. Prove that $\alpha(EFGH) = \frac{1}{2}\alpha(ABCD)$.

15. a. An area formula for a parallelogram, $A = hb$, is often found by finding a rectangle with equal area, as in Figure 11a. Explain this argument.

Figure 11a

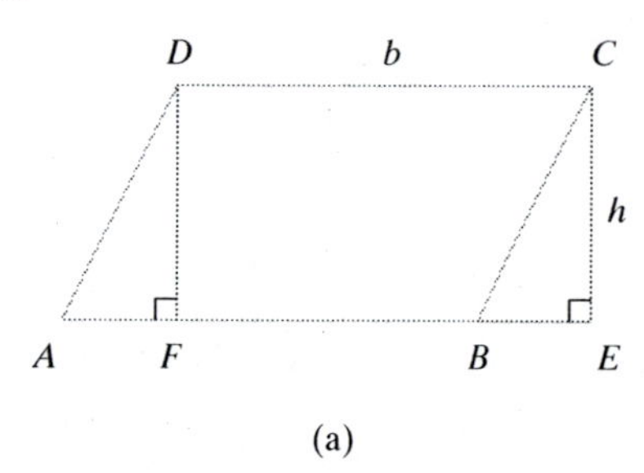

(a)

b. The argument of part **a** does not work for parallelograms like those in Figure 11b. Write a proof that $\alpha(ABCD) = hb$ for the situation in Figure 11b.

Figure 11b

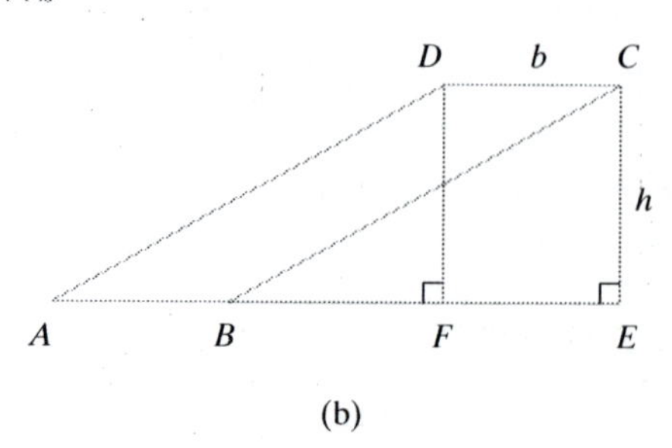

(b)

16. A regular n-sided polygon is inscribed in a unit circle.

a. Prove that its perimeter is $2n \sin \frac{180^\circ}{n}$.

b. Prove that its area is $\frac{n}{2} \sin \frac{360^\circ}{n}$.

c. Modify the formulas of parts **a** and **b** if the circle has radius r.

17. Deduce formulas similar to those found in Problem 16 for a regular n-gon that is circumscribed about a circle.

ANSWER TO QUESTION

By properties (1) and (2) of the definition of area function, $\alpha(EBFG) = \alpha(EADH) + \alpha(ABCD) + \alpha(DHGI) + \alpha(FCDI) = \alpha(EADH) + 2\alpha(ABCD) + \alpha(FCDI)$. Now use property (3) and substitute $(h + b)^2 = h^2 + 2\alpha(ABCD) + b^2$. Consequently, $h^2 + 2hb + b^2 = h^2 + 2\alpha(ABCD) + b^2$. Solving the equation for $\alpha(ABCD)$, we obtain $\alpha(ABCD) = bh$.

10.1.2 Area formulas for triangles

In Section 10.1.1, we deduced the familiar area formula $A = \frac{1}{2}bh$ for a triangle. However, in practice one may not know the length of any altitude of a triangle. We also arrived at a second formula, $A = \frac{1}{2}rp$, giving the area of a triangle in terms of its perimeter p and the radius r of its inscribed circle (its **incircle**). But the radius of the incircle is also unlikely to be known. More likely to be known are the sides and angles of the triangle. So we search for an area formula using only sides and angles.

Since congruent triangles have the same area, in theory we should be able to determine the area of a triangle given all three sides (SSS), two sides and an included angle (SAS), or two angles and the included side (ASA). Euclid was aware of this fact, but he knew of no such formulas. The first of these formulas, for the area of a triangle given its sides, was discovered by Archimedes about a half century after Euclid lived, but it is known today as *Hero's Formula*, after Hero the Elder of Alexandria, the same mathematician mentioned in Section 8.1.2 in connection with optics. It appeared in Hero's book *Geodesy* and was proved in two other books, *Dioptra* and *Metrica*. The proof is quite long. To help in understanding it, we split it into two parts and call the first part a lemma.[2]

Figure 12

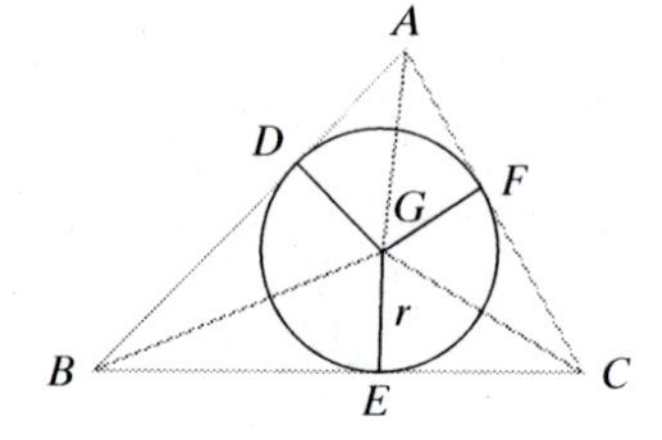

Lemma: Let D, E, and F be the points of tangency of the inscribed circle G of ΔABC and its sides $\overline{AB}$, $\overline{BC}$, and $\overline{AC}$, as shown in Figure 12. Let r be the radius of circle G and s be half the perimeter of ΔABC. Then

1. $\alpha(\Delta ABC) = sr$

2. $AD = AF = s - BC$
$BE = BD = s - AC$
$CF = CE = s - AB$.

[2] A **lemma** is a statement that is used in a proof of a theorem that immediately follows the lemma. The lemma is itself proved, but usually is not important enough to be labeled a theorem.

Proof: Let p be the perimeter of ΔABC. By Theorem 10.6 of Section 10.1.1,

$$\alpha(\Delta ABC) = r \cdot \frac{1}{2}p = rs.$$

Then

$$p = 2s = AD + DB + BE + EC + CF + FA.$$

$AD = AF$ because they are tangents from a point to the circle. ($\Delta ADG \cong \Delta AFG$ by HL Congruence.) Similarly, $BE = BD$ and $CF = CE$. So

$$2s = 2AD + 2BE + 2CF.$$

Dividing both sides by 2,

$$s = AD + BE + CF.$$

Now we can deduce any one of the three results by solving this equation for the desired distance. For instance,

$$\begin{aligned} AD &= s - BE - CF \\ &= s - BE - CE \\ &= s - BC. \end{aligned}$$

The other two parts follow in the same manner. ∎

The quantity s in the lemma is half the perimeter of ΔABC, so it is known as the **semiperimeter** of ΔABC. Notice that if the sides of ΔABC are named in the traditional manner as a, b, and c, then $AD = s - a$, $BE = s - b$, and $CF = s - c$. These lengths play major roles in Hero's Formula for the area of any triangle.

Theorem 10.7 **(Hero's Formula):** The area of a triangle with sides a, b, and c and semiperimeter s is $\sqrt{s(s-a)(s-b)(s-c)}$.

Proof 1 (synthetic): Let ABC be the triangle and let the inscribed circle for the triangle have center G and be tangent to ABC at points D, E, and F (Figure 13). Let $r = GD = GE = GF$. Because of part (1) of the lemma,

$$\alpha(\Delta ABC) = sr = (BE + EC + AD)r. \tag{1}$$

Now extend $\overline{CB}$ beyond B to H so that $BH = AD$.
Substituting into (1),

$$\begin{aligned} \alpha(\Delta ABC) &= (BE + EC + BH)r \\ &= CH \cdot r \end{aligned}$$

$$\alpha(\Delta ABC) = CH \cdot EG. \tag{2}$$

The goal now is to get some ratios involving CH and EG, and Hero does this through similar triangles. Let L be the intersection of the perpendiculars to $\overline{CG}$ at G and $\overline{CB}$ at B. $\angle CGL$ and $\angle CBL$ are right angles, so the midpoint of $\overline{LC}$ (not drawn) is equidistant from L, B, G, and C. This means that quadrilateral $LBGC$ is inscribed in a semicircle, and so its opposite angles CGB and CLB are supplementary. But also, $\angle CGB$ and $\angle AGD$ are supplementary (because the quadrilaterals $ADGF$, $CFGE$, and $BEGD$ are kites and their symmetry diagonals bisect the central angles). Consequently, $m\angle CLB = m\angle AGD$, and since these are acute angles in right triangles,

$$\Delta CBL \sim \Delta ADG. \tag{3}$$

Figure 13

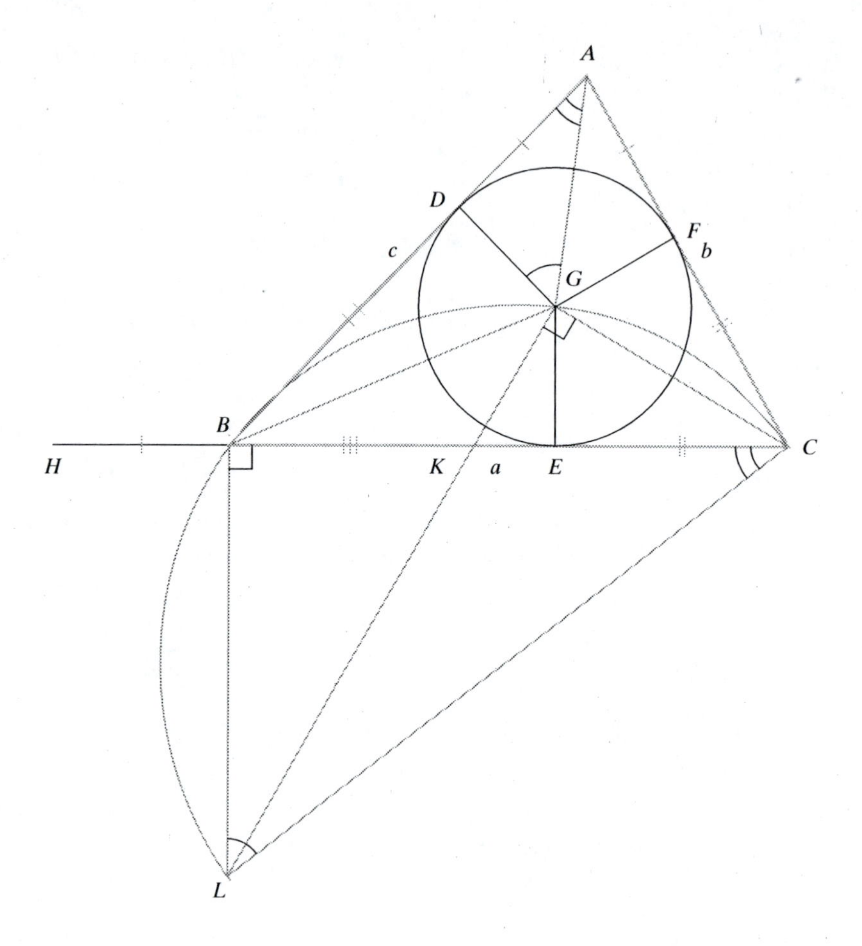

From (3) and recalling that $BH = AD$ and $DG = EG$,

(4) $$\frac{CB}{BL} = \frac{AD}{DG} = \frac{BH}{EG}.$$

From the left and right ratios in (4), $\frac{CB}{BH} = \frac{BL}{EG}$. With K being the intersection of $\overline{GL}$ and $\overline{BC}$, $\Delta GKE \sim \Delta LKB$, so that

(5) $$\frac{CB}{BH} = \frac{BL}{EG} = \frac{BK}{KE}.$$

Adding $\frac{BH}{BH}$ to the left and $\frac{KE}{KE}$ to the right ratios in (5),

(6) $$\frac{CH}{BH} = \frac{BE}{EK}.$$

With some multiplication by 1 and recognizing a geometric mean,

(7) $$\frac{CH^2}{CH \cdot BH} = \frac{BE \cdot EC}{EK \cdot EC} = \frac{BE \cdot EC}{EG^2}.$$

Again use the left and right ratios.

(8) $$\begin{aligned} CH^2 \cdot EG^2 &= CH \cdot BH \cdot BE \cdot EC \\ &= CH \cdot AD \cdot BE \cdot CF \end{aligned}$$

Now use part (2) of the lemma to rewrite the right side of (8) in terms of s, a, b, and c, and notice that $CH = s$. Rewrite the left side using (2) of this proof,

$$\begin{aligned}(9) \qquad \alpha(\Delta ABC)^2 &= CH \cdot AD \cdot BE \cdot CF \\ &= s(s-a)(s-b)(s-c).\end{aligned}$$

$$(10) \qquad \text{So} \quad \alpha(\Delta ABC) = \sqrt{s(s-a)(s-b)(s-c)}.$$

Hero's proof of the formula bearing his name is extraordinary because the result is algebraically quite complex and Hero had none of today's algebraic notation at his disposal. We have presented it to show that Hero's formula is geometrically derivable from the other area formulas, and also to exemplify the range of synthetic geometry proofs.

A shorter proof uses the Law of Cosines and quite a bit of algebraic manipulation.

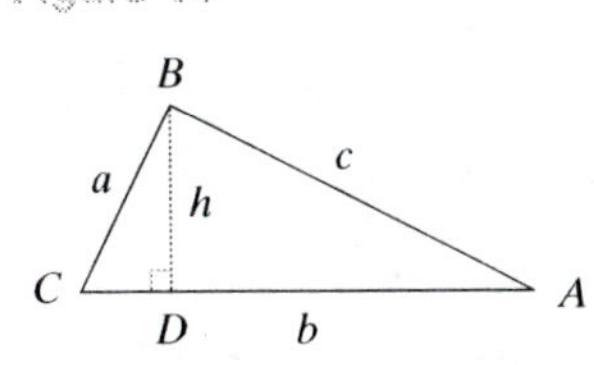

Figure 14

Proof 2 (analytic): The triangle ΔABC must have at least one acute angle. Let it be $\angle A$. Let D be the foot of the altitude from B to $\overline{AC}$ (Figure 14). Let $h = BD$. We know $\alpha(\Delta ABC) = \frac{1}{2}bh$. So we will search for a formula for h in terms of the sides a, b, and c. By the Law of Cosines,

$$a^2 = b^2 + c^2 - 2bc \cos A.$$

Now $AD = c \cos A$. Consequently,

$$a^2 = b^2 + c^2 - 2b \cdot AD.$$

We solve this for AD.

$$AD = \frac{b^2 + c^2 - a^2}{2b}$$

Using the Pythagorean Theorem, we can find an expression for h.

$$h^2 = c^2 - AD^2 = c^2 - \left(\frac{b^2 + c^2 - a^2}{2b}\right)^2 = \frac{4b^2c^2 - (b^2 + c^2 - a^2)^2}{4b^2}$$

Factor the difference of squares in the numerator of the last expression, rearrange the factors as differences of squares, and factor again to obtain

$$\begin{aligned} h^2 &= \frac{(2bc + b^2 + c^2 - a^2)(2bc - b^2 - c^2 + a^2)}{4b^2} \\ &= \frac{[(b^2 + 2bc + c^2) - a^2][a^2 - (b^2 - 2bc + c^2)]}{4b^2} \\ &= \frac{(b + c + a)(b + c - a)(a + b - c)(a - b + c)}{4b^2}. \end{aligned}$$

Now we introduce the semiperimeter s. If $s = \frac{1}{2}(a + b + c)$, then

$$\begin{aligned} b + c + a &= 2s \\ b + c - a &= 2(s - a) \\ a + b - c &= 2(s - c) \\ a - b + c &= 2(s - b). \end{aligned}$$

[These expressions are the equivalent of part (2) of the lemma before Theorem 10.7.] By substitution,

$$h^2 = \frac{[2s][2(s-a)][2(s-b)][2(s-c)]}{4b^2} = \frac{4s(s-a)(s-b)(s-c)}{b^2}.$$

So

$$h = \frac{2\sqrt{s(s-a)(s-b)(s-c)}}{b}.$$

Thus, $\alpha(\Delta ABC) = \frac{1}{2}bh = \sqrt{s(s-a)(s-b)(s-c)}$. ┘

Both Hero's Formula and the formula $A = \frac{1}{2}rp$ for the area of a triangle are *symmetric*, in that the three sides of the triangle each play the same role. They show that the area of a triangle does not depend on a particular side being singled out as special, and that regardless of which side might be called a base, the area of the triangle would be the same. This guarantees that, once a unit has been picked, the area of a triangle is unique.

Formulas when SAS or ASA is known

Hero's Formula produces the area of a triangle given SSS. There is a simple formula for the area of a triangle given SAS. This formula is often not encountered by students because area is studied before trigonometry.

Theorem 10.8 **(SAS Area Formula):** For all triangles ABC,

$$\alpha(\Delta ABC) = \frac{1}{2}ab \sin C.$$

Figure 15

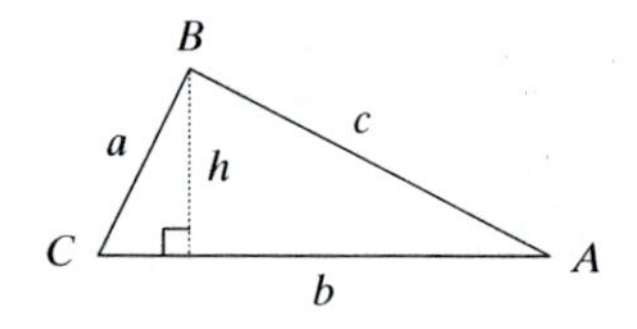

Proof: In ΔABC, let h be the altitude to side b (Figure 15). Then $\frac{h}{a} = \sin C$. Solving for h and substituting into the area formula $\alpha(\Delta ABC) = \frac{1}{2}bh$ yields this theorem. ┘

Corollary: The area of the parallelogram with adjacent sides a and b and included angle ϕ is $ab \sin \phi$.

The SAS area formula is not symmetric. Depending on which side lengths are known, we could have any of the following variants: $\alpha(\Delta ABC) = \frac{1}{2}ab \sin C = \frac{1}{2}bc \sin A = \frac{1}{2}ac \sin B$. We use these variants to develop a formula for the area of a triangle given two angles and an included side.

Theorem 10.9 **(ASA Area Formula):** For all triangles ABC,

$$\alpha(\Delta ABC) = \frac{1}{2}a^2 \frac{\sin B \cdot \sin C}{\sin(B+C)}.$$

Proof: Our goal is to obtain a formula with one side a and the included angles B and C. From Theorem 10.8,

$$\alpha(\Delta ABC) = \tfrac{1}{2}ac \sin B \cdot \frac{\frac{1}{2}ab \sin C}{\frac{1}{2}bc \sin A}$$

$$= \tfrac{1}{2}a^2 \sin B \cdot \frac{\sin C}{\sin A}.$$

Since for all x (measured in degrees), $\sin x = \sin(180° - x)$, and since $m\angle A + m\angle B + m\angle C = 180°$, $\sin A = \sin(B + C)$. Substitution of $\sin(B + C)$ for $\sin A$ in (2) yields the desired formula.

We have now shown five formulas for the area of a triangle. Are there others? Of course there are. A formula is possible using any segments or angles whose lengths and measures determine a unique triangle. Problem 16 shows a formula involving the radius of the circumcircle of a triangle. New formulas may be derived by reworking existing formulas. For instance, each line in the proof of Theorem 10.9 yields a different formula for the area of a triangle. The existence of so many ways of calculating the area of a triangle is useful not only because it enables calculation from a variety of given information, but also because the areas of other figures are often calculated by adding areas of triangles.

10.1.2 Problems

1. In the proof of the lemma in this section, demonstrate that $BE = s - AC$.

2. Modify Hero's Formula to yield a formula for the area of a triangle in terms of its sides a, b, and c, and its perimeter p.

3. Refer to the synthetic proof of Hero's Formula.
   a. To arrive at step (3), the statement is made that because $LBGC$ is inscribed in a semicircle, its opposite angles are supplementary. Why is this true?
   b. Why is $\Delta GKE \sim \Delta LKB$?
   c. Where is the geometric mean in step (7)?

4. a. Use Hero's Formula to find the area of a 3-4-5 right triangle.
   b. Find the area of the triangle with sides 13, 14, and 15.
   c. Use your answer to part **b** to determine the lengths of the three altitudes of the triangle with sides 13, 14, and 15.
   d. Your answers to parts **a** and **b** should be integers. Find another triangle whose sides are consecutive integers and whose area is an integer.

5. Use Hero's Formula to prove that if T is a similarity transformation with magnitude k, then, for any triangle ABC, $\alpha(T(\Delta ABC)) = k^2 \cdot \alpha(\Delta ABC)$.

6. Identify all the cyclic quadrilaterals shown in Figure 13 and explain why they are cyclic.

7. **Brahmagupta's Formula.** The following area formula for cyclic quadrilaterals was known to the Indian mathematician Brahmagupta (c. 598–c. 665) and is named after him. Let Q be a cyclic quadrilateral. Then

$$\alpha(Q) = \sqrt{(s-a)(s-b)(s-c)(s-d)},$$

where the semiperimeter $s = \frac{a+b+c+d}{2}$.
   a. Use a geometric construction program to construct a dynamic confirmation of Brahmagupta's Formula for any circle C and any convex quadrilateral Q inscribed in C.
   b. Prove that Brahmagupta's Formula does not hold for all quadrilaterals.
   c. Explain why Hero's Formula is a special case of Brahmagupta's Formula.[3]

8. If a circle can be inscribed in a cyclic quadrilateral (i.e., all sides of the cyclic quadrilateral are tangent to the circle), then the quadrilateral is called **cyclic-inscribable**. The area of a cyclic-inscribable quadrilateral with sides w, x, y, and z is $\sqrt{wxyz}$. Find a cyclic-inscribable cyclic quadrilateral and show that the formula works for it.

9. Deduce the Law of Sines (Theorem 9.6) using Theorem 10.8.

10. Suppose the area and length of one side of a triangle are fixed. Determine the conditions under which the sum of the lengths of the other two sides is smallest.

11. Suppose the length of one side and the sum of the lengths of the other two sides of a triangle are known. Determine the conditions under which the area is maximized.

12. Prove: Of all triangles with a given perimeter, the equilateral triangle has the greatest area.

13. Use the result of Problem 11 to prove: Of all polygons with a given perimeter, the one with greatest area is equilateral.

14. Find the area of a parallelogram with adjacent sides a and b and *non*-included angle ϕ.

15. a. Find the area of the triangle with consecutive vertices $(0, 0)$, (a, b), and (c, d).
   b. Three parallelograms have three vertices (not necessarily consecutive) at (a, b), $(0, 0)$, and (c, d). Find the fourth vertex and area of each parallelogram.

16. Let R be the radius of the circumcircle of ΔABC. Find and write a proof that $\alpha(\Delta ABC) = \frac{abc}{4R}$. (See, for example, Nathan Altschiller-Court, *College Geometry: An Introduction to the Modern Geometry of the Triangle and the Circle*, 2nd edition. New York: Barnes and Noble, 1952.)

[3] A proof of Brahmagupta's Formula can be found in Howard Eves, *An Introduction to the History of Mathematics with Cultural Connections*, sixth edition (New York: Saunders College Publishing, 1983).

10.1.3 Extended analysis: the line through a given point minimizing area

We devote this section to the analysis of the following problem, based on a problem found in a high school precalculus text.

> Of all lines through the point (5, 2), find the line that cuts off the triangle of smallest area in the first quadrant (Figure 16).

Figure 16

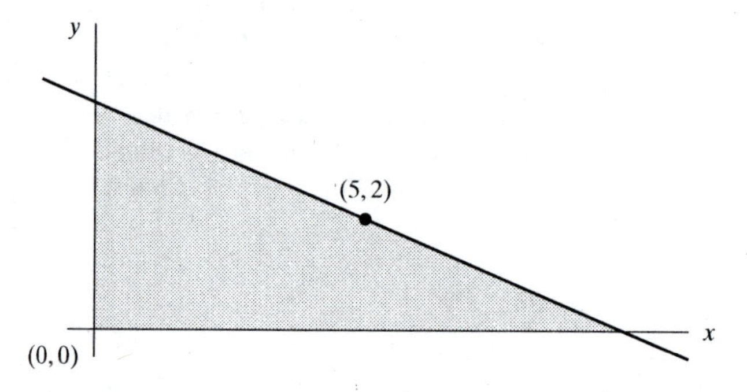

Doing the analysis yourself

You can get a rough feeling about this problem situation by sketching some different lines through the point (5, 2). Notice that a very steep line (Figure 17a) will cut off a large area. Similarly, a very flat line (Figure 17c) will also cut off a large area. (See Problem 1.) It makes sense that there is a line somewhere in the middle range of steepness that cuts off the smallest area.

Figure 17

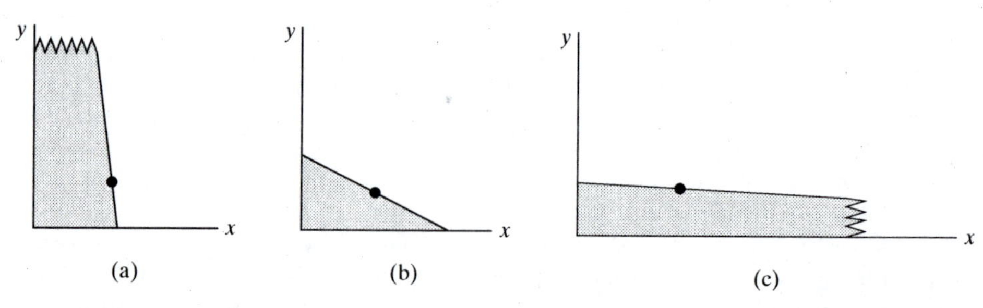

Before reading beyond this page, do as much as you can of the following.

a. Solve this problem using any method, and justify your solution.

b. Generalize your solution so that, starting with any point in the first quadrant, you could quickly find the line through that point cutting off minimum area.

c. Generalize your solution so that it would apply to a situation where the axes are not perpendicular.

d. Give an alternate approach to the problem that is fully geometrical. (Here a geometrical approach is an approach that does not rely on coordinates to identify the point.)

e. Suggest ways your result could be generalized even further, perhaps in the form of conjectures that you cannot yet see how to prove or disprove.

In typical school work, this sort of problem ends at (a). Yet there is a surprising mathematical richness in this situation that remains hidden if we limit ourselves to (a). The purpose of this section is to analyze this problem from an advanced standpoint, generalizing it and extending it along the lines of (a) to (e). Reading the rest of this section will make far more sense if you have at least tried (a)–(e).

The generalizations necessary to do parts (b) and (c) are aided by *parameterization.* We parameterize all the lines through the point $(5, 2)$ by choosing one of the parameters of a line as a variable. The interaction of variables, constants, and parameters is a primary feature of the analysis. The generalizations are also aided by *finding minimal values.* To prove that a particular line through $(5, 2)$ cuts off the smallest area, we use a simple method. We show geometrically that rotating the line a small amount produces a line that cuts off greater area.

Parameterizing the lines

There are several ways to parameterize the lines through $(5, 2)$. Three ways are suggested by Figures 18a, 18b, and 18c.

Figure 18

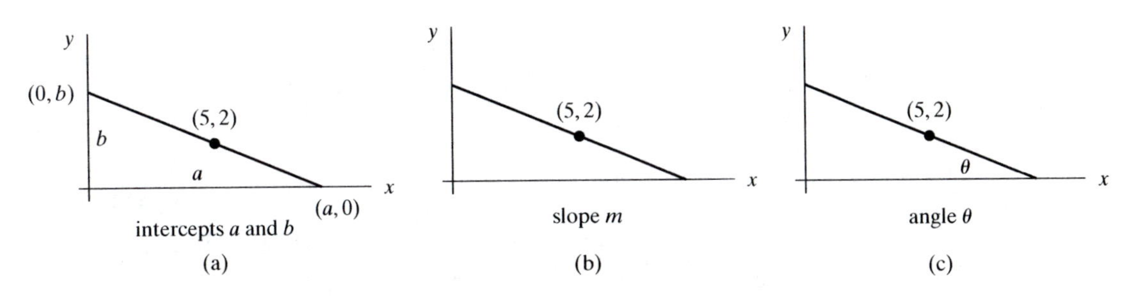

In Figure 18a, the lines are parameterized by their intercepts a and b. But notice that a and b are not independent. The fact that the line passes though $(5, 2)$ means that a and b are related by

$$\frac{2-b}{5} = \frac{2}{5-a}. \tag{1}$$

Question 1: Why is (1) true?

The parameterization of Figure 18b, using slope, is discussed in Problem 2. The parameterization using the angle θ, suggested in Figure 18c, is the subject of Problem 1. We show an analysis using Figure 18a.

Representing the area

Using the given parameters in Figure 18a, the area of the triangle is $\frac{1}{2}ab$. From the result in (1) we can express a in terms of b. Hence we can represent the area of the triangle in terms of the single parameter b.

$$\text{area} = f(b) = \frac{5b^2}{2(b-2)} \tag{2}$$

The problem is now reduced to finding the minimum value of the function f. One way to do this is to graph f and look for its minimum value (see Figure 19). From the graph it seems that the minimum of the function occurs when $b = 4$.

Figure 19

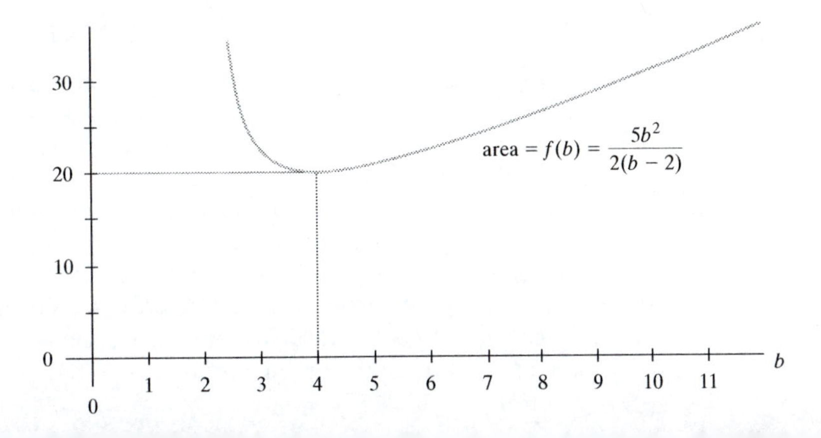

Alternatively, the function (2) can be analyzed using calculus. (See Problem 3.) Doing this verifies the graphical analysis. The minimum occurs precisely when $b = 4$.

We now have an answer to the original problem (a). The line cutting off minimal area has y-intercept $b = 4$. The area it cuts off is $f(4) = 20$. Figure 20 is a diagram to scale.

Figure 20

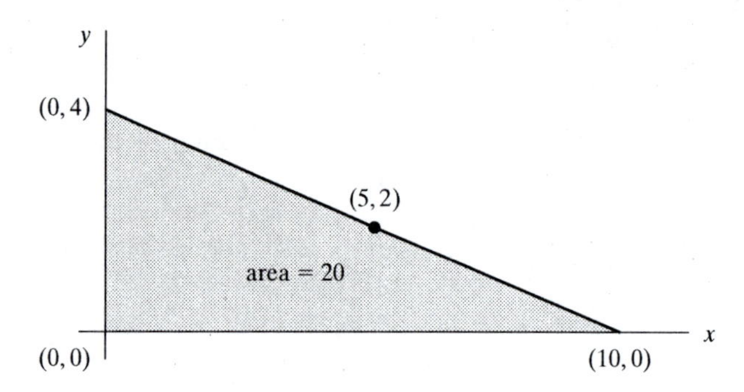

We have shown that the answer to the original problem is that the line through $(5, 2)$ cutting off the least area has x-intercept 10 and y-intercept 4. This answer is correct, but as it stands it has limited usefulness. For example, given another point, say $(3, 7)$, in the first quadrant and asked to find the line through this point cutting off minimum area, you would have to start the analysis over. At this point, if you have not done part (b) of the original problem, try to do so.

Interpreting the initial result

You may have noticed that there is a simple relationship between the numbers 10 and 4 and the given point $(5, 2)$. We can express this relationship in a general way using the slope.

(3) Of all lines through the point (p, q) in the first quadrant, the slope of the line that cuts off the triangle of smallest area is $-\frac{q}{p}$.

At this point (3) is only a conjecture. However, we can verify this conjecture readily using calculus. The relationships in (1) and (2) based on the specific point $(5, 2)$ can be rewritten in terms of a general point (p, q) as

$$\frac{b-q}{p} = \frac{q}{a-p} \tag{4}$$

and

$$\text{area} = g(b) = \frac{pb^2}{2(b-q)} \tag{5}$$

The area function g in (5) can be differentiated with respect to b as easily as the function f in (2). The function g has zero derivative when $b = 2q$. At this point $a = 2p$, and the area is $2pq$. Further, the slope of the line is $-\frac{q}{p}$. This analysis shows immediately that conjecture (3) is correct.

Without calculus (by finding graphically the minimum point of a function, as we did in Figure 19), you could not verify (3) directly. However, you could try a few other points and see if the conjecture holds for these points. (It will.) This would strengthen faith in the conjecture, but not prove it. Fortunately, there is a way to prove the conjecture geometrically without using calculus. Before showing this way, we proceed to part (c) of the original problem.

Generalization to angles other than right angles

In (c) you are asked to generalize further by looking at a point in any angle, not necessarily a right angle. You should try this problem if you have not done so already.

Here is a way to formulate (c) as a more general problem, which has been posed by Polya.

(6) Given a point P in the interior of an angle, what line through this point forms the triangle with minimum area?

In Figure 21, we have replaced the perpendicular axes with axes meeting to form $\angle ACB$. The line segment $\overline{FG}$ shown through P cuts off a region in this angle, forming ΔCFG. Problem (6) asks what line ℓ through P minimizes the area of ΔCFG.

Figure 21

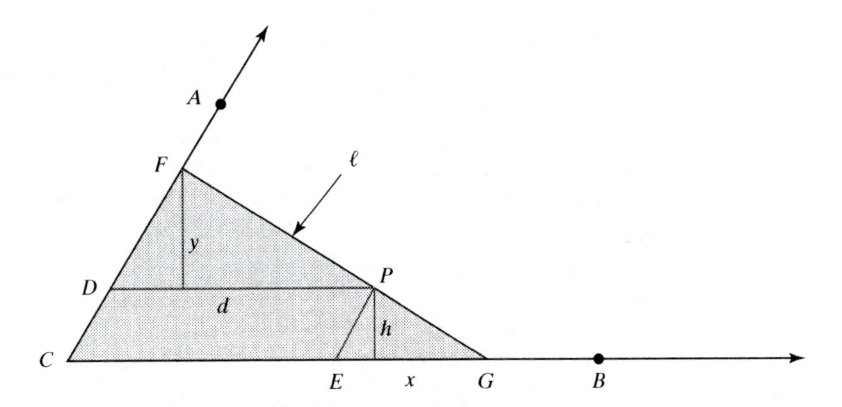

To proceed we need to identify in some way the location of the point P in the interior of $\angle ACB$. One way to do this is to use the parameters d (= DP) and h (= the distance from P to the side $\overline{CB}$ of the angle), where $\overline{DP} // \overline{CB}$, and $\overline{EP} // \overline{CA}$.

We also need to parameterize the line ℓ. Using x (= EG) and y (the distance from F to $\overline{DP}$) as parameters leads to a very simple analysis. The triangles DFP and EPG are similar. (Why?) Therefore, $\frac{y}{h} = \frac{d}{x}$, so the parameters x and y are not independent.

(7) $$y = \frac{dh}{x}$$

$\alpha(\Delta FCG) = dh + \frac{1}{2}dy + \frac{1}{2}hx$. Using (7), we can represent the area as a function of the parameter x.

(8) $$\alpha(\Delta FCG) = g(x) = h\left(d + \frac{1}{2}\frac{d^2}{x} + \frac{1}{2}x\right).$$

In the function defined by (8), d and h are constants. The variable is x, and varying x varies the position of the line ℓ through P. (This is what it means for the line ℓ to be parameterized by x.) A simple analysis using calculus shows that (8) has a minimum when $x = d$. At this value, $y = h$.

We now know that the answer to (6) is that the line minimizing area has $x = d$ and $y = h$. But this is a cumbersome description. Still, if we are alert we can see that the triangles DFP and EPG are not only similar, but *congruent*. (Why?) This means that the point P is *the midpoint* of the line minimizing area! Our analysis has led to a fairly general statement.

(9) Given a point P inside $\angle ACB$, the line through P that forms the triangle FCG of minimal area, with F on $\overrightarrow{CA}$ and G on $\overrightarrow{CB}$, is the line such that P is the *midpoint* of $\overline{FG}$.

The "midpoint" condition of (9) represents a significant result that, as you will see, is even more general than its formulation here.

Notice that as we generalized this problem, we moved from a result (3), which is described in terms of numbers (coordinates and slopes), to a result (9), which is described in purely geometrical terms. In other words, although the initial formulation and proofs used analytic concepts such as functions and graphs (and calculus), what we discovered is a result about geometry.

Seeing this as purely a geometry problem

Now we complete step (d) of the original problem by looking for a geometric proof of (9). Consider Figure 21. Our goal is to prove: *Of all line segments through P, the particular segment that has P as its midpoint cuts off the minimal area.*

Let ℓ' be another line through P that connects points on two sides of the angle. The two possible locations for ℓ' are as shown in Figures 22a and 22b. We try to prove that ℓ' cuts off more area than ℓ. The method that we use is a general and powerful technique for constructing proofs of results about optimization. Here is the major step:

(10) The two shaded triangles are congruent.

Question 2: Why is (10) true?

From (10), we see that the area enclosed by line ℓ' is greater than the area enclosed by line ℓ. In fact, it is greater by exactly the area of region R.

Figure 22

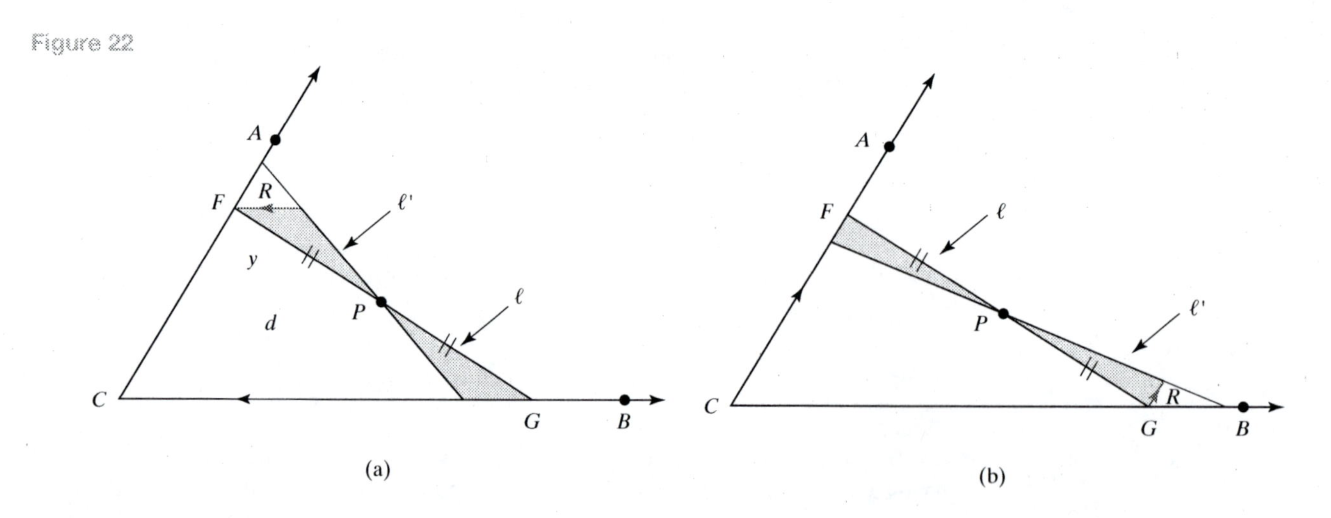

Summary of what we have done

In summary, we have constructed an extended analysis of the original high school problem following the steps of (a) to (d). The analysis resulted in significant generalizations. In particular, we generalized beyond particular numbers (the point $(5, 2)$), to a general point (p, q). We also generalized beyond particular shapes (a right angle) to any angle. The analysis progressed away from the original mathematical subject area of the problem (coordinate systems, graphs, and functions) to one more natural for the problem at hand. Finally, the more general analysis is no more difficult to carry out than the original very specific one.

The last feature is especially noteworthy. Although (9) is a very general result, the geometric proofs illustrated in Figures 22a and 22b are easier than the earlier analysis in terms of graphs and functions.

What about step (e) of the original problem? As an example, based on (9) we might conjecture that the line ℓ through a point P inside a parabola that cuts off the smallest area is the line bisected by P. (See Project 1 at the end of this chapter.)

What to look for in advance in general solutions

There is an important further perspective on what we have done in our analysis. In briefest terms, we have shown that the question asked in Figure 23a is answered by the particular line described in Figure 23b.

Figures 23a, b

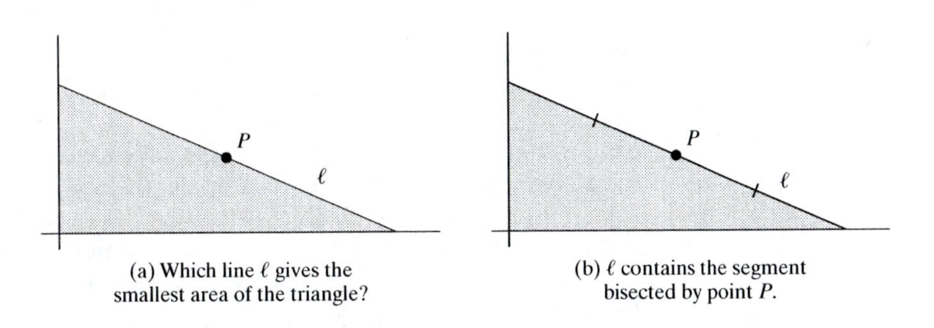

(a) Which line ℓ gives the smallest area of the triangle?

(b) ℓ contains the segment bisected by point P.

Let us call the line described in Figure 23b a "special" line. It is special in the sense that it has a simple geometric characterization in terms of the given point P. In Figures 23c through 23i, we indicate seven other conditions on lines through P.

Figures 23c–f

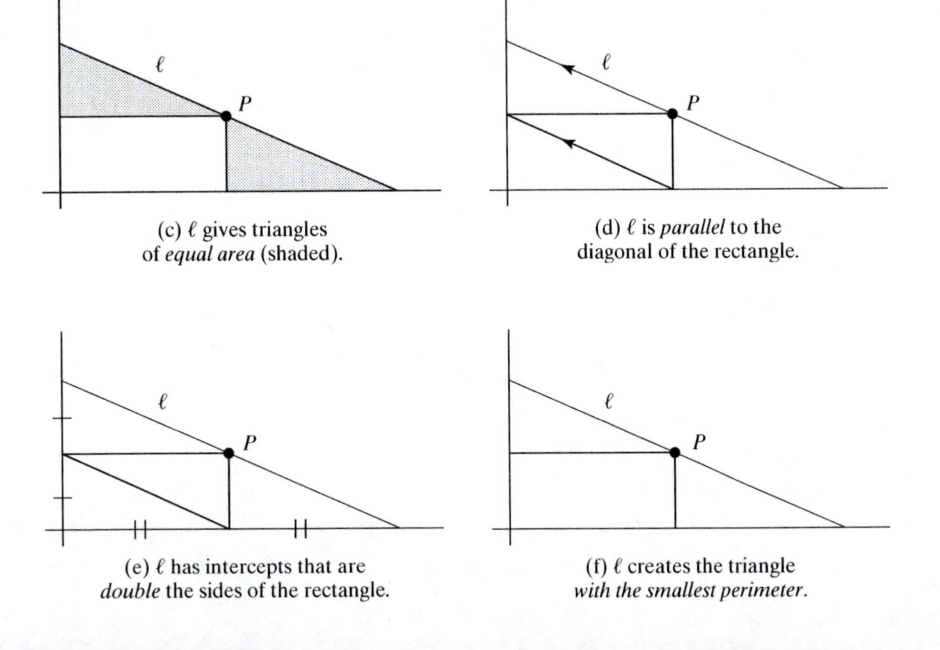

(c) ℓ gives triangles of *equal area* (shaded).

(d) ℓ is *parallel* to the diagonal of the rectangle.

(e) ℓ has intercepts that are *double* the sides of the rectangle.

(f) ℓ creates the triangle *with the smallest perimeter*.

Figures 23g–i

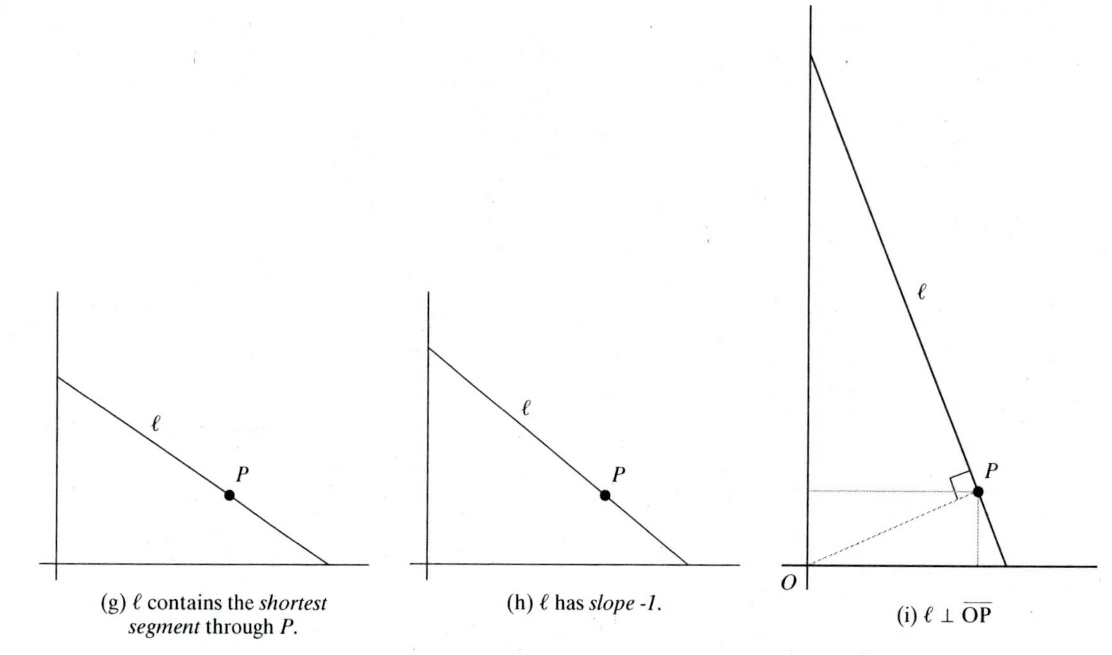

(g) ℓ contains the *shortest segment* through P.

(h) ℓ has *slope -1*.

(i) $\ell \perp \overline{OP}$

In short, Figure 23a–23i give geometric ways of characterizing special lines through a point. We can think of the "solution" to our original problem as being, possibly, one of these characterizations. Thinking in this way gets us away from focusing on numbers only and prompts us to look for more intrinsic features of our solutions.

Our analysis showed that Figures 23a and 23b are equivalent. This is the substance of result (9). What about the other "special" lines of Figure 23? It turns out that the five characterizations in Figures 23a to 23e are all equivalent: If a line has one of these properties, it has them all. (See Problem 7.)

In some cases, when people are asked to use intuition to guess what line gives minimum area, they come up with the properties described in Figures 23f through 23i. These are reasonable guesses, but further analysis shows them to be not correct. Figures 23f and 23g describe two other minimal properties, this time not minimal *area*, as in (a), but of minimal *length*. We might conjecture that one or the other of these lines is the same as the line giving minimal area. But this is not the case. (See Problems 7 and 12.)

10.1.3 Problems

1. Use Figure 24.

a. Prove that there is no upper limit to the area cut off by line ℓ through point $(5, 2)$ as the angle θ increases toward $\frac{\pi}{2}$.

b. Prove that there is no upper limit to the area cut off by line ℓ through point $(5, 2)$ as the angle θ decreases toward 0.

c. Give a rough sketch of the graph of a function that represents the area of ΔOAB as a function of θ.

Figure 24

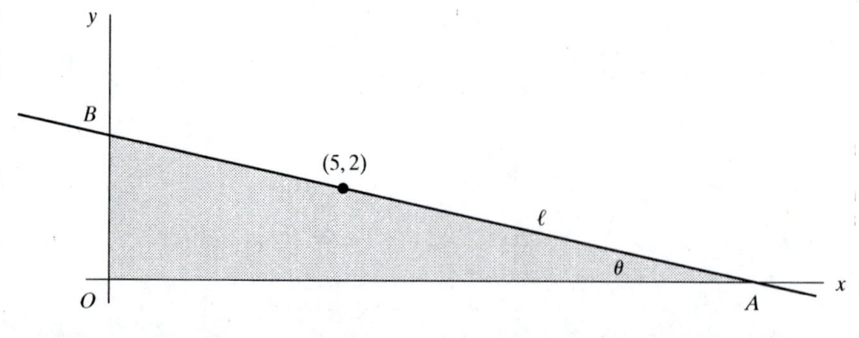

Figure 25

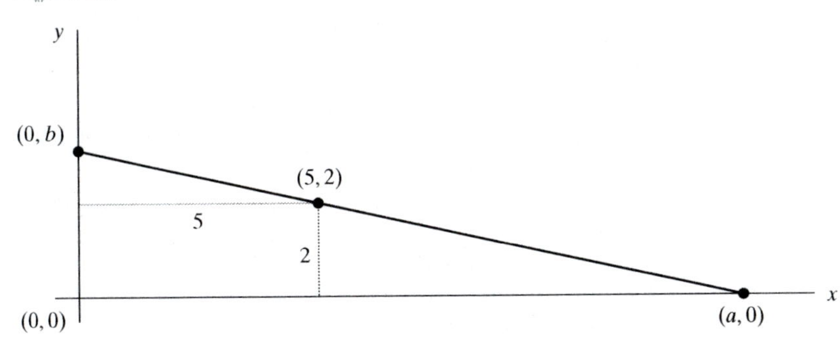

2. A nonvertical line in $\mathbf{R}^2$ is associated with three basic parameters: its slope m, its y-intercept b, and its x-intercept a. (See Figure 25.)

a. Show that any *two* of these parameters suffice to specify any nonvertical line.

b. Show that any *one* of these parameters suffices to specify any nonvertical line through the particular point $(5, 2)$.

c. Represent the area of the triangle cut off by a line through $(5, 2)$ in terms of the *slope* m of the line.

d. Represent the area of the triangle cut off by a line through $(5, 2)$ in terms of the *y-intercept* b of the line.

e. Represent the area of the triangle cut off by a line through $(5, 2)$ in terms of the *x-intercept* a of the line.

3. a. Give the details of the calculus proof that the function $f\colon b \to \frac{5b^2}{2(b-2)}$ of Figure 19 obtains its minimum at $b = 4$.

b. Give the details of the calculus proof that the more general function $f\colon b \to \frac{pb^2}{2(b-q)}$ obtains its minimum at $b = 2q$.

c. Use polynomial division to find an equation for the oblique asymptote of the graph of the function $f\colon b \to \frac{5b^2}{2(b-2)}$.

d. Answer the question of part **c** for the more general function of part **b**.

4. We solved the problem that opens this section using the intercepts a and b to parameterize the lines through the point $(5, 2)$. A somewhat simpler analysis results by parameterizing the lines in terms of the lengths u and v shown in Figure 26. Carry out this analysis. That is, show how u and v are related, express the area in terms of v alone, find the minimum of this function, and answer the original problem of this section.

Figure 26

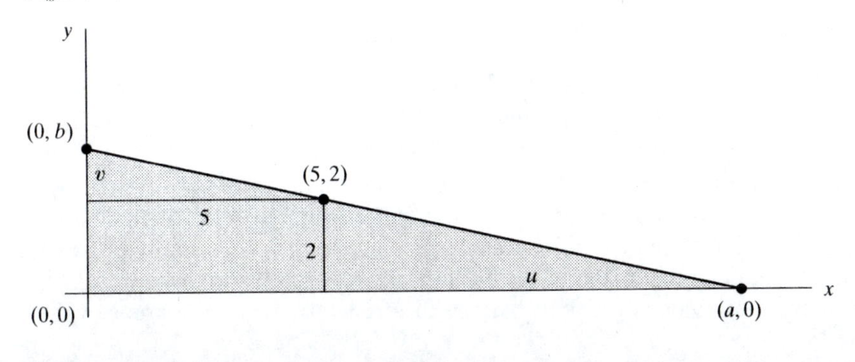

5. We have given a simple characterization of the segment through a point P that cuts off minimal area in the first quadrant: It is the segment bisected by P. Show that the other characterizations of the minimal area in Figures 23a to 23e are equivalent.

6. Suppose the point (p, q) in relationship (3) of this section is in the 2nd quadrant. Which, if any, of relationships (3) to (5) in this section are no longer true?

7. Show that the properties described in Figures 23g to 23i do not solve the original problem.

8. The function that we have minimized in Problem 4 involves the expression $10 + \frac{5}{2}v + \frac{10}{v}$. This is the sum of a constant and a variable and its reciprocal, with constant factors. Such expressions appear often in max-min problems. For example, such a function appears in (8). This problem asks that you show that there is a general answer to all such problems:

a. Consider a function $f(x) = Ax + \frac{B}{x}$, where $A > 0$ and $B > 0$ are constants. Find the minimum value of such a function, and find where it achieves this minimum. (Use calculus.)

b. Show that the minimum value of $Ax + \frac{B}{x}$ occurs at the value of x where the graph of $y = Ax$ intersects the graph of $y = \frac{B}{x}$.

c. By analyzing the graphs in part **b**, give a noncalculus argument for part **a**. (*Hint*: Show that where the graphs intersect, the slopes are opposites of each other. Use that and the concavity of the graph of $y = \frac{B}{x}$ to show that moving in either direction from the intersection will increase the value of the sum.)

d. Use the Arithmetic-Geometric Mean Inequality ($\frac{a+b}{2} \geq \sqrt{ab}$ for all real a and b) to do parts **a** and **b**.

*9. In this section, we *verified* that the line through point P cutting off minimum area has the midpoint property. But we had already conjectured the midpoint property. Show how the method we used can also be used in a more powerful way to actually *derive* the midpoint property. (*Hint*: Use the principle that if a line ℓ does cut off minimal area, then the "increments" of area, one positive and one negative, created by rotating this line a small amount about point P are approximately equal.)

10. A line ℓ passing through point $P = (1, 1)$ makes an angle θ with the horizontal. (See Figure 27.) Let $A(\theta)$ be the shaded area below ℓ in the square.

Figure 27

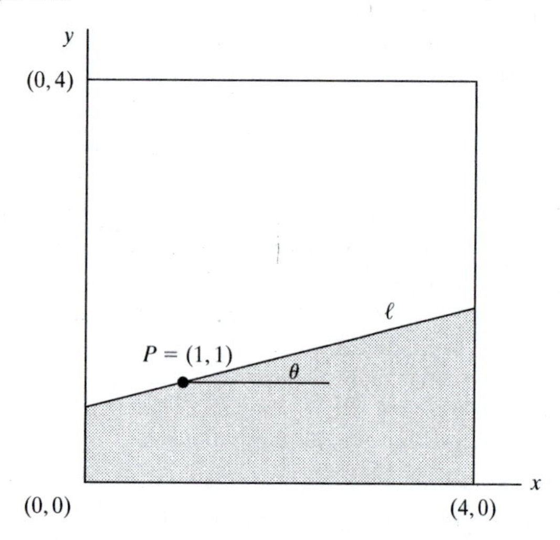

a. Sketch a graph of $y = A(\theta)$ as θ ranges from 0 to 2π.

b. Find a formula for $A(\theta)$ and graph this function. (There may need to be several formulas, depending on which sides of the square the line ℓ intersects.)

*11. In this section, we extended the initial problem about a point in a right angle to a problem about a point in any angle. In this problem you are asked to generalize the initial problem further. Prove or disprove the following results.

a. Given a point P inside a convex polygon C, there is at least one line ℓ through P whose portion inside C is bisected by P. Call such a line "P-centered".

b. Each line through P divides C into two convex polygons, and we can plot the area of one of them as we rotate the line through 360° (keeping it pivoted at P). This area function has a local minimum or local maximum for a certain line ℓ if ℓ is P-centered.

12. Given a point (m, n) in the first quadrant,

a. Use calculus to show that the *shortest* line segment in the quadrant through (m, n) has slope $\left(\frac{n}{m}\right)^3$.
(*Hint*: Parameterize the lines by the angle they make with the x-axis.)

b. Relate this problem to the problem of finding the shortest ladder that will touch the wall, the floor, and a box of height n and width m placed against the wall.

c. Relate this problem to the problem of finding the *longest* ladder that will fit around a corner where two hallways of widths m and n meet.

*d. Use a geometric argument to derive the property stated in part **a**.

13. Let P be a fixed point in the interior of a parabola. Let ℓ be a line containing P and intersecting the parabola at points A and B. What position of ℓ minimizes the area of the region bounded by ℓ and the parabola? (*Hint*: Because all parabolas are similar, answering the question for one parabola essentially answers the question for all parabolas.)

*14. Investigate the situation of the line ℓ through a fixed point P in the first quadrant outlining, with the axes, the triangle of minimal perimeter. (See Figure 23f.)

ANSWERS TO QUESTIONS

1. Each side of the equation is the slope determined by $(5, 2)$ and one of the intercepts.

2. In each case the midpoint P determines a pair of congruent sides, a pair of angles are vertical angles, and another pair of angles are alternate interior angles formed by two parallel lines. So the triangles are congruent by ASA congruence.

10.1.4 From polygons to regions bounded by curves

The area function defined in Section 10.1.1 applies only to regions that are unions of a finite number of triangular regions. This means that it does not immediately apply to circles or other regions that have curves as boundaries. However, from ancient times, mathematicians have extended the definition of an area function α from polygonal regions to curved regions by squeezing the curved region in between smaller and larger unions of triangular regions. This squeezing was first formally done by the Greek mathematician Eudoxus of Cnidus around 400 B.C. Eudoxus's strategy is called the "method of exhaustion" because the difference between the areas of the smaller and larger unions is exhausted in the sense that it approaches zero. It is akin to the way the rational numbers are extended to the real numbers via nested intervals. We add the following to our definition of area to enable the method of exhaustion to be applied to the common geometric figures with curved boundaries.

Definition (of area function, continued):

4. Let C be a region in E^2. Let s_i be a sequence of unions of triangular regions in E^2 with no interior points in common, such that each s_i is a subset of C. Let S_i be a sequence of unions of triangular regions in E^2 such that C is a subset of each S_i. If the least upper bound of the $\alpha(s_i)$ equals the greatest lower bound of the $\alpha(S_i)$, then this bound is $\alpha(C)$.

Approaching the area of a circle from above and below

We use property (4) to obtain the area of a circle. Figure 28a shows inscribed and circumscribed squares in a circle. The area of the circle is between the areas of the squares. If the radius of the circle is r, then Figure 28a shows that the area must be between $2r^2$ and $4r^2$. Figure 28b uses inscribed and circumscribed octagons, whose areas are $2\sqrt{2}r^2$ and $8(\sqrt{2} - 1)r^2$, that is, between approximately $2.8284r^2$ and $3.3137r^2$.

Figure 28

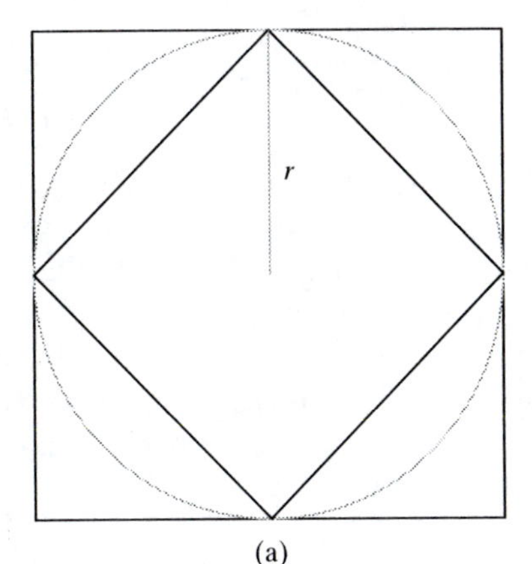

(a)

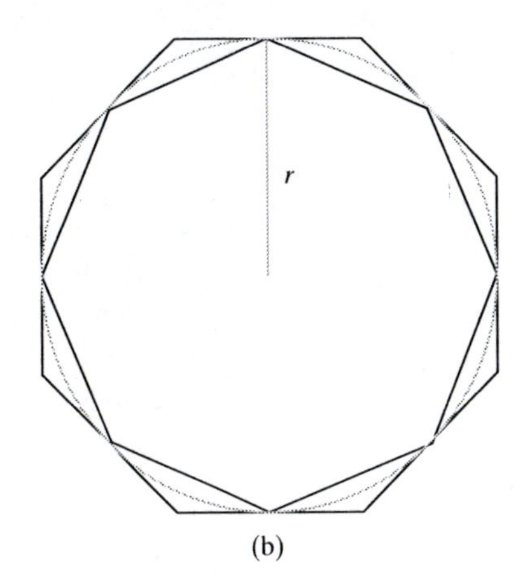

(b)

The calculations in Table 2 confirm that these areas continue to get closer to each other as the number of sides of the polygons is doubled again and again. In Table 2, s_i is a regular inscribed polygon with 2^i sides (a 2^i-gon) in a circle of radius 1, where $i \geq 2$, and S_i is the corresponding circumscribed polygon. Thus each polygon in both sequences has twice the number of sides of the preceding term of that sequence. Because each inscribed polygon can be thought of as connecting the points of tangency of the circumscribed polygon, for all i, $\alpha(s_i) \leq \alpha(S_i)$.

Table 2 Areas of Inscribed and Circumscribed 2^i-gons

i	2^i	$\alpha(s_i)$	$\alpha(S_i)$	$\alpha(S_i) - \alpha(s_i)$
2	4	$2r^2$	$4r^2$	$2r^2$
3	8	$2.82842711\ldots r^2$	$3.31370849\ldots r^2$	$.4852813\ldots r^2$
4	16	$3.06146745\ldots r^2$	$3.18259787\ldots r^2$	$.1211304\ldots r^2$
5	32	$3.12144515\ldots r^2$	$3.15172490\ldots r^2$	$.0302797\ldots r^2$
6	64	$3.13654849\ldots r^2$	$3.14411838\ldots r^2$	$.0075698\ldots r^2$
7	128	$3.14033115\ldots r^2$	$3.14222363\ldots r^2$	$.0018924\ldots r^2$

Figure 29

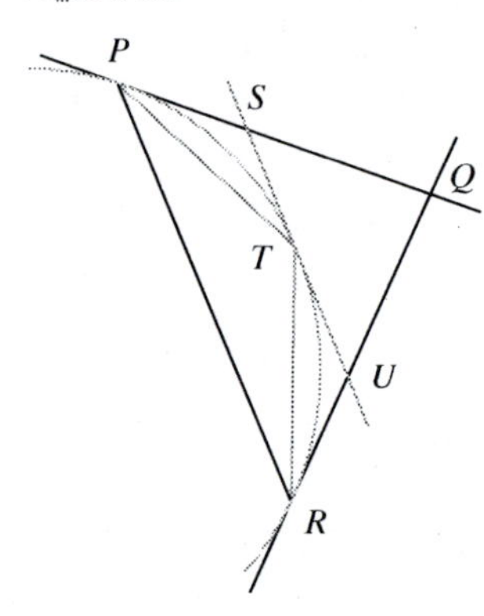

In Figure 29 we show parts of regular inscribed and circumscribed 2^i-gons and 2^{i+1}-gons. $\overline{PR}$ is a side of a regular inscribed 2^i-gon and $\overline{PQ}$ and $\overline{QR}$ are halves of sides of a regular circumscribed 2^i-gon. $\overline{PT}$ and $\overline{TR}$ are sides of a regular inscribed 2^{i+1}-gon, while $\overline{PS}$, $\overline{UR}$, and $\overline{SU}$ are two halves and a full side of a regular circumscribed 2^{i+1}-gon. The figure guides us to see that, as the number of sides is doubled, $\alpha(s_{i+1}) > \alpha(s_i)$ because $\alpha(\Delta PTR)$ is added to the area, while $\alpha(S_{i+1}) < \alpha(S_i)$ because $\alpha(\Delta SQU)$ is taken away. Thus we have, for each i,

$$\alpha(s_i) < \alpha(s_{i+1}) < \alpha(S_{i+1}) < \alpha(S_i).$$

So the closed intervals $I_i = [\alpha(s_i), \alpha(S_i)]$ form a nested sequence.

Figure 29 also shows us how the lengths of the intervals I_i go to zero as i increases. The length of I_i is $\alpha(S_i) - \alpha(s_i)$, the difference between the areas of the circumscribed and inscribed 2^i-gons. From Figure 29, which shows only $\frac{1}{2^i}$ of the circle,

$$\alpha(S_i) - \alpha(s_i) = 2^i(\alpha(\Delta PQR))$$

and

$$\begin{aligned}\alpha(S_{i+1}) - \alpha(s_{i+1}) &= 2^{i+1}(\alpha(\Delta TUR))\\ &= 2^i(\alpha(\Delta TUR) + \alpha(\Delta TSP)).\end{aligned}$$

Consequently,

$$\frac{\alpha(S_{i+1}) - \alpha(s_{i+1})}{\alpha(S_i) - \alpha(s_i)} = \frac{\alpha(\Delta TUR) + \alpha(\Delta TSP)}{\alpha(\Delta PQR)}.$$

The two regions identified in the numerator of the fraction on the right side are parts of the denominator. We leave it to you to show that their sum is less than half the area of trapezoid $PSUR$, and thus the value of the fraction is less than $\frac{1}{2}$. This means that, as i increases, the length of I_i is decreasing faster than a geometric sequence with common ratio $\frac{1}{2}$. (In fact, the right column of Table 2 shows that the ratio is nearer $\frac{1}{4}$.) Thus that difference goes to zero. By the completeness property of the real numbers, there is a unique real number between all the $\alpha(s_i)$ and all the $\alpha(S_i)$. This number is the area of the circle.

π and πr^2

But how do we know that the area of a circle with radius r is πr^2? This is perhaps the easiest part of the entire argument. We define π in its usual way, as the ratio of the circumference C of the circle to its diameter. That is, $\pi = \frac{C}{d}$. From this, $C = \pi d = 2\pi r$.

Theorem 10.10 The area of a circle with radius r is πr^2.

Proof: When a regular 2^n-gon circumscribes a circle of radius r, the circle is inscribed in the polygon. By Theorem 10.6, the area of this polygon is given by $A = \frac{1}{2}rp$. As n increases, the perimeter p is getting closer and closer to the perimeter of the inscribed 2^n-gon, and in between these perimeters is the circumference of the circle. So the area of the circle is between two areas, each of which is approaching the value $\frac{1}{2}rC$, that is, $\frac{1}{2}r \cdot 2\pi r$, which is πr^2. ┘

Thus the area formula for a circle is an extension of the area formula for polygons that can be circumscribed about a circle. The number π enters the formula because π is defined in terms of the "perimeter" of the circle, that is, in terms of the circumference.

Figure 30

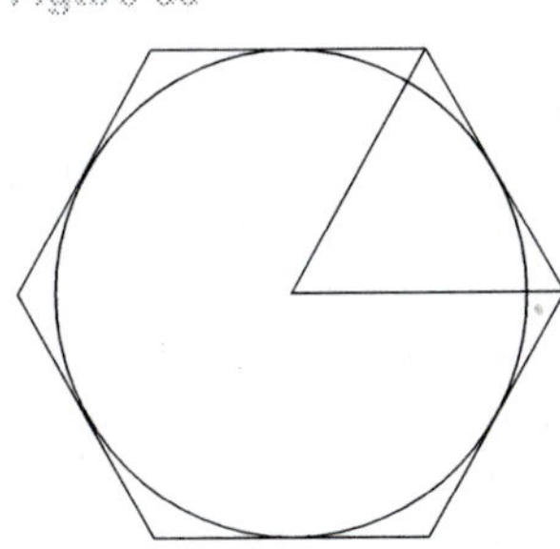

The connection between circumference and area, as mentioned in the proof of Theorem 10.10, was first made by Archimedes using the method of exhaustion. In doing so, Archimedes also was able to find a rather good approximation to π. Here is what he did. Instead of starting with inscribed and circumscribed squares as we did, he began with the triangle at the center of the circle that is $\frac{1}{6}$ of a circumscribed regular hexagon (see Figure 30). He successively bisected the central angle, compared ratios, took away parts of irrational square roots, and arrived at the conclusion that the circumference of a circle is less than $3\frac{1}{7}$ times the diameter.

Then Archimedes considered inscribed regular polygons of 6, 12, 24, 48, and 96 sides. He found the perimeter of each polygon and concluded that the circumference of a circle is more than $3 + \dfrac{1}{7 + \frac{1}{10}}$ times the diameter. In this way, Archimedes provided the first good numerical approximation to π: $3\frac{10}{71} < \pi < 3\frac{1}{7}$.

A brief history of π

Since the time of Archimedes, the calculation of better and better numerical approximations to π has occupied the attention of mathematicians throughout the world. At least two entire books are devoted to this history— Beckmann's *A History of Pi*, and Berggren and the Borweins *Pi: A Source Book*. We give only the briefest of histories of this calculation.

At first, the theory developed very slowly. The best approximation before 1600 that is known to us was by Viéte, who in 1576 found a value of π correct to 9 decimal places (in today's notation—decimal fractions were not invented until 1585!). Viéte used Archimedes' method with polygons of $6 \cdot 2^{16}$, or 393,216 sides. With decimals, using Archimedes' method with polygons of 2^{62} sides, Ludolph van Ceulen in 1610 was able to obtain a value of π correct to 35 decimal places, having spent most of his life on the calculations required for this task.

The first use of the Greek letter π to represent the ratio of a circle's circumference to its diameter seems to have been in the textbook *Synopsis Palmariorium Mathesios*, written by William Jones in 1706. He chose π because it was the first letter of the Greek word "perimetrog", meaning "surrounding perimeter".

After van Ceulen, most mathematicians started using methods of analysis and infinite series to approximate π. By 1844, Zacharias Dase, a German of prodigious calculating ability, computed π to 200 decimal places. And in 1873, with 15 years of work, William Shanks computed π to 707 places, of which the first 527 were correct. This was the most accurate calculation before the days of machine calculation.

Shanks's error was found in 1948 by mathematicians using a desk calculator. The next year, one of the very earliest computers, ENIAC, calculated π to 2,037 decimal places. By 1967, π had been calculated to 500,000 places. Two million places were calculated by Kazunori Miyoshi and Kazuhiko Nakayama in 1981 using the trigonometric identity

$$\pi = 32 \tan^{-1}\left(\frac{1}{10}\right) - 4 \tan^{-1}\left(\frac{1}{239}\right) - 16 \tan^{-1}\left(\frac{1}{515}\right).$$

Within five years, the billion-place standard was reached by David Bailey and Jonathan and Peter Borwein using a formula discovered by Ramanujan in 1910. By the year 2000, more than 51 billion decimal places of π were known.

The Riemann integral

In the 17th century, mathematicians refined the method of exhaustion in their development of calculus. If a curve can be described by a formula, calculus often enables the lengthy calculations of the method of exhaustion to be replaced by the relatively simple calculation of definite integrals.

In calculus, the area property (4) shows one way to obtain the area between the graph of a sufficiently well-behaved function (e.g., continuous or monotone) $y = f(x)$, the x-axis, and the lines $x = a$ and $x = b$. Each s_i is the total area of a union of rectangles whose total area is a *lower Riemann sum*, and each S_i is a union of rectangles whose area is an *upper Riemann sum*. In Figure 31, the area of the shaded region is a lower Riemann sum while the total area of the shaded and unshaded rectangles is an upper Riemann sum.

If we increase the number of partitions into which the interval $\overline{PQ}$ is decomposed, the difference between the sum of the areas of the outside rectangles and the sum of the areas of the inside rectangles becomes smaller and smaller, so that by taking a sufficiently large number of partitions, we can make this difference as small as we please. Since the area under the curve lies between these two sums, it is the limit toward which the sum of the outside or inside rectangles tends as the number of partitions is definitely increased, and the determination of this limit is accomplished by what we know from calculus as integration. In other words, the common bound on these sums for a monotone increasing function $y = f(x)$ is the desired area, the value of the definite integral $\int_a^b f(x)\,dx$.

Figure 31

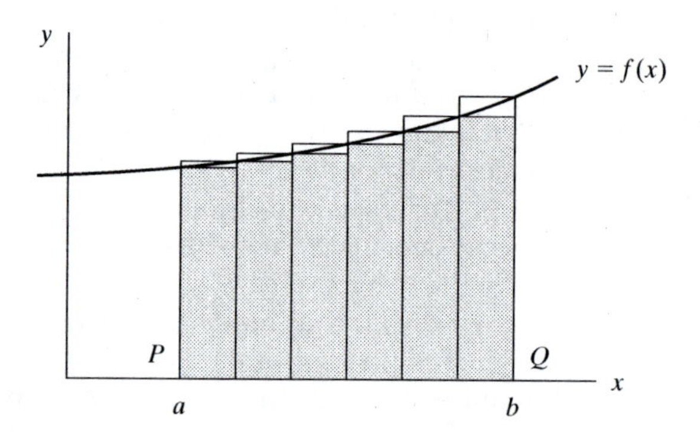

10.1.4 Problems

1. a. Find the exact areas of regular circumscribed and inscribed hexagons about a circle of radius r (Figure 32).

Figure 32

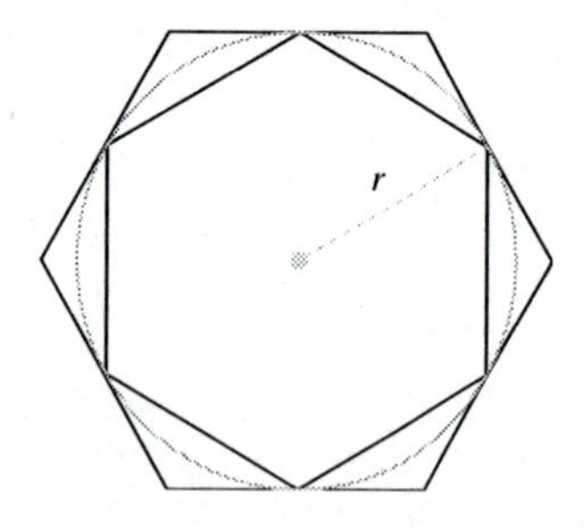

b. If a side of a regular n-gon inscribed in a circle of radius 1 has length x, determine the length of a side of a regular $2n$-gon inscribed in the same circle in terms of x.

*c. Use parts **a** and **b** and a calculator or computer to calculate the perimeter of regular polygons of 12, 24, 48, 96, and 192 sides inscribed in a circle of diameter 1. (Because of the unwieldy number system used at the time, Archimedes could go only so far as 96 sides.)

2. **The area of a sector.** A **sector** is the region bounded by two radii $\overline{OA}$ and $\overline{OB}$ and arc $\widehat{AB}$ of circle O (Figure 33). Assume that the area of a sector is proportional to the measure of the arc. Use this assumption to deduce the formula $A = \frac{r^2\theta}{2}$ for the area of a sector bounded by an arc of radian measure θ in a circle of radius r.

Figure 33

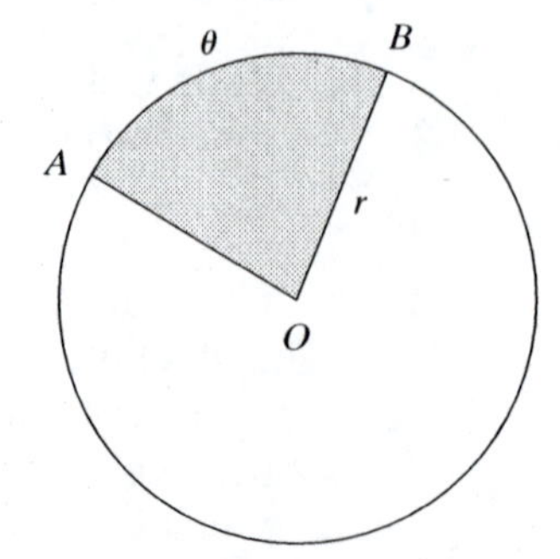

Figure 34

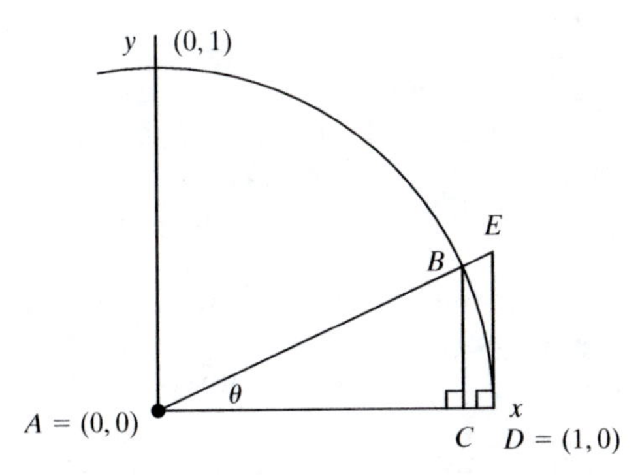

3. Figure 34 shows a part of the unit circle in the first quadrant.

a. Use the result of Problem 2 to prove that $\sin\theta < \theta < \tan\theta$, for θ in radians.

b. Explain why $\lim_{\theta\to 0} \frac{\sin\theta}{\theta} = \lim_{\theta\to 0} \frac{\tan\theta}{\theta} = 1$.

4. Find formulas for the area and perimeter of a Reuleaux triangle of width w. (See Problem 6 of Section 10.1.1.)

5. **The area of an ellipse.** Let T be the transformation in $\mathbf{R}^2$ defined by $T(x, y) = (ax, by)$.

a. Show that if $JKLM$ is any square in $\mathbf{R}^2$, and $T(JKLM) = J'K'L'M'$, then $\alpha(J'K'L'M') = ab\alpha(JKLM)$.

b. Show that the image of the unit circle under T is the ellipse with equation $\left(\frac{x}{a}\right)^2 + \left(\frac{y}{b}\right)^2 = 1$.

c. Use parts **a** and **b** to derive a formula for the area of any ellipse in terms of its semimajor axis a and its semiminor axis b.

6. **Area under a parabola.** Consider the region R bounded by the parabola $y = x^2$, the x-axis, and the line $x = 10$.

a. Calculate $L = \sum_{i=0}^{9} i^2$ and $U = \sum_{i=1}^{10} i^2$. Explain why $L < \alpha(R) < U$.

b. Find sums that provide a smaller interval containing $\alpha(R)$.

c. Use calculus to determine $\alpha(R)$ exactly.

7. **Estimation of π using perimeters.** In this section, π was estimated using the areas of inscribed and circumscribed regular polygons. The following theorem enables the estimation of π by calculating successive harmonic and geometric means of the perimeters of these polygons.

> **Theorem:** Let p and P be the perimeters of inscribed and circumscribed regular n-gons in a circle. Let p' and P' be perimeters of inscribed and circumscribed regular $2n$-gons in the same circle. Then
>
> (1) $$P' = \frac{2pP}{p + P} = H(p, P)$$
>
> (2) $$p' = \sqrt{pP'} = G(p, P').$$

That is, suppose the perimeters of the inscribed and circumscribed n-gons of a circle are known. Then the perimeter of the circumscribed $2n$-gon is the harmonic mean of these perimeters, and the perimeter of the inscribed $2n$-gon is the geometric mean of the perimeter of the circumscribed $2n$-gon and the inscribed n-gon. These two recursion relations enable π to be estimated as close as desired (given the computational wherewithal).

Figure 35

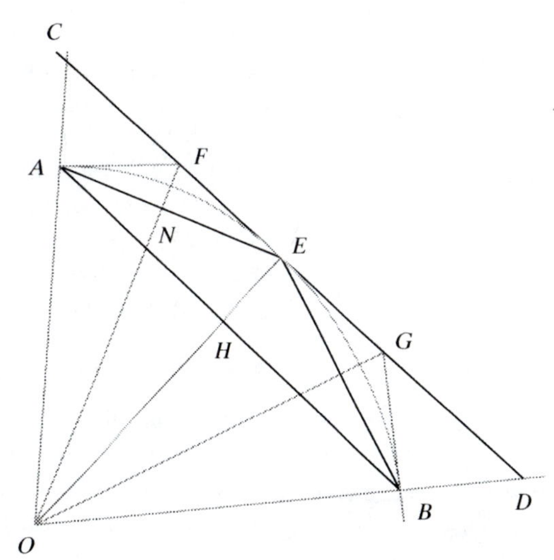

In Figure 35, $\overline{AB}$ is a side of a regular n-gon inscribed in circle O and $\overline{CD}$ is a side of the circumscribed regular n-gon to circle O, tangent to the arc $\widehat{AB}$ at its midpoint E. Then $\overline{AE}$ and $\overline{BE}$ are sides of a regular $2n$-gon inscribed in circle O. Tangents at A and B intersect $\overline{CD}$ at F and G, respectively. Then $\overline{FG}$ is a side of a circumscribed regular $2n$-gon in circle O. From these constructions, by definition of *perimeter*, $p = n \cdot AB$ and $P = n \cdot CD$. Also, $p' = 2n \cdot AE$ and $P' = 2n \cdot FG$.

a. Calculate p and P if $n = 2$ and the radius of the circle is 0.5.

b. Use the results of part **a** to calculate successive perimeters of regular inscribed and circumscribed polygons with 2^i-sides until π is estimated correct to 6 decimal places.

c. Prove both parts of the theorem by justifying each of these conclusions.

i. $\frac{P}{p} = \frac{CF}{FE}$ ii. $\frac{P+p}{2p} = \frac{CE}{FG}$ iii. $\frac{P}{P'} = \frac{CE}{FG}$

iv. $\frac{P+p}{2p} = \frac{P}{P'}$, from which (1) follows by solving for P'

v. $\frac{p}{p'} = \frac{AH}{AE}$ vi. $\frac{p'}{P'} = \frac{EN}{EF}$

vii. $\Delta ENF \sim \Delta AHE$, from which $\frac{AH}{AE} = \frac{EN}{EF}$

viii. $\frac{p}{p'} = \frac{p'}{P'}$, from which (2) follows by solving for p'.

10.1.5 The problem of quadrature

In Section 10.1.1, we proceeded from the area of a square to the area of any triangle, and then used areas of triangles to obtain area formulas for quadrilaterals with parallel sides (trapezoids) or perpendicular diagonals, and for polygons that can be circumscribed about a circle. In Section 10.1.4, we applied the last of these formulas to obtain a formula for the area of a circle and applied areas of trapezoids to obtain formulas for areas under curves.

The ancient Greeks utilized this process, but also reversed it. They were particularly interested in the problem of **quadrature**, that is, of constructing a square with the same area as a given figure. The reason for the importance of quadrature is simple: If a square can be constructed with the same area as a given region F, then you can be certain of the area of F in square units.

In this short section, we consider the question of quadrature. Our general scheme begins as follows. (1) Construct a square with the same area as any given rectangle. (2) Construct a rectangle with the same area as any given triangle. Then, using (2) and then (1), we can construct a square with the same area as any given triangle. The discussion provides a review of some of the ideas in earlier sections and also a few surprises.

Quadrature of the rectangle

Perhaps the first surprise is that we use ideas from similar triangles to construct a square with the same area as a given rectangle.

Theorem 10.11 A square can be constructed with the same area as a given rectangle.

Proof: Suppose a rectangle has dimensions x and y as in Figure 36a. Then a square with side s and the same area as the quadrilateral has area $s^2 = xy$, so $s = \sqrt{xy}$. Thus s is the geometric mean of x and y. A segment of length s can be constructed from segments of lengths x and y using ideas from Section 8.3.1. Specifically, place segments of lengths x and y on the same line next to each other, with point P in common, as shown in Figure 36b. Construct a circle whose center is the midpoint of the new segment of length $x + y$ and whose radius is $\frac{x+y}{2}$. Then construct the perpendicular at P. The length of the half-chord $\overline{PQ}$ from P to the circle has length $\sqrt{xy}$ and so is the length of a side of the desired square. ∎

Figure 36

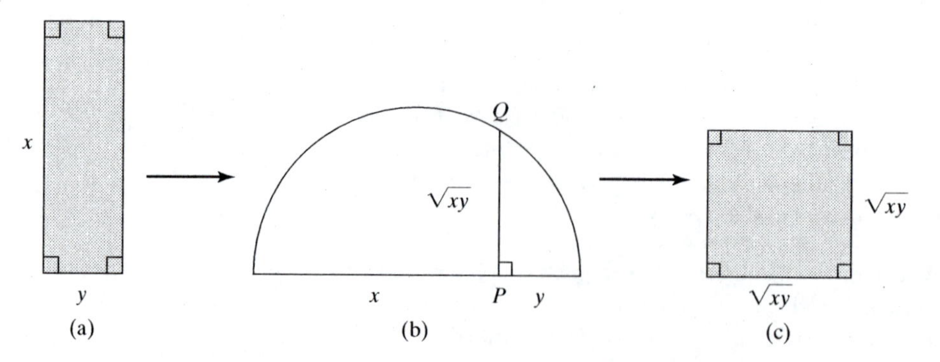

Question: Why does $\overline{PQ}$ have length $\sqrt{xy}$?

We say that Theorem 10.11 allows us to "square a rectangle". Theorem 10.11 is exceedingly important in the theory of quadrature because it is easy to construct a rectangle with the same area as any triangle.

Theorem 10.12 A rectangle can be constructed with the same area as a given triangle.

Proof: Suppose ΔABC is given. Let h be the altitude to side $\overline{BC}$. Then $\alpha(\Delta ABC) = \frac{1}{2}h \cdot BC$, so a rectangle with consecutive sides of lengths $\frac{1}{2}h$ and BC has the same area as ΔABC. Such a rectangle is easy to construct (Figure 37).

Figure 37

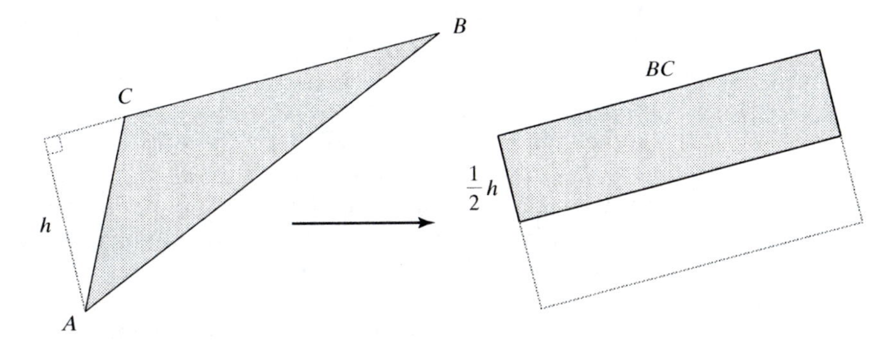

From Theorems 10.11 and 10.12, in two steps we can construct a square with the same area as any triangle. That is, we can "square a triangle". But a polygonal region may be split into two or more triangles. So the next question is: Suppose a region is the union of two triangles. Can a single square be found whose area equals the sum of the areas of the two triangles? The answer is an immediate and surprising consequence of a well-known theorem.

Theorem 10.13 Let F be the union of two triangular regions with no interior points in common. Then a square can be constructed with the same area as F.

Proof: Let the two triangular regions be F_1 and F_2 (Figure 38). By Theorems 10.11 and 10.12, a square region of side s_1 can be constructed with area $\alpha(F_1)$, and a square region of side s_2 can be constructed with area $\alpha(F_2)$.

Figure 38

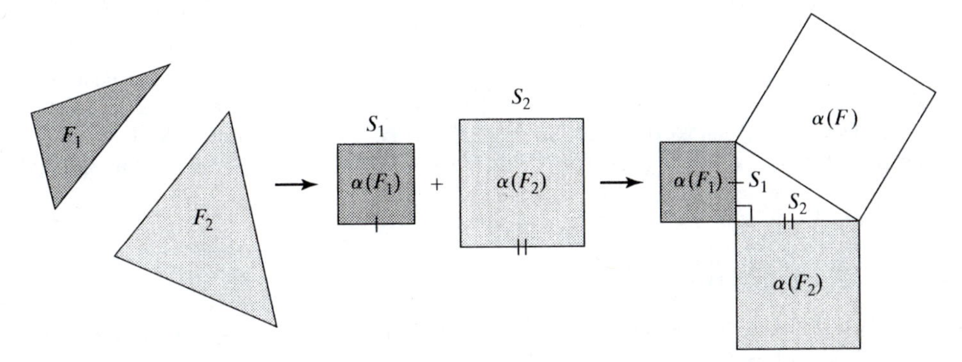

Now construct a right triangle with legs s_1 and s_2. By the Pythagorean Theorem (!), the square on the hypotenuse has area equal to $\alpha(F_1) + \alpha(F_2)$, so that square has area $\alpha(F)$.

The following corollary can be proved by mathematical induction. It shows that every polygonal region can be "squared". We leave its proof to you.

Corollary: Let F be the union of n triangular regions, no two of which have any interior points in common. Then a square can be constructed with the same area as F.

Armed with the knowledge of all the theorems in this section, it was natural for the Greeks to wonder if a square could be constructed with the same area as that of a given circle. But they were unable to solve the problem of "squaring a circle".

Not until the 19th century was the reason for their difficulty established. To square a circle, a side s must be constructed with $s^2 = \pi r^2$, where r is known. That implies that $s = r\sqrt{\pi}$. The rules of ruler-and-compass construction the Greeks had laid out enabled them only to construct finite combinations of sums, differences, products, quotients, and square roots of lengths of given segments. It follows that every length that can be constructed is an *algebraic* number, a number that can be the solution to a polynomial equation with integer coefficients. When in 1882 Ferdinand Lindemann proved that π is a *transcendental* number—one that cannot be the solution to a polynomial equation with integer coefficients—he simultaneously was showing that a length π could not be constructed. This implies that $\sqrt{\pi}$ cannot be constructed either. So s could not be constructed, and so a circle cannot be squared.

10.1.5 Problems

1. Trace the rectangle shown in Figure 39. Using a straightedge and compass, construct a square with the same area.

Figure 39

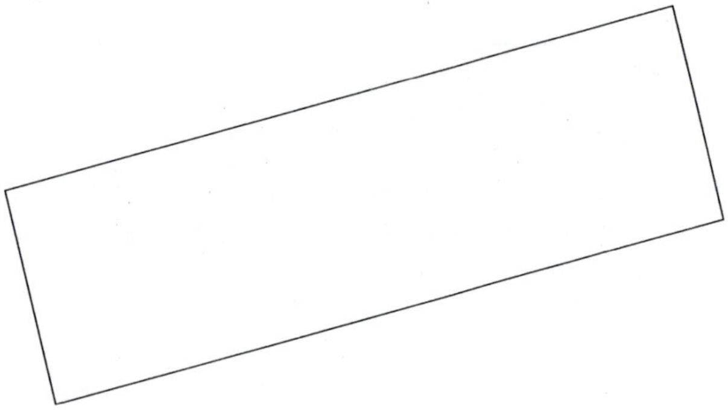

2. A triangle has vertices at $(0, 0)$, $(5, 0)$, and $(-6, 7)$. When this triangle is "squared", what is the length of a side of the square?

3. A trapezoid has vertices at $(0, 0)$, $(a, 0)$, (b, c), and (d, c). When this trapezoid is "squared", what is the length of a side of the square?

4. Prove the Corollary to Theorem 10.13.

5. A reasonable way to try to square a given circle is as follows. Circumscribe a square about the circle. Inscribe a second square in the circle. Let a third square have a side length equal to the arithmetic mean of the sides of the inscribed and circumscribed squares.

   a. Compare the area of the third square to the area of the circle.

   b. Is this a better or worse approximation to the area of the circle than if the areas of the first two squares are averaged?

6. Archimedes succeeded in the quadrature of the parabola. That is, he found a square equal to the area of a region bounded by the parabola and a line parallel to the parabola's directrix. Why can a parabola be squared while a circle cannot?

ANSWER TO QUESTION

$\overline{PQ}$ is the altitude to the hypotenuse of the right triangle whose vertices are Q and the endpoints of the diameter. So PQ is the geometric mean of the segments of the hypotenuse (Theorem 8.27).

10.1.6 Area as representing probability

If a probability experiment has n possible mutually exclusive outcomes O_1 to O_n, and these outcomes have probabilities $p(O_1)$ to $p(O_n)$, then $\sum_{i=1}^{n} p(O_i) = 1$. The probabilities can be pictured in a histogram. Figure 40 shows a histogram for the probability of each possible sum when 3 fair dice, each with 1 to 6 on its faces, are tossed.

Figure 40

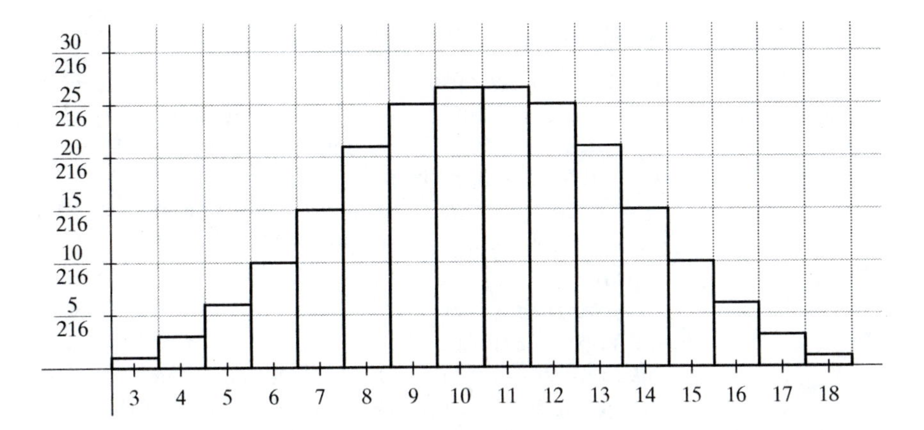

As the outcomes become more numerous, histograms like those in Figure 40 resemble curves. For instance, SAT scores were renormalized some years ago so that 500 is their mean and 100 is their standard deviation. Since any multiple of 10 from 200 to 800 is a possible score, there are 61 possible scores. Rather than draw a histogram with 61 bars, it is easier to draw a smooth (bell-shaped) curve connecting the tops of the bars, as in Figure 41. The area of the curve between $x = 445$ and $x = 605$ yields the probability that a randomly chosen test taker will score between 450 and 600, with the difference between the values of x and the scores due to the need to take into account rounding.

Figure 41

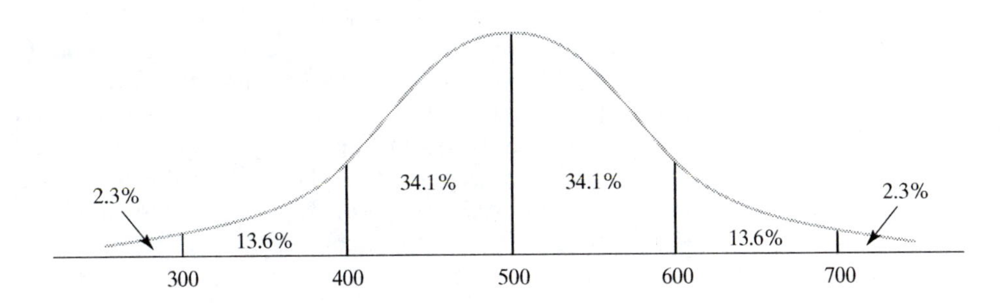

Using probability to determine area

In the preceding example, area represents probability. A class of techniques called *Monte Carlo methods* allow this idea to be turned around, for they use probability to determine area.

Monte Carlo methods are numerical methods that involve sampling from random numbers. They arose during World War II and owe their name to the similarities that can be made between statistical simulations and games of chance which are associated with the European gambling mecca Monte Carlo in the tiny country of Monaco. The Monte Carlo method can be used to simulate a Bingo game, simulate complex physical phenomena such as subnuclear processes in high-energy physics experiments, and study the flow of traffic in a city. There are also many applications of Monte Carlo methods in economics and computer science.

The idea of the Monte Carlo method is that a numerical problem in analytic form can be replaced by a problem in probability so that the numerical answer to the probability problem is the same as the numerical answer to the original problem. The new probability problem is solved using a computer program.

Figure 42

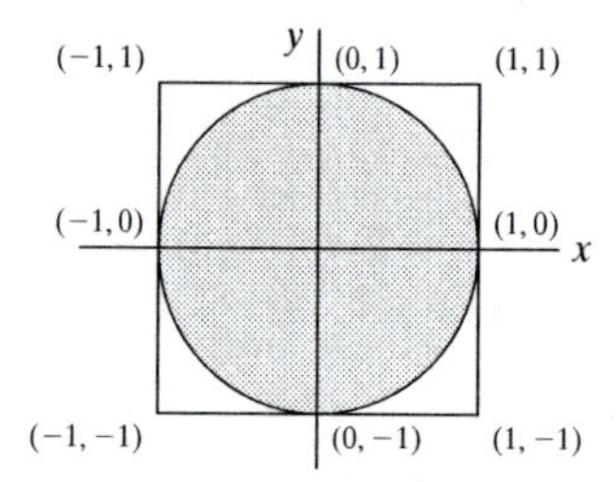

Approximating π using a Monte Carlo simulation

Monte Carlo methods allow us to investigate a complex system by sampling it in a number of random configurations, and then using the results of the sampling to describe the system. The connection with area will become apparent as we apply the method here to approximate π.

In $\mathbf{R}^2$, we inscribe a unit circle with center at $(0, 0)$ inside a square with vertices, as shown in Figure 42. The ratio $\frac{\alpha(\text{circle})}{\alpha(\text{square})} = \frac{\pi}{4}$. Now we use this ratio of areas to approximate π.

We pick a random point $A = (x, y)$ in such a way that the values for both x and y are between 1 and -1, that is, such that $|x| < 1$ and $|y| < 1$. The probability that this random point lies inside the unit circle and not in the space between the circle and the square is simply the ratio of the area of the circle to the area of the square.

$$P(x^2 + y^2 < 1) = \frac{\pi}{4}$$

Suppose we performed this experiment n times, and it turned out that x of those times produced a point inside the circle. Then we could estimate the probability to be $\frac{x}{n}$. As n approaches infinity, this probability estimate becomes arbitrarily close to $\frac{\pi}{4}$ and so we can write

$$\lim_{n\to\infty} \frac{x}{n} = \frac{\pi}{4} \quad \text{or} \quad \pi = 4 \lim_{n\to\infty} \frac{x}{n}.$$

How precise is this formula for π? The precision (number of digits) depends on how many times you perform the experiment. The greater x and n are, the more correct digits you are likely to get.

But this is not a particularly quick way to estimate π. If the experiment were performed 1,000,000 times, and 785,398 times the point was inside the circle, the estimate for π would be $\frac{4 \cdot 785{,}398}{1{,}000{,}000}$, or 3.141592 exactly, which is correct only to six places. Each deviation of 1 from the most likely but quite improbable result of 785,398 would cause a result correct only to 5 places, and a deviation of 10, still very likely, would mean the result was correct only to 4 places. Also, note that because π is irrational, the estimate—a rational number—can never be exact. When precision is needed, a Monte Carlo experiment must be repeated so often that computers are necessary.

For these reasons, Monte Carlo methods tend to be used when exact methods of calculation are unavailable or too unwieldy. For instance, if a function f had an equation for which the exact definite integral from $x = a$ to $x = b$ could not be calculated, then the graph of the function could be fit inside a rectangle with sides $x = a$ and $x = b$. By randomly selecting points in the rectangle and determining how many are under the graph of f, the definite integral could be estimated, as shown in Figure 43. Although this process is relatively tedious for calculating simple integrals, the Monte Carlo method is quite useful for calculating complicated integrals in n dimensions ($n \geq 3$), where other methods are relatively slow and expensive.

Figure 43

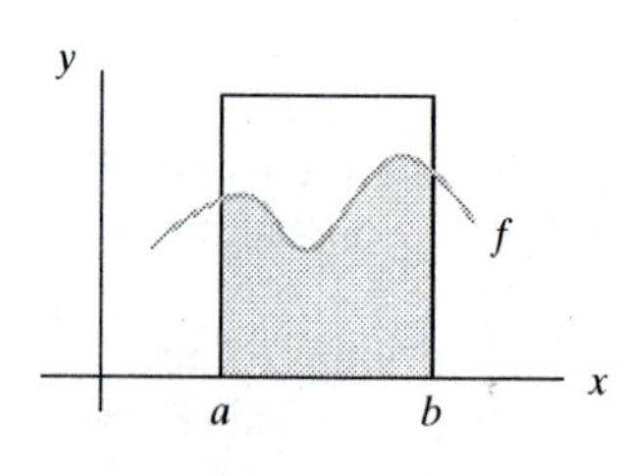

10.1.6 Problems

1. Write a scenario of a trip and a corresponding rate function and represent the total distance traveled as an area.
2. Determine the 16 probabilities in the histogram of Figure 40.
3. A general formula for a normal curve is $y = \frac{1}{\sqrt{2\pi}\sigma} e^{-(x-\mu)^2/(2\sigma^2)}$ where μ is the mean and σ the standard deviation. What is a specific formula for the normal curve in Figure 41?

4. Use the idea of Figure 42 with at least 100 random points to estimate π.

5. Imagine an election that is even between two candidates. Suppose you conducted a survey of 500 people. How likely are your survey results to be within 2% of the actual (even) situation? Study this question by repeatedly picking 500 random numbers between 0 and 1 and counting how many of the 500 are less than 0.5.

6. Prove: If a quarter has diameter equal to the distance between sides in a square lattice of infinite extent (Figure 44), then the probability that a quarter that is randomly thrown at the lattice covers a lattice point is $\frac{\pi}{4}$.

Figure 44

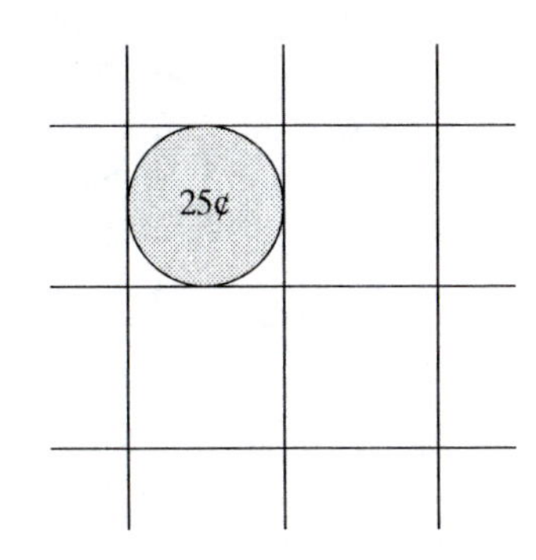

Unit 10.2 Volume

Ancient mathematicians needed to calculate volume to determine how much grain and other foodstuffs were stored in a particular location, or how much material would be needed to build structures or make other items. The Babylonians knew formulas for the volumes of boxes and, more generally, for the volume of a right prism with a trapezoidal base. They (incorrectly) found the volume of a truncated cone as the product of an altitude and half the sum of the areas of the bases. The *Rhind Papyrus* (which was a handbook for scribes) shows that the Egyptians computed the volumes of cylindrical granaries by multiplying the area of the circular base by the height. They were able to calculate the inclination of oblique planes, and used this calculation to find the volume of a pyramid. The Chinese investigated the calculation of volume using decomposition methods, and obtained formulas for the volumes of spheres and pyramids. Archimedes anticipated and influenced the integration methods of calculus used today for calculating volume. He considered a solid as being composed of a very large number of thin parallel layers. He envisioned these layers suspended at one end of a given lever in such a way as to be in equilibrium with a figure whose volume was known. He used these methods to conjecture relationships between the volumes of spheres, cones, and cylinders, and then supplied proofs of the relationships.

Can we develop a theory of volume by means of decomposition as we did with area? David Hilbert posed this question in 1900 when he encouraged mathematicians to investigate whether a definition of volume was possible for polyhedra analogous to that of area for polygons. Max Dehn (1878–1952) responded in that same year by showing that, although we can decompose two arbitrary polygonal regions of equal area into pairs of congruent triangles, it is not possible to decompose two arbitrary polyhedral regions of equal volume into pairs of congruent tetrahedra. He showed that decomposition is not always possible because figures as simple as a regular tetrahedron and a cube of equal volume (Figure 45) cannot be decomposed into congruent tetrahedral pieces.

Figure 45

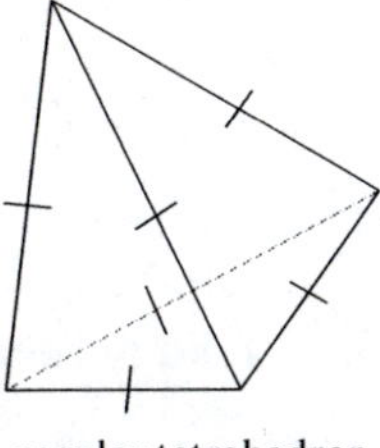
regular tetrahedron

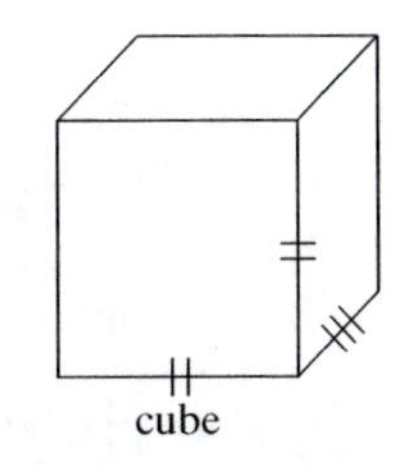
cube

As a consequence, like the areas of figures bounded by curves, infinite processes of some kind are needed to define volume for polyhedral regions. The volume of any solid bounded by plane polygons may be defined as the greatest lower bound of the sum of the volumes of nonoverlapping cubes, which together completely cover the solid. So methods of calculating volumes are analogous to those for calculating areas of regions bounded by curved lines.

In this unit, we investigate some figures of E^3, exploring different methods of calculating their volumes. We revisit some familiar formulas you probably learned in high school. Our goal is as it was with area, to derive these formulas from basic principles, show how they are related to each other, and illustrate them with applications.

10.2.1 What is volume?

We wish to define a volume function that will provide the means for obtaining the volumes of common 3-dimensional figures. You may wish at this time to refer back to the definition of area function found at the beginning of Section 10.1.1 and include the fourth part found at the start of Section 10.1.4.

Just as we think of area either as measuring the 2-dimensional space *occupied by* a region, or *contained in* the region's boundary, volume can be thought of either as a measure of the 3-dimensional space contained in a closed surface, or occupied by a solid that has that surface as a boundary. A **solid** is the set of points on a closed surface or in its interior.

Tetrahedrons

We began the study of area in Section 10.1.1 by considering the domain of the area function α to be the set of unions of triangular regions in E^2. This suggests that the domain of the volume function v might start from an analogous set in E^3. Such a set is the set of unions of solid *tetrahedrons* (an alternative plural is *tetrahedra*). Given four points not all in the same plane, a **tetrahedron** is the union of the four triangular regions (faces) determined by these points. That is, if the four given points are the noncoplanar points A, B, C, and D,

$$\text{tetrahedron } ABCD = \Delta ABC \cup \Delta ABD \cup \Delta ACD \cup \Delta BCD.$$

Figure 46

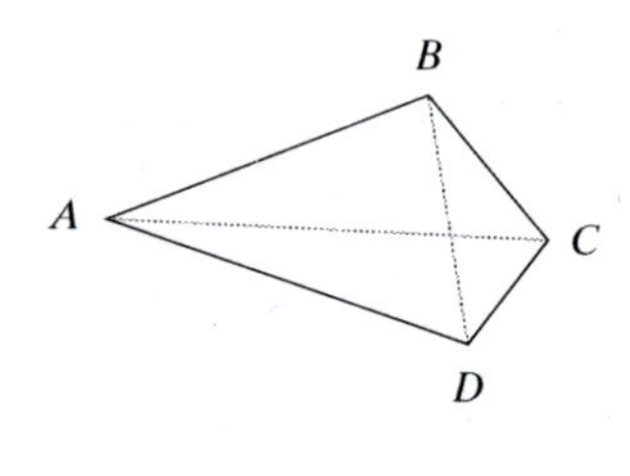

Such a tetrahedron is shown in Figure 46.

Tetrahedra are the spatial analogue of triangles. A triangle has three vertices, three sides, and three angles. A tetrahedron has four vertices, four faces, and four *solid angles*, one at each vertex. The solid angle at each vertex is the union of all points on and interior to the three plane angles of the tetrahedron at that vertex. Notice that these solid angles are not the twelve *plane angles* in the faces of the tetrahedron, nor are they the six *dihedral angles* that are formed by the union of the two half-planes that intersect at an edge of the tetrahedron.

The following sets associated with a triangle all describe the same region:

1. all points that lie between points on the sides of the triangle
2. the union of the sides of the triangle with the intersection of the interiors of any two of its angles
3. the intersection of all convex sets that contain the vertices of the triangle

The region described is the *triangular region* associated with the triangle.

Analogously, the following sets associated with a tetrahedron all describe the same **tetrahedral region** or **solid tetrahedron**.

1. all points that lie between points on the faces of the tetrahedron

2. the union of the faces of the tetrahedron with the intersection of the interiors of any two of its solid angles
3. the intersection of all convex sets that contain the vertices of the tetrahedron

The points in the tetrahedral region but not on any faces of the tetrahedron constitute the **interior** of the tetrahedron.

Defining properties of the volume function

In the following definition are four properties from which the volumes of solid figures can be derived. The first three of these defining properties of a volume function are analogous to those for area. Property (4) of this definition extends volume to figures that are not unions of a finite number of tetrahedral regions. It owes its name in the west to Bonaventura Cavalieri (1598–1647), a Jesuit priest who was a student of Galileo and who was the first western mathematician to realize its importance. But the first individuals to have used this principle to obtain volume seem to have been the Chinese mathematician Zu Chongzhi (429–500) and his son Zu Geng.

Definition

Let F be the union of tetrahedral regions in E^3. A **volume function** v is a function that assigns to each such F a positive real number $v(F)$ such that:

1. If $F_1 \cong F_2$, then $v(F_1) = v(F_2)$. (Congruence property)
2. If the tetrahedral regions making up F_1 and F_2 have no interior points in common, then $v(F_1 \cup F_2) = v(F_1) + v(F_2)$. (Additive property)
3. If F is a cube with edge of length x, then $v(F) = x^3$. (Volume of cube)
4. If F and a solid S lie between parallel planes a and b, and for each plane c parallel to and between a and b, $c \cap S$ is a region whose area is known from the properties of the area function, and $\alpha(c \cap F) = \alpha(c \cap S)$, then $v(F) = v(S)$. (Cavalieri's Principle)

A **cross section** of a surface or solid is the intersection of a plane and the surface or solid. The intersections $c \cap F$ and $c \cap S$ in Cavalieri's Principle are cross sections.

To apply Cavalieri's Principle, we think of a 3-dimensional solid as being made up of parallel cross sections, as we do in calculus. If all the pairs of cross sections of two solids F and S made by parallel planes have the same area, then the solids have the same volume. You may think of a deck of cards F that has been slanted into the position S, as in Figure 47.

Figure 47

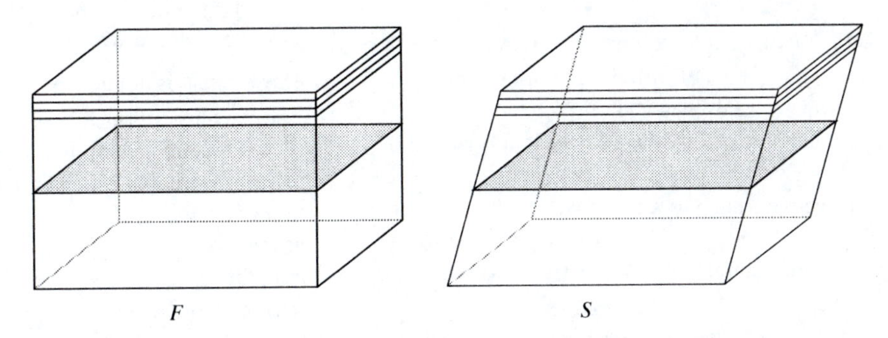

We will apply Cavalieri's Principle first when F and S are both unions of tetrahedral regions, and later when S is a cylinder, cone, sphere, or other figure with a curved boundary.

Volume is not area

In Section 10.1.1, we began with the area formula for any square, yet we defined the area of any polygonal region in terms of triangular regions. We could do this because it is possible to decompose any polygonal region into triangular regions, which can then be arranged to form a square, as we saw in Section 10.1.5. We might say that we can "square" any polygon.

Can we approach the volume of polyhedra in an analogous way? To do so we need to address two issues, first, the decomposition of polyhedra, and second, the possibility of comparing polyhedra with equal volume in terms of dissections.

First let us address the notion of dissection. Is it possible to decompose any polyhedral region into tetrahedral regions analogous to the decomposition we have seen of polygonal regions into triangular regions?

It is easy to decompose any convex polyhedron P into tetrahedra. Triangulate all the faces of P and select a point A in the interior of P. Then the tetrahedra with vertex A and the three vertices of each triangle form a decomposition of P. If the polyhedron is not convex, like the one shown in Figure 48, where segments $\overline{AE}$, $\overline{BF}$, and $\overline{CD}$ are in the exterior of the polyhedron, then the situation is more complex. More than one point in the interior must be used for the decomposition.

Figure 48

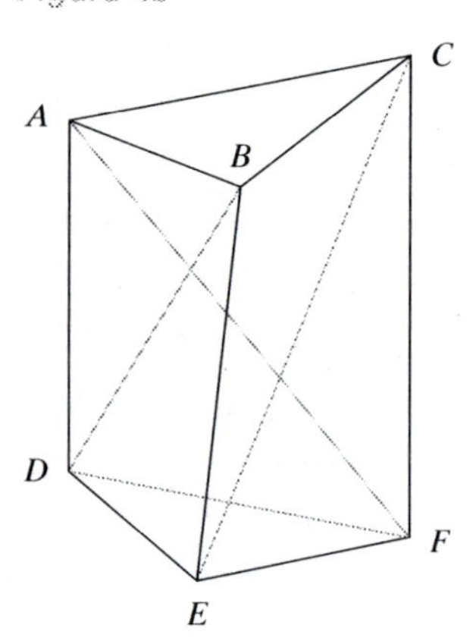

We next address the issue of defining volume as we did area. Recall that we defined polygons to be equal in area if they possess dissections into corresponding congruent polygonal pieces. Can we define polyhedra to be equal in volume if they possess an analogous property? It turns out that we cannot make such a definition because in 1900 Max Dehn came up with a counterexample. Dehn showed that there is no way to decompose a regular tetrahedron and a cube of equal volume into equal numbers of tetrahedra that are congruent in pairs. We might say, as a result, that we cannot "cube" every polyhedron.

This limitation on the decomposition of polyhedra is why we cannot construct a theory of volumes analogous to that of area. However, Cavalieri's Principle does allow us to overcome this difficulty. We can find volume formulas for some solids not by decomposing them, but by equating them, cross section by cross section, with the volume of a known solid.

10.2.1 Problems

1. Prove that any convex solid must contain a tetrahedron.
2. A square pyramid is the union of at least how many tetrahedrons?
3. A cube is the union of at least how many tetrahedrons?
4. Which part(s) of the definition of volume relate to Archimedes' principle of displacement found in physics?
5. a. State an analogue of Cavalieri's Principle for an area function.

   b. Use your principle from part **a** to prove that the transformation $T: (x, y) \to (x + ky, y)$ preserves area. This transformation is an example of a *shear* transformation of magnitude k.

   c. Every parallelogram with base b and height h can be placed on a coordinate plane so that its vertices are $(0, 0)$, $(b, 0)$, (c, h), and $(c - b, h)$. What magnitude shear of part **b** maps this parallelogram onto a rectangle with area bh?

   d. Prove that with composition, the set of shears of the type shown in part **b** forms a group.
6. Consider the transformation T in $\mathbf{R}^3$ defined by $T(x, y, z) = (x + ky, y, z)$, where k is fixed.

   a. Show that T is not an isometry.

   b. Use Cavalieri's Principle to show that T preserves volume. (T is a 3-dimensional shear transformation.)
7. Demonstrate that each of the following is a possible cross section of a cube.

   a. a single point

   b. a segment

   c. a triangular region

   d. a rectangular region

   e. a square region

   f. a pentagonal region

   g. a hexagonal region

10.2.2 From cubes to polyhedra

Definitions of figures in space can be simplified by extending to E^3 some of the language we used earlier in E^2. Two figures α and β in E^3 are **congruent** if and only if there is a distance-preserving transformation T with $T(\alpha) = \beta$. One congruence transformation in E^3 is the translation associated with the 3-dimensional vector $\overrightarrow{AB}$. In R^3, if $\overrightarrow{AB} = (h, k, j)$, then the **translation associated with the vector (h, k, j)** has the formula $T(x, y, z) = (x + h, y + k, z + j)$. It is distinguished by the fact that all segments connecting points on a preimage with their images are parallel and of equal length.

Translations make it easier to define prisms and cylinders. Let G' be the translation image of a polygonal region G in E^3, where G' and G are not in the same plane. The **prismatic solid** with **bases** G and G' is the set of all points on any segment connecting a point of G with its translation image point on G' (see Figure 49). A **prism** is the boundary of this solid. The **altitude** or **height** of the prism is the (perpendicular) distance between the planes of G and G'.

Figure 49

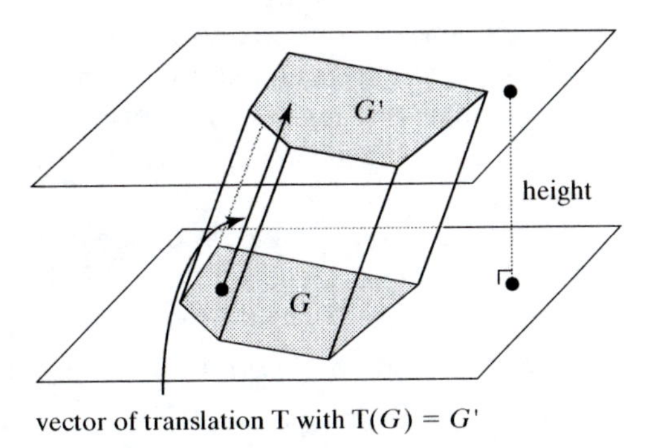

Types of prisms

A prism is classified as **square, rectangular, triangular, quadrangular**, etc., as its base is a square, rectangle, triangle, quadrilateral, etc. In a prism, its bases are congruent (since translations are isometries) and the plane section formed by the prism's intersection with any plane parallel to and between its bases is a region congruent to the bases. Therefore, all these sections have the same area.

When the direction of the translation mapping one base to the other is perpendicular to the base planes, the prism is a **right prism**. Otherwise, it is called **oblique**. Figure 50a shows an oblique prism. Figure 50b shows a right prism.

Figure 50

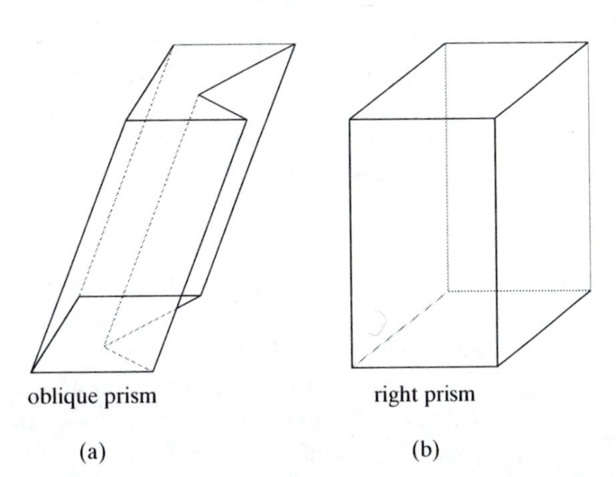

When a prism's base is a parallelogram, the prism is called a **parallelepiped** (Figure 51a). A right parallelepiped whose bases are rectangles is called a **right rectangular parallelepiped or box** (Figure 51b). A **cube** is a right rectangular parallelepiped whose faces and bases are squares.

Figure 51

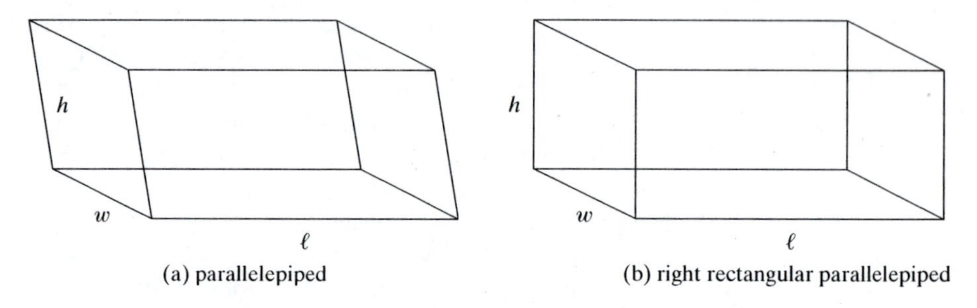

(a) parallelepiped (b) right rectangular parallelepiped

In developing the formulas for areas of polygonal regions, we first used areas of squares to obtain a formula for the area of any rectangle. The spatial analogue is to use volumes of cubes to obtain a formula for the volume of any box. The proof has two parts, which we split here into a lemma and the theorem. We doubt that the proof is original, but we have not seen it elsewhere.

Lemma: If a box has dimensions x, y, and $x + y$, then its volume is the product of its dimensions.

Proof: Consider a solid cube with edge $x + y$. It can be split into eight parts, shown in Figure 52.

Figure 52

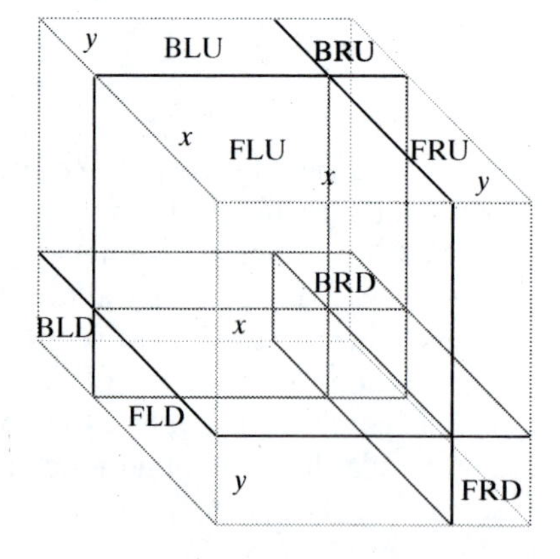

FLU (the front, left, upper part) and BRD (the back, right, lower part) are cubes with volumes x^3 and y^3. FRD, BLD, and BRU have dimensions x, y, and y, and FLD, BLU, and FRU have dimensions x, x, and y. These last six can be grouped to form three congruent boxes FRD ∪ FRU, BRU ∪ BLU, and BLD ∪ FLD, each with dimensions, x, y, and $x + y$. The volume of each of those boxes is $\frac{1}{3}((x + y)^3 - x^3 - y^3)$, or $xy(x + y)$. ┘

The proof of the main formula is similar to the proof of the lemma. We split a cube into pieces, some of which are smaller cubes. The rest we group together into congruent sets of boxes. The boxes in some of these sets have dimensions of form of the lemma, so their volumes are known. The remaining boxes have the dimensions we want, and their volumes can now be determined by subtracting the known volumes from the volume of the big cube.

Theorem 10.14 The volume of a right rectangular parallelepiped (box) with dimensions x, y, and z is xyz.

Proof: Consider a solid cube with edge $x + y + z$. It can be split into 27 parts, some of which can be seen in Figure 53.

Figure 53

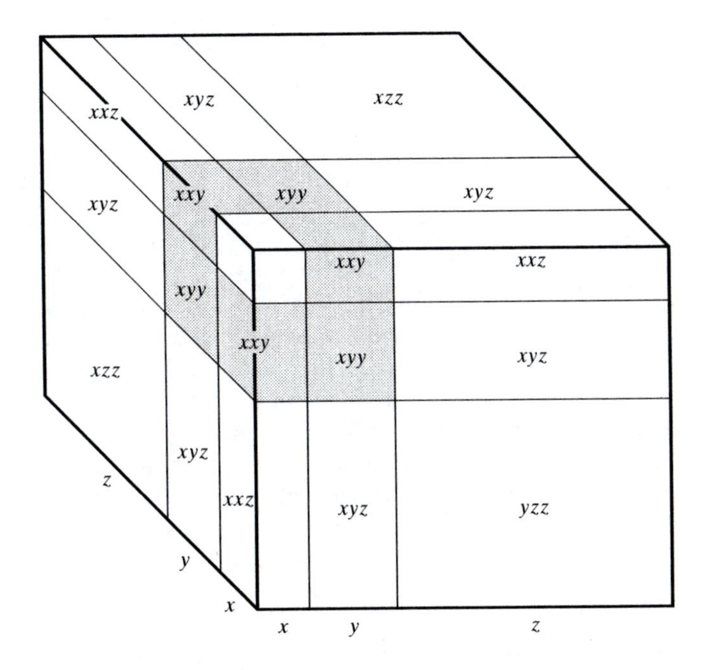

Of the 27 parts, 3 are cubes with volumes x^3, y^3, and z^3. (Only the cube with volume x^3 can be seen in Figure 53.) Eighteen of the others can be grouped as follows:

Three have dimensions x, y, y and three have dimensions x, x, y. These join to form 3 boxes with dimensions $x, x + y$, and y, so by the lemma their joint volume is $3xy(x + y)$. (These are blue in Figure 53.)

Three have dimensions x, z, z and three have dimensions x, x, z. These join to form 3 boxes with dimensions $x, x + z$, and z, so by the lemma their joint volume is $3xz(x + z)$.

Three have dimensions y, z, z and three have dimensions y, y, z. These join to form 3 boxes with dimensions $y, y + z$, and z, so by the lemma their joint volume is $3yz(y + z)$.

The remaining 6 parts are boxes with dimensions x, y, and z. The volume of each is $\frac{1}{6}((x + y + z)^3 - x^3 - y^3 - z^3 - 3xy(x + y) - 3xz(x + z) - 3yz(y + z))$, which is xyz. ⌟

Corollary: The volume of a right rectangular parallelepiped is the product of the area of the base and the corresponding altitude of the parallelepiped.

From this corollary, by using Cavalieri's Principle, we can prove the following more general theorem.

Theorem 10.15 The volume of any prism with base area B and altitude h is Bh.

Figure 54

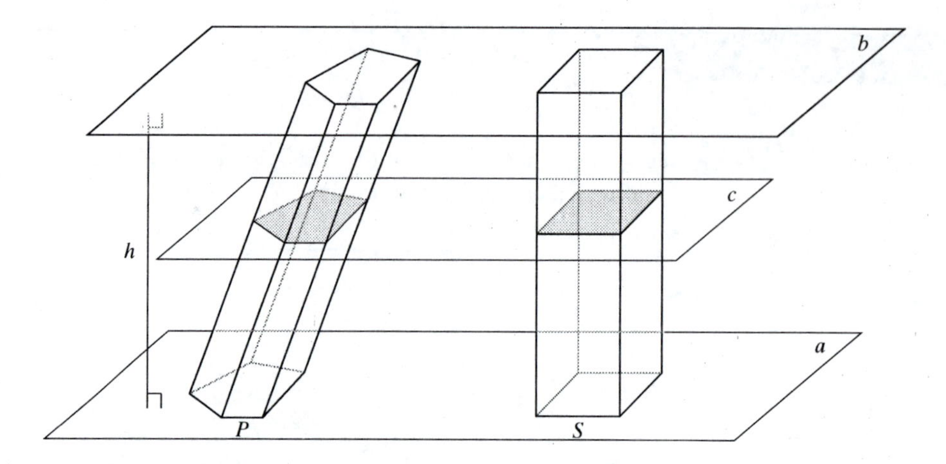

Proof: Let P be a prism whose bases are in planes a and b, and let P have base area B and altitude h. Construct a right square parallelepiped S with bases in planes a and b that also has base area B (Figure 54). This can be done because of the corollary to Theorem 10.13. S also has altitude h, so by the corollary to Theorem 10.14, $v(S) = Bh$. Now let c be any plane parallel to a or b and between them. The intersection of c and P is congruent to the bases of P, so has area B. The intersection of c and S is congruent to the bases of S, so also has area B. Thus, by Cavalieri's Principle, P and S have the same volume, and so $v(P) = Bh$. ┘

Pyramids

Although the previous arguments have obtained the volumes of prisms, in theory we have not yet established that prisms have a volume, for we have not shown a prismatic solid to be a union of tetrahedral regions. For this, we need to consider pyramids.

Given a polygonal region F in a plane, a **pyramidal solid** is the set of points on line segments connecting points of F (its **base**) with a point A (its **apex** or **vertex**) not in that plane. A **pyramid** is the boundary of this solid; it is union of F and the sets of points connecting the polygon boundary of F to A. The distance from the apex to the plane of the base is the **altitude** or **height** of the pyramid. Like prisms, pyramids are classified by their bases as triangular, quadrangular, pentagonal, etc. (Figure 55). From this definition, we can see that a triangular pyramid is a tetrahedron as defined in Section 10.2.1.

Figure 55

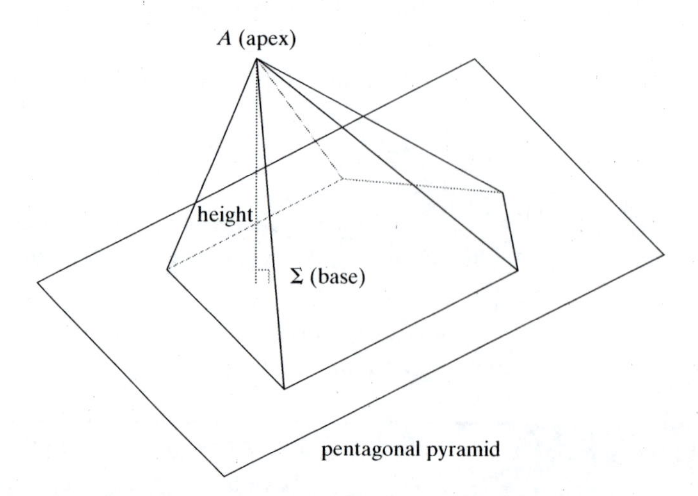

The word "pyramid" originates in the Greek word for the pyramids found in Egypt. These pyramids, and their counterparts built by the Mochica in Peru and the Maya in Central America, are *regular* square pyramids. A **regular pyramid** is a pyramid whose base F is a regular polygon and in which the segment connecting the apex to the center of F is perpendicular to the plane of F. That is, in a regular pyramid, if the base is horizontal, the apex is directly above the center of the base.

Figure 56

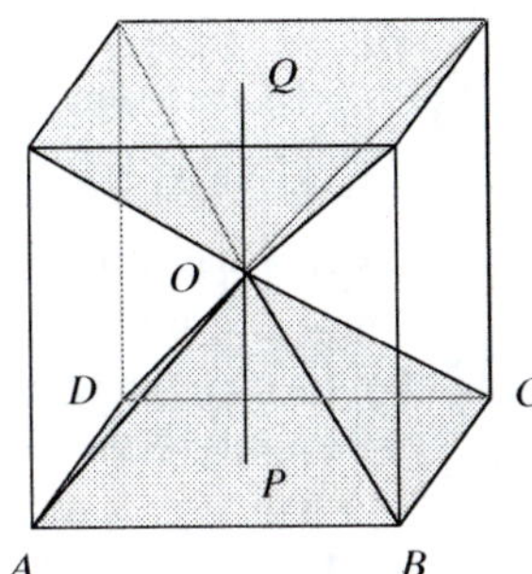

In Figure 56, we have dissected a cube into 6 regular square pyramids, each with apex at the center O of the cube. Furthermore, each of them can be further dissected into tetrahedra. For instance, the solid square pyramid $OABCD$ is the union of the solid tetrahedra $OABP$, $OBCP$, $OCDP$, and $ODAP$. Consequently, the volume function defined in Section 10.2.1 includes these square pyramids in its domain. Due to the reflection symmetry of the cube through any plane that is a perpendicular bisector of an edge, these six regular square pyramids are congruent. Thus, by property (1) of the volume function, the six of them have equal volume.

By property (3) of the volume function, the volume of the cube equals the cube of one of its edges, which in this case equals the area of the base times twice the height of any of the square pyramids. Since each of the six square pyramids have equal volume, the volume of one of them must equal $\frac{1}{6} \cdot 2h \cdot$ (area of base). Therefore, the volume of the square pyramid is the area of its base times $\frac{1}{3}$ its altitude. This is the formula we would like to derive for any pyramid. But the preceding argument does not apply to all pyramids, because not all square pyramids can be put together to form a cube. To achieve a proof of the desired formula, we need to know that pyramids with the same altitude and bases of equal area have equal volumes. For this, we use a size change in space.

Two figures α and β in E^3 are **similar** if and only if there is a distance-multiplying transformation T with $T(\alpha) = \beta$. One similarity transformation in E^3 is the **size change with center O and magnitude k > 0**. As in the two-dimensional case, the image of any point P under this size change is the point P' such that P' is on ray $\overrightarrow{OP}$ and $\frac{OP'}{OP} = k$.

Now we prove a lemma about the relationship between the area of a cross section of a pyramid parallel to its base and the area of the base itself.

Figure 57

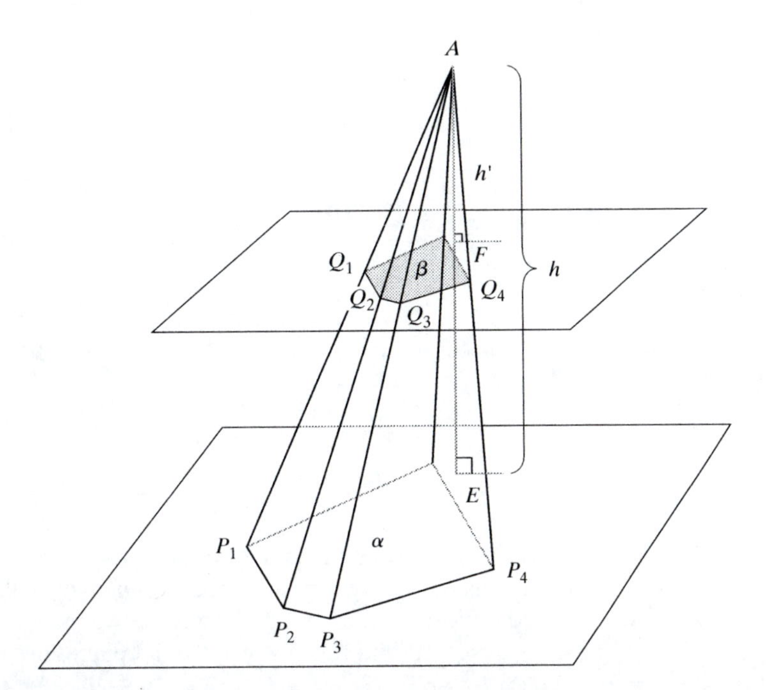

Lemma: Let P be a solid pyramid with altitude h and base α that has area B. Let Q be a pyramid with the same apex as P, and such that its base β is a section of P contained in a plane that is parallel to the plane of α. Let the area of the base of Q be B' and the altitude of Q be h'. Then $\frac{B'}{B} = \left(\frac{h'}{h}\right)^2$.

Proof: Let α be the polygon $P_1P_2 \ldots P_n$ and let A be the common apex of P and Q (Figure 57). Let β be the polygon $Q_1Q_2 \ldots Q_n$, so that for each i, A, P_i, and Q_i are collinear. Let ℓ be the line through A that is perpendicular to the plane of α at E and to the plane of β at F. Then $AE = h$ and $AF = h'$. Let $k = \frac{h'}{h}$.

Let T be a size change in space of magnitude k, center A. Then $T(E)$ is the point E' on $\overrightarrow{AE}$ such that $AE' = kAE$. Since $k = \frac{AF}{AE}$, $AE' = AF$. This implies that $E' = F$ since F is also on $\overrightarrow{AE}$. Since $T(E) = F, T$ also maps the plane α (which is perpendicular to ℓ at E) onto the plane perpendicular to ℓ at F. But this is the plane of β. Therefore, every point X of the plane of α is mapped onto the point in the plane of β at which $\overrightarrow{AX}$ intersects the plane. Thus T maps $P_1, P_2, \ldots, P_n$ to $Q_1, Q_2, \ldots, Q_n$, respectively. And so the base of P is mapped onto the base of Q. These bases are similar and k is the ratio of similitude. The area of $Q_1Q_2 \ldots Q_n = k^2$ times the area of $P_1P_2 \ldots P_n$. Thus $B' = k^2B$, and the lemma follows. ∎

We now use Cavalieri's Principle to show that the volume of a pyramid depends only on its base area and altitude.

Theorem 10.16 If two pyramids have the same altitude and same base areas, then they have the same volume.

Proof: Suppose P is any pyramid with a base in plane b. Any other pyramid P^* with the same height h and base area as P is congruent to a pyramid P' with height h whose base is in the same plane as the base of P and that lies on the same side of that plane (Figure 58).

Figure 58

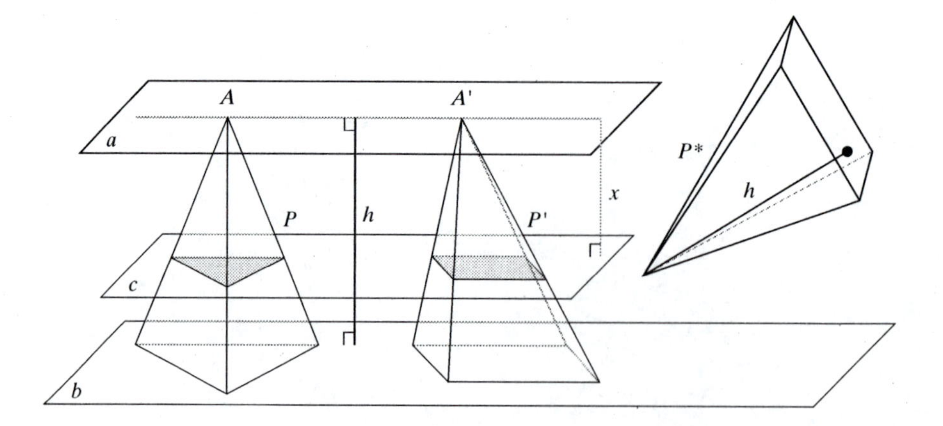

Let a be the plane through A and A' parallel to b. Since the altitude of each pyramid is h, the distance between these planes is h. Now let c be any plane parallel to a and b and between them. If the distance between A and this plane is x, so is the distance between A' and this plane. (Parallel planes are everywhere equidistant.)

If B is the area of the base in plane b, then, by the lemma, the area of the intersection of c and P has area $\left(\frac{x}{h}\right)^2 B$. By the same argument, the area of the intersection of c and P' is $\left(\frac{x}{h}\right)^2 B$. Since the areas of all parallel cross sections are equal, we can conclude by Cavalieri's Principle that the volumes of P and P' are equal. ⌟

From tetrahedra to pyramids

In the definitions of the area function (Section 10.1.1) and the volume function (Section 10.2.1), triangles and tetrahedra play analogous roles. Now we shall see that they also play the analogous roles in the derivation of formulas. That is, just as we put triangular regions together to get formulas for the areas of some polygons, we now put volumes of tetrahedral regions together to obtain formulas for the volumes of some pyramids. We first show that the volume of a tetrahedron is equal to one-third the area of its base times its altitude. We then generalize the formula to all pyramids.

Theorem 10.17 The volume of a tetrahedron with base area B and altitude h is $\frac{1}{3}Bh$.

Figure 59

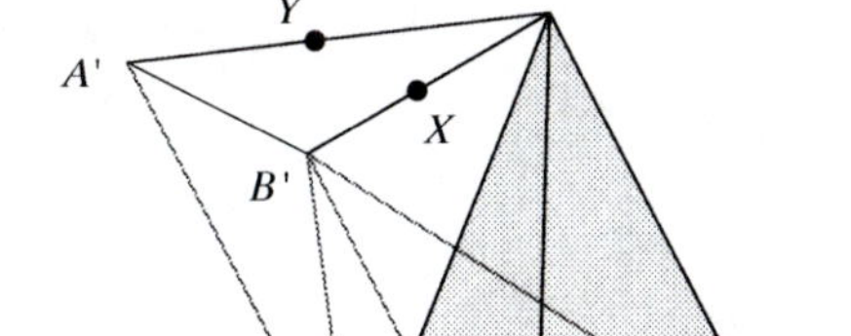

(a)

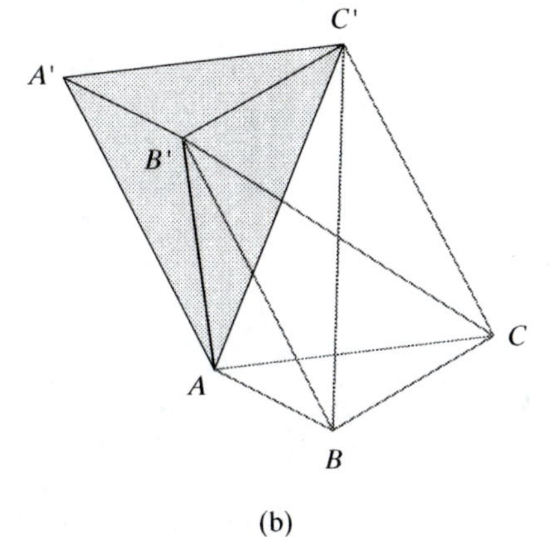

(b)

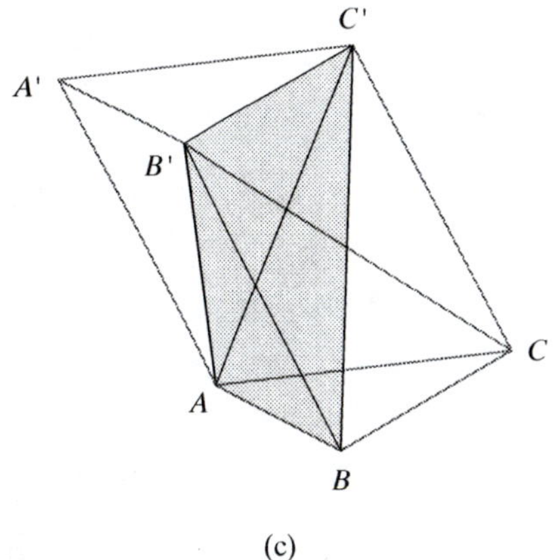

(c)

Proof: Let P be the triangular pyramid whose volume we want. Let the base of P be triangle ABC and let its apex be C'. Now we construct a triangular prism with base ΔABC (see Figure 59a). Let $\overrightarrow{C'X}$ be the ray in the direction of $\overrightarrow{CB}$ and let B' be the point on this ray such that $C'B' = CB$. Let $\overrightarrow{C'Y}$ be the ray in the direction of $\overrightarrow{CA}$ and let A' denote the point on this ray such that $C'A' = CA$. Notice that $CBB'C'$ and $CAA'C'$ are parallelograms. Thus, $\overline{AA'}$ and $\overline{BB'}$ are both parallel to $\overline{CC'}$ and are therefore parallel to each other. Also $\overline{AA'}$, $\overline{CC'}$, and $\overline{BB'}$ are congruent, so $ABB'A'$ is a parallelogram. Thus $\Delta ABC \cong \Delta A'B'C'$ by SSS Congruence (corresponding sides are opposite sides of parallelograms).

ΔABC and $\Delta A'B'C'$ are the bases of a triangular prism Q. Let h be the distance between the planes of ΔABC and $\Delta A'B'C'$. Then h is the altitude of both pyramid P and prism Q and, by Theorem 10.15, the volume of $Q = h \cdot \alpha(\Delta ABC)$.

Now consider the triangular pyramid P' with base $\Delta A'B'C'$ and with apex A (Figure 59b). Because P and P' have the same altitude and congruent bases, their volumes are equal. Note that P and P' have no interior point in common as their interiors are on opposite sides of the plane containing $\Delta AB'C'$. Next let P'' be the triangular pyramid with base $\Delta BC'B'$ and apex A (Figure 59c). Let h' denote the distance of A from the plane containing $\Delta BCB'$. We can also describe the pyramid P as a pyramid with base $\Delta C'BC$ and apex A. But $\Delta BC'B' \cong \Delta C'BC$ by SSS, and therefore these triangles have the same area. From this, using Cavalieri's Principle, it follows that the volume of P is equal to the volume of P''. Again, P and P'' have no interior point in common as they are on opposite sides of the plane of $\Delta ABC'$. Nor do P' and P'' have a point in common, since they are on opposite sides of the plane of $\Delta AB'C'$.

Since the triangular prism Q is the union of P, P', and P'', and no two of these tetrahedra have an interior point in common, the volume of the prism Q is the sum of the volumes of the tetrahedra. Thus the volume of Q is three times the volume of P, and the theorem follows. ∎

Theorem 10.18 The volume of a pyramid with base area B and altitude h is $\frac{1}{3}Bh$.

Figure 60

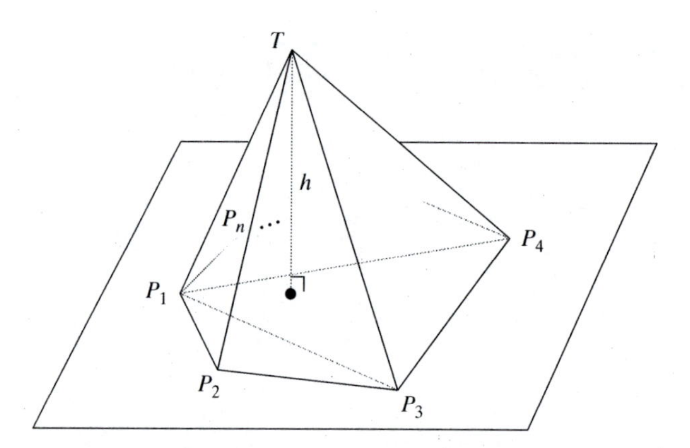

Proof: Let $P_1P_2 \dots P_n$ be the polygonal base of the pyramid P, and let T be its apex and h its altitude (Figure 60). Then each of the triangles at P_1: $\Delta P_1P_2P_3$, $\Delta P_1P_3P_4, \dots, \Delta P_1P_{n-1}P_n$ is the base of a triangular pyramid with apex T and altitude h. The pyramid P is the union of these $n - 2$ pyramids. No two of these pyramids have an interior point in common. Therefore, the volume of P is simply the sum of the volumes of these triangular pyramids, and the theorem follows. ∎

10.2.2 Problems

1. In 1999, the human population of Earth passed 6 billion. Could all people on Earth then fit into a cube with edges 1 mile long?

2. a. The diagonals of the faces of a box have lengths 3, 4, and 6. What is the volume of the box?

   b. Find a formula for the volume of a box whose face diagonals have lengths a, b, and c.

3. Prove the following theorem due to Legendre: Given a parallelepiped P, a rectangular parallelepiped can be constructed that has the same volume, same height, and same base area as P.

4. A **median of a tetrahedron** $ABCD$ is a segment from one vertex (say A) of a tetrahedron to the centroid of the opposite face (ΔBCD). (Recall that the centroid of a triangle is the point of concurrency of the medians of the triangle.) Prove that any plane passing through a median of a tetrahedron and containing a second vertex of the tetrahedron bisects the volume of the tetrahedron.

5. In Figure 56, if the volume of tetrahedron $OABC$ is V, what is the volume of the cube?

6. A regular octahedron is an 8-sided polyhedron whose faces are all equilateral triangles, as shown in Figure 61. What is the volume of a regular octahedron with edge e?

Figure 61

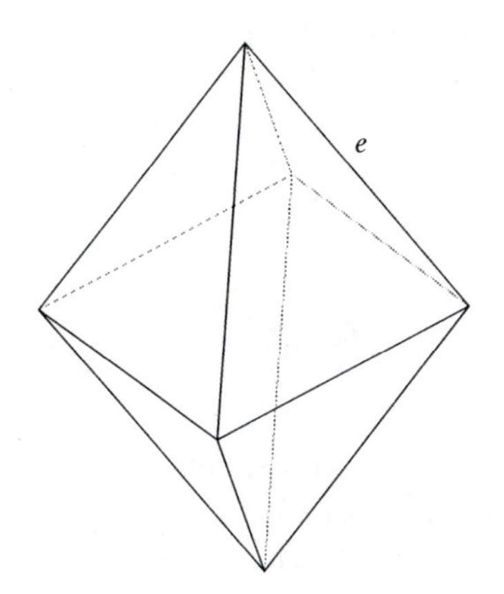

7. The pyramid of Khufu at Ghiza outside of Cairo is the largest of the Egyptian pyramids. It is a square pyramid with sides about 230 m long and had an original height of 147 m. The height is now about 137 m due to the loss of its outer stones.
   a. Estimate its original volume.
   b. Estimate its current volume.
   c. The pyramid contains about 2.3×10^6 stone blocks, each weighing about 2.75 tons. What is the volume of each block (assuming they are the same size)?

8. Is there a 3-dimensional analogue to a Reuleaux triangle (See Problem 6, Section 10.1.1)? That is, is there a solid R that is not a sphere such that, for any pair of parallel planes tangent to R, the distance between the plane is constant? If so, describe R. If not, explain why not.

10.2.3 From polyhedra to spheres

Cylindric solids and *conic solids* are generalizations of prismatic solids and pyramidal solids, respectively. They arise from allowing a base to be any planar region to which an area function applies. They provide a stepping stone in the development of volume formulas from prisms and pyramids to spheres.

From prisms to cylinders

Let G be any planar region to which the area function applies. A **cylindric solid** is the set of all points connecting a point of G with its translation image G' in a different (parallel) plane. A **cylinder** is the boundary of this solid (Figure 62a). Thus prisms are special kinds of cylinders.

The language of prismatic solids is used for the more general cylindric solids. G and G' are the **bases**. The distance between the bases is the **altitude** or **height** of the cylinder. If the translation connecting the bases of a cylinder is perpendicular to the base, then a **right cylinder** is formed (Figure 62b). Thus right cylinders generalize right prisms. Otherwise, the cylinder is called **oblique**. When the bases G and G' are circles, then a **circular cylinder** is formed.

Figure 62

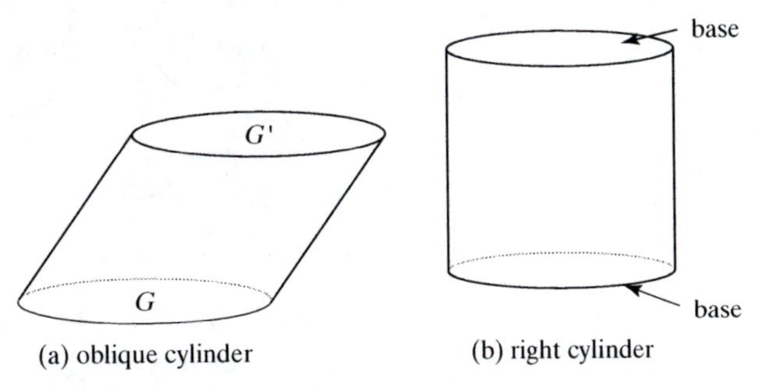

(a) oblique cylinder (b) right cylinder

The volume of any cylinder can be found if the area of its base is known.

Theorem 10.19 The volume of a cylinder with base area B and altitude h is Bh.

Proof: Let C be a cylinder with bases G_1 and G_2, height h, and let $B = \alpha(G_1) = \alpha(G_2)$. Because each cross section of the cylinder parallel to the base is congruent to the base, any cross section has area B. Now let P be a prism with rectangular bases of dimensions 1 by B in the same planes as G_1 and G_2. (We do not show a figure; you should draw one.) The area of each base of the prism P is B and its height is h. The area of any cross section of P parallel to the base is also B because all parallel cross sections are congruent to the base. Consequently, the prism P and cylinder C satisfy the given conditions of Cavalieri's Principle. Since the volume of the prism is Bh, the volume of the cylinder is also Bh. ┘

Corollary: The volume of a circular cylinder with base radius r is $\pi r^2 h$.

Notice that the volume formula for cylindrical solids applies both to right cylinders and oblique cylinders. This is akin to the area formula for a parallelogram applying both to right-angled parallelograms (i.e., rectangles) and those that have no right angles. This tends to violate many people's intuition about the volumes of these figures, which is affected by the different boundaries—surface area for the cylinders and perimeter for the parallelograms.

From pyramids to cones

The same principles that generalize prisms to cylinders also generalize pyramids to cones. Given a connected region G in a plane, a **conic solid** is the set of all points on line segments connecting a point of G (its **base**) with a single point (its **apex**) in a different plane (Figure 63a). The conic solids of main interest are those in which G is a circle. A **circular cone** is the boundary of a conic solid whose base is a circle. If the segment connecting the apex of the cone to the center of the base is perpendicular to the base, then a **right circular cone** is formed (Figure 63b). Right circular cones correspond to regular pyramids.

Figure 63

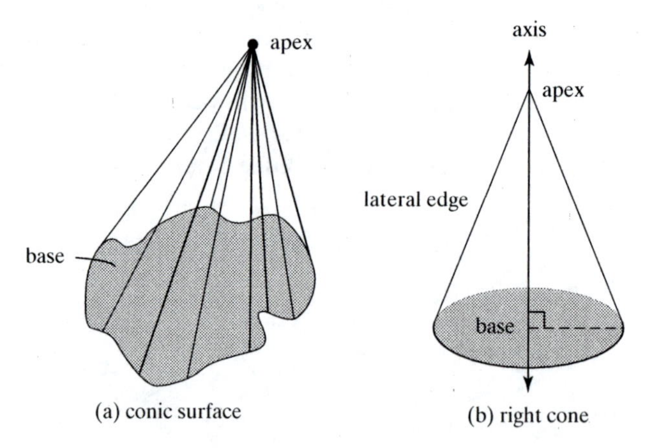

(a) conic surface (b) right cone

The *lines* containing the segments that join the apex to the points of the circular base in a right circular cone form a figure of infinite extent that is open on each side of the apex. This figure is called a **two-napped cone** (Figure 64a), and from it the conic sections arise (Figures 64b–64e). This provides another way of arriving at a right circular cone. Form a two-napped cone by rotating one of two intersecting lines in space about the other line at the point of intersection. Then cut off one of the nappes by a plane perpendicular to the line that is fixed to form a right circular cone.

This gives the word "cone" two closely related but different meanings, one meaning used when studying area and volume, the second used when studying the conic sections.

Figure 64

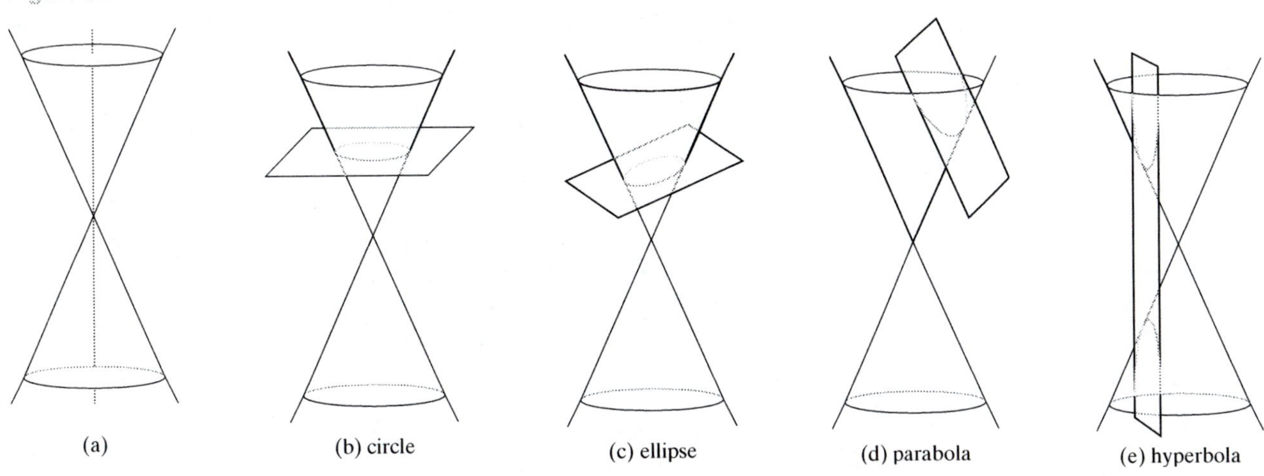

The derivation of a volume of a cone is slightly more complex than for a cylinder but also involves Cavalieri's Principle.

Theorem 10.20 The volume of a cone with base area B and altitude h is $\frac{1}{3}Bh$.

Proof: We begin as in the proof of Theorem 10.19. Let C be a cone with apex A, base G in plane b, and height h, and let $B = \alpha(G)$. Now let P be a pyramid with a rectangular base of dimensions 1 by B in the plane b, and apex in the plane a through A parallel to the plane of the base (Figure 65). Then the height of P is also h. Let c be a plane between a and b at a distance x from a, creating a cross section of P. (This argument should by now be rather familiar.) Because the cross section can be thought of as the image of the base under a size change with center A and magnitude $\frac{x}{h}$, the area of the cross section is $B\left(\frac{x}{h}\right)^2$.

Figure 65

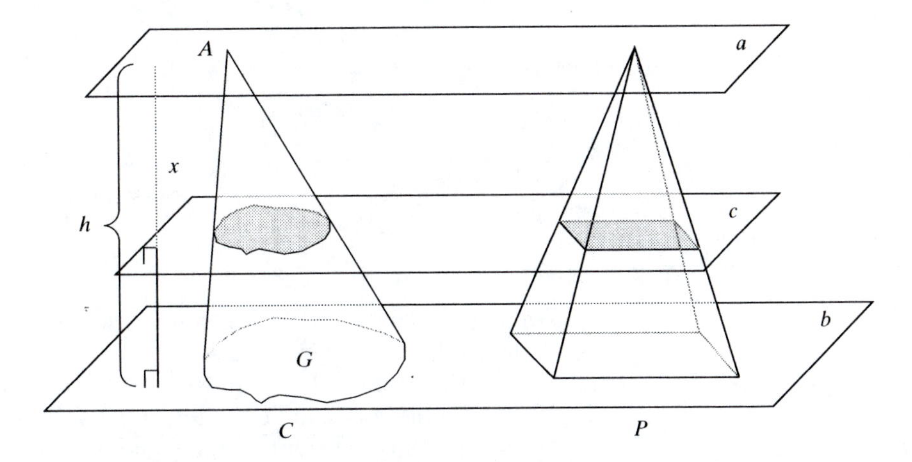

The intersection of the plane c with the cone C is a cross-section that is also similar to the base, being the image of the base under a size change with center at the cone's apex and magnitude $\frac{x}{h}$. The cone's base area B is the sum of areas of triangles or the greatest upper bound of such sums. The cross-section's area is the sum of areas of the size change images of these triangles. Because each triangle in calculating the cone's cross-sectional area has area $\left(\frac{x}{h}\right)^2$ times the corresponding triangle in the base, the cross-section has area $B\left(\frac{x}{h}\right)^2$. Consequently, the pyramid P and cone C satisfy the given conditions of Cavalieri's Principle. Since the volume of the pyramid is $\frac{1}{3}Bh$, the volume of the cone is also $\frac{1}{3}Bh$.

Corollary: The volume of a circular cone with radius r and height h is $\frac{1}{3}\pi r^2 h$.

From cylinders and cones to spheres

A **sphere** is the set of all points in E^3 that are the same distance from a given point. It is the 3-dimensional analogue to the circle, and the terms *center, radius, diameter, chord, secant,* and *tangent* are used in the same way with spheres as they are with circles. The solid figure bounded by a sphere is a **ball**. A ball is the set of all points in E^3 that are less than or equal to a fixed distance from a point. A ball is the 3-dimensional analogue to a 2-dimensional disk.

Our derivation of the formula for the volume of a ball or sphere involves surprising relationships between the volumes of cones, cylinders, and spheres, and an elegant application of Cavalieri's Principle.

Theorem 10.21 The volume of a sphere with radius r is $\frac{4}{3}\pi r^3$.

Proof: Let S be a sphere with radius r, and let a and b be planes tangent to the sphere at the endpoints of a diameter of S. To use Cavalieri's Principle, we construct a right circular cylinder with radius r and bases in the planes a and b. This cylinder has height $2r$, so (by the Corollary to Theorem 10.19) its volume is $2r \cdot \pi r^2$, or $2\pi r^3$. In Figure 66 we have shown the sphere, cylinder, and plane b. The plane a is not shown because it would hide helpful details.

Figure 66

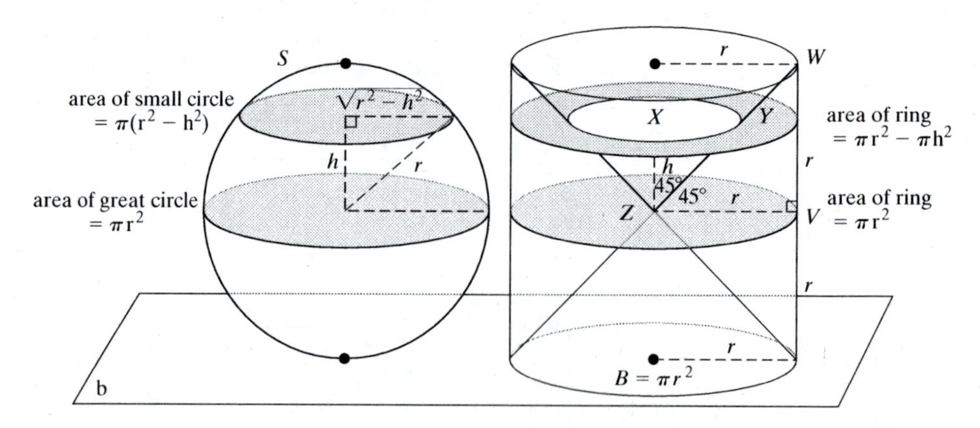

Now consider the two cones with apex at the center of the cylinder and whose bases are the bases of the cylinder. Each of these cones has height r and base area πr^2, so (by the Corollary to Theorem 10.20) each cone has volume $\frac{1}{3}\pi r^3$. Thus the region R between the cones and the cylinder has volume $2\pi r^3 - 2 \cdot \frac{1}{3}\pi r^3$, or $\frac{4}{3}\pi r^3$.

We now show that the volume of the sphere equals the volume of R. Consider a cross section parallel to a and b. If the cross section contains the center of the sphere, then its area is πr^2 in both the sphere and the cylinder. If the cross section of the sphere is at a distance h from the center, then, as Figure 66 shows, its area is $\pi(r^2 - h^2)$. The corresponding cross section of R is a ring. The outside radius of the ring is r. Notice that $WV = VZ = r$, so $m\angle WZV = 45°$. Consequently, $m\angle WZX = 45°$. Thus $XY = XZ$ and the inner radius of the ring is h. Consequently, the area of the ring is $\pi r^2 - \pi h^2$, the same as the area of the cross section of the sphere. Because the area of each cross section of the sphere equals the area of each cross section of R, by Cavalieri's Principle the volume of the sphere equals the volume of R. This proves the theorem.

Theorem 10.21 is found in Archimedes' work *On the Sphere and the Cylinder.* In that work Archimedes shows many other relationships between volumes and surface areas of cones, cylinders, and spheres. We have discussed the volume relationships. In the next section, we turn to surface area.

10.2.3 Problems

1. Show a drawing of the situation in the proof of Theorem 10.19.

2. Archimedes calculated the volume of a sphere by inscribing it in a cylinder. Suppose a cone is inscribed in a cylinder of the same size as that in which a sphere is inscribed.
   a. What is the ratio of the volume of the cone to the volume of the cylinder?
   b. What is the ratio of the volume of the sphere to that of the cone?

3. What common figure has a volume equal to $\frac{2}{3}hB$, where h is its height and B the area of its base?

4. Rain gauges are often made in the shape of cones. Figure 67 shows two such cones of the same height H but different radii r_1 and r_2.
   a. Show that if these rain gauges are placed in nearby locations in a rain storm, the water will reach the *same height* h in each of them, independent of their diameters.
   b. How does this height h vary with the amount d of rainfall?

5. Criticize the following explanation: To say that the volume of a sphere is 400 cubic feet means that the sphere contains 400 cubic feet, or that the amount of space is the same as in a 400-foot cube.

6. A **prismatoid** is a polyhedron with all of its vertices lying in two parallel planes. The faces in those planes are its bases and its altitude h is the distance between the bases. If the bases have areas B and B' and the cross section midway between the bases has area M, show that the volume of the prismatoid is given by $V = \left(\frac{1}{6}\right)(B + B' + 4M)h$. This formula is known as the **Prismoid** or **Prismoidal Formula**. (*Hint*: Pick any point P on the middle cross section, and find the volume of the two pyramids with P as apex and the bases of the prismatoid as their bases. Then partition the rest of the prismatoid into triangular pyramids with bases lying in the lateral faces of the prismatoid.)

7. Which theorems of Section 10.2.2 and this section can be considered as special cases of the Prismoidal Formula?

8. Find the volume of the largest right circular cone in which a sphere of radius 5 can be inscribed.

9. Derive the formula for the volume of a circular cone using calculus. (Consider the cone as a solid formed by rotating a right triangle in space about one of its legs.)

10. Derive the formula for the volume of a sphere using calculus. (Consider the sphere as a solid formed by rotating a semicircle about one of its diameters.)

11. Prove that there is exactly one sphere that contains the vertices of a tetrahedron. (*Hint*: Let A, B, C, and D be the vertices, and let E be the circumcenter of triangle ABC. Prove that if X belongs to a line ℓ that is perpendicular to the plane of ΔABC at E, then $AX = BX = CX$. Then prove that if Y is a point equidistant from A, B, and C, then Y belongs to ℓ.)

*12. Let P be a plane intersecting a ball S. A **spherical cap** is the union of $P \cap S$ and the set of points of S on one side of the plane. The *depth* of a spherical cap is the distance between P and the plane tangent to the cap parallel to P. Use Figure 66 and essentially the same proof as that of Theorem 10.21 to prove that the volume V of a spherical cap of depth h for a sphere of radius r is given by

$$V = \pi h^2\left(r - \frac{h}{3}\right).$$

Figure 67

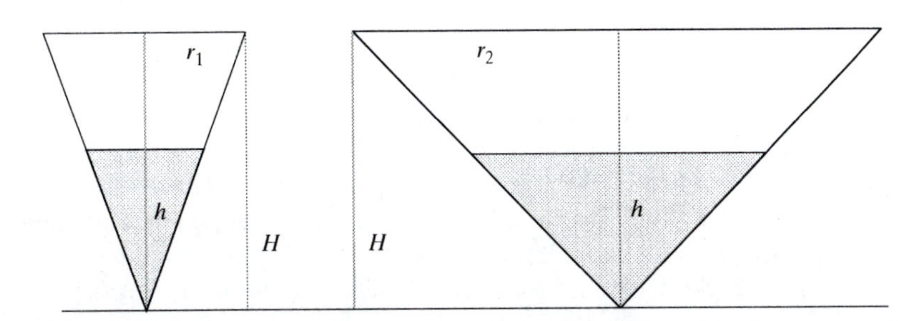

Unit 10.3 Relationships among Area, Volume, and Dimension

While perimeter, area, and volume measure different aspects of figures in different dimensions, and while knowing one of these seldom determines any other, these measures are related in many ways. For instance, every area formula involves the product of two lengths, and every volume formula involves the product of three lengths. Also, a rectangle with a given area can have as large a perimeter as one wants, but no smaller than the perimeter of a square with that area. In this unit we discuss a variety of other interesting relationships among length, area, and volume.

10.3.1 Surface area

The surface of any prism, pyramid, polyhedron, cylinder, or cone can be folded or rolled onto a plane. As a consequence, the surface areas of these 3-dimensional figures are found in the same way that 2-dimensional areas are found. When there are new formulas, it is only because the arrangement of the surface on the plane has regularity that allows the area to be calculated given certain dimensions of the original 3-dimensional surface. However, the use of an abbreviation such as S.A. for surface area misleads some students to believe that surface area is as different from area as volume is. The sphere is the only commonly studied solid whose surface cannot be rolled or folded to be a plane figure. Other common solids with this characteristic are ellipsoids and tori (doughnut-shaped solids).

Surface areas of prisms and pyramids

The part of the surface of a cylindric or conic solid that is not the base is called the **lateral surface** of the solid. The lateral surface of any prism can be unfolded to form a union of parallelograms. The total area of these parallelograms is the **lateral surface area (L.A.)** of the prism. The bases can then be attached at opposite sides of one of the parallelograms. The resulting figure is a **net** for the prism, and its area is the **total surface area (S.A.)** of the prism.

In school mathematics, surface areas of prisms are most often found for right prisms. Then the parallelograms of the net are rectangles and the lateral area is the product of the height of the prism and the perimeter of its base. An example is shown in Figure 68.

Figure 68

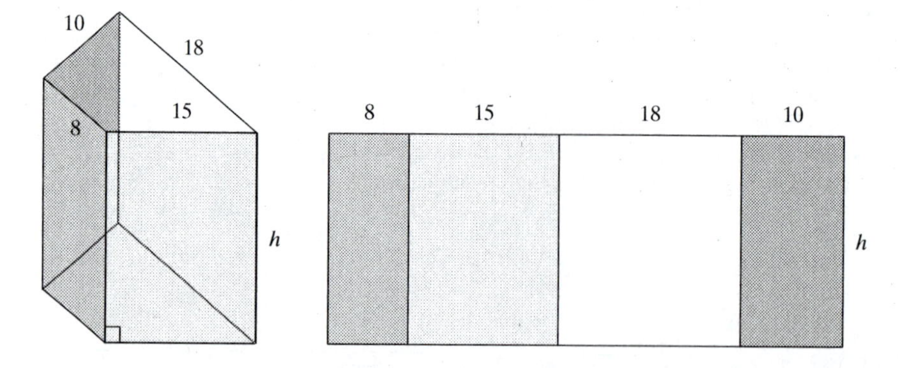

Whereas the lateral surface of a prism consists of parallelograms between two parallel lines, the lateral surface of a pyramid consists of triangles with a common vertex, as in Figure 69. For a regular pyramid, these triangles are isosceles and congruent, and the sum of their areas is the product of half an altitude of any of the triangles and the perimeter of the base. This altitude is called the **slant height of the pyramid** and is customarily denoted by the script letter ℓ. Thus L.A. $= \frac{1}{2}\ell p$. It seems as if this is a new formula, but it is simply an extension of the triangle area formula $A = \frac{1}{2}bh$.

Figure 69

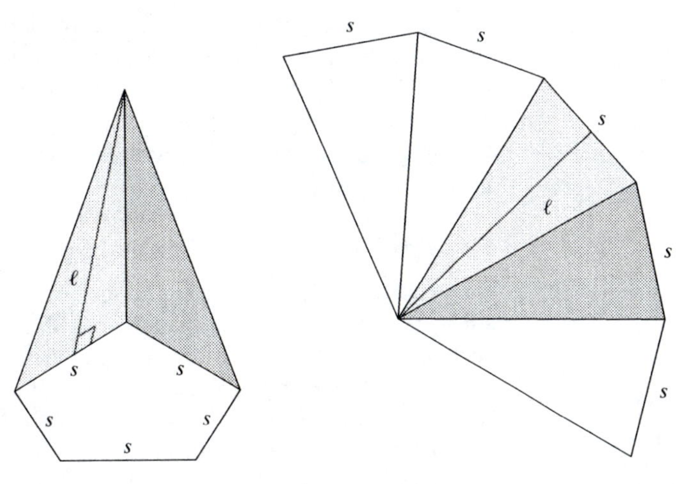

In Figure 70 we show two pyramids with congruent triangular bases ABC and $A'B'C'$ and the same height h. Notice that, whereas the volumes of the pyramids are equal, their surfaces areas are not equal. This is a 3-dimensional analogue to the fact that triangles with equal bases and altitudes can have different perimeters.

Figure 70

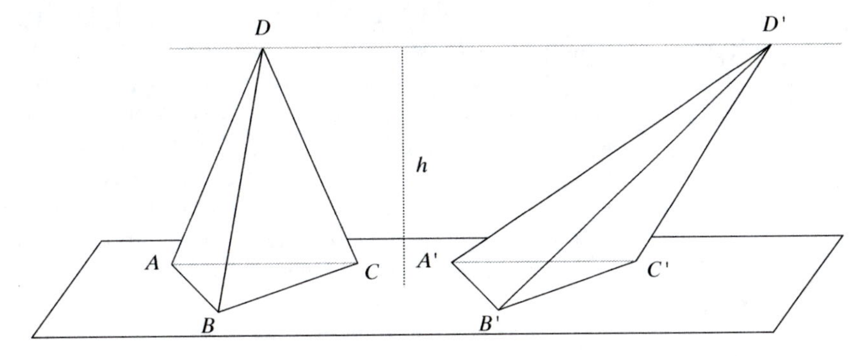

Surface areas of cylinders and cones

Unrolling the label from a soup can mimics the process used to determine the lateral area of a cylinder. When unrolled, the lateral surface of a right cylinder is a rectangle whose height is the height of the cylinder and whose width is the circumference of the base.

Unrolling a right circular cone is a little more interesting. The lateral surface is a sector of a circle. By thinking of the area of the sector as a limit of the sum of areas of triangles, the formula L.A. $= \frac{1}{2}\ell p$ for pyramids can be adapted. Here ℓ is the **slant height of the cone**, the length of any segment joining the apex to any point on the circle that is the boundary of the base; p is the perimeter of the base, or $2\pi r$, if the radius of the base is r. So one formula for the lateral area of a right circular cone is L.A. $= \pi r \ell$.

Figure 71

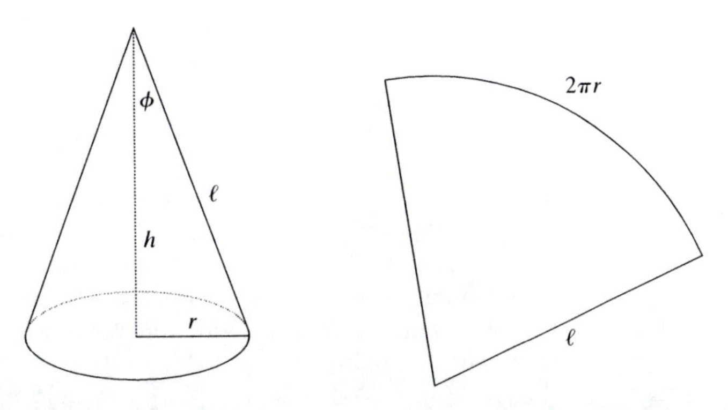

There is a very nice formula for the lateral area of a right circular cone in terms of its slant height and the angle between the segment from the apex perpendicular to the base and any segment whose length was used to measure the slant height. This angle can be thought of as measuring the opening of the cone. If ϕ is the measure of that angle (see Figure 71), then $\sin \phi = \frac{r}{\ell}$, from which $r = \ell \sin \phi$. Consequently, L.A. $= \pi \ell^2 \sin \phi$.

The surface area of a sphere

To the chagrin of mapmakers, the surface of a sphere cannot be unrolled onto a plane surface. This forces planar maps of Earth's surface to distort some aspect of the actual surface. It also forces a different approach for obtaining the surface area.

Our approach utilizes the volume of a sphere to obtain its surface area. While this may seem like a roundabout approach (no pun intended), it is not that much different from using the area of a circle to obtain a value for π and thus to obtain its circumference.

Theorem 10.22 The surface area of a sphere with radius r is $4\pi r^2$.

Figure 72

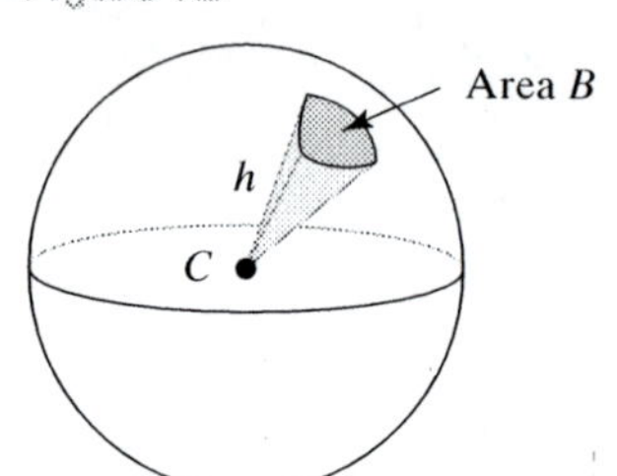

Proof: Consider a ball as being partitioned into "almost pyramids", each with an apex at the center of the sphere. One element of the partition is shown in Figure 72.

Each element of the partition is not exactly a pyramid because its base does not lie in a single plane. But as there are more and more elements of the partition, the base is closer and closer to lying in a single plane. If the areas of the bases of the partition are $B_1, B_2, \ldots, B_k$, then as k increases so that the largest B_k is made smaller, the volume of the ith element of the partition becomes closer and closer to $\frac{1}{3}B_i r$, where r is the radius of the sphere and also the height of the pyramid.

Let the volume of the sphere be V and its surface area be S.A.

$$\begin{aligned} V &\approx \frac{1}{3}B_1 r + \frac{1}{3}B_2 r + \cdots + \frac{1}{3}B_k r \\ &= \frac{1}{3}(B_1 + B_2 + \cdots + B_k)r \\ &\approx \frac{1}{3}\text{S.A.} \cdot r \end{aligned}$$

Now use the volume formula for a sphere (Theorem 10.21).

$$\frac{4}{3}\pi r^3 = \frac{1}{3}\text{S.A.} \cdot r$$

Solving this formula for S.A,

$$\frac{3 \cdot \frac{4}{3}\pi r^3}{r} = \text{S.A.}$$

Simplifying this fraction yields the theorem. ┘

Archimedes derived the formula for the surface area of a sphere by inscribing the sphere inside a circular cylinder with the same diameter and height as the sphere. He was so excited about this result that he directed the sphere in a cylinder diagram to be engraved on his tombstone. He noted (and we note) the following about a sphere with diameter d, which summarizes the formulas for both the circle and sphere.

Great circle: circumference πd, area $\frac{1}{4}\pi d^2$

Sphere: surface area πd^2, volume $\frac{1}{6}\pi d^3$

Notice that the coefficients of area and volume are proportional to the corresponding coefficients ($\frac{1}{2}$ and $\frac{1}{3}$) in formulas for areas of triangles and volumes of pyramids. These results suggest that it might be easier to have students learn the formulas for circles and spheres in terms of their diameters rather than in terms of their radii.

Relationships among formulas

Many calculus students notice that when $V = \frac{4}{3}\pi r^3$, then $\frac{dV}{dr} = 4\pi r^2 = \text{S.A.}$ That is, the derivative of the volume with respect to the radius seems to be the surface area. Also, when $A = \pi r^2$, then $\frac{dA}{dr} = 2\pi r = C$. Are these relationships calculational coincidences or are they the result of special properties of the circle and sphere?

To answer this question, recall that if g is a function of r, then

$$g'(r) = \lim_{h\to 0}\frac{g(r+h) - g(r)}{h}.$$

Figure 73

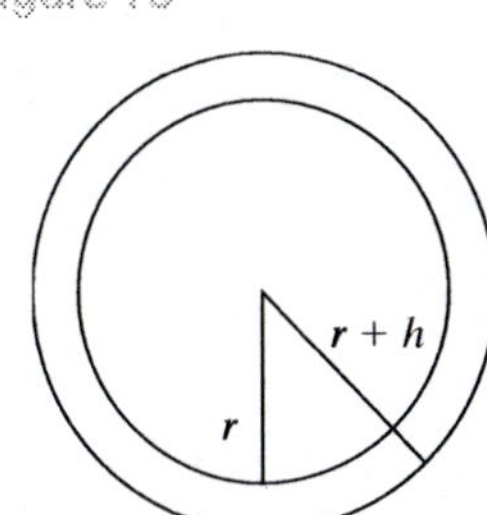

Let $A(r)$ be the area of a circle of radius r. Then

$$A(r+h) - A(r) = \pi(r+h)^2 - \pi r^2 = 2\pi rh + \pi h^2.$$

The quantity $A(r+h) - A(r)$ is the area of the ring-shaped region between two concentric circles of radii r and $r + h$ (Figure 73). As h approaches 0, it follows algebraically that

$$A'(r) = \lim_{h\to 0}\frac{A(r+h) - A(r)}{h} = \lim_{h\to 0}(2\pi r + \pi h) = 2\pi r = C(r).$$

Geometrically, as the width of the ring gets smaller, the relative change in area of the circle from r to $r + h$ approaches the circumference of the circle of radius r.

The equation $\frac{dV(r)}{dr} = 4\pi r^2 = S(r)$ can be interpreted in a similar way. Let $V(r)$ be the volume of a sphere of radius r. The quantity $V(r+h) - V(r) = \frac{4}{3}\pi(r+h)^3 - \frac{4}{3}\pi r^3$ is the volume of the spherical shell between the spheres of radii r and $r + h$. (A spherical shell is the 3-dimensional counterpart of a 2-dimensional ring.) This quantity can be simplified to

$$V(r+h) - V(r) = 4\pi r^2 h + 4\pi r h^2 + \frac{4}{3}\pi h^3.$$

The first term of this sum, $4\pi r^2 h$, is the approximation to the volume of this shell obtained by multiplying the area $4\pi r^2$ of the inner surface of the shell by the shell's thickness (an underestimate), while the term $4\pi r h^2 + \frac{4}{3}\pi h^3$ is the error resulting from this underestimate of the shell volume.

As h approaches 0, it follows algebraically that

$$V'(r) = \lim_{h\to 0}\frac{V(r+h) - V(r)}{h} = \lim_{h\to 0}\left(4\pi r^2 + 4\pi rh + \frac{4}{3}\pi h^2\right) = 4\pi r^2 = S(r).$$

Geometrically, as the thickness of the shell decreases to zero, the relative change in volume of the sphere between r and $r + h$ approaches the surface area of the sphere of radius r.

So these relationships are not coincidences, but explainable both geometrically and algebraically.

10.3.1 Problems

1. Interpret the formula L.A. $= \pi\ell^2 \sin\phi$ for the lateral area of a right circular cone when $\phi = 90°$.

2. Find a formula for the lateral area of a right circular cone in terms of the radius of its base and the angle between a lateral edge and the plane of its base.

3. a. Find the volume of a right circular cone in terms of the measure of the opening of the cone and its slant height.

   b. Use this formula to answer the following question. A sector is cut out of a disk, and the radii that are edges of the sector are made to coincide. Then the disk has become the lateral surface of a right circular cone. What is the measure of the central angle for which the cone's volume is maximized?

4. A rectangle with dimensions a and b can be rolled up in two ways to become the surface area of a cylinder.

   a. What are the volumes of the two cylinders so formed?

   b. Which way should an 8.5″ by 11″ sheet of paper be rolled up to obtain the cylinder of larger volume?

5. a. Suppose that T_r is an equilateral triangle and S_r is a square, both inscribed in a circle of radius r. Find formulas for the areas $A(r)$ and perimeters $P(r)$ of T_r and S_r in terms of r and show that it is not true that $A'(r) = P(r)$ for either of these figures.

   b. Suppose that, in part **a**, T_r and S_r were circumscribed about a circle of radius r. Is it true that $A'(r) = P(r)$ for either figure?

6. Let $A_n(r)$ and $P_n(r)$ be the area and perimeter of a regular n-sided polygon inscribed in a circle of radius r. Investigate the behavior of the derivative $A'_n(r)$ as n increases.

7. Prove: If a cylinder's base is the disk of a great circle of a sphere, and the cylinder's height is equal to the diameter of the sphere, then the cylinder's total surface area is $\frac{3}{2}$ the surface area of the sphere.

8. Show that the surface area $S_{h,r}$ of a spherical cap of depth h on a sphere of radius r is given by

$$S_{h,r} = 2\pi rh.$$

(*Hint*: Use the result of Problem 13 in Section 10.2.3 and an argument similar to that in the proof of Theorem 10.22.)

9. a. Figure 74 displays a cross-section of a sphere with center O by a plane that contains point P and the arc $\widehat{ACB}$ that is visible from P. Let r be the radius of the sphere, let a be the distance from P to the sphere, let H be the intersection of $\overline{AB}$ and $\overline{OP}$, and let $d = HC$. Explain why

$$OH = \frac{r^2}{r+a},$$

   and use this fact to show that $d = \frac{ra}{r+a}$.

Figure 74

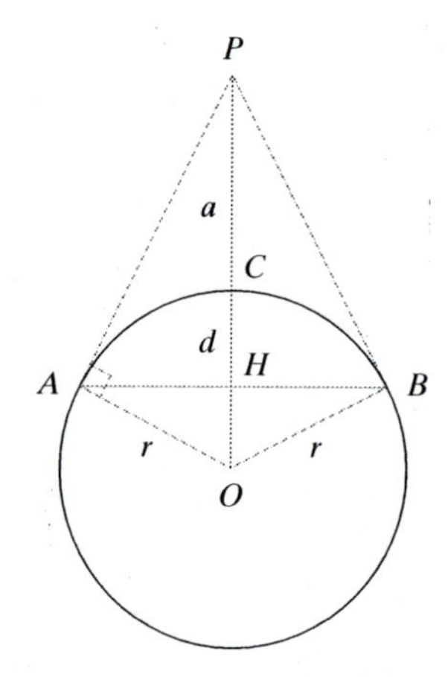

   b. Use part **a** and Problem 8 to conclude that the fraction $F_{a,r}$ of the surface area of a sphere of radius r that can be seen from a point P that is a units above the surface of the sphere is given by

$$F_{a,r} = \frac{a}{2(r+a)}.$$

   c. Assume that Earth is a sphere of radius 3960 miles. What fraction of Earth's surface can be viewed from a satellite in a circular orbit 1000 miles above the surface?

   d. Explain why 3 satellites placed in geosynchronous circular orbits 20,000 miles above the equator are insufficient to operate a communications network for the entire Earth.

   e. Based on the meaning of $F_{a,r}$, how do you expect $F_{a,r}$ to vary as a increases without bound for fixed r? Confirm your expectations mathematically using the formula for $F_{a,r}$ from part **b**.

10.3.2 The Isoperimetric Inequalities

Theorem 10.2 in Section 10.1.1 states a well-known result: Of all rectangles with a given perimeter, the square has the greatest area. The problem this theorem solves is an *isoperimetric problem*, that is, a max-min problem dealing with figures that have the same perimeter. That particular theorem relates to a situation where you have a fixed amount of fence and you wish to enclose the rectangular region with the largest area.

Dido's problem

If we do not restrict the shape of the region, then we are faced with the general problem for the plane. The problem of enclosing the largest possible area with a given perimeter is known as *Dido's problem*. The story of this problem is interesting.

Carthage was a port on the Mediterranean Sea in Northern Africa and one of the great cities of ancient times. A major foe of Rome, Carthage engaged Rome in three major wars in the 3rd and 2nd centuries B.C., being destroyed in the last of these. Carthage later was rebuilt and was important in the Roman empire. It was again destroyed around 450 A.D. and was depopulated after 698 A.D. The ancient site of Carthage is now in the suburbs of the city of Tunis in Tunisia. Archaeological excavations indicate that Carthage was founded around 750 B.C.

The story of Dido as told in the epic poem *Aeneid* by the Roman poet Virgil clearly involves many myths. According to legend (but possibly with some reality), Dido was the daughter of King Belus of the Phoenician city of Tyre. She fled to Africa with some devoted followers after her husband was murdered. She was offered only as much land as she could surround with a bull's hide. Determining the most efficient shape became Dido's problem. Her solution was to cut the hide into very thin strips and lay them out end to end to enclose the largest possible area. The enclosed region became the site of Carthage.

The next theorem states the solution to Dido's problem. A rigorous solution requires a very careful definition of the length of a *simple closed curve*, which we do not give. So we can present only a partial proof. This proof was first given by the Swiss mathematician Jakob Steiner (1796–1863). We assume that there exists a simple closed curve with the largest area for a given perimeter.

Theorem 10.23 Of all plane figures with the same perimeter, the circle has the largest area.

Proof: Let F be the simple closed curve with largest area for a given perimeter p.

1. First we show by an indirect proof that F must be convex. Suppose F is not convex. Then there exist two points A and B on F such that $\overline{AB}$ contains only points in the exterior of F (Figure 75). Then reflect the nonconvex part of F bounded by A and B over $\overleftrightarrow{AB}$. The new curve F' has the same perimeter as F and greater area. This contradicts F having the largest area for its perimeter.

Figure 75

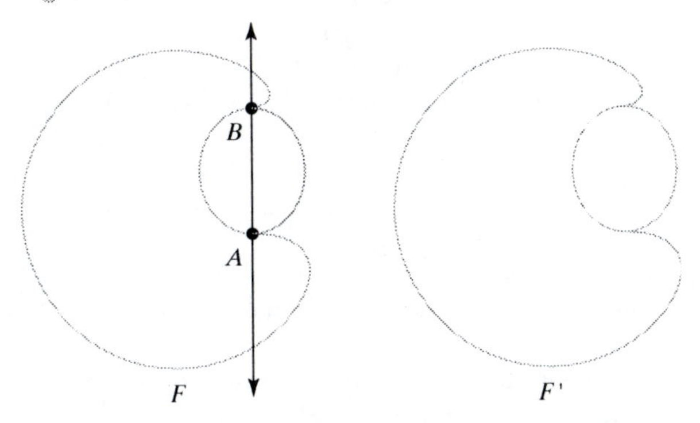

2. Now we show by indirect proof that there exist lines through any point of F that split the area of F into two equal halves. Suppose F cannot be split in this way. Let A be any point on F. Since F has perimeter p, there is a unique second point B on F whose distance from A along F is $\frac{p}{2}$. (B can be said to be

Figure 76

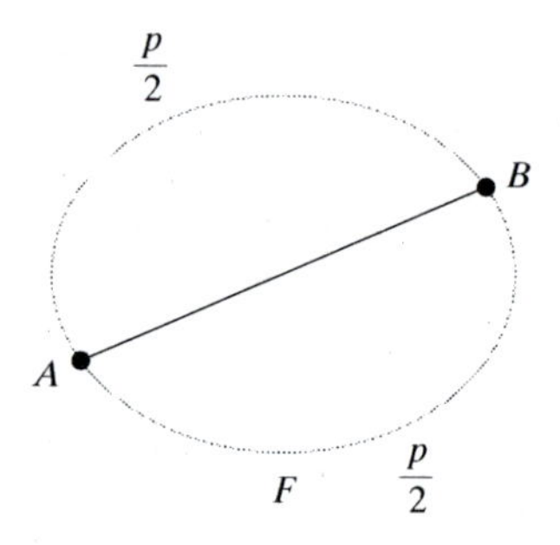

"halfway around" F from A.) Since from part (1) F is convex, $\overline{AB}$ splits F into two regions, each with the same perimeter (Figure 76). If these regions do not have the same area, then we can reflect the region with the larger area over $\overleftrightarrow{AB}$ and replace F by the new curve. But this means there is a curve with perimeter p enclosing a larger area than F, contrary to our assumption. So $\overline{AB}$ splits F into two regions with the same area.

3. Lastly, we show that each half of this split must be a semicircle. Let C be a point on F other than A or B. The half of the region containing C and bounded by $\overline{AB}$ consists of two regions R_1 and R_2 with areas A_1 and A_2 (Figure 77a) and ΔABC with area A_3. If $\angle ACB$ is not a right angle, then consider the new curve F' obtained by attaching regions R_1 and R_2 upon $\Delta A'C'B'$, where $A'C' = AC$, $B'C' = BC$, and $\angle C'$ is a right angle. Since $\alpha(\Delta ABC) = \frac{1}{2}AC \cdot BC \cdot \sin \angle ACB$, and $\alpha(\Delta A'B'C') = \frac{1}{2} \cdot AC \cdot BC$, the area of F' is greater than the area of F. This again contradicts F being the curve with maximal area for its perimeter, so $\angle ACB$ must be a right angle. Let M be the midpoint of $\overline{AB}$. Since ΔACB is a right triangle, M is equidistant from A, B, and C. So C lies on the circle with center M and passing through A and B. Thus all points of F on this side of $\overline{AB}$ are on this circle, and the same argument can be applied to the part of F on the other side of $\overline{AB}$.

Figure 77

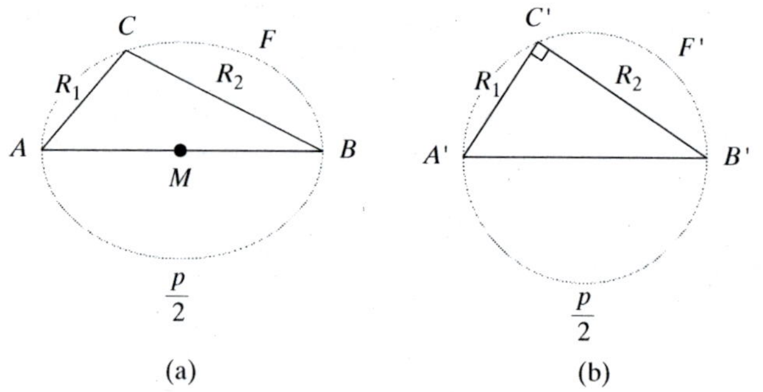

(a) (b)

This theorem may explain why tepees, hogans, igloos, yurts, and other structures of peoples throughout the world are circular in shape. These structures maximize the floor space that can be surrounded by a fixed amount of material. Since the materials that are used in these structures were valuable and sometimes scarce resources, people learned not to use more material than necessary. The same principle may explain why besieged pioneers in covered wagons would want to "circle the wagons"; the space enclosed by the wagons would then be maximized.

If the given perimeter is p, Theorem 10.23 allows us to calculate the maximum area of a figure with a given perimeter. All other figures must have less area. The result, an immediate consequence of Theorem 10.23, is an *isoperimetric inequality*.

Corollary 1 (Isoperimetric Inequalities for the Plane): Let a plane region have perimeter p and area A. Then $A \le \frac{p^2}{4\pi}$, or, equivalently, $p \ge 2\sqrt{\pi A}$, with equality occurring if the region is circular.

Proof: In the circle with perimeter (circumference) p, $p = 2\pi r$. Thus $r = \frac{p}{2\pi}$. Since $A = \pi r^2$, substituting for r, we have $A = \pi\left(\frac{p}{2\pi}\right)^2 = \frac{p^2}{4\pi}$. By Theorem 10.23, this is the maximum area figures with perimeter p can have. So for any other figure with area A and perimeter p, $A \le \frac{p^2}{4\pi}$. Solving for p yields the second inequality in the statement of the corollary.

The second inequality in the corollary provides a useful restatement of Theorem 10.23.

Corollary 2: Of all plane figures with the same area, the circle has the least perimeter.

It is not always desirable to have the smallest perimeter for a given area. Upscale suburban developments outside many cities include human-made lakes. Shoreline property is worth more per square foot than other property, so the developer may try to make the lake longer and thinner. Then the region covered by water will have a larger perimeter than a circle of equal area.

The 3-dimensional counterpart

Many igloos and yurts not only have circular bases but also have roughly hemispherical roofs. These roofs suggest that 3-dimensional counterparts to the 2-dimensional isoperimetric inequalities would involve hemispheres or spheres. In fact, the sphere plays the same role in these inequalities in three dimensions that the circle plays in two dimensions. However, the proof we have given for Theorem 10.23 does not generalize to three dimensions. A proof for the 3-dimensional counterpart was first found by Hermann Amandus Schwarz (1843–1921). We omit the proof.

Theorem 10.24 Of all solids with the same surface area, the sphere has the largest volume.

Theorem 10.24 has corollaries corresponding to those of Theorem 10.23.

Corollary 1 (Isoperimetric Inequalities for Space): Let a solid have surface area A and volume V. Then $V \leq \sqrt{\frac{A^3}{36\pi}}$, or, equivalently, $A \geq \sqrt[3]{36\pi V^2}$, with equality occurring if the region is spherical.

Proof: The proof follows the idea of the proof of Corollary 1 to Theorem 10.23 and is left to you. ┘

Theorem 10.24 can be restated in the language of minimums just as Theorem 10.23 could.

Corollary 2: Of all solids with the same volume, the sphere has the least surface area.

Corollary 2 suggests that containers should be spherical if they are to maximize their capacity for a given amount of material needed to form the container. So, for example, we might expect to have spherical milk cartons or spherical cereal boxes. One problem with such a solution is obvious: Spheres roll! A second problem is that they are not easily transported—even a bowling ball has holes!

Just as a 2-dimensional region can have a large perimeter and small area, a 3-dimensional solid can have a large surface area and small volume. Filters such as those found in heaters and air conditioners, water purifiers, and cigarettes are based on the idea that small particles can stick to surfaces. They have very large surface areas for their volumes.

10.3.2 Problems

1. An equilateral triangle, a square, and a regular hexagon each have perimeter p.

a. Give the area of each figure. How much larger in area is a circle with a circumference of p?

b. A lake has a surface area of 1000 km^2. What are the maximum and minimum lengths of beach this lake might have?

2. An equilateral triangle, a square, a regular hexagon, and a circle each have area A. Find the perimeter of each and show that the circle's perimeter is smallest.

3. Let the sides of a triangle be a, b, and c with c and $a + b$ constant. Under these conditions, prove that the triangle with maximal area is isosceles.

4. A boxing "ring" is actually square-shaped. Design a plan for 2000 people to have seats to view the ring that would allow maximum visibility for the greatest number of people.

5. Suppose a local zoning law requires a floor space of at least 600 square feet of living space for a summer cottage. How should the cottage be shaped to minimize the cost of materials for the walls?

6. A rancher has 300 yards of fencing. He wants to use the fencing to construct a rectangular pen and to divide the pen in two halves to separate the bulls from the cows.

a. Find the dimensions of the pen that provides the largest area for bulls and cows.

b. Show that neither splitting a big square pen in two congruent halves, nor using a pen with two square halves, yields the pen with the largest area.

7. A rectangular dog pen is to be built alongside a house. The house will form one side of the pen; fencing will form the other three sides. What shape pen gives the dog the most play area?

8. Prove Corollary 1 to Theorem 10.24.

9. When a spherical soap bubble lands on a flat surface so that the surface becomes one of the sides of the space enclosed by the bubble, the bubble will assume the shape of a hemisphere. (There is soap film on the surface when this happens.) If the radius of the original bubble was r, what is the radius of this hemisphere?

*10. Containers for many foods are circular cylinders. It is in the interest of food distributors to use cylinders with shapes that maximize volume for their surface areas.

a. What ratio of height to radius maximizes the volume of a circular cylinder with a given surface area S?

b. As the ratio of height to radius changes, by how much does the volume of a circular cylinder with a given surface area S change?

10.3.3 The Fundamental Theorem of Similarity

From the basic properties of similarity transformations, when two figures are similar with ratio of similitude k, corresponding angles have the same measure and corresponding distances are in the ratio k. But what happens to area and volume?

Consider a square with side s. Under a similarity transformation with ratio of similitude k, the image square has side ks. The original square, which had area s^2, gives rise to an image with area $(ks)^2$, or k^2s^2. The area of a figure that is not a square is determined by partitioning that figure into squares or parts of squares. If each square in the partition of a preimage figure has its area multiplied by k^2 in the image figure, then the area of the image is k^2 times the area of the preimage. Thus if two figures are similar with ratio of similitude k, then the ratio of their areas is k^2 (Figure 78).

Figure 78

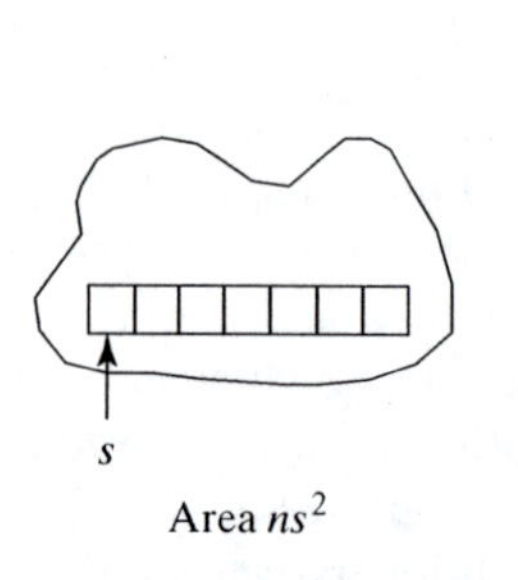

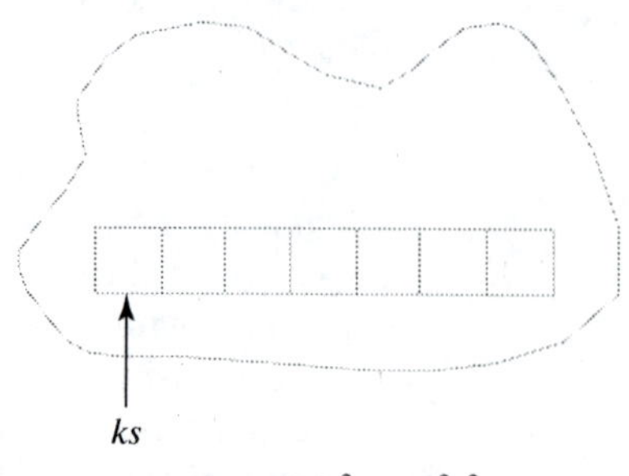

In 3-dimensional space, under a similarity transformation with ratio of similitude k, a cube with edge e gives rise to an image cube with edge ke. The original cube, which had volume e^3, gives rise to an image with volume $(ke)^3$, or k^3e^3. Using the same reasoning as in the 2-dimensional situation, if two figures are similar with ratio of similitude k, then the ratio of their volumes is k^3.

These properties are summarized in a theorem we call the Fundamental Theorem of Similarity because of its wide-ranging scope and applications.

Theorem 10.25 **(Fundamental Theorem of Similarity):** If β is the image of α under a similarity transformation of magnitude k, then

a. angle measures in β are equal to corresponding angle measures in α;

b. distances in β are k times corresponding distances in α;

c. areas in β are k^2 times corresponding areas in α; and

d. volumes in β are k^3 times corresponding volumes in α.

Figure 79

For instance, the ratio of similitude in Figure 79 is 1.75. So, while corresponding angle measures are equal, the lever in β is 1.75 times as long as the lever in α, and β occupies an area 1.75^2 times as great on the page as α does.

The parts of the Fundamental Theorem of Similarity are instances of an important pattern. Rewrite part (b) as *distances in* β *are* k^1 *times corresponding distances in* α. Then the exponent of k in parts (b), (c), and (d) signifies the dimension of the measure. Distance is 1-dimensional, area is 2-dimensional, and volume is 3-dimensional. This pattern also extends to angle measure. Part (a) can be written as *angle measures in* β *are* k^0 *times corresponding angle measures in* α. This suggests that angle measure is 0-dimensional, a property that is corroborated by examination of formulas that involve angle measures. One such formula is $s = r\theta$ for the length s of an arc with a central angle of θ radians in a circle with radius r. If θ had any dimension, then the lengths r and s would have different dimensions.

Another argument for angle measure being 0-dimensional is that angle measure is defined around a point, a 0-dimensional figure. From this perspective, the other dimensions fit in familiar ways. Distance and length are defined along a line, a 1-dimensional figure; area is defined on plane figures, which are 2-dimensional; and volume deals with the capacity of a figure in 3-dimensional space.

Applications to the strength of objects

Applications of the Fundamental Theorem of Similarity to the strength of natural objects, both animate and inanimate, and of manufactured objects as well, were first recognized by the great Italian scientist Galileo around the beginning of the 17th century and published by him in *On Two New Sciences* in 1638.

Until Galileo's time, it was commonly believed that larger and smaller objects would have the same strength as long as they were similar. However, the strength of an object is proportional to its cross-sectional area, while its weight is proportional to its volume. Through a dialogue somewhat like that found in Plato's writings and mathematical arguments like those found in Euclid's *Elements*, Galileo shows why larger objects made of the same materials and similar to smaller objects will be weaker. If the larger object is k times the smaller, then the weight of the larger object, being proportional to the volume, is k^3 times the smaller, while the cross-sectional area is only k^2 times the smaller. As a result, k times more pressure is placed on each point on the larger object. This is often enough to collapse the larger object.

For example, a model airplane made of balsa wood can fly, but a larger similar airplane made of the same balsa wood will collapse of its own weight. As another example, a fly has very thin legs compared to its body, but a larger animal could not support itself if its legs were proportional to the fly's. So the horse must have proportionally thicker legs than the fly, and the elephant must have thicker legs than the horse.

It is commonly said that an ant is proportionally stronger than a human because it can lift a leaf that is many times its weight. Birds are said to eat more than humans because they may eat many times their weight in a day, whereas a human eats only a small percent of his or her weight each day. These biological phenomena are not anomalies but due again to the fact that length, area, and volume are not proportional.

On the existence of giants

Suppose there were to be people on Earth similar in shape and substance to us but 5 times our size. By the Fundamental Theorem of Similarity, such people would have 125 times our weight, but that weight would be supported on only 25 times the area. As a result, each part of the body would have to support 5 times as much weight as our parts do. Even champion weightlifters seldom lift more than twice their body weight, and when they do, it is only for a few seconds. The bodies of such giants would collapse under their own weight.

Reality supports the theory. You may have seen television or movie pictures of people weighing well over 500 pounds. Even though their legs are wider, these people often can barely move. They cannot support their own weight.

Very heavy people tend not to have the same shape as lighter people. So they do not necessarily tell us how tall people can become without changing shape. Experience suggests that the answer is not much more than 1.5 times average height. The tallest man on record from anywhere in the world has been Robert Wadlow (1918–1940). Wadlow was born in Alton, Illinois, and was known from his youth as a giant, and so his condition was subject to medical study and there exist quite a number of pictures of him. He was growing his entire life, and on June 27, 1940, he was measured to have a height of 8 feet, 11.1 inches.

If you saw a picture of Wadlow without other people next to him, you would have little idea that he was not of average height. But he was so large that he could not support his weight without help and needed a leg brace for support. This tragically turned out to be the cause of his death. While getting out of a car—a difficult task for a man of such size—his brace cut a deep wound in his leg. Gangrene set in (penicillin was not in use at the time) and Wadlow died 18 days after his height was last measured.

The need for a brace by a man about 1.5 times the height of many other people shows that the human shape is meant to fall within a rather narrow range of heights. The shortest adult dwarfs in recorded history are just a little under two feet tall. The range of 2 feet to 9 feet belies giving any reality to humanlike miniatures or giants such as those found in fairy tales, cartoons, or movies.

Applications to formulas for area

The Fundamental Theorem of Similarity explains the structure of area and volume formulas. In Table 3 we arrange a variety of area formulas into two categories: those that involve one variable and those that involve two.

Table 3

One-Variable Area Formulas		Two-Variable Area Formulas	
Squares:	$A = s^2$	Rectangles:	$A = \ell w$
Equilateral triangles:	$A = \frac{s^2\sqrt{3}}{4}$	Triangles:	$A = \frac{1}{2}bh$
Circles:	$A = \pi r^2$	Ellipses:	$A = \pi ab$

The figures identified in the left column of Table 3 have the property that all members of the type are similar. For instance, all equilateral triangles are similar. This illustrates the following consequence from the Fundamental Theorem of Similarity.

Theorem 10.26 For every set of similar 2-dimensional figures, there is an area formula of the form $A = kL^2$, where L is a corresponding length on one of the figures.

Proof: We want a formula for all figures of a set S of similar figures in terms of a particular segment in S. (For instance, for the circle formula above, this segment is a radius.)

When this segment has length L, call the figure $F(L)$ and its area $A(L)$. Consider the figure $F(1)$. This is the figure in which $L = 1$. (For the circle, this is the unit circle.)

Every figure $F(L)$ in the set is the image of $F(1)$ under a similarity transformation with a magnitude L. So, by the Fundamental Theorem of Similarity, $A(L) = A(1)L^2$. ┘

In practice, L is usually a natural length on a figure, such as a diameter, height, or side.

The proof of Theorem 10.26 does more than prove the statement of the theorem. It shows that the constant k in the statement of the theorem is the area of the figure in the set when $L = 1$. For instance, the constant π in the expression πr^2 is the area of a circle when its radius is 1. The constant $\frac{\sqrt{3}}{4}$ in the formula for the area of an equilateral triangle is the area of an equilateral triangle with side $s = 1$. Thus, the formula for a set of similar figures in terms of a particular segment on those figures is entirely determined by the area when that segment has length 1.

Applications to formulas for volume

Volume formulas are slightly more complex, for they may involve one, two, or three variables. But the idea of Theorem 10.26 still holds. Examine the volume formulas shown in Table 4.

Table 4

One-Variable Volume Formulas	Two-Variable Volume Formulas	Three-Variable Volume Formulas
Cubes: $V = s^3$ Spheres: $V = \frac{4}{3}\pi r^3$	Circular cylinders: $V = \pi r^2 h$ Circular cones: $V = \frac{1}{3}\pi r^2 h$	Boxes: $V = \ell wh$

Theorem 10.27 For every set of similar 3-dimensional figures, there is a volume formula of the form $V = kL^3$, where L is a corresponding length on one of the figures.

We leave the proof of Theorem 10.27 to you. We also leave it to you to describe the quality of figures that separates those whose volume formulas involve two variables from those whose volume formulas require three variables.

10.3.3 Problems

1. a. A figurine weighs w kg. A similar figurine, made of the same materials, is twice the height of the first figurine. What is the weight of the larger figurine?

   b. Suppose a pail filled with sand weighs 5 lb. A similar pail, twice the height, is filled with sand. What is its weight?

   c. Generalize parts **a** and **b**.

2. According to 1998 weight guidelines from the National Institutes of Health, the middle weight of the healthy range for a 6-foot person of either sex is 158 lb.

   Use the ideas of this section to determine the corresponding weight for a similar person with the given height.

   a. 7 feet b. 5 feet c. h feet

3. Find the area of a circle with diameter 1. Use this result to find a formula for the area of a circle in terms of its diameter.

4. Find a formula for the area of a regular hexagon.

5. Find a formula for the area of a regular n-gon in terms of its side length s. (You will need trigonometry.)

6. In *On Two New Sciences*, Galileo proves the following theorem: The area of a circle is the geometric mean of the areas of two regular n-gons with the same number of sides, one circumscribed about the circle and one with the same perimeter as the circle.

   a. Prove this theorem for a particular value of n.

   b. Prove the general theorem.

7. Does Theorem 10.26 apply also to formulas for surface area? Explain why or why not.

8. Prove Theorem 10.27.

9. Find a formula for the volume of a sphere with diameter d.

10. Find a formula for the volume of a regular square pyramid with base of side s and height h.

11. Find a formula for the volume of a cone with height h and an elliptical base having axes of lengths a and b.

12. Give some volume formulas that involve two variables, and other volume formulas that involve three variables. What quality separates the figures in one group from the figures in the other?

10.3.4 Fractional dimension

Normally we think of dimension as a count of independent directions. For instance, if a figure extends in two directions, the figure is 2-dimensional. A direction such as "13° north of west" is not independent of "north" and "west" because we can go a certain distance north and a certain distance west to get to 13° north of west. But going north and west will never get us in an up or down direction, and so we view up or down as adding a 3rd dimension. If dimension is conceived as a count of directions, there can never be a fractional dimension.

However, recall that the Fundamental Theorem of Similarity (Theorem 10.25) has an interpretation closely linked to dimension. Its four parts can be combined into one statement: *If β is the image of α under a similarity transformation with magnitude k, then d-dimensional measures on β are k^d times corresponding d-dimensional measures on α.* In this form, the dimension d has a role as an exponent. Because exponents do not have to be integers, this form allows for the possibility of fractional dimension.

Another way to connect measures with dimension is found in unit conversion. For instance, we might begin with the conversion of feet to yards.

$$\begin{aligned} \text{1 yard} &= \text{3 feet} \\ \text{1 square yard} &= \text{9 square feet} \\ \text{1 cubic yard} &= \text{27 cubic feet} \end{aligned}$$

Let us rephrase these conversions using exponents.

$$\begin{aligned} \text{1 yard} &= 3^1 \text{ feet} \\ \text{1 square yard} &= 3^2 \text{ square feet} \\ \text{1 cubic yard} &= 3^3 \text{ cubic feet} \end{aligned}$$

Again the dimension of the unit is an exponent, in this case, an exponent in the conversion factor.

Might there be a kind of yard that does not convert to feet by an integer power of 3? The answer is that there is, and for this we consider a real problem.

The length of a coastline

Surveys measure lengths of coastlines. For instance, the National Oceanic and Atmospheric Administration (NOAA) of the U.S. Department of Commerce provides a length of coastline for every state bordering on one of the oceans. The four states with the most coastline are: Alaska, 5580 miles; Florida, 1350 miles; California, 840 miles; Hawaii, 750 miles. It is not easy to calculate the length of a coastline. How far in should you go inland, if at all, on a river? When should an inlet be counted as coastline? Should a small finger of land be included?

The NOAA figures were calculated in the following way: The general outline of the seacoast as found on charts as near the scale of 1:1,200,000 as possible was used. One inch on such a map is about 19 miles. Measurements were made with a unit measure of 30 minutes of latitude. Coastline of sounds and bays were included to a point where they narrow to a width of 30 minutes of latitude, and then the distance across at this point was used.

A key aspect of this description is that a unit of measure is selected. The unit of measure is critical because the smaller the unit, the more the measurer has to go into the narrower parts of sounds and bays. Fingers of land that might not be considered with a larger unit must be included with a smaller unit. Consequently, *when a smaller unit is used, the coastline is longer.*

This goes against the intuition that a coastline should have a definite length. But it agrees with another intuition, that measuring a coastline is difficult because a coastline really does not have a length at all. In between is a mathematical way of dealing with both intuitions: *coastlines have a measure but it is not 1-dimensional.*

The NOAA figures were calculated from maps, but suppose we were to actually go out to the coast to measure the coastline. Then we might proceed as follows. First we choose a unit, say 1500 meters. We put down a stake at a point O on the coast and draw an imaginary circle with O at the center and a radius of 1500 m. Assuming we are not on a small island, the circle intersects the coast at two points, one on each side of O. Let us call these points A_1 and B_1 (Figure 80). Now we go to A_1, put down a stake, and draw a second circle, this time with A_1 as center and with radius 1500 m. This circle intersects the coastline at O and a new point A_2. We continue this process with each new A_i as center intersecting the coast at A_{i-1} and a point A_{i+1}, until we have gone all the way around (if we are on a large island) or we reach the end of the territory whose coastline concerns us. If O were not at the edge of a territory, we would also need to use point B_1 and generate points B_i in the direction of B_1 just as we generated the points A_i.

Figure 80

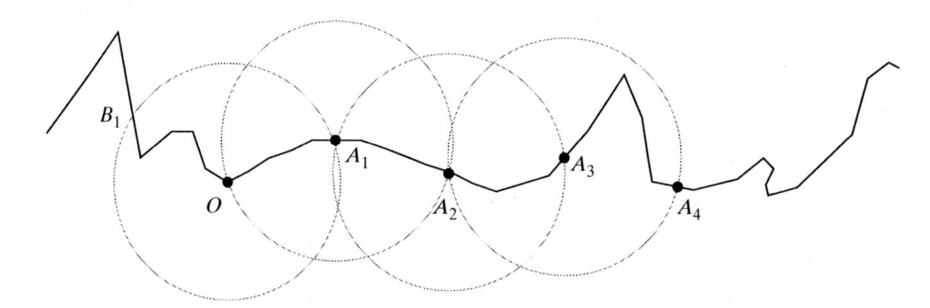

When we have finished covering the coastline, we count the number of spaces between stakes we have put down, and we arrive at a length of the coastline *for the unit of 1500 meters.*

But suppose we chose a unit of 300 meters. If we were measuring the length of a straight road, we would get 5 times the number of 300-meter intervals as we got for the 1500-meter interval. But a coastline is not a road. It goes in and out, and using 300-meter intervals we go in and out more than we did with the 1500-meter intervals. It would not take much of a disruption in the coastline for there to be six 300-meter intervals for each 1500-meter interval, on the average (Figure 81).

Figure 81

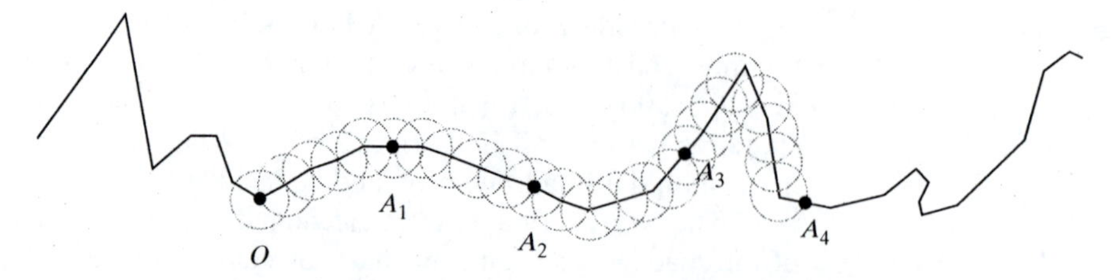

In such a case, if the coastline was 210 units long for the unit of 1500 meters, then it would be 1260 units long for each unit of 300 meters. With the unit 1500 meters long, the coastline would be $210 \cdot 1500$, or 315,000 meters long. But with the unit 300 meters long, the coastline would be $1260 \cdot 300$, or 378,000 meters long. Instead of

$$5 \text{ 300-meter units} = 1 \text{ 1500-meter unit}$$

we have

$$6 \text{ 300-meter coastline units} = 1 \text{ 1500-meter coastline unit.}$$

The dimension d of the coastline units is found by solving $5^d = 6$. The solution to this equation is $\log_5 6$, or $\frac{\log 6}{\log 5}$, or about 1.113. This hypothetical coastline would have dimension about 1.113.

The Koch curve

Study of the *Koch curve* enables us to connect the ideas of similarity, change of units, and dimension. The **Koch curve** is the limiting boundary of a sequence S_n of plane figures created by the following iterative procedure.

> S_1 is an equilateral triangle (Figure 82a).
>
> S_{n+1} is created from S_n in the following way. Divide each side of S_n into thirds. On the middle third, construct an equilateral triangle outward from the existing figure, then remove the middle third. The nonconvex polygon S_{n+1} created in this way has 4 times as many sides as S_n, and each side is $\frac{1}{3}$ the length of S_n. Figure 82b pictures S_2; Figure 82c pictures S_3.

Figure 82

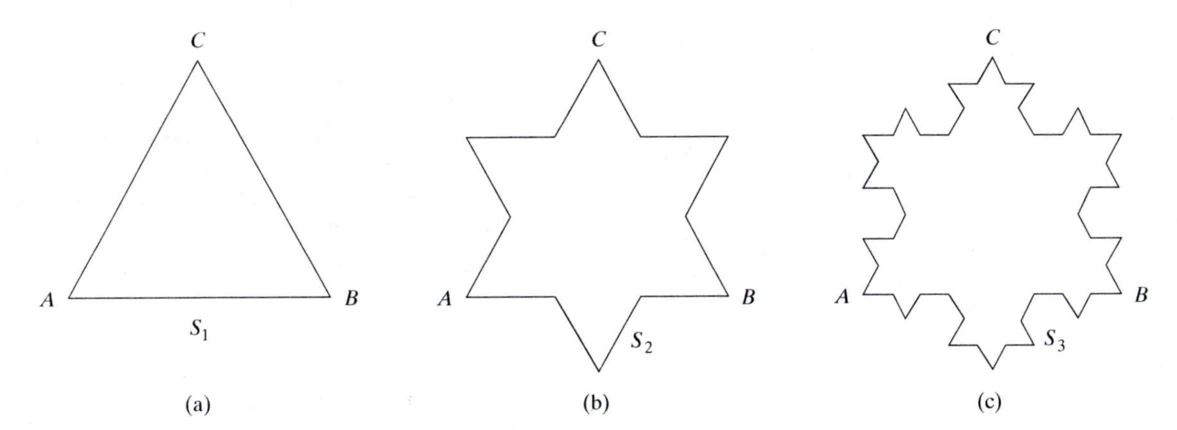

Viewed as a 1-dimensional object, the Koch curve has an interesting property. If S_1 has sides of length 1, then the perimeter of S_1 is 3. From the way that S_{n+1} is created from S_n, S_{n+1} has 4 times the number of sides as S_n, and each side of S_{n+1} is $\frac{1}{3}$ the length of a side of S_n. So the perimeter of S_{n+1} is $\frac{4}{3}$ the perimeter of S_n. The sequence of perimeters of $S_1, S_2, \ldots$ begins $3, 4, \frac{16}{3}, \frac{64}{9}, \ldots$. It is a geometric sequence with first term 4 and constant ratio $\frac{4}{3}$, and has nth term $3 \cdot \left(\frac{4}{3}\right)^{n-1}$. This sequence grows without bound, indicating that the limit, the Koch curve, has infinite perimeter. On the other hand, the Koch curve encloses a finite area that can be calculated (see Problem 1b).

The seemingly paradoxical situation of a curve enclosing a finite area but having an infinite length can be explained by the following argument that the Koch curve is not a 1-dimensional figure. Think of the Koch curve as a coastline whose dimension is not known. First use a unit of length 1 to measure the curve, starting at any vertex of the equilateral triangle. With this unit, the curve has length 3 because all of the infinitely many ins and outs are missed by such a large unit.

Next use a unit of length $\frac{1}{3}$. This unit picks up the ins and outs of S_1 but none of the ins and outs of any of the later curves. With this unit, the perimeter is 12, so the total length is 4. Continuing this process, each time picking a unit $\frac{1}{3}$ the size of the previous unit, the perimeter is multiplied by 4, instead of being multiplied by 3 as it would if this were a normal 1-dimensional curve. The situation can be thought of as follows:

$$3 \text{ small units of length} = 1 \text{ large unit of length}$$
$$4 \text{ small units of Koch curve} = 1 \text{ large unit of Koch curve.}$$

Thus the dimension of the Koch curve is the solution to the equation $3^d = 4$, from which $d = \frac{\log 4}{\log 3} \approx 1.26$.

The Koch curve, like many curves of fractional dimension, is *self-similar*, meaning that if a small part of the curve is blown up, the result cannot be distinguished from the original. Specifically, if a part of the Koch curve is blown up 3 times, then it looks like the (new) picture is merely covering 3 times as much as the original curve. When the whole curve is blown up 3 times, however, if its boundary is measured as a coastline is, then the boundary is found to be 4 times what it was. Again this tells us that the dimension d satisfies $3^d = 4$.

The Koch curve is an example of a *fractal*, a term coined by Benoit Mandelbrot in 1976. Since Mandelbrot's path-breaking book, a large literature about fractals has appeared. The study of fractals involves ideas from virtually every branch of mathematics: complex numbers, analysis, topology, geometry, random processes. While not always accessible with elementary mathematics, some of the pictures created using fractals are among the most beautiful ever created with mathematics.

10.3.4 Problems

1. a. In the sequence S of figures whose limit is the Koch curve, explain why the perimeter of S_{n+1} is $\frac{4}{3}$ the perimeter of S_n.

   b. In the sequence S of figures whose limit is the Koch curve, explain why $\alpha(S_{n+1}) = \alpha(S_n) + \frac{3\sqrt{3}}{16} \cdot \left(\frac{4}{9}\right)^n$. Then use this to determine the area enclosed by the Koch curve.

2. Suppose S_1 is a square, and S_{n+1} is created from S_n by dividing each edge of S_n into fifths and constructing a square outward from S_n on the 2nd and 4th fifth, then removing these fifths.

   a. Find $\lim_{n \to \infty}$ (perimeter of S_n).

   b. Find $\lim_{n \to \infty} \alpha(S_n)$.

   c. What is the dimension of S_n?

Chapter Projects

1. **Areas of regions with parabolic boundaries.** Consider a point P inside a parabola, and lines through this point. Each line cuts off a region inside the parabola. Does the midpoint principle (Section 10.1.3) apply to the problem of finding the line that cuts off the smallest area? (Use any method to explore this question.) Answer this question for points inside other curves: circle, ellipse, hyperbola.

2. **The method of exhaustion.** Investigate Archimedes' development of the method of exhaustion, and demonstrate its connection to integral calculus.

3. **Quadrature of the parabola.** Archimedes was able to find the area of a region between a segment and a parabola. This *quadrature of the parabola* was one of the developments that later mathematicians utilized in the evolution of calculus. Find a source that explains how Archimedes did this, and rewrite his method in your own words.

4. **Area under a cycloid.** In 1637, Gilles Persone de Roberval (1602–1675) devised an elegant geometric proof that the area under one arch of the cycloid

   (1) $x = a(t - \sin(t)) \qquad y = a(1 - \cos(t))$

   is $3\pi a^2$ by applying Cavalieri's Principle twice. Complete the following parts to see how Roberval's proof works.

   a. Use a graphing utility to plot the first half of the arch of the cycloid described in (1), the left half of the circle generating that cycloid,

   (2) $x(t) = -a\sin(t) \qquad y(t) = a(1 - \cos(t))$,

   and the curve with parametric equations

   (3) $x(t) = -at \qquad y(t) = a(1 - \cos(t))$,

   all for $a = 1$, on the same coordinate axes.

   (Roberval referred to the curve (3) as the *companion* of the cycloid (1).)

   b. At any given value t in the plotting interval $0 < t < \pi$, explain why the corresponding points P_s on the semicircle, P_c on the companion curve, and P_t on the cycloid are all on the same horizontal line L_t, and that the distance between P_s and the y-axis is equal to the distance between the points P_c and P_t.

   c. Use part **b** and Cavalieri's Principle to conclude that the area between the companion curve and the cycloid is $\frac{\pi a^2}{2}$.

d. Consider the rectangle R joining the points $(0,0)$, $(\pi a, 0)$, $(\pi a, 2a)$ and $(0, 2a)$. Prove that for any given t with $0 < t < \pi$, the point P_t is the same distance from the right side of R as the point $P_{\pi-t}$ is from the left side of R.

e. Use part **d** and Cavalieri's Principle to conclude that the companion curve (3) divides the rectangle R into two regions of equal area.

f. From parts **c** and **e**, show that the area under one arch of the cycloid (1) is $3\pi a^2$.

5. Squaring the circle. One of the classic problems of antiquity was "squaring the circle". Geometers sought to construct a square with area equal to that of a given circle, using straightedge and compass. It was not proved until the 19th century that, with those tools, the problem cannot be solved. Investigate the history of this problem, and write an essay explaining why it cannot be solved and then describing how it can be solved, if the restriction of using the tools of antiquity is removed.

6. A sphere inscribed in a cone.

a. Given a right circular cone of a given size and shape, what is the radius of a sphere inscribed in the cone?

b. How does the ratio of the sphere's volume to the cone's volume vary with the shape of the cone?

c. What shapes give the largest and the smallest ratios?

d. Sketch the cone with the largest ratio.

e. Prove that this is the shape of cone for which the inscribed sphere has diameter equal to exactly half the height of the cone.

7. Constructible numbers. The first three of Euclid's five postulates (Section 7.1.1) refer to geometric constructions. They establish the rules of constructions followed by Greek mathematicians.

1. Given two points P and Q in the plane, the line $\overleftrightarrow{PQ}$ can be constructed.

2. Given a point P and a given line segment $\overline{RS}$, a circle can be constructed with center P and radius RS.

The instruments used for these constructions are (1) the straightedge and (2) the compass. A point may be used as a given point in a construction if and only if it is given or is an intersection point of constructed figures, that is, an intersection point of lines and/or circles. (This means that in a classical Greek construction, you cannot merely open a compass to any radius.) A **Euclidean construction** of a geometric figure F is an algorithm that begins with given geometric objects (points, lines, line segments, triangles, circles, etc.) and proceeds in a finite number of allowable compass and straightedge constructions to the figure F. A real number c is **constructible** if and only if a line segment of length $|c|$ can be obtained from a given line segment of length 1 by a Euclidean construction.

a. Given that a and b are constructible numbers, prove that the numbers $a+b$, $a-b$, ab, $\frac{a}{b}$, and $\sqrt{|a|}$ are constructible.

b. Let E be the set of all constructible numbers. Prove that $\langle \mathrm{E}, +, \cdot \rangle$ is a field containing both the subfield $\langle \mathbf{Q}, +, \cdot \rangle$ of rational numbers and the subfield $\langle \mathbf{Q}(\sqrt{\mathbf{a}}), +, \cdot \rangle$ for any constructible number a that is not the square of a rational number.

c. Suppose that P and Q are points in the plane whose coordinates are in E and that r is in E. Prove that $\overleftrightarrow{PQ}$ and the circle with center P and radius r have equations whose coefficients are in E. Call such lines and circles *constructible lines* and *constructible circles*.

d. Prove that the coordinates of all points of intersection of two constructible lines, or a constructible line and a constructible circle, or two constructible circles are in E.

e. Explain why every constructible number is an algebraic number.

8. Duplicating the cube. Another classic problem of antiquity was "duplicating the cube". Geometers sought to construct a cube with volume equal to twice that of a given cube, using straightedge and compass. It was not proved until the 19th century that with those tools, the problem cannot be solved. Solutions to the problem, without the restriction to the tools of antiquity, were found by Erastosthenes, Menaechmus, and Nicomedes, among others. Investigate the history of this problem, and explain why it cannot be solved using straightedge and compass.

9. Measuring water in a container. The goal of this project is to construct three different scales on the side of an open cone-shaped container that measure in three different ways the amount of water in the container. For definiteness, begin with a cone with radius $r = 3.25$ cm, height $h = 21$ cm, and slant height $s = 21.25$ cm (Figure 83).

Figure 83

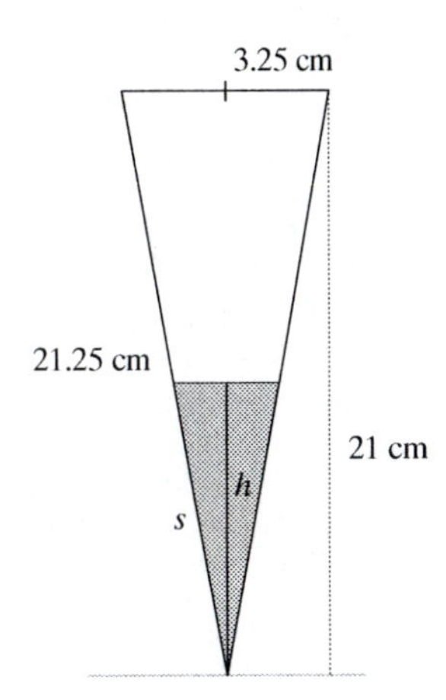

a. *Depth scale:* On a strip of paper 21.25 cm long, make a "depth scale" that can be pasted to the slanted side of the cone. The problem is to do this in such a way that a mark s on the scale measures the depth h of the water in the cone, with a mark for every centimeter of depth. Describe the functional relationship between s and h.

b. *Volume scale:* On a second strip of paper 21.25 cm long, make a "volume scale" that can be pasted to the slanted side of the cone. Do this in such a way that a mark s on the scale measures the volume V of the water in the cone, with a mark for every 10 mL of volume. Describe the functional relationship between s and V.

c. *Rainfall scale:* On a third strip of paper 21.25 cm long, make a "rainfall scale" that can be pasted to the slanted side of the cone. The idea here is that during a rainfall of d centimeters the cone will fill to a certain level. Make the scale in such a way that a mark s on the scale measures the amount d of rainfall, with a mark for every centimeter of rainfall. Describe the functional relationship between s and d.

10. The Mandelbrot set. Perhaps the most famous fractal is the Mandelbrot set, the graph of a set of complex numbers.

a. Explain how to determine whether a particular complex number is in the set, and give examples of numbers that are in the set and not in the set.

b. Explain why the Mandelbrot set is an example of a fractal.

c. Describe some of the properties of the Mandelbrot set.

Bibliography

Unit 10.1 References

Bailey, David H., Jonathan M. Borwein, and Peter B. Borwein. "Ramanujan, Modular Equations, and Approximations to Pi, *or* How to Compute One Billion Digits of Pi." *American Mathematical Monthly* 96(3):(1989) 201–219.

Bailey, David H., Jonathan M. Borwein, Peter B. Borwein, and S. Plouffe. "The Quest for Pi." *The Mathematical Intelligencer* 19(1):(1997) 50–57.
These two articles provide insight into the mathematics and the drama in the search to know ever more decimal places of π. The first article is reprinted in the Berggren reference below.

Beckmann, Petr. *A History of Pi*. Boulder, CO: Golem Press, 1971.
A broad and very readable detailed history.

Berggren, Lennart, Jonathan M. Borwein, and Peter B. Borwein. *Pi: A Source Book*. New York: Springer, 1997.
A compilation of 70 articles on the history and calculation and attraction of π, including most of the important original articles on the developments with respect to π, ranging from attempts by the ancients to estimate π to the most modern calculations of digits of π.

Fauvel, John, and J. Gray. *The History of Mathematics*. London: Macmillan, 1987.
This work examines some of the contributions of history's most renowned mathematicians. It includes proof and discussion of many of Euclid's propositions, including the construction of a parallelogram that has area equal to that of a triangle. Quadrature of circles and parabolas is also discussed.

Galilei, Galileo. *Dialogues Concerning Two New Sciences*. Translated from the Italian and Latin by Henry Crew and Alfonson de Salvi. New York: Macmillan, 1933.
This original text of Galileo discusses the relationships between height, weight, volume, and strength of similar-shaped objects.

Moise, Edwin E. *Elementary Geometry from an Advanced Standpoint*. Reading, MA: Addison-Wesley, 1963.
Chapter 13 defines an area function and then deduces well-known area theorems. Chapter 14 shows that the same area function can be deduced when the area theorems are taken as postulates.

Monte Carlo methods:
<http://csep1.phy.ornl.gov/mc/node1.html>

Unit 10.2 References

Altshiller-Court, Nathan. *Modern Pure Solid Geometry*. New York: Macmillan, 1935.
This text provides an in-depth look at properties of the tetrahedron. Of particular interest are several proven results on volume.

Courant, Richard, Herbert Robbins, and Ian Stewart. *What Is Mathematics*? New York: Oxford University Press. New Edition, 1996.
The text contains history and discussion of the problems of squaring the circle and duplicating the cube.

Unit 10.3 References

Emert, John, and Roger Nelson. "Volume and Surface Area for Polyhedra and Polytopes." *Mathematics Magazine* 70, December 1997, 365–371.
This article explores surface area as the derivative of volume with respect to inner radius. It is shown that this result generalizes to a wide range of n-dimensional figures.

Mandelbrot, Benoit B. *The Fractal Geometry of Nature*. San Francisco: W. H. Freeman, 1982.
This is the major work by the mathematician who coined the term "fractal" and helped to popularize the theory of fractals.

Stein, Sherman. *Archimedes: What Did He Do Besides Cry Eureka?* Washington, DC: Mathematical Association of America, 1999.
This book describes in some detail the work of the person many feel was the greatest mathematician of antiquity. Included are his work with the method of exhaustion and the quadrature of the parabola.

Project Reference

Peressini, Anthony L., and Donald R. Sherbert. *Topics in Modern Mathematics for Teachers*. New York: Holt, Rinehart and Winston 1971.
Chapter 9 of this book gives additional information on Euclidean constructions.

Chapter

11

AXIOMATICS AND EUCLIDEAN GEOMETRY

Before the work of Thales, Pythagoras, Euclid, Archimedes, and other mathematicians of the ancient Greek empire, existing records from Babylonia and Egypt show an understanding of *local deduction*, whereby a proposition is logically deduced from other propositions and principles. For instance, the Babylonians had derived a form of what we now call the Quadratic Formula by using mathematical principles not much different from those we would use today.

The Greek mathematicians were the first to move from local deduction to *global deduction*, in which every proposition is part of the same logical system. The global system that these mathematicians developed 2000 years ago is, aside from language, essentially the same system that we use today to study Euclidean geometry. We are only in recent decades beginning to understand the extent of trade and communication among peoples of Europe, Asia, and Africa before 1000 A.D. Geometry done later in Japan, China, and India seems to have been influenced by Greek geometry to such an extent that we can say that virtually all of today's formal geometry worldwide owes its roots to these mathematicians.

Every mathematical system emanates deductively from (1) *undefined terms*, (2) other terms that are defined from the undefined terms, and (3) *axioms*, assumed relationships among these terms. (In Euclid's *Elements*, both postulates and common notions are axioms.) The *theorems* of the system are the propositions that are logically deduced from the axioms.

In 1882 the German mathematician Moritz Pasch (1843–1930) gave the first axiomatic development of Euclidean geometry that would be considered rigorous by today's standards. By the early twentieth century, a multitude of similar, but different, axiom systems were produced for Euclidean as well as other geometries, by such noted mathematicians as David Hilbert (1862–1943), Mario Pieri (1860–1913), Giuseppe Peano (1858–1932), and Oswald Veblen (1880–1960), to name a few. The creation of new axiom systems for geometry continues even today. In this chapter we discuss the issues involved in creating an axiom system for Euclidean geometry from scratch.

Unit 11.1 Constructing Euclidean Geometry

In Section 7.1.1, we described **Euclidean geometry** as a mathematical system in which the statements that are assumed or can be deduced include all the axioms and propositions that are in Euclid's *Elements*. In trying to construct a rigorous development of Euclidean geometry, the first problem that faced mathematicians working in the era 1882–1910 was to choose undefined terms. Recall from Section 7.1.1 that Euclid had offered definitions for *all* important terms, so his work could not be used as historical precedent to help one decide which terms should be chosen. The only guide was that if a term T could be defined in terms of other terms $P_1, P_2, \ldots, P_n$ that were undefined or previously defined, then T did not need to be taken as undefined.

A quandary faced these mathematicians. How many terms needed to be undefined? Should geometry be separated from the rest of mathematics in the sense that no terms from outside the system (e.g., "number" or "function") would be assumed? Should undefined terms include logical terms such as "and" or "implies"? At this point in the history of mathematics, neither number nor logic had been placed on generally agreed-upon logical footings, so these mathematicians had major decisions to make.

Pasch chose *point, line segment, plane segment*, and *congruence of finite sets*. Peano worked from *point* and *line segment*. Pieri selected *point* and *rigid motion*. Veblen utilized *point* and *order*. Hilbert chose *point, line, plane, between, congruence of segments*, and *congruence of angles* as undefined terms in order to make his system a little more intuitive and easier to use.

In this and the next two sections we create a global system for Euclidean geometry. Our system is most like Hilbert's. We first establish our undefined terms as those pertaining to objects and those concerning relationships we associate with those objects. We take *point* and *line* as the basic *undefined objects* from which other objects of the geometry (segment, triangle, circle, angle, etc.) will be defined. Then we specify *undefined relations*, which, like our undefined objects, are implicitly defined through the axioms and will be used to define other relations. We want to be able to talk about a point being "on" a line or a circle (i.e., to specify whether a line or a circle contains that point). We should be able to identify some sort of order of points on a line, recognizing which points are "between" other points. Finally, we need to be able to compare the objects of our geometry by identifying which are equal or "congruent" to one another. To this end, we specify that *on, between*, and *congruent* are the undefined relations of our geometry, and these, along with the undefined objects, will be used to create the axioms and definitions we will need to prove theorems.

11.1.1 Axioms for incidence

We can classify the axioms that define Euclidean geometry into five categories: incidence, betweenness, congruence, continuity, and the parallel postulate. As you read, take the time to examine these axioms carefully, and use your intuition about the Euclidean geometry you studied in high school to convince yourself that they are indeed statements you can accept without proof. It is very helpful to draw diagrams to be sure that you understand exactly what each axiom allows you to say about the objects of geometry. We generally omit diagrams in this section to show that the propositions follow logically from the system without regard to any picture that may be drawn. We are careful only to use our undefined terms and relations (and any definitions derived from them) to create the axioms of our system. However, we do assume the familiar relations of sets and logic, including set membership, equality, implication, and the (natural) number of elements of a finite set.

Incidence axioms

The first axioms are called *incidence axioms* because they describe incidence properties of points and lines (i.e., what precisely it means for a point to be "on" a line, or for a line to be "on" a point). The relation of incidence is symmetric: If a point is incident with a line, then the line is incident with the point. In this case, we may say that "ℓ is on P" or "ℓ contains P" and that "P is on ℓ" or "P belongs to ℓ". The following incidence axioms for the Euclidean plane implicitly define the undefined relation *on*.

Axioms

Incidence:

I-1: There exist at least three distinct points.

I-2: For each two distinct points there exists a unique line on both of them.

I-3: For every line there exist at least two distinct points on it.

I-4: Not all points lie on the same line.

Notice the importance of Axiom I-1. Without it, we do not know if we have any points.

Question 1: Why do Axioms I-1 and I-2 guarantee the existence of (at least) one line?

Axiom I-2 can be reformulated with the familiar statement: Two points determine a line. If A and B are points on line ℓ, we may write $\overleftrightarrow{AB}$ for ℓ, since they determine ℓ.

Notice that Axiom I-3 is not a definition of a line, but simply a statement telling us one of its properties. From this axiom alone, we cannot assume, for example, that a line is infinite, or even that it contains three points. For that, we need another axiom, which we introduce in Section 11.1.2.

Axioms I-1 and I-2 together imply that there exists at least one line. However, it is possible that our geometry consists only of that line. For us to establish that there is more than one line, we need Axioms I-3 and I-4.

What kinds of theorems of Euclidean geometry can be proved using only these incidence axioms? One theorem is the following.

Theorem 11.1 Given a point, there exist at least two distinct lines on it.

Proof: Let P be a point (I-1). By I-4, there exists at least one line ℓ not on P. There are at least two distinct points on ℓ (I-3), call them A and B. So $\overleftrightarrow{AB}$ does not contain P. By I-2, A and P determine a line. $\overleftrightarrow{AP}$ is not $\overleftrightarrow{AB}$ because P is not on $\overleftrightarrow{AB}$. Similarly, B and P determine a line that is not $\overleftrightarrow{AP}$ or $\overleftrightarrow{AB}$. Consequently, there are at least two lines on P. ┘

When two or more lines are on the same point, we say that these lines *intersect* in that point.

Geometries satisfying the incidence axioms that are not Euclidean

Axioms I-1 to I-4 are sufficient to prove there are at least three lines in the geometry (see Problem 3b). Also, these axioms appear intuitively true for Euclidean geometry. Can we now conclude that with *point, line,* and *on* as undefined terms, and with

Figure 1

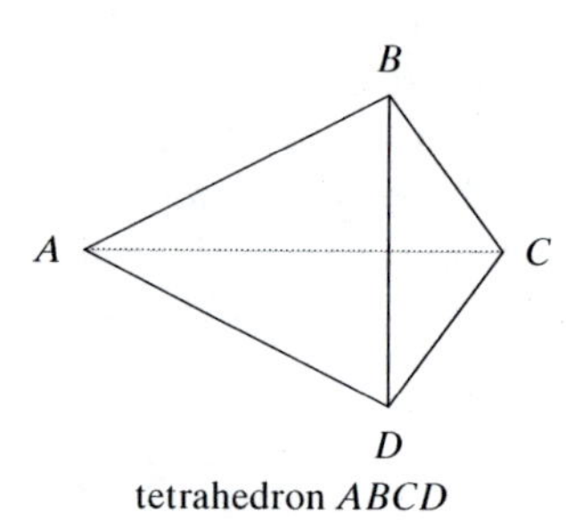

tetrahedron $ABCD$

the incidence Axioms I-1 to I-4, we will be able to deduce all the theorems of Euclidean geometry? The answer is no. One way we can confirm this is to create a *model* (or interpretation) of these four axioms that is not Euclidean.

One such model has the four vertices of a tetrahedron $ABCD$ as its *points*, and the six edges of the tetrahedron $\{A, B\}$, $\{A, C\}$, $\{A, D\}$, $\{B, C\}$, $\{B, D\}$, and $\{C, D\}$ as its *lines* (Figure 1).

(Convince yourself that Axioms I-1 to I-4 hold for this model.) Since for Euclidean geometry, we need infinitely many points, this model is not Euclidean.

Fano's geometry

Another model of Axioms I-1 to I-4 that is not Euclidean is called **Fano's plane geometry**, after the Italian mathematician Gino Fano (1871–1952).

Let **P** be the set of 7 points: $\{A, B, C, D, E, F, G\}$.

Let **L** be the set of 7 lines: $\{\{A, B, C\}, \{A, D, E\}, \{A, F, G\}, \{B, D, F\}, \{B, E, G\}, \{C, D, G\}, \{C, E, F\}\}$. We use set notation because no order to the points on a line is implied.

We now show that this *algebraic model* satisfies I-1 to I-4:

I-1: There exist at least three distinct points.
Since P contains 7 points, we can select any three to verify the axiom.

I-2: For each two distinct points there exists a unique line on both of them.
When we examine the lines we see that no two points belong to two different lines, and there is a line on every pair. To save space, we write ABC for the line $\{A, B, C\}$.

A, B ABC	B, C ABC	C, D CDG	D, E ADE	E, F CEF	F, G AFG
A, C ABC	B, D BDF	C, E CEF	D, F BDF	E, G BEG	
A, D ADE	B, E BEG	C, F CEF	D, G CDG		
A, E ADE	B, F BDF	C, G CDG			
A, F AFG	B, G BEG				
A, G AFG					

I-3: For every line there exist at least two distinct points on it.
This axiom is easily verified by observation. In fact, there are three distinct points on each line.

I-4: Not all points lie on the same line.
Since F does not belong to ABC, for example, the axiom is verified.

Question 2: Construct a geometric model of this system. Notice that there is nothing in the axioms that imply a line must be drawn straight! One geometric model of this system can be constructed using a "triangle" with an inscribed "circle".

These two models, the first consisting of 4 points and the second with 7 points, are examples of what are called *finite incidence geometries*. The study of finite incidence geometries was initiated by Fano in 1892, when he introduced a three-dimensional geometry with 15 points and 35 lines, in which each plane is on 7 points.

Question 3: What is the smallest finite geometry that can be created from the incidence axioms? Give an algebraic and geometric model.

All the models introduced in this section, as well as Euclidean geometry, are classified under the general title of **incidence geometries**, i.e., geometries consisting of points and lines with the relation *on*, and that satisfy Axioms I-1 to I-4.

These models show that Axioms I-1 to I-4 are not sufficient to prove that a line contains infinitely many points. We know this is a property of Euclidean lines. In the next section we introduce the required axioms to "fill in" the gaps between and beyond the points of the line given by the incidence axioms.

11.1.1 Problems

1. Create a model of a 3-point geometry that satisfies Axioms I-1 to I-4.

2. a. Create a model of a finite geometry that satisfies only two of Axioms I-1 to I-4.

b. Create a model of a finite geometry that satisfies only three of Axioms I-1 to I-4.

3. Prove the following theorems of Euclidean geometry using the incidence axioms alone.

a. The intersection of two distinct lines is exactly one point.

b. There exist three nonconcurrent lines (i.e., lines that are not on one common point).

c. For every line there is at least one point not on it.

4. Prove the following theorems of Euclidean geometry using the incidence axioms.

a. If a point A is not on a line ℓ containing distinct points B and C, then there is no line containing all three of them.

b. If there is no line containing all three points A, B, and C, then A, B, and C are distinct points.

5. Consider the following axioms, based on the undefined terms *point, line*, and *on*:

F-1: There exists at least one line.

F-2: There are exactly three points on every line.

F-3: Not all points are on the same line.

F-4: There exists exactly one line on any two distinct points.

F-5: There exists at least one point on any two distinct lines.

Show that Fano's 7-point geometry satisfies these axioms.

6. Rewrite the axioms for Fano's plane geometry (Problem 5) interchanging the words "point" and "line". Each of these new axioms is called the **dual** of its counterpart. Create a model for this new axiom set and determine if it also satisfies F-1 through F-5.

7. See Problem 5.

a. Construct a model for a geometry that satisfies F-1 and F-2, but not F-3.

b. Construct another model for a geometry that satisfies F-1 and F-3, but not F-2.

c. Determine whether each of your models is a model of incidence geometry. Explain why or why not.

8. a. Using the axioms for Fano's plane geometry (Problem 5), prove that each point is on exactly three lines.

b. Use Fano's model to answer the following question: A Rotary club with seven new members wants to plan meetings for the new members so they can get to know one another. The club wants each new member to meet with every other new member exactly once, and to have exactly three new members present at each meeting. How can such meetings be arranged, and how many of these meetings will each new member attend?

9. Let **P** be the set of 9 points $\{A, B, C, D, E, F, G, H, I\}$. Let **L** be the set of 12 lines $\{ABC, DEF, GHI, ADG, BEH, CFI, AEI, BFG, CDH, AFH, BDI, CEG\}$. Here, as in this section, we write ABC for $\{A, B, C\}$, etc. Show that this system, with the relation "on", satisfies axioms I-1 to I-4. This 9-point geometry is called **Young's geometry**. It is named after the famous mathematician John Wesley Young (1879–1932), who discovered it.

10. Consider the following model. The set of points is the interior of a given Euclidean circle. The set of lines consists of the open chords of the circle (an open chord is a chord minus its endpoints). Explain why this model satisfies or does not satisfy Axioms I-1 to I-4.

11. Consider the model in which the set of points are those inside a given Euclidean triangle, and the set of lines consists of the open segments joining two points lying on different sides of the triangle (an *open segment* is a segment without its endpoints). Explain why this model satisfies or does not satisfy Axioms I-1 to I-4.

12. Consider the follow axiom set, based on the undefined terms *point, line*, and *on*.

Axiom 1: There are exactly four points.

Axiom 2: For each two distinct points there is a unique line on them.

Axiom 3: There are exactly two points on every line.

Construct a model for this geometry and create and prove one theorem for it.

13. **Desargues's Theorem**, named after the French mathematician Girard Desargues (1596–1660), states that if ΔABC and ΔDEF are so situated that the lines $\overleftrightarrow{AD}$, $\overleftrightarrow{BE}$, $\overleftrightarrow{CF}$ all meet in a point, then the intersections of sides $\overleftrightarrow{AB}$ and $\overleftrightarrow{DE}$, $\overleftrightarrow{BC}$ and $\overleftrightarrow{EF}$, $\overleftrightarrow{CA}$ and $\overleftrightarrow{FD}$ are on the same line.

a. Draw a picture of this theorem in the Euclidean plane, choosing the vertices of ΔABC and ΔDEF so that no pair of corresponding sides is parallel.

b. This theorem is true in Fano's plane geometry of 7 points and 7 lines. Create two triangles in Fano's geometry (assuming a triangle is simply a set of three points not on the same line) that satisfy the hypothesis of Desargues's Theorem, and show that the conclusion holds.

ANSWERS TO QUESTIONS

1. There exist three distinct points by Axiom I-1. Call them A, B, C. Then, by Axiom I-2, there is at least one line $\overleftrightarrow{AB}$.
2. See Figure 2.

Figure 2

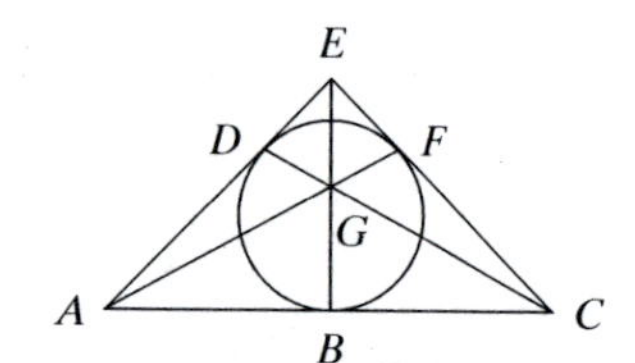

3. I-3 tells us that there exist at least three distinct points. Call them A, B, and C. By I-1 there exists a unique line on each pair of them. So there are at least three lines: $\overleftrightarrow{AB}$, $\overleftrightarrow{AC}$, and $\overleftrightarrow{BC}$. These lines satisfy the conditions of I-2: For every line there exist at least two distinct points on it. And I-4 is satisfied because B is not on $\overleftrightarrow{AC}$. So the smallest plane incidence geometry consists of the set of points $P = \{A, B, C\}$ and the set of lines $L = \{\overleftrightarrow{AB}, \overleftrightarrow{AC}, \overleftrightarrow{BC}\}$. A geometric model of this geometry consists of the vertices of triangle ABC.

11.1.2 Axioms for betweenness

We have seen that the incidence axioms alone do not imply that there are infinitely many points on a line. To achieve this, as well as to enable us to "order" points on a line, we introduce *betweenness* of points, and we specify the set of axioms that reveal the properties of this concept.

The betweenness axioms

Suppose a line contains four points A, B, C, and D. Assume A is between B and C, and C is between A and D. What can be concluded about A, B, and D? Suppose these four points were on a circle. Would the conclusion be the same? Euclid's axioms did not provide the means to determine the answers to these questions. The axioms that follow do. They also ensure that a geometry that satisfies them and the incidence axioms will contain infinitely many points. Notice that, just as with the incidence axioms, the only objects and relations we use are those we have accepted as undefined terms. You should draw diagrams to be sure you understand the properties of points that these axioms reveal.

Axiom B-1 is like Euclid's second postulate, which guarantees that a line can be extended "continuously". Axioms B-2 through B-4 ensure that there is an order to points on a line. For this reason some people call these "axioms of order". These axioms also enable us to distinguish lines from closed curves.

Axioms

Betweenness:

B-1: Let A and B be two distinct points. There exist points C, D, and E on line $\overleftrightarrow{AB}$ such that C is between A and B, B is between A and D, and A is between E and B.

B-2: If A, B, and C are points such that B is between A and C, then A, B, and C are distinct and on the same line.

B-3: If A, B, and C are points such that B is between A and C, then B is between C and A.

B-4: If A, B, and C are three distinct points on the same line, then exactly one of the following statements is true: B is between A and C; C is between A and B; or A is between C and B.

Question 1: What do Axiom B-1 and Axiom I-3 have in common?

Axiom B-2 makes the necessary connection between the incidence axioms and the betweenness axioms, relating the concept of *betweenness* to the *incidence* properties of point and line. Without it, we would have two separate theories—one for incidence, and one for betweenness.

Question 2: Axiom I-1 guarantees that there exist at least three distinct points. How many points do we obtain by Axioms B-1 and B-2?

From now on, because of Axioms B-1 and B-2, we do not need to be concerned that there are an insufficient number of points necessary for our theorems. Axiom B-3 says we can symmetrically permute A and C in the relation "B is between A and C", without destroying its validity, but Axiom B-4 says that if we apply certain other cyclic permutations, the validity of the relation is destroyed. Axiom B-4 indicates that all points on a line can be lined up so that, given any three points, one is between the other two.

At this juncture we are able to make certain definitions that would not have been possible before. It is important to realize that definitions are meaningful in a mathematical theory only if the axioms imply that the object being defined actually exists. With B-1 to B-4, for example, we are able to define *segment* because these betweenness axioms indicate that there are points between two given points, and others that are not. We need the concept of segment before we introduce our next betweenness axiom.

Definitions

Let A and B be two distinct points. **Segment AB**, written $\overline{\mathbf{AB}}$, is the set consisting of A and B and all points on the line $\overleftrightarrow{AB}$ that are between A and B. A and B are called the **endpoints** of $\overline{AB}$.

Notice we have defined segment using only the undefined terms of our system: "point", "line", "on", and "between".

Now we are free to incorporate the idea of segment into the definitions and axioms that follow. One idea that you have seen on occasion and that is defined in terms of segments is *convexity*. For instance, we spoke of convex and nonconvex kytes in Section 7.3.1 (Problem 9). Recall that a set S of points is **convex** if and only if whenever A and B are in S, so are all points between A and B. Since many elementary geometry books speak little if at all about convexity, many people think convexity is a frill or nonessential part of the study of geometry. Quite the contrary. A last betweenness assumption, Axiom B-5, involves convexity.

We can define the **plane** to be the set of all points and lines in this geometry. Our last betweenness axiom allows a plane to be partitioned into a line and two other parts, and is therefore described as a "plane separation" axiom.

Axiom

Plane Separation:

B-5: Every line ℓ partitions the plane into ℓ itself and two convex sets S_1 and S_2 such that if a point P is in S_1 and a point Q is in S_2, then $\overline{PQ}$ and ℓ have a point in common.

We say that S_1 and S_2 are the **sides** or **half-planes** of the line ℓ, and that they are **bounded** by ℓ. It is also common to say that S_1 and S_2 are **opposite sides of** ℓ. Recall the invalid "proof" in Section 7.1.2 that every triangle is isosceles. Our explanation of the fallacy relied on a careful definition of sides of a line that the plane separation axiom and the above definitions provide.

The **triangle with vertices A, B, C,** denoted **ΔABC**, is defined to be the union of the segments $\overline{AB}$, $\overline{BC}$, and $\overline{AC}$, its **sides**. With Axiom B-5, we are able to prove a statement based on an axiom introduced by Pasch in 1882. This statement essentially states that a line that goes into a triangle must come out.

Theorem 11.2 **(Postulate of Pasch):** If line ℓ different from $\overleftrightarrow{AB}$ intersects side $\overline{AB}$ of ΔABC in a point between A and B, then exactly one of the following holds: (1) ℓ contains C, (2) ℓ contains a point on $\overline{AC}$ between A and C, or (3) ℓ contains a point on $\overline{BC}$ between B and C.

Proof: Let line ℓ, different from $\overleftrightarrow{AB}$, intersect side $\overline{AB}$ in a point X between A and B. Then ℓ cannot also contain A or B because it would then equal $\overleftrightarrow{AB}$. Since ℓ intersects $\overleftrightarrow{AB}$ between A and B, A and B are on opposite sides of ℓ. Now, by the Plane Separation Axiom, either (1) C is on ℓ; (2) C is on the same side of ℓ as A, in which case C is on the opposite side of ℓ as B and so ℓ intersects $\overline{BC}$; or (3) C is on the same side of ℓ as B, in which case C is on the opposite side of ℓ as A, and so ℓ intersects $\overline{AC}$. ∎

Theorem 11.2 fills a significant gap in the deductive reasoning of the *Elements*. Euclid relied in many theorems on visual clues from diagrams, instead of using Pasch's axiom (or one equivalent to it) to validate his proofs.

The Plane Separation Axiom also allows us to order more than three points on a line. Theorem 11.3 and its corollary allow us to establish order for four points on a line. To help understand this theorem and its corollary, we introduce some notation. We write **A-B-C** if and only if B is between A and C.

Theorem 11.3 Let A, B, and C be points on a line with A-B-C. Suppose D is a fourth point on the line such that A-C-D. Then A-B-D and B-C-D.

Proof: Let A, B, C, and D be four distinct points on line ℓ such that A-B-C and A-C-D. There exists a point E not on ℓ. (Why?) Let $m = \overleftrightarrow{EC}$. Then $m \neq \ell$ (because E is on m and not on ℓ) and A, B, and D are not on m (because otherwise $m = \ell$). So $AD \neq m$. (Why?) Since A-C-D, $\overline{AD}$ intersects m at C. Consequently, A and D are on opposite sides of m. Now since A-B-C, $\overline{AB}$ does not intersect m (otherwise C is between A and B). Thus, by Plane Separation, A and B are on the same side of m. Consequently, again by Plane Separation, B and D are on opposite sides of m, and since $\overline{BD}$ intersects m at C, B-C-D.

To show that A-B-D, let $n = \overleftrightarrow{EB}$ and proceed in the same way as with m. We leave the details to you. ∎

A similar proof leads to this corollary. See Problem 4.

Corollary: (1): If A-B-C and B-C-D, then A-B-D and A-C-D.
(2): If A-B-D and B-C-D, then A-B-C and A-C-D.

When points A, B, C, and D satisfy a given condition of the corollary, we can write **A-B-C-D** because we can delete any one of these 4 points and the rest are in the same order.

The betweenness axioms provide us with the infinite lines and order of points on lines we need for Euclidean geometry. Does that mean that these axioms, in conjunction with the incidence axioms, are sufficient to define Euclidean geometry? Again, we can demonstrate they are not by constructing a model of these axioms that is not Euclidean.

Consider the *Beltrami-Cayley-Klein plane* model, which is named for Eugenio Beltrami (1835–1900), Arthur Cayley (1821–1895), and Felix Klein (1849–1929). Beltrami and Klein independently discovered it in 1871. However, it also appeared in an 1859 paper by Cayley.

Figure 3

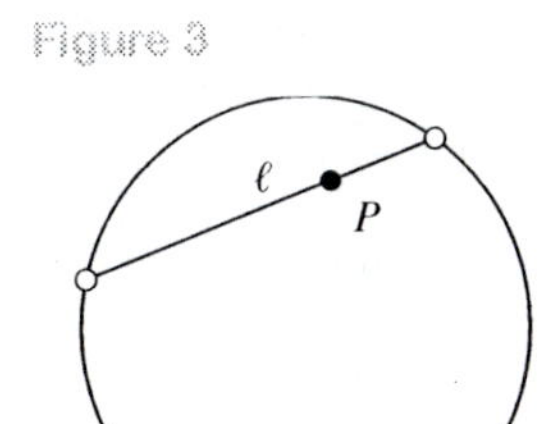

In the **Beltrami-Cayley-Klein plane model**, a point is interpreted as a point interior to a Euclidean circle **C**. A line is any open chord of **C**, i.e., a chord of the circle without its endpoints. Figure 3 shows a point P on a line ℓ in this model. Since the environment of this geometry is the interior of a Euclidean circle, it would appear that this could be a model of Euclidean geometry. We know that the incidence and betweenness axioms hold for Euclidean points and lines. Since chords can be viewed as subsets of lines, it makes intuitive sense that these axioms also hold in this model (see Problem 10 of Section 11.1.1).

Consider, for example, Axiom I-2. To show it holds in the Beltrami-Cayley-Klein plane, we need to show that for any two distinct points A and B in the interior of **C** there is a unique open chord containing them.

Proof of Axiom I-2 for the Beltrami-Cayley-Klein plane: Let A and B be interior to **C**. Let $\overleftrightarrow{AB}$ be the unique Euclidean line on them by Axiom I-2. This line intersects the boundary of **C** in two points. (This is a theorem of Euclidean geometry that ensures a line passing through the interior of a circle will intersect the circle in two distinct points. We discuss it in Section 11.1.3.) Call these points C and D. Then A and B lie on the open chord CD, which by Axiom I-1 for Euclidean geometry, is the only open chord on which they both lie. ∎

The other incidence axioms and betweenness axioms can be likewise verified for this model (see Problems 5 and 6).

Figure 4

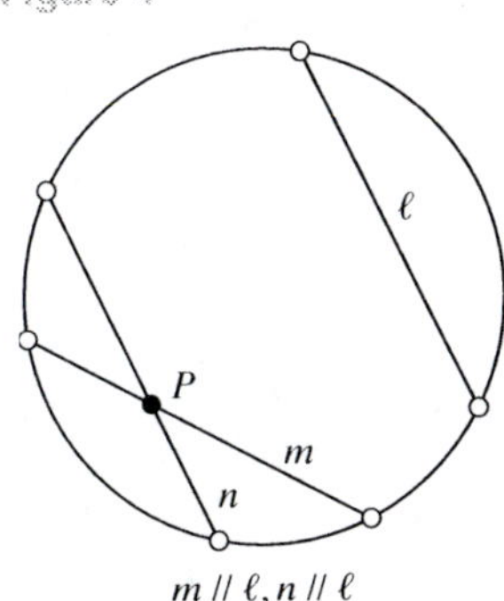

Once the incidence and betweenness axioms are verified, does this mean that the Beltrami-Cayley-Klein plane is a model of Euclidean geometry? No. It turns out that the geometry this model describes is not Euclidean! Here is one reason why (there are many others!): In Euclidean geometry there is only one line parallel to a given line through a point outside of that line. (By "parallel" we mean "nonintersecting".) We can see from Figure 4 that it is possible to have at least two such lines in the Beltrami-Cayley-Klein plane. In fact, there are infinitely many lines parallel to ℓ through P.

The Beltrami-Cayley-Klein plane is a model of a *non-Euclidean geometry*. Such geometries are infinite incidence geometries that share certain Euclidean axioms but that replace the Euclidean parallel postulate, which we discuss in Section 11.1.4, with an axiom of multiple parallels, or with an axiom that denies the existence of parallels. The Beltrami-Cayley-Klein plane is a model of the non-Euclidean geometry that is called *hyperbolic* or *Bolyai-Lobachevskian*, named for the Russian mathematician Nikolai Ivanovich Lobachevsky (1792–1856) and the Hungarian mathematician Jànos Bolyai (1802–1860), who first discovered it.

In hyperbolic geometry, the incidence and betweenness axioms I-1 to I-4 and B-1 to B-5 hold. For this reason, hyperbolic geometry is considered, as is Euclidean geometry, an *ordered incidence geometry*. But there must be certain axioms that distinguish hyperbolic from Euclidean geometry since it is non-Euclidean. From the discussion above, we can see that the Euclidean parallel postulate is false in hyperbolic geometry. Are there other Euclidean propositions that do not hold in hyperbolic geometry? Yes, in hyperbolic geometry, if two triangles are similar, then they are congruent. We next explore the congruence axioms of Euclidean geometry to determine if these are sufficient to categorize it, and perhaps further distance it (or not!) from hyperbolic geometry.

11.1.2 Problems

1. a. Which betweenness axiom guarantees that $\overline{AB}$ contains at least three points?
   b. Explain why $\overline{AB}$ represents the same set of points as $\overline{BA}$.
2. Rewrite Axioms B1 to B4 using the *A*-*B*-*C* notation for betweenness of points.
3. Complete the proof of Theorem 11.3.
4. Prove the Corollary to Theorem 11.3.
5. Verify incidence Axioms I-1, I-3, and I-4 in the Beltrami-Cayley-Klein model of hyperbolic geometry.
6. Consider the following analytic description of the Beltrami-Cayley-Klein plane: A point is an ordered pair (x, y) such that x and y are real numbers and $x^2 + y^2 < 1$. A line is any nonempty set of points (x, y) that satisfies some equation $ax + by + c = 0$, such that a and b are real numbers not both zero. Verify betweenness Axioms B-2 to B-5 for this model.
7. Let ℓ be an arbitrary line in a given plane. Prove that ℓ determines exactly two distinct half-planes.
8. Look up Euclid's proof of his Proposition 21 in Book I. Determine where in the proof there is a statement that requires Pasch's axiom.
9. Interpret points to be real numbers, and lines to be sets of real numbers.
   a. If the betweenness relation is interpreted in terms of $<$ (i.e., B is between A and C means $A < B < C$, or $C < B < A$), show that the betweenness Axioms B1–B4 are verified.
   b. If the betweenness relation is interpreted in the following way: B is between A and C means that $A \neq C$ and there exist positive real numbers x and y such that $B = Ax + Cy$, where $x + y = 1$, show the betweenness Axioms B1–B4 are verified.
10. Can the betweenness axioms hold in a finite incidence geometry? Support your answer by either showing they are verified in one of the finite incidence geometries discussed in Section 11.1.1, or explaining why they do not.

ANSWERS TO QUESTIONS

1. They are both existence axioms.
2. These axioms provide us with the means to generate an infinity of points on a line. We know by Axiom I-1 there exist three distinct points. Call them A, B, and C. By Axiom B-1, for distinct points A and B, we generate additional points D between A and B, E such that B is between A and E, and F, such that A is between F and B. These points are distinct from each other and from A and B by Axiom B-2. Now we can continue this process choosing distinct points in pairs and generating other points on the line.

11.1.3 Congruence and the basic figures

In Chapter 7, we detailed Euclid's treatment of congruence. We remarked that Euclid's approach to congruence assumed without proof that one figure could be superimposed onto another. To correct that weakness and also to allow the geometry of congruence to apply to all figures, we showed how congruence can be developed through transformations. We used synthetic, coordinate, and complex number descriptions of these transformations to deduce properties of congruence. Clearly, along the way we were utilizing a great deal of knowledge about Euclidean geometry, about functions, and about number.

A pure synthetic approach

We called one of our earlier approaches *synthetic* because it did not use coordinates or complex numbers. However, we did utilize numbers for distance and angle measure, and we defined congruence in terms of distance. So it was not a *pure* synthetic approach to congruence. The use of numbers in an approach which is otherwise synthetic is the most common current practice, and it is quite efficient and effective because it allows everything we know about numbers to be used. However, it leads many students to believe that geometry cannot be divorced from numbers.

A pure synthetic approach does not allow numbers at all. In this section we exhibit such an approach. We start from scratch about congruence, assuming only what we have explicitly mentioned in Sections 11.1.1 and 11.1.2. These axioms, together with the congruence axioms we introduce in this section, correct the defects in Euclid's approach to congruence and enable all his theorems about congruence to be deduced. Our approach here also shows how some terms previously defined in terms of number (e.g., supplementary angles and midpoints of segments) can be defined without number. We show how to examine two angles or two segments and determine that one is bigger than another without using numbers. We also treat triangle congruence without using numbers. Along the way, you will see many definitions of terms that you have been using in earlier chapters. Some are the same as in previous chapters and are repeated here so that this unit can be self-contained. Others are different due to the approach we are taking but are still not contradictory to definitions in earlier chapters.

Congruence axioms

Remember, in addition to *on* and *between*, we take the relation *is congruent to* (denoted "$\cong$") as an undefined term. In the following definitions and axioms, we develop the properties of congruent segments and angles. We first need to introduce the concept of *ray*. These definitions are the familiar ones.

Definitions

Given distinct points A and B, **ray AB**, written $\overrightarrow{\mathbf{AB}}$, is the set of points of the segment $\overline{AB}$ together with all points C on line $\overleftrightarrow{AB}$ such that B is between A and C. A is called the **vertex** (or **endpoint**) of the ray. $\overrightarrow{\mathbf{AB}}$ and $\overrightarrow{\mathbf{AC}}$ are **opposite rays** if A is between B and C.

Like a segment, a ray is convex. And, from their definitions, both segments and rays are subsets of the lines containing them. Rays can therefore be related to half-planes, or sides of a given line in the following way: If A is a point on a line ℓ and B is a point not on ℓ, then ray $\overrightarrow{AB}$ includes point A and exactly those points on line $\overleftrightarrow{AB}$ on the same side (half-plane) of ℓ as B. The only other points on $\overleftrightarrow{AB}$ are the points other than A on the ray $\overrightarrow{AC}$ opposite ray $\overrightarrow{AB}$. These points are on the opposite side of ℓ as B. We say that line ℓ **separates** line $\overleftrightarrow{AB}$ at A.

Segment congruence

Three axioms treat congruence of segments. Again you should draw diagrams to convince yourself that these axioms make sense to you.

Axioms

Segment Congruence:

C-1: Let A and B be distinct points. If C is any point, then for each ray r with vertex C, there exists a unique point D on r such that D is distinct from C and $\overline{AB}$ is congruent to $\overline{CD}$, written $\overline{AB} \cong \overline{CD}$.

C-2: Segment congruence is reflexive, symmetric, and transitive. That is, segment congruence is an equivalence relation.

C-3: Let B be between A and C, and E be between D and F. If $\overline{AB} \cong \overline{DE}$ and $\overline{BC} \cong \overline{EF}$, then $\overline{AC} \cong \overline{DF}$.

Question 1: Which of Euclid's common notions, if any, do each of the congruence axioms C-1, C-2, and C-3 replace?

Axiom C-1 gives justification for the familiar operation of "laying off" a segment on a ray. Intuitively, it tells us we can "move" the segment along the ray. Axiom C-3, in effect, tells us that if we "add" congruent segments, the sums are congruent. Notice that we are able to specify this axiom without introducing the concept of distance or length. From it we are able to prove the corresponding "segment subtraction" property.

Theorem 11.4 Let B be between A and C, and E be between D and F. If $\overline{AC} \cong \overline{DF}$ and $\overline{AB} \cong \overline{DE}$, then $\overline{BC} \cong \overline{EF}$.

Proof: Using Axiom C-1, let C' be the point on ray $\overrightarrow{BC}$ such that $\overline{BC'} \cong \overline{EF}$. Using this and the given $\overline{AB} \cong \overline{DE}$, $\overline{AC'} \cong \overline{DF}$ by Axiom C-3. But $\overline{AC} \cong \overline{DF}$ is given. So by Axiom C-2, $\overline{AC} \cong \overline{AC'}$. Since A-B-C, C is on $\overrightarrow{AB}$. We would like to show that $C = C'$. For this, we need to show that C' is on $\overrightarrow{AB}$. Since C is on $\overrightarrow{BC}$ either (1) B-C-C', (2) B-C'-C, or (3) $C = C'$. (1) If B-C-C', then by Theorem 11.3, A-B-C', so C' is on $\overrightarrow{AB}$; (2) if B-C'-C, then by the Corollary to Theorem 11.3, A-B-C', so C' is on $\overrightarrow{AB}$. So in all cases C' is on $\overrightarrow{AB}$. Because C and C' are on the ray $\overrightarrow{AB}$, by Axiom C-1, $C = C'$. Consequently, $\overline{BC} \cong \overline{BC'}$, and so using Axiom C-2 once again, $\overline{BC} \cong \overline{EF}$. ┘

With a suitable definition of "greater than", we can order segments by their size even without numbers!

Definition **Segment $\overline{CD}$ is greater than segment $\overline{AB}$**, written $\overline{CD} > \overline{AB}$, if and only if there exists a point E between C and D such that $\overline{CE} \cong \overline{AB}$.

We can now prove an important order relation for segments.

Theorem 11.5 If $\overline{CD} > \overline{AB}$ and $\overline{AB} \cong \overline{EF}$, then $\overline{CD} > \overline{EF}$.

Proof: If $\overline{CD} > \overline{AB}$, then by definition, there exists a point X between C and D such that $\overline{CX} \cong \overline{AB}$. If $\overline{AB} \cong \overline{EF}$ and $\overline{CX} \cong \overline{AB}$, then by Axiom C-2, $\overline{CX} \cong \overline{EF}$. Thus X is a point between C and D such that $\overline{CX} \cong \overline{EF}$. Therefore, by definition, $\overline{CD} > \overline{EF}$. ┘

We ask you to prove other order relations for segments in Problem 11.

With Axioms C-1 to C-3, we can now define some other important objects of Euclidean geometry in terms of congruence, and distance and length are not involved.

Definitions Let O and A be distinct points. The set of all points B such that $\overline{OB} \cong \overline{OA}$ is the **circle** with center O. Each of the segments $\overline{OB}$ is a **radius** of the circle.

Recall that Euclid explicitly postulated the existence of a circle (see Section 7.1.1), instead of making it the subject of a definition as we have done.

Question 2: How do we know the points B of the definition exist? How many of these points are necessary to determine a circle completely?

It turns out that although such points B exist, there are not yet sufficient axioms to actually prove that given three noncollinear points, a circle exists that contains them. You may be surprised to learn that the Euclidean parallel postulate is needed to prove this.

Angle congruence

From segments we proceed to angles. The concept of betweenness of points leads in a natural way to the concept of betweenness of rays, and that to the concept of angle. We can make the analogy that betweenness of rays is related to angles as betweenness of points is related to segments.

Definition

Let $\overrightarrow{AB}, \overrightarrow{AC}, \overrightarrow{AD}$ be distinct rays such that $\overrightarrow{AB}$ and $\overrightarrow{AD}$ are not opposite. Ray $\overrightarrow{AC}$ **is between rays** $\overrightarrow{AB}$ **and** $\overrightarrow{AD}$ if and only if there exist points X, Y, and Z, such that X belongs to $\overrightarrow{AB}$, Y belongs to $\overrightarrow{AC}$, Z belongs to $\overrightarrow{AD}$, and Y is between X and Z.

Because betweenness of rays is defined in terms of betweenness of points, each betweenness axiom for points B1–B4 has a counterpart in the betweenness of rays.

Question 3: What betweenness axiom allows us to conclude that betweenness of rays is symmetric?

It is also possible to prove that if $\overrightarrow{AC}$ is between $\overrightarrow{AB}$ and $\overrightarrow{AD}$, then, other than A,

i. each point of $\overrightarrow{AB}$ and each point of $\overrightarrow{AC}$ are on the same side of $\overleftrightarrow{AD}$, and likewise,
ii. each point of $\overrightarrow{AC}$ and each point of $\overrightarrow{AD}$ are on the same side of $\overleftrightarrow{AB}$,
iii. each point of $\overrightarrow{AB}$ and each point of $\overrightarrow{AD}$ are on opposite sides of $\overleftrightarrow{AC}$.

With the correspondence between betweenness of points and betweenness of rays in mind, we can use many of the results we achieved for betweenness of points to prove theorems about betweenness of rays.

Question 4: Can the following property of betweenness of points be extended to rays? If B is between A and C and C is between B and D, then B is between A and D.

Angles

We now turn to angles. The following definition should be familiar.

Definitions

An **angle** (written $\angle$) is the union of two distinct and nonopposite rays $\overrightarrow{AB}$ and $\overrightarrow{AC}$, called its **sides**. A is the **vertex** of the angle. The angle is denoted as $\angle BAC$ or $\angle CAB$. A point D is in the **interior of** $\angle$**BAC** if D is on the same side of $\overleftrightarrow{AC}$ as B and if D is also on the same side of $\overleftrightarrow{AB}$ as C.

Question 5: Describe the interior of an angle in terms of betweenness of rays and in terms of half planes.

It turns out that just as we can separate a plane by a line, we can separate a plane by an angle, into those points that are in its interior and those that are not.

Some developments of geometry allow *straight angles*, defined as angles that are the union of opposite rays, angles that in an analytic approach have measure 180°. Such angles, when they are allowed, can cause difficulties because they are identical to lines, any point on them can be a vertex, and either side of the line can be the interior of the angle. When Euclid wished to speak of the equivalent of 180°, he used the phrase "two right angles", as you can see from Proposition 32 in the list in Section 7.1.1. Some developments also allow *zero angles*, defined as angles that are the union of identical rays, angles that in an analytic approach have measure 0°. Such an angle would have no points in its interior and would be identical to a single ray. And we have pointed out in Section 7.2.2 that some books speak of *reflex angles*, angles with measures greater than 180°. Allowing reflex angles requires that the union of two rays determines two angles. All of these ideas are possible in a rigorous treatment of geometry, but because they tend to complicate the development, we do not employ them here.

The following axioms are the assumed properties of congruent angles. Notice that they do for angles what Axioms C-1, C-2, and C-3 do for segments. Axioms C-1 and C-4 convey the analytic idea that segments and angles have exactly one measure. In an analytic approach, we could replace Axioms C-2 and C-5 by assuming that congruent segments have the same measure and congruent angles have the same measure. Axioms C-3 and C-6 tell us that if we put together adjacent congruent segments and angles, the results are congruent.

Axioms

Angle Congruence:

C-4: Given $\angle BAC$ and ray $\overrightarrow{DE}$, there exist unique rays $\overrightarrow{DF}$ and $\overrightarrow{DG}$ on different sides of $\overleftrightarrow{DE}$ such that $\angle BAC \cong \angle EDF \cong \angle EDG$.

C-5: Angle congruence is reflexive, symmetric, and transitive. That is, angle congruence is an equivalence relation.

C-6: Let D be in the interior of $\angle ABC$ and E be in the interior of $\angle GHI$. If $\angle ABD \cong \angle GHE$ and $\angle DBC \cong \angle EHI$, then $\angle ABC \cong \angle GHI$.

From Axioms C-4 to C-6, we can deduce "angle subtraction" in a manner similar to the proof of Theorem 11.4.

Theorem 11.6 Let D be in the interior of $\angle ABC$ and E be in the interior of $\angle GHI$. If $\angle ABC \cong \angle GHI$ and $\angle ABD \cong \angle GHE$, then $\angle DBC \cong \angle EHI$.

Proof: The proof is left to you as Problem 12. ┘

The next definition enables us to be able to compare angles that are *not* congruent. It is analogous to the earlier definition in this section comparing segments.

Definition $\angle DEF$ **is greater than** $\angle ABC$, written $\angle DEF > \angle ABC$, if and only if there exists a ray $\overrightarrow{EG}$ such that $\angle ABC \cong \angle DEG$ and G is in the interior of $\angle DEF$.

Terms defined in earlier chapters without reference to number can be used here. Two angles with a common vertex form a pair of **vertical angles** if the sides of one are opposite to the sides of the other. Two angles form a **linear pair** if they have a common side and their noncommon sides are opposite rays. Two angles are **adjacent angles** if they are a linear pair, or if they have a common vertex and a common side that is in the interior of the angle formed by their noncommon sides.

These definitions enable us to define supplementary angles and right angles using congruence and with no reference to angle measure or other numbers. So these definitions are likely to be different from those you have seen before.

Definitions Two angles are **supplementary** and each is a **supplement** of the other if they are respectively congruent to the angles of a linear pair.

A **right angle** is an angle that is congruent to a supplement of itself.

Triangle congruence

Thus far, all of the congruence axioms have dealt with relationships along a line or around a point. We need an axiom which guarantees that congruence in one part of the plane is like congruence in every other part. An axiom that fills this void is *Side-Angle-Side (SAS) congruence for triangles.*

Axiom

Triangle Congruence Axiom (SAS Congruence):

C-7: Given two triangles, if two sides of one are congruent to two sides of the other, and the angles included by the congruent sides are congruent, then their third sides are congruent and their remaining corresponding angles are congruent.

When all six parts (sides and angles) of one triangle are congruent to the corresponding parts of another triangle, we call them **congruent triangles**. Thus, triangle congruence is defined in terms of the undefined terms *segment congruence* and *angle congruence*, and the properties of triangle congruence are known only by what can be deduced from postulates C-1 to C-7.

It follows from the definition of triangle congruence and Axioms C-2 and C-5 that triangle congruence is reflexive, symmetric, and transitive.

Notice that the statement of the SAS Congruence Axiom C-7 is very much like Euclid's statement of the SAS proposition. It does not mention congruent triangles! Rather, it goes directly to the usual reason for proving triangles congruent, namely, to obtain the congruence of the other sides and the other two pairs of corresponding angles.

The SAS Congruence Axiom is very powerful—it connects congruence of angle and congruence of segments, which until this point have, in a sense, been two separate theories. It also enables us to deduce all the other triangle congruence propositions.

You can see how the connection is made between congruent segments and congruent angles in the following theorem. Although its proof seems long, it is quite straightforward, simply involving repeated use of the SAS Congruence Axiom to achieve the result.

Theorem 11.7 Supplements of congruent angles are congruent.

Proof: It is sufficient to consider the case where the pairs of supplementary angles are linear pairs, since (by Axiom C-5) congruence of angles is an equivalence relation. Suppose then that $\angle AOB$ and $\angle AOC$, and $\angle DPE$ and $\angle DPF$ are two linear pairs, and that $\angle AOB \cong \angle DPE$. We show that $\angle AOC$ and $\angle DPF$ are congruent. We can assume that segments $\overline{OA}$, $\overline{OB}$, and $\overline{OC}$ are congruent respectively to $\overline{PD}$, $\overline{PE}$, and $\overline{PF}$ (Axiom C-1). (See Figure 5.) Then $\overline{AB} \cong \overline{DE}$ and $\angle OBA \cong \angle PED$ by SAS Congruence (Axiom C-7). Furthermore, we know that O is between B and C and that P is between E and F (definition of opposite rays). So $\overline{BC} \cong \overline{EF}$ by Axioms C-2 and C-3. Since $\angle ABC \cong \angle DEF$, then again by SAS Congruence, $\overline{AC} \cong \overline{DF}$ and $\angle ACB \cong \angle DFE$. Thus $\angle AOC \cong \angle DPF$ by SAS Congruence. ∎

Figure 5

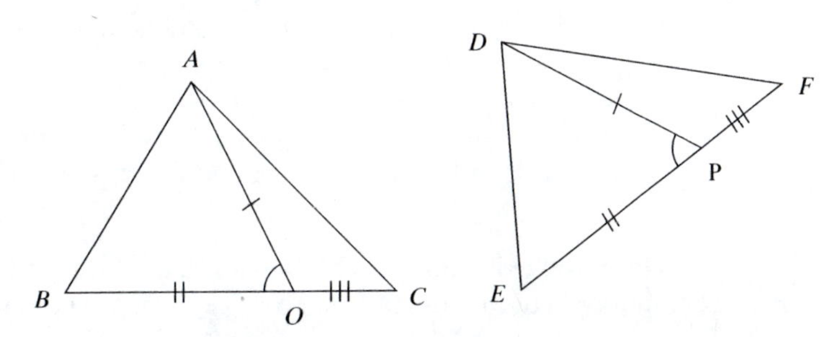

The need for an additional axiom

The incidence, betweenness, and congruence axioms allow the proving of many theorems about circles and polygons. But they do not enable rigorous proofs of some of the basic theorems that are in Euclid's geometry. As an example, consider Euclid's very first theorem and the proof he gave.

Euclid's Theorem 1 An equilateral triangle can be constructed on any segment $\overline{AB}$.

Proof: Draw a circle with radius $\overline{AB}$ and center A, and draw the circle with radius $\overline{AB}$ and center B (see Figure 6). (These can be drawn because of Euclid's Postulate 3.) Let C be a point of intersection of the circles. Now $\overline{AC} \cong \overline{AB}$ (Euclid's definition of "circle" applied to the circle with center A) and $\overline{BC} \cong \overline{AB}$ (Euclid's definition of "circle" applied to the circle with center B). By Common Notion 1, $\overline{AC} \cong \overline{BC}$. And so all sides of triangle ABC are congruent to each other, and the triangle is therefore equilateral. ∎

Figure 6

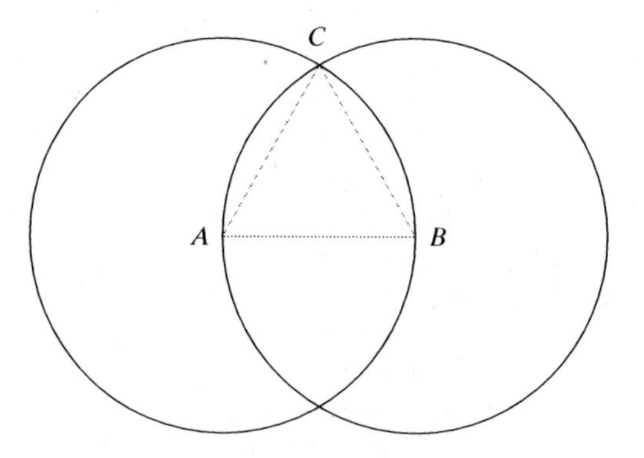

(We note that all of Euclid's proofs are written in paragraph form. The two-column form for writing proofs is a relatively recent invention, traceable back only to the 1890s.)

Question 6: Is Euclid's reasoning sound? Suppose we try to verify his theorem on a plane of points with rational coordinates. Let A be the origin $(0, 0)$ and $B = (0, 1)$. What happens?

Euclid's proof contains justifications for all its statements except one: Let C be a point of intersection of the circles. It happens that no postulate or theorem deducible from Euclid's postulates ensures that the circles intersect. An *axiom of continuity* fixes this gap. This axiom was first proposed by Dedekind.

Axiom **Axiom of Continuity:**

D-1: Let a line ℓ be partitioned into two nonempty subsets of points, S and T, in such a way that no point of S is between two points of T and no point of T is between two points of S. Then there exists a unique point O on ℓ with the following property: For any points A and B on ℓ distinct from O, O is between A and B if and only if A is in S and B is in T, or A is in T and B is in S.

Prior to Dedekind, most mathematicians had based their understanding of the continuity of a line on the fact that between any two points there exists another point

(Axiom B-1). But Dedekind realized that between any two rational numbers there exists a rational number, and yet the rational numbers do not form a continuum. The real numbers, on the other hand, do form a continuum and can be put into one-to-one correspondence with the points of a line.

Dedekind recognized that the continuity of a line can be based on a line separation property, i.e., the fact that a point can partition a line into two sets of points with the property that each point of the line belongs to one and only one set, and all the points of one set are to the left of all the points in the other set. Dedekind's axiom can be thought of as the converse to this line separation property: There is a unique point that creates that partition, separating the points of the line to the left and right of it.

The Axiom of Continuity guarantees that two circles have a point of intersection, but the argument is too long for us to give the details here. Instead, we outline a proof that is detailed in Heath's commentary on Euclid's *Elements*.

1. Call a ray an **interior ray** of an angle if its endpoint is the vertex of the angle and if it is between the sides of the angle. Consider the set I of all interior rays of an angle. Using the Axiom of Continuity, it can be proved that if a ray R in set I partitions I into itself and two subsets S and T so that all rays in S are on one side of the ray R, and all rays in T are on the other side of R, then R is either at the edge of S or at the edge of T.
2. If an angle is a central angle of a circle, the same partition of its interior rays also partitions the intersections of these rays with the circle itself. Consequently, the points of a minor arc can be partitioned just as the rays in (1) can be partitioned.
3. From (2) it can be proved that if a line has one point in the interior and one point in the exterior of a circle, then it has two points in common with the circle.
4. From (1), (2), and (3) it can be proved that if a circle A contains one point X that is in the interior of circle B, and if circle A contains a second point Y that is in the exterior of circle B, then circles A and B must intersect in two points. (See Problem 5.)

11.1.3 Problems

1. Draw a diagram illustrating each statement.

a. Axiom C-1

b. Axiom C-4

c. definition of "is greater than" for angles

2. Using only the language developed in Sections 11.1.2 and this section, define each term.

a. complementary angles

b. obtuse angle

3. How does the definition of triangle in this section differ from Euclid's definition shown in Section 7.1.1?

4. Each of the following terms is defined in this section. Reword that definition using the concepts of distance and/or angle measure.

a. circle

b. supplementary angles

c. right angle

5. a. Using the concept of betweenness and without using the concept of distance, define the interior of a circle and the exterior of a circle.

b. Use the axioms we have assembled so far to explain why there must exist two points X and Y that are in the interior of a circle with center O such that O is between X and Y.

6. Use Theorem 11.2 (the Postulate of Pasch) to prove that no line can contain points on all three sides of a triangle without containing a vertex of the triangle.

7. Use the triangle congruence axiom (C-7) to prove that the base angles of an isosceles triangle are equal. This is a famous theorem called *pons asinorum* or *Bridge of Asses*. (It receives its name from the diagram Euclid used in its proof.)

8. Using the betweenness axioms and the definitions of segment and ray, prove that the intersection of $\overrightarrow{AB}$ and $\overrightarrow{BA}$ is $\overline{AB}$.

9. A student argued that the Plane Separation Axiom B-5 can be converted into the line separation property merely by replacing the word "plane" by the word "line" and "line" by "point". Is the student correct? Why or why not?

10. Prove the following theorem: Suppose $\overline{AB} \cong \overline{CD}$. For any point E between A and B, there exists a unique point F between C and D such that $\overline{AE}$ is congruent to $\overline{CF}$.

11. Prove the following theorems using the segment congruence axioms and the definition of > (greater than) for segments.

a. Exactly one of the following conditions holds: $\overline{AB} > \overline{CD}$, $\overline{AB} \cong \overline{CD}$, or $\overline{CD} > \overline{AB}$.

b. If $\overline{AB} > \overline{CD}$ and $\overline{CD} > \overline{EF}$, then $\overline{AB} > \overline{EF}$.

With Theorem 11.5, these theorems provide for the ordering of segments.

12. Prove Theorem 11.6.

13. Prove the following theorems:

a. If P is in the interior of ΔABC, the line $\overleftrightarrow{AP}$ intersects the line segment $\overline{BC}$.

b. If P is in the interior of ΔABC, any line through P intersects two of the sides of the triangle.

14. Prove the following theorem: Suppose $\angle ABC \cong \angle DEF$. For any point X between A and C, there exists a unique point Y between D and F such that $\angle ABX \cong \angle DEY$. (*Hint:* Use Problem 10.)

15. Use Theorem 11.7 to prove the following theorems:

a. Vertical angles are congruent to each other.

b. An angle congruent to a right angle is a right angle.

16. Use the definition of > for angles to create and prove two theorems on the ordering of angles analogous to those for the ordering of segments in Problem 11.

ANSWERS TO QUESTIONS

1. Axiom C-2 replaces Euclid's first and fourth common notions. Axiom C-3 replaces his second common notion.

2. Axiom C-1 provides for them. We need three points.

3. Axiom B-3.

4. No. Let $\angle AOB$ be a right angle. Let $\angle COD$ be a right angle with $\overrightarrow{OC}$ not in the interior of $\angle AOB$, C and B on the same side of $\overleftrightarrow{OA}$, and C and D on opposite sides of $\overleftrightarrow{OA}$. Then $\overrightarrow{OB}$ is between $\overrightarrow{OA}$ and $\overrightarrow{OC}$, and $\overrightarrow{OC}$ is between rays $\overrightarrow{OB}$ and $\overrightarrow{OD}$, but $\overrightarrow{OB}$ is not between rays $\overrightarrow{OA}$ and $\overrightarrow{OD}$.

5. The interior of $\angle AOB$ is the set of points X such that $\overrightarrow{OX}$ is between $\overrightarrow{OA}$ and $\overrightarrow{OB}$. It can also be described as the intersection of two half planes: the half-plane of B bounded by $\overleftrightarrow{OA}$ and the half-plane of A bounded by $\overleftrightarrow{OB}$.

6. First construct a circle with center $(0, 0)$ and radius 1, then another circle using point $(0, 1)$ as the center and radius 1. These two circles do not intersect in a point with rational coordinates, so Euclid's proof breaks down on the given plane.

11.1.4 Geometry without the Parallel Postulate

All the axioms we have stated so far are valid propositions in both Euclidean and hyperbolic geometry. In the next section we add a final postulate that is not true in hyperbolic geometry and that allows us to define Euclidean geometry completely.

This postulate is Euclid's parallel postulate. However, we first examine some of the results of Euclidean (and hyperbolic) geometry that can be achieved without the parallel postulate. There are good reasons to do this. Euclid proved Propositions 1 through 28 in the first book of the *Elements* without using his parallel postulate. (See the list in Section 7.1.1.) By examining the body of theorems that do not depend on a parallel postulate we are better able to understand the role that the postulate plays in Euclidean geometry.

In particular we want to look at some important concepts and theorems of Euclidean geometry, see how they are related to each other, and determine if they require the parallel postulate. We first prove a familiar property of isosceles triangles. We call a triangle **isosceles** if and only if it has at least two congruent sides. The third side is called the **base** and the angles that include it are the **base angles** of the triangle. This proof of Theorem 11.8 is due to Pappus.

Theorem 11.8 Base angles of an isosceles triangle are congruent.

Proof: Let ΔABC be isosceles with $\overline{AB} \cong \overline{AC}$. We wish to show that $\angle ABC \cong \angle ACB$. Consider ΔABC and ΔACB. $\overline{AC} \cong \overline{AB}$ by Axiom C-2. Also, by Axiom C-5, $\angle CAB \cong \angle BAC$. Consequently, $\angle ABC \cong \angle ACB$ by the SAS Congruence Axiom C-7. ∎

Theorem 11.9 (SSS Congruence): If $\overline{AB} \cong \overline{DE}$, $\overline{BC} \cong \overline{EF}$, and $\overline{AC} \cong \overline{DF}$, then $\Delta ABC \cong \Delta DEF$.

Proof: By Axiom C-4, in the half plane of line $\overleftrightarrow{AB}$ that does not contain C there exists a ray $\overrightarrow{AG}$ such that $\angle BAG \cong \angle FDE$. By Axiom C-1, there is a point C' on $\overrightarrow{AG}$ such that $\overline{AC'} \cong \overline{DF}$, which by Axiom C-5 and the given $\overline{AC} \cong \overline{DF}$ implies $\overline{AC'} \cong \overline{AC}$. By Axiom C-7 (SAS Congruence), $\overline{BC'} \cong \overline{EF}$, which, with Axiom C-5 and the given $\overline{BC} \cong \overline{EF}$, implies $\overline{BC'} \cong \overline{BC}$.
By Axiom B-5, $\overline{CC'}$ intersects $\overleftrightarrow{AB}$ at a point P. There are now five cases.

Case 1: P is between A and B. Then $\Delta CAC'$ and $\Delta CBC'$ are isosceles. Consequently, by Theorem 11.8, $\angle ACC' \cong \angle AC'C$ and $\angle BCC' \cong \angle BC'C$. By Axiom C-6, $\angle ACB \cong \angle AC'B$. So $\Delta ABC \cong \Delta ABC'$ by SAS Congruence. But $\Delta ABC' \cong \Delta DEF$, and since triangle congruence is an equivalence relation, $\Delta ABC \cong \Delta DEF$, which was to be proved.

Case 2: A is between P and B. The argument for this case is like case (1), except that we use Theorem 11.6 instead of Axiom C-6.

Case 3: B is between P and A. The argument is just like case (2), with A and B changing places.

Case 4: $P = A$. Then C', P, and C are collinear. Thus $\Delta BCC'$ is isosceles. So, by Theorem 11.8, $\angle BCC' \cong \angle BC'C$. So $\Delta ABC \cong \Delta ABC'$ by SAS Congruence. The rest is identical to case (1).

Case 5: $P = B$. This argument is just like case (4), with A and B changing places. ∎

Theorem 11.10 (AAS Congruence): If $\overline{AB} \cong \overline{DE}$, $\angle ACB \cong \angle DFE$, and $\angle BAC \cong \angle EDF$, then $\Delta ABC \cong \Delta DEF$.

You are asked to prove this theorem in Problem 1.

We have not yet defined some simple ideas, among them *equidistance* and the *midpoint* of a segment.

Definitions A point P is **equidistant** from A and B if and only if $\overline{PA} \cong \overline{PB}$. M is the **midpoint** of $\overline{AB}$ if and only if M is on $\overleftrightarrow{AB}$ and equidistant from A and B.

Question 1: What is the significance of the word *the* in the definition of *midpoint*?

Although from Axiom B-1 we know that, given two different points A and B, there exists a point between them, we do not know automatically that there is one and only one point equidistant from them on the segment. Some geometers make the existence and uniqueness of a midpoint an axiom. Others deduce these properties from assumed metric properties of segments (when we postulate that a line can be coordinatized with real numbers, we have unique midpoints). Regardless of method, proofs of the existence and uniqueness of a midpoint often reveal interesting connections between it and other geometric figures.

You may have seen some proofs about unique midpoints based on the Euclidean parallel postulate or a postulate equivalent to it. But we can construct a proof without it, on the basis of axioms we have collected so far, with the addition of some definitions and theorems.

Theorem 11.11 Every segment has exactly one midpoint.

Proof: Let $\overline{AB}$ be a segment. Using Euclid's Theorem 1 and the Axiom of Continuity, construct equilateral triangles ABC and ABC' with C and C' on opposite sides of $\overleftrightarrow{AB}$. Thus $\overline{AC} \cong \overline{AB} \cong \overline{BC} \cong \overline{BC'} \cong \overline{AC'}$. Let M be the intersection of $\overline{CC'}$ and $\overleftrightarrow{AB}$. We want to show that M is a midpoint of $\overline{AB}$. $M \neq A$, for if $M = A$ then by the same argument starting with $\overline{BA}$ (which equals $\overline{AB}$), $M = B$, and so $A = B$, which is impossible since $\overline{AB}$ is a segment. So A and B do not belong to $\overline{CC'}$, and triangles $AC'C$ and $BC'C$ exist. $\angle AC'C \cong \angle BC'C$ by SSS Congruence. So $\angle AC'M \cong$ $\cong \angle BC'M$ (since M is on $\overline{CC'}$). Then $\overline{AM} \cong \overline{BM}$ by SAS Congruence. So M is a midpoint of $\overline{AB}$. Uniqueness is left to you as Problem 2. ∎

Question 2: Draw a diagram to illustrate this proof.

From the proof of Theorem 11.11, we see that the existence of midpoints is related to triangle congruence, incidence, betweenness, and separation properties of the plane. The parallel postulate is not required. To be certain of this, we would, however, need to examine the theorems (such as SSS and AAS) used, as well as, for example, whether anything besides continuity is required in the first proof to construct points C and C'.

It is often the case that proof reveals some surprising relationships between the concepts of geometry. For example, it turns out that the existence and uniqueness of a midpoint plays an important role in the proof of one of the famous theorems of Euclidean geometry—the exterior angle theorem.

Definition In ΔABC, if point D is on $\overleftrightarrow{BC}$ such that C is between B and D, then $\angle ACD$ is called an **exterior angle of ΔABC**.

Theorem 11.12 **(Exterior Angle Theorem):** In any triangle, an exterior angle is greater than either of its nonadjacent interior angles.

Figure 7

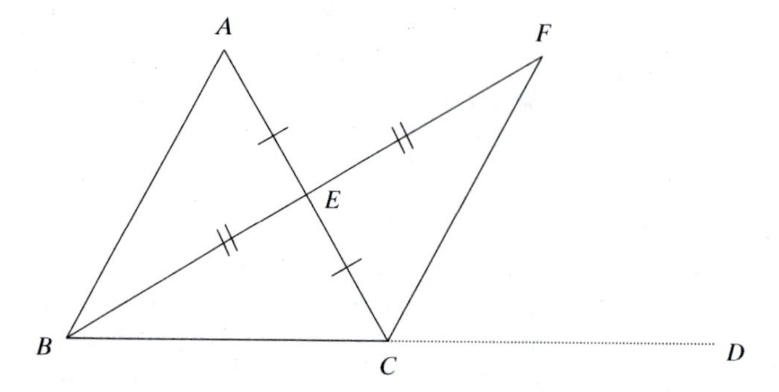

Proof: Suppose ΔABC has exterior angle ACD (Figure 7). Let E be the midpoint of side $\overline{AC}$. Extend $\overline{BE}$ to F so that $\overline{BE} \cong \overline{EF}$. The vertical angles AEB and CEF are congruent, and $\overline{AE} \cong \overline{EC}$, so $\angle ACF \cong \angle CAB$ by SAS Congruence (Axiom C-7). F is on the opposite side of $\overleftrightarrow{AC}$ from B, so it is on the same side of $\overleftrightarrow{AC}$ as D (Plane Separation Axiom). Also, since E is between B and F, and since $\overline{FB}$ intersects $\overleftrightarrow{CD}$

at B, F is on the same side of $\overleftrightarrow{CD}$ as E. Thus F is in the interior of $\angle ACD$. Consequently, $\angle ACD$ is greater than $\angle ACF$, so $\angle ACD > \angle CAB$. This same argument could be repeated starting with the midpoint of side $\overline{BC}$, so the theorem is proved. ┘

The exterior angle theorem is very powerful. It enables us to prove, for example, that given a line and a point not on it, there is a unique line on that point perpendicular to that line (see Problem 5).

Now we introduce parallelism. We adopt here Euclid's definition of *parallel*: **Parallel lines** are lines with no points in common. A sufficient condition for parallelism is found in an important theorem, which follows from the Exterior Angle Theorem. Before we explore this theorem, we again need some definitions:

Definitions Let lines ℓ, m, and n be distinct. Let B be on ℓ and n, and let C be on m and n such that B is distinct from C. Choose A on ℓ and D on m such that A and D are on opposite sides of n. Then $\angle ABC$ and $\angle BCD$ are a pair of **alternate interior angles** formed by the line n, which we call a **transversal**, and the lines ℓ and m.

Theorem 11.13 **(Alternate Interior Angle Theorem):** If two lines are intersected by a transversal so that alternate interior angles are congruent, then the lines are parallel.

Figure 8

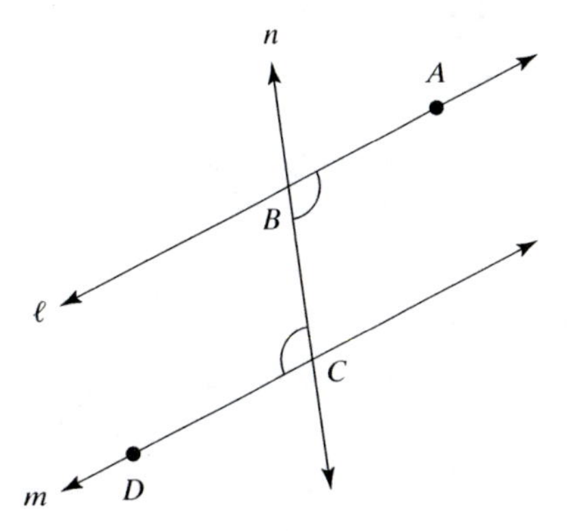

Proof: The proof is indirect. Refer to Figure 8. Assume that the alternate interior angles ABC and DCB are congruent but $\overleftrightarrow{AB}$ and $\overleftrightarrow{CD}$ intersect. Then find a contradiction to one of the theorems we have proved. The rest is left to you. ┘

Since Theorem 11.12 was used in the proof of Theorem 11.13, can we conclude that the exterior angle theorem is necessary to prove the alternate interior angle theorem, and therefore that the concept of midpoint is also necessary for its proof? No. Consider the following proof of the alternate interior angle theorem:

Proof: Let ℓ, m, n and points A, B, C, D be as in the preceding definition of alternate interior angles and $\angle ABC \cong \angle DCB$. Assume that ℓ and m meet at a point E, and that E is on the same side of n as D and the opposite side of n as A. There is a point F on $\overrightarrow{BA}$ such that $\overline{BF} \cong \overline{CE}$ (Axiom C-1). $\overline{CB}$ is congruent to itself (Axiom C-2) so in triangles BCE and CBF, $\angle EBC \cong \angle FCB$ (Axiom C-7). Since $\angle EBC$ is the supplement of $\angle FBC$, $\angle FCB$ is the supplement of $\angle ECB$ (Axiom C-4, Theorem 11.7). Then F is on m. Since both E and F are on ℓ and m, $\ell = m$ (Axiom I-1). This contradicts the given that ℓ and m are distinct lines. Therefore, ℓ is parallel to m. ┘

Question 3: Can we conclude from the Alternate Interior Angle Theorem that parallel lines exist?

We did not use the concept of midpoint in the previous proof. In fact, it turns out that we can use the alternate interior angle theorem to construct yet another proof of the existence of a midpoint, again based on the axioms we have collected so far.[1]

This discussion is intended to help you understand the power of proof in exposing relationships between geometric concepts. It also serves as a caution to you not to assume that simply because a result is used in the proof of a theorem, it is necessary for all proofs of that theorem. It might simply be sufficient. For example, there are proofs of the exterior angle theorem based on Euclid's parallel postulate, but, as we have shown, the parallel postulate is not necessary for the proof.

11.1.4 Problems

1. Prove Theorem 11.10 (AAS Congruence).
2. Prove the uniqueness of the midpoint of a line segment. (*Hint*: Assume there are two midpoints and derive a contradiction using AAS.)
3. Prove ASA Congruence using SAS Congruence.
4. Complete the proof of Theorem 11.13.
5. If $\angle AOB$ is a right angle, then lines $\overleftrightarrow{OA}$ and $\overleftrightarrow{OB}$ are **perpendicular** to each other at O.
   a. Prove that for every line ℓ and every point P there exists a line on P perpendicular to ℓ.
   b. Use part **a** to prove (without the parallel postulate or any equivalent statement) the following corollary to the Alternate Interior Angle theorem: If ℓ is any line and P is any point not on ℓ, there exists at least one line m through P parallel to ℓ.
   c. Prove that the perpendicular in part **a** is unique.
6. Use the following theorem (which can also be proved without the parallel postulate) to give an alternate proof of the corollary to the Alternate Interior Angle Theorem stated in Problem **5b**.
7. Prove that the alternate interior theorem, in the presence of the incidence, betweenness, and congruence axioms, is sufficient to show that parallel lines exist. You may assume that all right angles are congruent. (*Hint*: Use perpendicular lines.)

ANSWERS TO QUESTIONS

1. The word *the* signifies that if the midpoint exists, it is unique.
3. Yes, because congruent alternate interior angles can be formed at points B and C, the theorem is sufficient to prove the existence of parallel lines.

11.1.5 Euclid's Fifth Postulate

Although many theorems can be proved from the axioms we have so far stated, which fill in the gaps in the first four postulates of Euclid, not all the theorems of Euclidean geometry can be deduced from them. Like Euclid, when he was using only the first four of his postulates, we are one postulate short of those needed to prove all the theorems found in Euclid's *Elements*. That is, we need one more postulate in order to have a full set of postulates for Euclidean geometry. Here again is Euclid's fifth postulate.

Axiom **Euclid's fifth postulate (parallel postulate):**

P-E: If a straight line falling on two straight lines makes the interior angles on the same side less than two right angles, the two straight lines, if produced indefinitely, meet on that side on which are the angles less than the two right angles.

---

[1] See Marvin J. Greenberg, *Euclidean and Non-Euclidean Geometries* (Third Edition). San Francisco: W. H. Freeman, 1993, p. 137.

Question: Draw the figure described in the fifth postulate.

In the English translation (by Thomas Heath) that we have adapted for use here, the fifth postulate uses more words than Euclid's first four postulates put together. This also is the case in the original Greek, so it should not surprise you that, from the time of Euclid, the length and complexity of the fifth postulate led mathematicians to try to prove it from the other postulates. Examination of Euclid's first 28 propositions suggests that he also may have tried to prove his fifth postulate from the other four.

It was not until the 19th century that such attempts were revealed to be futile because models of geometry were described that satisfied the first four postulates but not the fifth. Still, much was learned from efforts to prove the fifth postulate, including the discovery of non-Euclidean geometries and the proofs of many statements equivalent to the parallel postulate of Euclid.

No original version of Euclid's *Elements* survives, and our knowledge of Euclid comes mainly from two detailed commentaries written by Greeks centuries later, Pappus (c. 320 A.D.) and Proclus (410–485 A.D.). Proclus was one of the first to advocate that Euclid's fifth postulate be a theorem, proved from a postulate he suggested:

1. There is only one parallel to a given line through a point not on that line.

This has become known as Playfair's version of the parallel postulate, after John Playfair (1748–1819), who suggested its use in 1795. We noted in Section 11.1.2 that Playfair's postulate is violated in the Beltrami-Cayley-Klein model of hyperbolic geometry. Thus once we include Euclid's parallel postulate (or any statement, such as Playfair's, equivalent to it), our axiom set can no longer describe hyperbolic geometry.

Other statements that have been proved to be equivalent to the fifth postulate include the following:

2. Any line that intersects one of two parallel lines intersects the other. (Proclus)
3. Given a triangle, another triangle can be constructed that is similar and not congruent to it. (John Wallis, 17th century)
4. The sum of the measures of the angles of a triangle equals 180°.

Euclid's fifth postulate does not mention the word "parallel", yet it is sometimes called Euclid's "parallel postulate" because it has implications for the behavior of parallel lines. Since Euclid had defined "angle" in terms of "inclination", he might have defined "parallel lines" as lines with the same inclination. But he never uses this word again and would have found it difficult to prove things about parallel lines had he adopted such a definition. He also could have defined parallel lines as lines that "go in the same direction", perhaps like chariot tracks. But he may have realized that the Earth is nearly a sphere (Greek mathematicians *did* know Earth was round) and that two lines drawn in the directions of due north and south would intersect at the poles. He might have defined parallel lines as lines that are equidistant, but this definition requires knowing that the set of points at a fixed distance from a line on one side of it is in fact a line. Instead, Euclid defines lines to be *parallel* if they are in the same plane and do not meet. Then he assumes the fifth postulate, which provides a sufficient condition for lines not to be parallel!

Figure 9

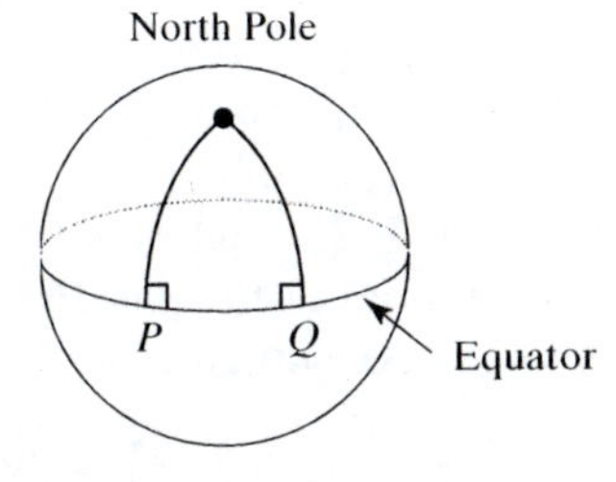

Euclid's fifth postulate shows once again that Euclidean geometry does not apply on the surface of the sphere. Recall the triangle NPQ from Section 10.1.1 (Figure 9) whose sides are the great circles of the sphere. Such a triangle includes two right angles, and so violates Euclid's fifth postulate.

The connection between the fifth postulate and parallelism comes through the logic of the **Principle of the Contrapositive**: If p and q are statements, then $p \Rightarrow q$ has the same truth value as its contrapositive not $q \Rightarrow$ not p.

Euclid's fifth postulate is of the form $p \Rightarrow q$, where

p is "A straight line falling on two straight lines makes the interior angles on the same side less than two right angles."

and q is "The two straight lines, if produced indefinitely, meet on that side on which are the angles less than the two right angles."

So the contrapositive of the fifth postulate is

not $q \Rightarrow$ not p: If two lines do not intersect on a side of a third line, then the interior angles on that side are either equal to or greater than two right angles.

But if the interior angles on one side are greater than two right angles, the interior angles on the other side of the line are less than two right angles, and (due to the postulate) the lines will meet there. So we can conclude: If two parallel lines are cut by a transversal, then the interior angles on the same side of the transversal equal two right angles. That is, (in modern language) they are supplementary.

This argument shows that the parallel postulate is also equivalent to the following statements:

5. If two parallel lines are cut by a transversal, then corresponding angles are congruent.
6. If two parallel lines are cut by a transversal, then alternate interior angles are congruent.

The converses of statements (5) and (6) are not equivalent to the parallel postulate. Theorem 11.13 shows that (6) can be deduced from our axioms without the parallel postulate.

Theorems requiring the Parallel Postulate

We can prove many theorems of Euclidean geometry without the parallel postulate, such as, for example, that two lines perpendicular to the same line are parallel to each other. However, the following theorem, which appears to be quite similar, requires it.

Theorem 11.14 Two distinct lines parallel to the same line are parallel to each other.

Proof: Given lines ℓ, m, n, such that ℓ is parallel to m and n is parallel to m, and lines ℓ and n are distinct. We prove that ℓ is parallel to n. Suppose not. Let A belong to ℓ and n. Since ℓ is distinct from n, there exist two distinct lines, ℓ and n, each on A and each parallel to m. But this contradicts the Euclidean parallel postulate (Playfair's version). Therefore, ℓ is parallel to n. ┘

You may be surprised at other theorems that require the parallel postulate. Because of their equivalence to the parallel postulate, statements (1) to (6) require the parallel postulate or some equivalent assumption in order to be proved. Because of statement (4), the sum of the measures of the angles of a quadrilateral will not be 360° without the parallel postulate. Thus the parallel postulate is needed to ensure the existence of rectangles (figures with four right angles) and the existence of squares. Consequently, the basic definition of area in Euclidean geometry (Section 10.1.1) requires the parallel postulate. Because of the equivalence of statement (3) to the parallel postulate, the existence of noncongruent similar figures also requires the parallel postulate. The Pythagorean Theorem, which relies either on area or similarity for its proof (Section 8.3.1), thus also relies on the parallel postulate.

With the parallel postulate, not only do we get all the above theorems, but we have an axiom set that defines Euclidean plane geometry. That is, using the 18 axioms I1–I4, B1–B5, C1–C7, D-1, and P-E, and the definitions we have given, all the theorems associated with Euclidean plane geometry can be deduced.

11.1.5 Problems

1. Consider this statement: If a line is perpendicular to one of two parallel lines, then it is perpendicular to the other. Prove that it is implied by Euclid's parallel postulate.
2. Prove that statement (5) in this section is equivalent to Euclid's parallel postulate.
3. Prove that the suggested alternates (1) and (2) to the parallel postulate mentioned on the first page of this section imply each other.
4. Prove that Playfair's parallel postulate (1) implies statement (6) in this lesson. (*Hint*: Use Theorem 11.10.)
5. Determine if the parallel postulate holds in the following geometries mentioned in Section 11.1.1. Support your answers.
   a. Fano's geometry
   b. Young's geometry (see Problem 9 of Section 11.1.1)
   c. three-point geometry (see Problem 1 of Section 11.1.1)
   d. four-point geometry (see Problem 11 of Section 11.1.1)
6. Consider the geometry consisting of 13 points $\{A, B, C, D, E, F, G, H, I, J, K, L, M\}$ and 26 lines: $\{ABC, BDF, CDG, DHI, EFK, FGM, GIK, HJL, ADE, BEI, CEJ, DJK, EGL, FIJ, AFH, BGH, CFL, DLM, EHM, AGJ, BJM, CHK, AIL, BKL, CIM, AKM\}$. Here ABC means the line $\{A, B, C\}$, etc.
   a. Does the Euclidean parallel postulate hold?
   b. Prove or disprove that this geometry satisfies the incidence axioms.
   c. Formulate a theorem in the above geometry.
   d. Does the postulate of Pasch (Theorem 11.2) hold in this geometry?
7. Answer the same questions as Problem 6 for the geometry consisting of 13 points $\{A, B, C, D, E, F, G, H, I, J, K, L, M\}$ and 13 lines $\{ABCD, AEFG, AHIJ, AKLM, BEHK, BFIL, BGJM, CEIM, CFJK, CGHL, DEJL, DFHM, DGIK\}$. Here $ABCD$ means the line with four points $\{A, B, C, D\}$.
8. a. Prove that the perpendicular bisectors of the three sides of a triangle have a point in common.
   b. Identify the place(s) in your proof where the parallel postulate is used.

ANSWER TO QUESTION

Figure 10 is one such figure.

Figure 10

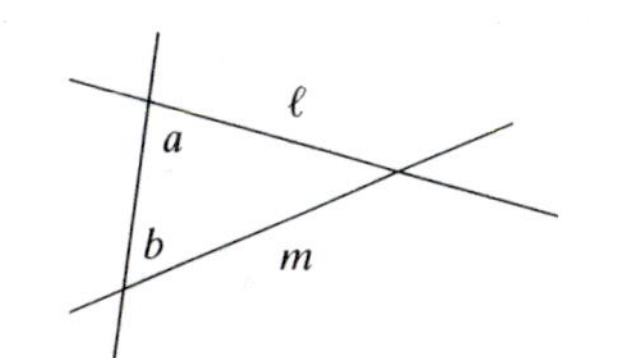

If $a + b < 180$, ℓ and m intersect.

Unit 11.2 The Cartesian Model for Euclidean Geometry

In this unit, we explore an approach to Euclidean geometry that is different from the synthetic approach exemplified by the axioms I1–I4, B1–B5, C1–C7, D-1, and P-E of Unit 11.1. This approach is analytic, and by examining it, we connect the lines that are studied in algebra with the lines in Euclidean geometry. It also justifies the analytic proofs that we have in this book for deducing some of the theorems of Euclidean geometry.

In this analytic approach to Euclidean geometry, we are led to ask questions about our axiom system for Euclidean geometry. Does our axiom system apply only to Euclidean geometry, or, like the incidence axioms, can it apply to different geometries?

11.2.1 The Cartesian coordinate system

Recall from Section 8.1.1 that the **Cartesian plane** is the set $\mathbf{R}^2$ of all ordered pairs of real numbers. Accordingly, a Cartesian **point** is an ordered pair of real numbers. A Cartesian **line** is the set of points (x, y) that satisfy an equation equivalent to a linear equation in the standard form $Ax + By + C = 0$, where A and B are constant real numbers that are not both zero. If $A = 0$, then the equation represents a **horizontal line**; if $B = 0$, a **vertical line**. While you can think of drawing the usual picture of such a system with the x-axis as horizontal and y-axis as vertical, all of the work that we do in this system in this section can be done with algebra alone, dependent not on the pictures but on the properties of real numbers.

Note, that the standard form $Ax + By + C = 0$ is not unique, since all multiples $kAx + kBy + kC = 0$, where k is a nonzero constant, represent the same line. However, a line is uniquely determined by its slope and y-intercept. So if the line is not vertical, then $B \neq 0$ and we can convert the standard form of the equation to the familiar slope-intercept form of an equation: $y = mx + b$, where m is the slope of the line and b is the y-intercept, where $m = -\frac{A}{B}$ and $b = -\frac{C}{B}$. If $B = 0$, we say the line has *no slope*, and the equation is of the form $x = h$, where $h = -\frac{C}{A}$. The equation $kAx + kBy + kC = 0$, when converted to slope-intercept form, yields the same slope and y-intercept as the original equation. Therefore, all such multiples can be treated as essentially the same equation.

We wish to demonstrate that the Cartesian plane with these points and lines is a *model* of Euclidean geometry, that is, an example or interpretation or manifestation of the axiom system of Euclid's geometry. To do this, we must show that each axiom in Euclidean geometry is satisfied in the model. Thus the axioms in our earlier development of Euclidean geometry become theorems in the model. Our first theorems are the four incidence axioms.

Theorem 11.15 **(Cartesian I-1):** There exist at least three distinct points.

Proof: We have an infinity of trios of Cartesian points from which to select. One trio is $(1, 1)$, $(0, -4)$, and $\left(\frac{3}{8}, 7\right)$. ┘

Theorem 11.16 **(Cartesian I-2):** For each two distinct points, there exists a unique line on them.

Proof: Let (x_1, y_1) and (x_2, y_2) be the two distinct points. Then, algebraically, the coordinates of these points must satisfy the equation of any line on them. Geometrically, the line passes through the two points.

An equation for the unique line on these two points is found by using the familiar two-point formula

$$y - y_1 = \frac{y_1 - y_2}{x_1 - x_2}(x - x_1),$$

which we can use when x_1 does not equal x_2. How do we know this is the *only* equation that describes the line on these two points? We convert our equation into standard form.

$$-(y_1 - y_2)x + (x_1 - x_2)y + (y_1x_2 - x_1y_2) = 0$$

Considering all multiples as representing the same line, we see that A, B, and C are uniquely determined by the coordinates of the given points. That is,

$$A = -(y_1 - y_2), \quad B = (x_1 - x_2), \quad \text{and} \quad C = (y_1x_2 - x_1y_2).$$ ┘

Question 1: Prove Cartesian I-2 for the case $x_1 = x_2$.

Theorem 11.17 **(Cartesian I-3):** For every line there exist at least two distinct points on it.

Proof: Let the line have equation $Ax + By + C = 0$. First assume A is zero, but B is not zero. Two possible points are $\left(1, -\frac{C}{B}\right)$ and $\left(2, -\frac{C}{B}\right)$. If B is zero, and A is not zero, our points can be $\left(-\frac{C}{A}, 13\right)$ and $\left(-\frac{C}{A}, 0\right)$, among others. If neither A nor B is zero, we could choose $\left(\frac{C}{A}, -\frac{2C}{B}\right)$ and $\left(-\frac{2C}{A}, \frac{C}{B}\right)$. ┘

Theorem 11.18 **(Cartesian I-4):** Not all points are on the same line.

Proof: Again, we have an infinity of points to select in order to demonstrate this axiom. The points we chose in the proof of Cartesian I-3 will suffice. ┘

Thus the Cartesian plane satisfies the incidence axioms of Euclidean geometry. What about the Euclidean parallel postulate? Let's check on this. We'll use the Playfair version: On a given point not on a given line, exactly one line can be drawn parallel to a given line. We must first specify what it means in our model for two lines to be parallel. We define parallel lines here as lines that do not intersect, so two lines in our model are parallel if there is no ordered pair that satisfies their respective equations.

Before we demonstrate the Playfair axiom, we first prove three lemmas.

Lemma 1 Every vertical line intersects every nonvertical line.

Proof: Any vertical line can be represented by the equation $x = A$, and any nonvertical line can be represented by the equation $y = mx + b$. The ordered pair that satisfies both these equations is $(A, mA + b)$. ┘

Lemma 2 Two lines are parallel if and only if (1) both are vertical lines; (2) neither is vertical and they both have the same slope.

Proof: Since Lemma 2 is an "if and only if" statement, we must prove both directions.

($\Leftarrow$) (1) If both distinct lines are vertical or (2) if neither of two distinct lines is vertical and both have the same slope, we must prove that the two lines are parallel.

(1) Let two distinct vertical lines be represented by the equations $x = A$ and $x = B$, where $A \neq B$. Since $A \neq B$, their respective solutions sets, $\{(A, y)\}$ and $\{(B, y)\}$ for all real numbers y have no common solution. Therefore, the lines are parallel.

(2) Let two distinct nonvertical lines with the same slope be represented by the equations $y = mx + b_1$ and $y = mx + b_2$, where $b_1 \neq b_2$. Solving these equations simultaneously produces a contradiction: $b_1 = b_2$, and so there is no common solution. The lines are parallel.

($\Rightarrow$) If two lines are parallel, we must prove that either (1) both lines are vertical or (2) if neither line is vertical, both have the same slope.

Let two lines be parallel. Then no ordered pair is a common solution to their respective equations. By Lemma 1, then either both lines are vertical or neither line is vertical. If both lines are vertical, then (1) holds. Therefore, it remains only to prove that if neither line is vertical, both must have the same slope in order to be parallel.

We proceed using an indirect proof. Assume two nonvertical lines have different slopes yet are parallel. Let the first line be represented by the equation $y = m_1x + b_1$. Let the second line be represented by the equation $y = m_2x + b_2$, where m_1 is not equal to m_2. Solving these equations simultaneously, we obtain

$$m_1x + b_1 = m_2x + b_2,$$

from which

$$x = \frac{b_2 - b_1}{m_1 - m_2}$$

and

$$y = m_1\left(\frac{b_2 - b_1}{m_1 - m_2}\right) + b_1.$$

So $(x, y) = \left(\frac{b_2-b_1}{m_1-m_2}, m_1\left(\frac{b_2-b_1}{m_1-m_2}\right) + b_1\right)$. These values for x and y satisfy the equations of both lines ($y = m_1x + b_1$ and $y = m_2x + b_2$), indicating that the lines intersect in the point represented by that ordered pair. This contradicts our hypothesis that the two lines are parallel, and so our assumption that they had different slopes must be false. ┘

Lemma 3 Given a point $P = (x_1, y_1)$ and a real number m, there is exactly one line that is on P and has slope m.

Proof: Lines with slope m can be represented by equations of the form $y = mx + b$. If such a line contains P, then the coordinates of P must satisfy its equation. Substituting x_1 and y_1 for x and y, respectively, we obtain $b = y_1 - mx_1$. Therefore, the unique line on P with slope m is $y = mx + (y_1 - mx_1)$. ┘

The lemmas we proved have established certain properties of vertical and nonvertical lines in the Cartesian plane. We are now ready to prove that the Playfair axiom holds in the Cartesian plane. In the proof below, we partition the lines of the Cartesian plane into vertical and nonvertical, treating each as a separate case.

Theorem 11.19 **(Cartesian Parallel Postulate):** On a given point not on a given line, exactly one straight line can be drawn parallel to a given straight line.

Proof: Given a line and a point $P = (x_1, y_1)$ not on it.

Case 1. The given line is vertical. Let this line be represented by the equation $x = A$. Then the unique vertical line on P is given by the equation $x = x_1$. Since both lines are vertical, by Lemma 2, they are parallel, and by Lemma 1 there is no nonvertical line parallel to our given line. Thus the Playfair axiom is verified for this case.

Case 2. The given line is not vertical. Let this line be represented by the equation $y = mx + b$. Then by Lemma 2, the only line on P parallel to the given line is a line on P with slope m. Using Lemma 3, this unique line is given by the equation $y = mx + (y_1 - mx_1)$. Thus the Playfair axiom is also verified for this case. ┘

We have thus established that the incidence axioms and the parallel postulate (Playfair version) are satisfied in the Cartesian plane. To complete the proof that the Cartesian plane is a model of Euclidean plane geometry as defined by our axiom system, we would have to show that all the other axioms of betweenness, congruence, and continuity are similarly verified. Some of these proofs are included as problems and others can be found in the literature.[2]

By demonstrating that all of the 18 axioms we gave in Unit 11.1 are true in the Cartesian plane, we allow coordinates to be used for proofs in Euclidean geometry.

11.2.1 Problems

1. For each of the cases in the proof of Cartesian I-3, find two points on the line $Ax + By + C = 0$ other than the two points shown.

2. Give an equation for the line through the two given points.

a. (a, b) and (c, d)
b. (x_0, y_0) and $(0, 0)$
c. $(a, 0)$ and $(0, b)$
d. (x_1, y_1) and (x_1, y_2)
e. $(a, -a)$ and $(-b, b)$

3. In the proof of Cartesian I-4, it is asserted that $(1, 1)$ $(0, -4)$, and $\left(\frac{3}{8}, 7\right)$ are not on the same line. Show that this is true with two different explanations.

4. Give an equation for the line parallel to the given line through the given point.

a. $\{(x, y)\colon Ax + By + C = 0\}; (0, 0)$
b. $\{(x, y)\colon x = 1\}; (2, 5)$
c. $\{(x, y)\colon ax + by = 1\}; (c, 0)$
d. $\{(x, y)\colon 4y = 2x + 3\}; \left(\frac{1}{2}, \frac{1}{3}\right)$

5. Provide an analytic proof of Euclid's Theorem 1 (Section 11.1.3), taking $A = (0, 0)$ and $B = (b, 0)$.

6. Let $A = (x_1, y_1)$ and $C = (x_2, y_2)$, with $A \neq C$. Define B to be between A and C if and only if there exists t such that $0 < t < 1$ and $B = ((1 - t)x_1 + tx_2, (1 - t)y_1 + ty_2)$. From this definition and properties of real numbers, deduce the indicated betweenness axiom from Section 11.1.2.

a. B-1 b. B-2 c. B-3 d. B-4

*7. Prove that the Postulate of Pasch (Theorem 11.2) is true in the Cartesian plane. (*Hint*: you must use the betweenness axioms and specify an interpretation of segment in terms of distance.)

8. Show that Theorem 11.1 of Section 11.1.1 is true in the Cartesian plane.

9. Consider the system of integers modulo 3. This system consists of the elements 0, 1, 2 with the definitions of multiplication and addition as in Figure 11 (see Unit 6.1).

Figure 11

×	0	1	2
0	0	0	0
1	0	1	2
2	0	2	1

+	0	1	2
0	0	1	2
1	1	2	0
2	2	0	1

All possible points (x, y) are given by $(0, 0)$, $(0, 1)$, $(0, 2)$, $(1, 0)$, $(1, 1)$, $(1, 2)$, $(2, 0)$, $(2, 1)$, and $(2, 2)$. We can put these nine points into one-to-one correspondence with the nine points of Young's geometry (see Problem 8 of Section 11.1.1). List all possible linear equations. Show that each line consists of exactly three points.

10. Use the axioms of Sections 11.1.1, 11.1.2, and 11.1.3 to construct the Cartesian coordinate system. You may assume certain principles, including the fact that the points on any line can be numbered so that number differences measure distances and the fact that we can assign direction. (*Hint*: Fix an arbitrary point as the origin (indicating what axiom justifies its existence). Choose an arbitrary direction (by convention we would choose from left to right), and an arbitrary unit of length to construct a real number line. Construct a second number line at right angles to the first, justifying its existence. From there, define the coordinates of a point, and deduce the straight line and two of its representations: the two-point form and slope-intercept form.)

ANSWER TO QUESTION

If $x_1 = x_2$, then $y_1 \neq y_2$ because the points are distinct. So (x_1, y_1) and (x_1, y_2) are on the line. The line containing these points has no slope; it is the vertical line $x = x_1$.

[2]See, for example, Edwin E. Moise, *Elementary Geometry from an Advanced Standpoint* (Reading, MA: Addison-Wesley, 1963), Chapter 26.

11.2.2 Verifying the definition of Euclidean geometry: the relationship between a mathematical theory and its models

The Cartesian plane is a model of Euclidean geometry. What does this tell us about Euclidean geometry, in general, and about other models of it? If we prove that certain statements hold in the Cartesian plane, does that mean that these statements are always true in Euclidean geometry? If there are other models of Euclidean geometry, what is their relationship to the Cartesian plane? If a statement is true in the Cartesian plane, will it be true in all other models of Euclidean geometry?

In Sections 11.1.1 to 11.1.5, we defined Euclidean geometry by means of a set of axioms. There are, however, other axiom sets that can define Euclidean geometry. The relationship of such axiom sets to the one we constructed is that each must produce the same body of knowledge (theorems). But how can we determine if another axiom set defines Euclidean geometry? If we accept our axiom system as a defining Euclidean geometry, we must be able to demonstrate that every one of its axioms is true (because it is either taken as an axiom or is provable as a theorem) in any other axiom set claiming to define Euclidean geometry.

Is there any other way to demonstrate that another set of axioms defines Euclidean geometry? Can we use the Cartesian coordinate system? It turns out that the answer is yes, because our axioms for Euclidean geometry enjoy a special property. They are *categorical.*

When an axiom set is **categorical**, it is possible to set up a one-to-one correspondence (isomorphism) between the objects of any two models of that axiom set in such a way that any property that holds for one model will hold for the other. Thus, *every* model of a categorical set of axioms exhibits the same properties.

The property of being categorical, which is applied both to a theory and its axiom sets, was first suggested by John Dewey (1859–1952). Oswald Veblen (1880–1960) introduced it in his systems of axioms for geometry in 1904. Since then, the term and the notion itself have been attributed to Veblen. However, the American mathematician Edward V. Huntington (1874–1952) is sometimes credited with being the first to state it clearly and use it. The first proof of what we today call categoricity, due to Dedekind in 1887, was about axioms for the real numbers.

A model becomes a very powerful tool when we are dealing with a categorical set of axioms. An important theorem in mathematical logic tells us that a statement is true in a mathematical theory if and only if it is true in *every* one of its models. Therefore, when a theory is categorical (where all of its models are essentially the same), we can determine what is provable in that theory on the basis of what we can verify for any single model of it!

Not all axiom systems for Euclidean geometry are categorical. The one we presented is because it includes Axiom D-1 (continuity), which provides sufficient points on a line to be put into 1-1 correspondence with the real numbers. There are other axiom systems for Euclidean geometry that are categorical because of the addition of axioms other than D-1 that allow the establishment of an isomorphism between the geometric line and the real numbers. Since our axiom system for Euclidean geometry is categorical, and we have demonstrated that the Cartesian coordinate system gives a model of this axiom system, we can say that all models of it are essentially the same as its analytic model based on the Cartesian coordinate system. This means that we can use the Cartesian model to make certain statements that we know will be provable from our axioms for Euclidean geometry, and will be valid for every model of it. For example, we can prove the following theorem of Euclidean geometry by verifying that this statement holds in the Cartesian plane.

Theorem 11.20 The altitudes of a Euclidean triangle are concurrent.

Proof (analytic version): A coordinate system can be located so that the triangle has vertices $P = (a, 0), Q = (b, 0)$, and $R = (0, c)$, with $abc \neq 0$. Then $\overleftrightarrow{QR}$ has slope $-\frac{c}{b}$, and $\overleftrightarrow{PR}$ has slope $-\frac{c}{a}$. The altitude of ΔPQR from P is perpendicular to $\overleftrightarrow{QR}$ and therefore has slope $\frac{b}{c}$. The altitude of the triangle from Q is perpendicular to $\overleftrightarrow{PR}$ and therefore has slope $\frac{a}{c}$. The equations of the altitudes from P and Q are respectively $bx - cy - ab = 0$ and $ax - cy - ab = 0$. Subtracting the second equation from the first, we obtain

$$\begin{array}{rl} bx - cy - ab &= 0 \\ ax - cy - ab &= 0 \\ \hline (b - a)x \quad\quad\quad &= 0 \end{array} \quad \Rightarrow x = 0 \text{ (since } a \neq b).$$

Thus the point of intersection of the altitudes from P and from Q lies on the y-axis. Since the altitude on R is the y-axis, all the altitudes pass through the same point. ∎

Figure 12 illustrates the proof.

Figure 12

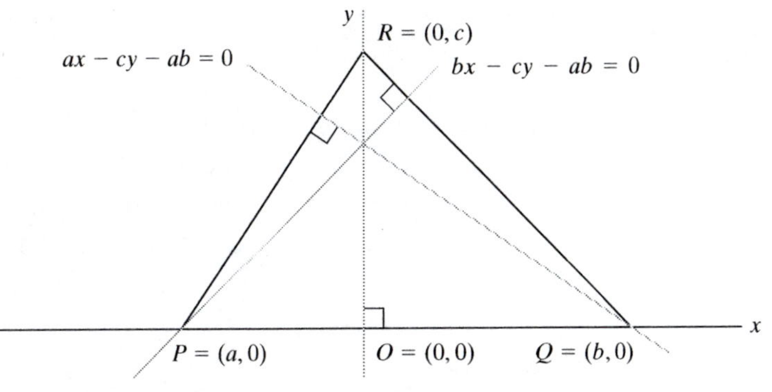

The point of concurrency of the altitudes of a triangle is called the **orthocenter** of the triangle.

The above proof of Theorem 11.20 is analytic since it makes use of the number properties of the Cartesian plane. Let's pause for a minute to analyze it. In constructing the proof, we assumed the following definitions and properties of the Cartesian plane:

1. The slope of a line on two points (x_1, y_1) and (x_2, y_2) is given by $\frac{y_1 - y_2}{x_1 - x_2}$.
2. Perpendicular lines have slopes that are opposite reciprocals of one another (see Problem 4, Section 7.2.2).
3. The line with slope m that passes through the point (x_0, y_0) is given by the equation $y - y_0 = m(x - x_0)$, which can be written in standard form.
4. To determine the point of intersection of two lines, we can solve their equations simultaneously to obtain the ordered pair which represents the point of intersection.
5. The y-axis consists of the points in the plane with coordinates $(0, y)$ where y is a real number.

Using these definitions and properties, we then translated the geometrical data of the problem into algebraic terms so we could develop the proof in our model. That is, we constructed a triangle using the Cartesian coordinate system. We used the laws of algebra to transform the data and draw algebraic conclusions from them. That is, we wrote the equations of the lines and solved these equations simultaneously. Then we translated the results back to geometric language to obtain the desired result for Euclidean geometry.

Since Theorem 11.20 holds for a model of Euclidean geometry, it is also provable directly from the axioms we took in Sections 11.1.1 to 11.1.5. Such a proof is *synthetic*. For a synthetic proof of Theorem 11.20, we first state some familiar definitions, and lemmas (which we ask you to prove in the problems) that we need to use.

Definitions A **polygon** is a plane figure consisting of the points on n line segments (its **sides**), such that each side intersects exactly one other side at each of its endpoints (its **vertices**). A **quadrilateral** is a polygon with four sides. A **parallelogram** is a quadrilateral with both pairs of opposite sides parallel.

Lemma 1: The opposite sides of a parallelogram are congruent.

Lemma 2: The perpendicular bisectors of the sides of a triangle meet in a point. (This point is called the **circumcenter** of the triangle.)

With these we are now ready to prove Theorem 11.20 synthetically.

Theorem 11.20 The altitudes of a triangle are concurrent.

Proof (synthetic version): Given triangle ABC. Through A, draw the line parallel to $\overleftrightarrow{BC}$, through B draw the line parallel to $\overleftrightarrow{AC}$, and through C, draw the line parallel to $\overleftrightarrow{AB}$. We know these lines exist by the Euclidean parallel postulate. In this way we construct a new triangle, which we call $A'B'C'$ (Figure 13).

Figure 13

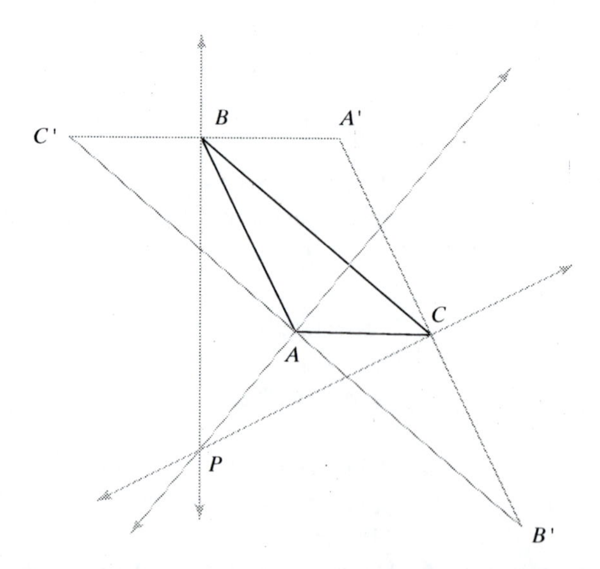

$ABA'C$ is a parallelogram, so be Lemma 1, $\overline{A'B} \cong \overline{AC'}$. Also, $BC'AC$ is a parallelogram, so $\overline{BC'} \cong \overline{AC}$. Thus $\overline{A'B} \cong \overline{BC'}$ (Axiom C-2). Since a line that is perpendicular to one of two parallel lines is perpendicular to the other (Problem 1 of Section 11.1.5), the altitude in ΔABC from B to $\overleftrightarrow{AC}$ is the perpendicular bisector of the side $\overline{A'C'}$ of $\Delta A'B'C'$. In similar fashion, the other two altitudes of ΔABC are the perpendicular bisectors of the other two sides of $\Delta A'B'C'$. By Lemma 2, the perpendicular bisectors of $\Delta A'B'C'$ are concurrent. Thus the altitudes of ΔABC are concurrent. ∎

You should compare the two proofs of Theorem 11.20—the analytic one constructed in the Cartesian model of Euclidean plane geometry, and the synthetic one deduced from the axioms and theorems of Euclidean plane geometry. The analytic proof requires a convenient position on the coordinate axis for the triangle. After that, the proof proceeds in a routine fashion, using the laws of algebra. The synthetic proof requires a little more ingenuity in the construction of the new triangle $A'B'C'$. Which method you prefer to employ is simply a matter of taste.

Another consequence from Euclidean geometry being a categorical system is that it enables us to determine whether a certain structure is a model of the Euclidean plane. One such structure is the Gaussian plane, named for Gauss, who studied it. The points in the **Gaussian plane** are the complex numbers $z = x + iy$, where x and y are real numbers, and i is $\sqrt{-1}$. A **Gaussian line** is the set of all points z that satisfy an equation $Dz + \overline{Dz} + E = 0$, where D is a nonzero complex number, $\overline{D}$ is its conjugate, and E is a real number.

We set up a 1-1 correspondence between the Cartesian plane and the Gaussian plane by mapping the real point (x, y) of the Cartesian plane into the complex point $x + iy$, and mapping the Cartesian line given by the equation $Ax + By + C = 0$ into the Gaussian line $Dz + \overline{Dz} + E = 0$, where $D = A - iB$, $\overline{D} = A + iB$, and $E = 2C$. We define the distance between two Gaussian points $z_1 = x_1 + iy_1$ and $z_2 = x_2 + iy_2$ to be $|z_1 - z_2| = |(x_1 - x_2) + i(y_1 - y_2)| = \sqrt{(x_1 - x_2)^2 + (y_1 - y_2)^2}$. With this definition, distance is preserved by the mapping. With these correspondences, the Cartesian plane and the Gaussian plane can be proved to be isomorphic, and as a result, the Gaussian plane is a model of Euclidean plane geometry. Thus, if a theorem can be proved in the Gaussian plane, then it is a theorem of Euclidean plane geometry.

We can also use a model to verify that a specific axiom (or its equivalent) is necessary to the definition of the theory in the presence of the other axioms, by showing that the reduced axiom set with that axiom removed does not define the categorical theory.

For example, we can use the Cartesian model to investigate the role of Dedekind's axiom (D-1) in our axiom set. Dedekind's axiom enables us to obtain sufficient points on a line so the points in the Cartesian plane can be put into one-to-one correspondence with ordered pairs of real numbers. Suppose that we remove Dedekind's axiom from our set. One model of the resulting reduced axiom set is known as the *surd plane*. We define a *surd* as a real number x with the following property: We can calculate x by a finite number of additions, subtractions, multiplications, divisions, and extractions of square roots, starting with 0 and 1. Certainly, every rational number is a surd. However, not all real numbers are surds. Therefore, the lines of the surd plane are full of holes. Furthermore, the surd plane is countably infinite, while the Cartesian plane is uncountable, so there is no way to establish a one-to-one correspondence between the points and lines of the Cartesian plane and the points and lines of the surd plane.

We can therefore conclude that the axiom set of Sections 11.1.1 to 11.1.5 with D-1 removed produces two different (nonisomorphic) models, the Cartesian plane and the surd plane. A theory for which there are nonisomorphic models is **noncategorical**. Thus, our axiom set with Dedekind's axiom removed is noncategorical and cannot define the Euclidean geometry of our axiom set, which we deemed categorical.

Are axiom sets for the most familiar mathematical structures categorical? No. For instance, the rational numbers and the real numbers with the usual operations of addition and multiplication satisfy the axioms for a field, but we cannot set up a one-to-one correspondence between these two models. Thus a theorem true for the rational numbers may not be true for the real numbers, and vice versa. But the axioms of a complete ordered field are categorical. So every theorem that holds for a complete ordered field also holds for the real numbers. (See Section 2.3.1.)

We have now come to the end of this book, but in a way we have returned to its beginning. Recall that in Chapter 2 we used the number line as a *geometric* model of the real numbers and as inspiration for properties of real numbers. In this unit, we have shown how real numbers enable us to develop an *algebraic* model for geometry. These models demonstrate the cross-fertilizations of algebraic and geometric ideas that permeate all of mathematics and lead to the richness of its theory and applications.

11.2.2 Problems

1. Prove Lemma 1: The opposite sides of a parallelogram are congruent. You may choose a synthetic or analytic proof.

2. Prove Lemma 2 both synthetically and analytically: The perpendicular bisectors of the sides of a triangle meet in a point, which is called the circumcenter of the triangle.

3. Prove Theorem 11.20 from Lemma 2, by means of the transformations discussed in Chapter 8. (*Hint*: Construct a size change that maps the altitudes of any triangle onto the perpendicular bisectors of the sides of a second triangle.)

4. Prove the following theorem synthetically and analytically: The diagonals of a parallelogram bisect each other.

5. Prove analytically and synthetically that any three non-collinear points lie on a circle.

6. An axiom is called **independent** in a set S of axioms if it cannot be proved or disproved using only the other axioms in S. An axiom system S is **complete** if it is impossible to add an independent axiom to it. By an indirect argument, prove that if an axiom system S is categorical, then it is complete.

7. In Section 11.1.1 we discussed a 7-point geometry that satisfies the incidence axioms. Our discussion was based on an algebraic model of the geometry. The following axioms *define* the geometry of that model (Fano's geometry):

F-1: There exists at least one line.

F-2: Every line has exactly three points on it.

F-3: Not all points are on the same line.

F-4: For each two distinct points, there exists exactly one line on both of them.

F-5: For each two lines there exists at least one point on both of them.

a. Using the axioms, prove that there are exactly 7 points in Fano's geometry, and therefore in every model of it. *Note*: You must not only show 7 points exist. You must also show an 8th point cannot exist.

b. Create a model for Fano's geometry that is different from the one given in Section 11.1.1.

c. Show that axiom F-5 of Fano's geometry is independent of the other four axioms. (*Hint*: Create a model for a geometry that satisfies all the axioms for Fano's geometry except for axiom F-5.)

d. Do you think Fano's geometry as defined by the above axioms is categorical? Explain why or why not.

Chapter Projects

1. Young's geometry.

a. Construct an axiom set (Y-1 to Y-5) for Young's geometry of Problem 9 in Section 11.1.1. Then prove that the incidence axioms (I-1 to I-4) are satisfied directly from the axioms you create.

b. Consider the following proof that every point in Young's geometry is on at least four lines.

Proof: Let P be any point, and let ℓ be any line that does not contain P. ℓ contains exactly three points. P and each point on ℓ must determine a distinct line; therefore, we have at least three lines. Now, there must be a line that contains P but contains no points on ℓ. Therefore, we have at least four lines on P. The theorem is proved.

Refer to the axioms you created in part **a** and justify every step of the above proof by means of an axiom. Then use your axioms to prove that Young's geometry must contain exactly nine points and exactly twelve lines.

c. Alter your axioms of part **a** for Young's geometry to assert that instead of three points, there are exactly two points on every line. How many points and lines would the new geometry have?

d. Alter your axioms for Young's geometry to assert that there are exactly four points on every line. How many points and lines would this new geometry have?

e. Generalize your results from parts **c** and **d** for the case where each line contains exactly n points (n = some positive integer). That is, consider a finite model satisfying the axioms for Young's geometry, assuming that instead of three points on a line, there are exactly n points on every line. How many points and lines would the geometry have?

f. Define parallel lines as lines with no points in common. Prove that two lines parallel to a third line are parallel to each other. Use the axioms to construct your proof. (*Hint*: First verify the statement by observing the model. Then construct your proof by assuming that two lines parallel to a third line are not parallel to each other and show that this assumption leads to a contradiction.)

2. Pappus's theorem. The following theorem was discovered and proved by Pappus of Alexandria about A.D. 340. If A, B, and C are three distinct points on one line, and if A', B', and C' are three different distinct points on a second line, then the intersections of $\overline{AC'}$ and $\overline{CA'}$, $\overline{AB'}$ and $\overline{BA'}$, and $\overline{BC'}$ and $\overline{CB'}$ are collinear.

a. Prove that Pappus's theorem holds in Young's geometry (Problem 9 of Section 11.1.1) for the six points on any pair of parallel lines.

b. It is possible to create a finite geometry from a geometric figure. For example, the finite geometry of Pappus arises from a figure resulting from his theorem. The figure, called the **Pappus configuration**, consists of nine points and nine lines, and inspires the definition of a new finite geometry with the following axioms:

P-1. There exists at least one line.

P-2. Every line has exactly three points.

P-3. Not all points are on the same line.

P-4. If a point is not on a given line, then there exists exactly one line on the point that is parallel to the given line.

P-5. If P is a point not on a line, then there exists exactly one point P' on the line such that no line joins P and P'.

P-6. With the exception of Axiom P-5, if P and Q are distinct points, then exactly one line contains both of them.

Prove the following theorems in this geometry:

a. Each point in the geometry of Pappus lies on exactly three lines.

b. There are exactly three lines on each point.

c. The geometry of Pappus has nine points and nine lines.

3. George E. Martin, in Chapter 5 of *The Foundations of Geometry and the Non-Euclidean Plane* (Springer-Verlag, New York, 1975), gives several models of Euclidean plane geometry. Choose one of these models, and write an essay describing it, showing how it is essentially the same as the Cartesian plane.

4. Investigate the work of Fano, in particular his three-dimensional finite incidence geometry. Design a talk for a local high school mathematics club, telling them about this mathematician and describing some characteristics of his three dimensional model.

5. Explore the interesting connection between Fano planes and Hamming error-correcting codes in *A Course in Modern Geometries* by J. Cederberg (New York: Springer-Verlag, 1989) or in *The Mathematical Theory of Coding* by I. F. Blake and R. C. Mullin (New York: Academic Press, 1975) or in texts in modern applied algebra. Prepare a lesson demonstrating this application of finite geometry.

6. The first model of plane hyperbolic geometry was given by the Italian mathematician Eugenio Beltrami in 1868. It is called a *pseudosphere* and is a surface formed by revolving a curve, called the *tractrix*, about an asymptote. Search the Internet or consult some books on non-Euclidean geometry for information about the pseudosphere. Draw a geometric model of the pseudosphere, and determine what are the points and lines of this model.

7. M. C. Escher's circular woodcuts (Circle Limits I, II, III, and IV) depict similar shapes (fish, angels, etc.) that diminish in size as they recede from the center and fit together to fill and cover a disk. I, II, and IV are based on a model of the hyperbolic plane that is owed to Henri Poincaré. Do some research on this and write an essay discussing how the Poincaré model was used in Escher's Circle Limit prints. Consult, for example, H. S. M. Coxeter; "The Trigonometry of Escher's Woodcut 'Circle Limit III'," *Mathematical Intelligencer* 18(4), 1996, or M. C. Escher: *Art and Science*, North-Holland, 1986 (edited by Coxeter and others).

8. Read the article "From Pappus to Today: The History of a Proof," *Mathematical Gazette* 74, 1990, 6–11, by Michael Deakin, which explores different proofs of the theorem that base angles of an isosceles triangle are equal. Select the proof that you believe has the greatest pedagogical advantage, and describe why you believe this to be so.

Bibliography

Unit 11.1 References

Prenowitz, Walter, and Meyer Jordan. *Basic Concepts of Geometry*. New York: Blaisdell, 1965.
The authors explore Euclidean and non-Euclidean geometries, as well as finite geometries, using incidence as a basic unifying idea.

Rosenfeld, B. A. *History of Non-Euclidean Geometry*. New York: Springer-Verlag, 1988.
This book investigates the mathematical and philosophical factors underlying the discovery of non-Euclidean geometry and the extension of the concept of space.

Unit 11.2 References

Henderson, David. *Experiencing Geometry on Plane and Sphere*. Upper Saddle River, NJ: Prentice Hall, 1996.
The author gives a series of problems designed to promote the learning of geometry by means of intuition and reasoning from experience.

Martin, George E. *The Foundations of Geometry and the Non-Euclidean Plane*. New York: Springer-Verlag, 1975.
The author provides an axiomatic development of the Euclidean and hyperbolic planes, providing historical aspects. Axiomatic systems of Euclid, Hilbert, and Pieri are discussed.

Millman, R. S., and G. D. Parker. *Geometry: A Metric Approach with Models*. New York: Springer-Verlag, 1981.
This book seeks to give the reader some intuition about Euclidean and hyperbolic geometry through an examination of analytic models.

Moise, Edwin E. *Elementary Geometry from an Advanced Standpoint*. Reading, MA: Addison-Wesley, 1963.
This classic geometry textbook gives a rigorous development of Euclidean and hyperbolic geometries, and attempts to reexamine familiar topics, providing valid definitions and proofs for concepts and theorems treated in more elementary courses.

Smith, James T. *Methods of Geometry*. New York: John Wiley & Sons, Inc., 2000.
This book builds on the mathematical knowledge of high school and lower division university courses to explore two- and three-dimensional geometry. It includes a wealth of nonroutine exercises and projects, an extensive annotated bibliography, and historical references.

Unit 11.3 References

Greenberg, Marvin J. *Euclidean and Non-Euclidean Geometries*. Third Edition. New York: W. H. Freeman and Co., 1993.
The author provides a rigorous treatment of the foundations of Euclidean geometry and an introduction to hyperbolic geometry, emphasizing its Euclidean models. This edition includes a discussion of the history of Euclid's parallel postulate, and the birth of non-Euclidean geometry.

Hartshorne, Robin. *Geometry: Euclid and Beyond*. New York: Springer-Verlag, 2000.
This book examines Euclid's *Elements*, explores questions that arise from it, and gives modern answers.

INDEX

(page numbers in **boldface** contain a definition of the term)